Functional Food Product Development

WILEY-BLACKWELL
A John Wiley & Sons, Ltd., Publication

Functional Food Science and Technology Series

Functional foods resemble traditional foods but are designed to confer physiological benefits beyond their nutritional function. Sources, ingredients, product development, processing and international regulatory issues are among the topics addressed in Wiley-Blackwell's new *Functional Food Science and Technology* book series. Coverage extends to the improvement of traditional foods by cultivation, biotechnological and other means, including novel physical fortification techniques and delivery systems such as nanotechnology. Extraction, isolation, identification and application of bioactives from food and food processing by-products are among other subjects considered for inclusion in the series.

Series Editor: **Professor Fereidoon Shahidi**, PhD, Department of Biochemistry, Memorial University of Newfoundland, St John's, Newfoundland, Canada.

Titles in the series

Nutrigenomics and Proteomics in Health and Disease: Food Factors and Gene Interactions
Editors: Yoshinori Mine, Kazuo Miyashita and Fereidoon Shahidi
ISBN 978-0-8138-0033-2

Functional Food Product Development
Editors: Jim Smith and Edward Charter
ISBN 978-1-4051-7876-1

Cereals and Pulses: Nutraceutical Properties and Health Benefits
Editors: Liangli Yu, Rong T Cao and Fereidoon Shahidi
ISBN 978-0-8138-1839-9

Functional Food Product Development

Edited by

Jim Smith and Edward Charter
Prince Edward Island Food Technology Centre
Charlottetown, Canada

WILEY-BLACKWELL
A John Wiley & Sons, Ltd., Publication

This edition first published 2010

Blackwell Publishing was acquired by John Wiley & Sons in February 2007. Blackwell's publishing programme has been merged with Wiley's global Scientific, Technical, and Medical business to form Wiley-Blackwell.

Registered office
John Wiley & Sons Ltd, The Atrium, Southern Gate, Chichester, West Sussex, PO19 8SQ, United Kingdom

Editorial offices
9600 Garsington Road, Oxford, OX4 2DQ, United Kingdom
2121 State Avenue, Ames, Iowa 50014-8300, USA

For details of our global editorial offices, for customer services and for information about how to apply for permission to reuse the copyright material in this book please see our website at www.wiley.com/wiley-blackwell.

Library of Congress Cataloging-in-Publication Data

Functional food product development / edited by Jim Smith and Edward Charter.
p. ; cm. – (Functional food science and technology)
Includes bibliographical references and index.
ISBN 978-1-4051-7876-1 (hardback : alk. paper) 1. Functional foods. 2. Food industry and trade.
I. Smith, Jim, 1953- II. Charter, Edward.
[DNLM: 1. Food Technology–methods. 2. Food–standards. 3. Food-Processing Industry–methods.
4. Nutritional Physiological Phenomena. WA 695 F9785 2010]
QP144.F85F853 2010
613.2–dc22

2009046210

A catalogue record for this book is available from the British Library.

Set in 10/12 pt Times by Aptara® Inc., New Delhi, India
Printed in Singapore by Markono Print Media Pte Ltd

1 2010

Contents

Preface

According to an August 2009 report from PricewaterhouseCoopers, the US market for functional foods in 2007 was US$27 billion. Forecasts of growth range between 8.5 and 20% per year or about four times that of the food industry in general. Global demand by 2013 is expected to be about US$100 billion. With this demand for new products comes a demand for product development and supporting literature for that purpose. There is a wealth of research and development going on in this area and much opportunity for commercialisation. This book provides a much-needed review of important opportunities for new products from many perspectives including those with in-depth knowledge of as yet unfulfilled health-related needs.

This book addresses functional food product development from a number of perspectives: the process itself, health research that may provide opportunities, idea creation, regulation; and processes and ingredients. It also features case studies that illustrate real product development and commercialisation histories. Written for food scientists and technologists, and scientists working in related fields, the book presents practical information for use in functional food product development. It is intended for use by practitioners in functional food companies and food technology centres and will also be of interest to researchers and students of food science.

Sections include New Technologies for Functional Food Manufacture, Functional Ingredients, Product Design and Regulation, Functional Foods and Health.

Within the text of the book, there are suggestions, ideas and clues for new functional food products; some are more obvious than others and some are closer to commercialisation than others, but numerous new products could result from the information contained herein. There is a large, growing market for unique functional food products that provide proven health benefits to consumers, and we hope that this book will play an important role in the creation of those products.

Jim Smith and Edward Charter

Contributors

S.L. Amaya-Llano
Programa de posgrado en Alimentos del Centro de la República, Universidad Autónoma de Querétaro, Querétaro, México

Juan-Carlos Arboleya
AZTI-Tecnalia, Food Research Division, Parque Tecnológico de Bizkaia, Astondo Bidea, Bizkaia, Spain

Mary Ann Augustin
Preventative Health, National Research Flagship, Food Science Australia, Werribee, Australia

Paula N. Brown
NHP Research Group, British Columbia Institute of Technology, Burnaby, British Columbia, Canada

Oscar Brunser
Laboratory of Microbiology and Probiotics, Institute of Nutrition and Food Technology, University of Chile, Santiago, Chile

M.P. Buera
University of Buenos Aires, Industry Department, School of Science, Ciudad Universitaria, Buenos Aires, Argentina

P.A. Ponce Cevallos
University of Buenos Aires, Industry Department, School of Science, Ciudad Universitaria, Buenos Aires, Argentina

Michael Chan
NHP Research Group, British Columbia Institute of Technology, Burnaby, British Columbia, Canada

Sylvia Cruchet
Laboratory of Microbiology and Probiotics, Institute of Nutrition and Food Technology, University of Chile, Santiago, Chile

Iñigo Martínez de Marañón
AZTI-Tecnalia, Food Research Division, Parque Tecnológico de Bizkaia, Astondo Bidea, Bizkaia, Spain

C. dos Santos
University of Buenos Aires, Industry Department, School of Science, Ciudad Universitaria, Buenos Aires, Argentina

Ans Eilander
Unilever R&D Vlaardingen, Vlaardingen, The Netherlands

B.E. Elizalde
University of Buenos Aires, Industry Department, School of Science, Ciudad Universitaria, Buenos Aires, Argentina

Jaouad Fichtali
Martek Biosciences Corporation, Winchester, KY, USA

P. Fustier
Agriculture and Agri-Food Canada, Saint-Hyacinthe, Quebec, Canada

Martin Gotteland
Laboratory of Microbiology and Probiotics, Institute of Nutrition and Food Technology, University of Chile, Santiago, Chile

Wlodzimierz Grajek
Department of Biotechnology and Food Microbiology, Poznan University of Life Sciences, Poznan, Poland

Peter J.H. Jones
Richardson Centre for Functional Foods and Nutraceuticals, University of Manitoba, Winnipeg, Manitoba, Canada

Jovana Kaludjerovic
Department of Nutritional Sciences, Faculty of Medicine, University of Toronto, Toronto, Ontario, Canada

Jerry W. King
Department of Chemical Engineering, University of Arkansas, Fayetteville, AR, USA

Daniel Lasa
Mugaritz Restaurant, Otzazulueta Baserria, Gipuzkoa, Spain

Beatrice Lau
Department of Nutritional Sciences, Faculty of Medicine, University of Toronto, Toronto, Ontario, Canada

Calvin London
Stirling Products Ltd, , Sydney, Australia

Christopher P.F. Marinangeli
Richardson Centre for Functional Foods and Nutraceuticals, University of Manitoba, Winnipeg, Manitoba, Canada

M.F. Mazzobre
University of Buenos Aires, Industry Department, School of Science, Ciudad Universitaria, Buenos Aires, Argentina

Anantha Narayana
Hindustan Unilever Research Centre, Bangalore, India

Idoia Olabarrieta
AZTI-Tecnalia, Food Research Division, Parque Tecnológico de Bizkaia, Astondo Bidea, Bizkaia, Spain

Anna Olejnik
Department of Biotechnology and Food Microbiology, Poznan University of Life Sciences, Poznan, Poland

Saskia Osendarp
Unilever R&D Vlaardingen, Vlaardingen, The Netherlands

Lech Ozimek
Department of Agricultural, Food and Nutritional Science, University of Alberta, Edmonton, Alberta, Canada

Federico Leighton Puga
Laboratorio de Nutricion Molecular, Centro de Nutricion Molecular y Enfermedades Cronicas, Facultad de Ciencias Biologicas, Universidad Catolica de Chile, Santiago, Chile

H.S. Ramaswamy
Agriculture and Agri-Food Canada, Saint-Hyacinthe, Quebec, Canada

Sandra M. Sacco
Department of Nutritional Sciences, Faculty of Medicine, University of Toronto, Toronto, Ontario, Canada

Luz Sanguansri
Preventative Health, National Research Flagship, Food Science Australia, Werribee, Australia

S.P.J. Namal Senanayake
Martek Biosciences Corporation, Winchester, KY, USA

Fereidoon Shahidi
Department of Biochemistry, Memorial University of Newfoundland, St. John's, Newfoundland, Canada

Anna Sip
Department of Biotechnology and Food Microbiology, Poznan University of Life Sciences, Poznan, Poland

Cai Song
Department of Biomedical Science, AVC, University of Prince Edward Island and NRC Institute for Nutrisciences and Health, Charlottetown, Prince Edward Island, Canada

Keerthi Srinivas
Department of Chemical Engineering, University of Arkansas, Fayetteville, AR, USA

A.R. Taherian
Agriculture and Agri-Food Canada, Saint-Hyacinthe, Quebec, Canada

Jyoti Kumar Tiwari
Unilever R&D Vlaardingen, Vlaardingen, The Netherlands

Inés Urquiaga
Laboratorio de Nutricion Molecular, Centro de Nutricion Molecular y Enfermedades Cronicas, Facultad de Ciencias Biologicas, Universidad Catolica de Chile, Santiago, Chile

Yanwen Wang
NRC Institute for Nutrisciences and Health, Charlottetown, Prince Edward Island, Canada

Wendy E. Ward
Department of Nutritional Sciences, Faculty of Medicine, University of Toronto, Toronto, Ontario, Canada

Jerzy Zawistowski
Food, Nutrition and Health, University of British Columbia, Vancouver, Canada

Part I

New technologies for functional food manufacture

1 Microencapsulation in functional food product development

Luz Sanguansri and Mary Ann Augustin

1.1 Introduction

Functional foods provide health benefits over and above normal nutrition. Functional foods are different from medical foods and dietary supplements, but they may overlap with those foods developed for special dietary uses and fortified foods. They are one of the fastest growing sectors of the food industry due to increasing demand from consumers for foods that promote health and well-being (Mollet & Lacroix 2007). The global functional food market, which has the potential to mitigate disease, promote health and reduce health care costs, is expected to rise to a value of US$167 billion by 2010, equating to a 5% share of total food expenditure in the developed world (Draguhn 2007).

Functional foods must generally be made available to consumers in forms that are consumed within the usual daily dietary pattern of the target population group. Consumers expect functional foods to have good organoleptic qualities (e.g. good aroma, taste, texture and visual aspects) and to be of similar qualities to the traditional foods in the market (Klont 1999; Augustin 2001; Kwak & Jukes 2001; Klahorst 2006). The demand for bioactive ingredients will continue to grow as the global market for functional foods and preventative or protective foods with associated health claims continues to rise. Over the last decade, there has been significant research and development in the areas of bioactive discovery and development of new materials, processes, ingredients and products that can contribute to the development of functional foods for improving the health of the general population.

New functional food products launched in the global food and drinks market have followed the route of fortification or addition of desirable nutrients and bioactives including vitamins, minerals, antioxidants, omega-3 fatty acids, plant extracts, prebiotics and probiotics, and fibre enrichments. Many of these ingredients are prone to degradation and/or can interact with other components in the food matrix, leading to loss in quality of the functional food products. To overcome problems associated with fortification, the added bioactive ingredient should be isolated from environments that promote degradation or undesirable interactions. This may be accomplished by the use of microencapsulation where the sensitive bioactive is packaged within a secondary material for delivery into food products. This chapter covers the microencapsulation of food components for use in functional food product formulations and how these components can be utilised to develop commercially successful functional foods.

Table 1.1 Food ingredients that have been microencapsulated

Types of ingredients
Flavouring agents (including sweeteners, seasonings and spices)
Acids, bases and buffers (e.g. citric acid, lactic acid and sodium bicarbonate)
Lipids (e.g. fish oils, milk fat and vegetable oils)
Enzymes (e.g. proteases) and microorganisms (e.g. probiotic bacteria)
Amino acids and peptides
Vitamins and minerals
Antioxidants
Polyphenols
Phytonutrients
Soluble fibres

1.2 Microencapsulation

Microencapsulation is a process by which a core, i.e. bioactive or functional ingredient, is packaged within a secondary material to form a microcapsule. The secondary material, known as the encapsulant, matrix or shell, forms a protective coating or matrix around the core, isolating it from its surrounding environment until its release is triggered by changes in its environment. This avoids undesirable interactions of the bioactive with other food components or chemical reactions that can lead to degradation of the bioactive, with the possible undesirable consequences on taste and odour as well as negative health effects.

It is essential to design a microencapsulated ingredient with its end use in mind. This requires knowledge of (1) the core, (2) the encapsulant materials, (3) interactions between the core, matrix and the environment, (4) the stability of the microencapsulated ingredient in storage and when incorporated into the food matrix and (5) the mechanisms that control the release of the core. Table 1.1 gives examples of cores that have been microencapsulated for use in functional food applications. The molecular structure of the core is usually known. However, information is sometimes lacking on how the core interacts with other food components, its fate upon consumption, its target site for action and in the case of a bioactive core, sometimes its function in the body after ingestion may also be unclear (de Vos *et al.* 2006).

1.2.1 Encapsulant materials

Depending on the properties of the core to be encapsulated and the purpose of microencapsulation, encapsulant materials are generally selected from a range of proteins, carbohydrates, lipids and waxes (Table 1.2), which may be used alone or in combination. The materials chosen as encapsulants are typically film forming, pliable, odourless, tasteless and non-hygroscopic. Solubility in aqueous media or solvent and/or ability to exhibit a phase transition, such as melting or gelling, are sometimes desirable, depending on the processing requirements for production of the microencapsulated ingredient and for when it is incorporated into the food product. Other additives, such as emulsifiers, plasticisers or defoaming agents, are sometimes included in the formulation to tune the final product's characteristics. The encapsulant material may also be modified by physical or chemical means in order to achieve the desired functionality of the microencapsulation matrix. The choice of encapsulant material is therefore dependent on a number of factors, including its physical and chemical

Table 1.2 Materials that have been used as encapsulants for food application

Encapsulant materials		
Carbohydrates	**Proteins**	**Lipids and waxes**
Native starches	Sodium caseinate	Vegetable fats and oils
Modified starches	Whey proteins	Hydrogenated fats
Resistant starches	Isolated wheat proteins	Palm stearin
Maltodextrins	Soy proteins	Carnauba wax
Dried glucose syrups	Gelatins	Bees wax
Gum acacia	Zein	Shellac
Alginates	Albumin	Polyethylene glycol
Pectins		
Carrageenan		
Chitosan		
Cellulosic materials		
Sugars and derivatives		

properties, its compatibility with the target food application and its influence on the sensory and aesthetic properties of the final food product (Brazel 1999; Gibbs *et al.* 1999).

The ability of carbohydrates to form gels and glassy matrices has been exploited for microencapsulation of bioactives (Reineccius 1991; Kebyon 1995). Starch and starch derivates have been extensively used for the delivery of sensitive ingredients through food (Shimoni 2008). Chemical modification has made a number of starches more suitable as encapsulants for oils by increasing their lipophilicity and improving their emulsifying properties. Starch that was hydrophobically modified by octenyl succinate anhydride had improved emulsification properties compared to the native starch (Bhosale & Singhal 2006; Nilsson & Bergenståhl 2007). Acid modification of tapioca starch has been shown to improve its encapsulation properties for β-carotene, compared to native starch or maltodextrin (Loksuwan 2007). Physical modification of starches by heat, shear and pressure has also been explored to alter its properties (Augustin *et al.* 2008), and the modified starch has been used in combination with proteins for microencapsulation of oils (Chung *et al.* 2008).

Carbohydrates used for microencapsulation of β-carotene, from sea buckthorn juice, by ionotropic gelation using furcellaran beads, achieved encapsulation efficiency of 97% (Laos *et al.* 2007). Interest in using cyclodextrins and cyclodextrin complexes for molecular encapsulation of lipophilic bioactive cores is ongoing, especially in applications where other traditional materials do not perform well, or where the final application can bear the cost of this expensive material. The majority of commercial applications for cyclodextrins have been for flavour encapsulation and packaging films (Szente & Szejtli 2004).

Proteins are used as encapsulants because of their excellent solubility in water, good gel-forming, film-forming and emulsifying properties (Kim & Moore 1995; Hogan *et al.* 2001). Protein-based microcapsules can be easily rehydrated or solubilised in water, which often results in immediate release of the core. Proteins are often combined with carbohydrates for microencapsulation of oils and oil-soluble components. In the manufacture of encapsulated oil powders, encapsulation efficiency was higher when the encapsulation matrix was a mixture of milk proteins and carbohydrates, compared to when protein was used alone (Young *et al.* 1993). Soy protein-based microcapsules of fish oil have been cross-linked using transglutaminase to improve the stability of the encapsulated fish oil (Cho *et al.* 2003). Protein-based hydrogels are also useful as nutraceutical delivery systems (Chen *et al.* 2006).

The release properties of protein-based hydrogels and emulsions may be modulated by coating the gelled particles with carbohydrates. A model-sensitive core, paprika oleoresin, was encapsulated in microspheres of whey proteins and coated with calcium alginate to modify the core's release properties (Rosenberg & Lee 2004). Whey protein-based hydrogels with an alginate coating altered the swelling properties of the gelled particles. The stability of these particles was increased at neutral and acidic conditions both in the presence and absence of proteolytic enzymes (Gunasekaran *et al.* 2007).

Lipids are generally used as secondary coating materials applied to primary microcapsules or to powdered bioactive cores to improve their moisture barrier properties (Wu *et al.* 2000). Lipids can also be incorporated in an emulsion formulation to form a matrix or film around the bioactive core (Crittenden *et al.* 2006).

The increasing demand for food-grade materials that will perform under the different stresses encountered during food processing has spurred the development of new encapsulant materials. Understanding the glass transition temperature of various polymers (e.g. proteins and carbohydrates) and their mixtures is also becoming important as this can influence the stability of the encapsulated core. The low water mobility and slow oxygen diffusion rates in glassy matrices can improve stability of bioactives (Porzio 2003). It is possible to exploit thermally induced interactions between proteins and polysaccharides and then to use the modified materials for encapsulation. Hydrogels formed by heat treatment of β-lactoglobulin – chitosan have been investigated, and it has been suggested that under controlled conditions these complexes may be useful for microencapsulation of functional food components (Hong & McClements 2007). Maillard reaction products formed by interactions between milk proteins and sugars or polysaccharides have been used as encapsulating matrices to protect sensitive oils and bioactive ingredients (Sanguansri & Augustin 2001).

1.2.2 Microencapsulation processes

Microencapsulation processes traditionally used to produce a range of microencapsulated food ingredients are listed in Table 1.3. A number of reviews give further details on microencapsulation technology in the food industry (Augustin *et al.* 2001; Gouin 2004). The choice of method used for microencapsulation depends on the properties of the core, the encapsulant materials and the requirements of the target food application. Figure 1.1 shows the structure of microencapsulated oil produced using three different microencapsulation processes.

Methods used for microencapsulation in the food industry have generally been adapted from technologies originally developed for the pharmaceutical industry. Mechanical processes use commercially available equipment to create and stabilise the microcapsules,

Table 1.3 Encapsulation processes used in the food industry

Mechanical processes	Chemical processes
Emulsification	Ionotropic gelation
Spray-drying	Simple coacervation
Fluidised-bed coating	Complex coacervation
Centrifugal extrusion	Solvent evaporation
Spinning disk	Liposomes
Pressure extrusion	Cyclodextrin complexation
Hot-melt extrusion	

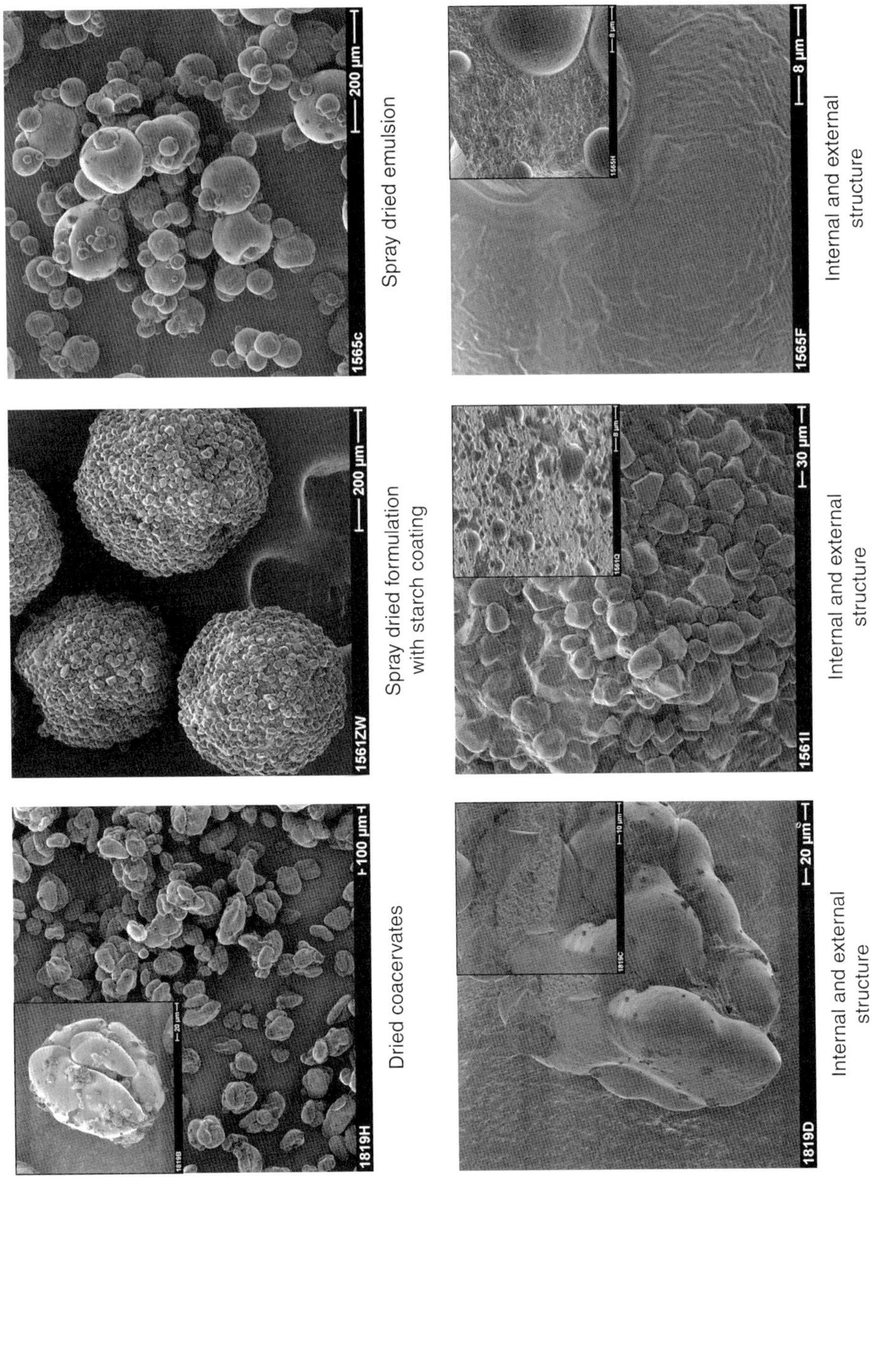

Fig. 1.1 Structure of microencapsulated oil produced using different microencapsulation processes.

whereas chemical processes capitalise on the possible interactions that can be promoted by varying the process conditions used to create the microcapsules.

Spray-drying is the most commonly used mechanical method for microencapsulation of bioactive food ingredients. Gharsallaoui *et al.* (2007) reviewed the use of spray-drying for the microencapsulation of food ingredients. It is efficient and cost-effective and uses unit processes and equipment readily available in most food processing plants. Spray-dried ingredients have reasonably good powder characteristics and good stability. Fluidised-bed coating is another mechanical process used for encapsulation of dry bioactive cores and ingredients. It consists of spraying an aqueous or solvent-based liquid coat onto the particles followed by drying. Dry particle coating of bioactive cores is an adaptation of the fluidised-bed coating technique that has been investigated by Ivanova *et al.* (2005) for microencapsulation of water-sensitive ingredients. The dry particle coating method avoids the use of aqueous or solvent-based coatings.

Of the different chemical microencapsulation processes available, only gelation and coacervation are widely used in the food industry. All current chemical methods are batch processes, although there is significant effort going into the development of continuous processes. Biopolymer–biopolymer supramolecular structures as complexes and coacervates may be formed under conditions where the two biopolymers carry opposite charges. These structures may have potential for controlled release and delivery of bioactives in foods (Turgeon *et al.* 2007; Livney 2008). The formation of native whey protein isolate-low methoxy pectin complexes by electrostatic interaction has potential for entrapment of water-soluble ingredients in acidic foods, as demonstrated by the entrapment of thiamine by Bedie *et al.* (2008).

Liquid emulsions may also be used as delivery systems in foods (Appelqvist *et al.* 2007; McClements *et al.* 2007). Oil-in-water emulsions are suitable for the delivery of lipids and lipid-soluble bioactives. Kinetically stable oil-in-water emulsions are made by homogenising a mixture of either an oil or an oil containing a lipid-soluble bioactive, with an aqueous solution containing the encapsulating material. Spontaneously formed, thermodynamically stable microemulsions may also be loaded with nutraceuticals and used as delivery systems. Garti and Amar (2008) have discussed the importance of understanding the nature of the microstructures and phase transitions in micro- and nanoemulsions for the effective delivery of nutraceuticals. Leal-Calderon *et al.* (2007) highlighted the need to understand the formulation and the design and characterisation of structured emulsions in order to control the release of bioactives in foods when ingested. Guzey and McClements (2006) explored ways of improving the release characteristics of conventional primary emulsions for controlled or triggered release delivery systems of bioactives, by developing multilayered emulsion formulations. Preparation of water-in-oil-in-water (w/o/w) emulsions by membrane filtration was explored by Shima *et al.* (2004) to encapsulate a model hydrophilic bioactive, with a view to protecting functional food ingredients for controlled release application.

1.2.3 Drivers for microencapsulation

The primary reasons for microencapsulation of food ingredient are to (1) protect the core from degradation during processing and storage, (2) facilitate or improve handling during the production processes of the final food application and (3) control release characteristics of the core, including its delivery to the desired site after ingestion.

Many bioactives (e.g. omega-3 oils, carotenes and polyphenols) need to be protected against degradation. For example, omega-3 oils are very susceptible to oxidation, leading to the development of off-flavours and off-odours. Microencapsulation protects the sensitive oils from exposure to oxygen, light and metal ions during processing and storage (Sanguansri & Augustin 2006). Protecting the core from degradation and from interactions with other food components can extend the shelf stability of the ingredient itself, as well as that of the final food product to which it is added.

Microencapsulation can facilitate or improve handling of ingredients during production processes used in the final applications. The conversion of a liquid ingredient into a powder offers significant convenience, as it is much easier to store, weigh and add a powdered ingredient, compared to its liquid version. Microencapsulation can aid in the addition and more uniform blending of bioactive ingredients into a food formulation. Bioactive ingredients in their pure or very concentrated forms are usually added in very low amounts (sometimes at ppm levels). Addition of a few milligrams or grams of ingredients into hundreds of kilos or tonnes of products can lead to uneven distribution within the food matrix, especially when ingredients are dry-blended. Microencapsulated forms with much lower payload can be used in these applications to facilitate a more homogeneous blending of these highly potent bioactive ingredients into food because the lower payloads provide a larger amount of the microencapsulated ingredient to be added to the mix into which it has to be blended.

Bioactive ingredients may require microencapsulation to improve or modify their functionality and release characteristics. Understanding the core, the final application, and the mechanism required to release the core is essential for effective design of the microcapsule's release characteristics. Different release characteristics can be achieved depending on the requirement, e.g. controlled, sustained or delayed release. Bioactive ingredients are often known to possess undesirable tastes and/or odours that require masking before they can be used successfully in food formulations. A significant challenge associated with nutraceutical ingredients is the need to mask bitterness and aftertaste (Anon 2006). With new developments in understanding the science of taste, the introduction of new bitterness blockers and sweetness potentiators in food formulations (McGregor 2004) can be combined within a microencapsulation system to allow controlled, delayed or sustained release of bioactives.

During the addition of bioactive ingredients into food, it is essential that both the bioactivity and the bioavailability are maintained to ensure that the bioactives achieve the desired function in the body. When direct addition of the bioactive could compromise its bioavailability, it needs to be protected by microencapsulation. The protection of the core from the acidic pH of the stomach during transit through the gastrointestinal (GI) tract may potentially enable more efficient delivery of the bioactive to the target site in the body and may also reduce the dosage required to achieve the heath benefits.

Advances in the development of microencapsulation technology for food applications have been driven by the need for (1) the core to be encapsulated, (2) new and alternative materials that are cost-effective encapsulants and (3) materials which will withstand the processes widely used in the food industry. More recent developments in microencapsulation technologies for food applications have focused on applying the technology to more cost-effective food-grade encapsulant materials and processes available in the food industry. The need for controlled release and delivery of bioactive food ingredients to target sites in the body continues to drive other new developments. Converting stable microcapsule formulations (emulsions, dispersions, suspensions, coacervates) into powders is still the preferred option for production of microencapsulated bioactive ingredients, as it offers more convenience and flexibility. An understanding of how these formulations will behave during the drying process and on reconstitution is critical to the success of powdered preparations.

With the primary reason or purpose of microencapsulation being clearly identified, other important factors need to be seriously considered to ensure proper selection of encapsulant materials and processes that are cost-effective, practical and scalable. Important considerations during the development of microencapsulated products include (1) core properties – e.g. chemical structure, solubility and stability, (2) product format – e.g. liquid or powder format depending on final application, (3) physical properties of the microencapsulated ingredient – e.g. particle size, bulk density and colour, (4) payload – i.e. amount of bioactive loading in the microcapsule, (5) release trigger mechanism – e.g. dissolution, pressure, heat and shear, (6) storage conditions and shelf-life requirements – e.g. refrigerated or ambient storage and (7) legal and regulatory requirements for addition into food in the country of its application. From a commercial perspective, there is the additional factor of material and production costs and whether the final food product can bear the additional cost of using a microencapsulated ingredient.

1.3 Microencapsulated food ingredients

There are several technical challenges in developing functional ingredients for incorporation into foods. They must satisfy the sensory demands of the consumers and ensure that the bioactive can be delivered to specific sites in the GI tract to exert the desired health benefit. Microencapsulation has been applied to a number of food ingredients to develop them into tailor-made bioactive ingredients (Augustin & Sanguansri 2008).

The increasing number of microencapsulated food ingredient launches has been the result of more creative translation and adaptation of microencapsulation techniques originally developed in the pharmaceutical industries. New encapsulant materials and more cost-effective formulations and processes have enabled the food industry to develop these new ingredients with added value and functionality. In more recent years, the addition of microencapsulated ingredients into a wider range of food products ensures that it does not significantly affect the cost of the final food product. This is a significant issue as food has very low profit margins compared to pharmaceuticals.

1.3.1 Vitamins and minerals

Fortification with vitamins and minerals is often challenging due to their susceptibility to degrade during processing and storage and to react with other components in the food system. Vitamins and minerals are generally sensitive to temperature, moisture, light and pH, and their potency is often compromised by their reaction with other ingredients or premature release.

Vitamins and minerals are added to a range of food products for the following reasons: (1) to replace those that are lost during processing and storage; (2) to meet special nutritional needs, e.g. for infants and elderly; and (3) to prevent disease in specific consumer or at-risk groups. Traditionally, higher levels than that are required in the end product have been added to overcome losses during processing and storage. These high overages may be avoided by using microencapsulated forms.

For water-soluble vitamins (e.g. vitamins B and C) and minerals (e.g. iron and calcium), spray-drying, spray chilling, fluidised-bed coating and spinning disk coating have been used to manufacture dry powder microcapsules. Where liquid microcapsule formats are preferred, microencapsulation in liposomal delivery systems can be used. There is also the possibility of entrapping water-soluble vitamins in double emulsions. Fechner *et al.* (2007) demonstrated that vitamin B12 in the inner phase of an oil/water/oil emulsion stabilised by caseinate–dextran conjugates, instead of pure protein, reduced the release of the vitamin under acidic conditions. For lipid-soluble vitamins (e.g. vitamins A, D, E and K) and provitamin A (β-carotene), stable emulsion formulations or spray-dried emulsions are commonly used as delivery systems. Emulsion-based systems are often used for delivery of lipid-soluble bioactives (McClements *et al.* 2007). However, where there are specific interactions between hydrophobic bioactives and a protein, an aqueous-based system may be exploited. Semo *et al.* (2007) demonstrated that casein micelles were useful for delivery of vitamin D2.

Microencapsulation has benefits when used for delivery of iron and calcium in foods. Direct addition of iron into foods may reduce its bioavailability through interaction with tannins, phytates and polyphenols. Free iron is also known to catalyse the oxidation of fats, vitamins and amino acids. These interactions can affect the sensory characteristics of the final food formulation, as well as decrease the nutritional value of the food due to iron-induced catalysis of deteriorative reactions. Many of these limitations of direct addition of iron may be overcome by microencapsulation. Other microencapsulation technologies used for encapsulation of iron include liposomal delivery systems and application of lipid coats by fluidised-bed coating (Xia & Xu 2005). Molecular inclusion of iron using cyclodextrins may also be used in its delivery (Leite *et al.* 2003).

The interaction of calcium with proteins can cause unwanted coagulation or precipitation of the protein, especially in calcium-fortified protein beverages. Calcium is naturally present in dairy products, but there is interest in fortifying other protein products with calcium, such as soy protein beverages. Calcium fortification of protein-based beverages may be achieved with the addition of calcium-chelating agents; however, this may result in an undesirable taste when high levels of calcium fortification are desired. Microencapsulation of calcium can prevent its negative interaction with other food components (e.g. soy proteins) in the food environment. A liposomal delivery system has also been examined for this application (Hirotsuka *et al.* 1984).

1.3.2 Functional fatty acids

Functional fatty acids, particularly docosahexaenoic acid, eicosapentaenoic acid, α-linolenic acid and conjugated linoleic acid, have attracted significant attention due to their potential health benefits (Ohr 2005). Emulsion-based technologies and spray-drying are currently the most common approaches employed for microencapsulation and delivery of functional fatty acids into food (Sanguansri & Augustin 2001; McClements *et al.* 2007).

Omega-3 fatty acids are highly susceptible to oxidation and have an inherent fishy taste and odour. Therefore, most food applications of omega-3 fatty acids require microencapsulation for protection from oxidation and to mask the fishy taste and odour. Significant research has been carried out on microencapsulation of omega-3 fatty acids. An increasing number of food companies are developing new functional food products containing omega-3 fatty acids. This increase in the number of food products launched containing omega-3 fatty acids has also been driven by the qualified health claims that were allowed by Food and Drug Administration (FDA) in 2004. Technologies that have been successfully used to encapsulate omega-3 oils include emulsification and spray-drying (Sanguansri & Augustin 2001), coacervation (Wu *et al.* 2005), cyclodextrin complexation and liposomal preparations (Tanouchi *et al.* 2007).

1.3.3 Probiotics

Probiotics are live microorganisms that must remain alive during processing, storage and gastric transit to fulfil their desired function in the body (Mattila-Sandholm *et al.* 2002). Much clinical data have been accumulated to support the role of probiotics in human health by benefiting the immune system, strengthening the mucosal barrier and suppressing intestinal infection (Saarela *et al.* 2002). This has driven interest in adding probiotics to a wider range of food products, other than traditional fermented dairy products such as yoghurt. As probiotics are sensitive to heat and moisture, keeping them alive during food processing and storage is not easy. Even in fermented dairy product applications, the survival of probiotics during storage still remains a challenge for the industry.

Processes that have been used to encapsulate probiotics include spray coating, spray-drying, extrusion, emulsification and gel particle technologies. Of these technologies, the technique most widely investigated by researchers involves the use of polysaccharides to form gelled particles (Krasaekoopt *et al.* 2003; Anal & Singh 2007). However, the use of gelled particles for microencapsulation of probiotics has not been widely adopted by commercial companies, as it is a batch process. The use of alginate–chitosan microcapsules has also been explored to improve the mechanical strength of the capsules to survive *in vitro* digestion (Urbanska *et al.* 2007). The application of a lipid coating by a fluid-bed technique has also been used for probiotic encapsulation (Lee & Richardson 2004). Probiotics encapsulated in lipid-based materials are used, in a limited range of food products, with varying degrees of success. The application of high-melting-point lipids and waxes allows protection of probiotics from high-moisture environments and thermal protection below the melting point of the coat. Starch-based encapsulation was also explored by Lahtinen *et al.* (2007), but their results showed no effect on improving the viability of *Bifidobacterium longum* strains.

Spray-drying has always been an attractive process for production of powdered food ingredients because it is a continuous, high-volume and cost-effective process. A number of researchers have explored spray-drying for production of probiotic microcapsules with varying degrees of success (Desmond *et al.* 2002; Ananta *et al.* 2005; Anal & Singh 2007; Su *et al.* 2007). The most important step still remains the selection and formulation of an encapsulant that can protect the probiotics during drying. A novel microencapsulation technology using protein–carbohydrate conjugate in the matrix provided significant protection to probiotic bacteria during spray-drying, during exposure to acidic pH and during non-refrigerated storage at low to intermediate water activity (Crittenden *et al.* 2006). The use of appropriate materials and process conditions applied during microencapsulation has the potential to enable the addition of probiotics to a much wider range of food products with intermediate water activity which do not require refrigeration.

1.3.4 Phytochemicals

Phytochemicals are biologically active plant chemicals, with increasing evidence that they can reduce the risk of chronic diseases (Hasler 1998). Ingredients claimed to be rich in phytochemicals are extracted from plant sources. Once isolated from their natural environment, these bioactive ingredients generally require microencapsulation to stabilise the active component and mask undesirable tastes, colours and odours. The phytochemicals of interest to the food industry include phytosterols, tocopherols, carotenoids, coenzyme Q10, curcumin, garlic extracts and polyphenols (e.g. resveratrol).

Resveratrol is a naturally occurring non-flavonoid polyphenolic compound present in plants such as grapes, berries and peanuts (Halls & Yu 2008), as well as in cocoa and chocolate

(Counet *et al.* 2006). Resveratrol is photosensitive and benefits from microencapsulation to maintain its stability when added to food products. Shi *et al.* (2008) have shown that encapsulation of resveratrol in yeast cells can offer protection and enhance its stability as an ingredient. The use of chitosan–alginate coacervates as an encapsulant has also exhibited potential for preparation of encapsulated powder ingredients from aqueous (water-soluble) antioxidant plant extracts (Deladino *et al.* 2008).

The use of natural fruit fibres as encapsulating agents for the microencapsulation and spray-drying of sticky bioactive extracts (*Hibiscus sabdariffa*) has been explored by Chiou and Langrish (2007). Extracts containing curcumin have been encapsulated using commercially available lecithin to form liposomes by homogenisation or microfluidisation (Takahashi *et al.* 2007). The delivery of curcumin through oil-in-water nanoemulsions has been shown to enhance its anti-inflammatory activity in animal tests (Wang *et al.* 2008). Szente *et al.* (1998) demonstrated that the stability of curcumin and carotenes is enhanced by molecular encapsulation using cyclodextrins.

1.3.5 Proteins, amino acids, peptides and enzymes

Proteins have traditionally been encapsulated for pharmaceutical applications (Putney 1998). The demand for more protein in food and beverages is on the rise (Sloan 2004). Whey, casein and soy proteins are commonly used in high-protein food formulations either in their native or hydrolysed forms.

Protein-derived peptides and amino acids are also being isolated from their source to enable addition at the correct dosage required for physiological health functions. The direct addition of these components into food and beverage formulations can result in an undesirable bitter taste and astringency. Encapsulation of casein hydrolysates in lipospheres has been found to reduce the bitterness (Barbosa *et al.* 2004).

Encapsulation may also be used to preserve the activity of enzymes. Components in garlic have also been shown to offer beneficial health effects (Gorinstein *et al.* 2007), and microencapsulation of garlic powder results in protection of alliinase activity, which improves the ratio of alliin to allicin conversion under *in vitro* conditions (Li *et al.* 2007).

1.3.6 Fibre

The trend of adding dietary fibres to food and beverage formulations that traditionally do not contain these fibres is increasing due to the increasing evidence of health benefits of high-fibre diets. Examples of dietary fibres for which the FDA has allowed health claims are β-glucan from oats and psyllium fibre. β-Glucan, a cholesterol-lowering soluble fibre, shown to reduce the risk of heart disease was allowed an FDA health claim in 1997. Later, in 1998, the FDA extended the health claim for soluble fibre to psyllium fibre. Other dietary fibres added to food and beverage formulations include indigestible gums, polysaccharides, oligosaccharides and lignins (Prosky 1999).

High levels of fibre need to be added in the final food formulation in order to make a health claim. The problems associated with the addition of high levels of dietary fibres to food and beverages are the unpalatability of the high-fibre ingredients and the significant effects they have on the viscosity of the final product. This has resulted in the development of expensive, refined fibre ingredients, e.g. polydextrose (Sunley 1998). Microencapsulation can minimise palatability problems as well as minimise water absorption during formulation and processing. Much cheaper sources (e.g. indigestible gums) can also be added at a much higher levels if the fibre in food formulations is encapsulated with materials that can reduce hydration and water absorption during processing. Chito-oligosaccharide, as a functional ingredient, offers a range of health benefits; however, direct addition to milk can affect its flavour and colour. Microencapsulation of chito-oligosaccharide with polyglycerol monostearate, as explored by Choi *et al.* (2006), reduced its adverse effects on the physicochemical or sensory properties when added to milk.

1.4 Development of microencapsulated ingredients

1.4.1 The approach

During the development of functional foods using microencapsulated food ingredients, the selection of ingredients and processes was traditionally based on empirical approaches. Ubbink and Kruger (2006) have suggested that an alternative concept is to use a retro-design approach that relies more on a fundamental understanding of the required performance of the ingredient in a complex food environment. This approach encompasses an understanding of the effects of processing and the factors controlling the chemical and physical events that govern the stability and release properties of a microencapsulated product; however, the test of whether a microencapsulation system is suitably tailored for its end product application is its acceptance in the marketplace. The route from concept to acceptance of functional foods by consumers has many stages and requires input from scientists, technologists, nutritionists and an understanding of the regulatory processes (Jones & Jew 2007). Our own program of research in designing microencapsulated ingredients has utilised multidisciplinary expertise, involving chemistry, physics, food science and process engineering, with the regulatory and market requirements in mind to minimize or avoid issues during scale-up and commercialisation. This approach ensures that both the food and ingredient manufacturers' requirements are met while consumers' demands are also considered during the development. The final product application must be the focus of the microencapsulated product development in order that the core is protected from various stresses during incorporation into the final product. It is important to ensure that when microencapsulation is used to deliver active ingredients into foods, it provides a simple, efficient and cost-effective solution compared to direct addition of bioactives.

1.4.2 Product and process developments

Understanding the fundamental science of the core, as well as a good knowledge of the materials and processes available, is a requirement for the process of developing a successful product. The stages in the development of a microencapsulated product from bench scale product concept to a commercial product acceptable to consumers in final food applications are shown in Figure 1.2. In designing a cost-effective and tailor-made microcapsule suitable for its intended use (i.e. the final food product application), the final product format (liquid or dry) and the market (size and value) need to be identified at the outset. These factors will significantly influence the choice of materials, formulation and process that can be employed. At this stage, the physical performance and characteristics, core stability and possible interactions with other ingredients during formulation and process should be tested. A few iterations of changes to the initial formulation may be required until reasonable product

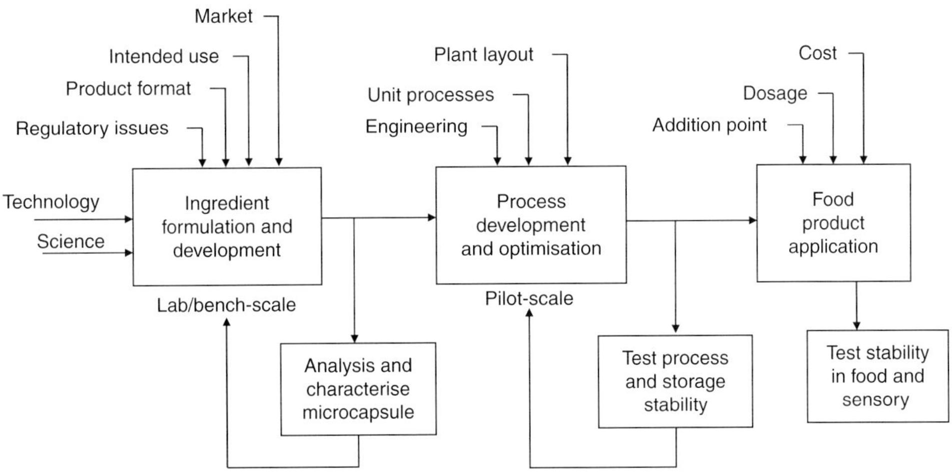

Fig. 1.2 Microencapsulated product development process.

properties are achieved at the laboratory scale. Once the formulation and desirable product properties are established, the next step is to develop a scalable process.

When considering processes for manufacture of microencapsulated food ingredients, the ability to use standard unit processes available in a conventional food processing operation is desirable. Their use will minimise future problems and assist in the commercial scale-up production of the microcapsules. During the scaling up of the process, the product specifications of the microencapsulated ingredient need to be clearly defined, as this will dictate the type of equipment and process conditions used during manufacture. For a powdered microencapsulated ingredient, these include the colour, particle size, bulk density, moisture content, payload, sensory aspects and other physical characteristics required in the final application. For a liquid (emulsion) microencapsulated ingredient, these include total solids concentration, viscosity, colour or clarity (if required), particle size, storage conditions and stability requirements. Sensory evaluation and storage stability trials of the final microencapsulated product need to be carried out during scale-up to assess consumer acceptability. Some minor formulation and process optimisation may be required at this final stage to achieve a product with the least production costs.

During scale-up, the final product performance during processing, the stability of the core and of the microcapsule under different processing conditions need to be fully established to define the conditions and the stage of addition during the manufacture of the final food application. The long-term stability of the microencapsulated ingredient itself also needs to be established to ensure that the ingredient stability equals or exceeds that of the final food product to which it is being added.

1.5 Delivery of microencapsulated ingredient into functional foods

1.5.1 Functional food product development

Diet has been a major focus of public health strategies aimed at maintaining optimum health throughout life stages. Nutrients and bioactive compounds (also called nutraceutical ingredients) which have shown potential in preventing or ameliorating the effect of major diseases (e.g. some types of cancer, cardiovascular disease, neurodegenerative disease and eye disorders) have driven the interest in developing functional foods for special health and dietary uses. The FDA's authorisation of qualified health claims for a number of ingredients, when used at specific levels, has helped accelerate the market for functional foods and to raise consumer awareness of several nutraceutical ingredients, e.g. omega-3 fatty acids, dietary fibre, plant sterols and soy protein. Microencapsulation technologies, through the use of appropriate formulations and processing strategies, have the potential to deliver a single bioactive or a cocktail of bioactives (Champagne & Fustier 2007). Functional food product launches with specific target health categories have continued to increase in the last decade. Functional health claims have been primarily focused on gut health, heart health, immune function, bone health and weight management.

Functional food ingredients designed to enhance GI tract health include probiotics, dietary fibre and prebiotics, and bioactive plant metabolites (e.g. phytochemicals such as polyphenols). Some of these ingredients have a role in gut fermentation, and by influencing the microflora composition and fermentation metabolites, they consequently contribute to both local and systemic effects in the body (Puupponen-Pimia *et al.* 2002). Other bioactive ingredients, such as fish oil (omega-3), polyphenols (resveratrol) and short-chain fatty acids (butyric acid), have been investigated and shown to be beneficial for gut health and as chemoprotective and chemopreventive agents against colon cancer (Schneider *et al.* 2000; Dwivedi *et al.* 2003; Orchel *et al.* 2005; Stehr & Heller 2006; Athar *et al.* 2007). The benefits of these gut health-promoting ingredients may be more effectively utilised by the general population if they are added into food products without affecting their shelf-life and sensory properties.

Microencapsulation has been used to assist the delivery of these ingredients into food, to stabilise and control their release during GI transit and to enhance their desired function in the body. A microencapsulation technology has been developed to protect these bioactives during processing and storage, as well as to target the release of the bioactive to specific sites in the GI tract (Augustin *et al.* 2005).

Heart health has been a major emphasis for many new products around the globe. As consumers continue to look for more ways to lower cholesterol and lessen their risk of heart-related illnesses, food manufacturers have continued to develop functional food products for this category. Dairy, beverage and bakery products are the top three categories with the addition of plant sterols, omega-3 fatty acids, peptides and whole grains being just a few examples of ingredient focus in heart-healthy food product developments.

Of the mainstream functional food product categories available commercially, dairy products accounted for about 40% of total functional food sales, followed by cereal products, beverages, fats and oils, soya products, bakery, eggs, and others (Watson *et al.* 2006). In this respect, where the consumption of functional foods is promoted as a fundamental way to proactively prevent or delay the onset of the disease, the ability to target the release and delivery of the bioactives to a specific site in the body and the bioavailability of the nutrients or bioactive compounds when they are released at the target site are more important than the amount originally present in the food (Parada & Aguilera 2007).

Microencapsulation is a logical solution for delivery of bioactives into functional foods as it can protect the bioactive during GI transit, until it reaches the target site in the body, as well as enhance its bioavailability when it is released. It also offers other advantages such as reduced dosage and overages during formulation, resulting in reduced ingredient cost during production.

1.5.2 Major food categories

Successful functional food product development in mainstream food categories requires special consideration as there is usually little room for reformulation and process modification as a result of adding the new active ingredient. This means that the ingredients used in the production of the microencapsulated ingredient must already be on the product label, and the microencapsulated ingredient must survive the processes that the product has to go through without affecting its sensory properties.

1.5.2.1 *Dairy products*

Functional dairy products account for 42.9% of the functional food market (Watson *et al.* 2006). Dairy products have been the most popular delivery vehicles for a number of functional and healthy ingredients, from vitamin and mineral fortification to addition of bioactives to promote health benefits. As milk and dairy products are a normal part of our daily diet, in all life stages, any new product launched can be expected to gain some market share. Much higher levels of vitamins and minerals have been added to dairy products in recent years. Omega-3 fatty acid fortification has also been popular despite the challenges in achieving acceptable flavour profiles in the final product. Addition of chito-oligosaccharide to milk has also been investigated by Choi *et al.* (2006).

1.5.2.2 *Cereal products*

Healthy bars and cereal products account for 19.4% of the functional food market (Watson *et al.* 2006). This category is the second most popular delivery vehicle in a number of functional ingredients for a number of reasons, e.g. market size, convenient format, easier to add to formulations and presence of ingredients that can mask unpleasant flavours.

1.5.2.3 *Beverages*

Functional beverages are the fastest growing product category for delivery of a range of functional ingredients. These currently account for 14.4% of the functional food and beverage market (Watson *et al.* 2006). The US market for fortified/functional beverages is expected to reach US$29 billion for standard beverages and US$815 million for dairy beverages by 2011 (Fuhrman 2007). Vitamin- and mineral-enriched drinks (e.g. with added calcium and vitamin C) are among the most popular, followed by weight-control beverages with added protein.

1.5.2.4 *Fats and oils products*

The fats and oils market accounts for 11.8% of the functional food market (Watson *et al.* 2006). In 2005, the global omega-3 ingredient market was worth over US$700 million (Haack 2007), and by 2010, the global market for omega-3 oils is expected to be worth US$1.2 billion (Lavers 2007). The development of spreads with cholesterol-lowering phytosterols, healthy oils, healthy spreads, sauces and dips with added nutraceutical ingredients is also increasing.

1.5.2.5 *Bakery products*

Bakery product launches containing functional ingredients account for about 1.7% of the functional food market (Watson *et al.* 2006); however, the use of microencapsulated ingredients in bakery products has applications beyond the addition of bioactive ingredients. Microencapsulated ingredients used for bakery applications include leavening agents, sweeteners, antimicrobial agents, dough conditioners and flavours. These ingredients are widely used in commercial baking operations where high volumes of dough and batter pre-mixes are prepared for further distribution. The development of microencapsulated ingredients for bakery applications has additional challenges, such as protection during high-shear and high-temperature processing. The coating materials used for bakery applications include fats and waxes. Processes used for bakery ingredient applications include hot-melt coating (fluid-bed technology), spray chilling and high-pressure congealing. New launches in functional bakery products have seen the addition of extra vitamins (vitamins A, C and E) and minerals (calcium and iron), long-chain polyunsaturated fatty acids (omega-3 and omega-6) and soluble fibres.

1.5.3 Factors affecting use of microencapsulation in the market

Success in translating research to commercial products has significant challenges, especially in stabilising and masking any undesirable tastes and odours of bioactive ingredients being added, as well as maintaining the overall sensory quality of the final food product (Hargreaves 2006). Microencapsulation has been employed as a technology that can minimise, if not solve, these challenges, and it also offers the possibility of developing tailor-made ingredients for specific applications. Important issues to consider for successful delivery of microencapsulated ingredients into commercial food products are shown in Table 1.4.

The trend of developing and using microencapsulated ingredients has increased significantly in the last decade as more cost-effective materials and production processes suitable for food applications have developed. Microencapsulated ingredients are used in functional food product formulations to improve nutritional content, to replace nutrients lost during processing (fortification) and to add other bioactive ingredients with known healthy benefits, without changing the sensory characteristics of the final food product.

1.6 Conclusion

Microencapsulation technology holds promise for the successful delivery of bioactive ingredients into functional foods, and has the potential to enhance the functionality of bioactive ingredients, thus maximising the health benefits available to consumers from these foods. Microencapsulation can offer significant advantages for improved delivery and protection of bioactive ingredients in food, which would not have been possible by direct addition.

New developments in a range of microencapsulation technologies continue to address different functionality challenges that occur when formulating bioactive ingredients into functional foods (Sunley 1998; Pszczola 2005). Opportunities for use of microencapsulation in the food industry continue to grow as greater demands are being made on the integrity of the capsules to control the release and delivery of the core material at a specific time during digestion and to a specified site in the body (Champagne & Fustier 2007). This often requires tailor-made microencapsulated ingredients that are fit for this purpose to be individually developed to take into account the final food application and format for delivery of the bioactive ingredients.

Table 1.4 Important issues to consider for successful delivery of microencapsulated bioactives into commercial food products

Important issues	Action or questions to ask
Regulatory standard	• Check the regulatory standards in each country for addition of bioactives. • Can the bioactive be added to the chosen food? • Are the ingredients used as encapsulants allowed in the chosen food? • What levels are required if there is to be a health claim?
Food product application format	• What is the format of the final product chosen? ○ For a powder, blending applications require good control of particle size, moisture and bulk density ○ For a liquid, rehydration and redispersion behaviour of powdered encapsulated bioactives is important
Protection and release characteristics	• What processing stresses has the ingredient to survive during incorporation into the food? • Under what conditions or in response to what trigger is the bioactive released?
Stage or point of addition	• Is the food manufacturing plant set-up automated? • At what point during production will the ingredient be added? • Will the ingredient be added using an automated process or manually?
Interaction with other ingredients	• Is there a need to avoid interaction of the bioactive with other ingredients in the final food during processing and storage?
Final product characteristics	• What are the storage conditions and shelf-life of the final food product? • What flavour characteristics are present in the food chosen? • Does the food product have a delicate flavour or strong flavour that can mask some undesirable taste and aroma?

Acknowledgements

The authors thank Christine Margetts for contributing to the sourcing of literatures and useful comments.

References

Anal, A.K. & Singh, H. (2007). Recent advances in microencapsulation of probiotics for industrial applications and targeted delivery. *Trends in Food Science and Technology*, **18**, 240–251.

Ananta, E, Volker, M. & Knorr, D. (2005). Cellular injuries and storage stability of spray dried *Lactobacillus rhamnosus* GG. *International Dairy Journal*, **15**, 399–409.

Anon (2006). Taste masking functional foods. *Nutraceutical Business and Technology*, **2**(1), 26–29.

Appelqvist, I.A.M., Golding, M., Vreeker, R. & Zuidam, N.J. (2007). Emulsion as delivery systems in foods. In: *Encapsulation and Controlled Release Technologies in Food Systems*. Lakkis, J.M. (ed.), Blackwell, Ames, IA, pp. 1–80.

Athar, M., Back, J.H., Tang, X., Kim, K.H., Kopelovich, L., Bickers, D.R. & Kim, A.L. (2007). Resveratrol: a review of preclinical studies for human cancer prevention. *Toxicology and Applied Pharmacology*, **224**, 274–283.

Augustin, M.A. (2001). Functional foods: an adventure in food formulation. *Food Australia*, **53**, 428–432.
Augustin, M.A. & Sanguansri, L. (2008). Encapsulation of bioactives. In: *Food Materials Science – Principles and Practice*. Aguilera, J.M. & Lillford, P.J. (eds), Springer, New York, pp. 577–601.
Augustin, M.A., Sanguansri, L. & Head, R. (2005). GI tract delivery systems. WO 2005/048998A1.
Augustin, M.A., Sanguansri, L., Margetts, C. & Young, B. (2001). Microencapsulation of food ingredients. *Food Australia*, **53**, 220–223.
Augustin, M.A., Sanguansri, P. & Htoon, A. (2008). Functional performance of a resistant starch ingredient modified using a microfluidiser. *Innovative Food Science and Emerging Technologies*, **9**, 224–231.
Barbosa, C.M.S., Morais, H.A., Delvivo, F.M., Mansur, H.S., De Oliveira, M.C. & Silvestre, M.P.C. (2004). Papain hydrolysates of casein: molecular weight profile and encapsulation in lipospheres. *Journal of the Science of Food and Agriculture*, **84**, 1891–1900.
Bedie, G.K., Turgeon, S.L. & Makhlouf, J. (2008). Formation of native whey protein isolate-low methoxy pectin complexes as a matrix for hydro-soluble food ingredient entrapment in acidic foods. *Food Hydrocolloids*, **22**, 836–844.
Bhosale, R. & Singhal, R. (2006). Process optimization for the synthesis of octenyl succinyl derivative of waxy corn and amaranth starches. *Carbohydrate Polymers*, **66**, 521–527.
Brazel, C.S. (1999). Microencapsulation: offering solutions for the food industry. *Cereal Foods*, **44**(6), 388–393.
Champagne, C.P. & Fustier, P. (2007). Microencapsulation for the improved delivery of bioactive compounds into foods. *Current Opinion in Biotechnology*, **18**, 184–190.
Chen, L., Remondetto, G.E. & Subirade, M. (2006). Food proteins-based materials as nutraceutical delivery systems. *Trends in Food Science and Technology*, **17**, 272–283.
Chiou, D. & Langrish, T.A.G. (2007). Development and characterisation of novel nutraceuticals with spray drying technology. *Journal of Food Engineering*, **82**, 84–91.
Cho, Y.H., Shim, H.K. & Park, J. (2003). Encapsulation of fish oil by an enzymatic gelation process using transglutaminase cross-linked proteins. *Journal of Food Science*, **68**, 2717–2723.
Choi, H.J., Ahn, J., Kim, N.C. & Kwak, H.S. (2006). The effects of microencapsulated chitooligosaccharide on physical and sensory properties of the milk. *Asian Australasian Journal of Animal Sciences*, **19**, 1347–1353.
Chung, C., Sanguansri, L. & Augustin, M.A. (2008). Effects of modification of encapsulant materials on the susceptibility of fish oil microcapsules to lipolysis. *Food Biophysics*, **3**, 140–145.
Counet, C., Callemien, D. & Collin, S. (2006). Chocolate and cocoa: new sources of trans-resveratrol and trans-piceid. *Food Chemistry*, **98**(4), 649–657.
Crittenden, R., Weerakkody, R., Sanguansri, L. & Augustin, M.A. (2006). Synbiotic microcapsules that enhance microbial viability during nonrefrigerated storage and gastrointestinal transit. *Applied and Environmental Microbiology*, **72**, 2280–2282.
de Vos, W.M., Castenmiller, J.J.M., Hamer, R.J. & Brummer, R.J.M. (2006). Nutridynamics – studying the dynamics of food components in products and in the consumers. *Current Opinion in Biotechnology*, **17**, 217–225.
Deladino, L., Anbinder, P.S., Navarro, A.S. & Martino, M.N. (2008). Encapsulation of natural antioxidants extracted from *Ilex paraguariensis*. *Carbohydrate Polymers*, **71**, 126–134.
Desmond, C., Stanton, C., Fitzgerald, G.F., Collins, K. & Ross, R.P. (2002). Environmental adaptation of probiotic lactobacilli towards improvement of performance during spray drying. *International Dairy Journal*, **12**, 183–190.
Draguhn, T. (2007). Health and nutrition cereal bars (global functional food market forecast). *Kennedy's Confection*, **March**, 12–14.
Dwivedi, C., Muller, L.A., Goetz-Parten, D.E., Kasperson, K. & Mistry, V.V. (2003). Chemopreventive effects of dietary mustard oil on colon tumor development. *Cancer Letters*, **196**, 29–34.
Fechner, C., Knoth, A., Scherze, I. & Muschiolik, G. (2007). Stability and release of double-emulsions stabilised by caseinsate-dextran conjugates. *Food Hydrocolloids*, **21**, 943–951.
Fuhrman, E. (2007). What can your beverage do for you? (US functional beverage market). *Beverage Industry*, **98**(8), 55–56.
Garti, N. & Amar, Y. (2008). Micro- and nano-emulsions for delivery of functional food ingredients. In: *Delivery and Controlled Release of Bioactives in Foods and Nutraceuticals*. Garti, N. (ed.), Woodhead, Cambridge, UK, pp. 149–183.
Gharsallaoui, A., Roudaut, G., Chambin, O., Voilley, A. & Saurel, R. (2007). Applications of spray-drying in microencapsulation of food ingredients: an overview. *Food Research International*, **40**, 1107–1121.

Gibbs, B.F., Kermasha, S., Alli, I. & Mulligan, C.N. (1999). Encapsulation in the food industry: a review. *International Journal of Food Science and Nutrition*, **50**, 213–224.

Gorinstein, S., Jastrzebski, Z., Namiesnik, J., Lentowicz, H., Leontowicz, M. & Trakhtenberg, S. (2007). The atherosclerotic heart disease and protecting properties of garlic: contemporary data. *Molecular Nutrition and Food Research*, **51**, 1365–1381.

Gouin, S. (2004). Microencapsulation: industrial appraisal of existing technologies and need. *Trends in Food Science and Technology*, **15**, 330–347.

Gunasekaran, S., Ko, S. & Xiao, L. (2007). Use of whey proteins for encapsulation and controlled delivery application. *Journal of Food Engineering*, **83**, 31–40.

Guzey, D. & McClements, D.J. (2006). Formation, stability and properties of multilayer emulsions for application in the food industry. *Advances in Colloid and Interface Science*, **128**, 227–248.

Haack, M. (2007). Supplying function (US and Swiss chocolate consumption and global omega-3 market). *Innova*, **5**(3), 23–25.

Halls, C. & Yu, O. (2008). Potential for metabolic engineering of resveratrol biosynthesis. *Trends in Biotechnology*, **26**(2), 77–81.

Hargreaves, C. (2006). Facing the functional challenge. *NutraCos*, **5**(1), 29–30.

Hasler, C.M. (1998). Functional foods: their role in disease prevention and health promotion. *Food Technology*, **52**(11), 63–70.

Hirotsuka, M., Taniguchi, H., Narita, H. & Kito, M. (1984). Calcium fortification of soy milk with calcium-lecithin liposome system. *Journal of Food Science*, **49**, 1111–1112, 1127.

Hogan, S.A., McNamee, B.F., O'Riordan, E.D. & O'Sullivan, M. (2001). Microencapsulating properties of whey protein concentrate 75. *Journal of Food Science*, **66**, 675–680.

Hong, Y.-H. & McClements, D.J. (2007). Formation of hydrogel particles by thermal treatment of β-lactoglobulin-chitosan complexes. *Journal of Agriculture and Food Chemistry*, **55**, 5653–5660.

Ivanova, E., Teunou, E. & Poncelet, D. (2005). Encapsulation of water sensitive products: effectiveness and assessment of fluid bed dry coating. *Journal of Food Engineering*, **71**, 223–230.

Jones, P.J. & Jew, S. (2007). Functional food development: concept to reality. *Trends in Food Science and Technology*, **18**, 387–390.

Kebyon, M.N. (1995). Modified starch, maltodextrin, and corn syrup solids as wall materials for food encapsulation. In: *Encapsulation and Controlled Release of Food Ingredients*. Risch, S.J. & Reineccius, G.A. (eds), American Chemical Society, Washington (ACS Symposium 590), pp. 42–50.

Kim, Y.D. & Moore, C.V. (1995). Encapsulating properties of several food proteins. *IFT Annual Meeting Poster*, p. 193.

Klahorst, S.J. (2006). Flavour and innovation meet. *World of Food Ingredients*, **June**, 26–30.

Klont, R. (1999). Healthy ingredients driving innovation. *World of Food Ingredients*, **March/April**, 43–23.

Krasaekoopt, W., Bhandari, B. & Deeth, H. (2003). Evaluation of encapsulation techniques of probiotics for yoghurt. *International Dairy Journal*, **13**, 3–13.

Kwak, N.S. & Jukes, D.J. (2001). Functional foods. Part 2: the impact on current regulatory terminology. *Food Control*, **12**, 109–117.

Lahtinen, S.J., Ouwehand, A.C., Salminen, S.J., Forssell, P. & Myllarinen, P. (2007). Effect of starch- and lipid-based encapsulation on the culturability of two *Bifidobacterium longum* strains. *Letters in Applied Microbiology*, **44**, 500–505.

Laos, K., Lougas, T., Mandmets, A. & Vokk, R. (2007). Encapsulation of β-carotene from sea buckthorn (*Hippophae rhamnoides* L.) juice in furcellaran beads. *Innovative Food Science and Emerging Technologies*, **8**, 395–398.

Lavers, B. (2007). Algae, plants and now krill. *Food Ingredients, Health and Nutrition*, **September–October**, 33–34.

Leal-Calderon, F., Thivilliers, F. & Schmitt, V. (2007). Structured emulsions. *Current Opinion in Colloid & Interface Science*, **12**, 206–212.

Lee, P.K. & Richardson, P.H. (2004). Controlled release encapsulated bioactive substances. US 6835397B2.

Leite, R.A., Lino, A.C.S. & Takahata, Y. (2003). Inclusion compounds between α-, β- and γ-cyclodextrins: iron II lactate: a theoretical and experimental study using diffusion coefficients and molecular mechanics. *Journal of Molecular Structure*, **644**, 49–53.

Li, Y., Xu, S.Y. & Sun, D.W. (2007). Preparation of garlic powder with high allicin content by using combined microwave-vacuum and vacuum drying as well as microencapsulation. *Journal of Food Engineering*, **83**, 76–83.

Livney, Y.D. (2008). Complexes and conjugates of biopolymers for delivery of bioactive ingredients via food. In: *Delivery and Controlled Release of Bioactives in Foods and Nutraceuticals*. Garti, N. (ed.), Woodhead, Cambridge, UK, pp. 234–250.
Loksuwan, J. (2007). Characteristics of microencapsulated β -carotene formed by spray drying with modified tapioca starch, native tapioca starch and maltodextrin. *Food Hydrocolloids*, **21**, 928–935.
Mattila-Sandholm, T., Myllarinen, P., Crittenden, R., Mogensen, G., Fonden, R. & Saarela, M. (2002). Technological challenges for future probiotic foods. *International Dairy Journal*, **12**, 173–182.
McClements, D.J., Decker, E.A. & Weiss, J. (2007). Emulsion based delivery systems for lipophilic bioactive components. *Journal of Food Science*, **72**, R109–R124.
McGregor, R. (2004). Taste modification in the biotech era. *Food Technology*, **58**(5), 24–30.
Mollet, B. & Lacroix, C. (2007). Where biology and technology meet for better nutrition and health. *Current Opinion in Biotechnology*, **18**, 154–155.
Nilsson, L. & Bergenståhl, B. (2007). Emulsification and adsorption properties of hydrophobically modified potato and barley starch. *Journal of Agricultural and Food Chemistry*, **55**, 1469–1474.
Ohr, L.M. (2005). Functional fatty acids. *Food Technology*, **59**(4), 63–65.
Orchel, A., Dzierzewicz, Z., Parfiniewicz, B., Weglarz, L. & Wilczok, T. (2005). Butyrate-induced differentiation of colon cancer cells is PKC and JNK dependent. *Digestive Diseases and Sciences*, **50**, 490–498.
Parada, J. & Aguilera, J.M. (2007). Food microstructure affects the bioavailability of several nutrients. *Journal of Food Science*, **72**, R21–R32.
Porzio, E.A. (2003). Encapsulation composition. US 6652895.
Prosky, L. (1999). Inulin and oligofructose are part of the dietary fiber complex. *Journal of AOAC International*, **82**(2), 223–226.
Pszczola, D.E. (2005). Making fortification functional. *Food Technology*, **59**(4), 44, 46, 48, 50, 52, 54, 56, 58–61.
Putney, S.D. (1998). Encapsulation of proteins for improved delivery. *Current Opinion in Chemical Biology*, **2**, 548–552.
Puupponen-Pimia, R., Aura, A.M., Oksman-Caldentey, K.M., Myllarinen, P., Saarela, M., Mattila-Sandholm, T. & Poutanen, K. (2002). Development of functional ingredients for gut health. *Trends in Food Science and Technology*, **13**, 3–11.
Reineccius, G.A. (1991). Carbohydrates for flavor encapsulation. *Food Technology*, **45**(3), 144–146, 149.
Rosenberg, M. & Lee, S.J. (2004). Water insoluble whey protein-based microsphere prepared by an all-aqueous process. *Journal of Food Science*, **69**, FEP50–FEP58.
Saarela, M., Lahteenmaki, L., Crittenden, R., Salminen, S. & Mattila-Sandholm, T. (2002). Gut bacteria and health foods – the European perspective. *International Journal of Food Microbiology*, **78**, 99–117.
Sanguansri, L. & Augustin, M.A. (2001). Encapsulation of food ingredients. WO 01/74175A1.
Sanguansri, L. & Augustin, M.A. (2006). Microencapsulation and delivery of omega-3 fatty acids. In: *Functional Food Ingredients and Nutraceuticals: Processing Technologies*. Shi, J. (ed.), Taylor and Francis, Boca Raton, FL, pp. 297–327.
Schneider, Y., Vincent, F., Duraton, B., Badolo, L., Gosse, F., Bergmann, C., Seiler, N. & Raul, F. (2000). Anti-proliferative effect of resveratrol, a natural component of grapes and wine, on human colonic cancer cells. *Cancer Letters*, **158**, 85–91.
Semo, E., Kesselman, E., Danino, D. & Livney, Y.D. (2007). Casein micelle as a natural nano-capsular vehicle for nutraceuticals. *Food Hydrocolloids*, **21**, 936–942.
Shi, G., Rao, L., Yu, H., Xiang, H., Yang, H. & Ji, R. (2008). Stabilization and encapsulation of photosensitive resveratrol within yeast cell. *International Journal of Pharmaceutics*, **349**, 83–93.
Shima, M., Kobayashi, Y., Fujii, T., Tanaka, M., Kimura, Y., Adachi, S. & Matsuno, R. (2004). Preparation of fine w/o/w emulsion through membrane filtration of coarse w/o/w emulsion and disappearance of the inclusion of outer phase solution. *Food Hydrocolloids*, **18**, 61–70.
Shimoni, E. (2008). Starch as an encapsulation material to control digestion rate in the delivery of active food components. In: *Delivery and Controlled Release of Bioactives in Foods and Nutraceuticals*. Garti, N (ed.), Woodhead, Cambridge, UK, pp. 279–293.
Sloan, A.E. (2004). The push for protein. *Food Technology*, **58**(12), 18.
Stehr, S.N. & Heller, A.R. (2006). Omega-3 fatty acid effects on biochemical indices following cancer surgery. *Clinica Chimica Acta*, **373**, 1–8.
Su, L.C., Lin, C.W. & Chen, M.J. (2007). Development of an oriental-style dairy product coagulated by microcapsules containing probiotics and filtrates from fermented rice. *Journal of the Society of Dairy Technology*, **60**, 49–54.

Sunley, N. (1998). Ingredients challenge. *Food Review*, **25**(12), 29, 31.
Szente, L., Mikuni, K., Hashimoto, H. & Szejtli, J. (1998). Stabilization and solubilization of lipophilic natural colorants with cyclodextrins. *Journal of Inclusion Phenomena and Molecular Recognition in Chemistry*, **32**, 81–89.
Szente, L. & Szejtli, J. (2004). Cyclodextrins as food ingredients. *Trends in Food Science and Technology*, **15**, 137–142.
Takahashi, M., Inafuku, K., Miyagi, T., Oku, H., Wada, K., Imura, T. & Kitamoto, D. (2007). Efficient preparation of liposomes encapsulating food materials using lecithins by a mechanochemical method. *Journal of Oleo Science*, **56**, 35–42.
Tanouchi, M., Takahashi, K., Fukunaga, K. & Murakawa, K. (2007). Functional material and functional food comprising useful phospholipid composition. WO 2007/046386A1.
Turgeon, S.L., Schmitt, C. & Sanchez, C. (2007). Protein-polysaccharide complexes and coacervates. *Current Opinion in Colloid and Interface Science*, **12**, 166–178.
Ubbink, J. & Kruger, J. (2006). Physical approaches for the delivery of active ingredients in foods. *Trends in Food Science and Technology*, **17**, 244–254.
Urbanska, A.M., Bhathena, J. & Prakash, S. (2007). Live encapsulated *Lactobacillus acidophilus* cells in yoghurt for therapeutic oral delivery: preparation and in-vitro analysis of alginate-chitosan microcapsules. *Canadian Journal of Physiology Pharmacology*, **85**, 884–893.
Wang, X., Jiang, Y., Wang, Y.W., Huang, M.T., Ho, C.T. & Huang, Q. (2008). Enhancing anti-inflammation activity of curcumin through o/w nanoemulsions. *Food Chemistry*, **108**, 419–424.
Watson, E., Carvalho, A., Green, R., Britton, S. & Scott, S. (2006). Functional ingredients. *Food Manufacture*, **81**(11), iv–xix.
Wu, K.G., Chai, X.H. & Chen, Y. (2005). Microencapsulation of fish oil by simple coacervation of hydroxypropyl methylcellulose. *Chinese Journal of Chemistry*, **23**, 1569–1572.
Wu, W.-H., Roe, W.S., Gimino, V.G., Seriburi, V., Martin, D.E. & Knapp, S.E. (2000). Low melt encapsulation with high laurate canola oil. US 6153236.
Xia, S. & Xu, S. (2005). Ferrous sulfate liposomes: preparation, stability and application in fluid milk. *Food Research International*, **38**, 289–296.
Young, S.L., Sarda, X. & Rosenberg, M. (1993). Microencapsulating properties of whey proteins. I. Microencapsulation of anhydrous milk fat. *Journal of Dairy Science*, **76**, 2868–2877.

2 Nanoencapsulation of food ingredients in cyclodextrins: Effect of water interactions and ligand structure

M.F. Mazzobre, B.E. Elizalde, C. dos Santos, P.A. Ponce Cevallos and M.P. Buera

2.1 Introduction

An important step in the development of functional ingredients is the design of formulation procedures for the stabilisation, solubilisation and delivery of the active components in the food product to which they are added. During the last few years, a new concept to prepare nanoparticles using amphiphilic cyclodextrin (CD) molecules for such purposes was considered (Duchene *et al.* 1999a,b; Memisoglu *et al.* 2002). CDs are cyclic oligomers of α-D-glucopyranose that can be produced through the transformation of starch by certain bacteria such as *Bacillus macerans* (Jeang *et al.* 2005; Qi & Zimmermann 2005; Qi *et al.* 2007). CDs have the ability to alter physical, chemical and biological properties of guest molecules through the formation of inclusion complexes. The widely used natural α-, β- and γ-cyclodextrins consist of six, seven and eight D-glucopyranose residues, respectively. Each cyclodextrin has its own ability to form inclusion complexes with specific guests, which depends on a proper fit of the guest molecule into the hydrophobic cyclodextrin cavity (Bender & Komiyama 1978; Saenger 1983; Szejtli 1998).

Cyclodextrins have become so important that there is now what we may call a 'cyclodextrin science', involving the search for improved properties and commercial applications of these molecules. Not only the relevant aspects of CDs are explored, but also applications of nuclear magnetic resonance (NMR), spectroscopy, calorimetry, computational chemistry and X-ray crystallography among other techniques are further developed for the elucidation of the structures of these compounds and their complexes.

In the food, cosmetics, toiletry and tobacco industries, CDs have been widely used either for the stabilisation of flavours, vitamins and natural colours or to eliminate off-flavours, microbiological contaminations and other undesired compounds (Singh *et al.* 2002; Del Valle 2004). In the chemical industry, CDs are also used as catalysts to improve the selectivity of reactions, as well as for the separation and purification of industrial-scale products (Hedges 1998). The most common applications of cyclodextrins are to enhance the solubility, stability and bioavailability of guest molecules. However, natural cyclodextrins have relatively low solubility, both in water and organic solvents, which thus limits their uses in formulations. Various kinds of cyclodextrin derivatives have been prepared to modify the physicochemical properties and inclusion capacity of natural cyclodextrins as novel drug carriers (Brewster & Loftsson 2007).

In an aqueous solution, the slightly polar CD cavity is occupied by water molecules and therefore they can be readily substituted by appropriate 'guest molecules', which are less polar than water. It is known that in order to form the complexes, water has to migrate from the interior of the cavity and to be replaced by the ligand. On the other side, once the complex is formed, the addition of water may cause the breakdown of the system. Thus, the interactions with water determine not only the ability to form the complexes, but also their stability. Under storage of the CD-encapsulated materials, water sorption properties are fundamental for defining the release of the guess molecule. It has been reported that the stability of the dried complexes is governed by the shape of the water sorption isotherms (Ponce Cevallos *et al.* 2009). Besides of polarity, the spatial structure of the ligand affects the degree of inclusion, its release and water adsorption. The aim of this chapter is to analyse the influence of water adsorption of CDs, their derivatives and their complexes on the chemical and physical stability of encapsulated compounds. These data are useful for selecting storage conditions for encapsulated nutraceuticals or for predicting the shelf-life of functional products formulated with nanoencapsulated compounds.

2.2 Brief history

In 1891, Villiers first reported the formation of some unidentified crystalline substance during starch fermentation. It was some sort of cellulose, and he named it 'cellulosine'. After their discovery, CDs were considered toxic substances and its capacity for complex formation was only considered a scientific curiosity. Later on, an Austrian microbiologist, Franz Schardinger, isolated a microorganism (*B. macerans*), which produced crystalline compounds when cultivated on starch-containing medium and he named them α- and β-dextrin (Szejtli 1998). Freudenberg and his co-workers elucidated the cyclic structure of these dextrins only in the second half of the 1930s. In the following years, the chemical structure of CDs, their general physicochemical properties and their abilities to form complexes were identified (Cramer *et al.* 1967). During this period, only small amounts of relatively impure CD were produced, which hampered industrial exploitation of these novel oligosaccharides. The biotechnological advances that occurred in the 1970s led to dramatic improvements in production of highly pure CDs, transforming them from expensive chemical additives to affordable industrial excipients (Brewster & Loftsson 2007). Further research on CDs proved that not only they are non-toxic but they can be helpful for different applications. In the last 30 years there was a progressive increase in the number of publications and patents related to the production of CDs. The purification of α- and γ-CDs increases considerably the cost of production; therefore, 97% of the CDs used in the market are β-CDs (Astray *et al.* 2009). They became accepted during 1980s as common ingredients for the food manufacture because they are not absorbed in the upper gastrointestinal tract and are completely metabolised by the colon microflora (Szente & Szejtli 2004).

CDs were also chemically or enzymatically modified. Chemical synthesis has produced a large number of derivatives. Nowadays, only α-CDs, β-CDs and γ-CDs, as well as some of their derivatives, are commercially available.

The regulatory status of CDs in foods differs among countries. In USA, α-, β- and γ-CDs are included in the Food and Drug Administration GRAS list (generally recognised as safe), while in Japan these CDs are recognised as natural products and their commercialisation in the food sector is only restricted by considerations of purity. The Joint FAO/WHO Expert Committee on Food Additives recommends a maximum level of 5 mg/kg per day of β-CD in foods. For α- and γ-CDs, an acceptable daily intake was not defined because of their innocuous profiles (Astray *et al.* 2009).

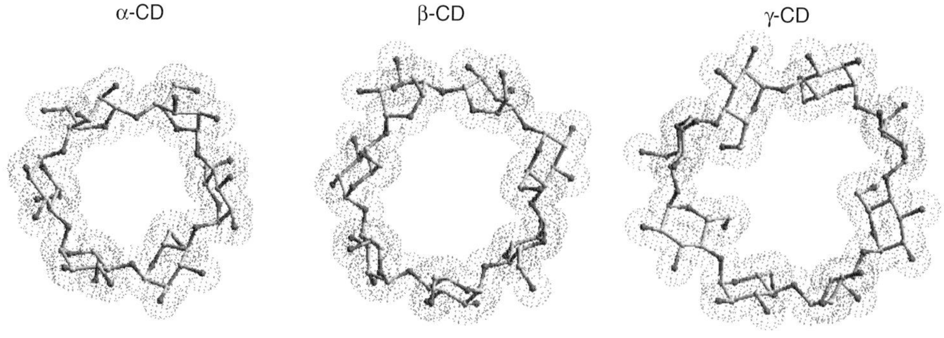

Fig. 2.1 Chemical structure of α-, β- and γ-cyclodextrins. All molecular models were obtained with *Hyperchem Professional, Release 7.5*. For a colour version of this figure, please see the colour plate section.

2.3 Structure and properties of cyclodextrins

Cyclodextrins are among the most important molecules studied in supramolecular chemistry, and the special structure of these materials has attracted biologists, chemists, physicists, engineers, food technologists and many others in attempts to take advantage of their properties (Villalonga *et al.* 2007).

The chemical structures of the most common CDs (α, β and γ) are shown in Figure 2.1, and some important physical properties and characteristics are summarised in Table 2.1. These cyclic oligomers, in which glucopyranose units are in the chair conformation are, crystalline and non-hygroscopic substances with non-reducing character that makes them behave as polyols. The ring that constitutes the CDs is a torus-like conical cylinder, which is frequently characterised as a truncated cone with a hydrophobic interior and a hydrophilic exterior. While the hydrophilic exterior helps CDs to be dissolved in water, the hydrophobic interior has the ability to form inclusion complexes with a wide variety of organic compounds. The internal cavity can include guest molecules, ranging from polar compounds such as alcohols, acids, amines and small inorganic anions to non-polar compounds such as aliphatic and aromatic hydrocarbons (Hapiot *et al.* 2006). This cavity is lined by the hydrogen atoms and the glycosidic oxygen bridges. The non-bonding electron pairs of the glycosidic oxygen bridges are directed towards the inside of the cavity, producing a high electron density, having similar characteristics to a Lewis base (Saenger 1983).

The C-2-OH group of one glucopyranose unit can form a hydrogen bond with the C-3-OH group of the adjacent glucopyranose unit (Bender & Komiyama 1978). In the CD molecule, a complete secondary belt is formed by these H-bonds; therefore, the β-CD has a rather rigid

Table 2.1 Physical properties of α-, β- and γ-cyclodextrins (adapted from Szejtli 1998)

	α-CD	**β-CD**	**γ-CD**
Glucose units	6	7	8
Internal diameter (nm)	0.47–0.53	0.60–0.65	0.75–0.83
Cavity depth (nm)	0.79	0.79	0.79
Solubility in water (mg/mL)	145	18.5	232
MW (Da)	972	1135	1297
Crystalline water (% w/w)	10.2	13.2	8.13–17.7

structure. This intramolecular hydrogen bond could be the explanation for the observation that β-CD has the lowest water solubility of all CDs. On the other side, the α-CD molecule has one glucopyranose unit in a distorted position, making the H-bond belt incomplete. Consequently, instead of the six possible H-bonds, only four can be fully established. The γ-CD is a non-coplanar, more flexible structure; therefore, it is the most soluble of the three CDs (Astray *et al.* 2009). The diameter of the cavity is larger on the side of the secondary hydroxyl groups than on the side of the primary hydroxyl groups since free rotation of the latter reduces its effective interior diameter. The approximate dimension of CDs is shown in Table 2.1.

NMR, infrared and optical rotary dispersion spectroscopy studies have demonstrated that D-glucopyranose units have the same conformation in both dimethyl sulfoxide and heavy water . The spectroscopic studies on CDs in aqueous solution suggest that the conformation of CDs in solution is almost identical to their conformation in the crystalline state (Bender & Komiyama 1978). β-CD has a perfect symmetry, while α- and γ-CD rings are slightly distorted, as can be observed in Figure 2.1.

During the past decade, series of the larger CDs have been isolated and studied. As the CD cavity diameter increases, the non-polar cavity can accommodate an increasing number of water molecules. The driving force for the complex formation (the substitution of the high enthalpy water molecules in the CD cavity) is weaker in the case of larger CDs. As a consequence, complex formation in such a system does not result in a significant gain on energy. Therefore, their utilisation as inclusion complexing agents will probably remain rather restricted. The larger CDs are not regular cylinder-shaped structures. They are collapsed and their real cavity is even smaller than in the γ-CD (Astray *et al.* 2009). To improve the solubility of natural CD or to obtain different physicochemical properties of this compounds, chemical-modified CDs have been prepared. Hydroxypropyl-β-cyclodextrin (HP β-CD) and methyl-β-cyclodextrin (M β-CD) are among of the most employed derivatives.

2.4 Formation and characterisation of the inclusion complexes

Lindner and Saenger (1978, 1982) noted that β-CD forms a dodecahydrated crystal, in which half the water molecules are placed on specific sites in the cavity, and the rest are distributed between the surface and the interstices of cyclodextrins. The water molecules in the cavity would be in an 'activated' state and could be easily removed by ligands capable of forming complexes with CDs. These observations are consistent with other works that predict hydrophobic interactions in the internal cavity and hydrophilic interactions on external surface (Koehler *et al.* 1988; Winkler *et al.* 2000).

In an aqueous solution, the non-polar cavity of CDs is occupied by water molecules. This conformation is energetically unfavourable due to the existence of polar–non-polar interactions, and therefore the water molecules can be readily substituted by appropriate 'guest molecules', which are less polar than water (Figure 2.2). The dissolved cyclodextrin is the 'host molecule'. A non-covalent 'inclusion complex' is formed with the guest molecules and this is the essence of 'molecular encapsulation'.

Depending on their size and shape, the guest molecules can be included totally or partially into the cyclodextrin cavity and also one, two or three cyclodextrins molecules can interact to contain one entrapped guest molecule. The formed inclusion complex can be isolated as a stable solid.

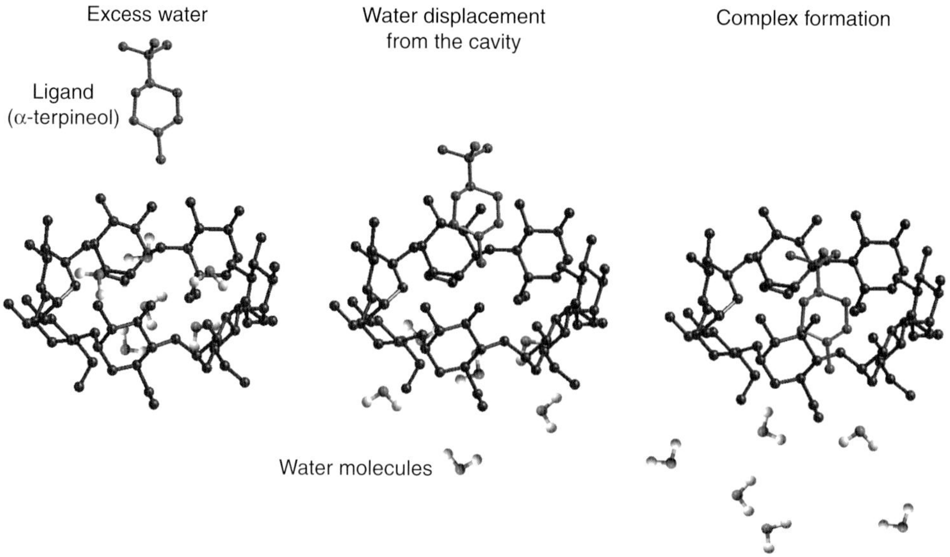

Fig. 2.2 Scheme showing the displacement of water during complex formation and of ligand displacement in excess water. Exemplified for α-terpineol as guest ligands and β-CD as a host. For a colour version of this figure, please see the plate section.

The 'driving force' of complexation is not yet completely understood, but it seems that it is the result of various effects:

- Substitution of water molecules from the inner cavity, which is energetically favoured;
- The release of CD ring strain when the complex is formed;
- Van der Waals forces and hydrogen bond interactions, which are established when the complex is formed.

The role of the hydrogen bonding is not universal and the stability of the complex increases with the electron–donor character of the substituents of the included molecule. To evaluate the van der Waals interactions in the system, it is important to establish the energy involved in the complex formation.

A dynamic equilibrium is established between the guest (G), the host (CD) and the complex, and this is expressed by the following equation:

$$nCD + mG \underset{k_d}{\overset{k_f}{\rightleftharpoons}} CD_n - G_m \tag{2.1}$$

where n is number of moles of CD and m is number of moles of the guest.

From this equilibrium, the constant of formation (k_f) and dissociation (k_d) can be calculated. The values of k_f and k_d are function of the chemical structure, the shape of the guest and also of the physicochemical properties of the environment.

The stability of the inclusion complex is associated to the complex stability constant (K_C), which is one of the most important characteristics of the encapsulation process.

K_C can be defined as follows:

$$K_C = \frac{k_F}{k_D}$$

For the particular case of CD–guest ratio 1:1, the expression of K_C is as follows:

$$K_{C(1:1)} = \frac{[CD - G]}{[CD][G]} \qquad (2.2)$$

The K_C of the complexes can be obtained from the phase-solubility diagrams according to the method developed by Higuchi and Connors (1965), where the concentrations of the guest, CD and the complex can be calculated, evaluating an appropriate physical or chemical property. Different methods can be used to achieve this objective: chemical reactivity, molar absorptivity and other spectrophotometric properties, NMR chemical shifts, pKa values and high-performance liquid chromatography retention times, among others (Chadha *et al.* 2004).

Phase-solubility studies also allow to calculate the thermodynamic parameters involved in the complex formation: free energy (ΔG^*), enthalpy (ΔH^*) and entropy (ΔS^*) by determining the K_C at different temperatures. In fact, the integrated form of the Van't Hoff equation (Equation 2.3) allows the calculation of the values of enthalpy and entropy changes (Tommasini *et al.* 2004):

$$\ln K_c = -\frac{\Delta H^*}{RT} + \frac{\Delta S^*}{R} \qquad (2.3)$$

Differential scanning calorimetry (DSC) is a routine method, commonly used to verify the formation of a complex in the solid state (Karathanos *et al.* 2007). The total or partial disappearance of thermal events (melting point) corresponding to guest molecules, when they are examined as CD complexes, is generally taken as a proof of complex formation (Williams *et al.* 1998; Pralhad & Rajendrekumar 2004).

Thermograms obtained by DSC for α-terpineol (Aterp), Aterp-HP β-CD and Aterp-β-CD complexes, both in a molar ratio Aterp-CD 1:1, are shown in Figure 2.3a. The thermogram corresponding to Aterp shows an endothermic peak at nearly 35°C corresponding to its melting point. As can be seen in Figure 2.3a, the curves corresponding to Aterp-β-CD and Aterp-HP β-CD complexes did not show any sharp endothermic peak in the range of the melting point of the pure compound (35°C). The disappearance of the α-terpineol melting is due to its encapsulation in the host β-CD or HP β-CD. A similar behaviour was observed for thymol complexed with β-CD (Figure 2.3b). The endothermic melting peak of thymol, nearly 50°C, disappears completely in the thymol-β-CD system when the complex is formed.

When the complexes of CD and a convenient guest are obtained by freeze-drying method (Karathanos *et al.* 2007), the whole guest will be in the final system, either free or encapsulated. The complete disappearance of the endothermic DSC signal is a strong evidence of the total inclusion of the ligand inside the cavity of CD. If the disappearance of the thermal signal in the thermograms is only partial, the inclusion may not be completed. Therefore, the ratio between this signal and the theoretical value of the melting enthalpy of the guest is an approximation to quantify the released amount of the guest from the system (dos Santos *et al.* 2007).

2.5 Water adsorption isotherms

The interactions with water define not only the formation of the complexes (Figure 2.2) but also their stability. It is known that in the presence of an enough amount of water the

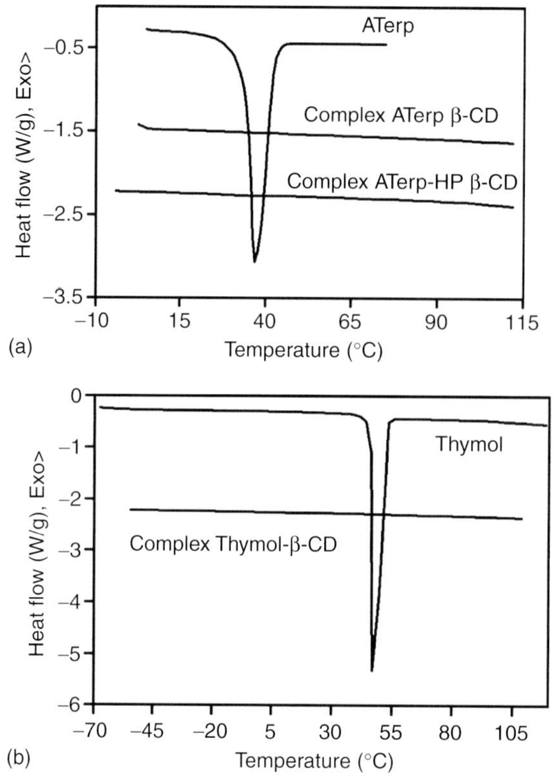

Fig. 2.3 Differential scanning calorimetry thermograms of (a) α-terpineol (ATerp) and the complexes ATerp-β-CD and ATerp-hydroxypropyl-β-CD (ATerp-HP β-CD) and (b) thymol and the complex thymol-β-CD. Scanning rate 10°C/minute.

equilibrium is displaced towards the single compounds, releasing the guest molecule. When the dry formulations of CD containing the nutraceutical-active components are stored, the analysis of the water adsorption behaviour becomes of fundamental importance to define the appropriate storage conditions. Water adsorption isotherms correlate the relative humidity (RH) of the environment to the corresponding equilibrium water content that a system will reach at a given temperature.

The water adsorption isotherms for freeze-dried β-CD and for its water-soluble derivatives, M β-CD and HP β-CD, are shown in Figure 2.4. It can be seen that at low relative humidities, the three analysed CDs have a similar water content. β-CD reaches a plateau at an RH close to 43% and its water content remains constant up to 85% RH. It has been reported that β-CD forms a stable crystalline hydrate and crystallizes from aqueous solutions as dodecahydrate or undecahydrate (Fujiwara *et al.* 1983). It is to be noted that the plateau observed in the adsorption isotherm of β-CD occurs when the molar ratio between water and β-CD is 12 (right axis in Figure 2.4), which is in accord with the formation of dodecahydrated crystals. Between 33 and 75% RH, M β-CD shows a sorption behaviour close to that observed for β-CD. However, M β-CD and HP β-CD did not show a plateau and the isotherms followed a sigmoid-type behaviour, which indicates that no hydrates were formed. At 95% RH, an interaction of about 35 water mol/CD mol was made possible in the modified CDs. It was observed that the freeze-dried modified CDs showed a structural collapse and volume

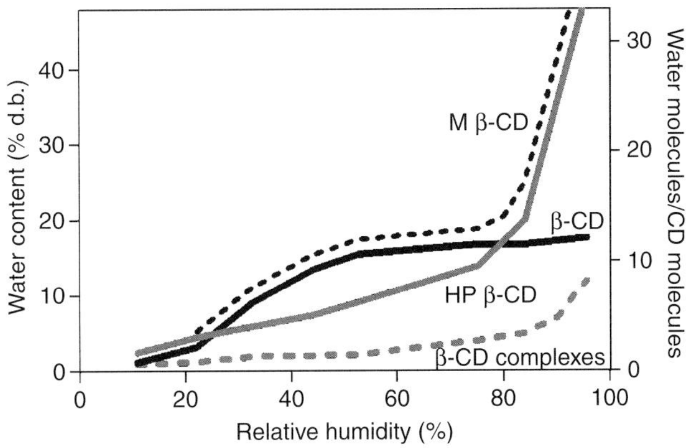

Fig. 2.4 Water adsorption isotherms of β-cyclodextrin (full black line), the complexes β-CD-α-terpineol (dashed line) and the derivatives HP β-CD and M β-CD (grey full line and dotted line, respectively).

shrinkage from 84% RH, while this phenomenon did not occur in β-CD at any of the studied RH. The isotherms shown in Figure 2.4 reflect that the β-CD modification greatly affects β-CD water interactions, mainly at high RH (>85%). The low solubility of the β-CD has been explained in terms of interactions between secondary hydroxyl groups adjacent to the cavity and water molecules. Therefore, any structural change in the cyclodextrin modifies its solubility and relationship with water (dos Santos *et al.* 2007; Astray *et al.* 2009).

The presence of hydrophobic ligands such as α-terpineol, cinnamaldehyde, thymol or myristic acid, which are constituents of essential oils used as nutraceuticals, greatly modified the β-CD sorption curves. The effect of complex formation on the water sorption properties of β-CD is also shown in Figure 2.4. Due to the presence of a host in the CD cavity, the β-CD complexes adsorbed less water than β-CD, and the plateau was not observed (Figure 2.4), indicating that the hydrated crystals were not formed. However, in the modified β-CDs the inclusion of the ligands occurred without changes in CD–water interactions. The sorption curves obtained for the methyl or hydroxypropyl β-CD complexes (data not shown) were similar to those shown in Figure 2.4 for the correspondent single CD. Only at RH of more than 90%, the water content was smaller in the complexed system. The modification of β-CD with hydroxypropyl or methyl moieties confers sorption, solubility, amorphicity properties and ability to form hydrates different from the original β-CD. The practical significance of the water adsorption characteristics of these systems is related not only to the physical structure, rehydration properties and microbial stability of the systems, but also to the release of the guest compound, as discussed in the following section.

2.6 Water and the stability and release of encapsulated nutraceuticals

In the study of CD encapsulation, it is not only important to evaluate the geometric compatibility between the guest and the CD cavity, but also the effect of the structure, the charge and polarity of the guest. The environment and the experimental conditions such as temperature, rate and time of stirring can also determine the success of the encapsulation procedure and thus the quality of the obtained products.

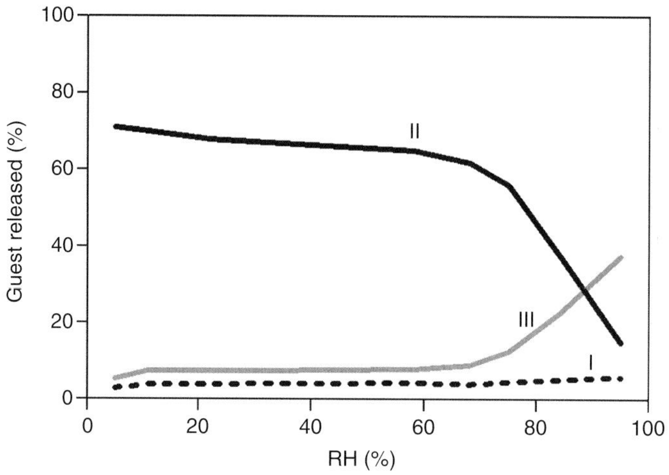

Fig. 2.5 Different types of responses of ligand release to RH according to the characteristics of ligands in systems stored during 60 days at 25°C. I, α-terpineol (ATerp)-β-CD; II, myristic acid-HP β-CD; III, cinnamaldehyde or thymol-β-CD.

The CD–guest molar ratio is also important to define the appropriate encapsulation conditions. For example, for the total encapsulation of the non-polar carboxylic acids, CD–guest molar ratios of 3:1, 4:1 or 2:1 were reported (Regiert 2007).

The stability of the complexes of β-CD and modified β-CDs with several types of compounds was analysed during storage at different RH at 25°C. According to calorimetric studies to determine complex stability, different types of behaviour could be proposed. In some systems such as α-terpineol encapsulated in β-CD at molar ratio 1:1 (represented by curve I in Figure 2.5), the ligand was completely encapsulated and remained included up to 60 days of storage, even at the higher RH (there was not any thermal signal of the melting of the α-terpineol in the correspondent thermograms of the complex). The molecular inclusion proposed for the complex is shown in Figure 2.6a. The compound remained encapsulated even increasing RH up to 95%.

The model proposed for myristic acid encapsulated in HP β-CD at 1:3 guest-CD molar ratio is represented by curve II in Figure 2.5. The inclusion of myristic acid was initially incomplete. The encapsulation of this non-polar aliphatic acid increased with increasing RH (Figure 2.5) and also with the time of storage at certain RH (the thermal signal of the endothermic melting of the ligand decreased both with storage time and RH, but it did not disappear completely) (data not shown). The presence of water in the molecular environment favoured the inclusion of myristic acid. According to the molar ratio for optimal encapsulation, a model of three CD molecules per acid molecule was proposed, as schematised in Figure 2.6b. The thermal signal of the endothermic melting of the ligand decreased during storage time, but it did not disappear completely.

A third type of behaviour was observed for thymol and cinnamaldehyde encapsulated in β-CD (curve III in Figure 2.5). These compounds were almost completely encapsulated at the optimal molar ratio 1:1 and remained included up to 75% RH. As a difference with α-terpineol encapsulated in β-CD (curve I, Figure 2.5), the amount of released ligands increased when the RH was higher than 75%. The molecular model for the complexes of thymol and cinnamaldehyde in β-CD is shown in Figures 2.6c and 2.6d, respectively. It is to

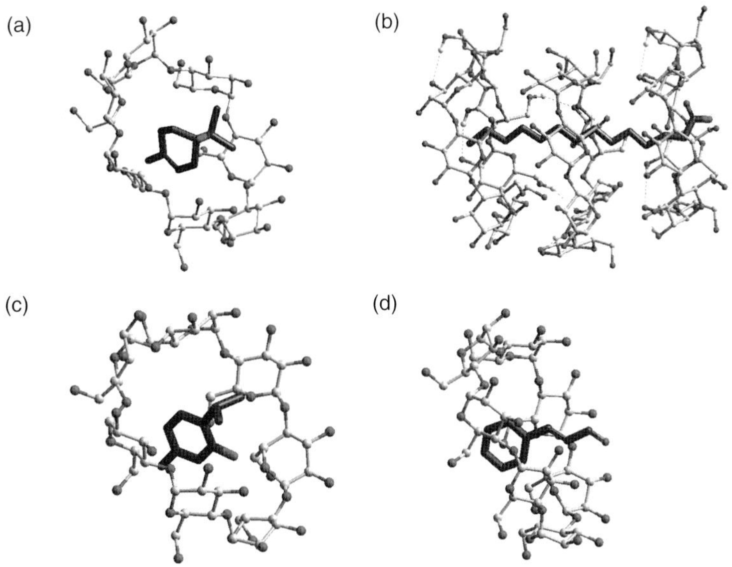

Fig. 2.6 Scheme of different cyclodextrin complexes: (a) α-terpineol (ATerp)-β-CD, (b) myristic acid-HP β-CD, (c) thymol-β-CD and (d) cinnamaldehyde-β-CD.

be noted that in both cases the release of the encapsulated compound occurred in the region of the water adsorption isotherm at which a sharp increase of water content occurs (Figure 2.4), which is coincident with the appearance of free water (Ponce Cevallos *et al.* 2009). Thus, the complex stability of these compounds is closely related to the shape of the water sorption isotherm.

2.7 Applications and future prospects

In the last 30 years, there was a progressive increase in the use of CD-assisted molecular nanoencapsulation in the food industry and also in the number of publications and patents related to the production of CDs. The widespread utilisation of CDs is reflected in pharmaceutical, food, chemical and other industrial areas. Its use has increased annually around 20–30%; most of that in food products (Samant & Pai 1991; Szejtli 1997; Hedges & McBride 1999; Cravotto *et al.* 2006).

Even though the CD molecule is composed of glucose subunits, α- and β-CDs do not taste sweet, while the γ-CD has only a slightly sweet taste. CDs and their complexes are usually colourless, non-hygroscopic powders and they are used in food industry with a variety of objectives. The main reported applications and the products in which they are employed are summarised in Table 2.2. They can be further described as follows:

- The solubility of many compounds (essential oils, flavours, vitamins, carotenoids and other colourants) can be improved by their encapsulation in cyclodextrins or modified cyclodextrins (Fereidon & Xiao-Qing 1993; Tommasini *et al.* 2004). Retinol–CD complexes have been prepared to protect the content of vitamin A and to improve its poor solubility in systems such as beverages or high-moisture foods (Loveday & Singh 2008).

Table 2.2 Some marketed products containing cyclodextrins and their role

Type of food product	Role of CDs
Low-cholesterol cheese	Cholesterol reducing
Flavoured sugar for baking	Flavour preservation on heating
Low-cholesterol eggs	Cholesterol reducing
Beer flavour standards	Preservation of standards of flavours
Chewing gum	Flavour stabilisation
Chocolate	Emulsifiers
Instant green tea	Colour stabilisation
Dietary fibre drink	Taste masking
Instant tea drink	Flavour preservation
Cranberry seed oil encapsulated in γ-CD	Free fatty acid protection
Low-cholesterol butter	Cholesterol reducing

- Encapsulation of monoterpenes in β-CD has been able to improve its solubility and to protect them against oxidation (Mourtzinos *et al.* 2008).
- The removal or reduction of unpleasant taste and odour has been used in reducing or eliminating unwanted flavours such as bitterness in fruit juices (Konno *et al.* 1981; Singh *et al.* 2002).
- Stabilisation of vitamins, flavour, essential and polyunsaturated oils. This property has been used in compounds such as antibiotics, unsaturated fats and pigments (Del Valle 2004).
- Achievement of controlled release of certain food ingredients. Interactions of cyclodextrin with the guest compounds produce a high-energy barrier that hinders an easy volatility. Menthol forms an odourless complex with β-cyclodextrin (Hedges 1998; Szente & Szejtli 2004). Soluble coffee powder containing β-CD results in an effective way to preserve the volatile compounds (Szente & Szejtli 2004).
- Technological aid for food modification processes. In the removal of cholesterol in animal products, such as milk (Lee *et al.* 1999; Kwak *et al.* 2004), whipping cream (Shim *et al.* 2003), mayonnaise (Jung *et al.* 2008), egg yolks (Mine & Bergougnoux 1998), butter (Jung *et al.* 2005) and lard, CD is removed at the end of the process (Szejtli 1998; Yamamoto *et al.* 2005). Another common use of cyclodextrins is the removal of free fatty acids, which may cause the formation of unwanted products in the frying process.

Other potential applications can be foreseen, related to their use in nutraceutical development, which include the following:

- Recently, many studies proved that CD acts as browning inhibitors when added to different fruit juices, such as apple, pear and banana (López-Nicolás *et al.* 2007a,b,c). CDs remove polyphenolic compounds, which cause enzymatic browning (Torres-Rivas *et al.* 2004; Alvarez-Parrilla *et al.* 2007). Thus, the polyphenol encapsulated in the cyclodextrin cavity is not available for the enzyme polyphenol oxidase (Del Valle 2004).
- The presence of β-CD considerably modifies the gelatinisation of starch wheat flour (Kim & Hill 1984) and denaturalisation of proteins since they are capable of forming complexes with hydrophobic amino acids (Mortensen & Jansson 2004).
- Many functional ingredients may be either employed to enhance their specific characteristics for flavour improvement or masked to avoid its sensorial perception of the food product where they are added. Due to their importance in consumer satisfaction and organoleptic

properties of final products, applications of CDs in flavour technology deserve a special attention. By forming inclusion complexes with cyclodextrins, the flavouring agents may be distributed evenly. This offers a great potential for the protection of volatile or labile flavouring materials present in food systems during technological processes such as freezing, thawing, microwaving, cooking and pasteurisation (Jouquand *et al.* 2004) and extrusion (Bhandari *et al.* 2001).
- The inclusion of the flavour into the CD improves the mixing process. It has been proved to be effective in protection against evaporation during heat treatment (Qi & Hedges 1995).
- The encapsulation of nutraceutical lipophilic and oxygen-sensitive components inside CDs protects them from oxidative processes. Essential polyunsaturated fatty acids may be protected against oxidation by formation of stable complexes with CDs. Various studies were carried out to determine the performance of CDs to obtain antioxidant effect on linoleic acid (Regiert 2007).
- Other new challenging application of CD is in the packaging of food materials, where they have shown to have two benefits: improved barrier properties of the packaging material and improved sensorial properties of the food systems (Wood 2001).

As noted previously, water significantly affects the formation of CD complexes and the release of encapsulated compounds. Thus, for any of the above-mentioned applications, the water relationships should be investigated for each particular system, regarding the environments at which the complexes will be exposed in order to develop the appropriate systems for functional ingredients, encapsulation.

Acknowledgements

The authors are grateful to Mr Nicolás Villagrán for the design of the molecular models (through Hyperchem Professional). Financial support from Universidad de Buenos Aires (UBACYT 024), CONICET (PIP 5799 and 00468) and ANPCYT (PICTs 20545 and 32916) is also acknowledged.

References

Alvarez-Parrilla, E., De la Rosa, L., Rodrigo-García, J., Escobedo-González, R., Mercado-Mercado, G., Moyers-Montoya, E., Vázquez-Flores, A. & González-Aguilar, G. (2007). Dual effect of β-cyclodextrin on the inhibition of apple polyphenol oxidase by 4-hexylresorcinol and methyl jasmonate. *Food Chemistry*, **101**, 1346–1356.

Astray, G., Gonzalez-Barreiro, C., Mejuto, J.C., Rial-Otero, R. & Simal-Gándara, J. (2009). A review on the use of cyclodextrins in foods. *Food Hydrocolloids*, **23**(7), 1631–1640.

Bender, M.L. & Komiyama, M. (1978). *Cyclodextrin Chemistry*. Springer, Berlin.

Bhandari, B., D'Arcy, B. & Young, G. (2001). Flavour retention during high temperature short time extrusion cooking process: a review. *International Journal of Food Sciences Technology*, **36**, 453–461.

Brewster, M. & Loftsson, T. (2007). Cyclodextrins as pharmaceutical solubilizers. *Advanced Drug Delivery Reviews*, **59**, 645–666.

Chadha, R., Kashid, N. & Saini, A. (2004). Account of analytical techniques employed for the determination of thermodynamics of inclusion complexation of drugs with cyclodextrins. *Journal of Scientific and Industrial Research*, **63**, 211–229.

Cramer, F., Saenger, W. & Spatz, H-C. (1967). Inclusion compounds. XIX. The formation of inclusion compounds of α-cyclodextrin in aqueous solutions. Thermodynamics and kinetics. *Journal of the American Chemical Society*, **89**(1), 14–20.

Cravotto, G., Binello, A., Baranelli, E., Carraro, P. & Trotta, F. (2006). Cyclodextrins as food additives and in food processing. *Current Nutrition and Food Science*, **2**, 343–350.
Del Valle, M. (2004). Cyclodextrins and their uses: a review. *Process Biochemistry*, **39**, 1033–1046.
dos Santos, C., Mazzobre, M.F., Elizalde, B. & Buera, M.P. (2007). Influencia de modificaciones de β-ciclodextrina sobre su interacción con el agua y ligandos de distinta estructura. *Alimentos Ciencia e Ingeniería de Alimentos*, **16**(2), 71–73.
Duchene, D., Ponchel, G. & Wouessidjewe, D. (1999a). Cyclodextrins in targeting: application to nanoparticles. *Advanced Drug Delivery Reviews*, **36**, 29–40.
Duchene, D., Wouessidjewe, D. & Ponchel, G. (1999b). Cyclodextrins and carrier systems. *Journal of Controlled Release*, **62**, 263–268.
Fereidon, S. & Xiao-Qing, H. (1993). Encapsulation of food ingredients. *Critical Reviews in Food Science and Nutrition*, **33**, 501–547.
Fujiwara, T., Yamazaki, M., Tomizu, Y., Tokuoka, R., Tomita, K., Matsuo, T., Suga, M. & Saenger, W. (1983). The crystal structure of a new form of β-cyclodextrin water inclusion compounds and thermal properties of β-cyclodextrin complexes. *Nippon Kagaku Kaishi*, **2**, 181–187.
Hapiot, F., Tilloy, S. & Monflier, E. (2006). Cyclodextrins as supramolecular hosts for organometallic complexes. *Chemical Reviews*, **106**, 767–781.
Hedges, A. (1998). Industrial applications of cyclodextrins. *Journal of the American Chemical Society*, **98**, 2035–2044.
Hedges, A.R. & McBride, C. (1999). Utilization of β-cyclodextrin in food. *Cereal Foods World*, **44**(10), 700–704.
Higuchi, T. & Connors, K. (1965). Phase-solubility techniques. *Advances in Analytical Chemistry and Instrumentation*, **4**, 117–120.
Jeang, C.L., Lin, D.G. & Hsieh, S.H. (2005). Characterization of cyclodextrin glysosyltranfesare of the same gene expressed from *Bacillus macerans*, *Bacillus sutilis* and *Scherichia colli*. *Journal of Agricultural and Food Chemistry*, **53**, 6301–6304.
Jouquand, C., Ducruet, V. & Giampaoli, P. (2004). Partition coefficients of aroma compounds in polysaccharide solutions by the phase ratio variation method. *Food Chemistry*, **85**, 467–474.
Jung, T.H., Ha, H.J., Ahn, J. & Kwak, H.S. (2008). Development of cholesterol-reduced mayonnaise with cross-linked β-cyclodextrin and added phystosterol. *Korean Journal for Food Science of Animal Resources*, **28**, 211–217.
Jung, T.H., Kim, J.J., Yu, S.H., Ahn, J. & Kwak, H.S. (2005). Properties of cholesterol-reduced butter and effect of gamma linolenic acid added butter on blood cholesterol. *Asian-Australasian Association of Animal Production Societies*, **18**, 1646–1654.
Karathanos, V., Mourtzinos, I., Yannakopoulou, K. & Andrikopoulos, N. (2007). Study of the solubility, antioxidant activity and structure of inclusion complex of vanillin with β-cyclodextrin. *Food Chemistry*, **101**, 652–658.
Kim, H. & Hill, R. (1984). Modification of wheat flour dough characteristics with cyclodextrins. *Cereal Chemistry*, **61**, 406–407.
Koehler, J.E., Saenger, W. & van Gunsteren, W.F. (1988). Conformational differences between α-cyclodextrin in aqueous solution and in crystalline form: a molecular dynamics study. *Journal of Molecular Biology*, **203**, 241–250.
Konno, A., Misaki, M. & Toda, J. (1981). Bitterness reduction of citrus fruits by beta-cyclodextrin. *Agricultural and Biological Chemistry*, **45**, 2341.
Kwak, H.S., Kim, S.H., Kim, J.H., Choi, H.J. & Kang, J. (2004). Immobilized β-cyclodextrin as a simple and recyclable method for cholesterol removal in milk. *Archives of Pharmacal Research*, **27**, 873–877.
Lee, D.K., Ahn, J. & Kwak, H.S. (1999). Cholesterol removal from homogenized milk with β-cyclodextrin. *Journal of Dairy Science*, **82**, 2327–2330.
Lindner, K. & Saenger, W. (1978). β-Cyclodextrin dodecahydrate: crowding of water molecules within a hydrofobic cavity. *Angewandte Chemie International Edition in English*, **17**, 694–695.
Lindner, K. & Saenger, W. (1982). Crystal and molecular structure of cyclohepta-amylose dodecahydrate. *Carbohydrate Research*, **99**, 103–115.
López-Nicolás, J. & García-Carmona, F. (2007b). Use of cyclodextrins as secondary antioxidants to improve the colour of fresh pear juice. *Journal of Agricultural and Food Chemistry* **55**, 6330–6338.
López-Nicolás, J., Nuñez-Delicado, E., Pérez-López, A., Sánchez-Ferrer, A. & García-Carmona, F. (2007a). Reaction's mechanism of fresh apple juice enzymatic browning in the presence of maltosyl-β-cyclodextrin. *Journal of Inclusion Phenomena and Macrocyclic Chemistry*, **57**, 219–222.

López-Nicolás, J., Pérez-López, A., Carbonell-Barrachina, A. & García-Carmona, F. (2007c). Kinetic study of the activation of banana juice enzymatic browning by the addition of maltosyl-β-cyclodextrin. *Journal of Agricultural and Food Chemistry*, **55**, 9655–9662.
Loveday, S. & Singh, H. (2008). Recent advances and technologies for vitamin A protection in foods. *Trends in Food Science and Technology* **19**, 657–668.
Memisoglu, E., Bochot, A., Sen, M., Charon, D., Duchene, D. & Hincal, A. (2002). Amphiphilic β-cyclodextrins modified on the primary face: synthesis, characterization, and evaluation of their potential as novel excipients in the preparation of nanocapsules. *Journal of Pharmaceutical Sciences*, **91**, 1214–1224.
Mine, Y. & Bergougnoux, M. (1998). Adsorption properties of cholesterol-reduced egg yolk low-density lipoprotein at oil-in-water interfaces. *Journal of Agricultural and Food Chemistry*, **46**, 2153–2158.
Mortensen, B. & Jansson, S. (2004). Complexes of cyclodextrins and carotenoids. Patent (WO/2004/005353).
Mourtzinos, I., Kalogeropoulos, N., Papdakis, S., Konstantinou, K. & Karathanos, V. (2008). Encapsulation of nutrceutical monoterpenes in β-cyclodextrin and modified starch. *Journal of Food Sciences*, **73**, 89–94.
Ponce Cevallos, P., Elizalde, B. & Buera, P. (2009). Water sorption properties and stability of inclusion complexes of thymol and cinnamaldehyde with β-cyclodextrins. In: *Water Properties in Food, Health, Pharmaceutical and Biological Systems*. Sajjaanantakul, T. & Reid, D. (eds), John Wiley & Sons, USA (in Press).
Pralhad, T. & Rajendrekumar, K. (2004). Study of freeze-dried quercetincyclodextrin binary systems by DSC, FT-IR, X-ray diffraction and SEM analysis. *Journal of Pharmaceutical and Biomedical Analysis*, **34**, 333–339.
Qi, Q., Mokhtar, M.N. & Zimmermann, W. (2007). Effect of ethanol on the synthesis of large-ring cyclodextrins by cyclodextrin glucanotransferases. *Journal of Inclusion Phenomena on Macrocyclic Chemistry*, **57**, 95–99.
Qi, Q. & Zimmermann, W. (2005). Cyclodextrin glucanotransferase: from gene to applications. *Applied from Microbiology and Biotechnology*, **66**, 475–485.
Qi, Z.H. & Hedges, A.R. (1995). Use of cyclodextrins for flavours. In *Flavour Technology: Physical Chemistry, Modification and Process*. Ho, C.T., Tan, C.T. & Tong, C.H. (eds), ACS Symposium Series 610, Washington, DC, pp. 231–243, American Chemical Society.
Regiert, M. (2007). Oxidation-stable linoleic acid by inclusion in α-cyclodextrin. *Journal of Inclusion Phenomena* and *Macrocyclic Chemistry*, **57**, 471–474.
Saenger, W. (1983). Sterochemistry of circularly closed oligosaccharides: cyclodextrins structure and function. *Biochemical Society Transactions*, **11**, 136–139.
Samant, S.K. & Pai, J.S. (1991). Cyclodextrins: new versatile food additive. *Indian Food Packer*, **45**, 55–65.
Shim, S.Y., Ahn, J. & Kwak, H.S. (2003). Functional properties of cholesterol-removed whipping cream treated by β-cyclodextrin. *Journal of Dairy Science*, **86**, 2767–2772.
Singh, M., Sharma, R. & Banerjee, U.C. (2002). Biotechnological applications of cyclodextrins. Research Review Paper. *Biotechnology Advances*, **20**, 341–359.
Szejtli, J. (1997). Utilization of cyclodextrins in industrial products and processes. *Journal of Materials Chemistry*, **7**, 575–587.
Szejtli, J. (1998). Introduction and general overview of cyclodextrin chemistry. *American Chemical Society*, **98**, 1743–1753.
Szente, L. & Szejtli, J. (2004). Cyclodextrins as food ingredients. *Trends in Food Science and Technology*, **15**, 137–142.
Tommasini, S., Raneri, D., Ficarra, R., Calabro, M.L., Stancanelli, R. & Ficarra, P. (2004). Improvement in solubility and dissolution rate of flavonoids by complexation with β-cyclodextrin. *Journal of Pharmaceutical and Biomedical Analysis*, **35**, 379–387.
Torres-Rivas, F., De la Rosa, L., Rodrigo-Garcia, J., Gonzalez-Aguilar, G. & Alvarez-Parrilla, E. (2004). Complexation of apple antioxidant by β-cyclodextrin as an approach to reduce its browning (proceeding). Food Science and Biotechnology in Developing Countries. International Congress, México, 20–23 de junio.
Villalonga, R., Cao, R. & Fragoso, A. (2007). Supramolecular chemistry of cyclodextrins in enzyme technology. *Chemical Reviews*, **107**, 3088–3116.
Villiers, A. (1891). Sur la fermentation de la fécule par l'action du ferment butyrique. *Compte Rendu de l Académie de Sciences*, **112**, 536–538.
Williams, R.O., Mahaguna, V. & Sriwongjanya, M. (1998). Characterization of an inclusion complex of cholesterol and hydroxypropyl-β-cyclodextrin. *European Journal of Pharmaceutics and Biopharmaceutics*, **46**, 355–360.

Winkler, R.G., Fioravanti, S., Ciccotti, G., Margheritis, C. & Villa, M. (2000). Hydration of β-cyclodextrin: a molecular dynamics simulation study. *Journal of Computer-Aided Molecular Design*, **14**, 659–667.
Wood, W.E. (2001). Improved aroma barrier properties in food packaging with cyclodextrins. TAPPI – Polymers, Laminations and Coatings Conference, pp. 367–377.
Yamamoto, S., Kurihara, H., Mutoh, T., Xing, X. & Unno, H. (2005). Cholesterol recovery from inclusion complex of β-cyclodextrin and cholesterol by aeration at elevated temperatures. *Biochemical Engineering Journal*, **22**, 197–205.

3 Supercritical carbon dioxide and subcritical water: Complementary agents in the processing of functional foods

Keerthi Srinivas and Jerry W. King

3.1 Introduction

Functional foods generally refer to foods that, when ingested, play a vital role in rejuvenating the physiological functioning of the body (beyond those of basic nutrition), and in this process, decreasing the risk of infections and disease. Commonly known as nutraceuticals, these functional foods aid in enhancing stamina and energy, control of body weight and providing essential nutrients to human health. These nutraceuticals are marketed as tablets or pills or as additives in conventional foods and can provide important health benefits. Common health-enhancing ingredients in functional foods present in natural products are carotenoids, fatty acids, dietary fibers, phenolic compounds, isoflavones and sterols. The antioxidant, antimicrobial and antiviral properties of these functional foods have been noted extensively in the literature (Hasler 2002).

Depending on the definition of a nutraceutical, the global market ranges of such products are estimated to be between US\$123.9 billion (2008) and US\$176.7 billion by 2013 (Sugla 2008). There is an estimated 14% average growth in the market primarily due to growth in such markets in Asia and Latin America over the past few years (AROQ 2004).

Traditional methods used in extraction of these nutraceutical ingredients from natural food sources involve the use of water and organic solvents. However, there has been growing concern due to the cumulative toxicity of the organic solvents as well as pesticide and other chemical residues, and in some cases, synthetic food preservatives. To overcome some of these problems, critical fluid extraction (CFE) using solvents such as carbon dioxide, ethanol and water has been advocated (King 2000). Apart from the low cost of these solvents, supercritical fluid extraction (SFE) can provide higher selectivities, shorter extraction times, a potential reduction in the processing steps and enhanced product quality.

The inception of supercritical fluid extraction in food processing technology dates back the early 1970s. The initial application of SFE in food processing was in the decaffeination of coffee beans (Zosel 1974) followed by a number of other applications related to the extraction of hops, aroma compounds, decaffeination of tea, and many more. By the 1980s, supercritical fluids were extensively used in many extraction and reaction schemes as well as in the chromatographic analysis of food products (Valcarcel & Tena 1997). However, it was not until the late 1990s that supercritical fluid processing technology was embraced as a 'green' processing platform. Nutraceutical products touting labels such as 'naturally processed' or 'supercritical CO_2 treated' have gained greater appeal among consumers and government agencies that are cognizant of food safety issues.

Table 3.1 Nutraceuticals extracted and enriched using critical fluids and their therapeutic utility

Nutraceutical	Therapeutic utility
Bixin	Hyperglycemia (diabetes)
Betulin	Anti-inflammatory, antimalarial
Lycopene	Anticarcinogenic (prostrate, gastrointestinal)
β-Carotene	Anticarcinogenic (lung, cardiovascular)
Limonene	Anticarcinogenic (mammary glands)
Terpenes	Skin
Flavonoids	Antioxidant
Lutein	Macular degeneration
Phospholipids	Cognitive
Isoflavones	Premenstrual syndrome, circulatory
Essential oils	Antioxidant, dermatological, circulatory
Phytosterols	Circulatory
Tocopherols	Antioxidant
Saponins	Anticarcinogenic, lower body cholesterol
Lipids	Anticarcinogenic, antioxidant
Carotenoid esters	Antioxidant
Hyperforin, adhyperforin (St John's wort)	Antidepression

Table 3.1 shows a number of nutraceutical ingredients processed using critical fluids and their function in human nutrition. There has been considerable research performed in the extraction of carotenoid pigments such as carotene (Spanos *et al.* 1993), cholesterol and lipids (Froning *et al.* 1989), tocopherols and sterols (Mendes *et al.* 2002) and polyphenolic compounds (Palma & Taylor 1999) using supercritical carbon dioxide (SC-CO_2) as a solvent. However, in most cases it was found that a single extraction step is not enough and further steps are required to improve the purity and composition of the extract from a natural product source. Hence, a number of sequential processing steps as shown in Figure 3.1, where the resultant extract from a supercritical fluid extraction step is enriched using fractionation or reaction using supercritical fluids, become necessary. Studies related to the concept of sequential unit processing with multiple fluids have been discussed in detail by King and Srinivas (2009).

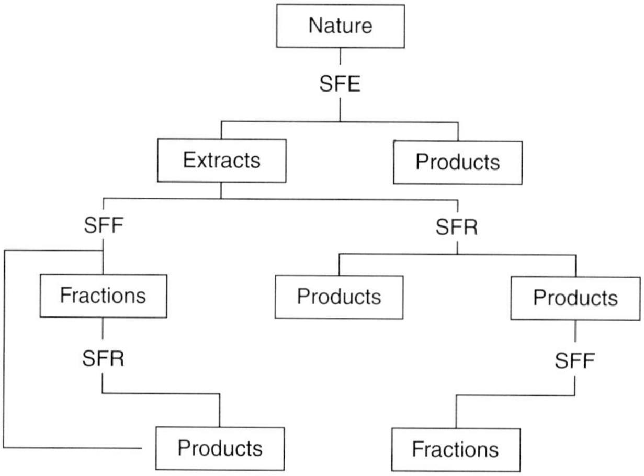

Fig. 3.1 Integration of multiple unit critical fluid processing for processing natural products.

In this chapter, we have reviewed some of the basic fundamentals of supercritical fluids and their use for the extraction and fractionation of food-related materials. The basic principles outlining the supercritical fluid extraction process, including factors such as solubility and phase equilibria of solutes in supercritical fluids, are discussed initially in this chapter. This is followed by a discussion at length on the application of supercritical fluids on both a laboratory and industrial scale.

3.2 Sub- and supercritical fluid solvents

Supercritical fluids such as carbon dioxide above its critical temperature (31°C) and pressure (73.8 MPa) exhibit physicochemical properties between those of a gas and a liquid. This allows supercritical fluids to dissolve compounds that may incompletely or poorly dissolve in the gas or liquid state. This dissolving power of the supercritical fluid is influenced by density, which is in turn dependent on the temperature and pressure conditions. These fluids also exhibit high diffusivity and low viscosity, which allows them to penetrate through the product matrix and dissolve the components from the raw material. The dissolved components can then be separated from the supercritical fluids by appropriately varying the pressure and temperature conditions (or low densities).

Table 3.2 shows the critical temperature and pressure of common sub- and supercritical fluids used for extraction and separation of nutraceutical ingredients, along with their solubility parameters calculated at room temperature. However, most studies have shown that SC-CO_2 tends to be selective toward low polarity and definitely non-polar compounds, while high polar and molecular weight compounds such as polysaccharides, amino acids, sugars, proteins, inorganic salts and flavonoid compounds have been found to be relatively insoluble in supercritical and liquefied carbon dioxide (del Valle & Aguilera 1999).

Since the early 1980s, studies have been performed to examine the effect of dense gases in terms of manipulating the solvent power of SC-CO_2. For gases such as nitrogen and helium, whose critical temperatures are lower than that of carbon dioxide, studies revealed a negligible effect of pressure on the solubility of various solutes in SC-CO_2. Actually, such gases when used as modifiers (i.e. antisolvents) reduce the solubility of caffeine (Gahrs 1984)

Table 3.2 Critical temperature, pressure and solubility parameter (at room temperature) for some common fluids

Fluid	Critical temperature T_C (°C)	Critical pressure P_C (MPa)	Solubility parameter δ^a ($MPa^{1/2}$)
Carbon dioxide	31.1	7.38	0.31
Ethylene	9.3	5.03	11.3
Ethane	32.3	4.88	11.6
Propane	96.7	4.24	12.7
Pentane	196.6	3.45	14.3
Hexane	234.2	3.09	14.8
Benzene	289.0	4.89	18.7
p-Xylene	343.1	3.52	17.9
Ethanol	240.8	6.30	26.2
Methanol	240.1	8.31	29.3
Water	374.2	22.04	48.1

[a] Solubility parameter values were obtained from Barton (1991).

and lipids (i.e. soybean oil) (King *et al.* 1995). Thus for SC-CO_2 extraction of caffeine, an increase in the concentration of nitrogen in CO_2 can dramatically reduce the solubility of the alkaloid.

The presence of dissolved gases in compressed dense fluids or liquids can also have a significant effect on solute extraction from natural products. For example, SC-CO_2 dissolved in pressurized water can facilitate extraction and/or a reaction due to the presence of carbonic acid (Toews *et al.* 1995). This concept has been explained by King and Srinivas (2009). Similar studies have been performed to study the solvent properties of SC-CO_2 with methanol and water, i.e. 'enhanced fluidity' liquids, as extraction agents (Wen & Olesik 2000).

By contrast, using cosolvents with critical temperatures greater than carbon dioxide, there is an increase in the solubility of solutes in SC-CO_2 with increasing modifier concentration in SC-CO_2. For example, the solubility of aromatic compounds such as benzoic acid, acridine, phthalic anhydride, 2-naphthol, 2-aminobenzoic acid and hexamethylbenzene using 3.5 mol percent methanol, acetone and octane as cosolvents is increased, permitting selective dissolution of some of these compounds at lower extraction pressures (Dobbs *et al.* 1987). Interestingly, these investigations have modeled this cosolvent-induced solubility enhancement using the solubility parameters of both the solutes and solvent mixtures. The solubility enhancement is primarily attributed to the higher polarity of methanol compared to the other mentioned cosolvents.

It is advantageous for the cosolvent to have intermediate volatility between that of the supercritical fluid and the compound to be extracted. Cosolvents utilized for SFE in the food industry are predominantly water and ethanol (King & Srinivas 2009). These solvents are considered as GRAS (generally regarded as safe) solvents having low toxicity. It has been suggested that by tuning the solvent power of a supercritical fluid by controlling factors such as temperature, pressure and density of the fluid, it is possible to recover a maximum yield of target product via a series of extraction steps (Osuna *et al.* 2005). Studies have indicated that with the rise of 'green' processing in various industries, the critical fluid technology platform using fluids such as water, carbon dioxide or ethanol, or a mixture of two or more of them in their supercritical or near-critical phase can enjoy widespread utilization (DeSimone 2002).

Table 3.3 shows the various multiple critical fluid processing options available for processing of natural products to yield nutraceutical components. We will study each of these cases in detail by using appropriate examples from the literature. Although abundant literature is available on processing using SC-CO_2, compressed water at temperatures and pressures above the boiling point of water has attracted increased interest especially for the extraction and separation of high polar compounds such as polysaccharides and flavonoids.

The phase diagram for water, as shown in Figure 3.2, is similar to that of carbon dioxide although the critical temperature and pressure of water are much higher ($T_C = 374°C$ and $P_C = 22.064$ MPa). As can be seen from Figure 3.2, the particular area of interest when using compressed water for the extraction of nutraceutical components is located between 100 and

Table 3.3 Multiple critical fluid processing options

Mode	Example
Single fluid	Carbon dioxide, water
Single fluid + cosolvent	Carbon dioxide + ethanol
Binary gases above their T_C	Carbon dioxide + hydrogen
Single fluid + dissolved gas	Water + carbon dioxide
Sequential fluids	Carbon dioxide, then water

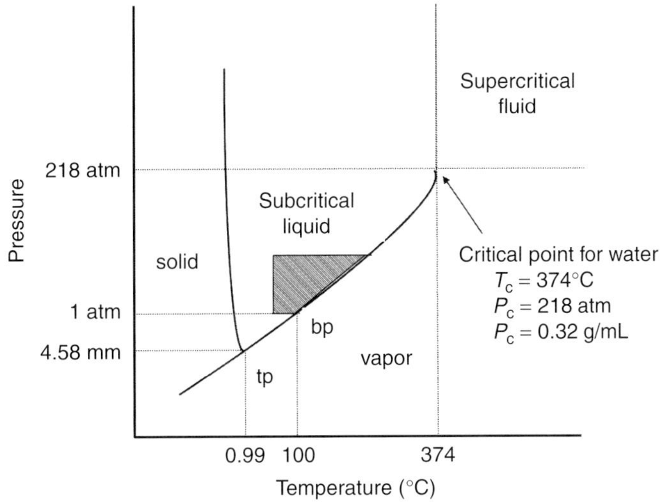

Fig. 3.2 Phase diagram of water.

200°C and at pressures less than 10 MPa (or the vapor pressure of water at the particular temperature). Studies have shown that subcritical water at these conditions is effective in extraction of antioxidant compounds from berry substrates (King *et al.* 2003), rosemary (Ibanez *et al.* 2003), oregano (Rodriguez-Meizoso *et al.* 2006) and winery by-products (Ju & Howard 2005; Garcia-Marino *et al.* 2006).

Concurrently, in the field of analytical chemistry, pressurized solvents are also used under subcritical conditions. However, these studies have indicated that the pressures used are far in excess to that required from the literature as vapor–liquid equilibrium curve given in Figure 3.2 (King 2006). Literature citations report that temperature is the key factor affecting the solvent power of water under subcritical conditions while pressure has but a minor effect on the solvent power (Hawthorne *et al.* 1999). Traditionally, this has been attributed to a decrease in the dielectric constant variation of water. We have attempted to explain the solvent behavior of carbon dioxide, water and ethanol in their sub- and supercritical conditions using an expanded solubility parameter theory (King *et al.* 2006, 2007). The solubility parameter, as defined by Hildebrand and Scott (1950), is the square root of the ratio of cohesive energy over the molar volume of the gas or liquid. However, in order to better understand the behavior of these compressed fluids, we extended the solubility parameter concept using the three-dimensional Hansen solubility parameter concept, which considers the contributions from the dispersion, polar and hydrogen-bonding forces between the solute and the sub-/supercritical solvent (Srinivas *et al.* 2008; King & Srinivas 2009).

Water does not attain an equivalent solvent power with organic solvents such as ethanol and methanol until very high temperatures and pressures. Figure 3.3 shows the three-dimensional solubility parameter components for water as a function of temperature. Note that the decrease in the solubility parameter of water with increasing temperature is primarily due to a reduction in hydrogen bonding as reflected by its hydrogen-bonding component, δ_h (Panayitou 1997). This concept is invoked again later in this chapter to estimate the solubility of structurally complex, biological solutes in sub- and/or supercritical fluids and predict optimized conditions for their extraction from natural product matrices.

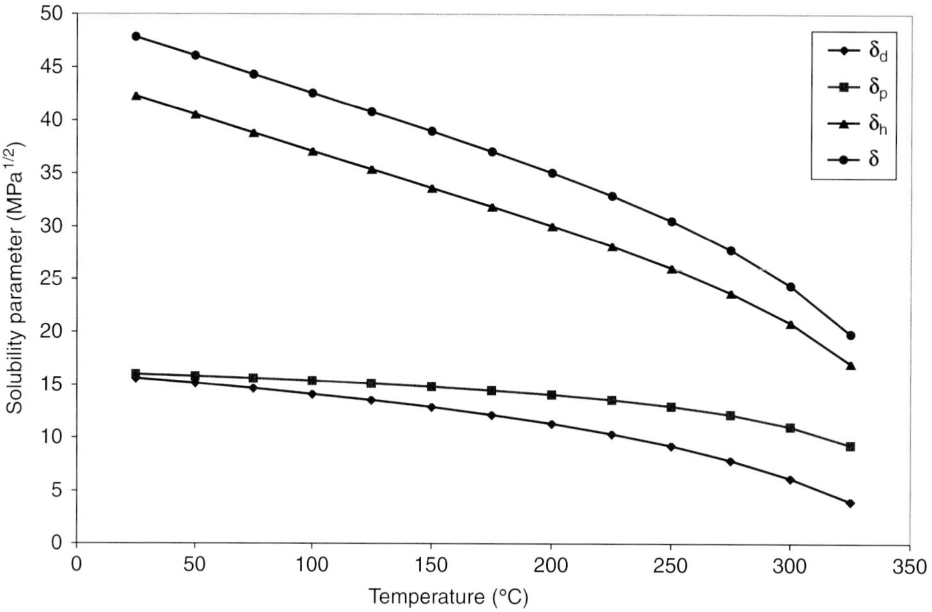

Fig. 3.3 Three-dimensional solubility parameters for water as a function of temperature (δ_d, δ_p and δ_h are the solubility parameters of water due to contributions of dispersive, polarity and hydrogen-bonding forces, respectively).

3.3 Sub- and supercritical fluid extraction

The utilization of sub- and supercritical fluids for the extraction of value-added compounds from natural products has been studied in detail. A typical extraction process design consists of a number of extraction vessels arranged with an appropriate valve sequence to permit several modes of operation. In such a system, carbon dioxide can be allowed to selectively enter any of the extraction vessels by using switching valves, and the extracts are collected through a back-pressure regulator or a micrometering valve by reducing the pressure. When one extraction vessel is in operation, the other extraction cells can be emptied or refilled. This is a basic design that can be even used on a pilot-plant scale by equipment manufacturers such as UHDE, Thar Designs, Applied Separations, Chematur and Separex (King 2004).

There has also been considerable progress in using expellers in supercritical fluid extraction operating in a continuous mode. Although this concept dates back to the mid-1980s where an auger-type screw press was used to assist in supercritical fluid extraction of oils from seed meal (Eggers 1996), only recently has this been realized on a plant production scale. In this system, the supercritical fluid contacts the natural product or seed matrix in the expeller barrel. The supercritical fluid phase is created due to the hydraulic compression of the feed matrix–fluid mixture, which results in an increase in the temperature and pressure in the expeller barrel. This compression process also enhances the fluidity of liquids being expelled from seeds in the expeller. This concept is being used commercially in the extraction of oil from seeds by Crown Iron Works in Minneapolis, USA. The commercial system used by Crown Iron Works is called the HIPLEX process and is due to initial research by Foidl

(1999). It has been reported that this process extracts close to 80–90% of oil from soybeans and 90% of available oil from canola.

Table 3.4 shows the supercritical fluid extraction of bioactive compounds from various natural product matrices with different temperature, pressure and cosolvent conditions. In all the studies shown in Table 3.4, it was found that using an alcoholic cosolvent, such as ethanol or isopropanol, a greater yield of bioactive compounds was achieved. For example, only about 30–40% of carotenes were recovered by extraction with SC-CO_2 at a temperature between 55 and 70°C and pressure of 34.5 MPa, but when using 10% ethanol as cosolvent, approximately 90% of the carotenes could be extracted at the same temperature and pressure conditions (Barth *et al.* 1995). In this study, it was established that pressure had a limited effect on the extraction of carotenes with supercritical CO_2 + 10% ethanol mixture. In a similar study by Vega *et al.* (1996), using response surface methodology models to quantify the carotene yield and using supercritical CO_2 + 10% ethanol mixtures at 50°C and 34.2 MPa, it was found that there was a considerable decrease in the amount of solvent and time required to extract carotenes using the supercritical fluid mixture in comparison with organic solvents used in traditional extraction process. However, in a study involving the extraction of carotenoids from leaf protein concentrates using SC-CO_2 in the absence of a cosolvent, it was found that no recovery of carotenoids occurred until around 30 MPa, above which approximately 90% of the β-carotene was recovered. It was also found in this study that for the extraction of lutein, a relatively polar carotenoid pigment, higher pressures (50–70 MPa) and SC-CO_2 volume were required for a maximum recovery of 70% (Favati *et al.* 1988).

In the processing of functional foods or nutraceuticals, application of subcritical water has been focused mainly on the extraction of polar compounds such as the antioxidants from natural product matrices. In order to study the effect of subcritical water as a solvent, however, it is also important to recognize its utility as a polar modifier for extraction of nutraceutical components using SC-CO_2. The effect of water and ethanol as cosolvents was evaluated in the treatment of tea leaves (Park *et al.* 2007), using extraction conditions of 50–80°C and 15–30 MPa. It was found that a greater amount of catechins was extracted in ethanol-modified SC-CO_2 in comparison with water-modified SC-CO_2 at these low cosolvent concentrations.

In another study, water was employed as a cosolvent for the SFE of sunflower leaves to yield bioactive compounds of nutraceutical value (Casas *et al.* 2007). In this study, both methanol and dimethyl sulfoxide were also used as modifiers. Water as a cosolvent yielded maximum extraction of bioactive compounds at 50°C and 50 MPa. Interestingly, lower extraction pressure and temperatures proved to be detrimental to the yield of bioactive compounds, especially when the sunflower leaf extract contained moisture. The maximum yield of bioactive compounds with water as a cosolvent at higher temperatures and pressures was rationalized in terms of higher density of water (when compared to SC-CO_2) under these extraction conditions, as well as the polar nature of the bioactive compounds being extracted.

Studies involving extraction of bioactive compounds from natural product matrices using subcritical water as a solvent are usually compared with traditional extraction techniques such as hydrodistillation and soxhlet extraction. Using subcritical water as solvent, bioactive compounds from kava roots (Kubatova *et al.* 2001) and anthocyanins from berry substrates (King *et al.* 2003; Ju & Howard 2005) have been extracted. All these studies reported a greater yield by a subcritical water extraction process in comparison with traditional extraction techniques. Continuous subcritical water extraction of nutraceutical components from laurel leaves has been reported by Fernandez-Perez *et al.* (2000). This continuous extraction system allows faster solvent flow rates at higher temperature and pressure conditions resulting in greater solvent flux through the substrate matrix, thereby resulting in faster extractions.

Table 3.4 Summary of studies on the supercritical carbon dioxide extraction of plant and botanical samples

Substrate name	Bioactive compound(s)	SFE conditions/ MPa/K/cosolvent	References
Avocado	Quercetin, isoquercetin, rutin, vitexin	10/333/EtOH	Takeuchi *et al.* (2006)
Basil	Eugenol	10–30/313	Leal *et al.* (2006)
Black pepper	Piperine, essential oils	20–32/318–338	Tipsrisukond *et al.* (1998) and Ferreira *et al.* (1999)
Carrots	β-Carotene	34.5/328–343/EtOH	Barth *et al.* (1995) and Vega *et al.* (1996)
Chamomile	α-Bisabolol, chamazulene	10–20/303–313	Povh *et al.* (2001)
Coriander	Volatile oil, phenolic compounds	20–30/298–331	Kraut *et al.* (2006)
Ginger	β-Pinene, *m*-diethyl benzene, o-diethyl-benzene, *ar*-curcumene, α-zingiberene, β-sesquiphellandrene	15–30/293–313	Zancan *et al.* (2002)
Grape seeds	Gallic acid, catechins, procyanidins	5–35/283–333; 30–40/288–323/EtOH; 20–30/313/EtOH	Gomez *et al.* (1996), Murga *et al.* (2000) and Cao and Ito (2003)
Jackfruit	Isoquercetin, vitexin	10/333/EtOH	Takeuchi *et al.* (2006)
Leaf protein concentrates	Lutein, β-carotene	10–70/313	Favati *et al.* (1988)
Onion	Essential oils	24.5/310	Dron *et al.* (1997)
Palm	Carotenoids, tocopherols, fatty acids, etc.	15–30/318–328	Franca and Meireles (1997) and Franca and Meireles (2000)
Paprika	Lipids, carotenoids and tocopherols	35/323; 15/313; 13.7–41.3/313	Vesper and Nitz (1997), Skerget *et al.* (1998), Jaren-Galan *et al.* (1999) and Illes *et al.* (1999)
Pitanga	Isoquercetin, rutin	10–40/273–333	Filho *et al.* (2008)
Rosemary	Camphor, carnosic acid, rosmarinic acids, phenolic diterpenes	10–30/303–313; 10–16/310–320	Ibanez *et al.* (1999) and Carvalho *et al.* (2005)
Turmeric	Circuminoids and terpenoids	20–30/303–318/ EtOH + IsoC3	Braga *et al.* (2005)
Tomato	Lycopene, β-carotene	17.2–27.6/313–353	Cadoni *et al.* (1999)
Vetiver roots	Zizanoic acid, α-vetivone, khusimol, bicyclovalencenol, isovalencenol	10–30/313	Talansier *et al.* (2008)

Bench-scale extractor units have been previously modified using booster-type pumps to allow for subcritical water extraction of natural products (King *et al.* 2003).

In order to maximize the yield of flavonoid components from natural products, hot water (under subcritical conditions) as a solvent was not sufficient. This was primarily due to a greater difference between the solubility parameter of water and the solute at the chosen operating temperature and pressure conditions. Subcritical water is still very dense and its solubility parameter is very high. Moderately polar modifiers such as methanol and ethanol are frequently used with subcritical water as reported in a particular study involving the extraction of flavonoids from grape pomace using subcritical water + ethanol mixtures between 40 and 140°C (Monrad *et al.* 2008).

3.3.1 Factors affecting sub- and supercritical fluid extraction

In order to better optimize the extraction process for maximizing the yield of functional food components from natural products, it is necessary to understand the thermodynamic and mass transfer parameters that affect the activity of structurally complex, biological solutes in sub- and/or supercritical fluids. The basic principle outlining the extraction of bioactive compounds using supercritical fluids is that the heated and pressurized solvent diffuses through the feed matrix where it dissolves the bioactive compounds and the mixture is transported out of the matrix. The various factors that affect the sub- and supercritical fluid extraction of such value-added compounds from natural products are as follows:

(a) Solubility of the bioactive compounds as a function of temperature and pressure
(b) Mass transfer or diffusion coefficients of the bioactive compounds in the sub- and supercritical fluids
(c) The chosen extraction or fractionation temperature, pressure, pH and solvent flow rate
(d) Substrate or matrix particle size
(e) Moisture content of the matrix
(f) Morphology of the sample matrix (flaked, ground, etc.)

3.3.2 Solute solubility in pressurized fluids

It is necessary to know the solubility maxima of the selected solutes in the sub- and supercritical solvents, as a function of temperature and pressure, to optimize extraction conditions of functional food components from natural product matrices (King 2000b). However, the solubilities of many nutraceutical compounds cannot be easily measured at high temperatures and pressures because of their tendency to thermally degrade. The CFE of these solutes from natural matrices is also dependent on the physical properties, particularly related to the molecular structure and temperature-dependent properties, of the solutes. Studies have been conducted to measure the solubility of some antioxidants in SC-CO_2 and SC-CO_2–cosolvent mixtures. The solubility of carnosic acid in SC-CO_2 decreased with an increase in temperature and a decrease in pressure (Chafer *et al.* 2005). Similar experiments were conducted to measure the solubility of catechin (Berna *et al.* 2001), epicatechin (Chafer *et al.* 2002) and quercetin (Chafer *et al.* 2004). In all the above-mentioned studies, it was found that the solubility of the solutes in SC-CO_2–ethanol mixtures was very low (of the order of 10^{-6} mole fraction) and this value increased with an increase in the concentration of ethanol. These low solubility values were primarily due to the presence of polar functional groups in the solute molecular structure. Hence, polar solvents such as water at high temperatures and pressures are an alternative and viable option to extract such compounds. However, there exist limited data on the solubility of such antioxidant compounds in, for example, water at its subcritical conditions.

Theoretical modeling of such solutes has been attempted to estimate the solubility of antioxidant compounds in sub- and supercritical solvents using cubic equations of state, group contribution methods such as the universal functional activity coefficient model and other models (Hartono *et al.* 2001; Diaz & Brignole 2009). A linear log–log relationship between the solute concentrations in the supercritical fluid phase and its density was reported by Chrastil (1982) for a number of food-related organic compounds. Solubility studies have focused on extraction curves, such as those reported for triglycerides from primrose seed oil in SC-CO_2 (King *et al.* 1999), soybean oil (Hong *et al.* 1990), sunflower oil (Perrut *et al.* 1997) and cotton seed oil (King 2000). These extraction curves are divided into three main sections (Sovova 2005): one, where the extraction rate is constant and is controlled by the thermodynamic equilibrium between the solute and the solvent; a second, non-lincar portion influenced by both internal and external mass transfer of the solutes from the matrix; and finally a section where the solute extraction rate is dependent on its internal mass transfer from the substrate. According to this model, most of the phase equilibria and solubility studies are made in the first portion of the curve where the effect of internal mass transfer is negligible.

Recently, we have studied the solubility of flavonoid-based antioxidants in pressurized water using the 'three-dimensional solubility parameter concept' (King & Srinivas 2009; Srinivas *et al.* 2009). A detailed review on the theory and principles behind the three-dimensional solubility parameter concept can be found in Hansen's tome (Hansen 2007). Initially, calculations of the Hildebrand solubility parameter for bioactive compounds were done using group contribution methods, such as that given by Fedors (1974). The temperature dependence of the solubility parameter of the bioactive compound was predicted using an equation given by Jayasri and Yaseen (1980). The solubility parameter of the solvents at sub- and/or supercritical temperature and pressure conditions was calculated using the law of corresponding states as given in Dunford *et al.* (2003). Hence, by plotting the solubility parameter of the solutes and the solvents as a function of temperature and pressure, we can estimate the solubility maxima of the solute in the solvent at a particular temperature and pressure. This prediction is based on the assumption that, when the solubility parameters of the solute and the solvent at a particular temperature and pressure condition are equal, the solute is completely miscible in the solvent at that particular temperature and pressure (Giddings & Keller 1969). A difference in the solubility parameter of the solute and the solvent of less than or equal to 2.5 Hildebrand units indicates miscibility of the solute in the solvent.

Figure 3.4 shows the solubility parameter of water and ethanol plotted as a function of temperature, at constant pressure of 5 MPa, along with the solubility parameter of betulin. Betulin is a natural antioxidant present in birch bark and is effective for treating cancers, viral and/or fungal infections. According to Figure 3.4, greater extraction of betulin should be achieved using ethanol at temperatures between 70 and 120°C than using water. This is based on the fact that the higher solubility of the solute in the solvent should occur when the solubility of the solute and solvent is closer to each other. This was substantiated by the higher yields of betulin obtained using ethanol at 90°C rather than hot water (Co *et al.* 2009).

The accuracy of such predictions can then be further extended using the 'three-dimensional solubility parameter concept'. In this model, the three-dimensional solubility parameters of the solutes were estimated using group contribution methods (Hansen 2007) and their variation with temperature predicted using an equation given by Jayasri and Yaseen (1980). However, it should be realized that the temperature dependence of the solubility parameter of the solute is dependent on the critical properties and the density of the biological solute, values which are very limited in the literature. Hence, group contribution methods such as those provided by Joback and Reid (1987) and/or Constantinou and Gani (1994) were

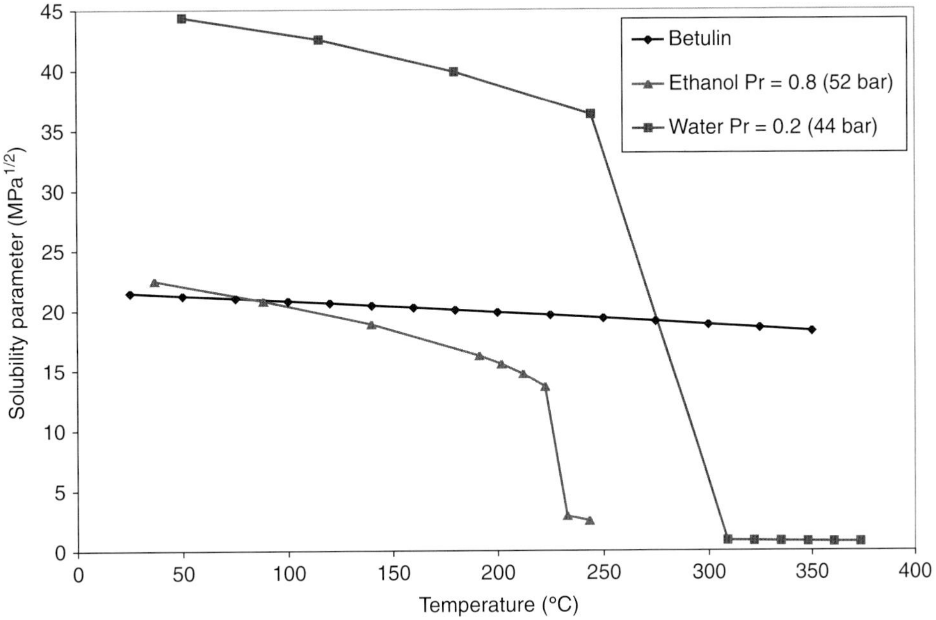

Fig. 3.4 Solubility parameter variation for betulin, water and ethanol with temperature.

utilized. The three-dimensional solubility parameters of the sub- and supercritical solvents were estimated using the set of equations provided by Williams *et al.* (2004), and the solvent densities at the specific temperature and pressure conditions were obtained from REFPROP (2007). The total solubility parameter (δ_T) of the solute or the solvent is given by:

$$\delta_T^2 = \delta_d^2 + \delta_p^2 + \delta_h^2 \qquad (3.1)$$

where δ_d, δ_p and δ_h are the solubility parameters of the solute or solvent due to contributions of dispersive, polarity and hydrogen-bonding forces, respectively.

The solubility parameter of carbon dioxide is around 0.31 $MPa^{1/2}$ at ambient temperature and pressure. Above the critical point, the solubility parameter of carbon dioxide increases to approximately 12.27 $MPa^{1/2}$. The variation in solubility parameter of carbon dioxide at different temperature and pressure combinations is shown in Figure 3.5. We can see from Figure 3.5 that the solubility parameter of CO_2 increases with an increase in pressure and a decrease in temperature. The solubility parameters of a selected group of carotenoids present in spices such as black pepper, cayenne, cinnamon, garlic, ginger, licorice, onion and chives are shown in Table 3.5. In order to optimize the SC-CO_2 extraction of such solutes from spice matrices, the solubility parameters of these carotenoids were compared with the solubility parameter curves of SC-CO_2 plotted as a function of pressure at 25, 40 and 60°C, as shown in Figure 3.6. We can see from Figure 3.6 that very high pressures are required for the extraction of these solutes from the spices. Mukhopadhyay (2007) suggests an optimum temperature range for SC-CO_2 extraction of carotenoids from spices as 35–65°C with pressures in excess of 60 MPa.

The addition of ethanol as a modifier to SC-CO_2 has shown promise in the extraction of nutraceutical components having a greater polarity than CO_2. Figures 3.7a and 3.7b show the variation of solubility parameter of SC-CO_2 and ethanol mixtures as a function of

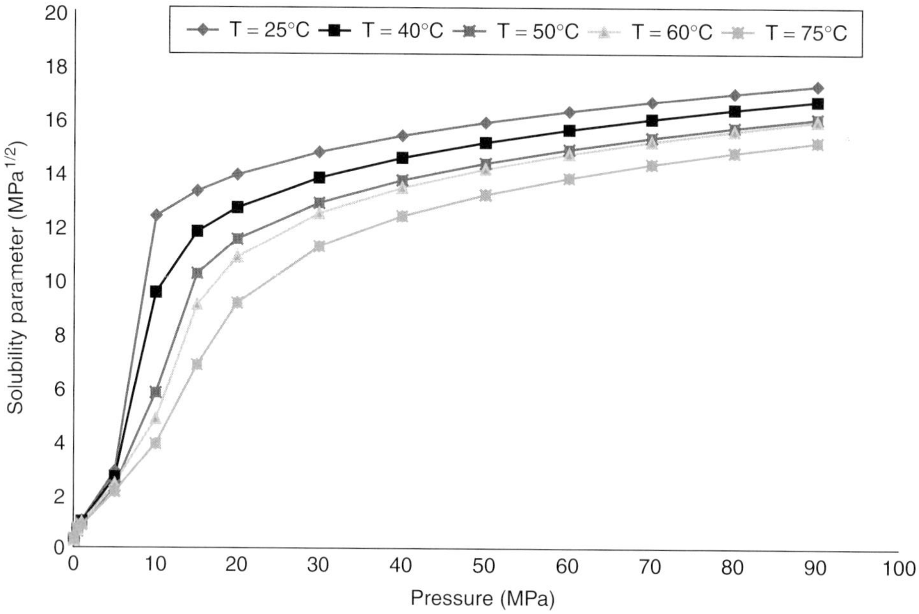

Fig. 3.5 Variation in solubility parameter of carbon dioxide at different temperature and pressures.

Table 3.5 Three-dimensional solubility parameters of certain spice oleoresin and their esters at room temperature

S. No.	Oleoresin	Dispersion solubility parameter, δ_d (MPa$^{1/2}$)	Polar solubility parameter, δ_p MPa$^{1/2}$	Hydrogen-bonding solubility parameter, δ_h (MPa$^{1/2}$)	Total solubility parameter, δ_T (MPa$^{1/2}$)
1	Curcumin	23.2	7.88	16.8	29.5
2	Dihydrocapsaicin	20.7	6.32	10.8	24.1
3	Capsaicin	19.6	6.01	10.5	22.9
4	Piperine	19.4	7.10	8.16	22.6
5	Capsanthin	17.5	4.64	10.2	20.6
6	Lutein	17.5	3.97	9.98	20.4
7	Violaxanthin laureate ester	18.0	4.83	7.79	20.2
8	Zeaxanthin laureate ester	17.8	4.52	7.53	19.7
9	Lutein stearate ester	17.6	3.78	7.14	19.3
10	Capsorubin	17.8	3.6	6.01	19.0
11	Lutein stearate diester	17.5	3.62	4.91	18.4
12	Cryptoxanthin laureate ester	17.4	3.50	5.40	18.4
13	β-Carotene	17.1	2.39	5.54	18.0

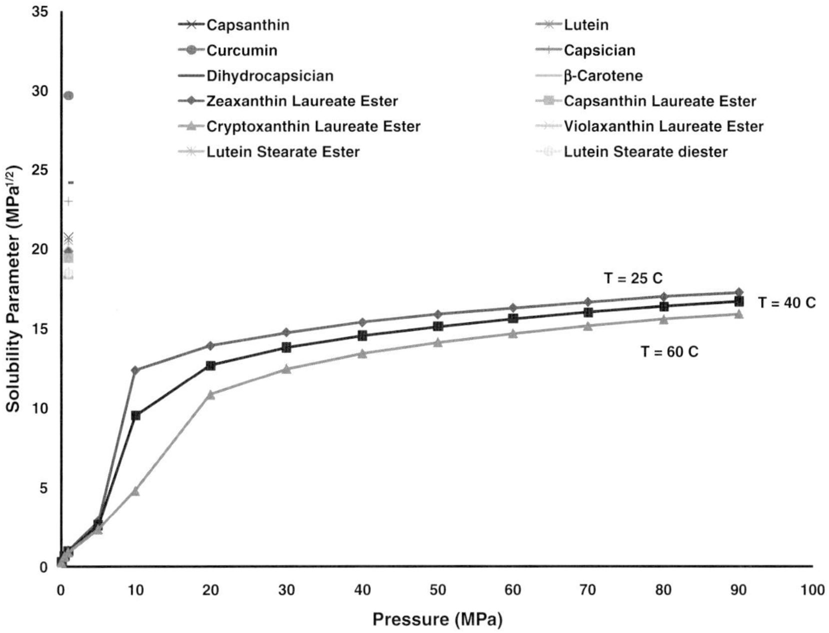

Fig. 3.6 Comparison of solubility parameters of certain solutes present in spices with the solubility parameter of carbon dioxide plotted as a function of temperature and pressure.

temperature at a pressure of 10 MPa, and with the mol percent composition of ethanol at a temperature of 35°C and pressure of 10 MPa, respectively. The pressure and temperature selection was made after reviewing experimental data. Figure 3.7a shows a change in the variation of total solubility parameter of carbon dioxide–ethanol mixtures with an increase in the concentration of ethanol and also temperature. The solubility parameter of SC-CO_2 decreases with temperature at a particular pressure, but we see in Figure 3.7a that for carbon dioxide–ethanol mixtures, the solubility parameter decreases until it reaches a minimum value at around 45°C and then starts to increase with temperature. This increase in the solubility parameter is due to an increase in the hydrogen-bonding solubility parameter with increasing concentration of ethanol in SC-CO_2, as predicted in Figure 3.7b. Such changes in the solubility parameter of carbon dioxide with addition of cosolvent at a particular temperature and pressure can be used to rationalize the increased extraction yield of solutes, for example carotenoids from spices as shown in Table 3.4.

Using the three-dimensional solubility approach coupled with computerized optimization using SPHERE and Hsp3D programs (Hansen 2007), we can plot Hansen spheres that show the miscibility region of the solvent in a three-dimensional axis. Using such an approach, the distance between the solute and the solvent inside the solubility envelope would determine the optimum conditions for dissolution of the solute in the solvent.

Figures 3.8a and 3.8b show Hansen solubility spheres plotted for betulin versus the solubility parameters of ethanol and water at different temperatures. The inverted open triangle on one side of the center of mass of the sphere is the solute (betulin, in this case) and the inverted triangles on the other side of the center of mass of the sphere are the temperature-dependent solvent points that lie within the Hansen solubility space, and a condition of miscibility between the pressurized solvent and betulin. The darkened black

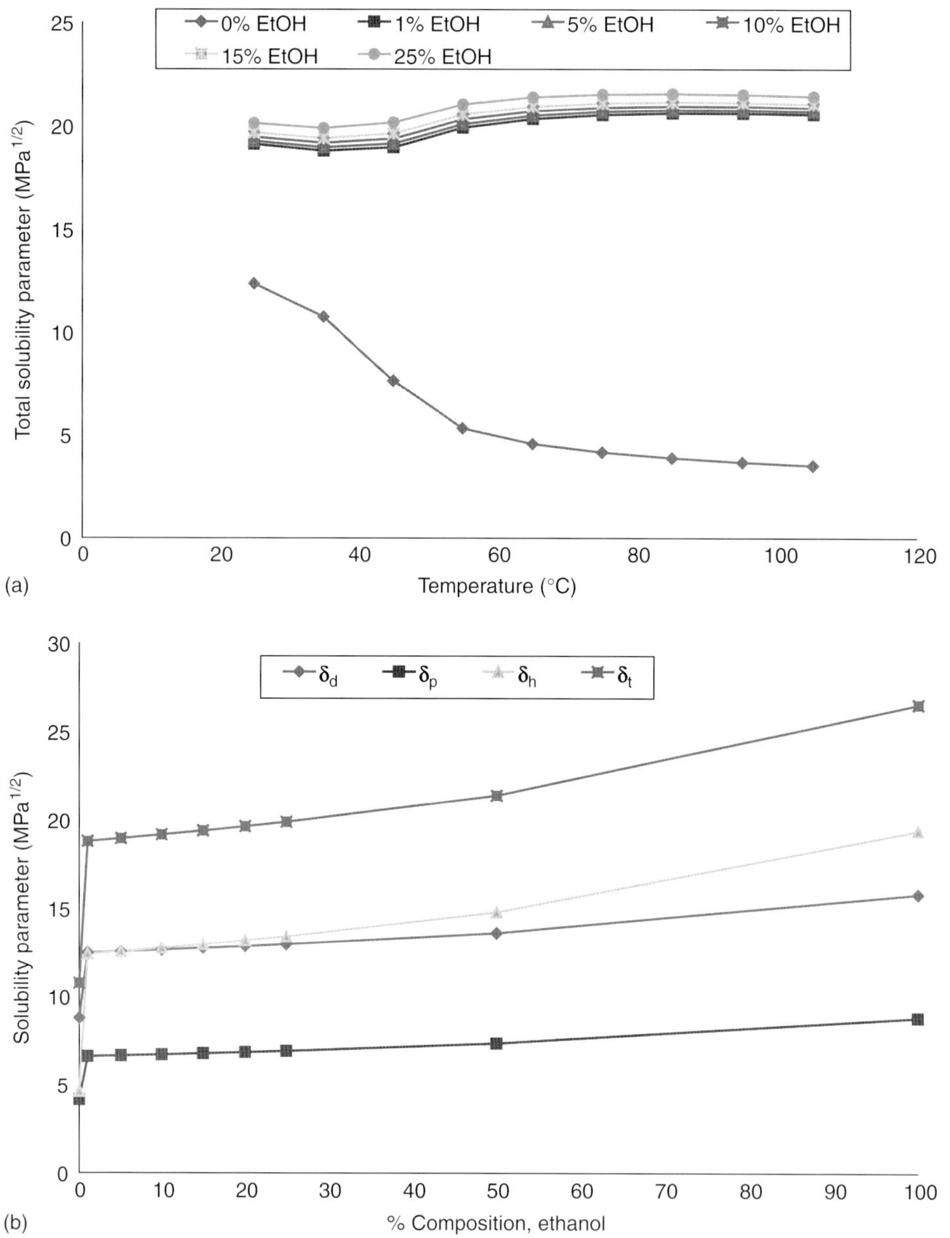

Fig. 3.7 (a) Variation in solubility parameter of CO_2 + ethanol mixtures with temperature at a pressure of 10 MPa; (b) variation in three-dimensional solubility parameter of CO_2 + ethanol mixtures with percent composition of ethanol at a temperature of 35°C and pressure 10 MPa.

triangles indicate immiscibility between the solute and the subcritical fluid at each of the specific temperatures. The relative energy difference (RED) is defined as follows:

$$\mathrm{RED} = \frac{R_\mathrm{a}}{R_\mathrm{o}} \tag{3.2}$$

where R_a shows distance between the solute or solvent and the center of mass of the sphere and R_o shows radius of the Hansen sphere. An RED value less than 1 indicates miscibility of the solute in the solvent, while RED value more than 1 indicates an immiscibility condition. An RED value equal to 1 indicates possible miscibility of the solute in the solvent.

It can be seen in Figure 3.8 that the radius of the Hansen sphere for the betulin–ethanol system is less than that for the betulin–water system, verifying that ethanol is a better solvent for betulin. The RED values indicate a maximum solubility of betulin in ethanol at 100°C although some miscibility between betulin and ethanol can be witnessed between 50 and 125°C. This is in agreement with the experimental data as cited above. A similar approach was used in predicting the extraction conditions of target solutes from natural product matrices: silymarins from milk thistle, vitamins B from brewer's yeast and anthocyanins from grape pomace (Srinivas *et al.* 2009).

It is evident from the above discussion that subcritical water is not as efficient for extracting such polar solutes from natural product matrices when compared with ethanol. This is primarily due to the greater hydrogen-bonding capacity of water in comparison with the alcohol. For example, in the extraction of anthocyanins from grape pomace, the solubility parameter of malvidin-3*O*-glucoside, a primary flavonoid present in grapes, is 33.8 $MPa^{1/2}$ at ambient conditions while that of water has been reported as 47.8 $MPa^{1/2}$ (Hansen 2007).

In order to efficiently extract malvidin-3-*O*-glucoside from grape pomace substrate, the use of an ethanol–water mixture as a solvent at temperatures between 25 and 245°C was modeled (Srinivas *et al.* 2008). Figure 3.9 shows the variation of the minimum radius of the Hansen solubility sphere for malvidin-3*O*-glucoside with the percent composition of ethanol in subcritical water. The minimum radius specified in Figure 3.9 refers to the smallest sphere formed by the computerized optimization method for complete miscibility of the solute in the solvent in a particular temperature range. The optimized temperature for the solubility maxima of malvidin-3*O*-glucoside in the solvent is based on the RED values obtained using the Hsp3D program. A study of the RED values for the malvidin-solvent system indicates maximum solubility in the water–ethanol mixture at temperatures between 25 and 75°C. There is an initial drop in the solubility parameter of water with addition of ethanol primarily due to the decrease in its hydrogen-bonding solubility parameter component. Note that although in Figure 3.9 the minimum interaction radius occurs at 80% ethanol concentration, there is not a significant difference with that recorded using only 10% ethanol concentration. This indicates a greater yield of anthocyanins from grape pomace using a lower concentration of ethanol in water at subcritical temperatures. Similarly, a greater yield of catechins from grape pomace was observed using 50% ethanol–water mixture at 65°C (Savova *et al.* 2007), which can be rationalized by this approach.

3.3.3 Other factors influencing sub- and supercritical fluid extraction

The mass transfer and the solubility of nutraceutical components in sub- and supercritical solvents are influenced by temperature, pressure, pH and solvent flow rates. Pressure plays the dominant role in supercritical CO_2 extraction processes, while temperature is the major factor in subcritical water extractions. The effect of pressure in extraction of essential oils from seed matrices was studied by Machmuddah *et al.* (2008) at 40–80°C and 15–49 MPa. SC-CO_2 was found to extract bioactive compounds, such as amygdalin from loquat seeds (which have applications in cancer therapy). It was found that a maximum yield of the

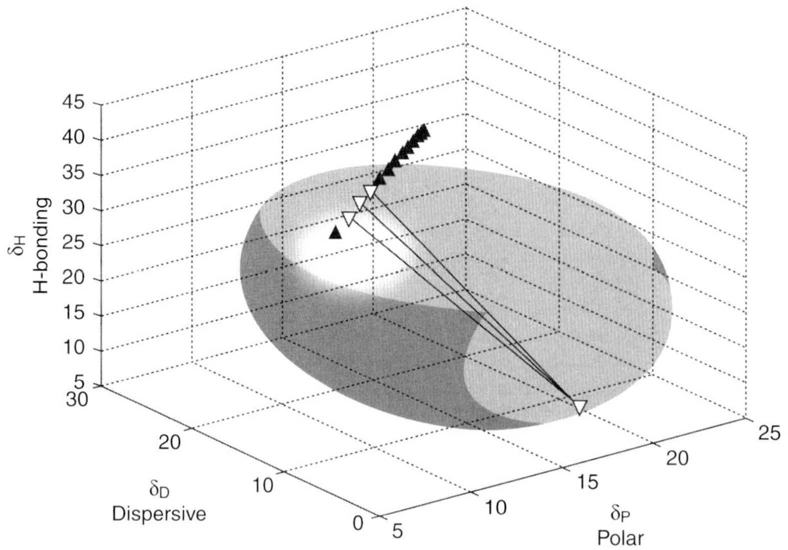

Betulin-Water System

Hansen sphere
Center of mass: D = 15.3, P = 12.0, H = 23.0, RAD = 16.53 $MPa^{1/2}$
True Volume = 9451.2 $MPa^{3/2}$, **True Radius** = 13.12 $MPa^{1/2}$

(a)

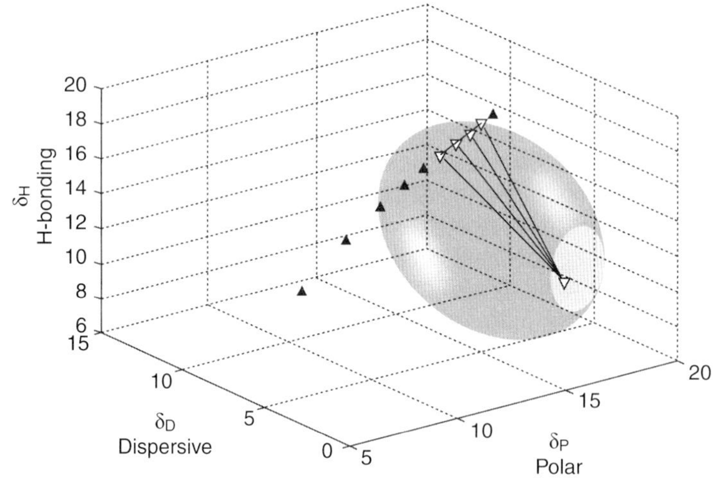

Betulin-Ethanol System

Hansen sphere
Center of mass: D = 15.4, P = 5.1, H = 12.7, RAD = 5.68 $MPa^{1/2}$
True Volume = 383.6 $MPa^{3/2}$, **True Radius** = 4.51 $MPa^{1/2}$

(b)

Fig. 3.8 Variation of Hansen three-dimensional solubility parameter sphere of betulin with (a) subcritical water and (b) ethanol at different temperatures.

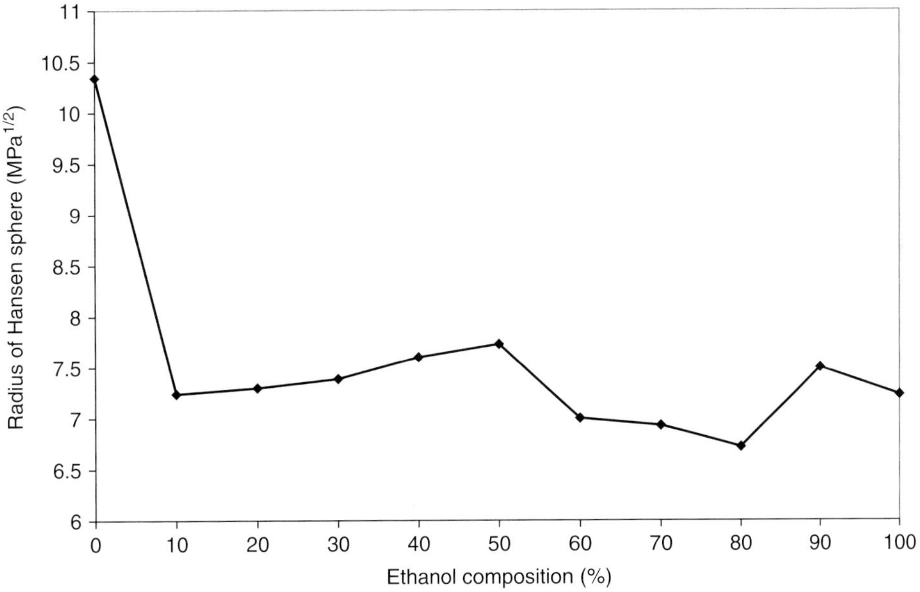

Fig. 3.9 Variation in the Hansen sphere radius (MPa$^{1/2}$) versus percent ethanol for malvidin-3O-glucoside in compressed water–ethanol extraction solvent.

bioactive compounds was obtained for lower residence times during the extraction from the seed matrix with an increase in extraction pressure. Higher temperatures proved a detrimental factor in this particular study due to the co-extraction of waxes and certain toxic substances, such as lectins, from the seed matrix.

When using water and ethanol as solvents (whose boiling points are higher than that of carbon dioxide), the solvent flow rate and the residence time of the organic solute in the extraction cell play a vital role in achieving maximum yield of such solutes from natural matrices. One particular study aimed at optimizing the extraction of flavonoid compounds from red onions using pressurized water with 5% ethanol as a modifier showed thermal degradation at higher temperatures within 8 minutes from the start of the extraction process (Petersson *et al.* 2009). Similar studies were conducted to evaluate the degradation kinetics in subcritical water extraction of silymarin compounds from milk thistle between 100 and 140°C (Duan *et al.* 2004). This study, conducted at a constant water flow rate at different temperatures, demonstrated an increase in thermal degradation of the silymarins with an increase in temperature and residence time in the extractor. When ethanol was added as a cosolvent to subcritical water, the degradation of silymarins was reduced significantly with increasing ethanol content (Duan 2005).

For continuous extractions using SC-CO_2, it was found that an increase in flow rate usually resulted in the achievement of maximum yield of nutraceutical components at shorter extraction times (King 2000b). However, it should also be noted that a very high flow rate would lead to a compaction of the feed matrix bed, thereby resulting in the channeling of fluid flow through the extraction system, which could affect the yield of desired products.

The control of solution pH can affect the equilibrium-based extraction of compounds when using sub- and supercritical fluids, as is the case for anthocyanins and similar flavonoid-based solutes (Clifford 2000). Traditionally, mineral acids have been used for controlling solution

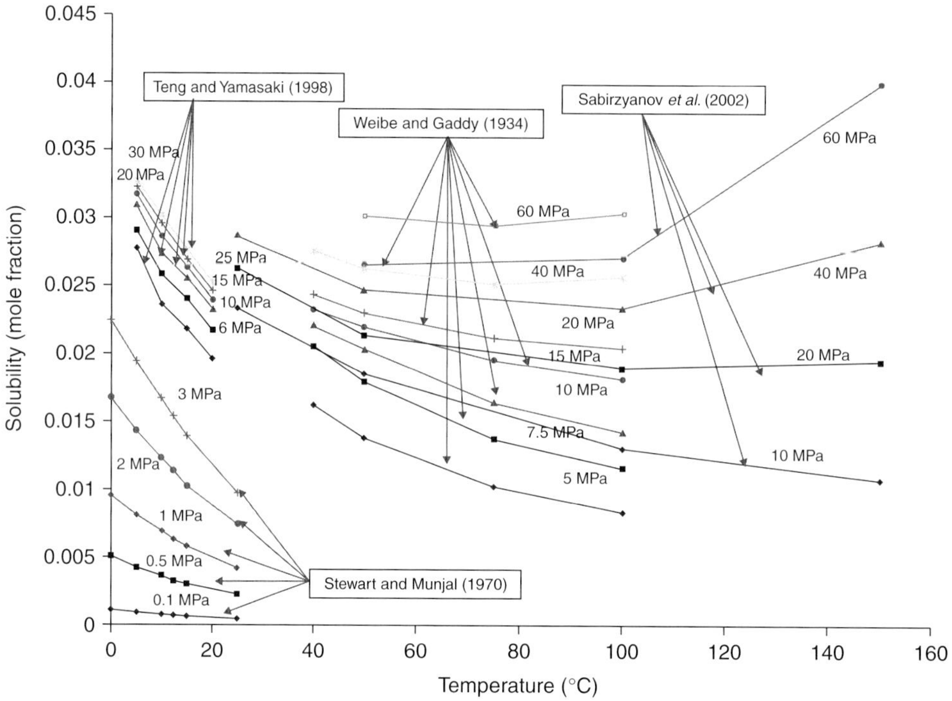

Fig. 3.10 Mole fraction solubility of CO_2 in water as a function of temperature and pressure. For a colour version of this figure, please see the plate section.

pH. However, for truly 'green' processing, the corrosive and toxic nature of these mineral acids is not always desired, nor their presence as a residue in the final processed foods. Another alternative source of pH control is dissolution of supercritical CO_2 in water. Figure 3.10 shows the solubility of carbon dioxide in water at different temperature and pressure conditions obtained from literature (Weibe & Gaddy 1934; Stewart & Munjal 1970; Teng & Yamasaki 1998; Sabirzyanov *et al.* 2002), indicating that some degree of acidic pH control is possible through the use of compressed CO_2 in water.

Other factors which influence the extraction process, as noted previously, are the moisture content and particle size of the substrate being extracted. Studies have shown the effect of particle size on extraction efficiency for SC-CO_2 (del Valle & Uquiche 2002) and subcritical water (Kubatova *et al.* 2001; Eikani *et al.* 2007). In general, grinding the matrix into fine particles allows for maximum extraction efficiency of functional food ingredients with sub- and supercritical fluids. However, there are case studies where an increase in particle size did not significantly affect the yield of the desired solutes.

In order to prevent degradation of carotenoids in red pepper extracts due to the presence of biologically active enzymes present in the red peppers, the water activity (a_w) in the matrix was decreased to 0.30 (Sun-Lee *et al.* 1992). However, decreasing water activity in a feed matrix involves utilization of drying operations, which can affect the thermal degradation of these bioactive compounds in the extracted substrate. Freeze drying is a viable option to control the water activity in natural products even though it requires a greater energy input than other dehydration methods. SC-CO_2 extraction of sweet potatoes at 48°C and 41.4 MPa showed a greater yield of carotenoid pigments with a freeze-dried sample in comparison with oven-dried samples (Spanos *et al.* 1993). A study on the effect of particle size and moisture content in supercritical fluid extraction of nutraceutical components from paprika revealed that a moisture content greater than 18% and particle size greater than 0.7 mm reduced the yield of the desired end product (Nagy & Simandi 2008). These numbers provide the reader with an idea of the effect that these factors have on extraction yield.

3.4 Tandem processing using sub- and supercritical fluids

Table 3.6 summarizes the various critical fluid-based options for processing of food and natural products using sub- and supercritical fluids as applied to essential oils. According to the processing options provided in Table 3.4, it is possible to efficiently separate and purify biological components from natural product matrices, using a combination of one or more unit operations with sub- and/or supercritical fluids as solvents.

There has been considerable progress in the bench-scale evaluation of methods for extracting nutraceutical components from food products. Such studies have been aimed at optimizing the extraction and/or reaction conditions for scaling up the process to pilot-plant scale and ultimately into full-scale industrial applications. Such scale-up issues involve mechanical complexities, safety issues, economic viabilities and production flexibility.

In order to increase extraction or fractionation efficiency when using supercritical fluid processing of nutraceutical components from food and natural products, it is necessary to use techniques such as columnar fractionation and chromatography. The need for these processes lies in the requirement for enriched or more purified forms of nutraceutical ingredients from the natural product matrix for incorporation into foods.

3.4.1 Supercritical fluid fractionation coupled with SFE

Fractionation columns can be operated in either a concurrent or countercurrent mode. An example of a thermal gradient supercritical fluid fractionation (SFF) column operating in concurrent mode is shown in Figure 3.11. Here, the supercritical CO_2-extracted mixture is subjected to a temperature gradient along the length of the fractionating tower, where each of the designated sections of the column has an increasing temperature in the sequence T1, T2, T3 and T4. As the mixture passes through each column section, the components in the mixture are separated, depending on their solubility in carbon dioxide, the specific column

Table 3.6 Coupled processing options for pressurized fluid extraction of essential oils from natural products

Process	(1) SFE (SC-CO_2)
	(2) SFF (SC-CO_2) – pressure reduction
	(3) SFF (SC-CO_2) – column deterpenation
	(4) SFC (SC-CO_2/cosolvent)
	(5) SFF – (subcritical H_2O deterpenation)
	(6) SFM – (aqueous extract/SC-CO_2)
Processing combinations	(1) + (2)
	(1) + (2) + (3)
	(1) + (4)
	(1) + (5)
	(1) + (5) + (6)

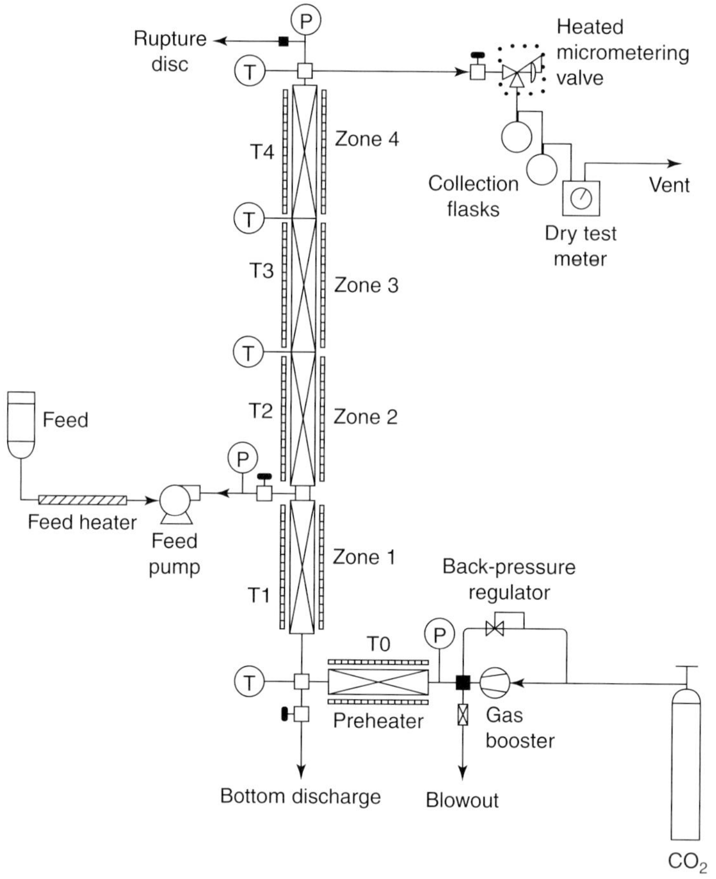

Fig. 3.11 Thermal gradient supercritical fluid fractionation column.

temperature and their respective vapor pressures. The variation in solubility parameter of carbon dioxide with temperature and pressure shown in Figure 3.4 can be used as an example of the observed solubility trend across the length of the fractionating tower. This type of fractionating tower approach has been used to deacidify olive oil, deterpenate citrus oils and fractionate fish oils or butterfat (King 2004). For example, by employing a thermal gradient fractionating column after preliminary extraction into SC-CO_2, glyceride mixtures could be separated by maintaining the four column stages between 60 and 90°C as shown in Figure 3.11 (King *et al.* 1997). In this case, a mixture of glycerides containing approximately 35–45 weight percent monoglyceride content is fed into the column right above the heated zone T1 (Figure 3.11). This glyceride mixture then makes contact with the rising SC-CO_2 stream and is fractionated along the length of the column, resulting in an enriched monoglyceride mixture at the top of the column (~95% enrichment). Studies have indicated that there are advantages to using the countercurrent mode of operation in place of a concurrent mode when using a fractionating column with supercritical fluids. Studies conducted using a countercurrent fractionation column include extraction of tocopherols from palm kernels (Zaidul *et al.* 2006) and carotenoid pigments from olive leaves (Tabera *et al.* 2006). In both the studies, it was found that higher temperatures and pressures yielded the maximum amount of desired solutes.

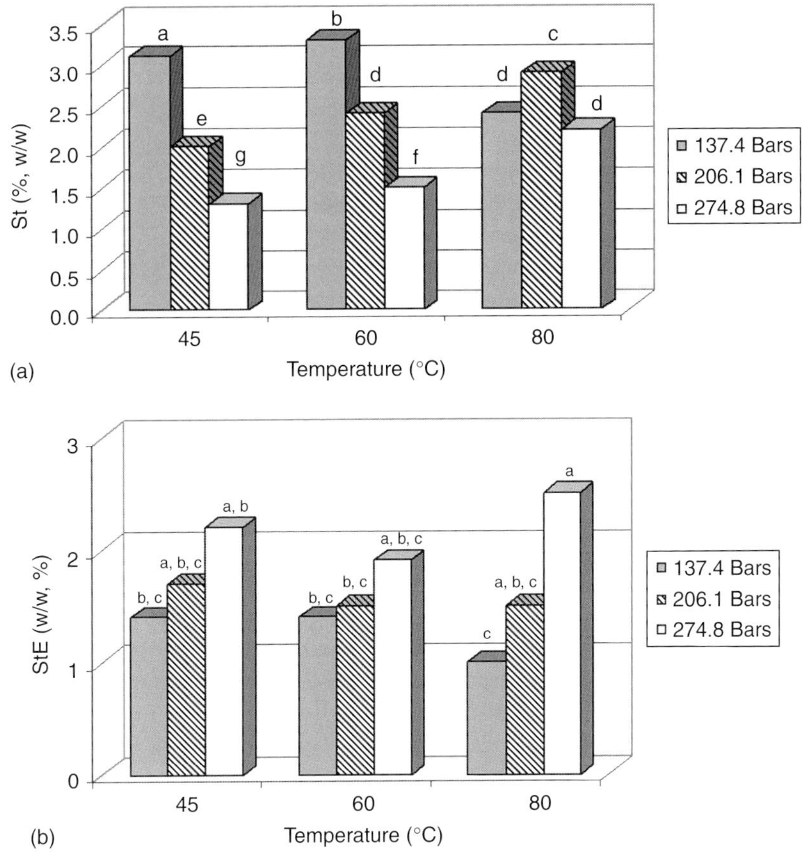

Fig. 3.12 Effect of temperature and pressure on the (a) phytosterol content (St) and (b) phytosterol esters (StE) content of SFF extract fractions (adapted from Dunford *et al.* 2003).

This countercurrent columnar mode technique was also used for the deacidification of crude rice bran oil (Dunford *et al.* 2003). This study was performed on a pilot-plant scale, involving a stainless steel fractionating column approximately 1.83 m in length and 10.2 cm outer diameter. The fractionation of crude rice bran oil, containing initial free fatty acid (FFA) content of approximately 5 weight percent, was performed using a columnar temperature gradient of 45–80°C and pressures of 13.8–27.5 MPa. Figures 3.12a and 3.12b show the effect of temperature and pressure on the fractionation of free phytosterols and fatty acid esters of free phytosterols. It was found that higher pressures and lower temperatures were detrimental in maximizing the removal of FFA from rice bran oil primarily due to loss of triglycerides under these conditions. From Figures 3.12a and 3.12b it can be seen that the maximum yield of free phytosterols was obtained at 60°C and 13.8 MPa, while maximum yield of fatty acid esters of the phytosterols was obtained at 80°C and 27.5 MPa. Oryzanol was only detected in the extract collected at 27.5 MPa and 45°C. Similar results were reported in a study on a pilot-plant scale process for removal of FFA from rice bran oil, with the extractor being maintained at 40–50°C and 24.1 MPa (Shen *et al.* 1997). Here, the fractionation efficiency was modeled by studying the fatty acid content in the raffinate and extract phase, showing an overall improvement in efficiency of a supercritical fluid fractionation scheme by coupling fractionation columns with SC-CO_2 for extracting nutraceutical components of commercial value.

In the studies described above, it can be summarized that the appropriate fractionation range for using a supercritical fluid can be established by varying the temperature and pressure in the column to establish differences in solute solubility in the supercritical fluid under a given set of operating conditions. This potential range to selectivity extract and fractionate solutes in a supercritical fluid frequently occurs in the 'crossover region' (Stahl *et al.* 1988) where the differences in solute solubility in a supercritical fluid are coupled with vapor pressure differences in the target solutes to selectively dissolve and fractionate botanical mixtures. This region was studied in the extraction of carotenoid pigments such as lycopene and rubixanthin from pitanga fruits using SC-CO_2 (Filho *et al.* 2008). Their studies showed that at 60°C and 25 MPa, almost 74% of rubixanthin and 78% of lycopene isomers were extracted. The yield increased by a factor of 66% when the system was maintained at 40 MPa in comparison with that resulting from extraction at 10 MPa and 40°C. It was found that the extraction yield increased with temperature at higher pressures while an inverse solubility trend was observed at lower temperatures. Modeling of the experimental data revealed a crossover region in terms of the yield of desired solutes as a function of pressure (15–20 MPa) – temperature exerting no effect on the overall yield.

Although fractionation studies (SFF) using SC-CO_2 have been performed, there are several disadvantages associated with this purification process. Fractionation of solute mixtures using a supercritical fluid requires significant changes in the polarity or molecular weight of the components to achieve complete separation. However, for complex natural products, such as those occurring in many nutraceutical products, SFF cannot often completely separate the natural product into its constituent components. However, the resulting product mixture can have application in the nutraceutical field (King 1991) since many existing nutraceutical products have low levels of bioactive ingredients in a mixture derived from a particular extraction process.

3.4.2 Supercritical fluid chromatography coupled with SFE

Figures 3.13a and 3.13b show processing schemes for coupled extraction and chromatographic separation of tocopherols (King *et al.* 1996) and phospholipids (PLs) (Taylor *et al.* 1999) from soybean flakes, respectively. In the first study, shown in Figure 3.13a, SC-CO_2 at 80°C and 25 MPa was used to extract a tocopherol–soybean oil mixture from soybean flakes. This can be achieved using a simplified apparatus, as shown in Figure 3.14. The process design comprises a packed bed column attached in series with an SFE unit. In this study, the supercritical fluid chromatography (SFC) column was maintained at 25 MPa and 40°C with silica gel as the sorbent. This allowed for further concentration of the tocopherol moieties, as shown in Figure 3.13a. It can be seen from the inset table in Figure 3.14 that the enrichment of tocopherols from soybean flakes was increased by a factor or 2–10 by using SFC in tandem with SFE.

As shown in Figure 3.13b, PLs have limited solubility in SC-CO_2, and hence, ethanol was used as a cosolvent. In this study, the soybean flakes were initially extracted with SC-CO_2, followed by sequential SFE–SFC utilizing SC-CO_2 – cosolvent mixtures on the lecithin-containing residue, which remain after the SC-CO_2 extraction. As shown in Table 3.7, the SFE–SFC step produced an extract containing a total of 43.7% by weight of the PLs. By using SFC with an alumina sorbent and SC-CO_2 modified with ethanol cosolvent (5–30% v/v),

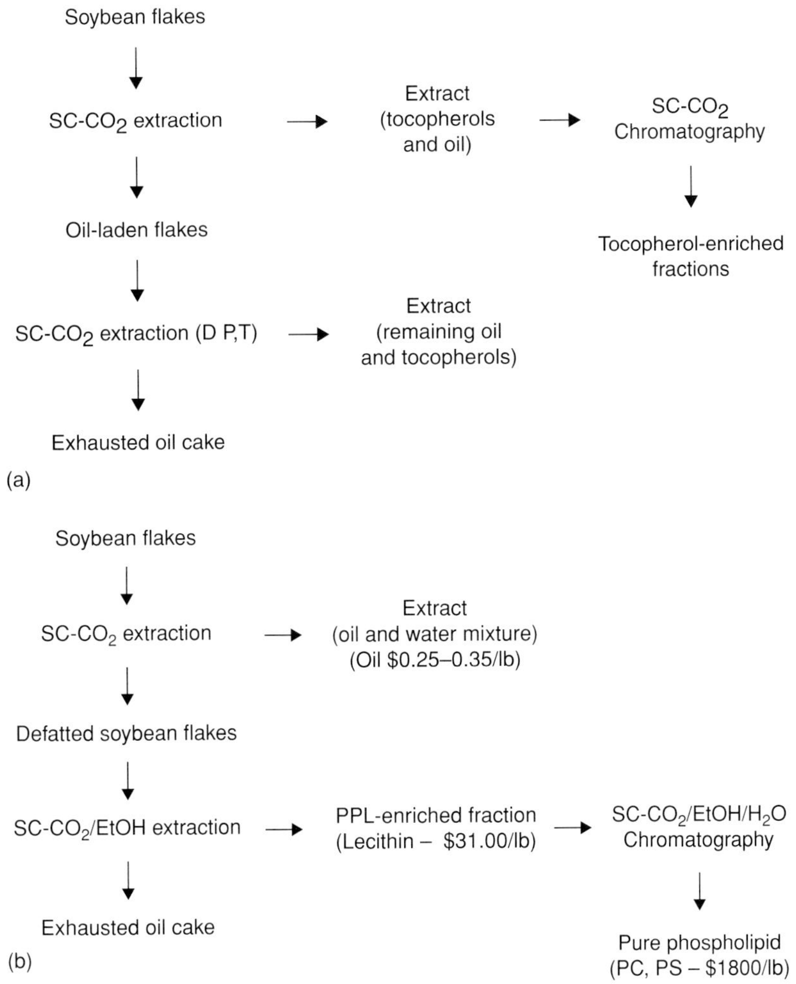

Fig. 3.13 (a) Tocopherol and (b) phospholipid enrichment/fractionation using coupled supercritical fluid-based processes.

and a small quantity of water in the eluent, enrichment of the PL fraction increased by a factor of 2–5 (Table 3.7). The purities of individual PLs, such as phosphatidylethanolamine and phosphatidylcholine, by SFC were approximately 75%. Note that only GRAS solvents are used in both the SFE and SFC steps.

In order to justify the incorporation of the SFC step, a cost analysis was performed as shown in Figure 3.13b, based on the selling price of the PLs. This cost analysis was based on the soybean oil market in the year 2004. The cost of soybean oil, due to an increase in food–fuel demand, has increased from US$25–35/lb in 2004 to US$55–60/lb in 2009. Using the crude soybean oil at US$25–35/lb, we can obtain a PL–lecithin-enriched fraction after SFE, using ethanol as cosolvent worth US$31/lb, and subsequently with the further use of SFC produce pure PL extracts that can retail for US$1800/lb.

Based on these results, a similar study was performed for isolation of sterols and phytosterol esters from corn bran and fiber (Taylor & King 2002). In this study, the extraction of corn bran

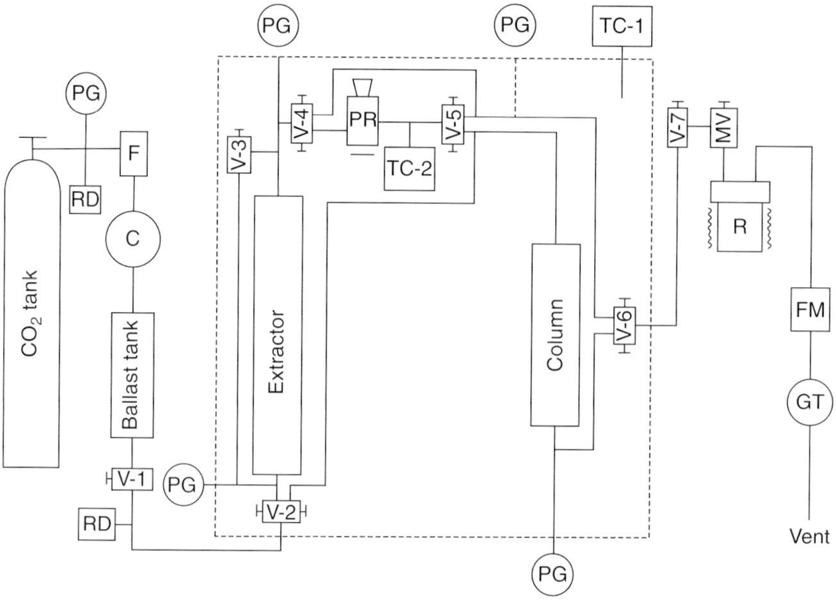

Enrichment of tocopherols from soybean flakes		
	SFE	SFE/SFC
α	4.33	12.1
β	1.83	2.4
γ	3.94	15.0
δ	3.75	30.8

Fig. 3.14 SFE–SFC process used in enriching tocopherol concentrates from soybean flakes (adapted from King *et al.* 1996).

and fiber was performed at 34.5 MPa and 40°C. The extracted mixture was then sent through a three-step SFC system using an aminopropyl sorbent. The first stage was maintained at 69 MPa and 80°C with neat CO_2 as the eluent, while the second and the third stages were maintained at 34.5 MPa and 40°C with CO_2 + 10 mol percent ethanol and CO_2 + 15 mol percent ethanol as eluents, respectively. The ethanol was removed from the enriched fractions

Table 3.7 Percentage amounts of phospholipids derived from SFE and SFC processing stages[a]

Phospholipid type	SFE stage[a]	SFC stage[a]
Phosphatidylethanolamine	16.1	74.9
Phosphatidylinositol	9.2	20.8
Phosphatidic acid	2.8	55.8
Phosphatidylcholine	15.6	76.8

[a] Relative to other eluting constituents (oil and unidentified peaks).

by heating to 60°C, and the sorbent column was reconditioned for reuse at 69 MPa and 80°C. The maximum amount of triglycerides was isolated in the first SFC stage, while the second stage yielded higher amounts of sterols and ferulate-phytosterol esters. The third stage yielded a low mass recovery containing 76% triglycerides, 2% sterols and 2% ferulate-phytosterol esters. Other studies utilizing an SFE–SFC system purification approach are Chandra and Nair (1996) for the extraction of isoflavones from soya-based products and Montanari *et al.* (1999) for the isolation from deoiled soy meal.

In contrast to using an elution-type batch SFC system, studies have been conducted on simulating moving bed–SFC (or SMB–SFC) systems. The simulated moving bed chromatography system has the advantages of a continuous countercurrent mode of operation with reduced solvent consumption. Instead of using a single sorbent column in elution type (batch) SFC, in an SMB system, there are multiple columns for separation of nutraceutical components, followed by a desorbent column at the end. The feed flow through the column can be controlled by the switching of valves through a central valve-switching unit. A typical SMB–SFC plant consisting of up to eight columns with dynamic axial compression and variable bed length is shown in Figure 3.15 (Depta *et al.* 1999). The principle behind the operation of an SMB–SFC is very similar to that of a thermal gradient fractionating column discussed previously in this chapter. In this system, a number of columns are connected to each other in a circular loop. A countercurrent flow between the stationary and mobile phases is simulated by consistent switching between the inlet and outlet ports in the direction of fluid flow. The SMB column can be divided into four different stages. The separation at each stage can be controlled not only by temperature and pressure conditions but also by controlling the packing bed length, column configuration and the type of packing used. In a mixture containing more than two components to be isolated, the least adsorbing component would travel through the system in the opposite direction to the fluid flow, toward the raffinate port. The four stages of the SMB–SFC system can be defined as desorbent, extract, feed and raffinate. The stationary phase is cleaned in stage 1 while the mobile phase (usually SC-CO_2 or SC-CO_2 – cosolvent mixture) is cleaned in stage 4. The number of stages in the SMB–SFC system can vary with the number of columns used. The pilot-plant schematic shown in Figure 3.15 houses eight columns, indicating eight stages of column deterpenations, using packed beds of varying length. The raffinate and the extract from the SMB plant are further separated by temperature control and pressure reduction to obtain solutes at near 100% purity, while the mobile phase can be recycled through an adsorbent filter (to remove any possible impurities) into the column stage.

These system configurations have been traditionally used in isolation and purification of pharmaceutical products (Depta *et al.* 1999). However, a study involving optimization of this system for enrichment of tocochromanols (Peper *et al.* 2007) has been carried out. In this study, unmodified silica gel (Kromasil Si 60–10) was used as sorbent column, with the separation of α-tocopherols conducted at 50°C and 20 MPa with CO_2 + 5% (w/w) isopropanol as the mobile phase. The study showed that at high separation factors, the specific productivity of the SMB column was same as that of a batch elution column. However, at lower separation factors, the SMB column had greater separation efficiency. Increased product costs for tocopherols isolated by an SMB–SFC system were higher than for tocopherols isolated by a batch-SFC system primarily due to greater capital costs for the chromatographic columns. However, further studies need to be performed to test the efficiency of an SMB–SFC system for various functional food applications.

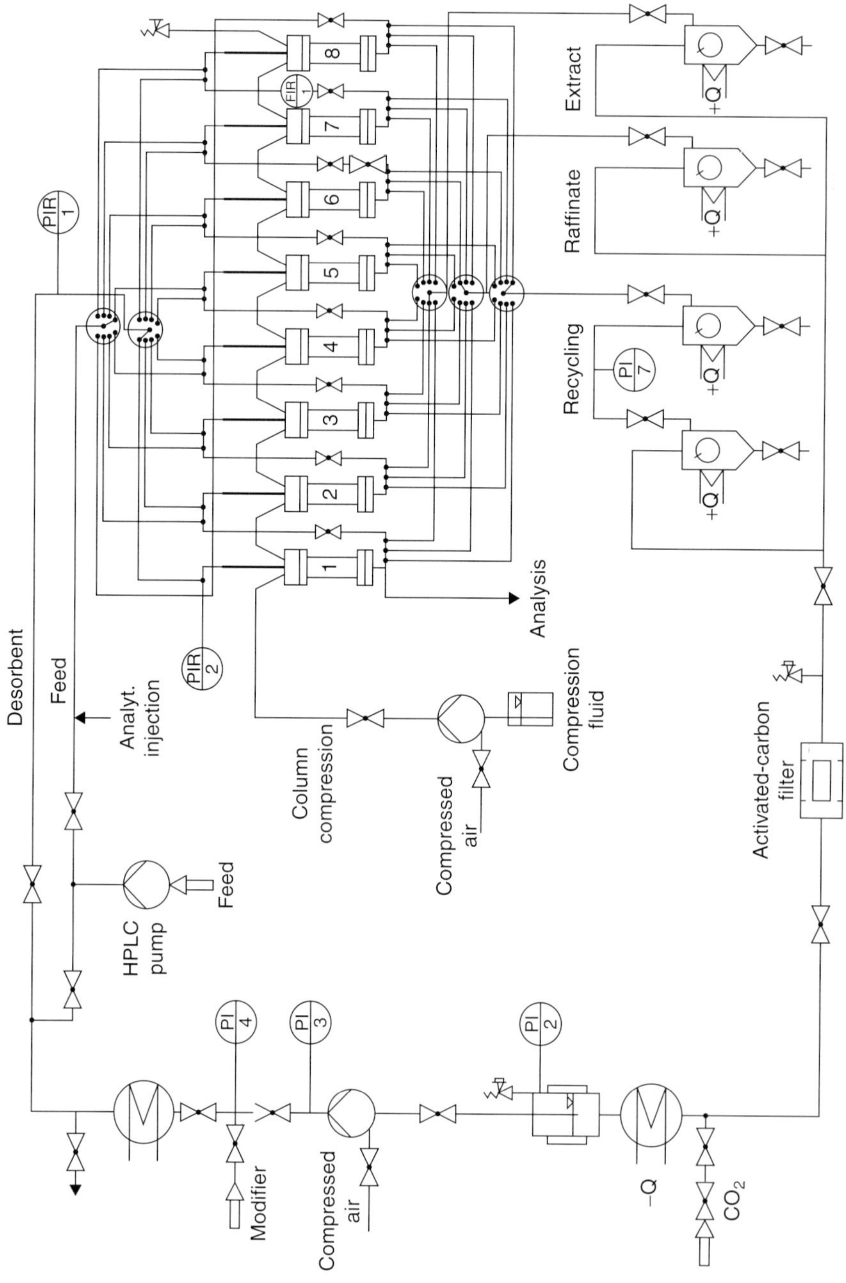

Fig. 3.15 A typical SMB–SFC pilot-plant processing scheme (adapted from Depta *et al.* 1999).

3.4.3 Supercritical fluid reaction coupled with SFE

Supercritical CO_2 has been proved as a versatile reaction medium in hydrogenation (Temelli *et al.* 1996) and enzyme-catalyzed reactions (Snyder *et al.* 1997). Supercritical CO_2 has been used in hydrogenation and transesterification reactions (in the presence of a Novozyme catalyst) to convert vegetable oils to fatty acid methyl esters (FAMEs) (Jackson & King 1997). Although not directly used in the nutraceuticals industry, FAMEs or similar esters can be used advantageously in SFE (Quancheng *et al.* 2004), columnar modes of SFF (Eller *et al.* 2008) and SFC (Pettinello *et al.* 2000). Soapstock and deodorizer distillate obtained as by-products from supercritical fluid reaction (SFR) of lipid-related products can be further separated using techniques such as SFF and SFC into squalene, tocopherols, fatty acid esters, sterols and triglycerides which have nutraceutical value. Approximately 70% of the FAMEs in deodorizer distillates, as well as tocopherols and sterols, can be extracted with SC-CO_2 (Fang *et al.* 2007). The use of supercritical fluids as a reaction medium in conversion of lipids and oleochemical mixtures to valuable by-products has been reviewed in detail by King (2004b).

3.4.4 Critical fluids coupled with membrane technologies

The initial coupling of membranes with supercritical fluid extraction and fractionations in the 1980s and 1990s was not intended to study the fractionation of product mixtures in SC-CO_2 but to recover excess carbon dioxide for sequestration and recycle (Schell *et al.* 1983; Semenova *et al.* 1992; Hsu & Tan 1993). Probably the first use of membrane technology in supercritical fluid processing of functional food components was by Robinson and Sims (1996), who devised a method employing microporous membranes to increase the fractionation efficiency of flavor and aroma components from fruit berry substrates using critical fluids. Such a critical fluid–membrane coupled system, known as the 'Porocrit' process, utilized the flow of liquefied CO_2 (or near-critical CO_2) through a concentric membrane tube arrangement concurrent to the flow of aqueous liquid feed mixture being pumped on the other side of the membrane material. It should be noted that this method was also applied in sterilization of citrus juices (Sims 1998). The use of membranes after extraction is justified on the basis of avoiding recompression costs on recycle of SC-CO_2 through the system, as well as lower energy requirements in comparison to a fractionation or chromatography system.

A number of studies have been conducted in tandem processing with membranes after an extraction and/or as a stand-alone separation system to increase the separation efficiency of components extracted into SC-CO_2. In one such study, reverse osmosis (RO) membranes were employed for the separation of essential oils from natural products such as orange, lemon grass and nutmegs after extraction using SC-CO_2 at 12 MPa and 40°C (Sarmento *et al.* 2004). In this study, close to 90% of the theoretical essential oils was obtained. Transmembrane pressure played a vital role in affecting the essential oil retention when using these membranes. In a similar study by Carlson *et al.* (2005), RO membranes were used in separation and recovery of carbon dioxide from limonene after extraction with SC-CO_2 at 12 MPa and 40°C. The RO membranes were aided in the recovery of approximately 70% of the carbon dioxide from the essential oils mixtures. Such studies involving coupling of SC-CO_2 extraction processes with nanofiltration membranes have shown advantages of good performance at high permeate fluxes mainly due to the low viscosity of SC-CO_2. Such integrated processing technology has been used to further fractionate the fats in foodstuffs such as butter or fish oil (Sarrade *et al.* 1999).

A study has been conducted to extract polyphenols from cocoa seeds using SC-CO_2, with or without ethanol as cosolvent, and to concentrate the extract using polymeric nanofiltration or RO membranes. The system, operating on a pilot scale at atmospheric pressure 8 and 15 MPa, at 40°C, produced a maximum yield of polyphenols of approximately 42.8% when a pressure of 8 MPa was maintained (Sarmento *et al.* 2008). Recently, antioxidant compounds from grape products have been recovered via the coupling of SFE and membrane technology (Beltran *et al.* 2008). A comparison study was performed to evaluate the separation efficiency of a mixture of antioxidant compounds, obtained by extraction with SC-CO_2, using SFF and/or membrane technology. The solubility of antioxidant compounds in SC-CO_2 is very low due to the presence of polar functional groups; hence, an ethanol modifier was required for the extraction process, which yielded a mixture of antioxidants and oils. SFF used in tandem with SFE yielded good separation efficiency of the antioxidant–oil mixture using a 'crossover' pressure around 15–20 MPa.

3.5 Integrated critical fluid processing technology

Previous discussions in this chapter have involved multiple unit processes such as SFE, SFF, SFC or a mixture of one or more processes to facilitate a 'green' platform for processing of functional food components from natural products. In this section, we intend to study the application of such processing combinations using multiple sub- and supercritical fluids to enhance the yield of the desired solutes from natural products.

Studies using tandem SC-CO_2–subcritical water extraction techniques are usually performed at lower temperature and pressure conditions to facilitate extraction of nutraceutical compounds from natural products. Mannila *et al.* (2002) sequentially applied SC-CO_2 followed by fractionating with subcritical water for extracting bioactive compounds such as hyperforin, adhyperforin, hypercin and pseudohypercin from St John's wort. Similar studies were performed for the extraction of essential oils and boldine, an alkaloid having pharmaceutical value, from boldo leaves with SC-CO_2, SC-CO_2 + ethanol mixtures and then with pressurized water (del Valle *et al.* 2005).

From the studies published in the literature, we can predict applications of integrated processing schemes on a number of substrates. Antioxidants from rosemary plants have been extracted using both supercritical CO_2 and pressurized water as solvents. In the first case, the SC-CO_2 extraction of rosemary extracts was performed in two stages, with maximum yield of essential oils obtained in the first stage at 10 MPa, 40°C, while a second set of extraction conditions at higher pressure and temperature conditions facilitated a more antioxidant-rich fraction (Ibanez *et al.* 1999). Using subcritical water extraction of the rosemary substrates yielded an even higher amount of antioxidants (carnosic acid) at temperatures in excess of 100°C (Ibanez *et al.* 2003). Increasing the temperature of the pressurized water increased the total antioxidant yield from 12.3% at 25°C to 48.6% at 200°C (antioxidant yield in a previous study using SC-CO_2 was only 1.0–1.8% based on rosemary substrate dry weight). Hence, coupling the two extraction processes under optimized conditions from each of the studies could lead to processing of rosemary plant substrates for the extraction and isolation of essential oils and antioxidants in a single integrated process. A similar approach in applying tandem SC-CO_2 and subcritical water extraction can be demonstrated for clove processing based on studies conducted by Clifford *et al.* (1999) and Rovio *et al.* (1999).

A particular substrate of interest to our research group is grape pomace or marc. Grapes form an important feed in the wineries and juice industries and a rich source of anthocyanins and procyanidins. SC-CO_2 has been used in extraction of oils from grape waste obtained in downstream processing of the wine and juice industries (Gomez *et al.* 1996). In this study, the extractor was maintained at pressures between 5 and 35 MPa, temperatures between 10 and 60°C and solvent flow rate between 0.5 and 2.0 L/minute. Optimal extraction yields close to 40% for grape seed oil were realized at 20 MPa, 40°C and 1.5 L/minute carbon dioxide flow rate. Further studies conducted by another research group (Cao & Ito 2003) to scale up a grape seed oil extraction process by 125 times used 30–40 MPa and 35–40°C for isolating the grape seed oil extract. Cao and Ito (2003) found it essential to use ethanol as a cosolvent to facilitate a greater removal of oil from the grape seeds. They also found that by increasing the concentration of ethanol in CO_2, there was a considerable increase in the recovery of gallic acid and procyanidin dimers and oligomers. Gallic acid and catechins were the main products after extraction, but by increasing the pressure from 20 to 30 MPa, close to 100% extraction of gallic acid was achieved. Similar studies have used methanol as a cosolvent; the methanol performed better as a cosolvent compared to ethanol (Murga *et al.* 2000).

As noted previously, a number of studies were performed using subcritical water in the extraction of anthocyanins and procyanidins from grape pomace. In such studies, an accelerated solvent extractor (ASE) was utilized. In one such study involving subcritical water and/or sulfured water as a solvent for the extraction of anthocyanins from grape skins, temperatures between 110 and 160°C were maintained in an ASE extractor using a very short static extraction time of 40 seconds. This resulted in a decrease in the amount of anthocyanins recovered as temperatures above 110°C were utilized (Ju & Howard 2005). In these studies, the effect of using sulfured water for the extraction of anthocyanin compounds was also studied. Similar studies performed with black currants showed an increased recovery of anthocyanin compounds and other phenolics when sulfur dioxide was added as a modifier to water at subcritical conditions (Cacae & Mazza 2002).

Additional studies were performed by Garcia-Marino *et al.* (2006) to extract procyanidin compounds and catechins from grape processing waste using subcritical water as the extraction solvent. Extractions conducted at 10.34 MPa and temperatures of 50, 100 and 150°C for static, short extraction times indicated an increase in the recovery of gallic acid and procyanidin dimers and corresponding oligomers with increasing temperature, the maximum recovery being obtained at 150°C (Garcia-Marino *et al.* 2006). These extractions were carried out in a single static extraction cycle using an ASE system, and the resultant recoveries were high in comparison with that obtained by a conventional extraction technique utilizing methanol/water in a 75:25 ratio as a solvent.

Sequential processing of grape seeds has been demonstrated with solvent extraction using methanol working in tandem with SFE (Ashraf-Khorassani & Taylor 2004). In the first step, SC-CO_2 at 65.5 MPa and 80°C flowing at 2 mL/minute contacts the grape seed matrix, resulting in the extraction of over 95% of the oil present in the seeds. After about 60 minutes, the remaining seed matrix was allowed to contact SC-CO_2 – 30–40% methanol mixture at the same temperature and pressure conditions. With a higher percentage of methanol in SC-CO_2, almost 80% of the catechins and epicatechins were separated from the seed matrix, which further contacts pure methanol at 13.8 MPa and 80°C in the third stage for further recovery of the catechins and the epicatechins. In this study, the SC-CO_2 extracted an oligomeric mixture of the catechins and epicatechins (with $n = 1–3$), but with addition of methanol, higher molecular weight flavonoids, such as procyanidins, could be extracted. In the above-mentioned study, it is possible to replace methanol with subcritical water in the second and third stages to allow for a 'green processing' platform.

Previous work conducted by one of the authors in extraction of anthocyanins from berry substrates using subcritical water as a solvent indicated a recovery yield of anthocyanins from elderberry pomace in excess of 90% in comparison with that obtained by using ethanolic solvents (King *et al.* 2003). A patent has been issued for the recovery of polyphenolics from fruits and vegetable substrates using subcritical water as the extraction solvent using the above-mentioned approach (King & Grabiel 2007). The above studies suggest that SC-CO_2 and subcritical water extraction can be coupled sequentially to recover an array of valuable products ranging from less polar essential oils to high molecular weight, high polarity polyphenolic compounds (King 2002; King *et al.* 2003). The use of ethanol as a cosolvent in the above-coupled process is favorable to increase the recovery yield of the polyphenolic compounds, while minimizing the amount of ethanol, resulting in cost savings and avoidance of regulatory issues concerning the mass balance of ethanol used in the process (King *et al.* 2003).

In conjunction with this potential reduction in the use of alcoholic solvents, numerous research groups are now practicing tandem SC-CO_2–subcritical water extraction. One such approach involves the application of a 'hybrid' SFE system with CO_2 and water (Yoo & Fukuzato 2006). This system consists of a tubular column in which the feed mixture, concentrated in the water phase, flows downward through an extraction column. This mixture contacts CO_2 (maintained at supercritical conditions) flowing concurrently to the direction of water flow.

Another patented technology involving the integration of sub- and supercritical technology is described for the extraction of nutraceutical components from palm oil (Brunner *et al.* 2007). In this system, fractions derived from palm oil are extracted with sub- or supercritical fluids in a single-stage or a multistage process and the enriched fractions separated on an adsorbent. After being carried by a continuous flow of the sub- or supercritical fluid along the length of the adsorption column, the separated fractions are sent through a desorbent for precipitation. Using such an approach, the tocochromanols were enriched from a concentration of only 500–1000 ppm in the first extraction stage to greater than 95% purified isomers in the last stage (desorption). Supercritical CO_2 was used to enrich tocochromanols, while carotenoid pigments were extracted and isolated using either SC-CO_2 or subcritical propane.

3.6 Production-scale critical fluid-based nutraceutical plants and commercial products

Irrespective of the high investment costs, there has been considerable application of critical fluid technology for extraction as applied to functional foods. These critical fluid-treated functional food products have gained considerable consumer acceptance. The initial applications of critical fluids, especially SC-CO_2, in the food industry were centralized on the decaffeination of coffee and tea leaves. Further developments were made by extending this technology to extraction of aroma compounds from hops and essential oils from varied flower and fruit substrates. There are close to 110 critical fluid-based plants concentrated mostly in Germany, United States, South Korea, France and Japan. Among plants conducting extractions and enrichments of nutraceutical-based ingredients are those owned by US Nutraceuticals, GreenTek 21, Hitex, Flavex, Newly Wed Foods and Fuji Flavors, among others, which have invested in utilizing critical fluid technology for varied applications inclusive of decaffeination, hops processing and/or extraction of flavors or aroma compounds from spices. Prominent organizations working toward setting up critical fluid-based plants worldwide for varied applications, but not necessarily restricted to functional foods, are Separex,

Table 3.8 Natex processing plants and their international locations

Country	Application
Germany	Tea decaffeination (3000 tons/year)
Italy	Coffee decaffeination (10 000 tons/year)
Czech Republic	Spices and herbs
India	Spices and herbs (3 × 300 liters; 2 × 600 liters)
Taiwan	Rice treatment (25 000 tons/year)
Poland	Hops extraction
Denmark	Wood impregnation (40 000–60 000 m^3 wood/year)
New Zealand	Hops and nutraceuticals
South Korea	Edible oils
Spain	Cork treatment (25 000 tons/year)

Source: Natex, Terpitz, Austria.

Natex, Uhde and Thar Technologies. Table 3.8 shows some of the critical fluid-based processing plants constructed by Natex in different countries and their intended applications. Of all the processing plants listed in the table, the Five King Cereals plant located in Taiwan is interesting because its purpose is not direct SFE. Established in 1999–2000 with an initial investment cost of US$10 million, this plant operates at 4 tons/hour through three 5800-liter supercritical CO_2 extractors. SC-CO_2 treatment here is principally used to facilitate a greater shelf-life, reduced cooking time, reduction in metal and pesticide residue content, as well as reduced waxy and fatty acid content – the overall result being a shelf-stable rice product exhibiting less rancidity and degradation over time. A similar plant was constructed by SNU-UMAX in South Korea, which produces a commercial 'diet' rice. The principle and the commercial product produced by both these plants are technically the same, with slight differences in rice texture when the rice is exposed to SC-CO_2.

There has been considerable growth in the number of SC-CO_2-based food processing plants in South Korea and India over the last few years. South Korea perhaps leads the world in using critical fluid-based technology partly due to an increase in health awareness in South Korea. Table 3.9 shows the various SC-CO_2-treated products available commercially in South Korea. Some of the leading organizations employing critical fluid processing in South Korea are Greentek21, Hanhwa, UMAX and Ottogi Food Corporation. Greentek21 has developed a process for the removal of fat content from processed instant noodles. Using SC-CO_2, this company states a 20% reduction in fat and a decrease in the sodium content. UMAX, a commercial producer of sesame oil in South Korea, is another company producing a product, which provides numerous health benefits.

Recently, India has also been providing a growing market for critical fluid-based processing plants. Initially restricted to pilot-scale plants at educational institutions such as IIT Mumbai, other organizations have come forward to provide production-scale units in India. As can be seen from Table 3.8, there are two Natex critical fluid-based spice processing plants in India. The most prominent feedstocks for these plants in India, apart from spices, are sandalwood and neem. Some of the functioning supercritical fluid-based processing units in India are as follows: SAMI industries, Novo Agritech, Prerna Biotech, Sat Group, Vedic Supercriticals, South East Agro, RKS Agro, Indo Global Spices and Nisarga Biotech.

CFE has been applied for some time now in the extraction of specialty oils, such as evening primrose, borage, black currant and flax. Recently, there has been considerable research in employing critical fluids to obtain oil from fungi or marine sources, such as spirulina, which

Table 3.9 Commercial supercritical fluid-treated products distributed in South Korea

Product name	**[a]INCI name (registered by GREENTEK21)**
SFE oil-soluble green tea extract	*Camellia sinensis* leaf extract
SFE 6 years ginseng extract	*Panax ginseng* root extract
SFE Korea Poongran	Orchid extract
SFE grape	*Vitis vinifera* (grape) fruit extract and squalene
SFE tomato	Tomato extract
SFE Jangnoisam Hanbang complex	*Panax ginseng* root extract
SFE Jaran	*Bletilla striata* extract
SFE *Castanea*	*Castanea sativa* bark extract and squalene
SFE rice	*Oryza sativa* (rice) extract
SFE mountain mushroom extract and phytosqualene	Squalene and *Tricholoma matsutake* (mushroom) extract
SFE somok	*Caesalpinia sappan* bark extract and squalene
SFE *Acorus* extract	*Acorus gramineus* extract
SFE *Camellia japonica* seed	*Camellia japonica* seed extract
SFE chrysanthemum flower	*Chrysanthemum coccineum* flower extract

[a] INCI: International Nomenclature Cosmetic Ingredient.

are devoid of cholesterol (Herrero *et al.* 2004). Other nutraceutical ingredients that can be obtained similarly include extracts from chamomile, paprika, feverfew and gingko biloba, garlic and ginger. Most of the common spices and mint oils, including a commercially available extract of rosemary, can be obtained via SC-CO_2 extraction. Some SC-CO_2-treated derived products are shown in Figure 3.16 and feature names involving 'supercritical' on their packaging.

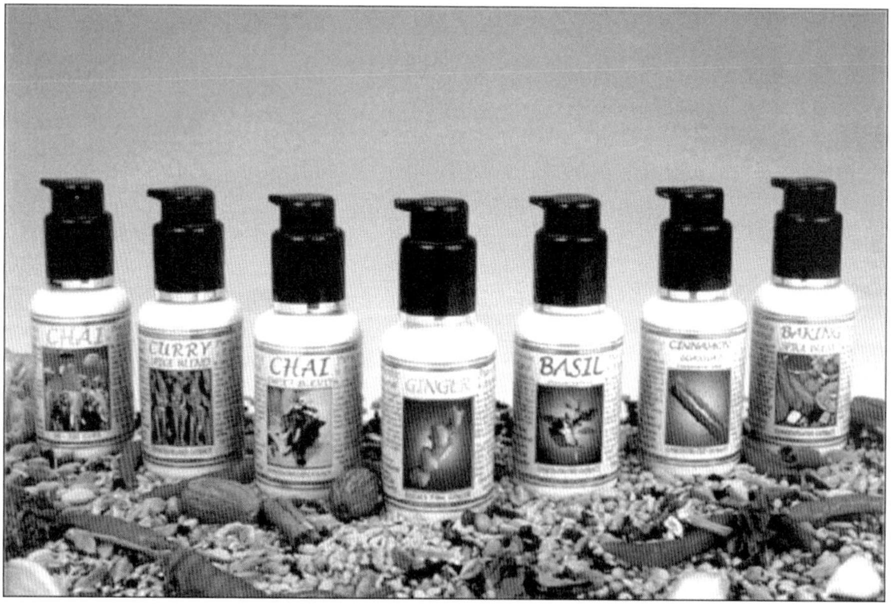

Fig. 3.16 Examples of commercially available supercritical fluid-treated nutraceutical and health products.

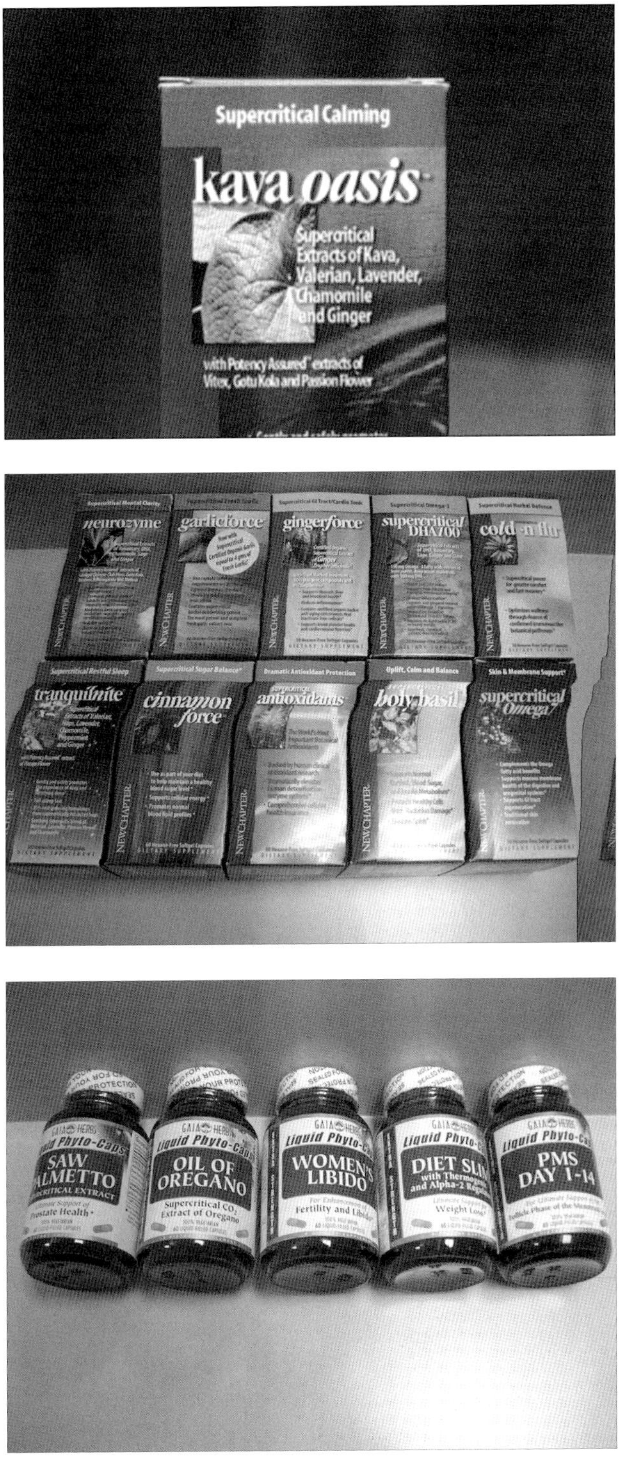

Fig. 3.16 *(Continued)*

References

AROQ Limited (2004). *Global Market Review of Functional Foods – Forecasts to 2010.* Worcestershire, England. Available online http://www.marketresearch.com (accessed 15 April 2009).

Ashraf-Khorassani, M. & Taylor, L.T. (2004). Sequential fractionation of grape seeds into oils, polyphenols, and procyanidins via a single system employing CO_2 based fluids. *Journal of Agricultural and Food Chemistry*, **52**, 2440–2444.

Barth, M.M., Zhou, C., Kute, K.M. & Rosenthal, G.A. (1995). Determination of optimum conditions for supercritical fluid extraction of carotenoids from carrot (*Daucus carota* L.) tissue. *Journal of Agricultural and Food Chemistry*, **43**(11), 2876–2878.

Barton, A.F.M. (1991). *CRC Handbook of Solubility Parameters and Other Cohesion Parameters*, 2nd edn. CRC Press, Boca Raton, FL.

Beltran, S., Sanz, M.T., Santanmaria, B., Murga, R. & Salazar, G. (2008). Recovery of antioxidants from grape products by using supercritical fluids and membrane technology. *Electronic Journal of Environmental, Agricultural and Food Chemistry*, **7**(8), 3270–3278.

Berna, A., Chafer, A., Monton, J.B. & Subirats, S. (2001). High pressure solubility data of system ethanol (1)+catechin (2)+CO_2 (3). *The Journal of Supercritical Fluids*, **20**(2), 157–162.

Braga, M.E.M., Leal, P.F., Sato, D.N. & Meirles, M.A.A. (2005). Comparison of yield, composition, and antioxidant activity of turmeric (*Curcuma longa* L.) extracts obtained using various techniques. *Journal of Agricultural and Food Chemistry*, **51**, 6604–6611.

Brunner, G., Gast, K., Chuang, M.H., Kumar, S., Chan, P. & Chan, W.P. (2007). Inventors: Carotech SDN.BHD, applicant. Process for production of highly enriched fractions of natural compounds from palm oil with supercritical and near critical fluids. Patent# WO 2007/090545.

Cacae, J.E. & Mazza, G. (2002). Extraction of anthocyanins and other phenolics from black currants with sulfured water. *Journal of Agricultural and Food Chemistry*, **50**, 5939–5946.

Cadoni, E., de Giorgi, M.R., Medda, E. & Poma, G. (1999). Supercritical CO_2 extraction of lycopene and β-carotene from ripe tomatoes. *Dyes and Pigments*, **44**(1), 27–32.

Cao, X. & Ito, Y. (2003). Supercritical fluid extraction of grape seed oil and subsequent separation of free fatty acids by high speed counter-current chromatography. *Journal of Chromatography A*, **1021**, 117–124.

Carlson, L.H.C., Bolzan, A. & Machado, R.A.F. (2005). Separation of D-limonene from supercritical CO_2 by means of membranes. *Journal of Supercritical Fluids*, **34**, 143–147.

Carvalho, R.N., Moura, L.S., Rosa, P.T.V. & Meirles, M.A.A. (2005). Supercritical fluid extraction from rosemary (*Rosmarinus officinalis*): kinetic data, extract's global yield, composition, and antioxidant activity. *Journal of Supercritical Fluids*, **35**, 197–204.

Casas, L., Mantell, C., Rodriguez, M., Torres, A., Macias, F.A. & de la Ossa, E.M. (2007). Effect of the addition of cosolvent on the supercritical fluid extraction of bioactive compounds from *Helianthus annuus* L. *Journal of Supercritical Fluids*, **41**, 43–49.

Chafer, A., Berna, A., Monton, J.B. & Munoz, R. (2002). High-pressure solubility data of system ethanol (1)+epicatechin (2) + CO_2 (3). *The Journal of Supercritical Fluids*, **24**(2), 103–109.

Chafer, A., Fornari, T., Berna, A. & Stateva, R.P. (2004). Solubility of quercetin in supercritical CO_2+ethanol as a modifier: measurements and thermodynamic modeling. *The Journal of Supercritical Fluids*, **32**(1–3), 89–96.

Chafer, A., Fornari, T., Berna, A., Ibanez, E. & Reglero, G. (2005). Solubility of solid carnosic acid in supercritical CO_2 with ethanol as cosolvent. *Journal of Supercritical Fluids*, **34**, 323–329.

Chandra, A. & Nair, M.G. (1996). Supercritical carbon dioxide extraction of daidzein and genistein from soybean products. *Phytochemical Analysis*, **7**, 259–262.

Chrastil, J. (1982). Solubility of solids and liquids in supercritical gases. *Journal of Physical Chemistry*, **86**, 3016–3021.

Clifford, A.A., Basile, A. & Al-Saidi, S.H. (1999). A comparison of the extraction of clove buds with supercritical carbon dioxide and superheated water. *Fresenius Journal of Analytical Chemistry*, **364**, 635–637.

Clifford, M.N. (2000). Anthocyanins – nature, occurrence and dietary burden. *Journal of the Science of Food and Agriculture*, **80**, 1063–1072.

Co, M., Koskela, P., Eklund-Akergen, P., Srinivas, K., King, J.W., Sjoberg, P.J.R. & Turner, C. (2009). Pressurized liquid extraction of betulin and antioxidants from birch bark. *Green Chemistry*, **11**, 668–674.

Constantinou, L. & Gani, R. (1994). New group contribution method for estimating properties of pure compounds. *AIChE Journal*, **40**(10), 1697–1710.
del Valle, J.M. & Aguilera, J.M. (1999). Review: high pressure CO_2 extraction. Fundamentals and applications in the food industry. *Food Science and Technology International*, **5**, 1–24.
del Valle, J.M., Rogalinsko, T., Zetzl, C. & Brunner, G. (2005). Extraction of boldo (*Peumus boldus* M.) leaves with supercritical CO_2 and hot pressurized water. *Food Research International*, **38**(2), 203–213.
del Valle, J.M. & Uquiche, E.L. (2002). Particle size effects on supercritical CO_2 extraction of oil containing seeds. *Journal of the American Oil Chemist's Society*, **79**(12), 1261–1266.
Depta, A., Giese, T., Johannsen, M. & Brunner, G. (1999). Separation of stereoisomers in a simulated moving bed-supercritical fluid chromatography plant. *Journal of Chromatography A*, **865**(1–2), 175–186.
DeSimone, J.M. (2002). Practical approaches to green solvents. *Science*, **297**(2), 799–803.
Diaz, M.S. & Brignole, E.A. (2009). Modeling and optimization of supercritical fluid processes. *The Journal of Supercritical Fluids*, **47**(2), 611–618.
Dobbs, J., Wong, J.M., Lahiere, R.J. & Johnston, K.P. (1987). Modification of supercritical fluid phase behavior using polar cosolvents. *Industrial Engineering and Chemistry Research*, **26**, 56–65.
Dron, A., Guyer, D.E., Gage, D.A. & Lira, C.T. (1997). Yield and quality of onion flavor oil obtained by supercritical fluid extraction and other methods. *Journal of Food Process Engineering*, **20**, 107–124.
Duan, L. (2005). *Extraction of Silymarins from Milk Thistile, Silybaum Marianum, Using Hot Water as Solvent*, Thesis (PhD). University of Arkansas, Fayetteville, AR.
Duan, L., Carrier, D.J. & Clausen, E.C. (2004). Silymarin extraction from milk thistle using hot water. *Applied Biochemistry and Biotechnology*, **114**(1–3), 559–568.
Dunford, N.T., King, J.W. & List, G.R. (2003). Supercritical fluid extraction in food engineering. In: *Extraction Optimization in Food Engineering*. Tzai, C. & Liadakis, G. (eds), Marcel Dekker, New York, pp. 57–93.
Eggers, R. (1996). Supercritical fluid extraction of oilseeds/lipids in natural products. In: *Supercritical Fluid Technology in Oil and Lipid Chemistry*. King, J.W. & List, G.R. (eds), AOCS Press, Champaign, IL, pp. 35–65.
Eikani, M.H., Golmohammad, F. & Rowshanzamir, S. (2007). Subcritical water extraction of essential oils from coriander seeds (*Coriandrum sativum* L.). *Journal of Food Engineering*, **80**(2), 735–740.
Eller, F.J., Taylor, S.L., Compton, D.L., Laszlo, J.A. & Palmquist, D.E. (2008). Counter-current liquid carbon dioxide purification of a model reaction mixture. *The Journal of Supercritical Fluids*, **43**(3), 510–514.
Fang, T., Goto, M., Wang, X., Ding, X., Geng, J., Sasaki, M. & Hirose, T. (2007). Separation of natural tocopherols from soybean oil byproduct with supercritical carbon dioxide. *The Journal of Supercritical Fluids*, **40**(1), 50–58.
Favati, F., King, J.W., Friedrich, J.P. & Eskins, K. (1988). Supercritical CO_2 extraction of carotene and lutein from leaf protein concentrates. *Journal of Food Science*, **53**(5), 1532–1536.
Fedors, R.F. (1974). A method for estimating both the solubility parameters and the molar volumes of liquids. *Polymer Engineering and Science*, **14**, 147–154.
Fernandez-Perez, V., Jimenez-Carmona, M.M. & de Castro, M.D.L. (2000). An approach to the static-dynamic subcritical water extraction of laurel essential oil: comparison with conventional techniques. *Analyst*, **125**, 481–485.
Ferreira, S.R.S., Nikolov, Z.L., Doraiswamy, L.K., Meirles, M.A.A. & Petenate, A.J. (1999). Supercritical fluid extraction of black pepper (*Piper nigrun* L.) essential oil. *The Journal of Supercritical Fluids*, **14**(3), 235–245.
Filho, G.L., de Rosso, V.V., Meirles, M.A.M., Rosa, P.T.V., Oliveira, A.L., Mercadante, A.Z. & Cabral, F.A. (2008). Supercritical CO_2 extraction of carotenoids from pitanga fruits (*Eugenia uniflora* L.). *The Journal of Supercritical Fluids*, **46**, 33–39.
Foidl, N. (1999). Device and process for the production of oils or other extractable substances. US Patent 5, 939, 571.
Franca, L.F. & Meirles, M.A.A. (1997). Extraction of oil from pressed palm oil (*Elase guineensis*) fibers using supercritical CO_2. *Ciencia e Tecnologia de Ailmentos*, **17**, 384–388.
Franca, L.F. & Meireles, M.A.A. (2000). Modeling the extraction of carotene and lipids from pressed palm oil (*Elaes guineensis*) fibers using supercritical CO_2. *Journal of Supercritical Fluids*, **18**, 35–47.
Froning, G.W., Wehling, R.L., Cuppett, S.L., Pierce, M.M., Niemann, L. & Siekman, D.K. (1989). Extraction of cholesterol and other lipids from dried egg yolk using supercritical carbon dioxide. *Journal of Food Science*, **55**(1), 95–98.

Gahrs, H.J. (1984). Applications of atmospheric gases in high pressure extraction. *Berichte der Bunsengesellschaft fuer Physikalische Chemie*, **88**(9), 894–897.

Garcia-Marino, M., Rivas-Gonzalo, J.C., Ibanez, E. & Garcia-Moreno, C. (2006). Recovery of catechins and proanthocyanidins from winery by-products using subcritical water extraction. *Analytica Chimica Acta*, **563**(1–2), 44–50.

Giddings, J.C. & Keller, R.A. (1969). *Advances in Chromatography*. CRC Press, Boca Raton, FL.

Gomez, A.M., Lopez, C.P. & de la Ossa, E.M. (1996). Recovery of grape seed oil by liquid and supercritical carbon dioxide extraction: a comparison with conventional solvent extraction. *The Chemical Engineering Journal and the Biochemical Engineering Journal*, **61**(3), 227–231.

Hansen, C.M. (2007). *Hansen Solubility Parameters: A User's Handbook*, 2nd edn, CRC Press, Boca Raton, FL.

Hartono, R., Mansoori, G.A. & Suwono, A. (2001). Prediction of solubility of biomolecules in supercritical solvents. *Chemical Engineering Science*, **56**(24), 6949–6958.

Hasler, C.M. (2002). Functional foods: benefits, concerns and challenges – a position paper from the American council on science and health. *Journal of Nutrition*, **132**, 3772–3781.

Hawthorne, S.B., Yang, Y. & Miller, D.J. (1999). Extraction of organic pollutants from environmental solid with sub- and supercritical water. *Analytical Chemistry*, **66**, 2912–2920.

Herrero, M., Ibanez, E., Senorans, J. & Cifuentes, A. (2004). Pressurized liquid extracts from *Spirulina platensis* microalga: determination of their antioxidant activity and preliminary analysis by micellar electrokinetic chromatography. *Journal of Chromatography A*, **1047**(2), 195–203.

Hildebrand, J.H. & Scott, R.L. (1950). Solubility of noneleectrolytes. *Annual Review of Physical Chemistry*, **1**, 75–92.

Hong, I.K., Rho, S.W., Lee, K.S., Lee, W.H. & Yoo, K.P. (1990). Modeling of soybean oil bed extraction with supercritical carbon dioxide. *Korean Journal of Chemical Engineering*, **7**(1), 40–46.

Hsu, J.H. & Tan, C.S. (1993). Separation of ethanol from aqueous solution by a method incorporating supercritical CO_2 with reverse osmosis. *Journal of Membrane Science*, **81**(3), 273–285.

Ibanez, E., Kubatova, A., Senorans, F.J., Cavero, S., Reglero, G. & Hawthorne, S.B. (2003). Subcritical water extraction of antioxidant compounds from rosemary plants. *Journal of Agricultural and Food Chemistry*, **51**(2), 375–382.

Ibanez, E., Oca, A., de Murga, G., Lopez-Sebastian, S., Tabera, J. & Reglero, G. (1999). Supercritical fluid extraction and fractionation of different prerpocessed rosemary plants. *Journal of Agircultural and Food Chemistry*, **47**, 1400–1404.

Illes, V., Daood, H.D., Biacs, P.A., Gnayfeed, M.H. & Meszaros, B. (1999). Supercritical CO_2 and subcritical propane extraction of spice red pepper oil with special regard to carotenoid and tocopherols content. *Journal of Chromatographic Science*, **37**, 345–352.

Jackson, M.A. & King, J.W. (1997). Lipase-catalyzed glycerolysis of soybean oil in supercritical carbon dioxide. *Journal of the American Oil Chemists' Society*, **73**(3), 353–356.

Jaren-Galan, M., Neinaber, U. & Schwartz, S.J. (1999). Paprika (*Capsicum annuum* L.) oleoresin extraction with supercritical carbon dioxide. *Journal of Agricultural and Food Chemistry*, **47**, 3558–3564.

Jayasri, A. & Yaseen, M. (1980). Nomograms for solubility parameter. *Journal of Coatings Technology*, **52**, 41–45.

Joback, K.G. & Reid, R.C. (1987). Estimation of pure-component properties from group contributions. *Chemical Engineering Communications*, **57**(1), 233–243.

Ju, Z. & Howard, L.R. (2005). Subcritical water and sulfured water extraction of anthocyanins and other phenolics from dried red grape skin. *Journal of Food Science*, **70**(4), S270–S276.

King, J.W. (1991). Supercritical fluid processing of cosmetic raw materials. *Cosmetic Toiletries*, **106**(8), 61–67.

King, J.W. (2000). Advances in critical fluid technology for food processing. *Food Science and Technology Today*, **14**, 186–191.

King, J.W. (2000b). Sub- and supercritical fluid processing of agrimaterials: extraction, fractionation and reaction modes. In: *Supercritical Fluids: Fundamentals and Applications*. Kiran, E., Debenedetti, P.G. & Peters, C.J. (eds), Kluwer Publishers, Dordrecht, The Netherlands, pp. 451–488.

King, J.W. (2004). Development and potential of critical fluid technology in the nutraceutical industry. In: *Drug Delivery and Supercritcal Technology*. York, P., Kompella, U.B. & Shekunov, B.V. (eds), Marcel Dekker, New York.

King, J.W. (2004b). Critical fluid technology for the processing of lipid-related natural products. *Comptes Redus-Chimie*, **7**, 647–659.

King, J.W. (2006). Pressurized water extraction: resources and techniques for optimizing analytical applications. In: *Modern Extraction Techniques: Food and Agricultural Samples*. Turner, C. (ed.), American Chemical Society, Washington, DC, pp. 79–95.

King, J.W., Cygnarowicz-Provost, M. & Favati, F. (1999). Supercritical fluid extraction of evening primrose oil kinetic and mass transfer effects. *Italian Journal of Food Scince*, **9**, 93–204.

King, J.W., Favati, F. & Taylor, S.L. (1996). Production of tocopherols concentrates by supercritical fluid extraction and chromatography. *Separation Science and Technology*, **31**, 1843–1857.

King, J.W. & Grabiel, R. (2007). Inventor: United States of America, assignee. Isolation of polyphenolic compounds from fruits or vegetables utilizing sub-critical water extraction. US Patent 7,208,181.

King, J.W., Grabiel, R.D. & Wightman, J.D. (2003). Subcritical water extraction of anthocyanins from fruit berry substrates. In: *Proceedings of the 6th International Symposium on Supercritical Fluids – Tome 1*. Versailles, France, 28–30 April, pp. 409–418.

King, J.W., Howard, L.R., Srinivas, K., Ju, Z.Y., Monrad, J. & Rice, L. (2007). Pressurized liquid extraction and processing of natural products. In: *Proceedings of the 5th International Symposium on Supercritical Fluids. Super Green*, Seoul, Korea, 28 November–1 December, CD proceedings #KL04.

King, J.W., Johnson, J.H. & Eller, F.J. (1995). Effect of supercritical carbon dioxide pressurized with helium on solute solubility during supercritical fluid extraction. *Analytical Chemistry*, **67**, 2288–2291.

King, J.W., Sahle-Demessie, E., Temelli, F. & Teel, J.A. (1997). Thermal gradient fractionation of glyceride mixtures under supercritical fluid conditions. *The Journal of Supercritical Fluids*, **10**, 127–137.

King, J.W. & Srinivas, K. (2009). Multiple unit fluid processing using sub- and supercritical fluids. *Journal of Supercritical Fluids*, **47**, 598–610.

King, J.W., Srinivas, K., del Valle, J.M. & de la Fuente, J.C. (2006). Design and optimization for the use of sub-critical fluids in biomass transformation, bio-fuel production, and bio-refinery utilization – 1. In: *Proceedings of the 8th International Symposium of Supercritical Fluids*, Kyoto, Japan, 5–8 November, CD proceedings #OC-2-17.

Kraut, S., Braga, M.E.M. & Meirles, M.A.A. (2006). Extraction of coriander (*Coriandrum sativum* L.) leaves oil by SFE: total phenolic content. In: *Proceedings of the 8th Conference on Supercritical Fluids and Their Applications*. Reverchon, E. (ed.), ISASF, SIchia, Italy, pp. 143–146.

Kubatova, A., Miller, D.J. & Hawthorne, S.B. (2001). Comparison of subcritical water and organic solvents for extracting kava lactones from kava root. *Journal of Chromatography A*, **923**(1–2), 187–194.

Leal, P.F., Chaves, F.C.M., Ming, L.C., Petenate, A.J. & Meirles, M.A.A. (2006). Global yields, chemical compositions, and antioxidant activities of clove basil (*Ocimum gratissimum* L.) extracts obtained by supercritical fluid extraction. *Journal of Food Process Engineerin*, **29**, 547–559.

Machmuddah, S., Kondo, M., Sasaki, M., Goto, M., Munemasa, J. & Yamagata, M. (2008). Pressure effect in supercritical CO_2 extraction of plant seeds. *Journal of Supercritical Fluids*, **44**, 301–307.

Mannila, M.H., Kim, H. & Wai, C.M. (2002). Supercritical carbon dioxide and high pressure water extraction of bioactive compounds in St. John's wort. In: *Proceedings of the First International Symposium on Supercritical Fluid Technology for Energy and Environmental Applications. Super Green*, Suwon, South Korea, 3–6 November, pp. 74–78.

Mendes, M.F., Pessoa, F.L.P. & Uller, A.M.C. (2002). An economic evaluation based on an experimental study of the vitamin E concentration present in deodorizer distillate of soybean oil using supercritical CO_2. *Journal of Supercritical Fluids*, **23**(3), 257–265.

Monrad, J.K., Howard, L.R., King, J.W. & Srinivas, K. (2008). Pressurized solvent extraction of anthocyanins from dried grape pomace. In: *Proceedings of the 2008 IFT Annual Meeting & Food Expo*. New Orleans, LA, 28 June–2 July.

Montanari, L., Fantozzi, P., Snyder, J.M. & King, J.W. (1999). Selective extraction of phospholipids with supercritical carbon dioxide and ethanol. *Journal of Supercritical Fluids*, **14**, 87–93.

Mukhopadhyay, M. (2007). Processing of spices using supercritical fluids. In: *Supercritical Fluid Extraction of Nutraceuticals and Bioactive Compounds*. Martinez, J.L.M. (ed.), CRC Press, Boca Raton, FL, pp. 337–366.

Murga, R., Ruiz, R., Beltran, S. & Cabezas, J.L. (2000). Extraction of natural complex phenols and tannins from grape seeds by using supercritical mixtures of carbon dioxide and alcohol. *Journal of Agricultural and Food Chemistry*, **48**, 3408–3412.

Nagy, B. & Simandi, B. (2008). Effects of particle size distribution, moisture content, and oil content on the supercritical fluid extraction of paprika. *The Journal of Supercritical Fluids*, **46**(3), 293–298.

Osuna, A.B., Serbanovic, A. & da Ponte, M.N. (2005). Fluid extraction: supercritical fluids. In: *Green Separation Processes: Fundamentals and Applications*. Afonso, C.A.M. & Crespo, J.G. (eds), Wiley-VCH Verlag, Washington, DC, pp. 207–218.

Palma, M. & Taylor, L.T. (1999). Extraction of polyphenolic compounds from grape seeds with near critical carbon dioxide. *Journal of Chromatography A*, **849**(1), 117–124.

Panayitou, C. (1997). Solubility parameter revisited: an equation-of-state approach for its estimation. *Fluid Phase Equilibria*, **131**(1–2), 21–35.

Park, H.S., Lee, H.J., Shin, M.H., Lee, K.W., Lee, H., Kim, Y.S., Kim, K.O. & Kim, K.H. (2007). Effects of cosolvents on the decaffeination of green tea by supercritical carbon dioxide. *Food Chemistry*, **105**, 1011–1017.

Peper, S., Johannsen, M. & Brunner, G. (2007). Preparative chromatography with supercritical fluids: comparison of simulating moving bed and batch processes. *Journal of Chromatography A*, **1176**(1–2), 246–253.

Perrut, M., Clavier, J.Y., Poletto, M. & Reverchon, E. (1997). Mathematical modeling of sunflower seed extraction by supercritical CO_2. *Industrial Engineering and Chemistry Research*, **36**(2), 430–435.

Petersson, E., Liu, J., Arapsitas, P. & Turner, C. (2009). Pressurized hot water extraction of anthocyanins from red onions: a study on extraction and degradation kinetics. In: *Proceedings of the 237th ACS National Meeting*. Salt Lake City, UT, 22–26 March.

Pettinello, G., Bertucco, A., Pallado, P. & Stassi, A. (2000). Production of EPA enriched mixtures by supercritical fluid chromatography: from the laboratory scale to pilot plant. *The Journal of Supercritical Fluids*, **19**(1), 51–60.

Povh, N.P., Marques, M.O.M. & Meirles, M.A.A. (2001). Supercritical CO_2 extraction of essential oil and oleoresin from chamomile (*Chamomilla recutita* [L.] *rauschert). The Journal of Supercritical Fluids*, **21**(3), 245–256.

Quancheng, Z., Guihua, S., Hong, J. & Moucheng, W. (2004). Concentration of tocopherols by supercritical carbon dioxide with cosolvents. *European Food Research and Technology*, **219**(4), 398–402.

REFPROP (2007). *[CD-ROM]* National Institutes of Science and Technology, Gaithersburg, MD.

Robinson, J.R. & Sims, M.J. (1996) Method and system for extracting a solute from a fluid using dense gas and a porous membrane. U.S. Patent 5,961,835.

Rodriguez-Meizoso, I., Marin, F.R., Herrero, M., Senorans, F.J., Reglero, G., Cifuentes, A. & Ibanez, E. (2006). Subcritical water extraction of nutraceuticals with antioxidant activity from oregano. Chemical and functional characterization. *Journal of Pharmaceutical and Biomedical Analysis*, **41**(5), 1560–1565.

Rovio, S., Hartonen, K., Holm, Y., Hiltunen, R. & Riekkola, M.L. (1999). Extraction of clove using pressurized hot water. *Flavor and Fragrances Journal*, **14**, 399–404.

Sabirzyanov, A.N., Il'in, A.P., Akunov, A.R. & Gumerov, F.M. (2002). Solubility of water in supercritical carbon dioxide. *High Temperature*, **40**(2), 203–206.

Sarmento, L.A.V., Machado, R.A.F., Petrus, J.C.C., Tamanimi, T.R. & Bolzan, A. (2008). Extraction of polyphenols from cocoa seeds and concentration through polymeric membranes. *Journal of Supercritical Fluids*, **45**, 64–69.

Sarmento, L.A.V., Spricigo, C.B., Petrus, J.C.C., Carlson, L.H.C. & Machado, R.A.F. (2004). Performance of reverse osmosis membranes in the separation of supercritical CO_2 and essential oils. *Journal of Membrane Science*, **237**(1–2), 71–76.

Sarrade, S., Carles, M., Perre, C. & Vignet, P. (1999). Process and installation for separation of heavy and light compounds by extraction using a supercritical fluid and nanofiltration. US Patent 5,961,835.

Savova, M., Kolusheva, T., Stourza, A. & Seikova, I. (2007). The use of group contribution method for predicting the solubility of seed polyphenols of *Vitis Vinifera* L. within a wide polarity range in solvent mixtures. *Journal of the University of Chemical Technology and Metallurgy*, **42**(3), 295–300.

Schell, W.J., Houston, C.D. & Hopper, W.L. (1983). Membranes can efficiently separate CO_2 from mixtures. *Oil and Gas Journal*, **8**, 53–56.

Semenova, S.I., Ohya, H., Higashijima, T. & Negishi, Y. (1992). Separation of supercritical CO_2 and ethanol mixtures with an asymmetric polyimide membrane. *Journal of Membrane Science*, **74**(1–2), 131–139.

Shen, Z., Palmer, M.V., Ting, S.S.T. & Fairclough, R.J. (1997). Pilot scale extraction and fractionation of rice bran oil using supercritical carbon dioxide. *Journal of Agricultural and Food Chemistry*, **45**, 4540–4544.

Sims, M. (1998). Porocritical fluid extraction from liquids using near-critical fluids. *Membrane Technology*, **97**, 11–12.

Skerget, M., Knez, Z., Novak, Z. & Bauman, D. (1998). Separation of paprika components using dense CO_2. *Acta Alimentaria*, **27**(2), 149–160.

Snyder, J.M., King, J.W. & Jackson, M.A. (1997). Analytical supercritical fluid extraction using lipase catalysis: conversion of different lipids to methyl esters and effect of moisture. *Journal of the American Oil Chemists' Society*, **74**(5), 585–588.

Sovova, H. (2005). Mathematical model for supercritical fluid extraction of natural products and extraction curve evaluation. *Journal of Supercritical Fluids*, **33**(1), 35–52.

Spanos, G.A., Chen, H. & Schwartz, S.J. (1993). Supercritical CO_2 extraction of β-carotene from sweet potatoes. *Journal of Food Science*, **58**(4), 817–820.

Srinivas, K., King, J.W. & Hansen, C.M. (2008). Prediction and modeling of solubility phenomena in subcritical fluids using an extended solubility parameter approach. In: *Proceedings of ACS-AIChE National Meeting*. New Orleans, LA, 6–10 April, Abstrach #174 h.

Srinivas, K., King, J.W., Monrad, J.K., Howard, L.R. & Hansen, C.M. (2009). Optimization of subcritical fluid extraction of bioactive compounds using Hansen solubility parameters. *Journal of Food Science*, **74**(6), E342–E354.

Stahl, E., Quirin, K.W. & Gerard, D. (1988). *Dense Gases for Extraction and Refining*. Springer-Verlag, Heidelberg, Germany.

Stewart, P.B. & Munjal, P. (1970). Solubility of carbon dioxide in pure water, synthetic sea water, and synthetic sea water concentrates at $-5°$ to 25°C and 10- to 45-atm pressure. *Journal of Chemical Engineering Data*, **15**(1), 67–71.

Sugla, S. (2008). Nutraceuticals: global markets and processing technologies. *Food and Beverage, BCC Research*, Report ID FODO13 C, Wellesley, MA. Available online http://www.bccresearch.com (accessed 16 April 2009).

Sun-Lee, D., Chung, S.K. & Yam, K.L. (1992). Carotenoid loss in dried red pepper products. *International Journal of Food Science and Technology*, **27**(2), 179–185.

Tabera, J., Guinda, A., Ruiz-Rodriguez, A., Senorans, F.J., Ibanez, E., Albi, T. & Reglero, G. (2006). Countercurrent supercritical fluid extraction and fractionation of high-added-value compounds from a hexane extract of olive leaves. *Journal of Agricultural and Food Chemistry*, **52**, 4774–4779.

Takeuchi, T.M., Braga, M.E.M., Benedetti, B.A., Orestes, T., Leonardi, V.V. & Meirles, M.A.A. (2006). Native and exotic plants from Brazil: leaves oil extraction by SFE and hydrodistillation. In: *Proceedings of the 8th Conference on Supercritical Fluids and Their Applications*. Reverchon, E. (ed.), ISASF, Ischia, Italy, pp. 147–150.

Talansier, E., Braga, M.E.M., Rosa, P.T.V., Paolucci-Jeanjean, D. & Meirles, M.A.A. (2008). Supercritical fluid extraction of vetiver roots: a study of SFE kinetics. *The Journal of Supercritical Fluids*, **47**, 200–208.

Taylor, S.L. & King, J.W. (2002). Preparative scale supercritical fluid extraction/supercritical fluid chromatography (SFE/SFC) of corn bran. *Journal of the American Oil Chemists' Society*, **79**(11), 1133–1136.

Taylor, S.L., King, J.W., Montanari, L., Fantozzi, P. & Blanco, M.A. (1999). Enrichment and fractionation of phospholipid concentrates by supercritical fluid extraction and chromatography. *Italian Journal of Food Science*, **12**, 65–76.

Temelli, F., King, J.W. & List, G.R. (1996). Conversion of oils to monoglycerides by glycerolysis in supercritical carbon dioxide media. *Journal of the American Oil Chemists' Society*, **73**(6), 699–706.

Teng, H. & Yamasaki, A. (1998). Solubility of liquid CO_2 in synthetic sea water at temperatures from 278 K to 293 K and pressures from 6.44 MPa to 29.49 MPa, and densities of the corresponding aqueous solutions. *Journal of Chemical Engineering Data*, **43**(1), 2–5.

Tipsrisukond, N., Fernando, L.N. & Clarke, A.D. (1998). Antioxidant effects of essential oil and oleoresin of black pepper from supercritical carbon dioxide extractions in groud pork. *Journal of Agricultural and Food Chemistry*, **46**, 4329–4333.

Toews, K., Shroll, R., Wai, C.M. & Smart, M.G. (1995). pH-defining equilibrium between water and supercritical CO_2. Influence on SFE of organics and metal chelates. *Analytical Chemistry*, **67**, 4040–4043.

Valcarcel, V. & Tena, M.T. (1997). Applications of supercritical fluid extraction in food analysis. *Fresenius Journal of Analytical Chemistry*, **358**, 561–573.

Vega, P.J., Balaban, M.O., Sims, C.A., O'Keefe, S.F. & Cornell, J.A. (1996). Supercritical carbon dioxide extraction efficiency for carotenes from carrots by RSM. *Journal of Food Science*, **61**(4), 757–759.

Vesper, H. & Nitz, S. (1997). Composition of extracts from paprika (*Capsicum annuum* L.) obtained by conventional and supercritical fluid extraction. *Advances in Food Science*, **19**(5–6), 172–177.

Weibe, R. & Gaddy, V.L. (1934). The solubility of carbon dioxide in water at various temperatures from 12 to 40° and at pressures to 500 atmospheres. Critical phenomena. *Journal of American Chemical Society*, **62**, 815–817.

Wen, D. & Olesik, S.V. (2000). Characterization of pH in liquid mixtures of methanol/H_2O/CO_2. *Analytical Chemistry*, **72**, 475–480.

Williams, L.L., Rubin, J.B. & Edwards, H.W. (2004). Calculation of hansen solubility parameter values for a range of pressure and temperature conditions, including the supercritical fluid region. *Industrial Engineering and Chemistry Research*, **43**(16), 4967–4972.

Yoo, K.P. & Fukuzato, R. (2006). Current status of commercial development and operation of SCF technology in China, Japan, Korea and Taiwan. In: *Proceedings of the 8th International Symposium of Supercritical Fluids*. Kyoto, Japan, 5–8 November.

Zaidul, I.S.M., Nik Norulaini, N.A., Mohd Omur, A.K. & Smith Jr, R.L. (2006). Supercritical carbon dioxide (SC-CO_2) extraction and fractionation of palm kernel oil from palm kernel as cocoa butter replacers blend. *Journal of Food Engineering*, **73**, 210–216.

Zancan, K.C., Marques, M.O.M., Petenate, A.J. & Meirles, M.A.A. (2002). Extraction of ginger (*Zingiber officinale* Roscoe) oleoresin with CO_2 and co-solvents: a study of the antioxidant action of the extracts. *Journal of Supercritical Fluids*, **24**, 57–76.

Zosel, K. (1974). Inventor; Studiengesellschaft Kohle, assignee. Process for recovering caffeine. US Patent 3,806,619.

4 Emulsion delivery systems for functional foods

P. Fustier, A.R. Taherian and H.S. Ramaswamy

4.1 Introduction

Although there is no internationally accepted definition for functional foods, consumer's demand for more natural and high-quality food products, presenting health benefits, has increased over the years. As a result of this trend, the emergence of dietary compounds with health benefits offers an excellent opportunity to improve public health. Known as bioactives (nutraceutical), this category of compounds has received much attention in recent years from the scientific community, consumers and food manufacturers. The list of nutraceutical compounds (e.g. vitamins, probiotics, bioactive peptides and antioxidants) is endless, and scientific evidence indicates growing support for health-promoting food ingredients (Wildman 2001).

Even though the exact nature of the involvement of nutraceutical substances in physiological functions is not fully understood, it is well recognized that their addition to food matrices decreases the incidence and risks associated with several diseases and improves the general well-being of subjects. Therefore, the scientific community should develop innovative functional foods with the potential to produce physiological benefits or reduce the long-term risk of developing diseases (Elliott & Ong 2002). For instance, food proteins show great promise for developing and engineering a range of new GRAS (generally recognized as safe) matrices with the potential to incorporate nutraceutical compounds and provide controlled release via the oral route. Clear advantages of food protein matrices include their high nutritional value, abundant renewable sources and acceptability as naturally occurring food components degradable by digestive enzymes. As vital macronutrients in food, proteins possess unique functional properties including their ability to form gels and emulsions, which allow them to be an ideal base for the encapsulation of bioactive compounds. Food proteins can also be used to prepare a wide range of multicomponent matrices in the form of emulsions, hydrogels, micro- or nanoparticles, all of which can be tailored for specific applications in the development of innovative functional food products (Chen *et al.* 2006).

In many food products, hydrocolloid gums occur together with proteins, and both types of food macromolecules contribute to the structure, texture and stability of food (Doublier *et al.* 2000; Maroziene & de Kruif 2000). While hydrocolloids are present as thickening and water-holding agents, the overall texture and stability of food products depends not only on the properties of proteins and polysaccharides but also on the nature and strength of protein–hydrocolloid interactions. As an example, pectin can prevent aggregation of casein micelles or be the cause of it. These phenomena are related to the interaction of pectin with the proteins; the outcome depends on whether the pectin adsorbs onto the casein particles or not (Maroziene & de Kruif 2000).

The protein–polysaccharide interactions in bulk solutions and at interfaces have an important influence on the stability properties of food dispersions. A binary mixture of proteins and polysaccharides in an aqueous solution can exhibit one of three different equilibrium situations: (a) miscibility, (b) thermodynamic incompatibility or (c) complex coacervation (or complexation) (de Kruif & Tuinier 2001). Thermodynamic incompatibility implies the separation into two distinct aqueous phases, one rich in protein and the other rich in polysaccharides (Dickinson 2003). Complex coacervation mainly occurs below the isoelectric point of protein as a result of net electrostatic interactions between the polymers carrying opposite charges and implies the separation of two distinct phases, one phase is rich in the two biopolymers and the other phase is depleted in both. Miscibility occurs commonly at low biopolymer concentrations. Either incompatibility or coacervation (or complexation) appears at high concentrations, depending on whether the protein–hydrocolloid interaction is a net repulsion or a net attraction, respectively.

4.2 Food emulsions

Many foods are sold in an emulsified state and include products such as ice cream, desserts, butter, salad dressing, meat emulsions, soups, margarine and beverages (Barbosa-Cánovas *et al.* 1996; McClements 2005). Emulsions are dispersions of one liquid phase in the form of fine droplets in another immiscible liquid phase. The immiscible phases are usually oil and water, so emulsions can be broadly classified as oil-in-water or water-in-oil emulsions, depending on the dispersed phase.

The oil phase (internal or dispersed phase) consists of some combination of vegetable oil, flavor oil, weighting agent, essential oil, oil-soluble vitamins, bioactive ingredients and antioxidants, whereas the water phase (external or aqueous phase) consists of water, various types of proteins, hydrocolloid gums, citric acid, preservatives, emulsifiers, sweeteners and salts (Chanamai & McClements 2000; Tan 2004).

Emulsions, however, are thermodynamically unstable systems that are prone to destabilization during storage. Emulsion destabilization may occur through a variety of different physiochemical processes, including gravitational separation, flocculation, coalescence and Ostwald ripening (Walstra 1993; McClements 2005).

4.2.1 Emulsion stability

Emulsion stability is a measure of the rate at which an emulsion destabilizes as a result of creaming, flocculation or coalescence. The rate of these changes can be measured by changes in rheological properties of component phases, particle size and distribution of the oil droplets and/or density difference between water and oil phases. A stabilizer can be defined as a single or mixed chemical component that confers long-term stability on emulsions. Stabilizers may operate by acting as emulsifiers or as texture modifiers. Emulsifiers are surface-active ingredients that are adsorbed at the surface of emulsion droplets and prevent them from aggregating. Texture modifiers are ingredients that increase the viscosity of the continuous phase of emulsions to slow down the gravitational separation of the droplets.

One of the most important and widely used methods for improving the stability of oil-in-water emulsions is the use of emulsifiers, which help to lower the interfacial tension between the oil and water. They also reduce droplet aggregation by generating repulsive forces between the droplets. Emulsifiers vary considerably in their ability to form and stabilize emulsions, as well as in their cost, ease of utilization, ingredient compatibility and environmental sensitivity. There is a growing trend within the food industry to replace synthetic emulsifiers with more natural and consumer-friendly ones, such as phospholipids, proteins and polysaccharides. Proteins and lecithin are good at producing small droplets but have relatively poor stability to environmental stresses, e.g. pH, salt, heating and freezing. Polysaccharides, on the other hand, provide good stability to the environmental stresses but are relatively poor at producing small emulsion droplets or are required in high concentrations (McClements 2003). Therefore, many studies have focused on improvement of the functionality of natural emulsifiers by utilizing emulsifier blends or conjugates as emulsifiers (e.g. protein–polysaccharide and hydrocolloid–polysaccharide) (Difits & Kiosseoglou 2003; Kim *et al.* 2003; Taherian *et al.* 2006, 2007a,b).

Polysaccharides are widely used as thickening, emulsifying and stabilizing agents in beverage emulsions due to their biocompatibility, biodegradability and non-toxicity. Polysaccharides are natural polymers, which can display different behaviors once dissolved in water owing to their molecular structure. In particular, their mechanical properties are influenced by the polysaccharide backbone and its side subsistent. The type of bond between sugar rings can result in random coil shapes (such as dextran solutions), semi-flexible chains (such as cellulose derivative solutions) or interrupted helical structures (such as those in amylose) (Lapasin & Pricl 1995; Morris *et al.* 1996).

Polysaccharides (hydrocolloid gums) are mostly hydrophilic polymers, which do not exhibit significant surface activity. However, as a stabilizer in food emulsions, some gums are found to migrate slowly to the air–water and oil–water interfaces and exhibit some surface and interfacial activities (Garti 1999). Hydrocolloid gums, although water-soluble, rigid and very hydrophilic, can precipitate/adsorb onto oil droplets and sterically stabilize emulsions against flocculation and coalescence (Taherian *et al.* 2008). There are different mechanisms by which these polymers can stabilize an emulsion. For example, the formation of an extended hydrogel network reflects into high viscosity of the continuous phase at low shear, thus slowing down the droplet motion. Such a polymeric structure surrounds the oil droplets, ensuring effective steric hindrance of their coalescence (McClements 2005; Taherian *et al.* 2007b).

Another contribution to stabilization is provided by a non-adsorbing depletion mechanism due to the pronounced hydrophilicity, low flexibility and low surface activity of these polymers (Garti 1999; McClements 2005). Finally, due to the presence of some impurities, such as hydrophobic groups or proteinic moieties, an additional stabilizing effect can derive from the formation of a viscoelastic-adsorbed layer. Polysaccharides and surfactants act through different stabilization methods once added to the continuous phase of an emulsion so that simultaneous use of these additives should improve the stability of these systems. Moreover, surfactants are also employed as rheology controllers of the polysaccharide hydrogel in order to tune the viscoelastic properties of the system and obtain the desired mechanical characteristics of the final product. Nevertheless, when all these additives are employed together, incompatibilities, preferential adsorption of one additive with respect to the others (Wilde 2000; McClements 2005) or the formation of polysaccharide–surfactant complexes can have a detrimental effect on the stability of the emulsion, inducing phase separation. To obtain an emulsion with good mechanical and stabilizing properties, it is necessary to consider all these aspects, and consequently, the choice of the constituent components should be made carefully.

4.2.2 Factors influencing the stability of emulsions

Some of the most important factors influencing the quality of emulsions are considered below.

4.2.2.1 *Oil droplet viscosity*

Unlike a rigid droplet, the liquid within an oil droplet can move when a force is applied to the surface of the droplet. This reduces the frictional forces that oppose the movement of a droplet and increase the creaming velocity as (Dickinson & Stainsby 1982):

$$\nu = \nu_{\text{Stokes}} \frac{3(\eta_1 + \eta_2)}{(3\eta_1 + 2\eta_2)}$$

If the viscosity of the droplet is much less than that of the water phase ($\eta_1 \ll \eta_2$), the creaming rate is 1.5 times faster than that predicted by Stokes' law. Conversely, when the viscosity of the oil droplet is much greater than that of the water phase ($\eta_1 \gg \eta_2$), the rate of creaming will be the same as Stokes' law (McClements 2005).

Since Stokes' law has been used to calculate the velocity of an isolated rigid spherical particle in an ideal liquid, it is then necessary to consider the droplet acting as rigid spheres by surrounding them with a viscoelastic interfacial layer, which prevents the fluid within them from moving (Walstra & Fennema 1996).

4.2.2.2 *Polydispersity*

This refers to droplet size range, which mostly depends on the concentration of droplets. In dilute systems, such as beverage emulsions, the average creaming velocity can be estimated from the mean droplet radius (Tan & Wu Holmes 1988).

4.2.2.3 *Electrical charge and zeta potential*

Emulsions containing charged droplets tend to move more slowly than uncharged particles. The reasons are (a) repulsive electrostatic interactions between similarly charged droplets do not allow them to get as close together as uncharged droplets; and (b) the cloud of counterions surrounding a droplet moves slower than the droplet itself. The potential which arises from the presence of charges in particles and the medium at the zone of shear is called 'zeta potential'. Zeta potential of less than –15 mV usually represents the onset of flocculation. Hydrocolloids present in the water phase have different zeta potentials (e.g. gum arabic zeta potential is −23 mV). However, one cannot categorically state that an emulsion will or will not be stable at a given zeta potential as some other factors such as density difference between the two phases and the droplet size should be taken into consideration (Tan & Wu Holmes 1988). Addition of minerals could also have a major effect on zeta potential. For instance, minerals increase the ionic strength of the aqueous phase, which reduces the electrostatic repulsion between droplets through electrostatic screening. Some minerals bind to oppositely charged groups on the surface of emulsion droplets, decreasing the magnitude of their zeta potential and thereby reducing the electrostatic repulsion between droplets (Hunter 1986). Ion binding can increase the short-range hydration repulsion between droplets because of the additional energy required to disrupt the sheath of water molecules associated with them.

At sufficiently high concentrations, minerals cause alterations in the structural organization of the water molecules, which alters the strength of the hydrophobic interactions between non-polar groups (Kulmyrzaev *et al.* 2000). It should be noted that zeta potential reflects both the electrolytes presence in the system and the dissociated ions accompanying the original colloid particles. Addition of cation electrolytes could neutralize the zeta potential and cause aggregation to occur due to van der Waals–London forces.

4.2.2.4 *Particle size distribution*

The properties of a food emulsion that most contribute to its perceived characteristics are particle size distribution and oil concentration. Direct determination of these parameters is difficult because droplets are often too small to be resolved by conventional light microscopy. Electron microscopy provides adequate resolution, but sample preparation is more disruptive and may generate artifacts. The most commonly used technique for the characterization of food emulsions is laser light scattering, but this is only suitable for very dilute systems ($\phi < 0.05$ wt%) and dilution of more concentrated emulsions may disrupt delicate aggregates. Nuclear magnetic resonance has been used to characterize concentrated food emulsions, but the apparatus is expensive and difficult to operate (Dickinson & McClements 1995; Coupland & McClements 2001). Stokes' law indicates that the velocity at which a droplet moves is proportional to the square of its radius. Therefore, the stability of an emulsion can be enhanced by reducing the droplet size. This is especially true for beverage emulsions, which have to be stable in both concentrated and diluted forms. Consequently, the size distribution of the droplets has a great effect on emulsion stability. For beverage emulsions, the determination of particle size distribution could serve two proposes: estimation of the quality of the emulsion concentrates and prediction of the stability of the emulsion in the finished product. For example, in a bottled beverage, a particle of 1 μm in diameter will travel upward 100 times faster than a particle of 0.1 μm in diameter (Tan & Wu Holmes 1988).

4.2.2.5 *Surface activity*

The interfacial region which separates the oil from the aqueous phase has a direct influence on the bulk physicochemical and sensory properties of food emulsions, including their formation, stability, rheology and flavor. Surface-active molecules such as proteins, polysaccharides, alcohols and surfactants, which can accumulate at the interface, are able to alter the properties of emulsions. An emulsifier forms a protective membrane that prevents the droplets from aggregating (flocculating and/or coalescing) with one another. In addition, an emulsifier reduces the oil–water interfacial tension, thereby facilitating the disruption of emulsion droplets during homogenization (Walstra 1993, 2003). In an oil-in-water emulsion, in the presence of an emulsifier, the direct contact between oil and water molecules is replaced by contact between the non-polar segments of the emulsifier and oil molecules and between the polar segments of the emulsifier and water molecules. This in turn causes a decrease in free energy and hence decreases interfacial tension. The reduction of interfacial tension by the presence of a surface-active molecule is referred to as surface tension, which is:

$$\pi = \gamma_{\text{o/w}} - \gamma_{\text{surface-active molecule}}$$

where $\gamma_{\text{o/w}}$ is the interfacial tension of a pure oil–water interface and $\gamma_{\text{surface-active molecule}}$ is the interfacial tension in the presence of the emulsifier. Therefore, a decrease in surface tension could be related to the activity of surface-active molecules present in the emulsion.

4.2.2.6 *Emulsion rheology*

The main reasons for oil or oil-soluble ingredients to be used in the form of an emulsion rather than in their original state is that a much wider range of rheological characteristics and consistencies can be achieved with an emulsion. The water phase of an emulsion contains polysaccharides and/or proteins, which can provide specific rheological properties, affecting stability of the emulsion. There are five major aspects, which determine emulsion rheology:

1. **Rheology of component phases:** The viscosity of an emulsion is directly proportional to the viscosity of the water phase, and so any alteration in the rheological properties of the water phase has a corresponding influence on the rheology of whole emulsion.
2. **Oil-phase volume fraction:** The viscosity of an emulsion increases with oil-phase volume fraction in a linear manner.
3. **Droplet size:** The mean droplet size and polydispersity have a significant influence on the rheology of a concentrated emulsion (McClements 2005). The effect of both droplet size and droplet size distribution on the rheology of an emulsion depends on the oil-phase volume fraction and the nature of colloidal interaction. At the same volume fraction, a polydispersed emulsion has a lower viscosity than monodispersed emulsions.
4. **Colloidal interactions:** Colloidal interaction governs whether emulsion droplets aggregate or remain as separate forms, as well as determine the characteristics of any aggregate formed. The rheological properties of an emulsion depends on the relative magnitude of attractive (van der Waals, hydrophobic and depletion) and repulsive (electrostatic, steric and thermal flocculation) interactions between the droplets. These properties can be controlled by manipulating the colloidal interactions between the droplets. For instance, the viscosity of an emulsion increases with the addition of biopolymers, which cause an increase in depletion attraction, bridging and flocculation (McClements 2005; Taherian *et al.* 2006).
5. **Particle charge:** When a charged droplet moves through a fluid, the cloud of counterions surrounding it becomes distorted and causes an attraction between charged droplets and cloud counterions. This attraction opposes the movement of the droplet and causes an increase in viscosity.

4.2.3 Kinetics of droplets coalescence

The stability of emulsions against oil droplet coalescence has also been determined by following the change with time of the average droplet size and droplet aggregation rate. The rate of coalescence of emulsion droplets (D_c) mainly follows first-order kinetics (Sherman 1983; Ye *et al.* 2004; Paraskevopoulou *et al.* 2005):

$$N_t = N_0 \exp(-D_c t)$$

where N_0 and N_t are the numbers of droplets per unit volume of emulsion initially, and time t, respectively, and D_c is the rate of droplets coalescence. In terms of average droplet size, the equation is given as (Sherman 1983; Taherian *et al.* 2007b):

$$\ln D_t = \ln D_0 + \frac{D_c t}{3}$$

where D_0 and D_t are the mean droplet sizes initially and at time t, respectively. Therefore, D_c can be determined by plotting $3(\ln(D_t/D_0))$ versus time (t).

4.3 Delivery systems for bioactive materials

Bioactive compounds are often extracted from their natural matrix. As a result, the stability of the bioactive ingredients is critical for successful incorporation into various food or pharmaceutical systems. Encapsulation of a food component consists of developing a continuous and thin coating around the compounds within a matrix such as a gel or a crystal. Mixtures of polysaccharides and proteins, two hydrophilic groups, have been essentially used as a shell material for encapsulation purposes. Encapsulation of nutraceuticals protects them against the surrounding conditions and improves the overall food product quality. It also limits the oxidation of bioactive compounds and provides delivery matrices for human nutrition.

Health-promoting compounds such as vitamins, probiotics, minerals, omega-3 fatty acids and phytosterols are sensitive to oxygen, light, heat and water. These factors limit shelf-life and bioavailability in product matrices (Champagne & Fustier 2007). If bioactive compounds form degradation products, off-flavors and off-colors, and even carcinogenic compounds are formed, then the shelf-life of the fortified product will be limited. The selection of the shell material for encapsulation is therefore important and depends primarily on the application: What is the bioactive molecule to be protected (hydrophilic, lipophilic, solid or liquid)? What kind of core material is to be selected? By which mechanism will the ingredient be released? What are the potential modes of release (melting, dissolution, diffusion, rupture or degradation)? Should the textural properties of the final products be improved? How will the shell material resist processing operations, storage and the digestive system, where enzymes and acidity play an important role? Are the shell materials GRAS? What are the particle size, density and stability requirements?

Microencapsulation consists of the entrapment of active molecules and ingredients such as omega-3 and omega-6 fatty acids, herbs and bioactives (probiotic, prebiotic, creatinine), food ingredients (enzymes, leavening agents, psyllium, acids, vitamins, minerals, flavors and minerals), oils and fats inside microparticles to immobilize, protect and release them (Flanagan & Singh 2006; Garti 2008).

In general, the aims of a delivery system for an emulsion are as follows:

(a) Minimize the reactivity of the core in relation to the outside environment (light, acidity, oxygen, water, heavy metals)
(b) Decrease the evaporation or transfer rate of the core material to the outside environment
(c) Control the release of the core material in order to achieve proper delay until the right stimulus in the digestive tract
(d) Mask some off-flavors and off-odors such as those associated with the omega-3 and omega-6 from fish oil
(e) Facilitate the handling of the core material

4.3.1 Major emulsion delivery systems for bioactive molecules

4.3.1.1 *Simple (conventional) emulsions*

Emulsions, oil/water (O/W) or water/oil (W/O), can be defined as microscopic dispersion of two immiscible liquids, one which forms the continuous phase of the system and the other the dispersed phase.

Owing to the immiscibility of these two components, the emulsion is a thermodynamically unstable system. Emulsification therefore involves adding a certain amount of energy to the system to create interfaces between the two media (direction 1 of the arrow) while, over time, the system tends to return to its thermodynamically stable state (direction 2 of the arrow)

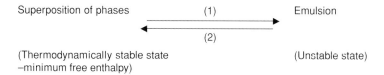

Superposition of phases (1) → Emulsion

← (2)

(Thermodynamically stable state –minimum free enthalpy) (Unstable state)

These emulsions are prepared by homogenizing both the oil phase and the aqueous phase together in the presence of an emulsifier. The nature and concentrations of the ingredients used to prepare these emulsions, the type of homogenization techniques (colloid mill, homogenizer, mixer, ultrasonic) and the processing steps used for their preparation will determine the physicochemical properties and stability of the emulsion (McClements 2003). The interfacial tension depends heavily on the use of small molecule surfactants (anionic, cationic), proteins and polysaccharides. The interfacial layer can vary from 1 to 10 nm thick, but can be increased significantly if biopolymer multilayers are formed around the oil droplets (Guzey & McClements 2007). The presence of a surface charge on the droplets due to the addition of an anionic or cationic surfactant can also contribute to form an electric barrier.

Oil is emulsified, as submicronized particles in an aqueous solution containing the encapsulating material, by a high-pressure homogenizer operating at 2500–20 000 psi and with one or several passes. The matrix-forming wall material can be either a single component (modified starch, acacia gum) or a combination of polymers (protein/polysaccharide).

Simple emulsions are mostly applied to essential oils as flavor systems, vegetable oils as clouding systems (Edris & Bergnstähl 2001; Buffo *et al.* 2002a,b; Tan 2004; Taherian *et al.* 2008), as well as some lipo- and hydrosoluble bioactive molecules. In the latter case, a bioactive lipid compound will be dispersed in the oil phase, whereas a hydrosoluble compound will be dispersed in the aqueous phase prior to the homogenization of the mixture. With crystalline lipid structures such as carotenoids, it is essential to warm the lipid phase to melt the crystals and also to control their saturation concentration in the oil phase prior to homogenization. Additionally, for bioactive materials sensitive to outside environmental conditions, it is critical to control the homogenizing conditions to inhibit their degradation. A few processes also involve survival of probiotics in simulated gastrointestinal tract. A probiotic suspension is blended in a mixture of sesame oil bodies in sesame oil. Compared to free cells, the bacteria entrapped in the emulsion demonstrate a significant increase in survival rate (Hou *et al.* 2003). Probiotics can also be encapsulated as freeze-dried or fresh cultures in water-insoluble microcapsules produced by the emulsion and/or spray drying using milk fat or denaturated whey protein (Picot & Lacroix 2004).

These diverse emulsions are generally easy to prepare, but they are also prone to physical instability due to droplet coalescence and to the diffusion of the water molecules from the internal phase to the bulk aqueous phase of food matrices (Garti 1997; Garti & Benichou 2004). Environmental stress such as pH extremes, elevated temperatures and high ionic strength are important destabilizing factors. Due to the small particle size of the oil droplets and interfacial layers, these conventional emulsions have limited potential to protect and control the release of the bioactive molecules. In order to minimize the oxidation of lipophilic bioactive molecules, it is possible to sequester the heavy metals of the environment by adding chelating agents into the aqueous phase and antioxidants such as tocopherols into the dispersed phase (Ribeiro & Shubert 2003; Hu *et al.* 2004). Augustin *et al.* (2006)

suggested the use of Maillard reaction products obtained by the reaction between protein and carbohydrate at elevated temperature to encapsulate oxidation-sensitive compounds such as fish oils. These conjugates are known to exhibit antioxidant activity and also form a robust shell around the emulsion droplets, and consequently, they are effective for protecting microencapsulated fish oils and other oils from oxidation.

O/W emulsions have found application in protecting many bioactive compounds in various food systems such as milk, juice, ice cream and yogurt. These bioactives include β-carotene (Santipanichwong & Suphantharika 2007), lycopene (Ribeiro & Shubert 2003) and omega-3 fatty acids (Chee *et al.* 2007). Polyunsaturated lipids can also be effectively protected against oxidation by selecting a milk protein, at low pH, as an emulsifier system. Lipid droplets become positively charged and as a result repel cationic metal ions that normally accelerate the oxidation mechanism. Protein (whey, sodium caseinate, soy, gelatin), carbohydrates (modified starch), gum (acacia gum) and small-molecule carbohydrates (maltodextrin, glucose, sucrose) are often used as encapsulating materials.

The encapsulating matrices must be able to stabilize the emulsions by forming a stable and dense film around the surface of the emulsion oil droplets and form a continuous wall matrix by acting as an oxygen barrier. Protein–carbohydrate combinations can also produce a wall system that is quite effective in stabilizing encapsulated oils, such as fish oil and other bioactive molecules during storage (Dickson & Galazka 1991; Benichou *et al.*, 2002). Carbohydrates in the amorphous state in the wall matrix, rather than a crystalline structure, are generally recommended for the protection of bioactive molecules against oxidation.

4.3.1.2 *Multiple emulsions*

Multiple emulsions are complex dispersion systems, often referred to as emulsions of emulsion or double emulsion. The most common multiple emulsion used in food matrices is W/O/W emulsion, but the O/W/O can also serve to protect lipophilic bioactive compounds. Compared to conventional emulsions, these are even more thermodynamically unstable with a tendency for coalescence, flocculation and creaming.

In multiple emulsions, the inner dispersed oil droplet is separated from the outer liquid phase by a layer of another phase (Garti & Benichou 2004), as shown in Figure 4.1.

W/O/W multiple emulsions have found application as potential vehicles for dispensing drugs (insulin) and cosmetics due to the prolonged release of active substances. They have the potential of combining incompatible compounds in the double compartments of the emulsion and the protection of sensitive compounds (Dahms 1999). Several articles highlight the potential of multiple emulsions as a delivery system for food application (Garti & Benichou 2004). Double emulsions have been used for the production of foods with lower fat and oil content (Labato-Calleros *et al.* 2006), low-calorie mayonnaise (Matsumoto 1985), protection of orange oil (Edris & Bergnstähl 2001; Cho & Park 2003), control of the aroma and flavor release (Gaonkar 1994), milk immunoglobulin IgG (Chen *et al.* 1999), proteins and amino acids (Su *et al.* 2006; Weiss *et al.* 2005), vitamin B (Fechner *et al.* 2007), vitamin A (Yoshida *et al.* 1999), vitamin C (Gallarate *et al.* 1999), β-carotene (Rodriguez-Huezo *et al.* 2006) and omega-3 fatty acids (Cournarie *et al.* 2004). No commercial food applications of these emulsions are known. The reasons for this are their inherent thermodynamic instability as well as the uncontrolled, fast release of the entrapped materials, such as electrolytes.

The release of the internal aqueous phase of this type of emulsion can be explained by the diffusion of the oily membrane and the break up of the oily globules, which leads to the loss of the multiple structure of the emulsion (Raynal *et al.* 1993; Grossiord 1996). For the

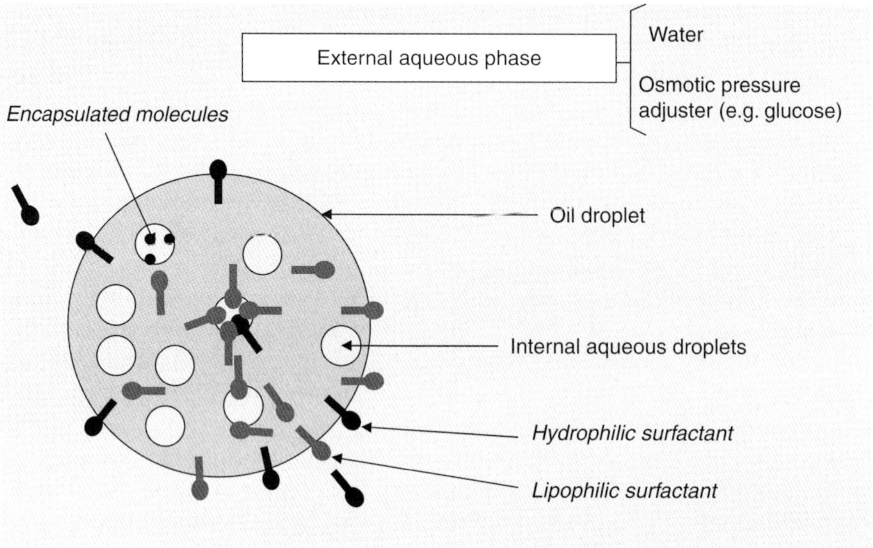

Fig. 4.1 Schematic representation of multiple emulsion. For a colour version of this figure, please see the plate section.

preparation of these emulsions, two surfactants of opposite solubility are used. Preparation involves a two-step procedure: first, water is mixed with an oily solution containing a lipophilic surfactant for the preparation of the primary emulsion (W/O), and then the primary emulsion is emulsified in an aqueous solution containing a hydrophilic surfactant, which will become the external aqueous phase of the multiple emulsion. Most of the surfactants used to prepare these emulsions are considered synthetic additives (Span, Tween and Admul). There is currently a need for identification of more friendly ingredients for the food industry. Milk proteins such as sodium caseinate and whey protein isolate (WPI) are among these emulsifiers; however, owing to their limited solubility due to thermal treatment and pH changes, their potential is limited. The Maillard conjugates prepared with proteins and polysaccharides where the amino groups of the proteins react with the carboxyl group of the sugars are very stable against change of pH value and ionic strength. The potential of these conjugates as food emulsifiers is promising, and they are an alternative to the monomeric and polymeric synthetic surfactants, which yield aftertaste in the food matrices at higher concentration (Benichou *et al.* 2002; Fechner *et al.* 2007). These Maillard-type protein–polysaccharide conjugates are easy to prepare and they have excellent emulsifying and steric stabilizing properties, especially under conditions where the protein alone is poorly soluble.

The advantage of these emulsions is their double-compartment structure. A W/O/W emulsion is generally recognized as more suitable for the delivery of bioactive molecules than a simple emulsion. In this case, hydrophilic molecules can be entrapped inside the inner droplet of the water phase and released at a controlled rate. They are also better protected from other water-soluble components that could promote their chemical degradation since the desired molecules are then isolated physically.

To produce a stable multiple emulsion, it is generally recommended to avoid severe mechanical stresses during the second homogenization step as it may disrupt the primary water in oil (W/O) emulsion droplets previously homogenized. This normally produces coarse W/O/W emulsion droplets with poor creaming stability. It is also suggested to incorporate a thickening or gelling polymer (xanthan, gelatin, sodium caseinate) within the aqueous phase of the primary emulsion in order to provide a gel network for entrapment of the emulsion droplets and reduce the mobility. Control of the osmotic pressure between the two aqueous phases is also recommended to ensure the physical stability of these emulsions and minimize the coalescence phenomenon. If required, these multiple emulsions can also be spray dried, resulting in a double-layered microcapsule, which provides better protection to sensitive materials such as citrus oil (Edris & Bergnstähl 2001). In this case, a triple-emulsion O/W/O/W is used to encapsulate orange oil, and then the continuous outer phase, which contains sodium caseinate and lactose as core material, is evaporated by spray drying. The powder consists of a double O/W/O emulsion, in which the inner orange oil is dispersed in an aqueous phase; this inner phase is dispersed in an oil phase encapsulated in the sodium caseinate and lactose.

4.3.1.3 *Multilayer emulsions*

Multilayer membranes are produced by sequential deposition of an oppositely charged emulsifier. A primary emulsion is first made by dispersing the oil in a solution of ionic emulsifier, such as lecithin, and then the emulsion is mixed with a solution of positively charged polysaccharides, such as chitosan. This procedure can be pursued by the addition of layers to the interfacial layers.

Layer-by-layer deposition produces cationic and also thick emulsion droplets. The characteristics of the interfacial layers surrounding the emulsified compounds such as permeability, electric charge, thickness, rheological behavior and the resistance to the layers to the outside environment can be easily modulated. It has been shown that the oxidative stability of tuna oil in water emulsion coated by the lecithin–chitosan multilayer membranes was superior to the emulsion droplets coated with lecithin alone (Ogawa *et al.* 2004; Klinkesorn *et al.* 2005).

This improved stability of this emulsion could be related to the cationic nature of the droplets that causes repulsion of the pro-oxidative metals coupled, and also the formation of a thicker membrane layer that reduces the interactions between lipids and water-soluble pro-oxidants. The same tendency in retarding lipid oxidation was observed with an emulsion stabilized with sodium dodecyl sulfate (SDS) combined with chitosan (Mun *et al.* 2005). An anionic interfacial membrane can be formed by the addition of pectin to form a third layer on the emulsion droplets, but the oxidative stability of this tertiary emulsion will be very similar to the SDS–chitosan emulsion, implying that the thickness increase of the interfacial membrane of the emulsion droplets can be controlled to minimize the pro-oxidative effect of an anionic emulsion droplet interface.

For application as delivery systems, multilayer emulsions present potential advantages over conventional emulsions for several reasons:

- Control of the thickness and properties of the interfacial layers which influence the release kinetics of the functional ingredients. This release can be controlled by modifying the thickness of the interfacial layer as a function of the ionic strength, pH, temperature, dilution and other environmental conditions.

- Oxidative stability of fatty acids can be reduced by minimizing the interaction between metal ions and the lipids via the adjustment of the interfacial charge and thickness (Klinkesorn *et al.* 2005)
- The composition and properties of the interfacial layer play an important role in the stabilization of the emulsion against factors such as pH, thermal treatment, mechanical stress and dehydration (Harnsilawat *et al.* 2006; Guzey & McClements 2007).

Multilayer emulsions have been used for encapsulating partially crystalline lipid droplets or compounds which tend to crystallize, namely, the carotenoids and phytosterols, which tend to aggregate (Walstra 2003). The formation of a thick blend of biopolymer around the crystalline lipid droplets prevents the aggregation.

4.3.1.4 *Coacervation*

The process of coacervation involves the separation of an aqueous phase of coating material from a polymeric solution via addition of another chemical or oppositely charged hydrocolloid. There are two methods of coacervation, a simple and a complex polymer coating.

In the simple polymer coating, the polymer materials which are initially soluble (e.g. proteins) can become insoluble by changing the pH and/or temperature of the solution. The encapsulation is initiated as a regular emulsion with the polymer adsorbed at the interface between the emulsified phase and a solvent. Then, the temperature or pH of the solution is modified so that the polymer becomes insoluble in the solvent and forms a separate phase that coats the emulsified phase. The removal of a solvent from a single colloid, dispersed in solution, by addition of a chemical component which has a greater affinity for the solvent than the hydrocolloid can also yield a coacervate. As a result, the colloid chains are coming closer together to form a coacervate. Gliadin, zein, albumin and gelatin microparticles have been prepared by such a procedure (Mauguet *et al.* 2002; Liu *et al.* 2005, 2006).

In the complex coacervation, two polymers are mixed together (one with negative charge and another with a positive charge) to form an insoluble complex on the surface of the oil droplets. Complexes of polysaccharides (arabic gum, xanthan gum, pectin, carrageenan and alginate) with food proteins (albumin, gelatin, casein and β-lactoglobulin) are generally recommended to form the microcapsules (de Kruif *et al.* 2004). For instance, a mixture of whey proteins and gum arabic can be used as coating material to encapsulate essential oils (Weinbreck *et al.* 2004). An emulsion of the essential oil-in-water in the presence of whey protein is first prepared. A concentrated solution of gum arabic is then added to the emulsion at a pH where both the protein and the gum are negatively charged. The pH value of the solution is changed below the isoelectric point of the whey protein, so the protein has a net positive charge, while the gum arabic remains negatively charged. This procedure induces a complexation of both polymers and the formation of a coacervate. An insoluble protective coating is thus formed that prevents the emulsion droplets from coalescing. There are many examples for using this procedure to deliver bioactive molecules, and different materials have been employed for such as purpose. Chitosan with β-lactoglobulin (Chen & Subirade 2005), chitosan and gelatin (Shu & Zhu 2002), gelatin and carrageenan (Jonganurakkun *et al.* 2006), pectin and chitosan (Chang & Lin 2000), β lactoglobulin and gum arabic (Schmitt *et al.* 1999), gelatin and gum arabic (Bungenberg de Jong & Kruyt 1924; de Kruif *et al.* 2004), and protein–polyelectrolyte (Da Silva *et al.* 2006) are among the complexes which have been studied.

4.3.1.5 *Filled hydrogel particles*

In this case, the oil droplets are incorporated within the hydrogel particles that are dispersed within an aqueous continuous phase. First a conventional oil-in-water (o/w) emulsion is prepared. The selection of an emulsifier (type and concentration) and the homogenization procedure will control both the particle size and charge of the droplets. The filled hydrogel particles are then prepared by mixing the oil-in-water emulsion with a biopolymer solution, and the environmental conditions are adjusted to facilitate the formation of these particles, which are also referred to as a type of oil-in-water-in-water emulsion. Several techniques are currently used to prepare these hydrogel particles. It involves the gelation of a biopolymer solution via coacervation, aggregative and segregative phase-separation procedures (McClements *et al.* 2007). The composition, thickness, permeability and the response to environmental conditions of the biopolymer layer can be controlled to minimize lipid oxidation, prevent instability related to the coalescence of the droplets and control the release of a bioactive component in the digestive tract.

Omega-3 fatty acids (Wu *et al.* 2005) and flavor oils (Weinbreck *et al.* 2004) have been incorporated in filled hydrogel particles based on aggregative phase-separation procedure, whereas complex coacervation using gelatin and gum acacia has been used to protect omega-3 fatty acids (Lamprecht *et al.* 2001).

4.3.1.6 *Nanoemulsion*

Nanoemulsions are defined as emulsions covering the size range of 50 to 200 nm (transparent) or up to 500 nm (milky appearance). Their small droplet size and high kinetic stability allow their use mainly in the pharmaceutical and cosmetic applications (Jafari *et al.* 2006).

Nanoemulsions can be prepared by low-energy emulsification methods or by phase inversion temperature or composition, but in most cases, they are prepared using high-pressure homogenization and microfluidization. Bioactive components can be incorporated within the oil droplets, the continuous phase or the interfacial region of the oil-in-water emulsion (Garti *et al.* 2005; Weiss *et al.* 2006).

Several publications deal with the use of nanoemulsions for bioactive molecules (Semo *et al.* 2007), essential oils (Parris *et al.* 2005), vitamins and antioxidants (Tan & Nakajima 2005), polypeptides (Were *et al.* 2003), phytosterols (Garti *et al.* 2005) and antimicrobial agents (Myc *et al.* 2001). However, there are still no food products using nanoemulsions for incorporating bioactive molecules. Lack of sufficient food-grade emulsifiers, lack of knowledge to construct phase diagrams, lack of sufficient protection of the bioactive components from environmental reactivity and the poor organoleptic quality of the nanoemulsion prepared via micelle solubilization may explain this trend (Jafari *et al.* 2006). Nutritional and safety issues related to the use of these emulsions remain today a topic to be addressed.

4.4 Encapsulation of polyunsaturated fatty acids – an example application

The nutritional benefits of omega-3 fatty acids make them excellent candidates as functional food ingredients if problems with oxidative rancidity can be overcome (Alamed *et al.* 2006). Successful incorporation of omega-3 fatty acids into processed foods would most likely be in the form of lipid dispersions. Lipid dispersions that consist of oil dispersed in an aqueous phase in the form of small spherical droplets are referred to as oil-in-water emulsions.

Owing to the evolution of oxidative degradation of oil, the level of oxidation at early stages of the process gives poor information about the later oil behavior. For this reason, oil stability against oxidation is considered even more important than the extent to which the oil is oxidized.

A study by McClements and Decker (2000) indicated that lipid oxidation in oil-in-water emulsions, believed to be duc to the interaction between lipid hydroperoxides located at the droplet surface and transition metals originating in the aqueous phase, is the most common cause of oxidative instability. Incorporating antioxidants into foods is one of the most effective means of retarding lipid oxidation. In oil-in-water emulsions, the most successful type of antioxidant is one that chelates transition metal ions. A chelate is a complex that results from the combination of a metal ion and a multidentate ligand such that the ligand forms two or more bonds with the metal, resulting in a ring structure that includes the metal ion (Miller 1996). Chelators that act as antioxidants can inhibit metal-catalyzed reactions by a variety of different mechanisms, including prevention of metal redox cycling, occupation of metal coordination sites, and steric hindrance of interactions between metals and lipid substrates (McClements & Decker 2000).

A study by Hu *et al.* (2004) indicates that the oxidative stability of oil-in-water emulsions can be improved by engineering the surface of the emulsion droplets to decrease transition metal–lipid interactions. This can be accomplished in protein-stabilized emulsions where the pH is less than the pI of the protein and the emulsion droplets are cationic. Combination of protein and polysaccharide, however, is another promising means to reduce lipid oxidation.

For instance, WPI is a surface-active globular protein and can be adsorbed to oil droplet surfaces in the form of a monolayer. Emulsification properties of WPI depend on many factors such as its structure and net charge, pH and ionic strength, and properties of the interfacial film formed (thickness, elasticity and viscosity). The major constituents of WPI are β-lactoglobulin and α-lactalbumin; they contain cysteyl residues, disulfide bonds and thiol functional groups, which can scavenge free radicals to inhibit lipid oxidation (Cayot & Lorient 1997). Therefore, WPI-stabilized emulsions may act as an antioxidant system.

On the other hand, xanthan gum (XG) is an anionic polysaccharide produced by the bacterium, *Xanthomonas campestris*. XG has a cellulose backbone with a trisaccharide side chain at the C-3 position. The terminal mannose residues are 4,6-pyruvated, which can chelate pro-oxidant metal ions and inhibit lipid oxidation (Morris & Foster 1994). The most important properties of XG are high low-shear viscosity and strong shear-thinning character. The relatively low viscosity at high shear rate makes it easy to mix, pour and swallow; its high viscosity at low shear rate gives it good suspension properties and lends stability to colloidal suspensions. The combination of protein and polysaccharide provides the system with high surface activity and high viscosity, and forms charged thick, gel-like adsorbed layers. The mechanical strength of the adsorbed layer as well as the electrostatic (repulsion of emulsion droplets carrying like charges) and steric (barrier of thick stabilizing layer) effects is the most important factors contributing to the kinetic stability of O/W emulsions.

4.5 Conclusions

The development of new food applications is particularly challenging since product safety, preparation mode and sensory quality of the product are not to be compromised by the incorporation of hydrosoluble, liposoluble and insoluble bioactive compounds. Several strategies to fortify individual categories of food products are already described by patents and publications. The type of molecule to be protected, the desired release profile of the bioactive molecule, the biodegradability and biocompatibility of the matrix, as well as the final properties of the food matrix, are among the requirements that need to be properly defined prior to the final selection of the microencapsulation technique. Each type of delivery system has its own specific advantages and disadvantages for the encapsulation and delivery of functional ingredients. A thorough approach with a precise definition of the goals of the microencapsulation process is therefore required prior to development of an application.

References

Alamed, J., McClements, D.J. & Decker, E.A. (2006). Influence of heat processing and calcium ions on the ability of EDTA to inhibit lipid oxidation in oil-in-water emulsions containing omega-3 fatty acids. *Food Chemistry*, **95**, 585–590.

Augustin, M.A., Sanguansri, L. & Bode, O. (2006). Maillard reaction products as encapsulants for fish oil powders. *Journal of Food Science*, **71**(2), E25–E32.

Barbosa-Cánovas, G.V., Kokini, J.L., Ma, L. & Ibarz, A. (1996). The rheology of semiliquid foods. *Advances in Food and Nutrition Research*, **39**, 1–69.

Benichou, A., Aserin, A. & Garti, N. (2002). Protein-polysaccharide interactions for stabilization of food emulsions. *Journal of Dispersion Science and Technology*, **23**(1–3), 93–123.

Buffo, R.A. & Reineccius, G.A. (2002b). Modeling the rheology of concentrated beverage emulsions. *Journal of Food Engineering*, **51**, 267–272.

Buffo, R.A., Reineccius, G.A. & Oehlert, O.W. (2002a). Influence of time–temperature treatments on the emulsifying properties of gum acacia in beverage emulsions. *Journal of Food Engineering*, **51**, 341–345.

Bungenberg de Jong, H.G. & Kruyt, H.R. (1924). Coacervation (partial miscibility in colloid systems). *Process Academia Science Amsterdam*, **32**, 849–856.

Cayot, P. & Lorient, D. (1997). Structure-function relationships in whey proteins. In: *Food Proteins and Applications*. Damoradan, S. & Parf, A. (eds), Marcel Dekker, New York, pp. 473–502.

Champagne, C.P. & Fustier, P. (2007). Microencapsulation for delivery of probiotics and other ingredients in functional dairy products. In: *Functional Dairy Products*, Vol. 2. Saarela, M. (ed.), VTT Biotechnology, Finland, pp. 404–422.

Chanamai, R. & McClements, D.J. (2000). Dependence of creaming and rheology of monodisperse oil-in-water emulsion on droplet size and concentration. *Colloids Surface*, **172**, 79–86.

Chang, K.L.B. & Lin, J. (2000). Swelling behaviour and the release of protein from chitosan-pectin particles. *Carbohydrate Polymers*, **43**(2), 163–169.

Chee, C.D., Djordjevic, D., Faraji, H., Decker, E.A., Hollender, R., McClements, D.J., Peterson, D.G., Roberts, R.F. & Coupland, J.N. (2007). Sensory properties of vanilla and strawberry flavoured ice cream supplemented with omega-3 fatty acids. *Milchwissenschaft*, **62**(1), 66–69.

Chen, C.C., Tu, Y.Y. & Chang, H.M. (1999). Efficiency and protective effect of encapsulation of milk immunoglobulin G in multiple emulsion. *Journal of Agricultural Food Chemistry*, **47**, 407–410.

Chen, L., Remondetto, G.E. & Subirade, M. (2006). Food protein-based materials as nutraceutical delivery systems. *Trends in Food Science and Technology*, **17**(5), 272–283.

Chen, L. & Subirade, M. (2005). Chitosan/β-lactoglobulin core shell nanoparticles as nutraceuticals carriers. *Biomaterials*, **26**, 6041–6053.

Cho, Y.H. & Park, J. (2003). Evaluation of process parameters in the O/W/O multiple emulsion method for flavour encapsulation. *Journal of Food Science* **63**(2), 534–538.

Coupland, J. & McClements, D.J. (2001). Droplet size determination in food emulsions: comparison of ultrasonic and light scattering methods. *Journal of Food Engineering*, **50**(2), 117–120.

Cournarie, F., Savelli, M.P., Rosilio, W., Bretez, F., Vauthier, C., Grossiord, J.L. & Seiller, M. (2004). Insulin loaded W/O/W multiple emulsion: comparison of the performances of systems prepared with medium chain triglycerides and fish oil. *European Journal of Pharmaceutics Biopharmaceutics*, **58**(3), 477–482.

Da Silva, F.L., Lund, M., Joensson, B. & Aakesson, T. (2006). On the complexation of proteins and polyelectrolytes. *Journal Physical Chemistry*, **110**(2), 4459–4464.
Dahms, G.H. (1999). Multiple phase emulsion – a carrier system for radical scavenger such as ascorbic acid and polyphenols. *SOFW*, **6**, 2–6.
de Kruif, C.G. & Tuinier, R. (2001). Polysaccharide protein interactions. *Food Hydrocolloids*, **15**, 555–563.
de Kruif, C.G., Weinberg, F. & Vries, R. (2004). Complex coacervation of proteins and anionic polysacharrides. *Current Opinion in Colloid and Interface Science*, **9**(5), 340–349.
Dickinson, E. (2003). Hydrocolloids at interfaces and the influence on the properties of dispersed systems. *Food Hydrocolloids*, **17**(1), 25–39.
Dickinson, E. & McClements, D.J. (1995). *Advances in Food Colloids*. Blackie & Academic Professional, Glasgow.
Dickinson, E. & Stainsby, G. (1982). *Colloids in Foods*. Applied Science Publisher, London.
Dickson, E. & Galazka, V.B. (1991). Emulsion stabilization by ionic and covalent complexes of β-lactoglobulin with polysacharrides. *Food Hydrocolloïds*, **5**, 281–296.
Difits, N. & Kiosseoglou, V. (2003). Improvement of emulsifying properties of soybean protein isolate by conjugation with carboxymethyl cellulose. *Food Chemistry*, **81**(1), 1–6.
Doublier, J.L., Garnier, C., Renard, D. & Sanchez, C. (2000). Protein–polysaccharide interactions. *Current Opinion in Colloid and Interface Science*, **5**, 202–214.
Edris, A. & Bergnstähl, B. (2001). Encapsulation of orange oil in spray dried double emulsion. *Nahrung/Food*, **45**(2), 133–137.
Elliott, R. & Ong, T.J. (2002). Science, medicine, and the future nutritional genomics. *British Medical Journal*, **324**, 1438–1442.
Fechner, A., Knoth, A., Scherze, I. & Muschiolik, G. (2007). Stability and release properties of double emulsions stabilized by caseinate-dextran conjugates. *Food Hydrocolloids*, **21**, 943–952.
Flanagan, J. & Singh, H. (2006). Microemulsions: apotential delivery system for bioactives in food. *Critical Reviews in Food Science and Nutrition*, **46**, 221–237.
Gallarate, M., Carlotti, M.E., Trotta, M. & Bovo, S. (1999). On the stability of ascorbic acid in emulsified systems for topical and cosmetic use. *International Journal of Pharmaceutics*, **188**, 233–241.
Gaonkar, A.G. (1994). Method for preparing a multiple emulsion. US patent 5322704.
Garti, N. (1997). Double emulsions – scope, limitations and new achievements. *Colloids and surface A: Physico-Chemical and Engineering Aspects*, **123–124**, 233–246.
Garti, N. (1999). Hydrocolloids as emulsifying agents for oil-in-water emulsions. *Journal of Dispersion Science and Technology*, **20**(12), 327–355.
Garti, N. (2008). *Delivery and Controlled Release of Bioactives in Foods and Nutraceuticals*. Woodhead Publishing in Food Science, Technology and Nutrition, CRC, Cambridge, UK, pp. 184–206.
Garti, N. & Benichou, A. (2004). Recent development in double emulsions for food applications. In: *Food Emulsions*. Friberg, S.E., Larsson, K. & Sjöblom, J. (eds), Marcel Dekker, New York, pp. 353–412.
Garti, N., Spernath, A., Aserin, A. & Lutz, R. (2005). Nano-sized self-assemblies of non-ionic surfactants as solubilization reservoirs and microreactors for food systems. *Soft Matter*, **1**, 206–218.
Grossiord, J.L. (1996). Les émulsions multiples: des systèmes vésiculaires d'intérêt industriel. *Oleagineux Corps Gras Lipides (OCL)*, **3**(3), 158–162.
Guzey, D. & McClements, D.J. (2007). Impact of electrostatic interactions on formation and stability of emulsions containing oil droplets coated with beta-lactoglobulin-pectin complexes. *Journal of Agriculture and Food Chemistry*, **55**(2), 475–485.
Harnsilawat, T., Pongsawatmanit, R. & McClements, D. (2006). Influence of pH and ionic strength on formation and stability of emulsions containing oil droplets coated by beta-lactoglobulin-alginate interfaces. *Biomacromolecules*, **7**, 2052–2058.
Hou, R.C., Lin, M.Y., Wang, M.M.C. & Tzen, J.T.C. (2003). Increase of viability of entrapped cells of *Lactobacillus delbrueckii bulgaricus* in artificial sesame oil emulsions. *Journal of Dairy Science*, **86**, 424–428.
Hu, M., McClements, D.J. & Decker, E.A. (2004). Impact of chelators on the oxidative stability of whey protein isolate-stabilized oil-in-water emulsions containing ω-3 fatty acids. *Food Chemistry*, **88**(1), 57–62.
Hunter, R.J. (1986). *Foundations of Colloid Science*, Vol. **1**. Oxford University Press, Oxford.
Jafari, S.M., He, Y. & Bandhari, B. (2006). Nanoemulsion production by sonification and microfluidization – a comparison. *International Journal of Food Properties*, **9**, 475–485.

Jonganurakkun, B., Nodasaka, Y., Sakairi, N. & Nishi, N. (2006). DNA-based gels for oral delivery of probiotic bacteria. *Macromolecular Bioscience*, **6**(1), 99–103.
Kim, H.-J., Choi, S.J., Shin, W.S. & Moon, T.W. (2003). Emulsifying properties of bovine serum albumin-galactomannan conjugates. *Journal of Agricultural and Food Chemistry*, **51**(4), 1049–1056.
Klinkesorn, U., Sophanadora, P., Chinachotti, P., McClements, D.J. & Decker, E.A. (2005). Increasing the oxidative stability of liquid and dried tuna oil-in-water emulsion with electrostatic layer-by-layer deposition technology. *Journal of Agricultural Food Chemistry*, **53**(11), 4561–4566.
Kulmyrzaev, A., Chanamai, R. & McClements, D.J. (2000). Influence of pH and $CaCl_2$ on the stability of dilute whey protein stabilized emulsions. *Food Research International*, **33**(1), 15–20.
Labato-Calleros, C., Rodriguez, E., Sandoval-Castilla, O., Vernon-Carter, E.J. & Alvarez-Raminez, J. (2006). Reduced fat white fresh cheese like products obtained from W/O/W multiple emulsion. *Food Research International*, **39**, 678–685.
Lamprecht, A., Schafer, U. & Lehr, C.M. (2001). Influences of process parameters on preparation of microparticles used as carrier system for omega-3 unsaturated fatty acid ethyl esters used in supplementary nutrition. *Journal of Microencapsulation*, **18**(30), 347–357.
Lapasin, R. & Pricl, S. (1995). *Rheology of Industrial Polysaccharides: Theory and Applications*. Blackie Academic and Professional, Glasgow.
Liu, L., Fishman, M.L., Hicks, K.B., Kende, M. & Ruthel, G. (2006). Pectin/zein beads for potential colon-specific drug delivery: synthesis and in vitro evaluation. *Drug Delivery*, **13**, 417–423.
Liu, X., Sun, Q., Wang, H., Zhang, L. & Wang, J. (2005). Microspheres of corn protein, zein, for the ivermectin drug delivery system. *Biomaterials*, **26**, 109–115.
Maroziene, A. & de Kruif, C.G. (2000). Interaction of pectin and casein micelles. *Food Hydrocolloids*, **14**, 391–394.
Matsumoto, S. (1985). Macro and micro-emulsions. In: *Theory and Applications*. Shah, D.O. (ed.), ACS Symposium Series, ACS, Washington, DC, pp. 272, 415.
Mauguet, M.C., Legrand, J., Brujes, L., Carnelle, G., Larre, C. & Popineau, Y. (2002). Gliadin matrices for microencapsulation processes by simple coacervation method. *Journal of Microencapsulation*, **19**, 377–384.
McClements, D.J. (2003). Role of hydrocolloids as food emulsifiers. In: *Gums and Stabilizers for the Food Industry 12*. Williams, P.A. & Phillips, G.O. (eds), The Royal Society of Chemistry, UK.
McClements, D.J. (2005). *Food Emulsions: Principles, Practice, and Techniques*, 2nd edn. CRC Press, New York.
McClements, D.J. & Decker, E.A. (2000). Lipid oxidation in oil-in-water emulsions: impact of molecular environment on chemical reactions in heterogeneous food systems. *Journal of Food Science*, **65**(8), 1270–1282.
McClements, D.J., Decker, E.A. & Weiss, J. (2007). Emulsion-based delivery systems for lipophilic bioactive components. *Journal of Food Science*, **72**(8), R109–R124.
Miller, D.D. (1996). Minerals. In: *Food Chemistry*, 3rd edn. Fennema, O.R. (ed.), Marcel Dekker, New York, pp. 617–649.
Morris, E.R. & Foster, T.J. (1994). Role of conformation in synergistic interactions of xanthan. *Carbohydrate Polymers*, **23**, 133–135.
Morris, E.R., Gothard, M.G.E., Hember, M.W.N., Manning, C.E. & Robinson, G. (1996). Conformational and rheological transitions of welan, rhamsan and acylated gellan. *Carbohydrate Polymers* **30**, 165–175.
Mun, S., Decker, E.A. & McClements, D.J. (2005). Influence of droplets characteristics on the formation of oil-in-water emulsion stabilized by surfactant chitosan layers. *Langmuir*, **21**(14), 6228–6234.
Myc, A., Vanhecke, T., Landers, J.J., Hamonda, T. & Baker, J.R. (2001). The fungicidal act emulsion ($X8W_{60}$ PC against clinically important yeast and filamentous fungi. *Mycopathologia*, **155**, 195–201.
Ogawa, S., Decker, E.A. & McClements, D.J. (2004). Production and characterization of the o/w emulsion containing droplets stabilized by lecithin-chitosan-pectin multilayered membranes. *Journal of Agricultural and Food Chemistry*, **52**(11), 3595–3600.
Paraskevopoulou, A., Boskou, D. & Kiosseoglou, V. (2005). Stabilization of olive oil – lemon juice emulsion with polysaccharides. *Food Chemistry*, **90**, 627–634.
Parris, N., Cooke, P.H. & Hicks, K.B. (2005). Encapsulation of essential oils in zein nanosperical particles. *Journal Agriculture and Food Chemistry*, **53**(12), 4788–4792.

Picot, A. & Lacroix, C. (2004). Encapsulation of bifibacteria in whey protein-based microcapsules and survival in simulated gastrointestinal conditions as yoghurt. *International Dairy Journal*, **14**(6), 505–515.

Raynal, S., Grossiord, J.L., Seiller, M. & Clausse, D. (1993). A typical W/O/W multiple emulsion containing several active substances: formulation, characterization and study of release. *Journal Control Release*, **26**, 129–140.

Ribeiro, H.S. & Shubert, H. (2003). Stability of lycopene emulsions in food systems. *Journal of Food Science*, **70**(2), E117–E123.

Rodriguez-Huezo, M.E., Pedroza-Islas, R., Prado-Barragan, L.A., Beristain, C.I., Vernon-Sanchez, C., Mekhloufi, G. & Renard, D. (2006). Complex coacervation between betalactoglobulin and acacia gum: a nucleation and growth mechanism. *Journal of Colloid and Interface Science*, **299**, 867–873.

Santipanichwong, R. & Suphantharika, M. (2007). Carotenoids as colorants in reduced-fat mayonnaise containing spent brewer's yeast beta-glucans as fat replacer. *Food Hydrocolloids*, **21**(4), 565–574.

Schmitt, C., Sanchez, C., Thomas, F. & Hardy, J. (1999). Complex coacervation between β-lactoglobulin and acacia gum in aqueous medium. *Food Hydrocolloids*, **13**, 483–496.

Semo, E., Kesselman, E., Danino, D. & Livney, Y.D. (2007). Casein micelle as a natural nano-capsular vehicle for nutraceuticals. *Food Hydrocolloids*, **21**(5/6), 936–942.

Sherman, P. (1983). Rheological properties of emulsions. In: *Encyclopedia of Emulsion Technology*, Vol. 1. Becher, P. (ed.), Marcel Deker Inc., New York, pp. 405–437.

Shu, X.Z. & Zhu, K.J. (2002). Controlled drug release properties of ionically cross linked chitosan beads: the influence of anion structure. *International Journal of Pharmaceutics*, **233**, 217–225.

Su, J.H., Flanagan, J., Hemar, Y. & Singh, H. (2006). Syergistic effects of polyglycerol ester of polyricinoleic acid and sodium caseinate on the stabilization of water-oil-water emulsion. *Food Hydrocolloids*, **20**(2–3), 261–268.

Taherian, A.R., Fustier, P., Britten, M. & Ramaswamy, H.S. (2008). Rheology and stability of beverage emulsions in presence and absence of weighting agents – a review. *Food Biophysics*, **3**(3), 279–286.

Taherian, A.R., Fustier, P. & Ramaswamy, H.S. (2006). Effect of added oil and modified starch on rheological properties, droplet size distribution, opacity and stability of beverage cloud emulsions. *Journal of Food Engineering*, **77**, 687–696.

Taherian, A.R., Fustier, P. & Ramaswamy, H.S. (2007a). Effect of added weighting agents and xanthan gum on stability and rheological properties of beverage cloud emulsions formulated using modified starch. *Journal of Food Process Engineering*, **30**, 204–224.

Taherian, A.R., Fustier, P. & Ramaswamy, H.S. (2007b). Steady and dynamic shear rheological properties, and stability of non-flocculated and flocculated beverage emulsions. *International Journal of Food Properties*, **10**, 1–20.

Tan, C.P. & Nakajima, M. (2005). Effect of polyglycerol esters of fatty acids on physicochemical properties and stability of β-carotene nanodispersions prepared by emulsification/evaporation method. *Journal Science Food Agriculture*, **85**, 121–126.

Tan, C.-T. (2004). Beverage emulsions. In: *Food Emulsions*. Friberg, S.E. & Larsson, K. (eds), 4th edn. pp. 485–524.

Tan, C.T. & Wu Holmes, J. (1988). Stability of beverage flavor emulsions. *Perfumer and Flavorist*, **13**, 23–41.

Walstra, P. (1993). Principles of emulsion formation. *Chemical Engineering and Science*, **48**, 333–350.

Walstra, P. (2003). *Physical Chemistry of Foods*. Marcel Dekker, New York.

Walstra, P. & Fennema, O.R. (1996). Dispersed systems: basic considerations. In: *Food Chemistry*, 3rd edn. Marcel Dekker, New York, pp. 133–151.

Weinbreck, F., Minor, M. & de Kruif, C.G. (2004). Microencapsulation of oils using whey protein/gum Arabic coacervates. *Journal of Microencapsulation*, **21**(6), 667–679.

Weiss, J., Scherze, F. & Muschiolik, G. (2005). Polysaccharide gel with multiple emulsion. *Food Hydrocolloids*, **19**(3), 605–615.

Weiss, J., Takhistou, P. & McClements, D.J. (2006). Functional materials in food nanotechnology. *Journal of Food Science*, **71**(9), R107–R115.

Were, L.M., Bruce, B.D., Davidson, P.M. & Weiss, J. (2003). Size, stability and entrapment efficiency of phospholipids nanocapsules containing polypeptide antimicrobials. *Journal of Agriculture and Food chemistry*, **51**(27), 8073–8079.

Wilde, P.J. (2000). Interfaces: their role in foam and emulsion behaviour. *Current Opinion in Colloid and Interface Science*, **5**, 176–181.

Wildman, R.E.C. (2001). *Handbook of Nutraceuticals and Functional Foods*. Wildman, R.E.C. (ed.), CRC Press, New York.
Wu, K.G., Chai, X.H. & Chen, Y. (2005). Microencapsulation of fish oil by simple coacervation of hydroxypropyl methylcellulose. *Chinese Journal of Chemistry*, **23**(11), 1459–1472.
Ye, A., Hemar, Y. & Singh, H. (2004). Enhancement of coalescence by xanthan addition to oil-in-water emulsions formed with extensively hydrolysed whey proteins. *Food Hydrocolloids*, **18**, 737–764.
Yoshida, K., Sekine, T., Matsuzaki, F., Yaniki, M. & Yamaguchi, M. (1999). Stability of vitamin A in oil-in water-in oil multiple emulsion. *Journal of American Oil Chemical Society*, **76**(2), 195–200.

Part II

Functional ingredients

5 Functional and nutraceutical lipids

Fereidoon Shahidi

Functional and nutraceutical lipids may originate from plants, animals and microorganisms. They may also be produced via modification processes for the purpose of concentration of specific fatty acids, elimination of certain components or enhancement of certain physicochemical characteristics. Interesterification is among processes whereby certain fatty acids may be included in triacylglycerol (TAG) or phospholipid (PL) molecules in order to impart desired properties to the lipid, and this process can also be used to prepare diacylglycerols (DAG) or monoacylglycerols (MAG) for specific applications. Lipids also serve as carriers of constituents such as fat-soluble vitamins, sterols, tocopherols, tocotrienols and other minor components that might influence the quality and stability of the products (Barrow & Shahidi, 2008; Shahidi, 2006).

This chapter provides a cursory account of specialty lipids, including their fatty acid constituents that may be used as ingredients in food and as natural health products to impart health benefits. Effect of processing and storage on the stability of oils and their minor components is also discussed.

5.1 Omega-3 fatty acids and products

Polyunsaturated fatty acids (PUFAs) are known to provide unique health benefits to consumers, but also present scientists and technologists with a difficult challenge in delivering foods, containing them without any off-flavor perception. Thus, interest in omega-3 fatty acids has expanded dramatically over the last decade or so.

Omega-3 fatty acids include α-linolenic acid (ALA; C18:3*n*-3) as their parent compound. ALA is found abundantly in flaxseed oil and is an essential fatty acid that may be converted to long-chain PUFAs such as eicosapentaenoic acid (EPA; C20:5*n*-3), docosahexaenoic acid (DHA; C22:6*n*-3) as well as docosapentaenoic acid (DPA; C22:5*n*-3) through a series of elongation and desaturation steps (see Figure 5.1). The conversion of ALA to EPA in humans is 2–5% (Aliam 2003). However, EPA and DHA are found in relative abundance in seafoods such as the flesh of fatty fish like herring, mackerel, anchovies, menhaden and salmon, or the liver of white lean fish such as cod and halibut, as well as the blubber of marine mammals like seals and whales (Ackman 2005; Shahidi & Zhong 2005; Shahidi 2007). While sea animals consume marine algae, which contain long-chain omega-3 fatty acids and thus serve as a source of omega-3 oils, algal oils such as DHASCO (DHA single cell oil), among others, may also be produced by fermentation processes (Senanayake &

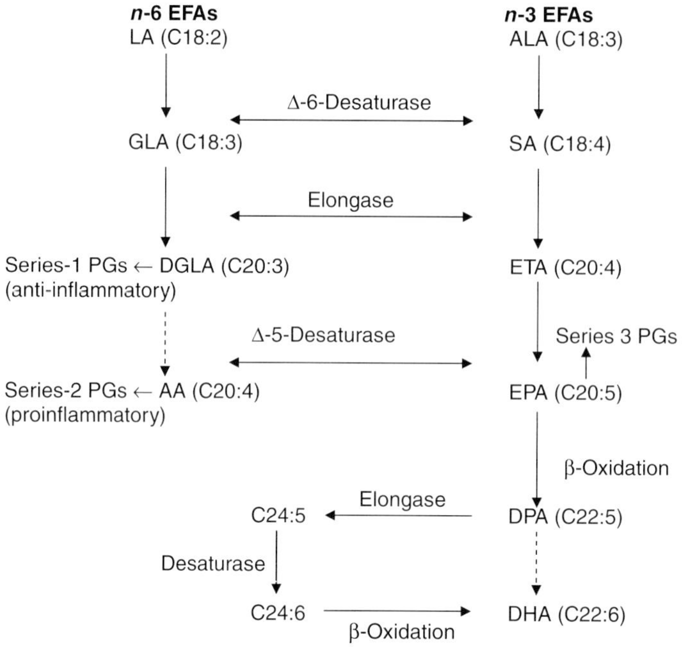

Fig. 5.1 The *n*-6 and *n*-3 fatty acids and their metabolites. EFA, essential fatty acids; LA, linoleic acid; ALA, α-linolenic acid; GLA, γ-linolenic acid; SA, stearidonic acid; PG, prostaglandin; DGLA, dihomo-γ-linolenic acid; ETA, eicosatetraenoic acid; AA, arachidonic acid; EPA, eicosapentaenoic acid; DPA, docosapentaenoic acid; DHA, docosahexaenoic acid.

Fichtali 2006). Table 5.1 summarizes the omega-3 fatty acid content of selected oils from plant and marine sources.

In most cases, crude fish oil is a by-product of the fish meal industry (Shahidi & Zhong 2007). The type of fish caught for the meal industry includes white anchovy, black anchovy, sardine, mackerel, capelin, blue whiting, herring, menhaden, sand eel and sprat. The extraction of the oil is carried out through a grinding process. Whole fish or heads and tails may be used for fish oil and meal production. The material is passed through a rotating grinder where

Table 5.1 Fatty acid composition of omega-3 oils[a]

Fatty acid	Flax oil	Menhaden	Cod liver oil	Seal blubber	Algal oil
14:0	—	9.4	2.7	3.7	8.7
16:0	5.3	16.0	12.4	6.0	22.2
16:1	—	12.8	5.0	18.0	—
18:0	4.1	2.9	6.1	0.9	0.7
18:1	20.2	10.4	16.3	20.1	0.7
18:2	12.7	2.1	3.0	1.5	0.5
18:3*n*-3	53.3	1.1	1.0	0.4	0
20:5*n*-3	—	11.4	13.2	6.4	1.4
22:5*n*-6	—	0.5	0	4.8	16.3
22:6*n*-3	—	11.3	15.4	7.6	41.1

[a] Seal blubber oil contains 12.2% of 20:1 and 2.0% of 22:1.

the oil is freed and pressed out. The resultant crude oil is generally stored in drums with nitrogen flushing. This oil may be stored for up to 3 years. The oil may be alkali refined, bleached and deodorized (RBD), similar to other edible oils. However, the deodorization process generally carried out for marine oils is molecular distillation, and employs a lower temperature. The RBD oil so produced is often stabilized using selected antioxidants and may also be encapsulated/microencapsulated (Shahidi & Han 1993; Shahidi & Wanasundra 1995).

The health benefits of omega-3 fatty acids are varied and range from those related to cardiovascular disease, cancer, type 2 diabetes, autoimmune and inflammatory disorders, brain health and vision as well as arthritis, among others (Shahidi & Wanasundara 1998; Shahidi & Finley 2001; Shahidi & Miraliakbari 2004, 2005, 2006). The mechanisms by which these effects are rendered are varied, but some involve the eicosanoid pathway. For cardiovascular health, the effects are positively related to a lowering of TAG and reducing the incidence of arrhythmias. For mental health, it is thought that EPA with its anti-inflammatory effect benefits patients with schizophrenia, depression and alike through its conversion to eicosanoids (Horrobin 1992; Keller 2002). However, DPA, which is found in relatively high abundance in blubber of marine mammals, may also play a major role in health promotion and disease risk reduction (e.g. Kanayasu-Toyoda *et al.* 1996). It is worth noting that ALA supplementation increased the concentration of ALA, EPA and DPA (but not DHA) in breast milk lipids (Francois *et al.* 2003).

While consumption of seafood is encouraged and daily intake of 0.6–1.0 g of long-chain omega-3 PUFA is considered beneficial, this is not the case for many populations and hence inclusion of omega-3 oils in processed foods or as dietary supplements is an alternate way for their delivery. Thus, omega-3 fatty acids may be included in bread, cereal-based products, bars, juices, dairy products, pastas, spreads and even in meat and seafoods, among others. However, a challenge for food manufacturers, as noted earlier, is related to the instability of omega-3 fatty acids as their autoxidation leads to the formation of a myriad of breakdown products with low off-flavor thresholds. Thus, microencapsulation of omega-3 oils is often considered. In this connection, several procedures are commonly used, among which is a coacervation technique that has been used to prepare microcapsules that remain intact in the products but are degraded in the gastrointestinal tract. This not only leads to the protection of the oil from autoxidation, it also facilitates its incorporation into different products. Currently, yogurts, juices, milk and a range of products containing MEG3, a commercial microencapsulated fish oil, are available in the marketplace. In addition, eggs containing omega-3 fatty acids may be purchased. The eggs are from hens fed on flaxseed or flaxseed oil or fish meal; in the latter case, they are then fed on a finishing diet devoid of fish meal in order to ensure their optimum flavor quality. Flaxseed oil has also been used in the formulation of a number of orange juice products in Canada. In addition, bread and other baked goods containing flaxseed are readily available in the marketplace (see Table 5.2).

The consumption of omega-3 fatty acids leads to an increased demand on the antioxidant system in the body; hence, marine and other highly unsaturated oils are expected to increase the body's burden on tocopherols (Wanasundara & Shahidi, 1998a). Manufacturers generally include mixed tocopherols from soybean oil processing by-products in the oils as stabilizers as well as for addressing the increased need of the body for antioxidants (Papas 2006). Natural antioxidants, other than tocopherols, may also be used to stabilize omega-3 oils. Thus, dechlorophillized green tea extracts (DGTEs) were found effective and their efficacy was similar to that of TBHQ (*tert*-butylhydroquinone) (Wanasundra & Shahidi, 1996).

The omega-3 fatty acids may also be used as dietary supplements as capsules, liquids or in other forms. When used as a liquid, flavorants may be added to the oil in order to mask any

Table 5.2 Different classes of food products in which omega-3 oils are used for fortification

Class	Items
Dairy	Milk, yogurt, yogurt-based drinks, butter
Grain based	Bread, cereals, pasta, bars
Confectionary	Sweets, candies, bars
Spreads	Margarine, spreads, etc.
Dressings	Salad dressing, mayonnaise, etc.
Juices	Orange juice, fruit juices, etc.
Muscle foods	Meat, fish and poultry products
Others	Specialty products, infant foods and formula

off-flavor that might be generated in the products on extended storage. For therapeutic purposes, the omega-3 fatty acid content of the oil may be concentrated. The omega-3 concentrates may generally be in the alkyl ester or in the TAG form. Thus, capsules of omega-3 oils, as such, or in the concentrated form, are available. The production of concentrated omega-3 fatty acids may commonly be carried out by the urea complexation process or by molecular distillation. In this manner, pure omega-3 fatty acids or different ratios of EPA to DHA oils may be produced, depending on their expected area of application. In addition, such oils, in the simple alkyl ester form, may then be subjected to reaction with glycerol to convert them to TAGs.

In the urea complexation process, straight-chain saturated fatty acids, and to a lesser extent monounsaturated fatty acids, are easily complexed with urea, while PUFAs do not participate in the complexation. Thus, TAGs are initially hydrolyzed into their constituent fatty acids via alkaline hydrolysis. The resultant fatty acids are subsequently mixed in an ethanolic solution of urea for complex formation. The urea complexed fraction, which includes saturated and monounsaturated fatty acid, is then removed by filtration. The liquid fraction is enriched with omega-3 fatty acids (Wanasundara & Shahidi 1999). By employing this procedure, we were able to enrich the DHA content of DHASCO from 42 to 97% (Wanasundara & Shahidi 1999).

Molecular distillation may also be employed for the removal of saturated and monounsaturated fatty acids from the hydrolyzed oils. In this manner, process conditions may be selected in such a way that enrichment provides varied ratios of EPA to DHA. Similarly, supercritical extraction may be used for the purpose of enriching omega-3 content of oils, which also provides the advantage of minimizing autoxidation of susceptible fatty acids. However, this procedure uses high pressures and has high initial costs.

Use of enzymes for concentration of specific fatty acids and other lipid reactions has been practiced for many years. For example, various vegetable and fish oils may be enriched with EPA and DHA by enzyme-catalyzed reactions (Wanasundara & Shahidi 1998a,b; Senanayake & Shahidi, 1999a,b, 2001) in which saturated and monounsaturated fatty acids are preferentially hydrolyzed, leaving behind acylglycerol products with enhanced EPA and DHA concentrations. Furthermore, as noted earlier, concentrated EPA and DHA or their corresponding simple alkyl esters may be esterified to glycerol to produce concentrates of omega-3 fatty acids in the acylglycerol form.

5.2 Monounsaturated fatty acids

Monounsaturated fatty acids (MUFAs), particularly oleic acid, have been reported to possess beneficial health effects similar to omega-3 fatty acids, but do not have the same level of fluidity, while being more stable against oxidative deterioration. MUFAs are known to have a positive effect in ameliorating the risk of breast cancer (Pala *et al.* 2001). In children, an association between the composition of MUFA oils and serum lipid profile has been reported (Sanchez-Bayle *et al.* 2008). In addition, minor components in such oils might play a major role in rendering their beneficial health effects, such as those in olive oil, a main component of the Mediterranean diet. The minor components present include phenolic compounds such as tyrosol, hydroxytyrosol, caffeic, *o*- and *p*-coumaric, cinnamic, ferulic, gallic, *p*-hydroxybenzoic, protocatechuic, sinapic, syringic and vanillic acid as well as oleorupin, among others (Shahidi 2006).

MUFAs are found abundantly in natural foods such as nuts, avocado, olive oil (50–83%), tree nut oils (generally >50%) and canola oil (~60%) as well as tea seed oil (~80%). New varieties of oils such as high-oleic sunflower oil have also been produced.

It has been recommended that 25–35% of total daily calories should be provided by dietary lipids, of which 20% should come from MUFA and 10% from PUFA. Thus, MUFAs are expected to contribute to approximately 60% of the recommended total dietary lipid intake (The National Heart, Lung and Blood Institute).

5.3 Medium-chain fatty acids and medium-chain triacylglycerols

Medium-chain fatty acids (MCFAs) are saturated fatty acids with 6–12 carbon atoms. MCFAs and medium-chain triacylglycerols (MCTs) are recognized for their health benefits. They are used for enteral and parenteral nutrition for a range of medical conditions for patients suffering from fat malabsorption, maldigestion, metabolic difficulties related to cystic fibrosis, Crohn's disease, colitis and enteritis (Kabara 2000; Pfeuffer & Schrezenmeir 2002). MCTs may also be incorporated into infant formula to help fat digestion and absorption of immature digestive systems of infants. Supplements of MCTs are sold for the purpose of increasing metabolic rate for weight reduction and providing additional energy for sport activities (Tsuji *et al.* 2001; Che Man & Abdul Manaf 2006; St-Onge & Bosarge 2008).

MCTs are procured from tropical fruit oils such as those of coconut and palm kernel where lauric acid (C12:0) is present in high amounts; other MCFAs present include caproic (C6:0), caprylic (C8:0) and capric (C10:0) acids. Canola oil is another example of an oil relatively rich in lauric acid. MCTs have a caloric value of 8.3 kcal/g, while MCFAs are more hydrophilic than their long-chain counterparts and thus do not need to be solubilized as micelles as a prerequisite for their absorption. MCT can pass discretely into the portal vein and readily oxidize in the liver to serve as a source of energy rather than being absorbed through the intestinal lymphatics. Since MCTs increase energy expenditure in humans, they may serve as weight-loss ingredients in foods. Thus, total energy expenditures were 48 and 65% greater in lean and obese individuals, respectively, after MCT compared to LCT (long-chain TAG) consumption (Scalfi *et al.* 1991). However, St-Onge and Jones (2003) reported that fat oxidation was greater in lean as opposed to obese individuals.

5.4 Conjugated linoleic acids and γ-linolenic acid

Conjugated linoleic acids (CLAs) refer to a group of fatty acids often found in the dairy products and meat from ruminants. During the process of biohydrogenation, anaerobic microorganisms in the rumen convert the methylene-interrupted *cis-cis* double bonds of the PUFAs to *cis-trans* isomers. Thus, isomers of linoleic acid with double-bond positions at 9 and 11, 10 and 12, and 11 and 13 are present. The proportion of different CLA isomers is different, but dairy products contain about 90% of 9-*cis*, 11-*trans* isomer. Cheeses are the richest source of CLA. CLAs have been reported to render health benefits due to their anticarcinogenic effect (Watkins & Li 2006). However, caution should be exercised when consuming artificially produced CLA, which is often dominant in 10-*trans*, 12-*cis* isomer and could have negative health effects. The latter is often used in supplements intended for weight loss/weight control purposes. It is worth noting that conjugated linolenic acids as well as conjugated eicosapentaenoic acids and conjugated docosahexaenoic acids might also be produced and these conjugated products have also been demonstrated to possess beneficial health effects (Tsuzuki *et al.* 2004; Yonezawa *et al.* 2005).

γ-Linolenic acid (GLA) is an omega-6 fatty acid and a desaturation product of linoleic acid. GLA is found in a relatively high proportion in seed oils of borage (20–25%), evening primrose (7–10%) and blackcurrant (15–19%) (Huang & Huang 2006). Algae such as *Spirulina* and various species of fungi also serve as desirable sources of GLA. GLA has been shown to possess therapeutic benefits in a number of diseases, notably atopic eczema, cyclic mastalgia, premenstrual syndrome, cardiovascular disease, inflammation, diabetes and cancer (Horrobin 1990), and may be used as a pharmaceutical product and as a food ingredient.

As there might be a problem in the enzymatic desaturation of linoleic acid to GLA, consumption of GLA may prove beneficial as its further conversion to arachidonic acid will be facilitated. GLA is generally available as an over-the-counter supplement and most frequently sold as capsules of evening primrose oil. The meal left behind after oil extraction from evening primrose is a source of oenothin B and other polyphenols with potential anticancer activity (Miyamoto *et al.* 1993; Wettasinghe & Shahidi 1999).

5.5 Diacylglycerol oils

DAGs are present as natural components in different oils. For example, cottonseed oil contains up to 10% DAG (Flickinger 2006). DAGs have been prepared in recent years, and products from canola oil are available in the market in Japan and in the United States. Such oils are produced via conversion of TAG to DAG to about 57%. Such oils contain mixtures of 1,3-DAG, 1,2-DAG and 2,3-DAG. In these cases, the actual benefit is due to the presence of 1,3-DAG. The absence of a fatty acid in the sn-2 position of DAGs does not allow the resynthesis of TAGs; hence, the latter cannot be deposited as subcutaneous fat, especially in the belly area.

5.6 Structured lipids

Structured lipids (SLs) are TAGs or PLs containing specific fatty acids located in the same glycerol molecule. SLs may be prepared by chemical or enzymatic processes in order to produce nutraceutical lipids for nutritional, pharmaceutical and medicinal applications.

The SL in the TAG form may be produced by the reaction between two TAG molecules (interesterification) or between a TAG and an acid (acidolysis). Thus, specific fatty acids may be incorporated into TAG in order to achieve specific effects or functionalities (Hamam &

Shahidi 2004, 2005a,b, 2006). As noted earlier, borage, evening primrose, black currant and fungal oils are good sources of γ-linolenic acid, which possesses beneficial effects in the treatment of atopic eczema, dermatitis, hypertension and premenstrual syndrome. On the other hand, *n*-3 PUFAs have potential for prevention of cardiovascular disease, among others. SLs containing both GLA and *n*-3 PUFA may be of interest because of their potential desired and possibly synergistic health benefits. We have successfully produced SLs containing GLA, EPA and DHA on the same glycerol backbone using borage and evening primrose oils as the main substrates (Senanayake & Shahidi 2002a,b). In this work, we noticed that the stability of SLs so prepared was less than that of the original substrate; a higher degree of unsaturation as well as reduction in the content of tocopherols was found to be responsible (Senanayake & Shahidi 2002c). In an attempt to introduce MCFA into a number of algal oils, we were able to successfully produce a number of SLs with potential application for patients with malabsorption and maldigestive problems, among others. Despite introduction of a saturated fatty acid in the molecules, the resultant products were found to be of lesser stability when compared to the original oils. We, however, were able to explain this apparent anomaly as the endogenous tocopherols were esterified to the free fatty acids which were present in the reaction medium, hence losing their antioxidant potency (Hamam & Shahidi 2006a,b). Nonetheless, it should be noted that esterified tocopherols will still exert their beneficial effects as a source of vitamin E and antioxidants once hydrolyzed on digestion.

5.7 Conclusions

This overview clearly demonstrates that fatty acids with beneficial health effect include certain saturated fatty acids such as MCFAs, MUFAs such as oleic acid and PUFAs belonging to both the omega-3 and omega-6 families as well as conjugated fatty acids, among others. These fatty acids may be present in naturally occurring oils or may have been included in specialty lipids on modification using different means and approaches.

References

Ackman, R.G. (2005). Fish oils. In: *Bailey's Industrial Oil and Fat Products*, 6th edn, Volume 3. Shahidi, F. (ed.), John Wiley & Sons, Hoboken, NJ, pp. 279–318.

Aliam, S.S.M. (2003). Long chain polyunsaturated fatty acids, nutritional and healthy aspects. Review article. *Rivista Italiana delle sostanze grasse*, **80**, 85–92.

Barrow, C. & Shahidi, F. (2008). *Marine Nutraceuticals and Functional Foods*. CRC Press, Boca Raton, FL.

Che Man, Y.B. & Abdul Manaf, M. (2006). Medium-chain triacylglycerols. In: *Nutraceutical and Specialty Lipids and Their Co-products*. Shahidi, F. (ed.), Taylor & Francis Group, Boca Raton, FL, pp. 27–56.

Flickinger, B.D. (2006). Diacylglycerols (DAGs) and their mode of action. In: *Nutraceutical and Specialty Lipids and Their Co-products*. Shahidi, F. (ed.), Taylor & Francis Group, Boca Raton, FL, pp. 181–186.

Francois, C.A., Conner, S.L., Bolewicz, L.C. & Conner, W.E. (2003). Supplementary lactating women with flaxseed oil does not increase docosahexaenoic acid in their milk. *American Journal of Clinical Nutrition*, **77**, 226–233.

Hamam, F. & Shahidi, F. (2004). Synthesis of structured Lipids via acidolysis of docosahexaenoic acid single cell oil (DHASCO) with capric acid. *Journal of Agricultural and Food Chemistry*, **52**, 2900–2906.

Hamam, F. & Shahidi, F. (2005a). Structured lipids from high-laurate canola oil and long-chain omega-3 fatty acids. *Journal of the American Oil Chemists Society*, **82**, 731–736.

Hamam, F. & Shahidi, F. (2005b). Enzymatic incorporation of capric acid into a single cell oil rich in docosahexaenoic acid and docosapentaenoic acid and oxidative stability of the resultant structured lipid. *Food Chemistry*, **91**, 583–592.

Hamam, F. & Shahidi, F. (2006a). Acidolysis reactions lead to esterification of endogenous tocopherols and compromised oxidative stability of modified oils. *Journal of Agricultural and Food Chemistry*, **54**, 7319–7323.

Hamam, F. & Shahidi, F. (2006b). Synthesis of structured lipids containing medium-chain and omega-3 fatty acids. *Journal of Agricultural and Food Chemistry*, **54**, 4390–4396.

Horrobin, D.F. (1990). Gamma-linolenic acid. *Reviews in Contemporary Pharmacotherapy*, **1**, 1–41.

Horrobin, D.F. (1992). The relationship between schizophrenia and essential fatty acids and eicosanoid metabolism. *Prostaglandins, Leukotrienes and Essential Fatty Acids*, **46**, 71–77.

Huang, Y.W. & Huang, C.Y. (2006). Gamma-linolenic acid (GLA). In: *Nutraceutical and Specialty Lipids and Their Co-products*. Shahidi, F. (ed.), Taylor & Francis Group, Boca Raton, FL, pp. 169–180.

Kabara, J. (2000). A health oil for the new millennium. *Inform*, **11**, 123–127.

Kanayasu-Toyoda, T., Morita, I. & Ad Murota, S.-I. (1996). Docosapentaneoic acid (22:5 n-3) and elongation metabolite of eicosapentaenoic acid (20:5 n-3), is a potent stimulator of endothelial cell migration on pretreatment *in vitro*. *Prostaglandins Leukotrienes Essential Fatty Acids*, **54**, 319–325.

Keller, J.R. (2002). Omega-3 fatty acids may be effective in the treatment of depression. *Topics in Clinical Nutrition*, **17**, 21–27.

Miyamoto, K.I., Nomura, M., Sasakura, M., Matsui, E., Koshiura, R., Murayama, T., Furukawa, T., Hatano, T., Yoshida, T. & Okuda, T. (1993). Antitumor activity of oenothein B, a unique macrocyclic ellagitannin. *Japanese Journal of Cancer Research*, **84**, 99–103.

Pala, V., Krogh, V., Muti, P., Chajès, V., Riboli, E., Micheli, A., Saadatian, M., Sieri, S. & Berrino, F. (2001). Erythrocyte membrane fatty acids and subsequent breast cancer: a prospective Italian study. *Journal of the National Cancer Institute*, **93**, 1088–1095.

Papas, A.M. (2006). Tocopherols and tocotrienols as byproducts of edible oil processing. Vitamin E: a new perspective. In: *Nutraceutical and Specialty Lipids and Their Co-products*. Shahidi, F. (ed.), Taylor & Francis Group, Boca Raton, FL, pp. 469–482.

Pfeuffer, M. & Schrezenmeir, J. (2002). Milk lipids in diet and health – medium chain fatty acids (MCFA). *Bulletin International Dairy Federation*, **377**, 32.

Sanchez-Bayle, M., Gonzalez-Requejo, A., Pelaez, M.J., Morales, M.T., Asensio-Anton, J. & Anton-Pacheco, E.A. (2008). Cross-sectional study of dietary habits and lipid profiles. The Rivas-Vaciamadrid study. *European Journal of Pediatrics*, **167**, 149–154.

Scalfi, L., Coltorti, A. & Contaldo, F. (1991). Postprandial thermogenesis in lean and obese subjects after meals supplemented with medium-chain and long-chain triglycerides. *American Journal of Clinical Nutrition*, **53**, 1130–1133.

Senanayake, S.P.J.N. & Fichtali, J. (2006). Single-cell oils as sources of nutraceutical and specialty lipids: processing technologies and applications. In: *Nutraceutical and Specialty Lipids and Their Co-products*. Shahidi, F. (ed.), Taylor & Francis Group, Boca Raton, FL, pp. 251–280.

Senanayake, S.P.J.N. & Shahidi, F. (1999a). Enzyme-assisted acidolysis of borage (*Borago officinalis* L.) and evening primrose (*Oenothera biennis* L.) oils: incorporation of omega-3 polyunsaturated fatty acids. *Journal of Agricultural and Food Chemistry*, **47**, 3105–3112.

Senanayake, S.P.J.N. & Shahidi, F. (1999b). Enzymatic incorporation of docosahexaenoic acid into borage oil. *Journal of the American Oil Chemists' Society*, **76**, 1009–1015.

Senanayake, S.P.J.N. & Shahidi, F. (2001). Modified oils containing highly unsaturated fatty acids and their stability. In: *Omega-3 Fatty Acids: Chemistry, Nutrition and Health Effects*. Shahidi, F. & Finley, J.W. (eds), ACS Symposium Series 788. American Chemical Society, Washington, DC, pp. 162–173.

Senanayake, S.P.J.N. & Shahidi, F. (2002a). Chemical and stability characteristics of structured lipids from borage (*Borago officinalis* L.) and evening primrose (*Oenothera biennis* L.) Oils. *Journal of Food Science*, **67**, 2038–2045.

Senanayake, S.P.J.N. & Shahidi, F. (2002b). Structured lipids via lipase-catalyzed incorporation of eicosapentaenoic acid into borage (*Borago officinalis* L.) and evening primrose (*Oenothera biennis* L.) oils. *Journal of Agricultural and Food Chemistry*, **50**, 477–483.

Senanayake, S.P.J.N. & Shahidi, F. (2002c). Oxidative stability of structured lipids produced from borage (*Borago officinalis* L.) and evening primrose (*Oenothera biennis* L.) oils with docosahexaenoic acid. *Journal of the American Oil Chemists' Society*, **79**, 1003–1014.

Shahidi, F. (2006). *Nutraceutical and Specialty Lipids and their Co-products*. CRC Press, Boca Raton, FL.

Shahidi, F. (2007). Omega-3 oils: sources, applications, and health effects. In: *Marine Nutraceuticals and Functional Foods*. Barrow, C. & Shahidi, F. (eds), Taylor & Francis Group, Boca Raton, FL, pp. 23–62.

Shahidi, F. & Finley, J. eds. (2001). *Omega-3 Fatty Acids: Chemistry, Nutrition, and Health Effects*. ACS Symposium Series 788, American Chemical Society, Washington, DC.

Shahidi, F. & Han, X.Q. (1993). Encapsulation of food ingredients. *Critical Reviews in Food Science and Nutrition*, **33**, 501–547.

Shahidi, F. & Miraliakbari, H. (2004). Omega-3 (n-3) fatty acids in health and disease: part 1 – cardiovascular disease and cancer. *Journal of Medicinal Food*, **7**, 387–401.

Shahidi, F. & Miraliakbari, H. (2005). Omega-3 fatty acids in health and disease: part 2–health effects of omega-3 fatty acids in autoimmune diseases, mental health, and gene expression. *Journal of Medicinal Food*, **8**, 133–148.

Shahidi, F. & Miraliakbari, H. (2006). Marine oils: compositional characteristics and health effects. In: *Nutraceutical and Specialty Lipids and their Co-products*. Shahidi, F. (ed.), Taylor & Francis Group, Boca Raton, FL, pp. 227–250.

Shahidi, F. & Wanasundra, U.N. (1995). Oxidative stability of encapsulated seal blubber oil. In: *Flavor Technology – Physical Chemistry, Modification, and Process*. Ho, C.T., Tan, C.T. & Tong, C.H. (eds), ACS Symposium Series 610, American Chemical Society, Washington, DC, pp. 139–151.

Shahidi, F. & Wanasundara, U.N. (1998). Omega-3 fatty acid concentrates: nutritional aspects and production technologies. *Trends in Food Science and Technology*, **9**, 230–240.

Shahidi, F. & Zhong, Y. (2005). Marine mammal oils. In: *Bailey's Industrial Oil and Fat Products*, 6th edn, Volume 3. Shahidi, F. (ed.), John Wiley & Sons, Hoboken, NJ, pp. 259–278.

Shahidi, F. & Zhong, Y. (2007) Antioxidants from marine by-products. In: *Maximising the Value of Marine By-products*. Shahidi, F. (ed.), Woodhead Publishing, Cambridge, UK, pp. 397–412.

St-Onge, M.P. & Bosarge, A. (2008). Weight-loss diet that includes consumption of medium-chain triacylglycerol oil leads to a greater rate of weight and fat mass loss than does olive oil. *American Journal of Clinical Nutrition*, **87**, 621–626.

St-Onge, M.P. & Jones, P.J. (2003). Greater rise in fat oxidation with medium-chain triglyceride consumption relative to long-chain triglyceride is associated with lower initial body weight and greater loss of subcutaneous adipose tissue. *International Journal of Obesity and Related Metabolic Disorders*, **27**, 1565–1571.

Tsuji, H., Kasai, M., Takeuchi, H., Nakamura, M., Okazaki, M. & Kondo, K. (2001). Dietary medium-chain triacylglycerols suppress accumulation of body fat in a double-blind, controlled trial in healthy men and women. *Journal of Nutrition*, **131**, 2853–2859.

Tsuzuki, T., Igarashi, M. & Miyazawa, T. (2004). Conjugated eicosapentaenoic acid (EPA) inhibits transplanted tumor growth via membrane lipid peroxidation in nude mice. *Journal of Nutrition*, **134**, 1162–1166.

Wanasundara, U.N. & Shahidi, F. (1996). Stabilization of seal blubber and menhaden oils with green tea catechins. *Journal of the American Oil Chemists' Society*, **73**, 1183–1190.

Wanasundara, U.N. & Shahidi, F. (1998a). Antioxidant and pro-oxidation activity of green tea extract in marine oils. *Food Chemistry*, **63**, 335–342.

Wanasundara, U.N. & Shahidi, F. (1998a). Concentration of omega-3 polyunsaturated fatty acids of marine oils using *Candida cylindracea*. Optimization of reaction conditions. *Journal of the American Oil Chemists' Society*, **75**, 1767–1774.

Wanasundara, U.N. & Shahidi, F. (1998b). Lipase-assisted concentration of T3-polyunsaturated fatty acids in acylglycerol forms from marine oils. *Journal of the American Oil Chemists' Society*, **75**, 945–951.

Wanasundara, U.N. & Shahidi, F. (1999). Concentration of omega-3 polyunsaturated fatty acids of seal blubber oil by urea complexation: optimization of reaction conditions. *Food Chemistry*, **65**, 41–49.

Watkins, B.A. & Li, Y. (2006). Conjugated linoleic acids (CLAs): food, nutrition and health. In: *Nutraceutical and Specialty Lipids and Their Co-products*. Shahidi, F. (ed.), Taylor & Francis Group, Boca Raton, FL, pp. 187–200.

Wettasinghe, M. & Shahidi, F. (1999). Evening primrose meal: a source of natural antioxidants and scavengers of hydrogen peroxide and oxygen-derived free radicals. *Journal of Agricultural and Food Chemistry*, **47**, 1801–1812.

Yonezawa, Y., Tsuzuki, T., Eitsuka, T., Miyazawa, T., Hada, T., Uryu, K., Murakami-Nakai, C., Ikawa, H., Kuriyama, I., Takemura, M., Oshige, M., Yoshida, H., Sakaguchi, K. & Mizushina, Y. (2005). Inhibitory effect of conjugated eicosapentaenoic acid on human DNA topoisomerases I and II. *Archives of Biochemistry and Biophysics*, **435**, 197–206.

6 The use of functional plant ingredients for the development of efficacious functional foods

Christopher P.F. Marinangeli and Peter J.H. Jones

6.1 Introduction

With few exceptions, most bioactive compounds characterized as nutraceuticals are derived from plants. Numerous bioactives isolated from legumes, cereals, grains, fruits and vegetables have been shown to be efficacious in reducing lipid and cholesterol levels, increasing bone mineral density and antioxidant status as well as possessing anticancer properties (Eskin & Tamir 2006). However, of the hundreds of plant-derived nutraceuticals that have been identified, few have been incorporated into common foods for habitual consumption.

A number of factors affect the suitability for identified bioactives to be integrated into manufactured foods as ingredients. First, the required dose to observe efficacy may render the use of a nutraceutical an inefficient means to disease prevention. Many studies that report efficacy regarding specific phytochemicals subject cell or animal models to very high doses and/or highly purified plant extracts (Espin *et al.* 2007). As a result, it would be difficult to incorporate efficacious doses of identified nutraceuticals into foodstuffs that could be consumed by humans as part of a normal dietary regimen. Second, plant-derived nutraceuticals can affect the sensory characteristics of foods by producing an undesirable texture, odor or taste. For example, legume extracts can produce a 'bean' flavor that some consumers find offensive (Fogliano & Vitaglione 2005). Third, the food matrices used for delivering the bioactive may affect a nutraceutical's stability and/or bioavailability (Fogliano & Vitaglione 2005), thus reducing the possibility of observing any effect on disease risk.

Taking the above facts into consideration, there are plant-derived nutraceuticals that have successfully made their way into various food matrices whereby efficacious doses have been achieved. The following chapter will use plant-derived bioactives such as soy proteins and isoflavones, plant sterols and fibers as paradigms for outlining criteria that need consideration when developing functional foods using plant-derived nutraceuticals. Themes that will be explored include proof of efficacy, dose, source, molecular structure/weight, food matrices for delivery of the bioactive of interest, and the impact of extraction and food manufacturing processes on the functional integrity of nutraceutical ingredients. By considering the above criteria, food manufacturers can be confident that they are developing food products which are efficacious and effective in lowering disease risk factors when consumed outside a clinical setting.

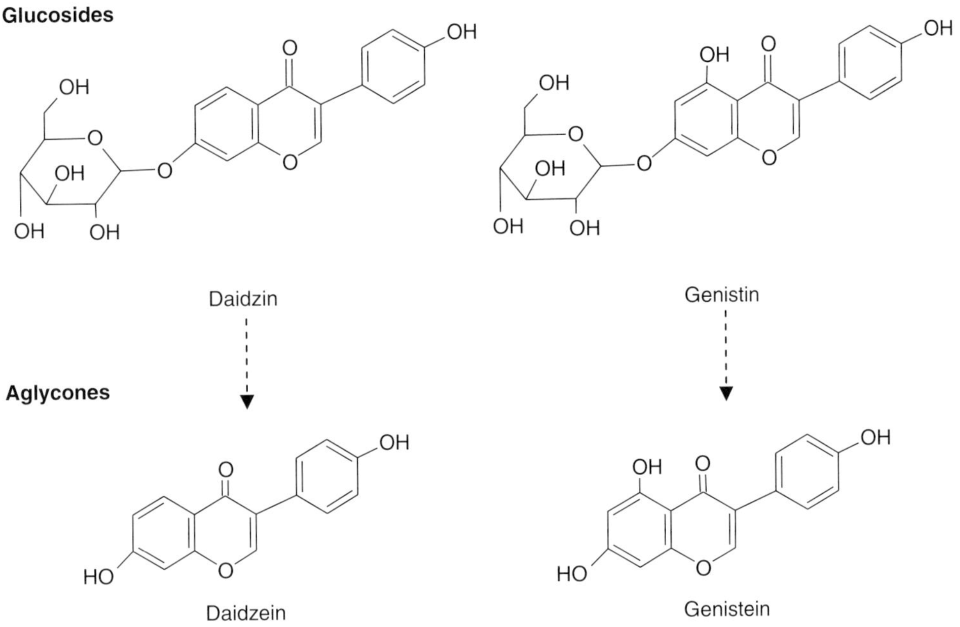

Fig. 6.1 Molecular structures of soy isoflavones showing glucosides, daidzin and genistin, and their hydrolyzed aglycone derivatives, daidzein and genistein.

6.2 Soy extracts

Soy extracts most commonly include isoflavones and soy-derived proteins. Isoflavones continue to be studied as possible measures for reducing risk factors associated with cardiovascular disease and cancer via reducing circulating lipid and cholesterol levels as well as increasing antioxidant capacity. However, results from human studies concerning efficacy are mixed, sometimes raising questions whether other soy-derived compounds, such as proteins, are the true bioactive ingredients found in soy products (Xiao *et al.* 2008).

6.2.1 Soy-derived isoflavones

Isoflavones in soybeans exist primarily as glucosides. The two major glucosides found in soy are genistin and daidzin. After consumption, genistin and daidzin are hydrolyzed by microbial and native intestinal β-glucosidases, producing the aglycones genistein and daidzein (Cassidy 2006; Nielsen & Williamson 2007) (Figure 6.1). Aglycones can be absorbed directly or after they are further metabolized into daughter compounds. Although glucosides need to be converted to their corresponding aglycone prior to absorption, there is no consensus regarding which compounds, glycosides or aglycones, are more bioavailable (Cassidy 2006; Nielsen & Williamson 2007).

6.2.2 Soy-derived isoflavones: The effect on antioxidant status

Efficacy regarding the use of soy isoflavones as antioxidants has been mixed. When volunteers were fed protein shakes containing 107 mg of soy-derived isoflavones, no change in plasma-conjugated dienes, oxidized low-density lipoprotein (LDL), ferric-reducing ability, oxygen radical absorbance capacity and perchloric acid-treated oxygen radical absorbance capacity were noted (Heneman *et al.* 2007). Similar results were observed by Vega-Lopez *et al.* (2005) when hypercholesterolemic men were fed diets whereby protein was derived from animal or soy sources containing either trace or substantial levels of isoflavones (50 mg/1000 kcal). Although total antioxidant capacity significantly increased 10% when subjects ingested soy protein with and without isoflavones, antioxidant endpoints including the production of protein carbonyls, F_2-isoprostanes, malondialdehyde and oxidized LDL did not change between treatments (Vega-Lopez *et al.* 2005).

On the other hand, some results regarding soy-derived isoflavones on antioxidant status have been positive. Pereira *et al.* (2006) investigated the effect of 2 g/day soy germ isoflavones on cholesterol oxides and electronegative LDL in hypercholesterolemic, postmenopausal women. Electronegative LDLs are characterized as having higher levels of conjugated dienes and malondialdehyde alongside lower levels of tocopherol (Avogaro *et al.* 1991). Soy germ significantly lowered both electronegative LDL protein and cholesterol oxides (Pereira *et al.* 2006). Furthermore, men and women taking either 50 or 100 mg/day of soy-derived isoflavones for 3 weeks noted significant reductions in oxidative DNA damage (Djuric *et al.* 2001). The above-mentioned studies offer some indication that soy isoflavones can positively affect antioxidant status.

6.2.3 Soy-derived isoflavones: Effects on circulating lipid levels

Efficacy is equally questionable when discussing the effect of soy isoflavones on circulating lipid levels. Supplementing postmenopausal women with 150–300 mg/day soy-derived isoflavones failed to elicit any significant change in plasma total cholesterol (TC), high-density lipoprotein cholesterol (HDL-C), low-density lipoprotein cholesterol (LDL-C) or triglyceride (TG) levels (Hsu *et al.* 2001; Dewell *et al.* 2002). A meta-analysis by Reynolds *et al.* (2006) suggested that soy isoflavones elicit only marginal reductions in cholesterol and TG, with efficacy depending largely on age, sex and initial circulating lipid levels (Zhan & Ho 2005). Nonetheless, a study supplementing volunteers with 62 mg/day soy-derived isoflavones for 4 weeks observed that TC and LDL-C were 10.6 and 15.6% lower than the placebo group, respectively (Uesugi *et al.* 2002).

6.2.4 Soy-derived proteins

Results from studies that tested the nutraceutical properties of soy suggest that bioactives other than isoflavones, such as proteins, are the true active ingredients found in soybeans. Results from Wang *et al.* (2004) noted a significant decrease in TG fractional synthesis rate of 13.3% alongside a significant 12.9% reduction in TG levels in men and woman consuming soy protein with and without soy-derived isoflavones. In the same study, soy protein significantly decreased TC and LDL-C by 4.4 and 5.7%, respectively. Vega-Lopez *et al.* (2005) showed an increase in total antioxidant status when volunteers consumed soy protein with and without soy-derived isoflavones. Similarly, McVeigh *et al.* (2006) observed that low (1.6 mg) and high (61.7 mg) doses of soy isoflavones equally improved the TC–HDL, LDL–HDL ratios, compared to milk protein. The above results raise questions whether isoflavones are the true bioactives regarding improvements in antioxidant and lipid levels (Vega-Lopez *et al.* 2005). Moreover, compared to soy-derived isoflavone studies, literature concerning *in vivo* efficacy of soy supplementation on cardiovascular risk factors in humans is more extensive when investigators use whole soy or extracted soy protein (Hoie *et al.* 2005; McVeigh *et al.* 2006; Allen *et al.* 2007; Azadbakht *et al.* 2007; Hoie *et al.* 2007; Maesta *et al.* 2007; Welty *et al.* 2007). Extraction of soy protein from a whole soy product could also capture bioactives other than isoflavones, which have a biological effect. When supplementing men and women with 2.5 g/day of the soy protein, β-conglycinin, significant reductions in TG and visceral body fat were noted (Kohno *et al.* 2006), indicating that specific soy proteins elicit health benefits.

6.2.5 Soy as a functional ingredient

The evidence regarding soy extracts as efficacious bioactives is enough for food companies to take advantage of the nutraceutical market share and include soy-derived extracts in manufactured food products. Literature examining the efficacy of soy isoflavones in food matrices indicates that soy-derived isoflavones are efficacious in increasing antioxidant status *in vivo*. At the same time, food containing soy proteins expresses significant lipid-lowering properties.

Human *in vivo* studies using soy-derived isoflavones have successfully demonstrated improvements in markers of antioxidant status and inflammation when incorporated into various food matrices (Tikkanen *et al.* 1998; Wiseman *et al.* 2000; Hall *et al.* 2005; Hallund *et al.* 2006; Clerici *et al.* 2007). When researchers incorporated 57 mg/day soy-derived aglycones into cereal bars, the lag time for copper-mediated LDL oxidation was significantly increased (Tikkanen *et al.* 1998). The cereal bars utilized in the present study contained soy protein, whose fraction could have also contained bioactives that would improve antioxidant status. However, when Wiseman *et al.* (2000) examined the effects of soy burgers containing low (1.9 mg) or high (56 mg) levels of soy aglycones, only isoflavone-enriched burgers significantly increased the LDL oxidation lag time, as well as significantly decreased plasma concentrations of F_2-isoprostanes. Furthermore, soy germ-enriched pasta, containing 33 mg of soy isoflavones and negligible amounts of soy protein (0.8 g), was shown to significantly decrease TC, LDL-C and F_2-isoprostane levels (Clerici *et al.* 2007). Significant increases in brachial flow-mediated dilation and decreases in C-reactive protein (CRP) concentrations, markers of vascular health and inflammation, respectively, were also observed (Clerici *et al.* 2007). Researchers also attributed the observed efficacy to β-glucosidases naturally found in semolina wheat (Clerici *et al.* 2007). Soy glucosides would be hydrolyzed into their corresponding aglycones during the manufacturing process, producing isoflavones that are readily available for absorption since they would not require hydrolysis via intestinal enzymes or microflora (Clerici *et al.* 2007). Efficacy regarding beneficial changes in endothelial-independent vasodilatation (Hallund *et al.* 2006) and CRP concentrations (Hall *et al.* 2005) has also been demonstrated in separate studies using low-protein cereal bars containing 40 and 50 mg soy-derived aglycones, respectively.

Soy protein-enriched breads and fortified beverages have been shown to significantly reduce TC levels by 10%, LDL-C levels by 12.5%, HDL-C levels by 12% (Ridges *et al.* 2001) as well as circulating testosterone levels by 19% (Goodin *et al.* 2007). The latter has been implicated in cancer prevention. Nonetheless, reductions in circulating lipid levels were not observed when hyperlipidemic men and women consumed cereals made with soy protein containing 168 mg/day soy isoflavones (Jenkins *et al.* 2000). In the same study, however, Jenkins *et al.* (2000) noted a 9.2% decrease in LDL-affiliated conjugated dienes, possibly indicating an increase in antioxidant status.

It is still not known whether it is soy protein, soy-derived isoflavones or a combination of both that best induce the functional properties of soy. Ultimately, manufacturing processes may explain why results among studies using soy extracts in food products are mixed. Many biological compounds can be easily modified with processing, secondary to changes in pH, temperature and pressure. Soy proteins have been shown to undergo property changes when subjected to changes in the above conditions (Huang *et al.* 1998; Maruyama *et al.* 1998, 1999; Puppo *et al.* 2004; Tsukada *et al.* 2006; Genovese *et al.* 2007), which may limit their suitability and more importantly their efficacy when used as ingredients in the creation of functional foods. Hoie *et al.* (2006) showed that the hypolipidemic effects of soy protein are diminished after thermal processing. Soy-derived isoflavones have also been shown to be sensitive to stresses associated with food processing (Xu *et al.* 2002; Ungar *et al.* 2003; Mathias *et al.* 2006; Stintzing *et al.* 2006; Uzzan *et al.* 2007). Typically, studies that report a favorable relationship between soy extract consumption and antioxidant status utilize food vehicles, i.e. cereal bars and vegetarian burgers, that can be produced in a manner that would limit the introduction of environmental stresses, which could facilitate the destruction of bioactives. Further research is required to determine not only which bioactives found in soy elicit the greatest health benefits, but also which compounds are best suited as food ingredients such that functional foods can be efficiently produced to deliver a health advantage to the population.

6.3 Plant sterols and stanols

Plant sterols can be found in virtually all plant materials and serve as structural components of cell walls. The biochemical structure of plant sterols is very similar to that of cholesterol with the addition of either a methyl or ethyl functional group. The most common plant sterols found in nature are sitosterol, stigmasterol and campesterol (Moreau *et al.* 2002; Ellegard *et al.* 2007). Saturated plant sterols are known as stanols (Figure 6.2) (Moreau *et al.* 2002). For the purpose of this chapter, plant sterols and stanols are used interchangeably and are abbreviated PS.

PSs have been shown to be efficacious in reducing circulating cholesterol levels in humans since 1954 (Best *et al.* 1954). Since then, an increasing number of studies and reviews have been published highlighting PS efficacy in producing *in vivo* reductions in circulating TC and LDL-C levels of up to 9 and 10%, respectively, in humans (Lees *et al.* 1977; Begemann *et al.* 1978; Mattson *et al.* 1982; Nguyen 1999; Hallikainen *et al.* 2000a; Nestel *et al.* 2001; Moreau *et al.* 2002; Berger *et al.* 2004; Noakes *et al.* 2005; Tikkanen 2005; Saito *et al.* 2006a; Ellegard *et al.* 2007; Ostlund 2007; Rudkowska & Jones 2007). Most studies regarding the use of PS as nutraceuticals have used PS as food ingredients rather than purified supplements. Reasons for using food as a vehicle for supplementation stem from the observation that free and esterified PSs are more soluble in the gastrointestinal tract as an emulsion rather than a crystalline formation (Ostlund 1999). When properly solubilized, free and esterified PSs possess similar cholesterol-lowering efficacy (Richelle *et al.* 2004).

6.3.1 Plant sterols and stanols: mechanism of action

Mechanistically, numerous studies have shown that PSs lower circulating TC and LDL-C levels by decreasing cholesterol absorption. Previously, it has been proposed that PSs

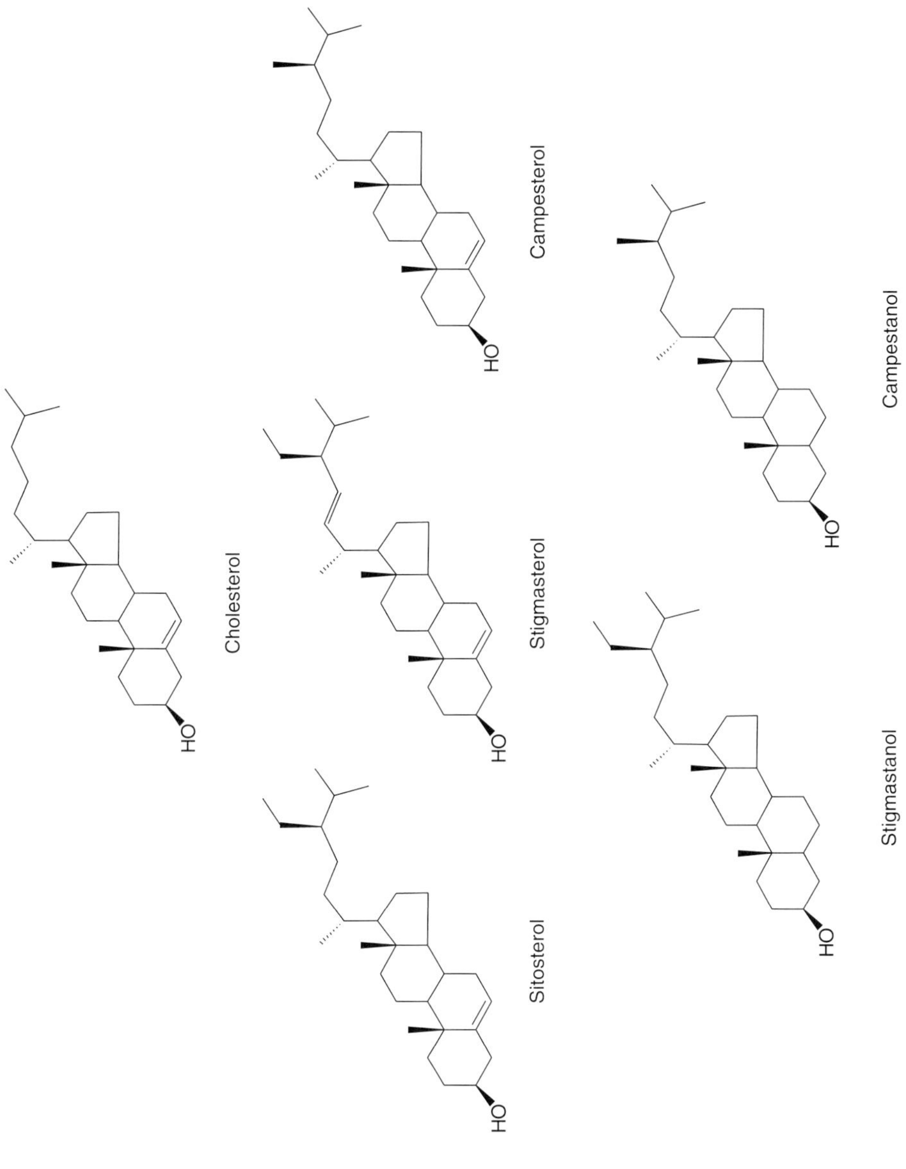

Fig. 6.2 Molecular structures of cholesterol and common plant sterols and stanols.

reduce circulating cholesterol levels by competitively inhibiting dietary and biliary cholesterol absorption during mixed micelle formation in the small intestine and colon, respectively (Mattson *et al.* 1982). However, evidence suggests that PSs absorbed by enterocytes upregulate protein transporters, which effectively shunt cholesterol back into the intestinal lumen, thus preventing cholesterol from being incorporated into chylomicrons (Plat & Mensink 2005; Marinangeli *et al.* 2006). Studies indicate that only 5% of dietary PSs are absorbed into the circulation (Heinemann *et al.* 1993) with the remainder shunted back into the intestinal lumen by adenosine triphosphate-binding cassette protein transporters (Allayee *et al.* 2000; Berge *et al.* 2000). A decrease in cholesterol absorption stimulates an increase in hepatic cholesterol synthesis, which activates transcription factors known as sterol response element-binding proteins. The activation of these transcription factors increases expression of hepatic LDL receptors, resulting in an increase in liver-mediated LDL clearance from the circulation (Brown & Goldstein 1997; Plat & Mensink 2005). Vanstone *et al.* (2002) indicated that both plant sterols and stanols equivalently decrease cholesterol absorption, TC and LDL-C levels, while increasing cholesterol synthesis rates.

6.3.2 Plant sterols and stanols: Efficacy for reducing cholesterol levels

Various foods have been used as vehicles to investigate the cholesterol-lowering properties of free and/or esterified PS since intestinal solubility can be problematic when crystalline forms are used. Efficacy studies that utilize PS supplements often use soft-gel formulations (Woodgate *et al.* 2006), whereby PSs are suspended in an oil. Studies using tablets add lecithin as an emulsifying agent (McPherson *et al.* 2005; Goldberg *et al.* 2006).

Food matrices that have been efficacious vehicles for administering PS include lipid-based spreads (Ntanios *et al.* 2002; Varady *et al.* 2004; Lau *et al.* 2005; Alhassan *et al.* 2006; Hallikainen *et al.* 2006; Chan *et al.* 2007; Varady *et al.* 2007; Clifton *et al.* 2008), milk (Clifton *et al.* 2004b; Richelle *et al.* 2004; Thomsen *et al.* 2004; Noakes *et al.* 2005; Goncalves *et al.* 2006; Hansel *et al.* 2007; Li *et al.* 2007; Madsen *et al.* 2007; Plana *et al.* 2008; Houweling *et al.* 2009), yogurt (Volpe *et al.* 2001; Clifton *et al.* 2004b; Noakes *et al.* 2005; Doornbos *et al.* 2006; Plana *et al.* 2008), mayonnaise (Saito *et al.* 2006b) and beef (Matvienko *et al.* 2002). Plant sterol-enriched cereals and breads have also been shown to be efficacious cholesterol-lowering foods (Clifton *et al.* 2004b). Collectively, the large amount of clinical research available is testament to the cholesterol-lowering efficacy regarding PS as a functional ingredient. Nonetheless, factors regarding dose, timing, source of PS, the lipid content of foods and safety need to be delineated so that manufacturers can produce PS-fortified foods that will bring forth meaningful reductions in cholesterol when consumed outside the clinical setting.

6.3.3 Plant sterols and stanols: Dose required for efficacy

Studies indicate that PSs are efficacious over a range of doses, from 0.3 to 8.0 g/day (AbuMweis *et al.* 2008). However, depending on the cholesterol-lowering response of individuals, certain doses could be more efficacious than others. Hence, it is important to establish a maximal dose such that the majority of people hoping to benefit from plant sterol supplementation have the best possible chance at lowering their cholesterol when using PS-fortified foods.

In a study by Hallikainen *et al.* (2000b), men and women consumed increasing doses of PS at 0.0, 0.8, 1.6, 2.3 and 3.0 g/day in margarine for 4 weeks. LDL-C was significantly decreased to 1.6, 7.0, 10.6 and 11.5%, respectively. Although LDL-C reductions increased with each subsequent increase in PS dose, 2.3 and 3.0 g/day failed to elicit any greater effect than 1.6 g/day (Hallikainen *et al.* 2000b; AbuMweis *et al.* 2008). Other studies have also shown 1.6 g/day to be the maximum dose response for cholesterol-lowering efficacy with PS supplementation (Hendriks *et al.* 1999; Clifton *et al.* 2008). Current recommendations suggest that hypercholesterolemic individuals consume 2.0–2.5 g/day PS to reduce circulating LDL-C levels (AbuMweis *et al.* 2008). The ability to incorporate efficacious doses of nutraceuticals into novel food products needs consideration when developing functional foods.

6.3.4 Plant sterols and stanols: Timing for efficacy

Factors related to physiology may render a nutraceutical most effective if taken at a certain time of the day. For instance, a bioactive may be most effective if ingested as a single dose in the morning, with sustained effectiveness, or in smaller doses throughout the day. Plant sterols, as nutraceuticals, provide a good example of the 'timing phenomenon' concerning their effectiveness as a functional ingredient in foods.

One of the mechanisms in which PSs have been proposed to induce a cholesterol-lowering effect is by reducing cholesterol absorption, causing an increase in hepatic cholesterol synthesis and hepatic LDL uptake (Marinangeli *et al.* 2006). Since hepatic synthesis of cholesterol has been shown to have a diurnal cycle, in theory, plant sterol intake would be optimal when hepatic synthesis is at its lowest point of the day, hence, forcing the liver to obtain cholesterol from circulating LDL-C levels. Studies investigating circadian rhythms indicate that the highest rates of endogenous cholesterol synthesis occur at night, while the lowest rates occur in the afternoon (Parker *et al.* 1982; Jones & Schoeller 1990; Cella *et al.* 1995). Hence, it could be more efficacious to ingest PS in the afternoon when cholesterol synthesis is lowest, possibly allowing for more flexibility for the liver to obtain cholesterol from circulating LDL. Results from AbuMweis *et al.* (2006) indicate that a single morning dose of PS failed to reduce LDL cholesterol levels. Other studies utilizing single morning bolus doses of PS note significant, but relatively small reductions in TC from non-significant to 8%, and LDL-C from 4.4 to 9% (Chan *et al.* 2007; Kassis *et al.* 2008). Comparatively, studies that distribute smaller doses of PS throughout the day observe greater cholesterol reductions from 8.2 to 13.1% and 8.8 to 13.4% for TC and LDL-C, respectively (Vanstone *et al.* 2002; Charest *et al.* 2004; Varady *et al.* 2004).

In light of the studies outlined above, PSs have been tested for efficacy in numerous food-derived vehicles, providing consumers with choice and ease when integrating them into their daily dietary regimen.

6.3.5 Plant sterols and stanols: Does source matter?

Tall, soybean and rapeseed oils are used as sources for PS. Although the major PS constituents remain the same, that is sitosterol and campesterol, the ratios of each can change drastically. For example, a study comparing the cholesterol-lowering efficacy of soybean oil, tall oil and combination of tall/rapeseed oil-derived PS, the percent β-sitosterol in each oil varied at 46.1, 77.9 and 63.5%, respectively (Clifton *et al.* 2008). Moreover, campesterol levels differed at 26.6, 9.5 and 23.1%, respectively (Clifton *et al.* 2008). Results from this study also indicated that PS derived from soybean, tall, and combination tall/rapeseed oil generated similar reductions in TC (7.7, 11.2 and 10.0%, respectively) and LDL-C (10.4, 14.7 and 11.8%, respectively) (Clifton *et al.* 2008). The above results indicate that PSs are effective cholesterol-lowering agents regardless of their source and composition.

6.3.6 Plant sterols and stanols: Matrices for delivery

Plant sterol solubility in a food matrix can greatly affect PS effectiveness in changing an ordinary food into a functional food. Since PSs are fat-soluble molecules, lipid is used to properly solubilize PS within food matrices. Studies using low-fat milk and yogurt have been successful at eliciting significant cholesterol reductions (Hallikainen *et al.* 2000a; Volpe *et al.* 2001; Noakes *et al.* 2005; Hansel *et al.* 2007; Madsen *et al.* 2007) probably because, although low, an appropriate amount of lipid is available to emulsify the added PS. The observation that low-fat functional foods are efficacious when fortified with PS is an important finding. Low-fat diets are often recommended to hypercholesterolemic individuals. Therefore, the availability of low-fat functional foods enriched with PS allows for more variety for the inclusion of PS into the diets of people for which these are targeted.

Studies using non-fat beverages have failed to produce significant reductions in TC or LDL-C levels when PSs were provided at 2.0 g/day (Jones *et al.* 2003). In the present study, PSs were directly added to a non-fat beverage without an emulsifier such as lecithin. The authors suggested that if an emulsifier had been added to the treatment, significant cholesterol reductions would have been observed (Jones *et al.* 2003). That being said, studies by Devarag *et al.* (2004, 2006) produced efficacious reductions in TC and LDL-C using PS (2 g/day) mixed with non-fat orange juice. Disparities in sample size, length of treatment and diet were thought to be reasons the orange juice/PS mixtures administered by Devarag *et al.* (2004, 2006) reduced cholesterol levels compared to other studies using non-fat PS formulations. However, a recent meta-analysis by AbuMweis *et al.* (2008) indicated that PSs are less effective at reducing LDL-C levels when incorporated into chocolate, orange juice, cheese, non-fat beverage, meats, croissants and muffins, oil in bread and cereal bars compared to fat spreads, mayonnaise, salad dressing, milk and yoghurt. Thus, emphasizing the food matrices for delivery of PS impacts efficacy.

6.3.7 Safety of chronic plant sterol and stanol ingestion

Given that PSs inhibit cholesterol absorption (Berger *et al.* 2004), it has been suggested that these bioactives could inadvertently reduce the absorption of fat-soluble vitamins. Richelle *et al.* (2004) gave volunteers 2.2 g/day of either free or esterified PS for 1 week. Coupled with a 60% reduction in cholesterol absorption, both free and esterified PSs significantly reduced the bioavailability of β-carotene and α-tocopherol by 50 and 20%, respectively (Richelle *et al.* 2004). However, the statistical model used for evaluating β-carotene and α-tocopherol levels after PS supplementation did not control for the observed reductions in cholesterol. Hence, a false-positive result was created regarding the effect of PS on the absorption of fat-soluble vitamins.

In a study by Korpela *et al.* (2006), 2.0 g/day PS suspended in low-fat dairy products for 6 weeks failed to significantly reduce levels of the fat-soluble vitamin K_1, retinol and vitamin D. While researchers observed a significant decrease in β-carotene and α-tocopherol after supplementation with PS, once statistical analysis controlled for reductions in TC, decreases in β-carotene and α-tocopherol were no longer significant. Other studies have shown similar results (Hendriks *et al.* 2003; Saito *et al.* 2006a). In a long-term, 52-week study, PS had no effect on levels of vitamin K, vitamin D, α-tocopherol, vitamin B_{12}, folic acid or zinc levels. Moreover, reductions in β-carotene and lycopene were no longer significant once a statistical correction for decreases in LDL-C was applied to the analysis (Hendriks *et al.* 2003). Nonetheless, statistical adjustments did not thwart observed reductions in carotenoids and lycopene, with 6.6 g/day PS over a 6-week period (Clifton *et al.* 2004a). Thus, these results emphasize the importance of adhering to the recommended plant sterol dose of 2.0–2.5 g/day. In addition, researchers have suggested that when supplementing the diet with food fortified with PS, a concurrent increase in fruit and vegetable consumption should be applied to decrease the risk of imposing a fat-soluble vitamin deficiency (Noakes *et al.* 2002; Clifton *et al.* 2004a).

6.3.8 Plant sterols and stanols as a functional ingredient

Plant sterols represent a solid success story in the realm of plant-derived functional ingredients. Numerous clinical trials have reported their efficacy as cholesterol-lowering agents using a broad range of food matrices. Nonetheless, PS provides a good example of the numerous other factors that must be considered before developing a successful functional food that will be efficacious outside the clinical setting. Using food matrices capable of delivering the appropriate dose, while considering the time of day, how a product should be consumed, i.e. single or multiple doses throughout the day, and safety are obviously important factors to consider. In addition, from where bioactives are derived could potentially affect whether a food product becomes functional. Due to the extensive research surrounding PS supplementation under numerous conditions, PSs have proved to be a success story as a functional food ingredient.

6.4 Fiber and its various components: β-Glucan and inulin

Fibers are characterized as the cell wall material isolated from plants. They are defined as indigestible carbohydrates that are fermented by bacteria in the colon. Fiber can be subdivided into three categories: water-insoluble, water-soluble and resistant starch. Water-insoluble fibers include celluloses, while water-soluble fibers include gums, psyllium, pectin, β-glucan and oligosaccharides. Resistant starches, on the other hand, are starch molecules that, for various reasons, are not digested in the small intestine and make their way to the large intestine for bacterial processing (Charalampopoulos *et al.* 2002).

For the most part, the public is aware that a diet high in fiber is healthy. As a result, fiber-based food claims are readily seen in supermarkets across North America. Although consumers mainly associate fiber with bowel regularity, in recent years, various components of water-soluble fibers have been implicated for possessing other physiological benefits including decreases in circulating lipid and cholesterol levels, improved glycemic responses to meals, satiety, bone health, anticancer and prebiotic effects. The latter has been further associated with improvements regarding cancer prevention and better-coordinated immune system responses. However, adding fiber to foods as functional ingredients can be difficult. Because different fibers have unique physical properties such as water retention, varying degrees of sweetness and viscosity, their addition to marketable food products as a means of boosting fiber content can be problematic. Hence, it is a challenge for food companies to not only fortify food products with fiber but also choose the right fiber to bring about the beneficial physiological changes mentioned above. Soluble fibers such as β-glucan and fructo-oligosaccharides, such as inulin, have successfully been added to marketable functional food products.

6.4.1 β-Glucan

β-Glucan is among the most researched fiber-derived functional ingredients. Primarily isolated from oats and barley, β-glucan is characterized by random repeating units of β(1→4) β-D-glucopyranose, linked together by single β(1→3) bonds (Lazaridou & Biliaderis 2007) (Figure 6.3). Researchers have begun evaluating the health-promoting properties of β-glucan as a functional ingredient in cookies (Casiraghi *et al.* 2006), crackers (Casiraghi *et al.* 2006), pasta (Bourdon *et al.* 1999), cereal (Davidson *et al.* 1991; Keenan *et al.* 2007), breads (Torronen *et al.* 1992) and juice (Keenan *et al.* 2007). Meaningful improvements in cardiovascular and diabetes risk factors including significant reductions in TC, LDL-C, postprandial glucose (Casiraghi *et al.* 2006; Nilsson *et al.* 2008; Shimizu *et al.* 2008), postprandial insulin (Casiraghi *et al.* 2006) and satiety (Nilsson *et al.* 2008) have been attributed to the capacity for β-glucan to produce a highly viscous mass in the gastrointestinal tract, slow gastric emptying and interfere with macronutrient absorption. Similar to soy and PS, some studies using β-glucan as a functional ingredient have failed to observe health-related efficacy (Keogh *et al.* 2003; Biorklund *et al.* 2005). A range of factors, including food preparation, the food matrix for β-glucan delivery, β-glucan extraction and/or processing, and dose need consideration when adding β-glucan to foods as bioactives.

6.4.1.1 *β-Glucan efficacy: The effect of the food matrix*

Cooking temperature, pressure, baking time and starch gelatinization have been shown to affect postprandial glycemic responses of different food products (Garcia-Alonso & Goni 2000; Granfeldt *et al.* 2000; Vervuert *et al.* 2003; Rizkalla *et al.* 2007). In a study comparing the postprandial glycemic response of cookies and crackers made with 3.5 g barley β-glucan, β-glucan crackers produced a 33 and 32% decrease in glycemic and insulin responses, respectively, compared to crackers made with wholewheat flour. On the other hand, cookies significantly reduced postprandial glucose and insulin responses 60 and 31%, respectively, compared to wholewheat cookies (Casiraghi *et al.* 2006). Casiraghi *et al.* (2006) suggested that differences in dough moisture, cooking temperature, pressure as well as the lower levels of simple sugars in β-glucan cookies may have accounted for better postprandial glucose responses compared to β-glucan crackers.

Using a relatively large dose, Torronen *et al.* (1992) attributed the ineffectiveness of 11 g/day of oat bran-derived β-glucan at lowering lipid levels in hypercholesterolemic men to the poor solubility of β-glucan in the bread matrix used to deliver the treatment. Similar results were noted in a study comparing high-carbohydrate foods and beverages fortified with β-glucan. The results indicated that only the high-carbohydrate foodstuffs enriched with β-glucan significantly decreased postprandial glycemic responses (Poppitt *et al.* 2007). Two factors were considered responsible for the inability of the β-glucan beverage to decrease

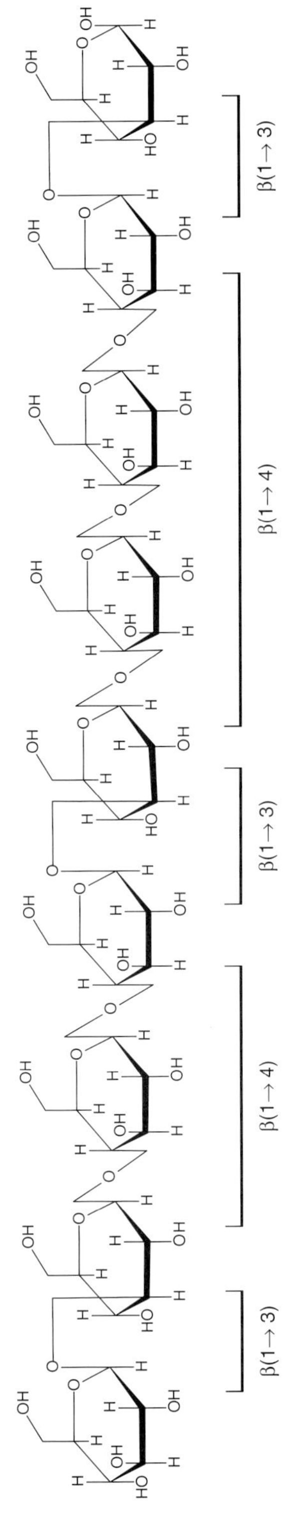

Fig. 6.3 Molecular structures of β-glucan, characterized by repeating units of β-D-glucopyranose connected by β(1→4) linkages, separated by β(1→3) bonds.

postprandial glycemia. First, the beverage presented carbohydrate to the gastrointestinal cells faster than the whole foods, possibly due to faster gastric emptying. Second, the beverage matrix did not provide the right conditions for β-glucan to produce a highly viscous mass (Poppitt *et al.* 2007). However, in another study, a beverage containing a single 5 g dose of oat-derived β-glucan lowered TC and postprandial glucose levels compared to the same beverage at 10 g β-glucan. Researchers suggested that because the solubility of β-glucan can decrease with storage over time, the process is expedited in highly concentrated solutions, thus reducing efficacy (Biorklund *et al.* 2005). Overall, the ability for β-glucan to improve health can be affected by food matrix.

6.4.1.2 *β-Glucan efficacy: The effect of extraction and processing*

Food processing and food preparation can make all the difference regarding the capacity for β-glucan to transform an ordinary food into a functional food. The ability for β-glucan to produce a viscous mass in the gastrointestinal tract ultimately depends on the solubility of β-glucan fractions. Different fractions or molecular weights of β-glucan confer different solubilities and contribute to viscosity (Jenkins *et al.* 1978; Wolever *et al.* 1978; Aman & Graham 1987; Tietyen *et al.* 1995; Lazaridou & Biliaderis 2007). Depending on the extraction procedure, the molecular weight of β-glucan can vary, and produce either a high-molecular-weight extract with higher viscosity or a low-molecular-weight product with lower viscosity but greater palatability (Redgwell & Fischer 2005).

The 11 g dose of β-glucan utilized by Torronen *et al.* (1992) was large compared to other studies that have reported efficacy. β-Glucan has been shown to be an efficacious cholesterol-lowering food ingredient, with as little as 3.5 g/day (Keenan *et al.* 2007). In a study that replaced wholewheat flour with barley (8 g/day) in pasta, bread, biscuits and muesli, volunteers noted a 6 and 7% reduction in TC and LDL-C levels, respectively. The authors suggested that the quality of β-glucan used in food products is more important than the dose, which is why their study and studies using even lower doses show health benefits when β-glucan is added as a bioactive (McIntosh *et al.* 1991). In a dose-dependent study comparing the effects of a 28, 56 or 84 g/day dose of oatmeal and oat bran on lipid levels, researchers noted that only the 84 g/day dose of oatmeal caused a significant decrease in LDL-C levels of 10.1% over 6 weeks. However, efficacious LDL-C reductions of 15.9 and 11.5% were reported with 56 and 84 g/day doses of oat bran, respectively. Fat content did not differ between treatments. Hence, differences in efficacy observed between either treatment were attributed to oat bran having higher levels of β-glucan compared to oatmeal (Davidson *et al.* 1991) which could be attributed to how oat products are processed.

Concerning the molecular weight of β-glucan, a direct relationship has been established between molecular weight, viscosity and efficacy (Wood *et al.* 2000; Lazaridou & Biliaderis 2007). In a retrospective analysis of previously published data regarding the effect of β-glucan on glucose tolerance, Wood *et al.* (2000) introduced molecular weight and viscosity as dependent variables. Results indicated that both viscosity and molecular weight contribute significantly to the ability of β-glucan to reduce postprandial glucose and insulin levels (Wood *et al.* 2000). A study indicated that low-molecular-weight β-glucan (50–400 kDa) with favorable sensory characteristics as well as bioactive properties had been isolated via a new extraction process (Keenan *et al.* 2007). In this study, men with hypercholesterolemia were given 3 and 5 g/day doses of high- and low-molecular-weight β-glucan in cereal and juice for 6 weeks. The 3 and 5 g/day doses of low- and high-molecular-weight β-glucan produced significant decreases in TC and LDL-C from 7 to 11% and 9 to 15%, respectively

(Keenan *et al.* 2007). The authors suggested that the low-molecular-weight β-glucan used in the present study has even greater potential as a nutraceutical ingredient since its molecular structure provides favorable sensory properties compared to equal doses of high-molecular-weight β-glucan.

Despite the above results, it would be premature to attribute the cholesterol and glycemic efficacy of β-glucan solely to molecular weight. The overall structure of β-glucan is complex and varies greatly between oats and barley. Thus, starting with a highly variable material, it can be hypothesized that different extraction procedures would break or even reform chemical bonds and produce highly variable structures with considerable diversity regarding structural branching. Thus, extracting and processing β-glucan one way versus another could produce similar molecules of similar weight, but confer different functional properties.

6.4.1.3 *β-Glucan efficacy: Does source matter?*

Finally, the choice of source of β-glucan as an ingredient could significantly contribute to whether β-glucan-fortified foods become functional. β-Glucan is most commonly derived from oats and barley. Not surprisingly, the physiochemical properties of β-glucan from either source can differ dramatically. For example, the β-glucan content in barley and oats has been shown to range between 3.0 and 6.9% and 2.2 and 4.2%, respectively. Nonetheless, 70–80% of the β-glucans found in oats are soluble compared to 15–54% in barley (Aman & Graham 1987; Lambo *et al.* 2005). Biorklund *et al.* (2005) examined the effect of 5 or 10 g/day β-glucan derived from either oats or barley in hypercholesterolemic men and women. Volunteers given the 5 g dose of oat-derived β-glucan noted a 7.4% reduction in TC levels as well as a significant decrease in glucose and insulin levels 30 minutes after a mixed meal compared to control. Barley-derived β-glucan produced no effect (Biorklund *et al.* 2005). The authors explained their results by acknowledging the principles outlined above regarding the importance of structural chemistry in order for β-glucan to improve health. β-Glucan derived from oats used in the present study had a molecular weight 200 000 kDa, which far surpassed that of β-glucan isolated from barley (40 000 kDa). Therefore, β-glucans from oats were likely highly soluble and better able to form highly viscous digesta in the gastrointestinal tract. The source from which β-glucan is derived must be considered when developing new functional food products since the physiochemical properties of β-glucan can differ greatly between sources. Ultimately, the physiochemical properties elicit efficacy, thus emphasizing that adding more of a functional ingredient does not necessarily confer a heightened effect.

6.4.2 Fructo-oligosaccharides: Inulin

The term 'fructo-oligosaccharide' is a broad term for all indigestible oligosaccharides containing fructose in their molecular structure. Because fructo-oligosaccharides are not absorbed, many of their health benefits are believed to come from their prebiotic activity in the colon. Fructo-oligosaccharides are fermented in the colon by resident microflora, promoting the growth and proliferation of advantageous bacteria, thus suppressing the expansion of detrimental bacteria. Studies demonstrate that maintaining a healthy milieu of certain strains of colonic microflora can promote local colonic as well as systemic health. Among many fructo-oligosaccharides, inulin is the main fructo-fiber that has made its way into mainstream food products as a functional ingredient.

6.4.2.1 *Inulin as a prebiotic: Efficacy*

Commercial inulin is most commonly extracted from chicory and Jerusalem artichoke (Roberfroid 2005). However, other sources such as burdock are becoming noticed (Li *et al.* 2008). Typically, two isoforms of inulin are available. The most common form of inulin consists of a D-glucopyranose molecule attached to repeating fructofructosyl units with a terminal β-D-fructofuranose molecule ($G_{py}F_n$). The second, and rarer inulin isoform, has a D-fructopyranose in place of the glucopyranose ($F_{py}F_n$) (Figure 6.4) (Stevens *et al.* 2001). The resistance of inulin to digestion is secondary to fructosyl–fructose glycosidic bonds with a β-(2←1) configuration. The degree of polymerization can vary between 2 and 70 units and can be influenced by the source of inulin, growth conditions and age of the plant (Stevens *et al.* 2001). Consumption of inulin has been shown to selectively promote the growth of colonic bacteria that are believed to promote health. In studies by Langlands *et al.* (2004) and Gibson *et al.* (1995), consumption of 7.5 and 15 g/day inulin, respectively, significantly changed the microflora population in the colon by selectively promoting the growth of *Bifidobacteria* and *Lactobacilli*. Animal studies suggest that dietary inulin may also promote the growth of beneficial strains of clostridium (Kleessen *et al.* 2001). Oral consumption of both isoforms of inulin ($G_{py}F_n$ and $F_{py}F_n$) has been shown to be efficacious prebiotics (Menne *et al.* 2000).

As science continues to investigate prebiotics, colonic microflora and health, an obvious symbiotic relationship between humans and resident bacteria is becoming obvious. Using compounds such as inulin to promote the growth of certain microflora inadvertently decreases levels of pathogenic and putrefactive bacteria as well as increases colonic absorption of dietary calcium and magnesium (Grizard & Barthomeuf 1999). Moreover, reviews by Pool-Zobel (2005), Kanauchi *et al.* (2003) and Pereira and Gibson (2002) suggest that the short-chain fatty acids produced by certain colonic bacteria, whose growth is promoted by fructo-oligosaccharides, prevent colon cancer and irritable bowel syndrome and reduce circulating cholesterol levels, respectively. Sauer *et al.* (2007) showed that the fermentation of inulin in human colonic microflora yielded short-chain fatty acids that significantly increased glutathione *S*-transferases, which are used to detoxify carcinogens. Supplementation with inulin strives to find a balance between increasing microbial short-chain fatty acid production while limiting gastrointestinal side effects.

6.4.2.2 *Inulin as a prebiotic: Efficacious dose*

The amount of daily inulin required to change colonic microbial populations has yet to be established. A study comparing high and low doses of inulin in men and women indicated that 5 and 8 g/day inulin for 2 weeks can significantly increase populations of bifidobacteria in the colon (Kolida *et al.* 2007). Researchers indicated that changes in *Bifidobacteria* are not necessarily dose-dependent. Instead, changes are dependent on initial levels of *Bifidobacteria*. That is, volunteers who began the study having lower levels of colonic *Bifidobacteria* noted greater changes than those with an already well-established colony (Kolida *et al.* 2007; de Preter *et al.* 2008). Nonetheless, adding more inulin to functional foods could increase the likelihood of observing a prebiotic effect. Conversely, higher levels of fermentable carbohydrate in the colon could translate into increased discomfort due to excess gas and bloating (Ghoddusi *et al.* 2007), a concern that food companies should keep in mind when producing functional foods containing fructo-oligosaccharides or other fermentable fibers.

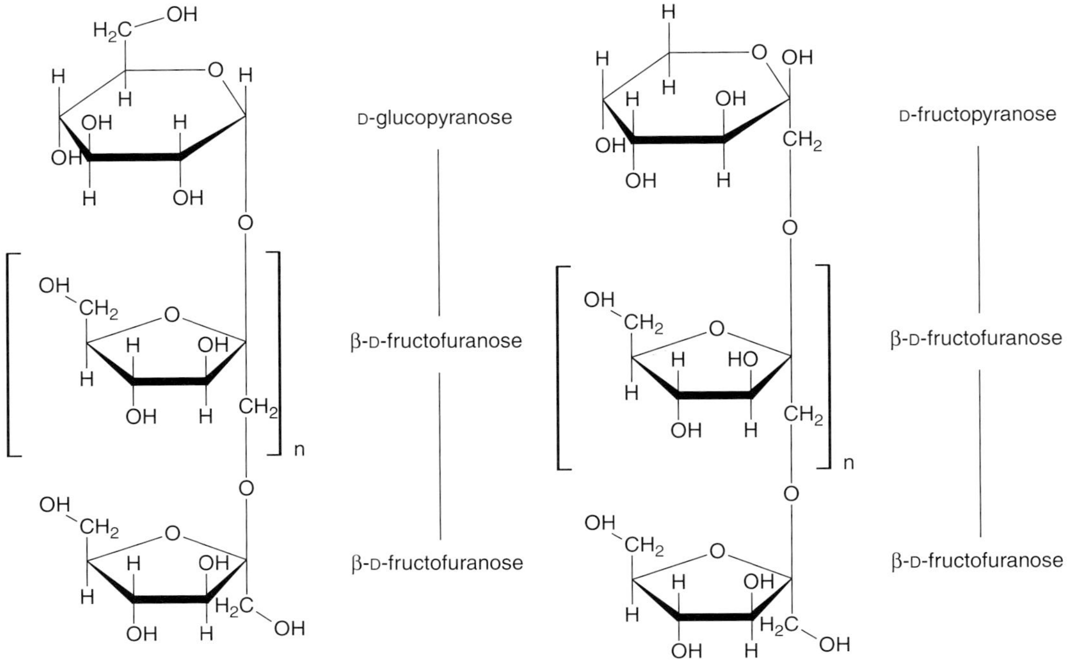

Fig. 6.4 Molecular structures of inulin isomers. The molecules differ by a D-glucopyranose ($G_{py}F_n$) and a D-fructopyranose ($F_{py}F_n$) at the reducing end. Each molecule has repeating units of β-D-fructofuranose linked by β(2←1) fructosyl–fructose bonds.

6.4.2.3 *Inulin as a prebiotic: The effect of molecular weight*

Like β-glucans, the effectiveness of inulin as a functional ingredient could be affected by molecular structure. The degree of chain length of added inulin has been shown to affect the sensory characteristics of fat-free yogurt. While short-chain inulins were sweeter, long-chain inulins resulted in a more acidic, bland-tasting product. Medium- and long-chain inulins had significantly less synergesis than control yogurt and yogurt containing short-chain inulin (Aryana *et al.* 2007). Rumessen and Gudmand-Hoyer (1998) demonstrated that gastrointestinal symptoms, such as gas and bloating, increase as inulin chain length decreases. Moreover, Kleessen *et al.* (2001) indicated that rats fed long-chain-length inulins demonstrated higher levels of fecal butyrate compared to rats fed short-chain inulin and the control. As outlined above, the production of bacteria-derived short-chain fatty acids is thought to elicit some of the positive health benefits secondary to prebiotic ingestion. Degree of polymerization of inulin fractions should be considered before adding inulin to food products. Depending on the food product being produced, the molecular structure of inulin used can influence sensory characteristics of the food, the potential for gastrointestinal discomfort and possibly efficacy.

6.4.2.4 *Successful inclusion of inulin into functional food products*

Inulin has successfully been incorporated into numerous foods such as yogurts (Aryana *et al.* 2007), breads (Seidel *et al.* 2007), snack bars (Kleessen *et al.* 2007) and milk (Sairanen *et al.* 2007). Inulin has also been added to commercial baby formulas to promote healthy colonization of colonic flora (Kim *et al.* 2007). Figure 6.5 illustrates an instant breakfast product produced by Nestlé© containing Prebio™, their own blend of inulin alongside other oligofructoses, added specifically for its prebiotic effects. Recently, studies have begun to investigate the utilization of prebiotic inulin with probiotics as a means of increasing the survival and subsequent colonization of probiotic bacteria in the colon. In a study using mice, inulin mediated the survival of probiotic *Lactobacillus casei* from 2 to 6 days (Su *et al.* 2007). However, a similar study in humans indicated that the addition of inulin to probiotic milk did not significantly increase the colonization of lactobacilli and bifidobacteria in the colon compared to probiotic alone. Instead, inulin combined with probiotics significantly increases gastrointestinal side effects among study volunteers (Sairanen *et al.* 2007). Overall, inulin is an efficacious functional ingredient that has been shown to have a prebiotic effect in a number of food matrices. However, undesirable gastrointestinal side effects may deter their consumption if dose and effects when combined with probiotics are not considered (Sairanen *et al.* 2007).

6.5 Conclusions

Most nutraceuticals are derived from plants. However, few have made their way into efficacious commercial functional food products. In many instances, the identified nutraceuticals have only been tested in *in vitro* and animal studies, whereby efficacious doses far exceed what could realistically be consumed by humans as a means of reducing disease risk factors. Moreover, the physical, chemical and sensory characteristics of plant-derived compounds can limit their suitability for product development. Nonetheless, efficacious and effective plant-derived nutraceuticals have been used to successfully create functional food products.

Fig. 6.5 Carnation® Breakfast Anytime® products produced by Nestlé© contain Prebio™. Prebio is Nestlé's proprietary blend of inulin and other oligofructoses, which are added to food products as a prebiotic for the promotion gastrointestinal health. Reproduced with permission of Nestlé Canada. For a colour version of this figure, please see the plate section.

Using soy proteins and isoflavones, PS and soluble fibers such as β-glucans and inulins as models, this chapter illustrated important factors that must be considered when developing foods that incorporate functional plant ingredients including proof of efficacy, dose, source, molecular structure/weight, and assurance that food matrices used to house bioactives are suitable for delivery and efficacy. Finally, prudence regarding the extraction and processing of nutraceuticals and food products is needed to ensure the functional integrity of the bioactive compound is not compromised during the manufacturing process. Consideration of these criteria will help to ensure that the incorporation of plant-derived nutraceuticals in food production will produce efficacious functional foods that effectively lower the risk of chronic disease.

References

AbuMweis, S.S., Barake, R. & Jones, P.J. (2008). Plant sterols/stanols as cholesterol lowering agents: a meta-analysis of randomized controlled trials. *Food and Nutrition Research*, **52**.

AbuMweis, S.S., Vanstone, C.A., Ebine, N., Kassis, A., Ausman, L.M., Jones, P.J. & Lichtenstein, A.H. (2006). Intake of a single morning dose of standard and novel plant sterol preparations for 4 weeks does not dramatically affect plasma lipid concentrations in humans. *Journal of Nutrition*, **136**, 1012–1016.

Alhassan, S., Reese, K.A., Mahurin, J., Plaisance, E.P., Hilson, B.D., Garner, J.C., Wee, S. & Grandjean, P. (2006). Blood lipid responses to plant stanol ester supplementation and aerobic exercise training. *Metabolism*, **55**, 541–549.

Allayee, H., Laffitte, B.A. & Lusis, A.J. (2000). Biochemistry. An absorbing study of cholesterol. *Science*, **290**, 1709–1711.

Allen, J.K., Becker, D.M., Kwiterovich, P.O., Lindenstruth, K.A. & Curtis, C. (2007). Effect of soy protein-containing isoflavones on lipoproteins in post-menopausal women. *Menopause*, **14**, 106–114.

Aman, P. & Graham, H. (1987). Analysis of total and insoluble mixed-linked (1-> 3)(1-> 4)-β-D-glucans in barley and oats. *Journal of Agricultural and Food Chemistry*, **35**, 704–709.

Aryana, K.J., Plauche, S., Rao, R.M., McGrew, P. & Shah, N.P. (2007). Fat-free plain yogurt manufactured with inulins of various chain lengths and *Lactobacillus acidophilus*. *Journal of Food Science*, **72**, M79–M84.

Avogaro, P., Cazzolato, G. & Bittolo-Bon, G. (1991). Some questions concerning a small, more electronegative LDL circulating in human plasma. *Atherosclerosis*, **91**, 163–171.

Azadbakht, L., Kimiagar, M., Mehrabi, Y., Esmaillzadeh, A., Padyab, M. & Hu, F.B. & Willett, W.C. (2007). Soy inclusion in the diet improves features of the metabolic syndrome: a randomized crossover study in post-menopausal women. *American Journal of Clinical Nutrition*, **85**, 735–741.

Begemann, F., Bandomer, G. & Herget, H.J. (1978). The influence of beta-sitosterol on biliary cholesterol saturation and bile acid kinetics in man. *Scandinavian Journal of Gastroenterology*, **13**, 57–63.

Berge, K.E., Tian, H., Graf, G.A., Yu, L., Grishin, N.V., Schultz, J., Kwiterovich, P., Shan, B., Barnes, R. & Hobbs, H.H. (2000). Accumulation of dietary cholesterol in sitosterolemia caused by mutations in adjacent ABC transporters. *Science*, **290**, 1771–1775.

Berger, A., Jones, P.J. & AbuMweis, S.S. (2004). Plant sterols: factors affecting their efficacy and safety as functional food ingredients. *Lipids in Health and Disease*, **3**, 5.

Best, M.M., Duncan, C.H., Van Loon, E.J. & Wathen, J.D. (1954). Lowering of serum cholesterol by the administration of a plant sterol. *Circulation*, **10**, 201–206.

Biorklund, M., van Rees, A., Mensink, R.P. & Onning, G. (2005). Changes in serum lipids and postprandial glucose and insulin concentrations after consumption of beverages with beta-glucans from oats or barley: a randomised dose-controlled trial. *European Journal of Clinical Nutrition*, **59**, 1272–1281.

Bourdon, I., Yokoyama, W., Davis, P., Hudson, C., Backus, R., Richter, D. Knuckles, B. & Schneeman, B.O. (1999). Postprandial lipid, glucose, insulin, and cholecystokinin responses in men fed barley pasta enriched with beta-glucan. *American Journal of Clinical Nutrition*, **69**, 55–63.

Brown, M.S. & Goldstein, J.L. (1997). The SREBP pathway: regulation of cholesterol metabolism by proteolysis of a membrane-bound transcription factor. *Cell*, **89**, 331–340.

Casiraghi, M.C., Garsetti, M., Testolin, G. & Brighenti, F. (2006). Post-prandial responses to cereal products enriched with barley beta-glucan. *Journal of the American College of Nutrition*, **25**, 313–320.

Cassidy, A. (2006). Factors affecting the bioavailability of soy isoflavones in humans. *Journal of AOAC International*, **89**, 1182–1188.

Cella, L.K., Van Cauter, E. & Schoeller, D.A. (1995). Diurnal rhythmicity of human cholesterol synthesis: normal pattern and adaptation to simulated 'jet lag'. *American Journal of Physiology*, **269**, E489–E498.

Chan, Y.M., Demonty, I., Pelled, D. & Jones, P.J. (2007). Olive oil containing olive oil fatty acid esters of plant sterols and dietary diacylglycerol reduces low-density lipoprotein cholesterol and decreases the tendency for peroxidation in hypercholesterolaemic subjects. *British Journal of Nutrition*, **98**, 563–570.

Charalampopoulos, D., Wang, R., Pandiella, S.S. & Webb, C. (2002). Application of cereals and cereal components in functional foods: a review. *International Journal of Food Microbiology*, **79**, 131–141.

Charest, A., Desroches, S., Vanstone, C.A., Jones, P.J. & Lamarche, B. (2004). Unesterified plant sterols and stanols do not affect LDL electrophoretic characteristics in hypercholesterolemic subjects. *Journal of Nutrition*, **134**, 592–595.

Clerici, C., Setchell, K.D., Battezzati, P.M., Pirro, M., Giuliano, V., Asciutti, S., Castellani, D., Nardi, E., Sabatino, G., Orlandi, S., Baldoni, M., Morelli, O., Mannarino, E. & Morelli, A. (2007). Pasta naturally enriched with isoflavone aglycons from soy germ reduces serum lipids and improves markers of cardiovascular risk. *Journal of Nutrition*, **137**, 2270–2278.

Clifton, P.M., Mano, M., Duchateau, G.S., Van Der Knaap, H.C. & Trautwein, E.A. (2008). Dose-response effects of different plant sterol sources in fat spreads on serum lipids and C-reactive protein and on the kinetic behavior of serum plant sterols. *European Journal of Clinical Nutrition*, **62**(8), 968–977.

Clifton, P.M., Noakes, M., Ross, D., Fassoulakis, A., Cehun, M. & Nestel, P. (2004a). High dietary intake of phytosterol esters decreases carotenoids and increases plasma plant sterol levels with no additional cholesterol lowering. *Journal of Lipid Research*, **45**, 1493–1499.

Clifton, P.M., Noakes, M., Sullivan, D., Erichsen, N., Ross, D., Annison, G., Fassoulakis, A., Cehun, M. & Nestel, P. (2004b). Cholesterol-lowering effects of plant sterol esters differ in milk, yoghurt, bread and cereal. *European Journal of Clinical Nutrition*, **58**, 503–509.

Davidson, M.H., Dugan, L.D., Burns, J.H., Bova, J., Story, K. & Drennan, K.B. (1991). The hypocholesterolemic effects of beta-glucan in oatmeal and oat bran. A dose-controlled study. *The Journal of the American Medical Association*, **265**, 1833–1839.

de Preter, V., Vanhoutte, T., Huys, G., Swings, J., Rutgeerts, P. & Verbeke, K. (2008). Baseline microbiota activity and initial bifidobacteria counts influence responses to prebiotic dosing in healthy subjects. *Alimentary Pharmacology and Therapeutics*, **27**, 504–513.

Devaraj, S., Autret, B.C. & Jialal, I. (2006). Reduced-calorie orange juice beverage with plant sterols lowers C-reactive protein concentrations and improves the lipid profile in human volunteers. *American Journal of Clinical Nutrition*, **84**, 756–761.

Devaraj, S., Jialal, I. & Vega-Lopez, S. (2004). Plant sterol-fortified orange juice effectively lowers cholesterol levels in mildly hypercholesterolemic healthy individuals. *Arteriosclerosis, Thrombosis, and Vascular Biology*, **24**, e25–e28.

Dewell, A., Hollenbeck, C.B. & Bruce, B. (2002). The effects of soy-derived phytoestrogens on serum lipids and lipoproteins in moderately hypercholesterolemic post-menopausal women. *Journal of Clinical Endocrinology and Metabolism*, **87**, 118–121.

Djuric, Z., Chen, G., Doerge, D.R., Heilbrun, L.K. & Kucuk, O. (2001). Effect of soy isoflavone supplementation on markers of oxidative stress in men and women. *Cancer Letters*, **172**, 1–6.

Doornbos, A.M., Meynen, E.M., Duchateau, G.S., Van Der Knaap, H.C. & Trautwein, E.A. (2006). Intake occasion affects the serum cholesterol lowering of a plant sterol-enriched single-dose yoghurt drink in mildly hypercholesterolaemic subjects. *European Journal of Clinical Nutrition*, **60**, 325–333.

Ellegard, L.H., Andersson, S.W., Normen, A.L. & Andersson, H.A. (2007). Dietary plant sterols and cholesterol metabolism. *Nutrition Review*, **65**, 39–45.

Eskin, N.A.M., & Tamir, S. (2006). Dictionary of nutraceuticals and functional foods. Taylor & Francis Group/CRC Press, Boca Raton, FL.

Espin, J.C., Garcia-Conesa, M.T. & Tomas-Barberan, F.A. (2007). Nutraceuticals: facts and fiction. *Phytochemistry*, **68**, 2986–3008.

Fogliano, V. & Vitaglione, P. (2005). Functional foods: planning and development. *Molecular Nutrition and Food Research*, **49**, 256–262.

Garcia-Alonso, A. & Goni, I. (2000). Effect of processing on potato starch: in vitro availability and glycaemic index. *Nahrung*, **44**, 19–22.

Genovese, M.I., Lopes Barbosa, A.C., Da Silva Pinto, M. & Lajolo, F.M. (2007). Commercial soy protein ingredients as isoflavone sources for functional foods. *Plant Foods for Human Nutrition*, **62**, 53–58.

Ghoddusi, H.B., Grandison, M.A., Grandison, A.S. & Tuohy, K.M. (2007). In vitro study on gas generation and prebiotic effects of some carbohydrates and their mixtures. *Anaerobe*, **13**, 193–199.

Gibson, G.R., Beatty, E.R., Wang, X. & Cummings, J.H. (1995). Selective stimulation of bifidobacteria in the human colon by oligofructose and inulin. *Gastroenterology*, **108**, 975–982.

Goldberg, A.C., Ostlund, R.E., Jr, Bateman, J.H., Schimmoeller, L., McPherson, T.B. & Spilburg, C.A. (2006). Effect of plant stanol tablets on low-density lipoprotein cholesterol lowering in patients on statin drugs. *American Journal of Cardiology*, **97**, 376–379.

Goncalves, S., Maria, A.V., Silva, A.S., Martins-Silva, J. & Saldanha, C. (2006). Phytosterols in milk as a depressor of plasma cholesterol levels: experimental evidence with hypercholesterolemic Portuguese subjects. *Clinical Hemorheology and Microcirculation*, **35**, 251–255.

Goodin, S., Shen, F., Shih, W.J., Dave, N., Kane, M.P., Medina, P., Lambert, G.H., Aisner, J., Gallo, M. & DiPaola, R.S. (2007). Clinical and biological activity of soy protein powder supplementation in healthy male volunteers. *Cancer Epidemiology, Biomarkers and Prevention*, **16**, 829–833.

Granfeldt, Y., Eliasson, A.C. & Bjorck, I. (2000). An examination of the possibility of lowering the glycemic index of oat and barley flakes by minimal processing. *Journal of Nutrition*, **130**, 2207–2214.

Grizard, D. & Barthomeuf, C. (1999). Non-digestible oligosaccharides used as prebiotic agents: mode of production and beneficial effects on animal and human health. *Reproduction, Nutrition, Development*, **39**, 563–588.

Hall, W.L., Vafeiadou, K., Hallund, J., Bugel, S., Koebnick, C., Reimann, M., Ferrari, M., Branca, F., Talbot, D., Dadd, T., Nilsson, M. & Dahlman-Wright, K. (2005). Soy-isoflavone-enriched foods and inflammatory biomarkers of cardiovascular disease risk in post-menopausal women: interactions with genotype and equol production. *American Journal of Clinical Nutrition*, **82**, 1260–1268; quiz 365–366.

Hallikainen, M., Lyyra-Laitinen, T., Laitinen, T., Agren, J.J., Pihlajamaki, J., Rauramaa, R., Miettinen, T.A. & Gylling, H. (2006). Endothelial function in hypercholesterolemic subjects: effects of plant stanol and sterol esters. *Atherosclerosis*, **188**, 425–432.

Hallikainen, M.A., Sarkkinen, E.S., Gylling, H., Erkkila, A.T. & Uusitupa, M.I. (2000a). Comparison of the effects of plant sterol ester and plant stanol ester-enriched margarines in lowering serum cholesterol concentrations in hypercholesterolaemic subjects on a low-fat diet. *European Journal of Clinical Nutrition*, **54**, 715–725.

Hallikainen, M.A., Sarkkinen, E.S. & Uusitupa, M.I. (2000b). Plant stanol esters affect serum cholesterol concentrations of hypercholesterolemic men and women in a dose-dependent manner. *Journal of Nutrition*, **130**, 767–776.

Hallund, J., Bugel, S., Tholstrup, T., Ferrari, M., Talbot, D., Hall, W.L., Reimann, M., Williams, C.M. & Wiinberg, N. (2006). Soya isoflavone-enriched cereal bars affect markers of endothelial function in post-menopausal women. *British Journal of Nutrition*, **95**, 1120–1126.

Hansel, B., Nicolle, C., Lalanne, F., Tondu, F., Lassel, T., Donazzolo, Y., Ferrieres, J., Krempf, M., Schlienger, J.L., Verges, B., Chapman, M.J. & Bruckert, E. (2007). Effect of low-fat, fermented milk enriched with plant sterols on serum lipid profile and oxidative stress in moderate hypercholesterolemia. *American Journal of Clinical Nutrition*, **86**, 790–796.

Heinemann, T., Axtmann, G. & von Bergmann, K. (1993). Comparison of intestinal absorption of cholesterol with different plant sterols in man. *European Journal of Clinical Investigation*, **23**, 827–831.

Hendriks, H.F., Brink, E.J., Meijer, G.W., Princen, H.M. & Ntanios, F.Y. (2003). Safety of long-term consumption of plant sterol esters-enriched spread. *European Journal of Clinical Nutrition*, **57**, 681–692.

Hendriks, H.F., Weststrate, J.A., van Vliet, T. & Meijer, G.W. (1999). Spreads enriched with three different levels of vegetable oil sterols and the degree of cholesterol lowering in normocholesterolaemic and mildly hypercholesterolaemic subjects. *European Journal of Clinical Nutrition*, **53**, 319–327.

Heneman, K.M., Chang, H.C., Prior, R.L. & Steinberg, F.M. (2007). Soy protein with and without isoflavones fails to substantially increase postprandial antioxidant capacity. *The Journal of Nutritional Biochemistry*, **18**, 46–53.

Hoie, L.H., Graubaum, H.J., Harde, A., Gruenwald, J. & Wernecke, K.D. (2005). Lipid-lowering effect of 2 dosages of a soy protein supplement in hypercholesterolemia. *Advances in Therapy*, **22**, 175–186.

Hoie, L.H., Guldstrand, M., Sjoholm, A., Graubaum, H.J., Gruenwald, J., Zunft, H.J. & Lueder, W. (2007). Cholesterol-lowering effects of a new isolated soy protein with high levels of nondenaturated protein in hypercholesterolemic patients. *Advances in Therapy*, **24**, 439–447.

Hoie, L.H., Sjoholm, A., Guldstrand, M., Zunft, H.J., Lueder, W., Graubaum, H.J. & Gruenwald, J. (2006). Ultra heat treatment destroys cholesterol-lowering effect of soy protein. *International Journal of Food Sciences and Nutrition*, **57**, 512–519.

Houweling, A.H., Vanstone, C.A., Trautwein, E.A., Duchateau, G.S. & Jones, P.J. (2009). Baseline plasma plant sterol concentrations do not predict changes in serum lipids, C-reactive protein (CRP) and plasma plant sterols following intake of a plant sterol-enriched food. *European Journal of Clinical Nutrition*, **63**(4), 543–551.

Hsu, C.S., Shen, W.W., Hsueh, Y.M. & Yeh, S.L. (2001). Soy isoflavone supplementation in post-menopausal women. Effects on plasma lipids, antioxidant enzyme activities and bone density. *Journal of Reproductive Medicine*, **46**, 221–226.

Huang, L., Mills, E.N., Carter, J.M. & Morgan, M.R. (1998). Analysis of thermal stability of soya globulins using monoclonal antibodies. *Biochimica et Biophysica Acta*, **1388**, 215–226.

Jenkins, D.J., Kendall, C.W., Vidgen, E., Vuksan, V., Jackson, C.J., Augustin, L.S., Lee, B., Garsetti, M., Agarwal, S., Rao, A.V., Cagampang, G.B. & Fulgoni, V. (2000). Effect of soy-based breakfast cereal on blood lipids and oxidized low-density lipoprotein. *Metabolism*, **49**, 1496–1500.

Jenkins, D.J., Wolever, T.M., Leeds, A.R., Gassull, M.A., Haisman, P., Dilawari, J., Goff, D.V., Metz, G.L. & Alberti, K.G. (1978). Dietary fibres, fibre analogues, and glucose tolerance: importance of viscosity. *British Medical Journal*, **1**, 1392–1394.

Jones, P.J. & Schoeller, D.A. (1990). Evidence for diurnal periodicity in human cholesterol synthesis. *Journal of Lipid Research*, **31**, 667–673.

Jones, P.J., Vanstone, C.A., Raeini-Sarjaz, M. & St-Onge, M.P. (2003). Phytosterols in low- and nonfat beverages as part of a controlled diet fail to lower plasma lipid levels. *Journal of Lipid Research*, **44**, 1713–1719.

Kanauchi, O., Mitsuyama, K., Araki, Y. & Andoh, A. (2003). Modification of intestinal flora in the treatment of inflammatory bowel disease. *Current Pharmaceutical Design*, **9**, 333–346.

Kassis, A.N., Vanstone, C.A., AbuMweis, S.S. & Jones, P.J. (2008). Efficacy of plant sterols is not influenced by dietary cholesterol intake in hypercholesterolemic individuals. *Metabolism*, **57**, 339–346.

Keenan, J.M., Goulson, M., Shamliyan, T., Knutson, N., Kolberg, L. & Curry, L. (2007). The effects of concentrated barley beta-glucan on blood lipids in a population of hypercholesterolaemic men and women. *British Journal of Nutrition*, **97**, 1162–1168.

Keogh, G.F., Cooper, G.J., Mulvey, T.B., McArdle, B.H., Coles, G.D., Monro, J.A. & Poppitt, S.D. (2003). Randomized controlled crossover study of the effect of a highly beta-glucan-enriched barley on cardiovascular disease risk factors in mildly hypercholesterolemic men. *American Journal of Clinical Nutrition*, **78**, 711–718.

Kim, S.H., Lee da, H. & Meyer, D. (2007). Supplementation of baby formula with native inulin has a prebiotic effect in formula-fed babies. *Asia Pacific Journal of Clinical Nutrition*, **16**, 172–177.

Kleessen, B., Hartmann, L. & Blaut, M. (2001). Oligofructose and long-chain inulin: influence on the gut microbial ecology of rats associated with a human faecal flora. *British Journal of Nutrition*, **86**, 291–300.

Kleessen, B., Schwarz, S., Boehm, A., Fuhrmann, H., Richter, A., Henle, T. & Krueger, M. (2007). Jerusalem artichoke and chicory inulin in bakery products affect faecal microbiota of healthy volunteers. *British Journal of Nutrition*, **98**, 540–549.

Kohno, M., Hirotsuka, M., Kito, M. & Matsuzawa, Y. (2006). Decreases in serum triacylglycerol and visceral fat mediated by dietary soybean beta-conglycinin. *Journal of Atherosclerosis and Thrombosis*, **13**, 247–255.

Kolida, S., Meyer, D. & Gibson, G.R. (2007). A double-blind placebo-controlled study to establish the bifidogenic dose of inulin in healthy humans. *European Journal of Clinical Nutrition*, **61**, 1189–1195.

Korpela, R., Tuomilehto, J., Hogstrom, P., Seppo, L., Piironen, V., Salo-Vaananen, P., Toivo, J., Lamberg-Allardt, C., Karkkainen, M., Outila, T., Sundvall, J., Vilkkila, S. & Tikkanen, M.J. (2006). Safety aspects and cholesterol-lowering efficacy of low fat dairy products containing plant sterols. *European Journal of Clinical Nutrition*, **60**, 633–642.

Lambo, A., Oste, R. & Nyman, M. (2005). Dietary fibre in fermented oat and barly β-glucan rich concentrates. *Food Chemistry*, **89**, 283–293.

Langlands, S.J., Hopkins, M.J., Coleman, N. & Cummings, J.H. (2004). Prebiotic carbohydrates modify the mucosa associated microflora of the human large bowel. *Gut*, **53**, 1610–1616.

Lau, V.W., Journoud, M. & Jones, P.J. (2005). Plant sterols are efficacious in lowering plasma LDL and non-HDL cholesterol in hypercholesterolemic type 2 diabetic and nondiabetic persons. *American Journal of Clinical Nutrition*, **81**, 1351–1358.

Lazaridou, A. & Biliaderis, C.G. (2007). Molecular aspects of cereal β-glucan functionality: physical properties, technological applications and physiological effects. *Journal of Cereal Science*, **46**, 101–118.

Lees, A.M., Mok, H.Y., Lees, R.S., McCluskey, M.A. & Grundy, S.M. (1977). Plant sterols as cholesterol-lowering agents: clinical trials in patients with hypercholesterolemia and studies of sterol balance. *Atherosclerosis*, **28**, 325–338.

Li, D., Kim, J.M., Jin, Z. & Zhou, J. (2008). Prebiotic effectiveness of inulin extracted from edible burdock. *Anaerobe*, **14**, 29–34.

Li, N.Y., Li, K., Qi, Z., Demonty, I., Gordon, M., Francis, L., Molhuizen, H.O. & Neal, B.C. (2007). Plant sterol-enriched milk tea decreases blood cholesterol concentrations in Chinese adults: a randomised controlled trial. *British Journal of Nutrition*, **98**, 978–983.

Madsen, M.B., Jensen, A.M. & Schmidt, E.B. (2007). The effect of a combination of plant sterol-enriched foods in mildly hypercholesterolemic subjects. *Clinical Nutrition*, **26**, 792–798.

Maesta, N., Nahas, E.A., Nahas-Neto, J., Orsatti, F.L., Fernandes, C.E., Traiman, P. & Burini, R.C. (2007). Effects of soy protein and resistance exercise on body composition and blood lipids in post-menopausal women. *Maturitas*, **56**, 350–358.

Marinangeli, C.P., Varady, K.A. & Jones, P.J. (2006). Plant sterols combined with exercise for the treatment of hypercholesterolemia: overview of independent and synergistic mechanisms of action. *The Journal of Nutritional Biochemistry*, **17**, 217–224.

Maruyama, N., Katsube, T., Wada, Y., Oh, M.H., Barba De La Rosa, A.P., Okuda, E., Nakagawa, S. & Utsumi, S. (1998). The roles of the N-linked glycans and extension regions of soybean beta-conglycinin in folding, assembly and structural features. *European Journal of Biochemistry*, **258**, 854–862.

Maruyama, N., Sato, R., Wada, Y., Matsumura, Y., Goto, H., Okuda, E., Nakagawa, S. & Utsumi, S. (1999). Structure-physicochemical function relationships of soybean beta-conglycinin constituent subunits. *Journal of Agricultural and Food Chemistry*, **47**, 5278–5284.

Mathias, K., Ismail, B., Corvalan, C.M. & Hayes, K.D. (2006). Heat and pH effects on the conjugated forms of genistin and daidzin isoflavones. *Journal of Agricultural and Food Chemistry*, **54**, 7495–7502.

Mattson, F.H., Grundy, S.M. & Crouse, J.R. (1982). Optimizing the effect of plant sterols on cholesterol absorption in man. *American Journal of Clinical Nutrition*, **35**, 697–700.

Matvienko, O.A., Lewis, D.S., Swanson, M., Arndt, B., Rainwater, D.L., Stewart, J. & Alekel, D.L. (2002). A single daily dose of soybean phytosterols in ground beef decreases serum total cholesterol and LDL cholesterol in young, mildly hypercholesterolemic men. *American Journal of Clinical Nutrition*, **76**, 57–64.

McIntosh, G.H., Whyte, J., McArthur, R. & Nestel, P.J. (1991). Barley and wheat foods: influence on plasma cholesterol concentrations in hypercholesterolemic men. *American Journal of Clinical Nutrition*, **53**, 1205–1209.

McPherson, T.B., Ostlund, R.E., Goldberg, A.C., Bateman, J.H., Schimmoeller, L. & Spilburg, C.A. (2005). Phytostanol tablets reduce human LDL-cholesterol. *Journal of Pharmacy and Pharmacology*, **57**, 889–896.

McVeigh, B.L., Dillingham, B.L., Lampe, J.W. & Duncan, A.M. (2006). Effect of soy protein varying in isoflavone content on serum lipids in healthy young men. *American Journal of Clinical Nutrition*, **83**, 244–251.

Menne, E., Guggenbuhl, N. & Roberfroid, M. (2000). Fn-type chicory inulin hydrolysate has a prebiotic effect in humans. *Journal of Nutrition*, **130**, 1197–1199.

Moreau, R.A., Whitaker, B.D. & Hicks, K.B. (2002). Phytosterols, phytostanols, and their conjugates in foods: structural diversity, quantitative analysis, and health-promoting uses. *Progress in Lipid Research*, **41**, 457–500.

Nestel, P., Cehun, M., Pomeroy, S., Abbey, M. & Weldon, G. (2001). Cholesterol-lowering effects of plant sterol esters and non-esterified stanols in margarine, butter and low-fat foods. *European Journal of Clinical Nutrition*, **55**, 1084–1090.

Nguyen, T.T. (1999). The cholesterol-lowering action of plant stanol esters. *Journal of Nutrition*, **129**, 2109–2112.

Nielsen, I.L. & Williamson, G. (2007). Review of the factors affecting bioavailability of soy isoflavones in humans. *Nutrition and Cancer*, **57**, 1–10.

Nilsson, A.C., Ostman, E.M., Holst, J.J. & Bjorck, I.M. (2008). Including indigestible carbohydrates in the evening meal of healthy subjects improves glucose tolerance, lowers inflammatory markers, and increases satiety after a subsequent standardized breakfast. *Journal of Nutrition*, **138**, 732–739.

Noakes, M., Clifton, P.M., Doornbos, A.M. & Trautwein, E.A. (2005). Plant sterol ester-enriched milk and yoghurt effectively reduce serum cholesterol in modestly hypercholesterolemic subjects. *European Journal of Nutrition*, **44**, 214–222.

Noakes, M., Clifton, P.M., Ntanios, F., Shrapnel, W., Record, I. & McInerney, J. (2002). An increase in dietary carotenoids when consuming plant sterols or stanols is effective in maintaining plasma carotenoid concentrations. *American Journal of Clinical Nutrition*, **75**, 79–86.

Ntanios, F.Y., Homma, Y. & Ushiro, S. (2002). A spread enriched with plant sterol-esters lowers blood cholesterol and lipoproteins without affecting vitamins A and E in normal and hypercholesterolemic Japanese men and women. *Journal of Nutrition*, **132**, 3650–3655.

Ostlund, R.E., Jr (2007). Phytosterols, cholesterol absorption and healthy diets. *Lipids*, **42**, 41–45.

Ostlund, R.E., Jr, Spilburg, C.A. & Stenson, W.F. (1999). Sitostanol administered in lecithin micelles potently reduces cholesterol absorption in humans. *American Journal of Clinical Nutrition*, **70**, 826–831.

Parker, T.S., McNamara, D.J., Brown, C., Garrigan, O., Kolb, R., Batwin, H. & Ahrens, E.H., Jr (1982). Mevalonic acid in human plasma: relationship of concentration and circadian rhythm to cholesterol synthesis rates in man. *Proceedings of the National Academy of Sciences of the United States of America*, **79**, 3037–3041.

Pereira, D.I. & Gibson, G.R. (2002). Effects of consumption of probiotics and prebiotics on serum lipid levels in humans. *Critical Reviews in Biochemistry and Molecular Biology*, **37**, 259–281.

Pereira, I.R., Faludi, A.A., Aldrighi, J.M., Bertolami, M.C., Saleh, M.H., Silva, R.A., Nakamura, Y., Campos, M.F., Novaes, N. & Abdalla, D.S. (2006). Effects of soy germ isoflavones and hormone therapy on nitric oxide derivatives, low-density lipoprotein oxidation, and vascular reactivity in hypercholesterolemic postmenopausal women. *Menopause*, **13**, 942–950.

Plana, N., Nicolle, C., Ferre, R., Camps, J., Cos, R., Villoria, J. & Masana, L. (2008). Plant sterol-enriched fermented milk enhances the attainment of LDL-cholesterol goal in hypercholesterolemic subjects. *European Journal of Nutrition*, **47**(1), 32–39.

Plat, J. & Mensink, R.P. (2005). Plant stanol and sterol esters in the control of blood cholesterol levels: mechanism and safety aspects. *American Journal of Cardiology*, **96**, 15D–22D.

Pool-Zobel, B.L. (2005). Inulin-type fructans and reduction in colon cancer risk: review of experimental and human data. *British Journal of Nutrition*, **93**(Suppl 1), S73–S90.

Poppitt, S.D., van Drunen, J.D., McGill, A.T., Mulvey, T.B. & Leahy, F.E. (2007). Supplementation of a high-carbohydrate breakfast with barley beta-glucan improves postprandial glycaemic response for meals but not beverages. *Asia Pacific Journal of Clinical Nutrition*, **16**, 16–24.
Puppo, C., Chapleau, N., Speroni, F., De Lamballerie-Anton, M., Michel, F., Anon, C. & Anton, M. (2004). Physicochemical modifications of high-pressure-treated soybean protein isolates. *Journal of Agricultural and Food Chemistry*, **52**, 1564–1571.
Redgwell, R.J. & Fischer, M. (2005). Dietary fibre as a versatile food component: an industrial perspective. *Molecular Nutrition and Food Research*, **49**, 521–535.
Reynolds, K., Chin, A., Lees, K.A., Nguyen, A., Bujnowski, D. & He, J. (2006). A meta-analysis of the effect of soy protein supplementation on serum lipids. *American Journal of Cardiology*, **98**, 633–640.
Richelle, M., Enslen, M., Hager, C., Groux, M., Tavazzi, I., Godin, J.P., Berger, A., Metairon, S., Quaile, S., Piguet-Welsch, C., Sagalowicz, L., Green, H. & Fay, L.B. (2004). Both free and esterified plant sterols reduce cholesterol absorption and the bioavailability of beta-carotene and alpha-tocopherol in normocholesterolemic humans. *American Journal of Clinical Nutrition*, **80**, 171–177.
Ridges, L., Sunderland, R., Moerman, K., Meyer, B., Astheimer, L. & Howe, P. (2001). Cholesterol lowering benefits of soy and linseed enriched foods. *Asia Pacific Journal of Clinical Nutrition*, **10**, 204–211.
Rizkalla, S.W., Laromiguiere, M., Champ, M., Bruzzo, F., Boillot, J. & Slama, G. (2007). Effect of baking process on postprandial metabolic consequences: randomized trials in normal and type 2 diabetic subjects. *European Journal of Clinical Nutrition*, **61**, 175–183.
Roberfroid, M.B. (2005). Introducing inulin-type fructans. *British Journal of Nutrition*, **93**(Suppl 1), S13–S25.
Rudkowska, I. & Jones, P.J. (2007). Functional foods for the prevention and treatment of cardiovascular diseases: cholesterol and beyond. *Expert Review of Cardiovascular Therapy*, **5**, 477–490.
Rumessen, J.J. & Gudmand-Hoyer, E. (1998). Fructans of chicory: intestinal transport and fermentation of different chain lengths and relation to fructose and sorbitol malabsorption. *American Journal of Clinical Nutrition*, **68**, 357–364.
Sairanen, U., Piirainen, L., Grasten, S., Tompuri, T., Matto, J., Saarela, M. & Korpela, R. (2007). The effect of probiotic fermented milk and inulin on the functions and microecology of the intestine. *Journal of Dairy Research*, **74**, 367–373.
Saito, S., Takeshita, M., Tomonobu, K., Kudo, N., Shiiba, D., Hase, T., Tokimitsu, I. & Yasukawa, T. (2006a). Dose-dependent cholesterol-lowering effect of a mayonnaise-type product with a main component of diacylglycerol-containing plant sterol esters. *Nutrition*, **22**, 174–178.
Saito, S., Tomonobu, K., Kudo, N., Shiiba, D., Hase, T. & Tokimitsu, I. (2006b). Serum retinol, alpha-tocopherol, and beta-carotene levels are not altered by excess ingestion of diacylglycerol-containing plant sterol esters. *Annals of Nutrition and Metabolism*, **50**, 372–379.
Sauer, J., Richter, K.K. & Pool-Zobel, B.L. (2007). Products formed during fermentation of the prebiotic inulin with human gut flora enhance expression of biotransformation genes in human primary colon cells. *British Journal of Nutrition*, **97**, 928–937.
Seidel, C., Boehm, V., Vogelsang, H., Wagner, A., Persin, C., Glei, M., Pool-Zobel, B.L. & Jahreis, G. (2007). Influence of prebiotics and antioxidants in bread on the immune system, antioxidative status and antioxidative capacity in male smokers and non-smokers. *British Journal of Nutrition*, **97**, 349–356.
Shimizu, C., Kihara, M., Aoe, S., Araki, S., Ito, K., Hayashi, K., Watari, J., Sakata, Y. & Ikegami, S. (2008). Effect of high beta-glucan barley on serum cholesterol concentrations and visceral fat area in Japanese men-a randomized, double-blinded, placebo-controlled trial. *Plant Foods for Human Nutrition*, **63**, 21–25.
Stevens, C.V., Meriggi, A. & Booten, K. (2001). Chemical modification of inulin, a valuable renewable resource, and its industrial applications. *Biomacromolecules*, **2**, 1–16.
Stintzing, F.C., Hoffmann, M. & Carle, R. (2006). Thermal degradation kinetics of isoflavone aglycones from soy and red clover. *Molecular Nutrition and Food Research*, **50**, 373–377.
Su, P., Henriksson, A. & Mitchell, H. (2007). Prebiotics enhance survival and prolong the retention period of specific probiotic inocula in an in vivo murine model. *Journal of Applied Microbiology*, **103**, 2392–2400.
Thomsen, A.B., Hansen, H.B., Christiansen, C., Green, H. & Berger, A. (2004). Effect of free plant sterols in low-fat milk on serum lipid profile in hypercholesterolemic subjects. *European Journal of Clinical Nutrition*, **58**, 860–870.
Tietyen, J.L., Nevins, D.L., Shoemaker, C.F. & Schneeman, B.O. (1995). Hypocholesterolemic potential of oat bran treated with an endo-beta-D-glucanase from *Bacillus subtilis*. *Journal of Food Science*, **60**, 558–560.
Tikkanen, M.J. (2005). Plant sterols and stanols. *Handbook of Experimental Pharmacology*, **170**, 215–230.

Tikkanen, M.J., Wahala, K., Ojala, S., Vihma, V. & Adlercreutz, H. (1998). Effect of soybean phytoestrogen intake on low density lipoprotein oxidation resistance. *Proceedings of the National Academy of Sciences of the United States of America*, **95**, 3106–3110.

Torronen, R., Kansanen, L., Uusitupa, M., Hanninen, O., Myllymaki, O., Harkonen, H. & Malkki, Y. (1992). Effects of an oat bran concentrate on serum lipids in free-living men with mild to moderate hypercholesterolaemia. *European Journal of Clinical Nutrition*, **46**, 621–627.

Tsukada, H., Takano, K., Hattori, M., Yoshida, T., Kanuma, S. & Takahashi, K. (2006). Effect of sorbed water on the thermal stability of soybean protein. *Bioscience, Biotechnology, and Biochemistry*, **70**, 2096–2103.

Uesugi, T., Fukui, Y. & Yamori, Y. (2002). Beneficial effects of soybean isoflavone supplementation on bone metabolism and serum lipids in post-menopausal Japanese women: a four-week study. *Journal of the American College of Nutrition*, **21**, 97–102.

Ungar, Y., Osundahunsi, O.F. & Shimoni, E. (2003). Thermal stability of genistein and daidzein and its effect on their antioxidant activity. *Journal of Agricultural and Food Chemistry*, **51**, 4394–4399.

Uzzan, M., Nechrebeki, J. & Labuza, T.P. (2007). Thermal and storage stability of nutraceuticals in a milk beverage dietary supplement. *Journal Food Science*, **72**, E109–E114.

Vanstone, C.A., Raeini-Sarjaz, M., Parsons, W.E. & Jones, P.J. (2002). Unesterified plant sterols and stanols lower LDL-cholesterol concentrations equivalently in hypercholesterolemic persons. *American Journal of Clinical Nutrition*, **76**, 1272–1288.

Varady, K.A., Ebine, N., Vanstone, C.A., Parsons, W.E. & Jones, P.J. (2004). Plant sterols and endurance training combine to favorably alter plasma lipid profiles in previously sedentary hypercholesterolemic adults after 8 wk. *American Journal of Clinical Nutrition*, **80**, 1159–1166.

Varady, K.A., Houweling, A.H. & Jones, P.J. (2007). Effect of plant sterols and exercise training on cholesterol absorption and synthesis in previously sedentary hypercholesterolemic subjects. *Translational Research*, **149**, 22–30.

Vega-Lopez, S., Yeum, K.J., Lecker, J.L., Ausman, L.M., Johnson, E.J., Devaraj, S., Jialal, I. & Lichtenstein, A.H. (2005). Plasma antioxidant capacity in response to diets high in soy or animal protein with or without isoflavones. *American Journal of Clinical Nutrition*, **81**, 43–49.

Vervuert, I., Coenen, M. & Bothe, C. (2003). Effects of oat processing on the glycaemic and insulin responses in horses. *Journal of Animal Physiology and Animal Nutrition (Berl)*, **87**, 96–104.

Volpe, R., Niittynen, L., Korpela, R., Sirtori, C., Bucci, A., Fraone, N. & Pazzucconi, F. (2001). Effects of yoghurt enriched with plant sterols on serum lipids in patients with moderate hypercholesterolaemia. *British Journal of Nutrition*, **86**, 233–239.

Wang, Y., Jones, P.J., Ausman, L.M. & Lichtenstein, A.H. (2004). Soy protein reduces triglyceride levels and triglyceride fatty acid fractional synthesis rate in hypercholesterolemic subjects. *Atherosclerosis*, **173**, 269–275.

Welty, F.K., Lee, K.S., Lew, N.S. & Zhou, J.R. (2007). Effect of soy nuts on blood pressure and lipid levels in hypertensive, prehypertensive, and normotensive post-menopausal women. *Archives of Internal Medicine*, **167**, 1060–1067.

Wiseman, H., O'Reilly, J.D., Adlercreutz, H., Mallet, A.I., Bowey, E.A., Rowland, I.R. & Sanders, T.A. (2000). Isoflavone phytoestrogens consumed in soy decrease F(2)-isoprostane concentrations and increase resistance of low-density lipoprotein to oxidation in humans. *American Journal of Clinical Nutrition*, **72**, 395–400.

Wolever, T.M., Jenkins, D.J., Leeds, A.R., Gassull, M.A., Dilawari, J.B., Goff, D.V., Metz, G.L. & Alberti, G.M. (1978). Dietary fibre and glucose tolerance importance of viscosity. *The Proceedings of the Nutrition Society*, **37**, 47A.

Wood, P.J., Beer, M.U. & Butler, G. (2000). Evaluation of role of concentration and molecular weight of oat beta-glucan in determining effect of viscosity on plasma glucose and insulin following an oral glucose load. *British Journal of Nutrition*, **84**, 19–23.

Woodgate, D., Chan, C.H. & Conquer, J.A. (2006). Cholesterol-lowering ability of a phytostanol softgel supplement in adults with mild to moderate hypercholesterolemia. *Lipids*, **41**, 127–132.

Xiao, C.W., Mei, J. & Wood, C.M. (2008). Effect of soy proteins and isoflavones on lipid metabolism and involved gene expression. *Frontiers in Bioscience*, **13**, 2660–2273.

Xu, Z., Wu, Q. & Godber, J.S. (2002). Stabilities of daidzin, glycitin, genistin, and generation of derivatives during heating. *Journal of Agricultural and Food Chemistry*, **50**, 7402–7406.

Zhan, S. & Ho, S.C. (2005). Meta-analysis of the effects of soy protein containing isoflavones on the lipid profile. *American Journal of Clinical Nutrition*, **81**, 397–408.

7 Dairy ingredients in new functional food product development

S.L. Amaya-Llano and Lech Ozimek

7.1 Historical aspects

For centuries, dairy products have been used as one of the principal foods for millions of people around the world. Milk is a natural food with demonstrated nutritional value. Recognition of the value of milk is reflected in the increasing interest in development programs focused on small-scale dairying in developing countries where malnutrition and poverty are the main challenge. Demand for milk in developing countries is expected to increase by 25% by 2025 (Delgado *et al.* 1999), partly due to population growth but also because disposable income is being spent on a greater diversity of food products to meet nutritional needs. The Food and Agriculture Organization projection scenario estimates that consumption of milk in developing countries will grow 3.3% per year since the early 1990s and until 2020. The corresponding developed world growth rates are 0.2% per year. By 2020, developing countries will consume 223 million metric tons more milk than they did in 1993, dwarfing developed country increases of 18 million metric tons for milk (Delgado *et al.* 1999). Designing and developing functionality in dairy-based products simply means modifying and/or enriching the healthy nature of the milk. The concept of functional food first came to light in Japan in the mid-1980s when an aging population and rising health care costs led the Ministry of Health and Welfare to initiate the regulatory approval of functional foods (Swinbanks & O'Brien 1993). As a result of a long decision-making process to establish a category of foods for potential enhancing benefits as part of a national effort to reduce the escalating cost of health care, the concept of foods for specified health use (FOSHU) was established in 1991. In the meantime, but mainly in the 1990s, a variety of terms, more or less related to the Japanese FOSHU, have appeared worldwide. In addition to functional foods, these include more exotic terms such as 'nutraceuticals', 'designer foods', 'pharmafoods', 'medifoods' and 'vitafoods', but also the more traditional 'dietary supplements' and 'fortified foods'. 'Functional food' appears as a quite unique concept that deserves a category of its own, a category different from nutraceutical, pharmafood, medifood, designer food or vitafood, and a category that does not include dietary supplement. It is also a concept that belongs to nutrition and not to pharmacology. Functional foods are and must be foods, not drugs, as they have no therapeutic effects. Moreover, their role regarding disease will, in most cases, be in reducing the risk of disease rather than preventing it (Roberfroid 2000). It is accepted that a functional food provides a health benefit that goes beyond a general nutritional benefit (Shortt *et al.* 2004). During the 1990s, functional foods and nutraceuticals emerged as the dominant trend for the food industry, both in the United

States and internationally. Consumer studies repeatedly indicate that there are segments in the population that have attitudes and lifestyles more consonant with the concept of foods for health. In the United States, typically, consumer research can segment about 40 million consumers who are 'health active,' meaning they act today to ensure good health when older, are concerned about family nutrition, regularly eat fruits, accept medications and exercise twice a week. In addition, research recognizes a somewhat similar group of consumers, known as 'health aware,' who are like the former 'active' group except they do not exercise twice a week. This group, also comprising consumers over age 18, is estimated to include 13 million consumers. These are sizable consumer markets (Childs 2007). Quantifying the value of the functional food market is a difficult task. Using a strict definition – food and drinks that make some kind of specific health claim on packaging or in advertising – the market in Europe, the United States, Japan and Australia has been valued at US$5.7 billion (Hilliam 2000). The European market for functional foods was estimated to be between US$4 and 8 billion in 2003 depending on which foods are regarded as functional (Menrad 2003). This value has increased to around US$15 billion in 2006 (Kotilainen *et al.* 2006). The current market share of functional food is still less than 1% of the total food and drink market. Germany, France, the United Kingdom and the Netherlands represent the most important countries within the functional food market in Europe (Mäkinen-Aakula 2006). According to the latest research from Euromonitor International, the Dutch market for fortified and functional foods surpassed the US$384.27 million in 2004, making the Netherlands the sixth largest market of functional food products in Europe (Benkouider 2005a; Siró *et al.* 2008). Euromonitor predicted that value sales for functional foods will rise moderately from 2005 to 2009 in the newly emerging markets of Hungary, Poland and Russia (Benkouider 2005b). Although these markets are still undeveloped, numerous new products have been introduced in the last few years. Furthermore, the demand for functional foods is high in these countries, especially among the higher income population. The value of Russian functional food market, for example, was estimated at US$75 million in 2004, and an annual growth of 20% is expected (Kotilainen *et al.* 2006; Siró *et al.* 2008). The functional food market in 2006 represented approximately 17% of the total food market in Spain; moreover, the predicted value for 2020 is around 40%. More than 50% growth was reported between 2000 and 2005 (Monár 2007; Siró *et al.* 2008).

7.2 Functional dairy product development

The concept of foods that could provide health-enhancing and disease-preventing properties was embraced by a growing number of consumers, increasingly documented by nutritionists and scientists, and legally endorsed by public policy and legislative mandates for food and dietary supplement labeling. Nutraceuticals and functional foods are clearly a twenty-first-century industry. They promise value-added opportunities in the food industry and new market opportunities for the pharmaceutical industry. They offer advances in public health as health claim marketing messages empower consumers to select healthier food choices (Childs 2007). Dairy products form the major part of functional foods. The literature on new product development is vast, and many approaches have been made to divide product development into specific phases (Cooper 1990; Cohen *et al.* 1998). Biströn and Nordström (2002) chose a five-phase system to analyze the functional dairy food development processes: (1) idea generation, (2) basic research, (3) development of the final product concept, (4) clinical testing of the final product and (5) marketing activities and product launch. When introducing a new component or ingredient in a dairy food, his new product might fit into generally accepted definition (in Europe and the United States) of the so-called functional, or enhanced, food. This term is often used to indicate a food that includes a health-promoting component beyond traditional biochemical composition. Functional foods have no therapeutic effect and their main mission must be to prevent the risk of diseases. They are foods, not drugs, even though the idea that a food must be modified before it can be considered functional is debatable. It can be argued that milk itself is a functional food, replete with bioactive peptides, antioxidants and other biologically active components (Berner & O'Donell 1998). Traditional dairy streams (milk, colostrum, whey) contain a multitude of constituents with potential as functional ingredients, both for inclusion into dairy and non-dairy food products. Once extracted or isolated, and their bioactivity substantiated, it is these dairy constituents that will provide the functional food sector with a range of ingredients with which to formulate new and novel functional foods, and provide the basis for any health claims (Pitts 1994).

7.3 Health and dairy functional ingredients

7.3.1 Milk

Milk from all mammalian species is a rich source of nutrients for the newborn, including protein, carbohydrate, lipids and minerals. Apart from these nutrients, milk represents an excellent natural source of a number of key vitamins, vitamin precursors and minerals important in disease prevention (Playne *et al.* 2003). It contains components that provide critical nutritive elements, immunological protection and biologically active substances to both neonates and adults (Warner *et al.* 2001). Milk contains high levels of immunoglobulins and other physiologically active compounds for warding off infection in the newborn (Séverin & Wenshui 2005). Milk proteins are currently the main source of a range of biologically active peptides even though other animal and plant proteins contain potential bioactive sequences. These peptides, which are encrypted within the sequence of the parent proteins, can be released by enzymatic proteolysis, for example, during gastrointestinal digestion or during food processing (Gobbetti *et al.* 2002). It is now well established that physiologically active peptides are produced from several food proteins during gastrointestinal digestion and fermentation of food materials with lactic acid bacteria (Korhonen & Pihlanto 2006). Once the bioactive peptides are liberated, they may act as regulatory compounds with hormone-like activity. The activity is based on the inherent amino acid composition and sequence. Bioactive peptides usually contain 3–20 amino acid residues per molecule (Pihlanto-Leppälä 2001). Although chemical and physical treatments may have an influence, proteolysis by naturally occurring enzymes in milk – by exogenous enzymes and by enzymes from microbial starters such as lactic acid bacteria – is mainly responsible for the generation of bioactive peptides during dairy processing, thereby enriching the dairy products. On the other hand, once produced, bioactive peptides may influence the biochemical activities of the microbial communities (Smacchi & Gobbetti 2000). Some of the functions ascribed to the milk proteins or their peptide fragments may be supported by other non-peptide components of milk. These can include milk lipids (Isaacs *et al.* 1995), glycolipids (Newburg 1996), sphingolipids (Dillehay *et al.* 1994; Merrill *et al.* 1995) and oligosaccharides (Ebner & Schanbacher 1974; Prieto *et al.* 1995; Newburg 1996), which contribute to the total probiotic activity as well as antibiotic defense against microbial and viral pathogens (Steinhoff *et al.* 1994; Harmsen *et al.* 1995). Components in cheese or milk such as protein (casein and whey), lipids, calcium and phosphorus may be partly responsible for the beneficial effects of these foods on oral health. Milk proteins may protect against caries by inhibiting the ability of cavity-causing bacteria to adhere to tooth surfaces. Milk proteins, particularly κ-casein, have been demonstrated to reduce the adherence of *Streptoccocus mutans* to the saliva-coated hydroxyapatite surfaces of teeth (Miller *et al.* 2000).

7.3.2 Casein

Casein is the main proteinaceous component of milk, where it accounts for approximately 80% of the total protein inventory. Until recently, the main physiological role of casein in the milk system was widely accepted to be a source of amino acids required by growth of the neonate. While no specific physiological property has been proposed for the whole casein system (or its individual fractions for that matter), various peptides hidden (or inactive) in the amino acid sequence have been the subject of increasingly intense studies. Much work regarding those peptides, which are known to possess bioactivities, is currently underway regarding their release via selective enzymatic hydrolysis. Functional peptides derived from casein, present in either milk or dairy products, have been shown to have effects in the cardiovascular system, mainly via antithrombotic and antihypertensive features (Silva & Malcata 2004).

7.3.3 Whey products

Whey is the portion of milk remaining after casein and fat are formed into cheese curd usually by acid, heat or rennet (Kosikowski & Mistry 1997). Whey is recognized for its high nutritional quality. This by-product of the cheese industry contains the water-soluble nutrients of milk, specifically lactose, non-casein protein (albumins and globulins), and some minerals and vitamins. There are two types of whey, sweet whey and acid whey. Sweet whey (pH greater than or equal to 5.6) is obtained from whole milk used in the manufacture of natural enzyme-produced cheeses such as cheddar cheese. Acid whey (pH less than or equal to 5.1) is obtained from non-fat milk used in the manufacture of cottage or similar cheeses. Fresh pasteurized liquid whey is rarely used as such for foods or feeds because of high transporting costs and susceptibility to deterioration during storage. Consequently, whey is processed to provide a wide range of products including condensed whey, dry whey and modified whey products, each with unique functional characteristics (e.g. whipping/foaming, emulsification, high solubility, gelation and viscosity). These whey products contain a high concentration of whey solids that are easily transported, have enhanced storage stability, blend well with other foods, and are economical sources of milk solids (Miller *et al.* 2000). They have a higher nutritional value, mostly because of their sulfur amino acid and lysine content, as well as molecules with a high added value (lactoferrin, lactoperoxidase). Although proteins do not constitute the major fraction of whey, it is the one that is most important from an economic and nutritional point of view, and as far as potential uses are concerned. The nitrogen substances of milk are retained in whey, with the exception of the caseins (Linden & Lorient 1999).

7.3.4 Calcium

Because more than 99% of the total calcium content of the body is found in the skeleton, it is not surprising that considerable interest lies in the role of calcium and vitamin D (which enhances calcium absorption) in bone health. The Food and Drug Administration has concluded that a lifetime of 'adequate calcium intake is important for maintenance of bone health and may help reduce the risk of osteoporosis particularly for individuals at greatest risk'. Although the 'optimal' calcium intake for skeletal health and prevention of osteoporosis is unknown, the Institute of Medicine, National Academy of Sciences (NAS) has released new calcium recommendations based on intakes consistent with desirable calcium retention, which in turn is associated with increased bone mass and reduced risk of osteoporosis. The NAS calcium recommendations are as follows: 800 mg/day for children 4–8 years, 1300 mg/day for children and adolescents 9–18 years, 1000 mg/day for adults 19–50 years, and 1200 mg/day for adults 51 years and over (Miller *et al.* 2000). Since the early 1980s, a considerable body of evidence has accumulated from investigations in experimental animals, epidemiological studies and clinical intervention trials in humans to support a beneficial role for calcium or calcium-rich foods such as milk and other dairy foods in blood pressure control (Reusser & McCarron 1994; Hamet 1995; Osborne *et al.* 1996); however, interactions among calcium and other nutrients or dietary components may contribute to inconsistent effects of calcium on blood pressure (Miller *et al.* 2000). Numerous epidemiological, experimental animal, *in vitro*, and clinical studies in humans have investigated the protective effect of calcium, vitamin D and dairy foods against colon cancer (Lipkin 1991; Van der Meer *et al.* 1998). In individuals at risk for colon cancer, hyperproliferation of colon epithelium is reduced toward normal by increased dietary calcium. Unfortunately, calcium intake in the United States is generally lower than recommended.

7.4 Galacto-oligosaccharides, lactulose, lactitol and lactosucrose

The enzymatic hydrolysis of lactose into its component monosaccharides (glucose and galactose) is of interest from both the nutritional and technological viewpoints. The resulting sugars are sweeter, more readily fermented and are absorbed directly from the intestine. More recently, interest in the reaction has been raised by observation that oligosaccharides may have beneficial effects as 'bifidus factors', promoting the growth of desirable intestinal microflora. Also, the transferase reaction can be used to attach galactose to other chemicals and consequently has potential applications in the production of food ingredients, pharmaceuticals and other biologically active compounds (Mahoney 1998). Lactulose, lactitol and lactobionic acid are compounds that can be produced from lactose (or whey). The molecular formula of lactulose is similar to that of lactose (disaccharide of galactose and glucose), the only difference being that in lactulose the glucose residue of lactose is isomerized to fructose (Strohmaier 1998). Lactulose is used in various types of food products (infant formula, baby food, confectionary, soft drink, milk products) and also pharmaceutically to improve hepatic encephalopathy and constipation (Mizota 1996; Strohmaier 1998). Besides lactulose, lactitol, a sugar alcohol consisting of galactose and sorbitol, is also used as a sugar substitute in foods such as sugar-free confections, chocolates, chewing gums, no-sugar-added baked goods and ice creams (Kummel & Brokx 2001), and medically to improve hepatic encephalopathy and constipation (Riggio *et al.* 1990; Petticrew *et al.* 1997; Saarela *et al.* 2003). Similarly to lactulose, lactitol is also fermented *in vitro* by several important intestinal bacteria including representatives of *Bacteroides*, *Clostridium*, *Lactobacillus*, *Enterococcus* and *Bifidobacterium* species (Saarela *et al.* 2003). Lactobionic acid is commercially produced by chemical oxidation of lactose. Lactobionate is a strong chelator of calcium and is used in calcium supplements in pharmaceuticals. The chelating properties of lactobionate also enable applications as ion sequestrant in detergent solutions (Gerling 1997). Moreover, lactobionate at a concentration of 100 mmol/L is a key component of solutions used for the cold storage of transplant organs (Gerling 1997; Gänzle *et al.* 2008). Lactosucrose is a trisaccharide produced enzymatically from a mixture of sucrose and lactose using β-fructofuranoside (invertase) obtained from *Arthrobacter* species. Its sweetness relative to sucrose is 0.3–0.6. Lactosucrose is not resorbed in the upper intestine and is, thus, available for hydrolysis and metabolism by the colonic microflora. Lactosucrose has a bifidogenic effect, and its consumption was reported to decrease fecal pH and to inhibit growth of colonic clostridia (Ogata *et al.* 1993; Gänzle *et al.* 2008).

7.5 Growth factors

7.5.1 Immunoglobulins

Immunoglobulins are glycoproteins, each comprising two glycosylated 'heavy' and two 'light' chain subdomains designed to bind antigens and elicit host defense processes, and thereby offer passive immunity to either calves or other potential consumers (Playne *et al.* 2003). Colostrum, which is the richest source of bovine immunoglobulins, is under intense development as a food ingredient, targeting lucrative niche markets, such as sports nutrition, which is large and growing rapidly (Playne *et al.* 2003). Food allergies occur in some individuals as the result of abnormal immunoglobulin responses to a particular food or food component, usually naturally occurring proteins (Hefle & Taylor 2004; Sampson 2004). Food allergies affect only a small percentage of the population; however, food allergic reactions can be quite severe and even life-threatening in some of individuals.

7.5.2 Lactoferrin

Milk contains between 20 and 200 mg/mL lactoferrin, whereas human milk contains over 2 mg/mL; bovine colostrum also contains appreciable quantities: between 2 and 5 mg/mL. Lactoferrin has an antitoxic activity. It is involved in conveying the biologically available iron. Thus, it has been possible to show that it is involved in conveying iron to the intestine, and that it interacts with the peritoneal macrophages and the hepatocytes. Lactoferrin is capable of delaying bacterial growth, and it is thought that, together with lysozyme and lactoperoxidase, it forms part of the primary defense system against bacterial infection. Lactoferrin has a strong inhibitory effect on bacteria that have a metabolic requirement for iron, as a consequence of its powerful competitive affinity for the iron (III) ion. The higher resistance of breastfed infants to intestinal infections in comparison with bottle-fed babies has been attributed to antimicrobial substances such as lactoferrin present in large quantities in human milk. As a consequence, this glycoprotein was offered as a supplement in milks for babies and young children (Linden & Lorient 1999).

7.5.3 Lactoperoxidase

Lactoperoxidase, which was the first enzyme to be discovered in milk, where it is found in appreciable quantities, has an antibacterial effect. It is generally found in a soluble form in the serum and has been identified with L2 lactenine. Lactoperoxidase, on its own, has no bacteriostatic or bacteriocidal effect. It catalyzes the oxidation of thiocyanate (SCN^-) using hydrogen peroxide (H_2O_2). The products obtained at the end of the reaction (sulfate, cyanate, etc.) have no effect, but the reaction intermediates have a powerful action. The hypothiocyanite ion ($OSCN^-$) is the active principle. One of the industrial advantages of this antibacterial system lies in the storage of raw milk. So, this treatment, combined with storage at low temperatures, can, for example, prevent the proliferation of psychotropic flora over a long period (Linden & Lorient 1999).

7.5.4 Lactose

This is the carbohydrate found in milk. It is predominantly found in the milk of ruminants, at a level of 5% in cows' milk. Anhydrous lactose and monohydrated lactose have little affinity for water, so this disaccharide can be used in many formulae without any risk of clotting (Linden & Lorient 1999). In the last decade, a number of novel dietary carbohydrates have been introduced for food applications. One important group is non-digestible oligosaccharides (NDOs) such as galacto-oligosaccharides derived from lactose, which are increasingly being added to foods, particularly in some European countries and Japan. Dietary carbohydrates that have escaped digestion in the upper gastrointestinal tract are principal substrates for bacterial growth in the colon next to mucus (Cummings 1983; Cummings *et al.* 1989). Several beneficial effects are claimed on the consumption of NDOs and since their average daily ingestion is lower than the level considered as safe at not over 15 g/day (Crittenden & Playne 1996). Tomomatsu (1994) mentions effective daily doses of NDOs in pure form of 2 ± 2.5 g of galacto-oligosaccharides. They are classified as prebiotics because they are not hydrolyzed or absorbed in the upper part of the gastrointestinal tract, and are claimed to beneficially affect the host by selectively stimulating the growth and/or activity of one or a limited number of bacterial species already resident in the colon and in addition may repress pathogen colonization, growth or virulence and induce systemic effects, which can be beneficial to health. NDOs have also been shown to inhibit development of pathogens in the intestines by inhibiting the attachment of bacterial pathogens to mucosal surfaces (Voragen 1998).

7.6 Specific lipids

7.6.1 Conjugated linoleic acid

Conjugated linoleic acid (CLA) is a collective term referring to a mixture of positional geometric isomers of linoleic acid (*cis*-9,*cis*-12-octadecadienoic acid) in which the double bonds are conjugated in either the *cis* or *trans* configuration. CLA has received considerable attention because it has been associated with various health-related benefits in animals due to its anticarcinogenic, antidiabetic, antiadipogenic and antiatherogenic properties as well as its immune modulation effects (Lee *et al.* 2005; Seo *et al.* 2008). In particular, it has long been recognized that CLA inhibits chemically induced carcinogenesis in tissues such as the mammary gland, skin, forestomach and intestine. Furthermore, CLA has been shown to inhibit the proliferation of hepatoma, lung, colon and breast cancer cells *in vitro* studies (Schonberg & Krokan 1995; Chujo *et al.* 2003; Liu & Sidell 2005).

7.7 The *n*-3 and *n*-6 polyunsaturated fatty acids

Over the last few years, dairy companies have launched a wide variety of innovative products in the European market, enhancing first the nutritional properties of dairy components, and then providing new components that offered new health benefits outside the intrinsic characteristics of the raw material as polyunsaturated fatty acids or PUFA. Milk enriched with PUFA (*n*-3, *n*-6) has been, and remains today, successful in Europe, especially in Italy, Spain and France. Daily intake of *n*-3 PUFA-supplemented milk, plus folic acid and B-vitamins, has favorable effects on heart health (Baro *et al.* 2003). Also, rheumatoid arthritis, a debilitating disease, is associated with osteoporosis and an increased risk of cardiovascular disease. In order to prevent this pathology, elder populations should consume dairy products supplemented with long-chain *n*-3 polyunsaturated fatty acids and antioxidants (Rennie *et al.* 2003; Sancho 2003).

Polar lipids, which constitute between 0.2 and 1% of total lipids, are essentially phospholipids (90–99%) with small quantities of ceramides (1–8%). The principal phospholipids in milk are phosphatidylcholine, phosphatidylethanolamine and sphingomyelin. They each represent between 19 and 35% of total phospholipids. Phosphatidylserine and phosphatidylinositol are present in small quantities (3–5%) (Linden & Lorient 1999).

7.8 Uses in food systems

Diet is thought to contribute to six of the ten leading causes of death. Up to 70% of certain cancers may be attributed to diet. There is an increasing demand by consumers for quality of life and this in part is fueling the nutraceutical revolution. The use of nutraceuticals in diets can be seen as a means to reduce escalating health care costs that will contribute not only to a longer lifespan, but more importantly to a longer healthspan (Belem 1999).

7.9 Regulations

In most countries there is no legislative definition of the term, and drawing a borderline between conventional and functional foods is challenging even for nutrition and food experts (Mark-Herbert 2004; Niva 2007). Regarding functional foods, claims associated with specific food products are the preferable means of communicating to consumers. In application of the fundamental principle, any claim must be true and not misleading; it must be scientifically valid, unambiguous and clear to the consumer. However, these basic principles should be safeguarded without becoming a disincentive to the production of functional foods or to their acceptance by consumers (Roberfroid 2000).

7.10 Future considerations

Biotechnology has a key role to play in this new industry. Traditionally, the application of biotechnology techniques in the food industry focused on the major energy-providing foods, such as bread, alcohol, fermented starch, yogurt, cheese, vinegar, and others. More recently, there has been increased interest in biologically active non-nutritive ingredients (or components) from natural products or foods. For those firms having the financial, scientific and technical resources, and management expertise necessary to steer a product through the drug evaluation and approval process, approval of a food or food product intended for use as a nutraceutical or functional food can, in fact, represent an opportunity. The opportunity lies in the strong probability that very few products recognized by consumers as 'functional foods' will ever come to market, in turn conferring a high degree of market exclusivity to those firms that do choose to 'go the drug route' with their food products (Belem 1999).

References

Baro, L., Fonolla, J., Peña, J.L., Martinez-Perez, A., Lucena, A., Jimenez, J., Boza, J.J. & Lopez-Huertas, E. (2003). N-3 fatty acid plus oleic acid and vitamin supplemented milk consumption reduces total and LDL cholesterol, homocysteine and levels of endothelial adhesion molecules in healthy humans. *Clinical Nutrition*, **22**, 175–182.

Belem, M.A.F. (1999). Application of biotechnology in the product development of nutraceuticals in Canada. *Trends in Food Science and Technology*, **10**, 101–106.

Benkouider, C. (2005a). Dining with the Dutch. *Functional Foods and Nutraceuticals*. Available online http://www.ffnmag.com/ASP/articleDisplay.asp?strArticleId=753&strSite=FFNSITE&Screen=CURRENTISSUE (accessed 30 July 2008).

Benkouider, C. (2005b). The world's emerging markets. *Functional Foods and Nutraceuticals*. Available online http://www.ffnmag.com/NH/ASP/strArticleID/770/strSite/FFNSite/articleDisplay.asp (accessed 28 July 2008).

Berner, L.A. & O'Donell, J.A. (1998). Functional foods and health claim legislation: applications to dairy foods. *International Dairy Journal*, **8**, 355–362.

Biströn, M. & Nordström, K. (2002). Identification of key success factors of functional dairy foods product development. *Trends in Food Science and Technology*, **13**, 372–379.

Childs, N.M. (2007). Marketing and regulatory issues for functional foods and nutraceuticals. In: *Handbook of Nutraceuticals and Functional Foods*, 2nd edn. Wildman, R.E.C. (ed.), CRC Press, Boca Raton, FL.

Chujo, H., Yamasaki, M., Nou, S., Koyanagi, N., Tachibana, H. & Yamada, K. (2003). Effect of conjugated linoleic acid isomers on growth factor-induced proliferation of human breast cancer cells. *Cancer Letters*, **202**, 81–87.

Cohen, L.Y., Kamienski, P.W. & Espino, R.L. (1998). Gate system focuses industrial basic research. *Research Technology Management*, **41**, 34–37.

Cooper, R.G. (1990). Stage-gate systems: a new tool for managing new products. *Business Horizons*, **33**, 44–54.

Crittenden, R.G. & Playne, M.J. (1996). Production, properties and applications of food-grade oligosaccharides. *Trends in Food Science and Technology*, **7**, 353–361.

Cummings, J.H. (1983). Fermentation in the human large intestine: evidence and implications for health. *Lancet*, **321**, 1206–1209.

Cummings, J.H., Gibson, G.R. & Macfarlane, G.T. (1989). Quantitative estimate of fermentation in the hindgut of man. *Acta Veterinaria Scandinavica*, **86**, 76–82.

Delgado, C., Rosegrant, M., Steinfeld, H., Ehui, S. & Courbois, C. (1999). Livestock to 2020: the next food revolution. *Food Agriculture and the Environment*. IFPRI/FAO/ILRI Discussion Paper 28:83.

Dillehay, D.L., Webb, S.K., Schmelz, E.-M. & Merrill, A.H., Jr. (1994). Dietary sphingomyelin inhibits 1,2-dimethylhydrazineinduced colon cancer in CFI mice. *Journal of Nutrition*, **124**, 615–620.

Ebner, K.E. & Schanbacher, F.L. (1974). Biochemistry of lactose and related carbohydrates. In: *Lactation: A Comprehensive Treatise*, Vol. 2. Larson, B.L. & Smith, V.R. (eds), Academic Press, New York, pp. 77–113.

Gänzle, M.G., Haase, G. & Jelen, P. (2008). Lactose: crystallization, hydrolysis and value-added derivatives. *International Dairy Journal*, **18**, 685–694.

Gerling, K.G. (1997). Large scale production of lactobionic acid – use and new applications. In: *Whey*. International Dairy Federation, Brussels, Belgium, pp. 251–261.

Gobbetti, M., Stepaniak, L., De Angelis, M., Corsetti, A. & Di Cagno, R. (2002). Latent bioactive peptides in milk proteins: proteolytic activation and significance in dairy processing. *Critical Reviews in Food Science and Nutrition*, **42**, 223–239.
Hamet, P. (1995). The evaluation of the scientific evidence for a relationship between calcium and hypertension. *Journal of Nutrition*, **125**, 311s–400s.
Harmsen, M.C., Swart, P.J., de Bétthune, M.P., Pauwels, R., De Clercq, E., The, T.H. & Meijer, D.K. (1995). Antiviral effects of plasma and milk proteins: lactoferrin shows potent activity against both human immunodeficiency virus and human cytomegalovirus replication *in vitro*. *The Journal of Infectious Diseases*, **172**, 380–388.
Hefle, S.L. & Taylor, S.L. (2004). Food allergy and the food industry. *Current Allergy and Asthma Reports*, **4**, 55–59.
Hilliam, M. (2000). Functional food – How big is the market? *The World of Food Ingredients*, **12**, 50–52.
Isaacs, C.E., Litov, R.E. & Thormar, H. (1995). Antimicrobial activity of lipids added to human milk, infant formula, and bovine milk. *The Journal of Nutritional Biochemistry*, **6**, 362–366.
Korhonen, H. & Pihlanto, A. (2006). Bioactive peptides: production and functionality. *International Dairy Journal*, **16**, 945–960.
Kosikowski, F.V. & Mistry, V.V. (1997). Cheese and fermented milk foods, Vol. 1. Origins and principles, Vol. II. *Procedures and Analysis*, 3rd edn. F.V. Kosikowski and Associates, Brooktondale, NY.
Kotilainen, L., Rajalahti, R., Ragasa, C. & Pehu, E. (2006). Health enhancing foods: opportunities for strengthening the sector in developing countries. Agriculture and Rural Development Discussion Paper 30.
Kummel, K.F. & Brokx, S. (2001). Lactitol as a functional prebiotic. *Cereal Foods World*, **46**, 424–429.
Lee, K.W., Lee, H.J., Cho, H.Y. & Kim, Y.J. (2005). Role of the conjugated linoleic acid in the prevention of cancer. *Critical Reviews in Food Science*, **45**, 135–144.
Linden, G. & Lorient, D. (1999). New ingredients in food processing. In: *Biochemistry and Agriculture*. Woodhead Publishing, Cambridge, England.
Lipkin, M. (1991). Application of intermediate biomarkers to studies of cancer prevention in the gastrointestinal tract: introduction and perspective. *American Journal of Clinical Nutrition*, **54**, 188S–192S.
Liu, J. & Sidell, N. (2005). Anti-estrogenic effects of conjugated linoleic acid through modulation of estrogen receptor phosphorylation. *Breast Cancer Research and Treatment*, **94**, 161–169.
Mahoney, R.R. (1998). Galactosyl-oligosaccharide formation during lactose hydrolysis: a review. *Food Chemistry*, **63**, 147–154.
Mäkinen-Aakula, M. (2006). Trends in functional foods dairy market. In: *Proceedings of the Third Functional Food Net Meeting*, Liverpool, UK.
Mark-Herbert, C. (2004). Innovation of a new product category – functional foods. *Technovation*, **24**, 713–719.
Menrad, K. (2003). Market and marketing of functional food in Europe. *Journal of Food Process Engineering*, **56**, 181–188.
Merrill, A.H., Jr, Schmelz, E.M., Wang, E., Schroeder, J.J., Dillehay, D.L. & Riley, R.T. (1995). Role of dietary sphingolipids and inhibitors of sphingolipid metabolism in cancer and other diseases. *Journal of Nutrition*, **125**, 1677s–1682s.
Miller, G.D., Jarvis, J.K. & McBean, L.D. (2000). *Dairy foods and nutrition*, 2nd edn. CRC Press, Boca Raton, FL.
Mizota, T. (1996). Lactulose as a growth promoting factor for *Bifidobacterium* and its physiological aspects. International Dairy Federation, Bulletin no. 313, 43–48.
Monár, J. (2007). The Spanish functional food market: present and future perspectives. Functional FoodNet (FFNet) network meeting, IATA-CSIC.
Newburg, D.S. (1996). Oligosaccharides and glycoconjugates in human milk: their role in host defense. *Journal of Mammary Gland Biology and Neoplasia*, **1**, 271–282.
Niva, M. (2007). All foods affect health: understandings of functional foods and healthy eating among health-oriented Finns. *Appetite*, **48**, 384–393.
Ogata, Y., Fujita, K., Ishigami, H., Hara, K., Terada, A., Hara, H., Fujimori, I. & Misuoka, T. (1993). Effect of a small amount of 4G-beta-D-galactosylsucrose (lactosucrose) on fecal flora and fecal properties. *Journal of Japanese Society of Nutrition and Food Science*, **46**, 317–323.
Osborne, C.G., McTyre, R.B., Dudek, J., Roche, K.E., Scheuplein, R., Silverstein, B., Weinberg, M.S. & Salkeld, A.A. (1996). Evidence for the relationship of calcium to blood pressure. *Nutrition Reviews*, **54**, 365–381.
Petticrew, M., Watt, I. & Sheldon, T. (1997). Systemic review of the effectiveness of laxatives in the elderly. *Health Technology Assessment*, **1**, 1–52.

Pihlanto-Leppälä, A. (2001). Bioactive peptides derived from bovine whey proteins: opioid and ace-inhibitory peptides. *Trends in Food Science and Technology*, **11**, 347–356.
Pitts, E. (1994). Dairy ingredients: new opportunities in functional foods. *The World of Food Ingredients*, **October/November**, 40–43.
Playne, M., Bennet, L.E. & Smithers, G.W. (2003). Functional dairy foods and ingredients. *The Australian Journal of Dairy Technology*, **58**, 242–264.
Prieto, P.A., Mukeiji, P., Kelder, B., Emey, R., Gonzalez, D., Yun, J.S., Smith, D.F., Moremen, K.W., Nardelli, C., Pierce, M., Li, Y., Chen, X., Wagner, T.E., Cummings, R.D. & Kopchick, J.J. (1995). Remodeling of mouse milk glycoconjugates by transgenic expression of a human glycosyltransferase. *Journal of Biological Chemistry*, **270**, 29515–29519.
Rennie, K.L., Hughes, J.L., Lang, R. & Jebb, S.A. (2003). Nutritional management of rheumatoid arthritis: a review of the evidence. *Journal of Human Nutrition and Dietetics*, **16**, 97–109.
Reusser, M.E. & McCarron, D.A. (1994). Micronutrient effects on blood pressure regulation. *Nutrition Review*, **52**, 367–375.
Riggio, O., Balducci, G., Ariosto, F., Merli, M., Tremiterra, S., Ziparo, V. & Capocaccia, L. (1990). Lactitol in the treatment of chronic hepatic encephalopathy – a randomised crossover comparison with lactulose. *Hepatogastroenterology*, **37**, 524–527.
Roberfroid, M.B. (2000). Defining functional foods. In: *Functional Foods, Concept to Product*. Gibson, G.R. & Williams, C.M. (eds), CRC Press, Boca Raton, FL.
Saarela, M., Hallamaa, K., Mattila-Sandholm, T. & Mättö, J. (2003). The effect of lactose derivatives lactulose, lactitol and lactobionic acid on the functional and technological properties of potentially probiotic Lactobacillus strains. *International Dairy Journal*, **13**, 291–302.
Sampson, H.A. (2004). Update on food allergy. *Journal of Allergy and Clinical Immunology*, **113**, 805–819.
Sancho, F. (2003). Combination of dairy components and other ingredients – exciting new opportunities. *The Australian Journal of Dairy Technology*, **58**, 153–155.
Schonberg, S. & Krokan, H.E. (1995). The inhibitory effect of conjugated dienoic derivatives (CLA) of linoleic acid on the growth of human tumor cell lines is in part due to increased lipid peroxidation. *Anticancer Research*, **15**, 1241–1246.
Seo, J.H., Moon, H.S., Kim, I.Y., Guo, D.D., Lee, H.G., Choi, Y.J. & Cho, C.S. (2008). PEGylated conjugated linoleic acid stimulation of apoptosis via a p53-mediated signaling pathway in MCF-7 breast cancer cells. *European Journal of Pharmaceutics and Biopharmaceutics*, **70**(2), 621–626.
Séverin, A. & Wenshui, X. (2005). Milk biologically active components as nutraceuticals: review. *Critical Reviews in Food Science*, **45**, 645–656.
Shortt, C., Shaw, D. & Mazza, G. (2004). Opportunities for health-enhancing functional dairy products. In: *Handbook of Functional Dairy Products*. Shortt, C. & O'Brien, J. (eds), CRC Press, Boca Raton, FL.
Silva, S.V. & Malcata, F.X. (2004). Caseins as asource of bioactive peptides. *International Dairy Journal*, **15**, 1–15.
Siró, I., Kápolna, E., Kápolna, B. & Lugasi, A. (2008). Functional food: product development, marketing and consumer acceptance – a review. *Appetite*, **51**(3), 456–467.
Smacchi, E. & Gobbetti, M. (2000). Bioactive peptides in dairy products: synthesis and interaction with proteolytic enzymes. *Food Microbiology*, **17**, 129–141.
Steinhoff, U.M., Senft, B. & Seyfert, H.M. (1994). Lysozyme-encoding bovine cDNAs from neutrophile granulocytes and mammary gland are derived from a different gene than stomach lysozymes. *Gene*, **143**, 271–276.
Strohmaier, W. (1998). Lactulose: status of health-related applications. International Dairy Federation, Bulletin no. 9804, 262–271.
Swinbanks, D. & O'Brien, J. (1993). Japan explores the boundary between food and medicine. *Nature*, **364**, 180.
Tomomatsu, H. (1994). Health effects of oligosaccharides. *Food Technology*, **10**, 61–65.
Van Der Meer, R., Bovee-Oudenhoven, I.M.J., Sesink, A.L.A. & Kleibeuker, J.H. (1998). Milk products and intestinal health. *International Dairy Journal*, **8**, 163–170.
Voragen, A.G.J. (1998). Technological aspects of functional food-related carbohydrates. *Trends in Food Science and Technology*, **9**, 328–335.
Warner, E.A., Kanekanian, A.D. & Andrews, A.T. (2001). Bioactivity of milk proteins: 1. Anticariogenicity of whey proteins. *International Dairy Journal*, **54**, 151–153.

8 Probiotics and prebiotics

Anna Sip and Wlodzimierz Grajek

8.1 Introduction

In the last two decades, knowledge on microorganisms and their active role in the maintenance of human health has increased considerably. The first studies on the presence of lactic acid bacteria (LAB) in the intestinal ecosystem were published in the early 1900s (Moro 1900; Beijerinck 1901). However, a real breakthrough was marked by studies by Metchnikoff, who indicated a relationship between the consumption of fermented dairy products and good health and longevity in humans (the Caucasians). A token of acknowledgment of the momentous significance of these studies was the Nobel Prize awarded to Metchnikoff (1908). The conviction of the beneficial effect of LAB on human health has become one of the canons of contemporary knowledge on human nutrition. Benefits resulting from the consumption of products containing LAB have been reported in numerous studies, which have shown that certain LAB strains have additional, unique properties that may have an effect on the functioning of the human organism. Bacteria with such properties are termed 'probiotic'. Probably the first researcher who introduced this term was Vergio (1954); however, the first definition of probiotics was proposed by Fuller (1989), who used this term for food products containing live microorganisms, which apart from their normal nutritive value benefit the health of consumers by promoting an advantageous balance of the microbial population of the gastrointestinal tract. Schrezenmeir and de Vrese (2001) defined probiotics as preparations or foodstuffs containing single or mixed cultures of live microorganisms, which when administered to humans or animals in appropriate amounts have a beneficial effect on their health. Currently, the definition given by the FAO/WHO is commonly adopted, which says that probiotics are 'live microorganisms which when administered in adequate amounts confer a health benefit on the host'.

The beneficial effect of probiotics on the human organism may consist of the improvement of metabolic or physiological processes, as well as medical effects resulting in a reduced risk of incidence of many diseases or their limited duration (Saarela *et al.* 2000; Holzapfel & Schillinger 2002; Saxelin *et al.* 2005; Shah 2007). Products improving human welfare may make enhanced-function claims, while those reducing the risk of disease incidence may make disease risk-reduction claims.

8.2 Probiotic strains

Based on detailed studies, it has been shown that probiotic characteristics are not connected with a specific species of microorganisms but rather they are attributes ascribed to specific and at the same time rare strains (Table 8.1). The most thoroughly investigated probiotic microorganisms are bacteria from the genus *Lactobacillus* (*L. acidophilus, L. casei, L. paracasei, L. rhamnosus, L. fermentum, L. johnsonii, L. reuterii*), *Bifidobacterium* (*B. animals* – present name *B. lactis, B. longum, B. breve*) and the yeast *Saccharomyces bulardii* (Holzapfel & Schillinger 2002; Shah 2007). Most probiotic strains of the above-mentioned microorganisms have already been introduced in the market by dairy companies.

Table 8.1 Examples of microorganisms used as probiotics (Holzapfel & Schillinger 2002; Playne *et al.* 2003; Shah 2004, 2007; Saxelin *et al.* 2005)

Species	Strains
Lactobacillus	
L. acidophillus	La-1/La-5 (Chr. Hansen), NCFM (Rhodia), La1 (Nestle), DDS-1
L. bulgaricus	(Nebraska Cultures), LAFTI®L10
L. casei	Lb12
L. fermentum	Immunitals (Danone), Defensis DN 114 001 (Danone), Shirota
L. helveticus	(Yakult)
L. johnsonii	RC-14 (Urex Biotech), KLD
L. reuterii	B02, L89
L. rhamnosus	LA1 (Nestle)
L. paracasei	ING1
L. plantarum	GG (Valio), HN001
L. reuteri	33 (Uni-President Enterprises Corp.) CRL 431 (Chr. Hansen)
L. rhamnosus	(Probi AB), 299v, Lp01, ATTC 8014 (Valio)
L. salivarius	SD2112 (also known as MM2)
	271 (Probi AB), GR-1 (Urex Biotech), LB21 (Essum AB)
	UCC118
Bifidobacterium	
B. adolescentis	ATTC 15703, 94-BIM
B. amimals (lactis)	Bb-12 (Chr. Hansen), Lafti^{TM}, B94 (DSD), DR 10/HOWARU
B. bifidus	(Danisco), HN019
B. breve	Bb-11
B. essensis	Yakult
B. infantis	Danone (Bioactivia)
B. longum	Shirota, Immunitass, 744, 01
	UCC 35624 (UCCork), SBT 2928, B6, BB536
Other LAB	
Carnobacterium divergens	V41, AS7
Enterococcus fecalis	Unspecified strain
Enterococcus faecium	SF68, M-74
Streptococcus thermophilus	CCRC 14079, CCRC 14085, F4, V3
Streptococcus intetmedicus	Unspecified strain
Non-lactic bacteria	
Bacillus subtilis	Unspecified strain
Propionibacterium freudenreichii ssp. *shermanii*	SJ (Valio)
Yeast	
Saccharomyces boulardii	Unspecified strain

However, intensive studies conducted in recent years are rapidly extending this list of probiotic microorganisms.

Selected strains of *Lactobacillus* and *Bifidobacterium* have found the widest applications in the production of probiotic food. As a rule, it is required for a probiotic strain to be isolated from the human organism. This results from the conviction that 'human' strains are best attached to the intestinal epithelium and most effectively colonize the gut. Moreover, it is suggested that they efficiently reduce the growth of pathogens and act as immunomodulators. However, this opinion has not been sufficiently confirmed and there are known animal-origin strains, which also have a beneficial effect on the human organism, such as *Bifidobacterium animalis (lactis)*. They easily survive the intestinal passage and adhere well to cells of the intestinal epithelium (Playne *et al.* 2003; Shah 2007). It also needs to be stressed that in Japan, human-origin cultures have been used in the production of dairy products for over 40 years, while in Germany for at least 20 years.

The best-known probiotic strain is *Lactobacillus rhamnosus* GG (ATCC 53013). It colonizes the alimentary tract, reduces the activity of fecal enzymes, protects the host against antibiotic-associated diarrhea, reduces the duration of therapy, prevents or treats rotavirus-associated diarrhea, traveler's diarrhea and acute diarrhea, treats Crohn's disease and juvenile rheumatoid arthritis, as well as exhibits antagonistic properties in relation to bacteria causing dental caries (Guandalini *et al.* 2000; Marteau *et al.* 2001; Nase *et al.* 2001; Szajewska *et al.* 2001; Femia *et al.* 2002; Hatakka *et al.* 2003; McFarland 2006).

Another probiotic strain widely discussed in the literature is *L. casei* Shirota. This organism maintains the balance of intestinal microflora, protects the host against intestinal disorders, treats rotavirus-associated diarrhea, reduces the activity of fecal enzymes, protects the organism against food mutagens, is used in adjunctive treatment of bladder cancer and supports the immune system at early stages of colon cancer (Kato *et al.* 1999; Cats *et al.* 2003).

In recent years, products containing *L. casei* Defensis DN 114 001 have been introduced in the market. These products, thanks to the presence of probiotic bacteria exhibiting high survival rates in the stomach and the duodenum, stimulate the immune system, prevent and treat gastrointestinal infections, and reduce the incidence and duration of acute diarrhea in children (Meyer *et al.* 2006). Another thoroughly investigated probiotic microorganism is the bacterium *Lactobacillus johnsonii* (La1) (NCC533). This strain stabilizes intestinal microflora, stimulates the immune system, is effective in the treatment of gastroenteritis, and is antagonistic against *Helicobacter pylori*. Moreover, it is also characterized by strong adhesion to intestinal cells (Granato *et al.* 2004).

Probiotic strains are also found among bifidobacteria. For instance, *Bifidobacterium lactis* DN 173010 exhibits high survival rates in the stomach and the duodenum. It has a positive effect on the shortening of intestinal passage, especially in the elderly (Chouraqui *et al.* 2004). In turn, the strain *Bifidobacterium breve* Yakult protects against food mutagens, maintains balance in intestinal microflora and prevents diarrhea (Shimakawa *et al.* 2003). Probiotic action of those probiotic microorganisms which have been most thoroughly described in the literature is comprehensively presented in Table 8.2.

8.3 Functional properties of probiotics

In-depth and long-term studies are required to prove probiotic properties of a strain, and this process consists of many stages. Usually, basic research is conducted on *in vitro* models with the use of an artificial alimentary tract as well as animal and human cell cultures.

Table 8.2 Beneficial effects of selected probiotics documented in clinical trials (Holzapfel & Schillinger 2002; Saarela *et al.* 2002; Guarner & Malagelada 2003; Saxelin *et al.* 2005; Shah 2007)

Probiotic strains	Health benefits
Lactobacillus rhamnosus GG (Valio)	• Effective in reducing the incidence of diarrheal diseases (rotavirus diarrhea, antibiotic-associated diarrhea, *C. difficile* diarrhea and traveler's diarrhea) • Reduced risk of gastrointestinal disorders such as inflammatory bowel diseases (pouchitis and Crohn's disease) • Reduced incidence of *H. pylori* infections • Reduced incidence of atopic diseases • Prevention of atopic diseases • Inhibition of generation of carcinogenic products by reducing the activity of microbial enzymes • Normalization of intestinal permeability • Reduced risk of respiratory infections • Reduced risk of dental caries
Lactobacillus casei Shirota (Yakult)	• Effective in treatment of rotavirus diarrhea • Reduced risk of gastrointestinal disorders • Reduced activity of procarcinogenic enzymes • Induces production of INF-γ
Lactobacillus casei Defensis DN 114 001 (Danone)	• Shortens the diarrheal phase in children with rotavirus infection • Effective in treatment and prevention of gastrointestinal infections • Maintains constant urease activity • Stimulation of immune system
Lactobacillus acidophilus NCFM (Rhodia)	• Improved lactose metabolism • Reduced risk of colon cancer by limiting DNA damage in colon cells, reducing the activity of procarcinogenic enzymes and binding mutagens • Reduced serum cholesterol level • Prevention of urogenital infections
Lactobacillus johnsonii La1 (Danone)	• Reduced incidence of *H. pylori* infections • Reduction of inflammation • Stimulation of immune system
Lactobacillus plantarum 299v (Probi AB)	• Relief in inflammatory bowel symptoms, e.g. enterocolitis and pouchitis • Reduced recurrence of *C. difficile* enterocolitis • Stimulation of immune system
Lactobacillus reuterii (Stoneyfield, Biogaia)	• Reduced incidence of rotaviral diarrhea in children • Shortens the duration of acute gastroenteritis
Bifidobacterium animalis DN 173 010 (Danone)	• Alleviation of symptoms of atopic eczema in infants with milk hypersensitivity • Reduces the duration of increased stool output in children with diarrheal illnesses
Bifidobacterium bifidum	• Reduces the incidence of rotavirus infections in children and stimulates rotavirus-specific antibody response • Improved lactose digestion
Saccharomyces boulardii	• Reduced incidence of *C. difficile* diarrhea • Shortens the duration of acute gastroenteritis

Table 8.3 A relationship between activity of probiotics and their health effects (Holzapfel & Schillinger 2002; Saarela *et al.* 2002; Guarner & Malagelada 2003; Saxelin *et al.* 2005; Shah 2007)

Activity of probiotics	Physiological effects
Intestinal microbiota consumption	• Reduced incidence of rotavirus-associated diarrhea • Reduced incidence of antibiotic-associated diarrhea • Reduced incidence of traveler's diarrhea • Reduced incidence of bacterial diarrhea • Control of irritable bowel syndrome • Control of inflammatory bowel diseases, e.g. pouchitis and Crohn's disease • Reduced incidence of *H. pylori* infections and complications • Prevention of intestinal infections
Metabolic effects	• Reduced serum cholesterol level • Improvement of lactose tolerance • Reduction in risk factors for colon cancer • Reduced risk of osteoporosis • Improved bioavailability
Immunomodulation	• Alleviation of food allergy symptoms • Alleviation of atopic disease symptoms in infants • Control of inflammatory bowel diseases, e.g. pouchitis and Crohn's disease • Strengthened innate immunity

These studies aim to determine a relationship between a microorganism and the course of a selected physiological process (Table 8.3). On this basis, the first hypotheses are posed on the mechanism of action for a given probiotic strain. The next step comprises studies with an *in vivo* animal model. Laboratory rats and mice are typically used for this purpose, while other animal species are used only occasionally. Results of these studies usually make it possible to develop the final hypothesis concerning the mechanism of action for probiotic bacteria, which is next verified in trials on volunteers where all restrictions applicable to clinical trials are followed (Shah 2007). The above-mentioned multistage studies have to include several aspects related to consumer safety, medical, as well as technological aspects, connected with the production of probiotics and the manner of their administration. It needs to be stressed that a given product may be termed probiotic only on the basis of positive results of clinical trials (Playne *et al.* 2003). In studies on functional food, including those on probiotic microorganisms, a key role is played by the properly assumed and validated biological markers definitely related with the investigated functional characteristic.

The list of known probiotic properties of such bacteria is long (Reid *et al.* 2003; Shah 2004, 2007):

- Synthesis and secretion of important digestive enzymes (e.g. β-galactosidase)
- Production of antibacterial substances inhibiting the pathogen's growth
- Protective effect on and rebuilding of microflora after such gastrointestinal disorders as infection-related diarrhea (traveler's diarrhea, acute viral diarrhea in children), antibiotic-associated diarrhea and radiotherapy-associated diarrhea
- Reduction of cholesterol levels
- Stimulation of the immune system
- Increased colon peristalsis and prevention of constipation
- Reduction of adherence and colonization by pathogens

- Anti-allergenic activity
- Reduction of activity of fecal enzymes responsible for the transformation of procarcinogens into carcinogens (protection against colon cancer)
- Maintenance of continuity of intestinal mucosa

Most of the above-mentioned probiotic properties result from the activity of metabolites produced by probiotic bacteria. In this respect several examples may be presented, such as bacteriocins responsible for the elimination of pathogenic bacteria, peptides for reduction of blood pressure, proteolytic enzymes for the elimination of food allergies, β-galactosidase for lactose digestion, cholesterol oxidase for the reduction of cholesterol, linolic acid conjugates for increasing antioxidant potential, enzymes for eliminating carcinogens and adhesins on the surface of bacterial cells acting as immunostimulants (Fooks & Gibson 2002; Saxelin *et al.* 2005; Shah 2007).

A product may be considered functional and introduced in the market with a specific health claim on condition it contains cells of a microorganism exhibiting at least one of the above-mentioned and clinically confirmed probiotic properties (FDA/WHO 2002).

8.4 Medical applications

Probiotic bacteria are also used for therapeutic purposes. The mechanism of their action consists of the inhibition of pathogenic and toxin-forming bacterial growth, their effect on metabolism and on immunomodulation (Table 8.3). In recent years, intensive clinical and epidemiological trials have been conducted by numerous research institutions and international teams, with the results of their studies being published in prestigious scientific periodicals. Thus, reliable documentation has been collected presenting the therapeutic potential of probiotics. Therapeutic potential of probiotics in relation to different diseases and disorders of the alimentary tract is given below (Table 8.4).

8.5 Gastrointestinal infections of different etiology

8.5.1 Rotavirus and bacterial diarrhea

Diarrhea is most typically the effect of infections of the alimentary tract caused by viruses or bacteria. Among viral diarrhea cases, those caused by rotavirus and calcivirus predominate (Mrukowicz *et al.* 1999). In turn, the most frequent cases of bacterial diarrhea are infections caused by *Campylobacter jejuni*, *Salmonella (S. typhimurium* and *S. enteritidis)*, *Shigella*, *Yersinia*, as well as enterotoxigenic *Escherichia coli*. Diarrhea may also be a consequence of the administration of antibiotics (the so-called antibiotic-associated and antibiotic-dependent diarrhea). Diarrhea associated with antibiotic therapy is usually mild. An exception in this respect may be diarrhea caused by *Clostridium difficile*, which may lead to pseudomembranous colitis (Kyne & Kelly 2000; Barbut *et al.* 2001).

From an epidemiological and economic perspective, the biggest problem is posed by viral diarrhea, especially rotavirus-associated diarrhea. Rotaviral diarrhea is frequently acute and leads to damage to mature enterocytes, which may result in lactose intolerance (Isolauri *et al.* 1994; Ciarlet & Esters 2001). The application of certain probiotics in combination with fluid therapy makes it possible to markedly shorten the duration of diarrhea, particularly

Table 8.4 Selected examples of diseases in which prevention and treatment probiotics may be applied (Holzapfel & Schillinger 2002; Saarela *et al.* 2002; Guarner & Malagelada 2003; Saxelin *et al.* 2005; Shah 2007)

Reduced disease risk	Probiotics
Diarrhea in children, mainly rotavirus-associated diarrhea	*L. rhamnosus* GG *L. reuteri* *L. acidophilus* *L. casei* subsp. *rhamnosus* (Lacidophilus) *L. delbrueckii* subsp. *bulgaricus* (Yalaclu) *S. thermophilus* +*B. bifidum* *B. bifidum* + *B. infantis* *S. boulardii*
Antibiotic-associated *C. difficile* diarrhea	*L. rhamnosus* GG *S. boulardii*
Traveler's diarrhea	*L. rhamnosus* GG *L. fermentum* KLD *L. acidophilus* (unspecified strain) *L. acidophilus* + *L. bulgaricus* *L. bulgaricus* + *B. bifidum* + *S. thermophilus* *S. boulardii*
Bacterial diarrhea	*L. rhamnosus* GG *L. acidophilus* *L. plantarum* *B. bifidum*
Bacterial gastroenteritis	*L. rhamnosus* GG *L. reuteri* *E. faecium* SF68 *S. boulardii*
Inflammatory bowel diseases, e.g. pouchitis and Crohn's disease	*L. rhamnosus* GG VSL#3 (containing four strains of lactobacilli, three strains of bifidobacteria and one strain of *S salivarius* subsp. *thermophilus*) *L. reuterii* *L. salivarius* UCC118 *B. longum infantis* UCC35624 *S. boulardii*
Irritable bowel syndrome	*L. plantarum* 299 V
Hypercholesterolemia	*L. acidophilus* (unspecified strain) *L. plantarum*
Food allergies and atopic diseases	*L. rhamnosus* GG *L. paracasei* F19 *B. lactis* Bb-12
Lactose intolerance	Different LAB starters
Colon cancer	*L. acidophilus* *L. casei* Shirota *L. rhamnosus* GG *B. longum* *Propionibacterium* sp.
H. pylori infections and complications	*L. acidophilus* La1 *L. johnsonii* *L. gasseri* LG21 *L. reuterii* *L. salivarius* *L. casei* *Bifidobacterium*

acute rotavirus-related diarrhea in infants and young children (Saavedra *et al.* 1994). To date, the highest efficacy in the treatment of acute rotavirus-related diarrhea has been reported in young children when using *L. rhamnosus* GG and *B. lactis* Bb-12 (formerly referred to as *B. animals* Bb-12). For instance, supplementing orally administered rehydration fluids with a probiotic *L. rhamnosus* GG reduced the duration of rotavirus-related diarrhea by 18–32 hours and reduced its intensity (Szajewska & Mrukowicz 2001; Van Niel *et al.* 2002). The beneficial action of *L. rhamnosus* GG resulted from increased synthesis of secreted immunoglobulin IgA in the Peyer's patch.

Because many probiotic strains inhibit enteropathogenic growth and/or reduce their adhesion to the intestinal epithelium, probiotics may also be useful in treatment of bacterial diarrhea. However, the efficacy of their action is lower than in the case of viral diarrhea. For example, *L. rhamnosus* GG reduces only by 14–26 hours the duration of bacterial diarrhea in children, connected with the excretion of stools containing blood, pus and mucus, and moderately alleviates its symptoms (Guandalini *et al.* 2000). To date, the effectiveness of probiotics in the treatment of infectious diarrhea has not been confirmed in adults. Moreover, there are no reliable results concerning the role of probiotics in the treatment of traveler's diarrhea. Most authors did not observe any effect of the administration of probiotics containing *L. fermentum*, *L. acidophilus* or *Lactobacillus bulgaricus* on the course of traveler's diarrhea. Only a study by Hilton *et al.* (1997), in which Danish and Finnish tourists were examined during their 2-week stay in Egypt, showed that the consumption of lyophilized preparations containing live bacteria *L. acidophilus, B. animals, L. delbrueckii* ssp. *bulgaricus, Streptococcus thermophilus* at a concentration of approximately 10^9 colony-forming units (cfu)/day and preparations of *L. rhamnosus* GG reduced the incidence of diarrhea.

Probiotics are also effective in the treatment of diarrhea caused by *C. difficile*. These bacteria are found in small numbers in the alimentary tract of healthy humans. The application of antibiotics, especially clindamycin, aminopenicillin and cephalosporins, causes changes in the composition of intestinal microflora and leads to increased proportions of toxin-forming *C. difficile*. Toxins produced by *C. difficile* (enterotoxin A and cytotoxin B) cause severe diarrhea (the so-called antibiotic-associated diarrhea) and colon mucosal edema (Barlett 2002; Beaugerie *et al.* 2003). Symptoms of *C. difficile* infection also appear occasionally in patients who had not been treated with antibiotics and in such cases they are usually the effect of hospital infections. Probiotics play an important role in the treatment of diarrhea caused by *C. difficile* infection. In clinical trials, a beneficial effect of preparations containing *Saccharomyces boulardi* and *L. rhamnosus* GG was shown. The application of these probiotics markedly improved physiological parameters in patients with symptoms related to the activity of *C. difficile* (McFarland *et al.* 1994; Surawicz *et al.* 2000). Antidiarrheal action was also observed for other probiotic strains, such as *L. reuteri*, *L. acidophilus* LB, *L. plantarum*, *L. casei* DN-114 001, *Bifidobacterium bifidum* and *S. thermophilus* (Cremonini *et al.* 2002; Wullt *et al.* 2003; Plumer *et al.* 2004; McFarland 2006). Methodological irregularities in trials with the use of the above-mentioned probiotics, as well as highly diverse populations of examined patients, make it impossible to formulate definite conclusions concerning their role in the treatment of diarrhea caused by *C. difficile*.

Some authors have suggested a potential application of probiotics in the prevention of diarrhea, especially associated with antibiotic therapy. Currently available data may not be considered as fully justifying such an application for probiotics.

Despite numerous clinical trials, the mechanism of action of probiotics in the treatment of diarrhea with different etiology has not been thoroughly clarified. However, it is believed that the antidiarrheal action may result from (1) the production of antimicrobial substances, especially bacteriocins and organic acids, (2) competition for nutrients, (3) blockage of adhesion sites, (4) inhibition of toxin formation and blockage of their receptors, (5) cytoprotective effect on the intestinal mucosa, (6) stimulation of synthesis of immunoglobulin IgA and/or (7) stimulation of type Th1 response, by the stimulation of formation of cytokinins Il-12, Il-2 and particularly INF-γ (Levy 2000; Isolauri *et al.* 2001).

8.6 Colitis

8.6.1 Inflammatory bowel disease – ulcerative colitis and Crohn's disease

In the last decade, the number of patients suffering from chronic enteritis, particularly ulcerative colitis and Crohn's disease, has increased considerably. These diseases, in approximately 20% of cases, already start during childhood. Despite numerous studies, it still has not been clarified whether the above-mentioned diseases are separate disease entities or only two forms of one disease, with different locations and rate of changes in the intestinal wall. Probably, both genetic factors (chronic inflammatory conditions occur most frequently in individuals genetically predisposed to such disorders) and the composition of intestinal microflora are responsible for the incidence of the above-mentioned diseases and their course (Shanahan 2000; Marteau *et al.* 2001; Swidsinski *et al.* 2002). To date, however, no pathogens specific to ulcerative colitis have been identified. In contrast, there is an extensive body of scientific evidence indicating a relationship between the presence of *Mycobacterium avium* subsp. *paratuberculosis* in the intestines and Crohn's disease. Immunological properties of the intestinal mucosa have a strong effect on the intensity of symptoms accompanying Crohn's disease. In healthy individuals, the immune system tolerates the commensal intestinal microflora and the appearance of pathogenic microflora triggers a defense reaction. In the case of non-specific enteritis, mechanisms regulating the response of the mucosa to intestinal microflora are disturbed, and this in turn leads to inflammations in the intestinal wall. It was shown that the immune response of the intestinal mucosa to the presence of intestinal microflora depends on the profile of secreted cytokinins. In Crohn's disease, this means T_H1-associated cytokinins, i.e. interferon-γ, tumor necrosis factor INF-α and internalin 12 (IL-12). In turn, in the case of ulcerative enteritis, an enhanced response of T_H2 was observed, connected with the production of internalins 5 (IL-5) and 10 (IL-10). Activated immunocytes, together with cytokinins, also produce numerous mediators of the inflammation process (leukotrienes, thromboxanes, oxygen radicals, nitrite oxide). As a consequence of their action, tissue destruction and fibrosis occur, which is especially apparent in Crohn's disease (Brandtzaeg *et al.* 1989; Dianda *et al.* 1997; Shanahan 2002). Because there is a growing body of evidence confirming a strong relationship between intestinal microflora, immune response and the occurrence of inflammations (i.e. intestinal microflora induces the incidence of inflammations), attempts have been made to apply probiotics in the treatment of non-specific enteritis (Camilleri 2006).

The first such trials were conducted on animals. McCarthy *et al.* (2003) showed the anti-inflammatory action of bacteria *Lactobacillus salivarius*. In mice administered with a probiotic, intestinal inflammatory lesions were observed to decrease. Similar results were reported by Madsen *et al.* (2001). These researchers, as a consequence of the administration of probiotic VSL#3, found a reduction in the secretion of pro-inflammatory cytokinins, TNF-α and INF-γ, and observed enhanced integrity of the mucosa, as well as an improved histological picture of the intestine in mice with a deficit of the Il-10 gene. The administered probiotic VSL#3 was a combination of three *Bifidobacterium* strains (*B. longum, B. breve and B. infantis*), four *Lactobacillus* strains (*L. casei, L. plantarum, L. acidophillus, L. bulgaricus*) as well as one strain of *Streptococcus salivarius*. This preparation was more effective than the individual strains it contained when acting independently, probably due to the potential summation of the anti-inflammatory action of its components. The suitability of probiotic VSL#3 in the treatment of recurrent enteritis was also confirmed in clinical trials on humans. When administering probiotic VSL#3, remission of ulcerative enteritis was achieved for a period of 1 year in 17 of 20 patients. The administration of this probiotic also limited the recurrence of pouchitis. In feces of individuals treated using probiotic VSL#3, throughout the entire treatment period, the presence of bacteria typical of the applied probiotic was detected. In turn, no marked changes were observed in the counts of the other intestinal microflora (Hart *et al.* 2003).

Another method to affect the composition of intestinal microflora, which may prove suitable in the treatment and prevention of remissions of enteritis, is to use prebiotics.

8.7 Functional bowel disorders

8.7.1 Constipation

Approximately 20% of patients coming to gastrological outpatient clinics report problems with bowel movements. In over 90% of cases, constipation is functional in character and results from low dietary fiber, carbohydrate and liquid intake levels. Retention of fecal mass in the intestines disturbs the balance of the intestinal microflora (leading to an increase in *Clostridium* counts), and increases the intensity of putrefaction processes and synthesis of sulfides, which are toxic for the colon epithelium. Restoration of balance in the intestinal microflora, particularly the proportion of LAB from genus *Lactobacillus* and *Bifidobacterium*, is a key factor in the treatment of constipation. LAB produce short-chain fatty acids (SCFAs), which improve intestinal motor activity (stimulating muscle contractions in the intestinal wall) and as a result accelerate the removal of chyme. Some SCFAs are absorbed by the intestinal mucosa, while some are used by LAB for the formation of biomass. An increase in the LAB population in fecal deposits results in an increase in their mass. This also leads to the loosening of fecal structure. Such an effect is believed to stem from a high water content (~80%) in LAB biomass and the ability of these bacteria to synthesize gases, which facilitate feces evacuation by making it spongier. The easiest method to increase counts of *Lactobacillus* and *Bifidobacterium* in the intestines, having a beneficial effect in the case of constipation, is to supplement the diet with probiotic products, such as bio-yogurt. Studies have shown that the application of a diet enriched with bio-yogurt reduces the time of intestinal passage (reduces intervals between bowel movements) (Ouwehand *et al.* 2002a; Koebnick *et al.* 2003). Efficacy of constipation treatment may also be enhanced by the application of prebiotics. Some of them (e.g. lactulose or oligosaccharides) have a strong laxative effect. They increase osmolarity of feces (by binding water) and as a result contribute to feces loosening.

8.7.2 Irritable bowel syndrome

Irritable bowel syndrome (IBS) is a functional disorder of the colon, accompanied by abdominal pain, bloating and problems with bowel movements (alternately constipation and diarrhea). In developed, industrialized countries, this problem is found in as many as 10%

of inhabitants (Camilleri 2001). Although IBS is neither a threat to human life nor promotes the development of other serious diseases, this disorder markedly deteriorates the quality of life. IBS is caused by changes in intestinal motor activity and disorders in the functioning of neurotransmitters, especially an excessive reaction to stimuli, caused by an abnormal functioning of the intestines as well as infections (Horwitz & Fisher 2001). The above-mentioned factors are directly connected with changes in the intestinal microflora (especially excessive proliferation of certain groups of bacteria in the small intestine) and the course of 'abnormal' fermentation, resulting in the production of considerable amounts of gases (King *et al.* 1998; Noback *et al.* 2000). Symptoms of IBS are also enhanced by one's tendency to swallow air (Haderstorfer *et al.* 1989). Due to a lack of effective and safe medication for IBS and the relationship between the accompanying disorders and the intestinal microflora, modification of intestinal microflora using selected probiotic strains is an interesting therapeutic option for patients suffering from this syndrome. Many authors through clinical trials have already confirmed the efficacy of the application of probiotics and their effect on the normalization of intestinal microflora composition and activity (evidence found in reports from 19 randomized controlled clinical trials, conducted on 1628 patients). Very promising results were obtained in clinical trials with the use of the probiotic strain *L. plantarum* 299v. It was shown that this strain much more effectively than other *L. plantarum* strains colonizes the mucosa of the small intestine. As a result, this markedly normalizes the rhythm of bowel movements, and reduces abnormal fermentation and related gas production. In 95% of patients, after 4 weeks of oral administration of a probiotic *L. plantarum* 299v, a definite improvement of physiological parameters was observed, especially in terms of a reduced incidence of bloating connected with breaking wind. This effect was also maintained for as long as 12 months after the completion of probiotic administration (Nobaek *et al.* 2000; Niedzielin *et al.* 2001).

8.8 Disorders in lipid metabolism

8.8.1 Atherosclerosis, dyslipidemia, hypertension

Disorders in lipid metabolism leading to hypercholesterolemia are the main causes of atherosclerosis, considered to be the primary civilization-related disease. One of the possible methods to prevent atherosclerosis is to apply a low-calorie and low-fat diet. A strict low-fat diet is troublesome and not always, especially in terms of the taste (sensory perception), accepted by consumers. A much more convenient solution, making it possible to reduce levels of cholesterol and triacylglycerols, i.e. compounds being the primary causes of atherosclerosis, is to apply a diet comprising probiotic products. Numerous clinical trials have confirmed the beneficial effect of probiotics on lipid metabolism and stressed its complex, frequently multifaceted (multifactorial) character. They have suggested that a reduction of cholesterol concentration, particularly that of LDL cholesterol, may be associated with its increased excretion with feces, a possible cause of this phenomenon being the capacity of probiotics to inhibit the formation of readily digestible lipid micelles. This suggestion comes from studies indicating increased cholesterol concentration in feces of individuals fed diets high in probiotic products (Fukushima & Nakano 1995; Ouwehand *et al.* 2002b). Some authors have also suggested that probiotic strains of *L. acidophilus* may be capable of anaerobic cholesterol metabolism (Liong & Shah 2005). However, there is a lack of studies in the available literature which would explicitly explain the mechanism of probiotic action on the metabolism of cholesterol and bile acids. The role of prebiotics in lipid metabolism has been documented much more thoroughly. To date, most conclusions concerning the mechanism of prebiotic action on lipid metabolism have been formulated on the basis of clinical trials on animals and humans. For example, in rats, after a 5-week inulin administration, a considerable reduction was observed in triacylglycerol concentrations (Delzenne *et al.* 1993). Analogous results (reduction of triacylglycerol concentration by 27% and cholesterol concentration by 5%) were reported when administering a 4-week inulin-rich diet to humans (Canzi *et al.* 1995). In turn, Schafsma *et al.* (1998) showed that total cholesterol and LDL cholesterol levels could be lowered in humans as a result of administering milk fermented by *L. acidophilus* with the addition of fructo-oligosaccharides. In order to explain the causes for the reduction of triacylglycerol concentrations, studies were conducted on the biochemical and physiological levels. Results of these studies suggest that prebiotics have an effect on hepatic metabolism and inhibit synthesis of lipogenic enzymes, i.e. acetyl-CoA carboxylase, acetyl-CoA synthetase and glucose 6-phosphate dehydrogenase. This is confirmed by the results of *in vitro* hepatocyte cultures, in the course of which reduced utilization of palmitic and acetic acids for lipid synthesis was observed under the influence of prebiotic administration (Kok *et al.* 1996). In the reduction of lipid levels, the catabolic mechanism may also play a significant role. For example, oligofructose accelerates lipid catabolism and thus exhibits a hypovolemic action. Moreover, the administration of prebiotics also reduces glycemia and insulinemia (Kok *et al.* 1998). This may be connected with the effect of prebiotics on the kinetics of carbohydrate adsorption. Since products of prebiotic metabolism include SCFA, e.g. acetic and propionic acids, it is assumed that these compounds may also affect lipid metabolism. It was shown that propionic acid inhibits the synthesis of fatty acids, whereas acetic acid has a lipogenic action (Nishina & Freeland 1990).

Moreover, information may be found in available literature indicating a potential effect of probiotics on blood pressure. For instance, a marked reduction of systolic blood pressure, as well as LDL cholesterol, insulin and leptin levels, was observed in smokers after a 6-week administration of a preparation containing probiotic bacteria *L. plantarum* (Naruszewicz *et al.* 2001). Due to the effect of probiotics and prebiotics on lipid metabolism, some authors have also suggested that a diet containing synbiotics may successfully limit the effects of obesity. To date, such a diet has been shown to exhibit a hepatoprotective action. In genetically obese mice, subjected to a 10-week diet treatment using products enriched with oligofructoses, hepatic triacylglycerol level was reduced by as much as 57% (Daubioul *et al.* 2000).

8.9 Disorders of calcium and phosphate metabolism

8.9.1 Bone tissue defects, lowered bone strength, osteoporosis

Maintenance of an appropriate calcium balance is especially important during growth, menopause and aging. Calcium deficits in the diet lead to increased resorption of bone tissue. Disturbed balance between resorption and osteogenesis in favor of resorption results in irreversible changes in bone structure and leads to their excessive fragility (osteoporosis). Consumption of foodstuffs containing calcium even in considerable amounts is not always sufficient to ensure an adequate calcium balance due to the low bioavailability of this element. One of the methods to enhance bioavailability and calcium absorption is to consume products rich in oligosaccharides (prebiotic products). It has been shown that SCFAs, such as acetic, propionic or butyric acids, as well as hydroxy acids – primarily lactic acid, formed in the large intestine as a result of oligosaccharide fermentation – lower pH in the large intestine and thus result in the dissolution of calcium–phosphate–magnesium complexes formed during chyme passage through the small intestine. Thus, the amount of ionized calcium, i.e. its absorbable form, is increased. Oligosaccharides also have a beneficial effect on the process of calcium absorption both via both passive and active transport. This increases the size of intestinal crypts, the number of epithelial cells per crypt, blood flow in the intestinal circulation and stimulates the expression of calcium-binding protein in the course of active transport (Scholz-Ahrens *et al.* 2001).

In the literature there are many studies confirming the positive effect of diet supplementation with oligosaccharides (i.e. inulin, oligofructose, lactulose and transoligosaccharides) on the absorption and retention of calcium, both in intensively growing boys (aged 14–16 years), young men (~21 years old) and post-menopausal women (Coudrary *et al.* 1997; van den Heuvel *et al.* 2000; Scholz-Ahrens *et al.* 2001). An advantageous dependence was observed between the amount of oligosaccharides introduced to the diet and their positive effect on calcium absorption and retention (an increase in calcium absorption by 12–58%). For instance, the consumption of products containing 1000 mg calcium and 10 g lactulose may increase the amount of absorbed calcium by approximately 50 mg/day. This amount is sufficient to improve the negative calcium balance in post-menopausal women and as a consequence to inhibit the development of osteoporosis (van den Heuvel *et al.* 1999). Many authors have also stressed that oligosaccharide supply does not enhance excretion of calcium with urine, which may suggest intensified incorporation of this element into bones and/or indicate inhibited resorption of the bone tissue (van den Heuvel *et al.* 2000).

Since oligosaccharides improve calcium bioavailability, and as a consequence also enhance the density of bone material, bone strength and breaking strength, prebiotic products may be recommended as factors eliminating disorders connected with negative calcium balance in the human organisms.

In contrast, to date, no probiotic effect has been shown in clinical trials on the application of probiotics on calcium and phosphate metabolism and on bone metabolism.

8.10 Food allergy

Allergic reactions to food, of which the pathophysiological basis is a specific immune mechanism, are found in 5.4–9% infants and 1.4–2.4% adults. Probiotics may also be useful in the treatment of food allergies due to their effect on the immune system and intestinal microflora. Clinical trials conducted by many researchers on infants suffering from atopic dermatitis have shown that oral administration of probiotics *Lactobacillus* strain GG and *B. lactis* for at least 2 months reduced the intensity and extent of atopic dermatitis assessed using the SCORAD (score for atopic dermatitis) scale. Irrespective of the type of feeding (breast or bottle feeding), in every second child treated with probiotics for a period of 2 months the SCORAD value dropped by at least 10 points (Majamaa & Isolauri 1997; Isolauri *et al.* 2000).

Moreover, it was shown that some probiotics have not only a therapeutic but also a preventive effect, i.e. they prevent food allergies. Long-term preventive administration of preparations containing *L. rhamnosus* GG to pregnant women with a positive family history of atopy, and then to their children, reduced the risk of atopic dermatitis in children (Kalliomakki *et al.* 2001). Moreover, the administration of *L. rhamnosus* GG and *B. lactis* Bb-12 reduced the risk of allergy to cow milk protein (Majamaa & Isolauri 1997; Isolauri *et al.* 2000).

To date, the mechanism of probiotic action on food allergy has not been fully clarified. Probably, it is the effect of the action of many factors, among which the following are mentioned most commonly: (1) reduced permeability of the intestinal mucosa (Isolauri *et al.* 1993), (2) reduced adhesion of intestinal pathogens to the intestinal epithelium resulting in reduced incidence of inflammation in the intestines (Isolauri *et al.* 1994) and (3) stimulation of synthesis of secretory immunoglobulin IgA in the Peyer's patch, participating in the elimination of food allergens penetrating from the mucosa to the intestinal lumen (Kalia *et al.* 1992). Moreover, other mechanisms of probiotic action have also been investigated, including their effect on changes in immunogenicity of food allergens, stimulation of Th1-type response, inhibition of Th2-type response, stimulation of production of interleukin-10, antagonistism toward IgE and enhancement of food tolerance by probiotics, thanks to the stimulation of TGF-β synthesis. The latter mechanism may play a key role in allergy prevention. It is believed that an insufficient TGF-β production in the neonatal period increases susceptibility to sensitization by low allergen doses. Moreover, TGF-β inhibits Th2-type response (Kirjavainen *et al.* 1999; Pessi *et al.* 2000; Isolauri *et al.* 2001; Kalliomakki *et al.* 2001).

8.11 Metabolic disorders

8.11.1 Lactose intolerance

In most cases, lactose intolerance, i.e. impaired lactose absorption (metabolism), is a genetically determined disease and its symptoms (cramp-like abdominal pain, bloating and/or breaking excess wind, diarrhea) appear gradually and are intensified with age. Lactose intolerance (the so-called secondary lactose intolerance) may also occur as a result of damage or decline in brush border erythrocytes, which are responsible for the synthesis of β-galactosidase (lactase), an enzyme essential for lactose hydrolysis. Damage to the small intestine epithelium is usually caused by alimentary infections (acute or chronic infection-associated diarrhea, parasitic infections, especially giardiases and cryptosporidioses) or inflammations of different etiology. Unabsorbed lactose increases osmotic pressure in the intestinal lumen, which stimulates the transfer of water and electrolytes to the gut, leading to diarrhea, while hydrogen produced in the intestines causes bloating (Shah *et al.* 1992; Shah 1993). In order to eliminate the effects of lactose intolerance, it is recommended to eliminate milk and dairy products from the diet. However, this solution is connected with a reduced supply of protein and calcium. In turn, lactose may be fermented by LAB. In the available literature there is an extensive body of evidence, confirming a potential reduction of lactose level by its partial fermentation. Moreover, many authors also stress that the consumption of fermented dairy drinks, especially those containing probiotic LAB (e.g. *L. acidophilus*), reduces the intensity of lactose intolerance symptoms (Shah 2000). Patients with lactose intolerance may thus consume dairy products, containing partly digested lactose, such as yogurts, hard cheeses or milk chocolate. Increased intestinal proportions of LAB capable of β-galactosidase synthesis, thanks to the consumption of probiotic products, also alleviate the effects of lactose intolerance (Shah *et al.* 1992).

8.12 Cancer

The development of cancer is the outcome of genetic and environmental factors. Moreover, the role of intestinal microflora in carcinogenesis is being increasingly stressed, including its direct relationship with neoplasm formation, especially colon cancer. Treatment of the intestinal microflora as a potential source of neoplasms is connected with its capacity for the synthesis of enzymes that are precursors of carcinogenic substances (Commane *et al.* 2005). Probiotics are more and more frequently listed among factors potentially resulting in a reduced activity of enzymes initiating carcinogenesis. In studies on animals exposed to the action of carcinogens, a beneficial effect of *L. acidophilus* and *Bifidobacterium longum* was shown (Rafter 1995; Pool-Zobel *et al.* 1996; Singh *et al.* 1997; Wollowski *et al.* 2001). It was found that *L. acidophilus* and *Bifidobacterium* spp. reduce the activity of β-glucuronidase, nitroreductase and azoreductase, enzymes responsible for the activation of procarcinogens, and as a consequence reduce the risk of cancer development (Ling *et al.* 1992, 1994). Substances responsible for this process include the SCFAs that they produce. Moreover, probiotic bacteria regulate the production of interferon-γ – they activate macrophages and natural killer cells, and thanks to this, they normalize the composition of intestinal microflora (they reduce the population of bacteria producing procarcinogenic enzymes). What is more, a synergistic action was found for *Bifidobacterium* and oligofructoses, which in turn may suggest the application of a synbiotic based on these components in the prevention of colon cancer (Gallaher & Khil 1999). Carcinogenesis may also be modified by prebiotics themselves. In trials on rats it was shown that oligofructose and inulin stimulate the growth of *Bifidobacterium*, while components of their cell walls inhibit the development of the tumor and limit the formation of aberrative crypt foci (Gibson *et al.* 1995). The potential reduction of hepatic lipogenesis, thanks to the administration of inulin and oligofructose, may also be important in the treatment of colon cancer. For their development, malignant cells require the endogenous synthesis of fatty acids.

8.13 Other disease entities

8.13.1 Gastritis type B, gastric and duodenal peptic ulcers

The bacterium *H. pylori* is an etiological factor in gastritis type B as well as gastric and duodenal ulcers (Armuzzi *et al.* 2001; Sakamoto *et al.* 2001). Antibiotic therapy effectively destroys these bacteria, but its application may also cause undesirable symptoms (bloating, diarrhea, taste disturbances) and initiate the formation of antibiotic-resistant microorganisms. The antibiotic therapy may be supplemented with probiotics, as *L. johnsonii* La1 and *Lactobacillus gasseri* OLL2716 control the colonization by *H. pylori* and reduce inflammation (Felley *et al.* 2001). The capacity to inhibit the growth of *H. pylori* has also been observed in the bacteria *L. casei* Shirota and *L. acidophilus* (Cats *et al.* 2003). These probiotics do not destroy *H. pylori*, but in patients after antibiotic therapy they reduce bacterial load and as a consequence minimize gastrointestinal side effects.

8.13.2 Hereditary angiovascular edema

Hereditary angiovascular edema is a disease caused by the defect of protein regulating the initial stage of the C1-inhibitor complement cascade. Symptoms of this disease are transient, non-itchy edema of the skin, larynx, alimentary tract and brain. Due to the danger of suffocation during larynx edema, this disease constitutes a life threat for humans. Because probiotics, especially probiotic LAB, may induce the synthesis of interferon-γ, which in turn stimulates the synthesis of C1-inhibitor, attempts have been made to apply them in the treatment of hereditary angiovascular edema. To date, promising results have been obtained when administering yogurt containing probiotic bacteria *L. acidophillus* to patients over a period of 6 months. Clinical symptoms of angiovascular edema were found to decrease and serum concentrations of complement C4 and C1 inhibitor were recorded. Thus, these results suggest potential applications of products containing probiotic LAB as adjunctive factors in the treatment of hereditary angiovascular edema (Jaworska *et al.* 2002).

8.14 Selection of probiotic strains

Selection of strains exhibiting probiotic properties is a highly complex and time-consuming procedure (Klaenhammer & Kullen 1999; Saarela *et al.* 2000; Mishra & Prasad 2005). First, it is necessary to determine whether among the isolated microorganisms there are strains exhibiting any of the known probiotic properties. Next, the biological mechanism, from which a given trait originates, is identified. Moreover, safety measures need to be considered and finally the probiotic strain has to meet specific technological requirements for the probiotic product to be easily produced and attractive for the consumer.

Among crucial selection criteria, we need to take into consideration is competitiveness in relation to microflora colonizing the intestinal ecosystem. This is a necessary condition for the introduced probiotic strain to remain in the alimentary tract of the host over a longer period. Another desirable feature is the capacity for survival, growth and metabolic activity in its destination, typically the colon. This is connected with the capacity to produce antimicrobial substances.

In the course of the passage of a food product through the alimentary tract, bacteria it contains are exposed to numerous destructive factors, such as low pH in the stomach and high pH in the colon, the presence of digestive enzymes, including proteases, as well as contact with bile salts, which are strong emulsifiers. Thus, the basic selection criteria in the case of probiotics include tests of resistance to low pH, digestive enzymes and bile salts (Mishra & Prasad 2005).

An essential role is played by the adhesion properties of isolated strains. Strong adhesion to the intestinal epithelium extends the time during which a given strain remains in the alimentary tract and offers a better chance for humans to be protected against contact with pathogens. A long-lasting colonization of the alimentary tract by probiotic microorganisms is also promoted by their resistance to bacteriocins, acids and other antagonistic compounds produced by the endogenous microflora colonizing the intestinal ecosystem (Fooks & Gibson 2002).

The most advantageous trait of probiotic bacteria is their antagonistic activity toward such pathogens as *Salmonella* sp., *Shigella*, *Listeria monocytogenes*, *C difficile*, *Staphylococcus aureus* and *H. pylori*. This characteristic is especially useful in medical applications (Fooks & Gibson 2002).

Studies on the selection of probiotic strains and confirmation of their probiotic properties comprise three important stages:

1. *In vitro* trials conducted in terms of both their safety and their probiotic effects
2. *In vivo* trials on animal models in order to explain the mechanism of their probiotic action
3. Clinical trials on selected groups of volunteers (Reid *et al.* 2003)

Procedures for selecting probiotic strains were described in the Report of the Joint FAO/WHO Expert Consultation on Evaluation of Health and Nutritional Properties of

Probiotics in Food Including Powdered Milk with Live Lactic Acid Bacteria (Cordoba, Argentina, 1–4 October 2001) and in the Report of the Joint FAO/WHO Working Group on Drafting Guidelines for the Evaluation of Probiotics in Food (London, Ontario, Canada, 30 April, 1 May, 2002).

The first step in the selection procedure comprises a thorough identification of a candidate probiotic strain. It has to take into consideration both phenotypic and genetic tests. In the determination of genus affiliation, phenotypic trait evaluation is usually applied, i.e. morphology, API tests (miniaturized rapid biochemical tests) and culture on selective media. Such analyses should be verified using molecular methods with the application of genus-specific DNA–DNA probes. Next, the species affiliation is determined using DNA–DNA hybridization (recommended) and the analysis of 16S rRNA region sequences (for the characteristics of individual strains the recommended procedure is pulse field gel electrophoreses or RAPD – less reproducible than pulsed field gel electrophoresis). On this basis, the taxonomic affiliation of a given strain is determined (Gardiner *et al.* 2002; FDA/WHO 2002). The name of the identified probiotic strain needs to be consistent with the accepted list of bacterial nomenclature available at http://www.bacterio.cict.fr. In recent years, safety and non-pathogenicity of a new strain are considered to be of special importance. It is required for a probiotic strain to have a 'history of safe use' (Salminen *et al.* 1999).

The second step in studies on probiotic strains includes *in vitro* trials confirming the safety of application of a given strain and suggesting the mechanism of its probiotic activity.

In the course of these trials, tests should be conducted in order to determine:

- survival of a given strain at low pH and when exposed to bile salts,
- adhesability to mucin and cells of the intestinal epithelium,
- antagonistic activity toward pathogenic bacteria,
- capacity to reduce adhesion of pathogen cells to the intestinal epithelium,
- bile salts resistance (probiotics do not deconjugate bile salts because deconjugation would be a negative trait in the small bowel) (Dunne *et al.* 1999; FDA/WHO 2002).

In these studies, models of artificial alimentary tracts are used along with *in vitro* cultures of intestinal epithelium cell lines. Caco-2 and HT-29 cell lines are applied most commonly when investigating cell adhesion (Blum *et al.* 1999).

The third step in the identification of probiotic bacteria comprises trials on probiotic safety. Theoretically, probiotic microorganisms may be responsible for four undesirable effects:

1. Systemic infections
2. Harmful metabolic activity
3. Side effects of immunomodulation
4. Transfer of undesirable genes (Marteau 2001; FDA/WHO 2002)

When examining the safety of application for a given probiotic strain, the history of its safe application is also investigated, including its non-pathogenicity. Information on links with infectious diseases is searched for; it is determined whether the strain is capable of bile acid cleavage (an undesirable characteristic), whether it produces toxins and harmful metabolites, or exhibits hemolytic properties, or causes infections in humans and animals with impaired immunity. The potential transfer of genes coding antibiotic resistance is also studied. It is inadmissible for genes of antibiotic resistance to be found on mobile elements of the genome (plasmids, transposons) (FDA/WHO 2001).

In vitro tests and animal trials do not make possible a comprehensive evaluation of the essence of the probiotic action, and in many cases they are insufficient for a given microorganism to be considered probiotic. The probiotic action has to be confirmed in clinical trials on volunteers. Clinical trials constitute the most important stage in the investigation of functional traits of probiotic strains. They aim at the confirmation of a statistically and biologically significant health benefit and improved quality of life, or a reduction of risk for a disease, or potentially more rapid recovery after the administration of probiotics (Reid *et al.* 2003). Clinical trials are also crucial when evaluating safety of probiotic products.

8.15 Technological aspects and production of probiotic foods

Probiotic cultures are produced on the commercial scale separately and introduced to the foodstuff in the form of ready-to-use preparations. They are typically either in a dried or frozen form. Probiotic cultures may be composed of single strains or a mixture of several strains. Most probiotic cultures are produced in a concentrated form, with cell density of over 10^{10} cfu/g.

Many of the produced probiotic cultures grow poorly in food products. This is true especially of bifidobacteria, which not only require special media and anaerobic culture conditions, but also are characterized by very slow growth. Due to differences in growth rate and different nutrient requirements, they may not be cultured in the foodstuff together with commercial strains of LAB. For this reason they are produced in a strongly concentrated form so that their counts after being added to the foodstuff are sufficient to provide a probiotic effect. A large numbers of probiotic cultures are produced in the direct vat set (DVS) form.

Production technology for probiotic preparations is identical to that of starter cultures and comprises several basic stages:

- Preparation of medium and inoculum
- Culture in a bioreactor, preferably a membrane bioreactor
- Isolation of cells from medium
- Preparation of cell suspension to be dried or frozen
- Packaging in bulk and unit containers (Lacroix & Yildirim 2007)

Many of these processes are run in a way that may be considered stressful for cultured microorganisms. For bacterial cells, a stress is connected already with the transfer from their habitat (the large intestine) to a bioreactor, with its completely different conditions, especially in terms of medium and gas atmosphere composition. Adverse environmental conditions include extreme temperatures (drying, freezing), osmotic stress, mechanical stress resulting from shear forces of moving elements of processing equipment, strong acidity of the medium caused by lactic acid, as well as nutrient-deficit stress, caused by the deficit of certain medium components, particularly toward the end of culture. These difficult conditions for commercial culture require resistant strains, which should retain adequate viability, metabolic activity and genetic stability (Girgis *et al.* 2003).

The culture process in the case of probiotic microorganisms is run in several stages by culture passage to tanks of increasing size. During culture, it is necessary to maintain medium temperature and pH at optimal levels for the growth of cultured microorganisms. Lactic acid secreted to the medium by LAB, being main components of probiotic preparations, has to be constantly neutralized by the automatic addition of calcium or soda lye. Cell viability is significantly affected by their acid formation capacity. It is advisable for probiotic strains to exhibit moderate capacity to synthesize lactic acid. The required acid neutralization during culture leads to an increase in osmotic pressure, which may inhibit cell propagation. Moreover, excessive acid formation capacity of LAB is also undesirable in foodstuffs since it has an adverse effect on their sensory attributes. Generally, bacteria from genus *Lactobacillus* are more tolerant of environmental acidity than bifidobacteria. After culturing, the probiotic microorganisms are isolated by centrifugation, and less commonly by membrane microfiltration. In the course of these procedures, some cells are killed as a result of mechanical stress they are exposed to in the centrifuge or pump attached to the microfilter module. Next, the isolated cell biomass is suspended in a mixture of a neutral carrier and protectants, and then preserved.

In order to enhance resistance of cells to stress factors, they are exposed to in the course of preservation, stress resistance genes may be activated, for example, by subtle stressogenic treatment. For this purpose, toward the end of culture the cell biomass is exposed to mild stress by a slight increase in osmotic pressure or an abrupt elevation of temperature. Under the influence of such a shock, cells initiate the synthesis of intracellular protectants, including heat shock protein, protective amino acids and trehalose. Exposure of cells to one of the shock factors frequently results in resistance to many other shock factors (Desmond *et al.* 2001). In this way, so-called cross-resistance is acquired. In order for probiotic microorganisms to adapt to extreme temperatures, components may be introduced to the medium, changing the saturation ratio of fatty acids found in cell membranes. When applying low-temperature preservation, membrane fluidity should be increased, while during high-temperature drying the membranes need to be made rigid. However, it needs to be remembered that adaptation to stress is inductive and transient, and is found only in the generation subjected to adaptation.

Cultures of probiotic microorganisms are produced in the form of frozen concentrates, either freeze- or spray-dried (Holzapfel *et al.* 2001). In the case of spray-drying, the cell suspension is subjected to the action of hot air, with an inlet air temperature of 160–200°C, and an outlet air temperature of 70–80°C. The lower the drying temperature, the higher the survival rate of cells. The application of such a high-temperature drying results in a suspension droplet temperature in the drying chamber of 60°C. Special additives may be applied to protect dried cells, such as skim milk, soluble fiber, gum acacia, granular starch, adonitol, trehalose or a mixture of these substances. However, many authors have stressed that among factors causing cell death, the most harmful factor is osmotic shock, caused by the condensation of cytoplasmic components, with high temperature ranking second (Selmer-Olsen *et al.* 1999; Conrad *et al.* 2000; Desmond *et al.* 2002).

Probiotic microorganism biomass is most commonly preserved by freeze-drying. The freeze-drying process is run in two stages. First, the cell suspension is frozen to a temperature ranging from −25 to −50°C, followed by the initiation of the drying process itself based on sublimation. The frozen slurry is subjected to controlled heating so that an addition of energy facilitates evaporation of water with an omission of the liquid phase. The drying process is run while maintaining high vacuum in the drying chamber and constant water vapor removal. Freeze-drying requires high vacuum, amounting to 0.01 torr, which may be obtained when the condenser temperature is below −50°C. For this reason, condensers cooled with dry ice are commonly used. Freeze-drying time is approximately 7–8 hours. Optimum water content in the freeze-dried material is 2–6%. As a result of freeze-drying a powder is produced, which is easy to use and highly stable (Garcia De Castro *et al.* 2000).

Freezing is also a highly efficient method of probiotic microorganism preservation. In order to increase survival at low temperatures, cryoprotectants are added, such as glycerol, sorbitol, skim milk, glucose, starch, sucrose, fructose, lactose, monosodium glutamate, whey protein, milk protein or mixtures of these substances. Cryoprotectant concentrations are usually 5–20%. In the case of deep-frozen cultures, cell concentration in the preparation is up to 10^{10} cfu/g, while in freeze-dried preparations it is up to 10^{11} cfu/g. In commercial practice when inoculating a food product with a probiotic bacterial culture, approximately 70 g of deep-frozen culture or 20–25 g of powdered culture is added per 1000 liters of milk.

When probiotic microorganisms are preserved by drying, especially freeze-drying, their nutritional environment is crucial. Media rich in nutrients enhance cell resistance to stress, including freezing. These media also facilitate an increased synthesis of glycogen substances, lipids and polypeptides, which serve an important role during freezing and thawing. For example, the addition of sodium oleate to the medium for *L. bulgaricus* increases the content of unsaturated fatty acids in the phospholipid cellular membrane, making it more elastic, which in turn results in high survival rates of bacterial populations during freezing.

The age of a culture is another factor determining the capacity of probiotic cells to survive the stress caused by freezing. Cells which reach the stationary growth phase are preserved most efficiently. Younger cells, collected from the logarithmic growth phase, survive storage much less successfully. The survival rate of probiotic microorganisms in the course of freezing also depends on the density of their population in a positive manner. This is explained by the protective action of autolyzed cells on the other live cells.

One of the methods applied to reduce cell death rates in the course of preservation processes is encapsulation (Anal & Singh 2007). This procedure consists of the mixing of a cell suspension with an appropriate carrier, the formation of a capsule or gel globule, followed by drying or freezing. Alginates, carrageen, cellulose derivatives (acetate, phthalate), chitosan, gelatin and starch are applied most commonly as capsule-forming substances. When encapsulating the cells, prebiotics may be introduced to the capsule matrix, such as starch or fructo-oligosaccharides (Sultana *et al.* 2000). The basic encapsulation method is polymer cross-linking (the bead method with the application of alginate) and emulsifying. In the former case, polymer cross-linking takes place via divalent metal ions. The best-known example of such a procedure is drop-wise addition of a sodium alginate solution, containing bacterial cells, to a calcium chloride solution. In the other above-mentioned case, two-phase systems are formed in which a mixture of polymer solution and cells is introduced to oil. A water-in-oil emulsion is formed, followed by the formation of cross-linkage with the polymer, and then a gel in oil is formed.

Additionally, microcapsules may be covered with a lipid membrane using waxes, glycerol, organic esters, soybean oil, palm oil or long-chain fatty acids. Capsules may also be formed in the course of spray- or freeze-drying (Anal & Singh 2007).

Encapsulation provides higher survival rates during storage under various conditions (temperature, product matrix) and improves microorganism survival in the alimentary tract after the consumption of the product. It also extends the scope of probiotic application (e.g. milk chocolates, muesli, snacks, raisins in chocolate–*Bifidobacterium* coating).

The effectiveness of probiotic preservation is next assessed in terms of reproduction ability, as well as stability of functional and genetic properties. Reproduction ability is assessed by quantitative growth measurements (turbidity, colony formation on plates). This supplies information on the effectiveness of a given preservation method. One of the basic criteria here is the ability to recreate a normal population. This is not equivalent to the counts of surviving colony-forming cells. There are many cells for which the genome was damaged in the course of preservation and storage. Such damage is repaired and these cells are still alive. However, they are no longer capable of growing on minimal media, although they may grow on complete media (auxotrophic mutants). Another symptom of irregularities is reduced salt tolerance.

In order to appropriately evaluate survival rates of stored cultures, selective media need to be applied. Prior to such tests, biochemical and morphological properties of a given parental strain need to be defined. Survival is not a sufficient criterion when assessing the effectiveness of a given method of microorganism preservation. When the culture is used commercially without the stage of preliminary propagation, the selection of criteria other than survival rate is a key issue. One such criterion may be determination of its selected functional properties, e.g. adhesion to the intestinal epithelium or antagonistic activity toward pathogens. Moreover, growth assays on media containing antibiotics or inhibitors may be used as criteria.

Genetic stability of culture is also very important criteria for choosing the best method of cell preservation. Genetic stability is usually evaluated on the basis of observed behavior of the entire offspring population. The primary criteria used when evaluating genetic stability include changes in pigmentation, temperature requirements and fermentation reactions. It needs to be stressed that auxotrophic mutants are much less stable than prototrophs.

Following production, the probiotic preparation has to be appropriately packaged. The packaging is required to form an effective barrier isolating cells from oxygen and light. For this purpose, air-tight aluminum foil packaging or dark glass containers are usually used.

Most probiotics require low temperatures during transport and storage, while only dry products tolerate room temperature. It needs to be stressed that storage temperature is a key factor determining the stability of a probiotic preparation. Cold storage conditions are recommended, whereas freezing of probiotic cultures reduces cell viability.

Another important stage affecting a probiotic culture is its introduction to the foodstuff matrix. Cell survival rates in the product are determined by the chemical composition of the matrix (media, inhibitors), its structure, available oxygen, water activity and pH.

Growth of probiotic microorganisms in non-fermented products is usually undesirable, while in fermented products it may be advantageous and lead to a reduction of their production costs. In milk-based matrices, the probiotic effect is affected by pH, lactic and acetic acid concentrations, as well as oxygen content. In order to improve growth conditions for probiotic bacteria, glucose, yeast extract, antioxidants, minerals and vitamins are usually introduced to the food product. Interactions with starter microorganisms in fermented products are also very important. Starter bacteria may produce inhibitory metabolites, i.e. lactic acid, hydrogen peroxide and bacteriocins. Probiotic culture may be a component of starter culture on condition it exhibits resistance to the above-mentioned metabolites.

When comparing survival rates of probiotic cultures in fermented and raw milk, it was found that it is markedly higher in raw milk.

When adding probiotic cultures to sweet products, they need to be cooled in order to inhibit bacterial growth. A factor, which also needs to be considered when developing probiotic products, is the ability of probiotic microorganisms to survive in the matrix of the developed (final) food product, usually in the presence of other microorganisms and components differing from the components of the culture medium, often at a reduced water activity. Thus, when selecting probiotic strains, their technical properties need to be investigated. These properties include growth rate, survival rate during culture and in the course of drying processes, ability to acidify products, tolerance to low pH, survival during transport and storage, effect on sensory attributes of the product, lack of antagonism toward commercial strains, resistance to bacteriophages and several other characteristics. An important character of probiotic strains is also their enzyme profile. The presence of some enzymes may lead to undesirable changes in product texture or the formation of off-flavors.

8.16 Probiotic products

Currently, the market for functional food is dominated by probiotic dairy products, primarily fermented drinks and milk, with the addition of probiotic microorganisms. They are particularly popular in Europe, where they account for over two-thirds of the entire market for this category of functional food. The range of probiotic fermented dairy drinks is being constantly expanded. For example, studies have been conducted on probiotic cheese (Stanton *et al.* 1998), probiotic ice cream (Salem *et al.* 2005) and snacks. In order to provide probiotic effects, the product must contain an adequate count of active probiotic microorganisms. It is assumed that functional food needs to contain at least 10^9 cells in a unit container to be used as a single meal. This generally means that the minimum cell density in the product, throughout its entire shelf-life should be 10^6–10^8 cells/mL or g (Lee & Salminen 1997).

Products containing probiotics are found in two forms:

1. Traditional products containing valuable nutrients and cultures of probiotic bacteria, such as yogurt, kefir and cottage cheese
2. Enriched or fermented food, constituting a source of probiotic bacteria or their metabolites

Producers of probiotic food focus on the sensory attributes of final products. High nutritive value and probiotic effects have to be combined with sensory attractiveness. Generally, fermented dairy drinks with added probiotic bacteria have a modified taste – they are milder in taste and less aromatic. Thus, taste and aroma additives are frequently introduced to such products. Yogurts of mild acidity are sold in Germany as mild yogurts or bio-yogurts. In the United States, mainly acidophilus milk, containing *L. acidophilus*, has been developed as a probiotic dairy drink. In the last decade, the market for products containing probiotic bacteria has been growing very rapidly (Table 8.5)

A considerable problem in the production of probiotics is the standardization of their quality. The most frequent defect is too low a count of bacterial cells in relation to the

Table 8.5 The market for probiotic products (report of JHNFA conference, 2003)

	Market of functional foodstuffs (sales in billions of yens)		
Use of product	**1997**	**1999**	**2001**
Tooth care products	0	0.4	18.7
Mineral nutrients	9.2	4.5	11.4
Diabetes	0.7	0.5	18.4
Hypertension	1.4	7.2	10.0
Obesity	0	7.0	15.2
Cholesterol	0	0.4	2.8
Gastrointestinal disorders			
Oligosaccharides	10.4	9.1	5.6
Dietary fiber	11.9	11.6	12.8
Probiotic bacteria	97.9	186.3	317.1

JHNFA – Japan Health Food and Nutrition Food Association.

declared level. Very often, instead of a density of 10^{10} cells/mL, products may contain less than 10^5 cells/mL.

When producing probiotics, active strains are typically introduced to the formulation of starter cultures to initiate fermentation. Probiotic strains are included in the formulation of fermented dairy drinks, cottage cheese, ice cream and frozen dairy desserts (Shah 2007). Fermented drinks containing, in addition to typical commercial strains, probiotic strains, are referred to as second-generation drinks, while they are considered third-generation drinks if they contain probiotic bacteria of intestinal origin.

Most producers launching probiotic products add the 'bio' prefix to their names to stress their natural origin and probiotic character (bio-yogurt, biodrink, biogarde, biobest) or the 'Acti' prefix (Actimel, Activia, Actifit). Names of certain products refer to the bacterial species used in their production (e.g. acidophilous milk).

In Europe, one of the most popular probiotic products is probiotic yogurt produced by Nestlé. It contains the strain *L. johnsonii* La1 (NCC533), claimed to strongly stimulate the immune system. Another well-known example of a probiotic product is Actimel, produced by Danone. It contains the active strain *L. casei* DN 114 000.

Apart from dairy products, probiotics are also introduced to baby foods, confectionery such as chocolate, as well as fruit and vegetable products. In the case of confectionery, the bacteria are introduced into the filling. This requires a careful selection of strains in terms of their tolerance toward a low water activity medium, in the presence of oxygen and at room temperature, since non-dairy products such as cereal bars, breads, biscuits, cookies, chocolate, tablet candy, powdered soup, infant formulae and muesli-type products are stored over extended periods in warehouses and on shop shelves without refrigeration. In this case, the application of encapsulated cultures is recommended.

Moreover, probiotic fruit and vegetable juices (Yoon *et al.* 2007) and probiotic fermented meat products (Työppönen *et al.* 2003) have been gaining in popularity.

8.17 Prebiotics

Obtaining and preserving balance in the intestinal microflora may be done in two ways, either by consumption of products containing probiotic bacteria or products containing nutrients stimulating the growth of specific, beneficial groups of autochthonous microorganisms. The latter products are referred to as prebiotics. They are defined as indigestible ingredients of food, which selectively stimulate the growth and/or activity of one or a limited number of microorganisms in the large intestine, thus improving the health state of humans (Shah 2004). They were first introduced into the market in Japan.

Prebiotic components need to meet the following criteria:

- They should be indigestible in the stomach and the small intestine by human digestive enzymes.
- They should stimulate growth of selected groups of bacteria beneficial for human health, especially bacteria, such as bifidobacteria in adult humans, and have an indirect regulatory effect on the microbial equilibrium in the alimentary tract.
- Their metabolism should have a beneficial effect, including the production of SCFAs and organic acids, reducing the pH of intestinal contents.
- They should be safe for human health (Gibson & Roberfroid 1995; Roberfroid 2001).

Long-term studies have shown that prebiotic properties are exhibited first of all by oligosaccharides, which although they are digested neither in the stomach nor in the small intestine, still selectively stimulate growth or activity of groups of microorganisms which promote good health (Rivero-Urgell & Santamaria-Orleans 2001). This has been confirmed especially in relation to bifidobacteria. In addition, prebiotics include certain polysaccharides, protein, peptides and fats. The most thoroughly investigated oligosaccharides are fructo-oligosaccharides (FOS), galacto-oligosaccharides, isomalto-oligosaccharides, malto-oligosaccharides, isomalto-oligosaccharides, glucosylsucrose, palatinose oligosaccharides, soy oligosaccharides, lactose sucrose, xylo-oligosaccharides, arabinogalactan, raffinose and stachyose. The above-mentioned saccharides reach the large intestine and reach the cecum, where they constitute a medium for bifidobacteria, undergoing fermentation to SCFAs (Shah 2004).

From a practical point of view, the most important source of fructo-oligosaccharides is inulin. It is a polymer of β-D-fructofuranose, having β-(2,1)-glycoside bonds, rarely β-(2,6)-glycoside bonds. Glucose is found at the reducing end of the polyfructose chain. Inulin is broken down by an enzyme, inulinase, to oligosaccharides with a degree of polymerization of 2–6. Inulin hydrolysates strongly stimulate the growth of bifidobacteria, while they inhibit growth of *Clostridium*, *Fusobacterium*, *Salmonella* and *Escherichia*. Satisfactory results have been observed when consuming 2–4 g of inulin derivatives daily. At present, it is the most popular prebiotic worldwide.

Apart from chicory, inulin is found in artichokes, onion, garlic, leek, asparagus, tomatoes, wheat germ, barley and bananas. By controlled enzymatic hydrolysis, commercial preparations are produced from chicory, containing fructo-oligosaccharides with different degrees of polymerization.

Plant origin materials are also sources of other oligosaccharides. From lupine seeds we may obtain galactanes, while from soybeans, soy oligosaccharides. Mannanoligosaccharides are isolated from yeast cell walls. Oligosaccharides are also produced by enzymatic synthesis. For example, oligofructosides are synthesized from sucrose, while galacto-oligosaccharides, from lactose. The synthesis of oligosaccharides is catalyzed by transferases attaching single sugar groups to the formed oligosaccharide chain (Johnson 1999).

Metabolism of FOS in the large intestine leads to their transformation to volatile fatty acids (50%; acetic acid, lactic acid, propionic acid and butyric acid), gaseous products (10%; carbon dioxide, hydrogen and methane) and to the formation of bacterial biomass (40%). Gaseous products and cell biomass are excreted, while SCFAs are used by the intestinal epithelium. Their presence contributes to a reduction of triacylglycerol, phospholipid and cholesterol levels. From the metabolism of one FOS molecule, a total of 14 ATP molecules are produced, i.e. relatively large amounts of energy.

Prebiotic properties have also been observed in some polysaccharides, such as resistant starch. They are considered to be a special type of dietary fiber, purposely added to foodstuffs.

Consumption of FOS with the diet provides similar benefits to the consumer as the consumption of probiotics. For instance, a daily dose of FOS required for the bifidogenous effect is 2–10 g/day. Probiotic effects resulting from FOS consumption include the following:

- Reduced calorie content of products
- Modification of intestinal microflora composition, by favoring growth of *Bifidobacterium*, *Lactobacillus* and *Eubacterium* spp.
- Inhibition of intestinal inflammations
- Easier bowel emptying
- Alleviation of intestinal disorders following antibiotic therapy

- Inhibition of osteoporosis
- Regulation of lipid metabolism
- Improved absorption of calcium and phosphorus
- Prevention of cancer development
- Reduced absorption of cholesterol (Holzapfel & Schillinger 2002)

Products of this type are also recommended in the diet of diabetics.

The best-known prebiotic preparation produced using inulin from chicory is Raftilose. Another popular prebiotic, containing oligofructose, is a line of Raftiline preparations. Both above-mentioned preparations have been deemed as fit for human consumption as additives to dairy products. Other known commercial preparations containing FOS include NutraFlora, Actilight, Neosugar and Meioligo. However, it needs to be remembered that the daily consumption of prebiotics should be limited to several grams. Excessive consumption of these preparations has a laxative effect and causes strong bloating.

8.18 The application of prebiotics

Prebiotics, due to their neutral character in relation to the foodstuff matrix, have found many more applications than supporting probiotics. They generally exhibit considerably lower sweetness than sucrose, are stable and practically imperceptible in the food product. Thanks to these properties, they may be introduced to a large range of products such as cookies, bread, soups, ready-to-eat dinner dishes, puff snacks, chocolate products and food concentrates (Shah 2007).

A commercial example of a prebiotic is Actilight, produced by the French company Vivis from beets, and added to products such as cookies and soups. The Japanese company, Beghin Meiji Industries, produces milk enriched with a soluble fraction of dietary fiber, and Bauer (Germany) produces the fermented product Probiotic Plus Oligofructose, which contains two probiotic bacterial strains and the prebiotic Raftilose.

8.19 Synbiotics

Taking into consideration the health benefits, it is recommended to consume probiotics and prebiotics simultaneously. This has encouraged many producers to launch so-called synbiotic products, combining both types of health-promoting components. In this way, a multifaceted effect is obtained, stimulating both the growth of probiotics and the presence of bacteria with a defined probiotic property (Roberfroid 2000).

A highly convenient form for synbiotic products includes fermented dairy drinks with the addition of fruits, such as fruit-flavored yogurts. Added preparations usually contain sucrose, sucrose syrup, inverted sugar, oligofructose, and taste and aroma additives such as peach, strawberry or blueberry flavor. Moreover, the formulation of such products includes thickening agents (such as modified starch, gelatin, xanthan gum or pectin), acidity regulators (such as ascorbic or citric acids), as well as coloring agents (such as β-carotene, carrot juice, riboflavin, chokeberry, black elder or black currant extracts, or curcumin). Added fruit material containing the above-mentioned substances accounts for approximately 10–25% of the product weight.

Generally, the same probiotic strains are used in synbiotic products as in the case of probiotics. For example, *B. longum* and fructo-oligosaccharides have been introduced into the Japanese and then the world markets. Typically, oligosaccharides are introduced into the matrices of foodstuffs and drinks. However, it needs to be stressed that synbiotic preparations may also be offered in the form of tablets, usually two-thirds bacteria and one-third FOS. The formulation of tablets usually includes calcium carbonate, fructo-oligosaccharides, fructose, microcrystalline cellulose, powdered yogurt, encapsulated culture of probiotic bacteria, calcium stearate, ascorbic acid, filling agent and natural vanilla flavor. Each tablet contains over 10^9 cells, and their stability is over 1 year.

8.20 Conclusions

The effectiveness of probiotic action is determined by the microflora found in the alimentary tract, the dose of pro- and prebiotics, frequency of consumption, and the daily diet and health state of the consumer. Relationships between intestinal bacteria and the human organism are highly complex; however, the presence of probiotic bacteria has a modulating effect on the immune system.

In recent years, research has brought immense advances in the methods used to identify bacteria, to investigate their physiology, determine their genomes, as well as evaluate their probiotic properties. Research on the mechanism of probiotic action has been considerably accelerated, thanks to the rapid development of tests based on models of cell lines conducted *in vitro*, and on experimental trials on animals conducted *in vivo*. More and more effort is dedicated to the search for biological markers, making it possible to explain the action of probiotics and prebiotics in the human organism.

Studies conducted on functional properties aim at a more precise definition of targets for specific consumer groups. A trend may be observed, consisting of focusing on individual functional targets – one claim per product, and searching for appropriate biomarkers for a given target. An increasing number of probiotics are addressed to specific groups based on age, sex or increased risk for a particular condition or disease.

Further advances in the launching of probiotics on the functional food market will depend to a large extent on progress in clinical trials run according to pharmaceutical standards. These standards require such studies to be conducted by research networks with adequate coordination and appropriate scope of investigated parameters. Currently, such research projects are being intensively developed in the European Union.

Studies on probiotics require interdisciplinary cooperation, combining studies on the physiology of the alimentary tract with microbiological, genetic, immunological and nutritional research. Moreover, there is a marked trend consisting in the development of an increasing range of probiotic, prebiotic and synbiotic products, including dry products, such as cereal and fruit and vegetable products.

References

Anal, A.K. & Singh, H. (2007). Recent advances in microencapsulation of probiotics for industrial applications and targeted delivery. *Food Science and Technology*, **18**, 240–251.

Armuzzi, A., Cremonini, F. & Bartolozzi, F. (2001). The effect of oral administration of *Lactobacillus* GG on antibiotic-associated gastrointestinal side-effects during *Helicobacter pylori* eradication therapy. *Alimentary Pharmacology and Therapeutics*, **15**, 163–169.

Barbut, F., Richatd, A., Hamadi, K., Chomette, V., Burghoffer, B. & Petit, J.C. (2001). Epidemiology of recurrences of reinfections of *Clostridium difficile*-associated diarrhea. *Journal of Clinical Microbiology*, **38**, 2386–2388.

Barlett, J.G. (2002). Clinical practice. Antibiotic-associated diarrhea. *New England Journal of Medicine*, **346**, 334–339.

Beaugerie, L.A., Flahault, F., Barbut, P., Atlan, P., Lalande, V., Cousin, P., Cadilhac, M. & Petit, J.C. (2003). Antibiotic-associated diarrhea and *Clostridium difficile* in the community. *Alimentary Pharmacology and Therapeutics*, **17**, 905–912.

Beijerinck, M.W. (1901). Sur les ferments de lactique de l'industrie. *Des Sciences Exactes et Naturelles*, **6**, 212–243.

Blum, S., Reniero, R., Schiffrin, E.J., Cittenden, R., Mattila Sandholm, T., Ouwehand, A.C., Salminen, S., Wright, A., von Saarela, M., Saxelin, M., Collins, K. & Morelli, L. (1999). Adhesion studies for probiotics: reed for validation and refinement. *Trends in Food Science and Technology*, **10**, 405–410.

Brandtzaeg, P., Halstensen, T.S., Kett, K., Krajci, P., Kvale, D., Rognum, T.O., Scott, H. & Sollid, L.M. (1989). Immunobiology and immune-pathology of human gut mucosa: humoral immunity and intraepithelial lymphocytes. *Gastroenterology*, **97**, 1562–1584.

Camilleri, M. (2001). Management of the irritable bowel syndrome. *Gastroenterology*, **120**, 652–663.

Camilleri, M. (2006). Is there a role for probiotics in irritable bowel syndrome? *Digestive and Liver Disease*, **38**(Suppl.), 266s–269s.

Canzi, E., Brighenti, F., Casiraghi, M.C., Del Puppo, E. & Ferrari, A. (1995). Prolonged consumption of inulin in ready to eat breakfast cereals: effects on intestinal ecosystem, bowel habits and lipid metabolism. *Cost 92. Workshop on Dietary Fiber and Fermentation in the Colon*, Helsinki, Finland.

Cats, A., Kuipers, E.J., Bosschaert, M.A., Pot, R.G., Vandenbroucke-Grauls, C.M & Kusters, J.G. (2003). Effect of frequent consumption of *Lactobacillus casei*-containing milk drink in *Helicobacter pylori*-colonized subjects. *Pharmacology and Therapeutics*, **17**, 429–435.

Chouraqui, J.P., Van Ergoo, L.D. & Fichot, M.C. (2004). Acidified milk formula supplemented with *Bifidobacterium lactis*: impact infant diarrhea in residential care settings. *Journal of Pediatric Gastroenterology and Nutrition*, **38**, 288–292.

Ciarlet, M. & Esters, M.K. (2001). Interactions between rotavirus and gastrointestinal cells. *Current Opinion in Microbiology*, **4**, 435–441.

Commane, D., Hughes, R., Shortt, C. & Rowland I. (2005). The potential mechanisms involved in the ant-carcinogenic action of probiotics. *Mutation Research*, **591**, 276–289.

Conrad, P.B., Miller, D.P., Cielenski, P.R & de Pablo, J. (2000). Stabilization and preservation of *Lactobacillus acidophilus* in saccharide matrices. *Cryobiology*, **41**, 17–21.

Coudrary, C., Bellanger, J., Castiglie-Delavand, C., Remesy, C., Vermorel, M. & Rayssignuier, Y. (1997). Effect of soluble and partly soluble dietary fibers supplementation on absorption and balance of calcium, magnesium, iron, zince in healthy young men. *European Journal of Clinical Nutrition*, **51**, 375–380.

Cremonini, F., Di Caro, S., Nista, E.C., Bartolozzi, F., Capelli, G., Gasbarrini, G. & Gasbarrini, A. (2002). Meta-analysis: the effect of probiotic administration on antibiotic-associated diarrhea. *Alimentary Pharmacology and Therapeutics*, **16**, 1461–1467.

Daubioul, C.A., Taper, H.S., Wispelaere, L.D. & Delzenne, N.M. (2000). Dietary oligofructose lessens hepatic steatosis, but does not prevent hypertriglicerydemia in obsese Zucker rats. *Journal of Nutrition*, **130**, 1314–1319.

Delzenne, N.M., Kok, N., Fiordaliso, M.F., Deboyser, D.M., Goethals, F.M. & Roberfroid M.B. (1993). Dietary fructooligosaccharides modify lipid metabolism in rats. *American Journal of Clinical Nutrition*, **69**(1), 64–69.

Desmond, C., Ross, R.P., O'callaghan, E., Fitzgerald, G. & Stanton, C. (2002). Improved survival of *Lactobacillus paracasei* NFBC 338 in spray-dried powders containing gum acacia. *Journal of Applied Microbiology*, **93**, 1003–1011.

Desmond, C., Stanton, C., Fitzgerald, G.F., Collins, K. & Ross, R.P. (2001). Environmental adaptation of probiotic lactobacilli towards improvement of performance during spray drying. *International Dairy Journal*, **11**, 801–808.

Dianda, L., Hanby, A.M., Wright, N.A., Sebesteny, A., Hayday, A.C. & Owen, M.J. (1997). T cell receptor-alpha beta-deficient mice fail to develop colitis in the absence of microbial environment. *American Journal of Pathology*, **150**, 91–97.

Dunne, C., Murphy, L., Flynn, S., O'Mahony, L., O'Halloran, S., Feeney, M., Morrissey, D., Thornton, G., Fitzgerald, G., Daly, C., Kiely, B., Quigley, E.M.M., O'Sullivan, G.C., Shanahan, F. & Kevin, J. (1999). Probiotics: from myth to reality. Demonstration of functionality in animal models of disease and human clinical trials. *Antonie van Leeuwenhoek*, **76**, 279–292.

FDA/WHO (2001). Health and nutrition properties of probiotics in food including power milk with lactic acid bacteria. Report of a joint FDA/WHO expert consultation on evaluation of health and nutritional properties of probiotics in food including power milk with lactic acid bacteria. World Health Organization, Cordoba, Argentina.

FDA/WHO (2002). Guidelines for the evaluation of probiotics in food. Report of a joint FDA/WHO working group on drafting guidelines for the evaluation of probiotics in food. World Health Organization, London, Ontario, Canada.

Felley, C.P., Corhesy-Theulaz, I., Rivero, J.L., Sipponen, P., Kaufmann, M., Bauerfeind, P., Wiesel, P.H., Brassart, D., Pfeifer, A., Blum, A.L. & Michetti, P. (2001). Favorable effect of an acidified milk (LC-1) on *Helicobacter pylori* gastritis in man. *European Journal of Gastroenterology and Hepatology*, **13**, 25–29.

Femia, A.P., Luceri, C., Dolara, P., Giannini, A., Biggeri, A., Salvadori, M., Clune, Y., Collins, K.J., Paglierani, M. & Caderni, G. (2002). Antitumorigenic activity of the prebiotic inulin enriched with oligofructose in combination with the probiotics *Lactobacillus rhamnosus* and *Bifidobacterium lactis* on azoxymethane-induced colon carcinogenesis in rats. *Carcinogenesis*, **23**, 1953–1960.

Fooks, L.J. & Gibson, G.R. (2002). In vitro investigations of the effect of probiotica and prebiotics on selected human intestinal pathogens. *FEMS Microbiology Ecology*, **39**, 67–75.

Fukushima, M. & Nakano, M. (1995). The effect of probiotic on fecal and liver lipid classes in rats. *British Journal of Nutrition*, **73**, 701–710.

Fuller, R. (1989). Probiotics in man and animals. *Journal of Applied Bacteriology*, **66**, 365–378.

Gallaher, D.D. & Khil, J. (1999). The effect of synbiotics on colon carcinogenesis in rats. *Journal of Nutrition*, **129**(Suppl.), 1483s–1497s.

Garcia De Castro, A., Bredholt, H., Strom, A.R. & Tunnacliffe, A. (2000). Anhydrobiotic engineering of gram-negative bacteria. *Applied and Environmental Microbiology*, **66**, 4142–4144.

Gardiner, G.E., Heinemann, Ch., Baroja, M.L., Bruce, A.W., Beuerman, D., Madrenas, J. & Reid, G. (2002). Oral administration of the probiotic combination *Lactobacillus rhamnosus* GR-1 and *L. fermentum* RC-14 for human intestinal applications. *International Dairy Journal* **12**, 191–196.

Gibson, G.R., Beatty, E.R., Wang, X. & Cummings, J.H. (1995). Selective stimulation of bifidobacteria in the colon by oligofructose and inulin. *Gastroenterology*, **108**, 975–982.

Gibson, G.R. & Roberfroid, M.B. (1995). Dietary modulation of the human colonic microbiota, introducing concept of prebiotics. *Journal of Nutrition*, **125**, 1401–1412.

Girgis, H.S., Smith, J., Luchansky, J.B. & Klaenhammer, T.R. (2003). Stress adaptation of lactic acid bacteria. In: *Microbial Stress Adaptation and Food Safety*. Yousef, A.E. & Juneja, V.K. (eds), CRC Press, Boca Raton, FL, pp. 159–211.

Granato, D., Bergonzelli, G.E., Pridmore, R.D., Marvin, L., Rouvet, M. & Corthesy-Theulaz, I.E. (2004). Cell surface-associated elongation factor Tu mediated the attachment *Lactobacillus johnsonii* NCC511 (La1) to human intestinal cells and mucin. *Infection and Immunity*, **72**, 2160–2169.

Guandalini, S., Pensabene, L., Zikri, M.A., Dias, J.A., Casali, L.G. & Hoekstra, H. (2000). *Lactobacillus* GG administrated in oral rehydratation solution to children with acute diarrhea: a multicenter European trial. *Journal of Pediatric Gastroenterology and Nutrition*, **138**, 361–365.

Guarner, F. & Malagelada, J.R. (2003). Gut flora in health and disease. *Lancet*, **361**, 512–519.

Haderstorfer, B., Whitehead, W.E. & Schuster, M.M. (1989). Intestinal gas production from bacterial fermentation of undigested carbohydrate in irritable bowel syndrome. *American Journal of Gastroenterology*, **84**, 375–378.

Hart, A.L., Stagg, A.J. & Kamm, M.A. (2003). Use of probiotics in the treatment of inflammatory bowel diseases. *Journal of Clinical Gastroenterology*, **36**, 111–116.

Hatakka, K., Martini, J., Korpela, M., Herranen, M., Poussa, T. & Laasanen, T. (2003). Effect of probiotic on the activity and activation of mild rheumatoid arthritis. *Scandinavian Journal of Rheumatology*, **32**, 211–215.

Hilton, E., Kolakowski, P., Smith, M. & Singer, C. (1997). Efficiency of *Lactobacillus* GG as diarrhoeal prevention in Travelers. *Journal of Travel Medicine*, **4**, 41–43.

Holzapfel, W.H., Haberer, P., Geisen, R., Bjorkroth, J. & Schillinger, U. (2001). Taxonomy and important features of probiotic microorganisms in food and nutrition. *American Journal of Clinical Nutrition*, **73**, 365–373.

Holzapfel, W.H. & Schillinger, U. (2002). Introduction to pre- and probiotics. *Food Research International*, **35**, 109–116.

Horwitz, B.J. & Fisher, R.S. (2001). Current concepts: the irritable bowel syndrome. *New England Journal of Medicine*, **344**, 1846–1850.

Isolauri, E., Arvola, T., Sutas, Y., Moilanen, E. & Salminen, S. (2000). Probiotics in the management of atopic eczema. *Clinical and Experimental Allergy*, **30**, 1604–1610.

Isolauri, E., Kaila, M., Arvola, T., Majamaa, H., Rantala, I., Virtanen, E. & Arvilommi, H. (1993). Diet during rotavirus enteritis effect jejuna permeability to macromolecules in suckling rats. *Podiatry Research*, **33**, 548–553.

Isolauri, E., Kaila, M., Mykkanen, H., Ling, W.H. & Salminen, S. (1994). Oral bacteriotherapy for viral gastroenteritis. *Digest Diseases Science*, **39**, 2595–2600.

Isolauri, E., Sutas, Y., Kankaanpaa, P., Arvilommi, H. & Salminen, S. (2001). Probiotics effect on immunity. *American Journal of Clinical Nutrition*, **73**(Suppl.), 444s–450s.

Jaworska, H., Gregorek, H., Nowicka, E. & Madaliński K. (2002). Estimation of the assay to treat hereditary andioedema with the use of probiotics. *Wspolczesna Pediatria, Hepatologia i Zywienie Dziecka*, **1**, 96.

Johnson, K.F. (1999). Synthesis of oligosaccharides by bacterial enzymes. *Glycoconjugate Journal*, **16**, 141–146.

Kalia, M., Isolauri, E., Soppi, E., Virtanen, E., Laime, S. & Arvilommi, H. (1992). Enhancement of the circulating antibody secreting cell response in human diarrhea by a human *Lactobacillus* stain. *Pediatric Research*, **32**, 141–144.

Kalliomakki, M., Salminen, S., Arvilommi, H. & Isolauri, E. (2001). Probiotics in primary prevention of atopic disease: a randomized placebo-controlled trial. *Lancet*, **357**, 1076–1079.

Kato, I., Tanaka, K. & Yokokura T. (1999). Lactic acid bacterium potentially induces the production of interleukin-12 and interferon-gamma by mouse splenocytes. *International Journal of Immunopharmacology*, **21**, 121–131.

King, T.S., Elia, M. & Hunter, J.O. (1998). Abnormal clonic fermentation in irritable bowel syndrome. *Lancet* **352**, 1187–1189.

Kirjavainen, P.V., Apostolou, E., Salminen, S.J., & Isolauri, E. (1999). New aspects of probiotics – a novel approach in the management of food allergy. *Allergy*, **54**, 909–915.

Klaenhammer, T.R. & Kullen, M.J. (1999). Selection and design of probiotics. *International Journal of Food Microbiology*, **50**, 45–57.

Koebnick, C., Wagner, I., Leitzmann, P., Stern, U. & Zunft, H.J. (2003). Probiotic beverage containing *Lactobacillus casei* Shirota improves gastrointestinal symptoms in patients with chronic constipation. *Canadian Journal of Gastroenterology*, **17**, 655–659.

Kok, N., Roberfroid, M. & Delzenne, N. (1996). Involvement of lipogenesis in the lower VLDL secretion induced by oligofructose in rats. *British Journal of Nutrition*, **76**, 881–890.

Kok, N., Taper, H. & Delzanne, N. (1998). Oligofructose modulates lipid metabolism alteration induced by a fat-rich diet in rats. *Journal of Applied Toxicology*, **18**, 47–53.

Kyne, L. & Kelly, C.P. (2000). Recurrent *Clostridium difficile* diarrhea. *Gut*, **49**, 152–153.

Lacroix, C.H. & Yildirim, S. (2007). Fermentation technologies for the production of probiotics with high viability and functionality. *Current Opinion in Biotechnology*, **2**, 176–183.

Lee, C.H. & Salminen, S. (1997). The coming age of probiotics. *Trends in Food Science and Technology*, **6**, 241–245.

Levy, J. (2000). The effect of antibiotics on gastrointestinal function. *American Journal of Gastroenterology*, **95**(Suppl.), 8–10.

Ling, W.H., Hanninen, O., Mykkanen, H., Heikura, M., Salminen, S. & von Wright, A. (1992). Colonization and fecal enzymes activities after oral *Lactobacillus* GG administration in elderly nursing home resident. *Annals of Nutrition and Metabolism*, **36**, 162–166.

Ling, W.H., Korpella, H., Mykkanen, H., Salminen, S. & Hänninen, O. (1994). *Lactobacillus* strain GG supplementation decreases colonic hydrolytic activity and reductive enzyme activities in healthy female adults. *Journal of Nutrition*, **124**, 18–23.

Liong, M.T. & Shah, N.P. (2005). Acid and bile tolerance and cholesterol removal ability of Lactobacillus strains. *Journal of Dairy Science*, **88**, 55–66.

Madsen, K., Cornish, A., Soper, P., McKaigney, C., Jijon, H., Yachimec, C., Doyle, J., Jewell, L. & De Simone, C. (2001). Probiotic bacteria enhance mureine and human intestinal epithelial barrier function. *Gastroenterology*, **121**, 580–591.

Majamaa, H. & Isolauri, E. (1997). Probiotics: a novel approach in the management of food allergy. *Journal of Allergy and Clinical Immunology*, **99**, 179–185.

Marteau, P. (2001). Safety aspects of probiotic products. *Scandinavian Journal of Nutrition*, **45**, 22–24.

Marteau, P., Pochart, P., Dore, J., Bera-Maillet, C., Bernalier, A. & Corthier, G. (2001). Comparative study of bacterial grups within the human cecal and fecal microbiota. *Applied and Environmental Microbiology*, **67**, 4939–4942.

McCarthy, J., O'Mahony, L., O'Callaghan, L., Sheil, B., Vaughan, E.E., Fitzsimons, N., Fitzgibbon, J., O'Sullivan, G.C., Kiely, B., Collins, J.K. & Shanahan, F. (2003). Double blind, placebo controlled trial of two probiotic strains in interleukin 10 knockout mice and mechanistic link with cytokine balance. *Gut*, **2**, 975–980.

McFarland, L.V. (2006). Meta-analysis of probiotics for the prevention of antibiotic associated diarrhea and the treatment of *Clostridium difficile* disease. *American Journal of Gastroenterology*, **101**, 812–822.

McFarland, L.V., Surawicz, C.M., Greenberg, R.N., Fekety, R., Elmer, G.W., Moyer, K.A., Melcher, S.A., Bowen, K.E., Cox, J.L., Noorani, Z., Harrington, G., Rubin, M. & Greenwald, D. (1994). A randomised placebo-controlled trial of *Saccharmomyces bulardii* in combination with standard antibiotics for *Clostridium difficile* disease. *Journal of American Medical Association*, **271**, 1913–1918.

Meyer, M.L., Micksche, M., Herbacek, I. & Elmadfa, I. (2006). Daily intake of probiotic as well as conventional yogurt has a stimulating effect on cellular immunity in young healthy women. *Annals Nutrition and Metabolism*, **50**, 282–289.

Mishra, V. & Prasad, D.N. (2005). Application in vitro methods for selection of *Lactobacillus casei* strains as potential probiotics. *International Journal of Food Microbiology*, **103**, 109–115.

Moro, E. (1900). Uber den *Bacillus acidophilus* n. spec. Ein Beitrag zur Kenntnis der normalen Darmbakterien des Säuglings (*Bacillus acidophilus* n. spec.). *Jahrbuch für Kinderheilkunde*, **52**, 38–55.

Mrukowicz, J., Krobicka, B. & Duplaga, M. (1999). Epidemiology and impact of rotavirus diarrhoea in Poland. *Acta Pediatrica*, **426**(Suppl. 1), 53–60.

Naruszewicz, M., Johansson, M.L., Zapolska-Downar, D. & Bukowska, H. (2001). Effect of *Lactobacillus plantarum* 299v on cardiovascular risk factors in smokers. *Czynniki ryzyka*, **19**, 3–4.

Nase, L., Hatakka, K., Savilathi, E., Saxelin, M., Ponka, A. & Poussa, T. (2001). Effect of long-term consumption of a probiotic bacterium of *Lactobacillus rhamnosus* GG, in milk on dental caries and caries risk in children. *Caries Research*, **35**, 412–420.

Niedzielin, K., Kordecki, H. & Birkenfeld, B. (2001). A controlled double-blind, randomized study on the efficiency of *Lactobacillus plantarum* 299v in patients with irritable bowel syndrome. *European Journal of Gastroenterology and Hepatology*, **13**, 1143–1147.

Nishina, P. & Freeland, R. (1990). Effect of propionate on lipid biosynthesis in isolated hepatocytes. *Hepatology*, **16**, 1350–1356.

Nobaek, S., Johansson, M.L., Molin, G., Ahrné, S. & Jeppsson, B. (2000). Alteration of intestinal microflora is associated with reduction in abdominal bloating pain in patients with irritable bowel syndrome. *American Journal of Gastroenterology*, **95**, 1231–1238.

Ouwehand, A.C., Lagström, H., Suomalainen, T. & Salminen, S. (2002a). Effect of probiotics on constipation, fecal azoreductase activity and fecal mucin content in the elderly. *Annals of Nutrition and Metabolism*, **46**, 159–162.

Ouwehand, A.C., Salminen, S. & Isolauri, E. (2002b). Probiotics: an overview of beneficial effects. *Antonie Van Leeuwenhoek*, **82**, 279–289.

Pessi, T., Sutas, Y., Hurme, M. & Isolauri, E. (2000). Interleukin-10 generation in atopic children following oral *Lactobacillus rhamnosus* GG. *Clinical and Experimental Allergy*, **30**, 1804–1808.

Playne, M., Bennet, L.E. & Smithers, G.W. (2003). Functional dairy foods and ingredients. *The Australian Journal of Dairy Technology*, **58**, 242–264.

Plumer, S., Weaver, M.A., Harris, J.C., Dee, P. & Hunter, J. (2004). *Clostridium difficile* pilot study: effects of probiotic supplementation on incidence of *C. difficile* diarrhea. *International Microbiology*, **7**, 59–62.

Pool-Zobel, B.L., Neudecker, C., Domizlaff, I., Ji, S., Schillinger, U., Rumney, C., Moretti, M., Vilarini, I., Scasellati-Sforzolini, R. & Rowland, I.R. (1996). Lactobacillus-and bifidobacterium-mediated antigenotoxicity in colon of rats. *Nutrition and Cancer*, **26**, 365–380.

Rafter, J.J. (1995). The role of lactic acid bacteria in colon cancer prevention. *Scandinavian Journal of Gastroenterology*, **30**, 497–502.

Reid, G., Jass, J., Sebulsky, M.T. & McCormick, J.K. (2003). Potential uses of probiotics in clinical practice. *Clinical Microbiology Reviews*, **16**, 658–672.

Rivero-Urgell, M. & Santamaria-Orleans, A. (2001). Oligosaccharides: application in infant food. *Early Human Development*, **65**, 43–52.

Roberfroid, M.B. (2000). Prebiotics and probiotics: are they functional foods? *American Journal of Clinical Nutrition*, **71**(Suppl.), 1682s–1690s.

Roberfroid, M.B. (2001). Prebiotics: preferential substances for specific germs? *American Journal of Clinical Nutrition*, **73**(Suppl.), 406s–409s.
Saarela, M., Mogensen, G., Fonden, R., Matto, J. & Mattila-Sand- Holm, T. (2000). Probiotic bacteria: safety, functional and technological properties. *Journal of Biotechnology*, **84**, 197–215.
Saavedra, J., Bauman, N.A., Oung, I., Perman, J. & Yolken, R. (1994). Feeding of *Bifidobacterium bifidum* and *Streptococcus thermophilus* to infants in hospital for prevention of diarrhea and shedding of rotavirus. *Lancet*, **344**, 1046–1049.
Sakamoto, S., Igarashi, M., Kimura, K., Takagi, A., Miwa, A. & Koga, Y. (2001). Suppressive effect of *Lactobacillus gasseri* OLL2716 (LG21) on *Helicobacter pylori* infection in humans. *Journal of Antimicrobial Chemotherapy*, **47**, 709–710.
Salem, M.M.E., Fathi, F.A. & Awad, R.A. (2005). Production of probiotic ice cream. *Polish Journal of Food and Nutrition Science*, **14**, 267–271.
Salminen, S., Ouwehand, A., Benno, Y. & Lee, Y.-K. (1999). Probiotics: how should they be defined? *Trends Food Science and Technology*, **10**, 107–110.
Saarela, M., Lähteenmäki, L., Crittenden, R., Salminen, S. & Mattila-Sandholm, T. (2002). Gut bacteria and health foods – the European perspective. *International Journal of Food Microbiology*, **78**, 99–117.
Saxelin, M., Tynkkynen, S., Mattila-Sanholm, T. & de Vos, W.M. (2005). Probiotic and other functional microbes: from markets to mechanisms. *Current Opinion in Biotechnology*, **16**, 204–211.
Schafsma, G., Meuling, W.J.A., Dokkum, W. & Bouley, C. (1998). Effect of a milk product, fermented by *Lactobacillus acidophilus* with fructo-oligosaccharides added, on blood lipids in male volunteers. *European Journal of Clinical Nutrition*, **52**, 436–440.
Scholz-Ahrens, K.E., Schaafsma, G., van den Heuvel, E.G.H.M. & Schrezenmeir, J. (2001). Effects of prebiotics on mineral metabolism. *American Journal of Clinical Nutrition*, **73**(Suppl.), 459s–464s.
Schrezenmeir, J. & de Vrese, M. (2001). Probiotics, prebiotics, and synbiotics – approaching a definition. *American Journal of Clinical Nutrition*, **73**(Suppl. 2), 361S–364S.
Selmer-Olsen, E., Sorhaug, T., Birkeland, S.E. & Pehrson, R. (1999). Survival of *Lactobacillus helveticus* entrapped in Ca-alginate in relation to water content, storage and rehydratation. *Journal of Industrial Microbiology and Biotechnology*, **23**, 79–85.
Shah, N.P. (1993). Effectiveness of dairy products in alleviation of lactose intolerance. *Food Australia*, **45**, 268–270.
Shah, N.P. (2000). Effects of milk-derived bioactives: an overview. *British Journal of Nutrition*, **84**(Suppl.), 3s–10s.
Shah, N.P. (2004). Probiotics and prebiotocs. *Agro Food Industry HiTech*, **15**, 13–16.
Shah, N.P. (2007). Functional cultures and health benefits. *International Dairy Journal*, **17**, 1262–1277.
Shah, N.P., Fedorak, R.N. & Jelen, P. (1992). Food consistency effects of quarq in lactose absorption by lactose intolerant individuals. *International Dairy Journal*, **2**, 257–269.
Shanahan, F. (2000). Immunology. Therapeutic manipulation of gut flora. *Science*, **289**, 1311–1312.
Shanahan, F. (2002). Crohn's disease. *Lancet*, **359**, 62–69.
Shimakawa, Y., Matsubara, S., Yuki, N., Ikeda, M. & Ishikawa, F. (2003). Evaluation of *Bifidobacterium breve* strain Yakult-fermented soymilk as a probiotic food. *International Journal of Food Microbiology*, **81**, 131–136.
Singh, J., Rivenson, A., Tomita, A., Shimamura, S., Ishibashi, N. & Reddy, B.S. (1997). *Bifidobacterium longum*, a lactic acid-producing intestinal bacterium inhibits colon cancer and modulates the intermediate biomarkers of colon carcinogenesis. *Carcinogenesis*, **18**, 833–841.
Stanton, C., Gardiner, G., Lynch, P.B., Collins, J.K., Fitzgerald, G. & Ross, R.P. (1998). Probiotic cheese. *International Dairy Journal*, **8**, 491–496.
Sultana, K., Godward, G., Reynolds, N., Arumgaswamy, R., Peiris, P. & Kailaspathy, K. (2000). Encapsulation of probiotic bacteria with alginate-starch and evaluation of survival in simulated gastrointestinal conditions and in yoghurt. *International Journal of Food Microbiology*, **62**, 47–55.
Surawicz, C.M., McFarland, L.V., Greenberg, R.N., Rubin, M., Fekety, R., Mulligan, M.E., Garcia, R.J., Brandmarker, S., Bowen, K., Borjal, D. & Elmer, G.W. (2000). The search for better treatment for recurrent *Clostridium difficile* disease: use of high-dose vancomycin combined with *Saccharmomyces bulardii*. *Clinical Infectious Diseases*, **31**, 1012–1017.
Swidsinski, A., Ladhoff, A., Pernthaler, A., Swidsinski, S., Loening-Baucke, V., Ortner, M., Weber, J., Hoffmann, U. & Schreiber, S. (2002). Mucosal flora in inflammatory bowel disease. *Gastroenterology*, **122**, 44–54.

Szajewska, H., Kotowska, M., Mrukowicz, J.Z., Armanska, M. & Mikolajczyk, W. (2001). Efficacy *of Lactobacillus* GG in prevention of nosocomial diarhea in infants. *Journal of Pediatric*, **138**, 361–365.
Szajewska, H. & Mrukowicz, J.Z. (2001). Probiotics in the treatment and prevention of acute infectious diarrhea in infants and children: a systematic review of Publisher randomized, double-blind, placebo-controlled trials. *Journal of Pediatric Gastroenterology and Nutrition*, **33**(Suppl. 2), 17s–25s.
Työppönen, S., Petäjä, E. & Mattila-Sandholm, T. (2003). Bioprotectives and probiotics for dry sausages. *International Journal of Food Microbiology*, **83**, 233–244.
van den Heuvel, E.G.H.M., Muijs, T., van Dokkum, W. & Schaafsmam, G. (1999). Lactulose stimulates calcium absorption in postmenopausal women. *Journal of Bone Mineral Research*, **14**, 1211–1216.
van den Heuvel, E.G.H.M., Schoterman, M.H. & Muijs, T. (2000). Transgalactooligosaccharides stimulate calcium absorption in postmenopausal women. *Journal of Nutrition*, **130**, 2938–2942.
Van Niel, C.W., Feudtner, C., Garrison, M.M. & Christakis, D.A. (2002). *Lactobacillus* therapy for acute infectious diarrhea in children: a meta-analysis. *Pediatrics*, **109**, 678–684.
Vergio, F. (1954). Anti- und Probiotika. *Hippocrates*, **25**, 116–119.
Wollowski, I., Rechkemmer, G. & Pool-Zobel, B.L. (2001). Protective role of probiotics and prebiotics in colon cancer. *American Journal of Clinical Nutrition*, **73**(Suppl.), s451–s545.
Wullt, M., Hagslatt, M.L. & Odenholt, I. (2003). *Lactobacillus plantarum* 299v for the treatment of recurrent *Clostridium difficile*-associated diarrhea: a double-bind, placebo-controlled trial. *Scandinavian Journal of Infectious Diseases*, **35**, 365–367.
Yoon, K.Y., Woodams, E.E. & Hang, Y.D. (2007). Production of probiotic cabbage juice by lactic acid bacteria. *Bioresource Technology*, **7**, 1427–1430.

9 The influence of food processing and home cooking on the antioxidant stability in foods

Wlodzimierz Grajek and Anna Olejnik

9.1 Introduction

Antioxidants present in food undergo chemical changes during technological processing (Pokorny & Schmidt 2001). Literature devoted to this is generally limited to losses in described antioxidants during single processing events, without assessing changes in their biological activity. Most research is based on model experiments where synthetic antioxidants are added to a product.

Technological processing may include many drastic thermal and hydrothermal processes such as pasteurisation, sterilisation, blanching, concentrating by evaporation, drying, extrusion, microwave heating as well as culinary processing such as cooking, roasting, stewing and frying. New processing methods such as high-pressure processing and high-intensity pulsed electric field treatment are also considered drastic (Knorr 1999).

Technological processes mentioned above may lead to inconvenient changes in a product matrix (Table 9.1) and in antioxidant stability (Table 9.2). Among them are antioxidant oxidation, complexing with other food compounds, enzymatic modifications and changes from active antioxidant into pro-oxidative form.

Most changes in food compounds, especially lipids and proteins, are caused by oxidative reactions. These reactions are especially dangerous when a product is subjected to contact with oxygen at increased temperature and in the presence of light. In oxidative stress conditions, reactive oxygen species and free radicals are produced by numerous biochemical pathways. Antioxidants, as they are especially sensitive to oxidation, have been commonly accepted as indicators of processing damage. Oxidation of food compounds concerns both types of compounds: those soluble in water and those soluble in lipid phase. Also, microbiological and enzymatic processing during fermentation may lead to changes in bioactive compounds. However, keeping the high antioxidant activity of antioxidants is a complex problem and the analysis is hard to perform. Negative effects due to food processing have led to the concept of food reconstitution. The idea is to add natural antioxidants, usually of the same origin, to a product to cover losses that are a result of processing. It is generally assumed that processed products have a lower nutritive value and lower health-protecting capacity than fresh ones.

Oxidative processes in fat are especially dangerous to antioxidants and their stability. They may lead to chain reactions started by lipid radicals $ROO^{\bullet}$ or $RO^{\bullet}$, lipid hydroxides ROOH and lipid dioxylans. Singlet or triplet oxygen may be the reason for oxidative reactions. The latter changes phenols into inactive quinones. The best-known example of such reaction is tocopherol oxidation by hydrogen peroxide in the presence of Fe^{2+} ions. Tocopherols change

Table 9.1 Effect of the processing on the food matrix

Treatment	Changes in food matrix
Peeling, cutting, slicing, mechanical disintegration	Mechanical damages, exposition of hypodermal tissue to oxygen and microorganisms, surface dehydration, moisture loss, high respiratory activity, loss of firmness
Washing	Dry matter loss, hydration, lowering the product temperature, reduction of surface contamination
Blanching	Structure looseness, leaching of soluble compounds from the surface layer, partial inactivation of enzymes, loss of firmness
Pasteurisation	Denaturation and precipitation of protein, inactivation of enzymes, loss of firmness
Precooking	Loss of dry matter, denaturation of membrane proteins, inhibition of some enzymes, increase of the resistance to softening during cooking
Cooking	Structure looseness, tissue disintegration, denaturation of membrane protein, inhibition of enzymes, leaching of soluble compounds, loss of firmness, softening
Concentration by evaporation	Concentration of dry matter, precipitation of some colloids
Caning, jarring	Structure looseness, protein denaturation, leaching of soluble compounds, inactivation of enzymes
Roasting, frying	Volume reduction, porosity reduction, saturation with fat, hydrolysis of glycerols, high resistance against destructively high temperature (180°C), oxidation of lipids affecting flavour, taste, colour and aroma, thermal polymerisation of secondary oxidation products, darkening polymer formation
Air-drying	Volume reduction, porosity reduction, denaturation of protein, contraction of surface layer, temperature increase
Microwave heating	Volume reduction, porosity reduction, inhibition of enzymes, protein denaturation
Puffing	Increase of volume of dried products
Freeze-drying	Plant tissue damages by ice crystals
Freezing and refrigerated storage	Plant tissue damages by ice crystals
High-pressure treatment	Disruption of tissue structure of matrix, homogenisation, denaturation of protein, inactivation of enzymes
Electric field pulses	Concentration of dry matter, volume reduction
Gamma-irradiation	Increase of cell permeability
Fermentation	Increase of moisture, structure looseness

into tocopherols and dimerise with unsaturated fatty acid residues (Yamauchi *et al.* 1995). The strongest radical scavengers, which abolish lipid oxidation, are compounds that contain hydroxyl groups (phenolics), amino groups (proteins, uric acid and spermine) and sulfhydryl (histidine, cysteine, lysine and glutathione). Effectiveness of these antioxidants increases in an environment deprived of oxygen. It is helpful when food products contain, in addition to antioxidants, metal chelators. These are proteins such as serum albumin (Cu), ferritin (Fe), lactoferrin (Fe), carnosine (Cu), histidine (Cu) and transferrin (Fe). Some reducing agents (e.g. ascorbate and superoxide anions) may release ferrous ions from proteins which usually lead to pro-oxidative reactions.

Table 9.2 Effect of the processing on the antioxidant stability in the food matrix

Treatment	Changes in antioxidant stability
Peeling, cutting, slicing, mechanical disintegration	Oxidation, quantitative losses due to leakage
Blanching	Inactivation of oxidising enzymes, losses of antioxidants due to thermal treatment and leaching, isomerisation of carotenoids
Pasteurisation	Thermal degradation, leaching
Cooking	Losses depending on boiling time and water quantity, antioxidant leaking, thermal degradation, release of some carotenoids, small reduction in total antioxidant capacity, significant loss of vitamin C
Concentration by evaporation	Large losses, thermal degradation of some antioxidants, oxidation
Caning, jarring	Large losses, thermal degradation of antioxidants, leaching of antioxidants
Baking, frying	Large losses, thermal degradation, oxidation, dramatic losses in vitamin C, evaporation of some antioxidants, increase of flavonoids at baking, polymerisation of secondary oxidation products
Air-drying	Losses depending on temperature and time of treatment, oxidation, thermal degradation, interaction of polyphenols with proteins and carbohydrates, formation of Maillard reaction products with pro- and antioxidant activity, sometimes increase of total antioxidant capacity, *cis*-isomerisation of carotenoids, dramatic decrease of vitamin C and tocopherols with increase of temperature over 80°C, increase in β-carotene content
Microwave heating	Losses depending on temperature and time of treatment, oxidation, thermal degradation, formation of Maillard reaction products with pro- and antioxidant activity, decrease of vitamin C and tocopherols with increase of temperature over 80°C, increase in β-carotene content
Puffing	Small losses, partial oxidation, thermal degradation, evaporation
Freeze-drying	Oxidation, evaporation
Freezing and refrigerated storage	Oxidation, leaching after thawing, precipitation of flavanones, interaction with insoluble matrix compounds
High-pressure treatment	Slight losses due to partial degradation and depolymerisation, improvement of extractability of carotenoids
Electric field pulses	Small thermal degradation of antioxidants
Gamma-irradiation	Antioxidant alterations, oxidation, partial degradation
Fermentation	Antioxidant losses, glycoside hydrolysis, ring scission, enzymatic oxidation, enzymatic degradation, enzymatic polymerisation of simple polyphenols, uptake by microorganisms

High losses of antioxidants in oxidative processes show how important it is to protect products from oxygen. Up-to-date technologies allow us to process food in an atmosphere of lowered oxygen content or even fully deprived of it. These processes include the following:

- Vacuum concentration and drying
- Drying in an atmosphere of inert gases
- Processing with superheated steam, although in this case the high temperature of processing may be a limit

Oxygen presence may also be limited by encapsulation of the components that contain antioxidants, covering products in polymer filters and packing in an atmosphere of decreased oxygen concentration (Schreiner *et al.* 2003). Research by Ramesh *et al.* (1999) on inert gas processing of vegetables showed that short initial steam blanching followed by fast cooling of chopped vegetables and drying at 60°C in the atmosphere of N_2 reduced losses of carotenoids in carrots by 15% and vitamin C in paprika by 13% in comparison to conventional air-drying. The authors additionally emphasise the apparent increase in carotenoids after blanching, which was due to the leaching of soluble solids.

Taking biological activity of antioxidants into consideration, the most important changes to the chemical nature of these compounds are those taking place during technological and culinary processing (Table 9.2) (Boekel & Jongen 1997; Nicoli *et al.* 1997b, 1999). These processes may have both negative and positive effects. Production of new chemical compounds that have antioxidant properties is advantageous. There are many examples of this, for example non-enzymatic browning, especially Maillard reactions (Manzocco *et al.* 2001; Yilmaz & Toledo 2005). These reactions take place during technological processing as well as during long-term storage and food preparation in the home environment. The main role in production of colourful compounds is played by enzymatic phenol oxidation and so-called non-enzymatic browning. The latter takes place mainly during heat processing. Next to Maillard reactions, carbohydrate caramelisation, chemical oxidation of phenols and maderisation should be mentioned. Heat treatment of plant materials liberates phenolic compounds and thus increases antioxidant activity of products. During this process, several low-molecular-weight phenolic compounds may form (Kim *et al.* 2006).

Reactions of polyphenol and other antioxidant oxidation usually lead to losses in the antioxidant potential of these compounds. Lately, though, there have been publications that state that partly oxidised polyphenols may have even higher antioxidant potential than their non-oxidised counterparts (Cheigh *et al.* 1995; Manzocco *et al.* 1998; Nicoli *et al.* 2000). The increase of antioxidant potential of red wine due to polyphenol oxidation may be an example (Manzocco *et al.* 1999). In Maillard reactions, brown melanoidins that usually have a high antioxidant potential are produced. These reactions are influenced by temperature, heating time, initial pH, oxygen presence, carbohydrate and amino acid content in processed material as well as by the proportions of reacting compounds. One of the most effective melanoidin-producing processes is roasting (e.g. coffee and nut roasting). Heating for too long, though, may lead to a rapid decrease in antioxidant potential of a product due to melanoidin and polyphenol breakdown (Yilmaz & Toledo 2005). There is usually a positive correlation between the intensity of brown colour and the antioxidant potential of a product. But it should be mentioned that our knowledge on how the browning substances are produced and their properties is very limited. These compounds are produced in many different oxidation reactions, as well as reduction and condensation of aldols. There are also many complicated interactions between compounds mentioned above, which makes them even harder to study. Most research is led with the use of model systems with basic carbohydrates and single amino acids. In food products, the situation is far more complicated and additionally the increase of antioxidant potential during browning reactions is being masked by losses of other antioxidants present in a product. Besides carbohydrates, there are also other compounds that may react with amino acids. These are compounds that contain carbonyl groups, e.g. phenols, ascorbic acid and lipid oxidation products. For example, incorporation of ascorbic acid into Maillard reactions may lead to a decrease in antioxidant potential. Maillard reaction products (MRPs) may also react with lipids and pro-oxidants. As a result, the increase in browning is not always linearly correlated to antioxidant potential of a product. In relation to the reaction between sugars and amino acids at elevated temperatures, it should be mentioned that these conditions also lead to formation of genotoxic compounds such as acylamide (Jägerstad & Skog 2005).

9.2 Mechanical processing

Mechanical processing is responsible for significant losses in antioxidants. During peeling, cutting, slicing and shredding, the plant tissue is subjected to direct contact with oxygen. Additionally, many antioxidants are lost because of removal of the outer parts of fruit, seeds and roots that contain the highest antioxidant concentration. This has been confirmed by research showing a decrease of antioxidant potential of peeled apples and cucumbers (Wu *et al.* 2004). Mechanical processing is also the main reason for antioxidant losses in minimally processed foods. These are mainly fruit and vegetable salads prepared from fresh materials. These products are getting more and more popular among consumers because of their freshness, high nutritional value and convenience. Fresh-cut tissues are primarily submitted to oxidative stress altering the content and composition of antioxidant compounds, resulting in changes of total antioxidant capacity of the product. Decrease of the antioxidant capacity for fresh-cut spinach, mandarin oranges and tomatoes has been observed (Gil *et al.* 1999; Piga *et al.* 2002; Lana & Tijskens 2006). Losses of antioxidants and other important compounds due to oxidation may be slowed by processing at low temperature (below 5°C). There is also an important role of tight packaging, especially in boxes that contain single-meal portions. These are very popular in Northern Europe.

There are some interesting results of investigations of the antioxidant properties of tomatoes after home processing (Sahlin *et al.* 2004). Tomato fruits, sliced and soaked in a vinegar marinade of olive oil and white vinegar showed an unexpectedly significant reduction in the ascorbic acid, total phenolic and antioxidant capacity. Treatment with oil significantly reduced the amount of lycopene that could be extracted from the tomatoes, while incubation with vinegar had no effect.

Citrus fruits sold as peeled segments or fresh juices are other examples of minimally processed foods. They can be cold-stored for up to 12 or 15 days, respectively. In addition to vitamin C, these fruits contain water-soluble flavanone glycosides, limonoids, carotenoids, flavones and phenylpropanoids. Research by Del Caro *et al.* (2004) showed that minimal processing had almost no effect on the main chemical constituents. Losses of antioxidants rather occur during lengthy storage. The biggest losses affected ascorbic acid and differed on the basis of species and cultivars of citrus fruits as well as type of preparation. Fruit segments retain their health-promoting properties due to increase of flavonoid content in some cases after minimal processing. There are similar results from research by Piga *et al.* (2002). The authors studied changes in overall antioxidant capacity of water-soluble fraction of some mandarin and Satsuma segments and juices packed in plastic trays and glass bottles. They assumed that during storage, antioxidant potential due to vitamin C and polyphenols decreased more in fruit segments packed in trays because of higher oxygen availability. These losses, though, were low.

9.3 Drying

Dehydration of raw materials and food products is one of the oldest methods of preservation and ensures excellent stability at room temperature, convenience, good sensory properties, reduced volume and weight of product, high versatility, and a good technological experience. Air-drying of plant materials runs usually at 60–110°C for 2–10 hours, at airflow rates of 0.5–2.0 m/second, and results in a final product moisture content of less than 15%. Long treatment at high temperatures and in the presence of oxygen causes oxidative heat damage to the product. Depending on initial humidity and chemical composition of the dried product, it can undergo different Maillard reactions, chemical oxidation and enzymatic reactions. Some of these changes are positive, but most of them lead to a decrease of antioxidant potential of the product. The range of thermal damage is directly proportional to the temperature, time of drying and final water activity in the dried products. Thermal destruction of antioxidants is usually observed on the surface of these products in a solid state. Drying also ensures good conditions for lipid oxidation. Lowering of water quantity and uncovering of polymers enable oxygen to penetrate easily. High concentration of lipid radicals leads to rapid oxidation of antioxidants and their inactivation. These processes are also observed during sublimation drying, microwave drying and long-term storage of dry products. Microwave heating of solid substances may additionally lead to vaporisation of some antioxidants. Portions of ascorbic acid and polyphenols may be used in Maillard reactions and result in pro-oxidative compound production.

In the literature there are many descriptions of the influence of drying conditions on losses of plant antioxidants. Generally, if drying of vegetables is performed at low temperature and within a short time, the retention of thermolabile antioxidants is usually high. Ascorbic acid is one of the most thermolabile antioxidants and is used as a good indicator to assess the drying process. Zanoni *et al.* (1999) described the influence of drying on oxidative heat damage of lycopene and ascorbic acid of tomato halves. They observed a marked loss of ascorbic acid largely dependent on temperature. After 430-minute drying at 80°C, losses of vitamin C reached about 92%, while at 110°C absolute decay of ascorbic acid was observed after 350 minutes. In the same trials, losses of lycopene were either minimal or not detectable. During drying at 80°C, lycopene was intact, while at 110°C losses of lycopene reached less than 12%. In extreme temperatures of drying, the number of *cis*-isomers increases (Shi *et al.* 1999; Schieber & Carle 2005). What is interesting is that lipophilic extracts of dried tomatoes had an identical antioxidant potential to those of fresh tomatoes, while hydrophilic extracts of dried tomatoes had higher antioxidant potential compared to those of fresh tomatoes. The nature of this phenomenon is still not known (Giovanelli *et al.* 2002). High losses of vitamin C and good stability of carotenoids during drying, especially at temperatures in the 70–80°C range, are confirmed by other authors (Dewanto *et al.* 2002; Toor & Savage 2006). Comparative studies on freeze-drying and hot air-drying of tomatoes showed that freeze-drying retained high levels of antioxidant compounds (8–10% loss), whereas high-temperature treatment caused a tremendous decrease in the content of antioxidants (56–61% loss) (Chang *et al.* 2006). Interestingly, the total phenolic and flavonoid contents in both freeze- and hot-air-dried tomatoes were significantly higher than in fresh material. Different changes appeared in lycopene content. In freeze-dried tomatoes, lycopene content was reduced by 33–48%; however, the amounts of lycopene in hot-air-dried tomatoes increased 152–197%, probably due to breaking of cell walls and weakening of the binding forces between lycopene and the tissue matrix.

Air-drying also substantially influences paprika quality and causes high losses in antioxidants. The air-drying process is usually performed using two methods: natural drying under ambient conditions and forced air-drying under ambient conditions or heated air. Dehydration of paprika at ambient temperature maintained the colour, but dramatically decreased the tocopherol and ascorbic acid content, while carotenoid content decrease was lower under these conditions (Carbonell *et al.* 1986; Daood *et al.* 1996). Retention of ascorbic acid and tocopherol was significantly lower in naturally dried (46%) than in forced air-dried (37%) paprika. α-Tocopherol *in vivo* synthesis during natural drying was continued to the moment when paprika humidity was lowered to the 32–47% range. At the final stage of drying, a drastic decrease in tocopherol was observed. Tocopherol losses were significantly higher in forced air-drying due to oxidation. Additionally, it was observed that changes in vitamin C and tocopherol strongly depended on paprika breed. In contrast, β-carotene content of forced air-dried paprika was significantly higher than that of naturally dried. In spray-drying, when the air temperature of dryer is very high (often above 200°C) and the temperature of liquid drops reaches 80–100°C (Mujumdar 2006), losses of carotenoids are very high. This drying method leads to strong oxidation and degradation of lycopene and β-carotene (Goula *et al.* 2005).

Air-drying also leads to antioxidant decrease in herbs. Capecka *et al.* (2005) dried peppermint, lemon balm and oregano at 25–32°C for 10 hours. The authors investigated total antioxidant activity, radical scavenging activity (DPPH) and ascorbic acid, carotenoid and total phenols in fresh and dry herbs. Drying resulted in significant loss of vitamin C (83–94%). Carotenoid losses reached on average 50%. Total phenolic content of peppermint increased by 32% and of oregano by 58%. DPPH value in fresh and dried material was almost the same, but total antioxidant capacity in dried material was a little lower in all dried herbs. In general, during drying of vegetables significant losses of β-carotene are observed due to thermal degradation and isomerisation (Hiranvarachat *et al.* 2008).

Methakhup *et al.* (2005) compared the quality of Indian gooseberry flake after vacuum drying and low-pressure superheated steam drying (LPSSD). The drying was carried out at 65 and 75°C and absolute pressure of 7–13 kPa. It was found that LPSSD retained the colour better than a vacuum-drying system because the degree of ascorbic acid and chlorophyll degradation of LPSSD was much lower than that of the vacuum-drying system. Depending on the vacuum system, the retention of ascorbic acid in dried material obtained in the vacuum-drying process varied from 78 to 94%, while LPSSD value reached approximately 95%. Such high retention of ascorbic acid in dried products indicates that vacuum drying, especially LPSSD, is much better than traditional air-drying and should be performed to obtain food products with high antioxidant potential.

During last few years there has been much research on microwave drying of vegetables and fruits as an alternative to hot air-drying (Lin *et al.* 1998; Feng *et al.* 1999; Zanoni *et al.* 1999; Nindo *et al.* 2003; Soysal 2004; Wang & Xi 2005; Soysal *et al.* 2006). Fast development allowed new hybrid solutions like microwave–hot air-drying, microwave–vacuum drying, microwave–spouted bed drying and microwave–halogen lamp drying. These methods allow reduced drying time and maintenance of the high nutritive quality of products. Comparative studies on antioxidant potential of green asparagus especially for use as an ingredient in instant foods or as a nutraceutical product were performed by Nindo *et al.* (2003). They investigated tray drying, spouted bed drying, combined microwave and spouted bed drying, Refractance Window (RW) drying and freeze-drying. Asparagus contains flavonoids, mainly rutin, and other phenolic compounds, resulting in high antioxidant activity. Research results show that the highest antioxidant potential was observed in product dried in RW and freeze-drying. In the RW drying, pureed product heats up very fast taking heat from a plate heated with hot water at 95°C. This leads to intensive release of phenolic compounds. Due to high steam tension over the dried product surface, oxygen availability is lowered, which protects antioxidants. This method also reduces loss of vitamin C. These observations confirm the results of research on strawberry puree and carrot puree, where losses of vitamin C and β-carotene during RW drying reached 6 and 9.9%, respectively (Abonyi *et al.* 2002).

Combined microwave–spouted bed drying was the fastest method to dry asparagus, but the antioxidant potential of dried product was lower than in RW drying.

In other comparative research, Lin *et al.* (1998) investigated vacuum microwave, air-drying and freeze-drying for carrot slices drying. Carrots were blanched to limit enzymatic oxidation of vitamin C by ascorbic acid oxidase. In this part of the process, vitamin C amount decreased by about 42%. In air-drying further decomposition of vitamin C was observed, and in the final product there was only, 38% of the amount that was measured after blanching. Total losses of α- and β-carotene during air-drying were 19.2% and for vacuum–microwave drying only 3.2%. In freeze-drying, no losses of carotenoids were observed. Almost unnoticeable changes of carrot quality were also reported by Sumnu *et al.* (2005), who dried carrots in a microwave oven additionally equipped with halogen lamps. Halogen lamp heating provides near-infrared radiation. Far-infrared (FIR) radiation was also used to process peanut hulls after antioxidant extraction (Lee *et al.* 2006). Total phenolic content of water extract from treated hulls increased from 72.9 to 141.6 μM after 60 minutes of FIR radiation, while during heat treatment at 150°C there was an increase in total phenolic content to 90.3 μM. FIR radiation increased radical scavenging activity from 2.34 to 48.33%, which was twice as much as after hot air-drying. In another study, microwave–vacuum drying of carrot slices using 400 W, up to residual water contents below 0.1 g/g dried material, maintained all-*trans*-lycopene and all-*trans*-β-carotene stable (Mayer-Miebach *et al.* 2005). Using a combined microwave power program (600 W for 75 minutes followed by 240 W) leads to significant losses of carotenoids because of their thermal degradation. The authors showed that heating carrot homogenates at temperatures above 100°C initiated isomerisation of all-*trans*-lycopene under non-oxidising conditions, and a significant increase of 9-*cis*-lycopene was observed.

9.3.1 Hydrothermal processing

Hydrothermal processing is the basis for such processes as blanching, pasteurisation, sterilisation, cooking and steaming. It is also the main method of food preparation at home. It should generally be noted that a water environment enhances fast heat transfer to the whole product volume, which leads to equal heating of the product. These conditions support Maillard reactions (Yilmaz & Toledo 2005). In the first steps of these reactions, sometimes a decrease of antioxidant potential is observed. But with further heating this potential actually increases rapidly. The reason for this may be pro-oxidative compound production. Slow and mild heating is a factor that enhances, and high-temperature treatment one that decreases, the content of pro-oxidative compounds. High-temperature processing, therefore, may lead to thermal destruction of antioxidants. Due to this, long cooking times and sterilisation are considered antioxidant-destructive.

Changes connected to mild hydrothermal processing (<100°C) are usually advantageous. Due to heating, oxygen is removed from solutions, oxidoreductases are denaturated and heteroglycosides are hydrolysed to aglycones. On the other hand, increased temperature may lead to higher losses because a portion of water-soluble antioxidants are extracted. Losses in water-soluble vitamins are a good indicator of the antioxidant potential decrease of a given food product. Blanching, where solid material is in direct contact with hot steam or hot water, effectively inactivates oxidative enzymes and due to that losses of antioxidants. If the process is performed at too low a temperature, though, it may be ineffective and lead to polyphenol oxidation by polyphenoloxygenase. During treatment, a portion of antioxidants leach into the water, which decreases the antioxidant potential of plant materials (Lin & Chang 2005;

Amin *et al.* 2006; Wachtel-Galor *et al.* 2008). Blanching of vegetables is usually an initial step before further processing or freezing (Ewald *et al.* 1999; Nilsson *et al.* 2004).

Long-term cooking of a product is usually negative to its antioxidant potential. This is observed, for example, in the process of liquid concentration by vaporisation. In such conditions thermolabile vitamins are destroyed and oxidative reactions occur. To limit the losses, oxygen should be removed and vacuum concentration performed. Research on processing of carrots for catering showed that in water and steam cooking, high losses of α- and β-carotene occur (Sant'Ana *et al.* 1998). It was stated that the lowest losses of carotenoids were observed during water cooking without pressure treatment and next in steam cooking, shredded raw, water cooking with pressure and moist/dry cooking. These observations were confirmed by Gayathri *et al.* (2004), who deeply investigated the effect on β-carotene during pressure cooking and open pan boiling with the addition of acidulants (tamarind and citric acid) and plant preparations rich in antioxidants (turmeric and onion powder). The loss of carotene ranged from 27 to 71% during pressure cooking and 16 to 67% during boiling for 4 hours, depending on the plant source. It should be stressed that to limit losses of antioxidants, the same portion of water should be used many times. The study on the pressure cooking of common bean showed that a significant portion of phenolic compounds are released to cooking water (Rocha-Guzman *et al.* 2007). Therefore, in the case of home preparation, vegetables should be consumed in the form of vegetable soups to minimise such losses. Wachtel-Galor *et al.* (2008) observed that antioxidant content was highest in steamed > boiled > microwaved *Brassica* vegetables and decreased with longer cooking time.

The combination of antioxidant and acidulant spices improves provitamin retention. It should be noted that some antioxidants that are poorly soluble in water, i.e. lycopene and β-carotene, are relatively heat stable even during prolonged cooking or sterilisation. Under drastic processing conditions, decomposition of β-carotene into *cis*-isomers or into fragmentation products may occur. Isomerisation of β-carotene takes place during blanching and sterilisation at 130°C, whereas pasteurisation or sterilisation at 121°C causes only minor isomerisation (Marx *et al.* 2003). This process is additionally accelerated in the presence of plant oils that are able to dissolve β-carotene. The production of tomato paste from fresh tomatoes involves mechanical homogenisation and heat treatment. In this process, bioavailability of β-carotene is enhanced, but other labile antioxidants are destroyed. The increase in carotenoids is due to enzymatic degradation, weakening of protein–carotenoid aggregates and concentration of dry matter during dehydration.

Comparative studies on antioxidant activity in vegetables – fresh, frozen, jarred and canned – showed that all methods mentioned led to some degree of antioxidant loss. Even during ambient or chilled storage, antioxidant capacity declined (Hunter & Fletcher 2002). Blanching and freezing of peas and spinach reduced water-soluble antioxidant capacity by 30 and 50%, respectively, but further storage at −20°C did not change the antioxidant level. The lowest antioxidant capacity was observed in canned and jarred vegetables that underwent the dramatic process of sterilisation. These products retained only 20–30% of initial antioxidant potential. The authors also showed that short-term cooking and microwave processing led to relatively low losses in antioxidants while long-term cooking resulted in high losses. Another example of comparative studies is the research by Lombard *et al.* (2005) who assessed the influence of sautéing, baking and boiling on quercetin content in onion. They showed that baking and sautéing increased the concentration of flavonols by 7–25% compared to raw onions, while boiling produced an 18% decrease in quercetin concentration due to leaching of flavonols into cooking water. Leaching of the monoglucoside over the diglucoside form was favoured.

Some hydrothermal processing techniques are performed to inactivate harmful enzymes. As an example, there are some harmful enzymes thermally inactivated in ground paprika and chili (Schweiggert *et al.* 2005), and lipase and lipoxidase inactivation leads to a decrease in lipid oxidation in cereals. In the latter example, the activity of these enzymes in non-germinated grains under normal conditions is very low, but during slicing and grinding they are strongly activated and lipids are oxidised, thus negatively influencing product sensory quality. To prevent this, grains are heated. For example, oat groats are heated to 90–100°C by introducing a flow of steam (Lehtinen *et al.* 2003). It has been shown that intense heat treatment of whole groats reduces the degree of lipid hydrolysis in moist material sixfold. However, the lower the residual lipase activity in the whole kernels, the higher the oxidation of lipids during prolonged storage of the dry fraction. Heating of grains and fractions leads to formation of hexanal, which is linked to the oxidation of polar lipids. The oxidation of polar lipids induces disintegration of membrane structures and inactivation of heat-labile antioxidants, mainly tocopherols.

Fruit juice production is an example of combining mechanical, thermal and enzymatic processing. Initial steps of technological processing depend on fruit shredding and juice squeezing. Often there are enzymes added that digest the plant cell wall, which increases juicing efficiency. Thermal processing is used to denaturise oxidative enzymes and to inactivate microorganisms. Usually, pasteurisation is used to achieve this aim. In juice production from concentrates, the range of thermal processing is wider and additionally includes concentration of juice. All these processes lead to decomposition of thermolabile compounds, which include antioxidants.

Extraction of antioxidants from a fruit matrix is an important processing step that determines the initial antioxidant capacity. During juice production, most antioxidants stay in the fruit pulp while only a portion are transferred to the juice. Research by Dekker *et al.* (1999) showed that in apple juice there are only 5–10% of the flavonoids that are present in the fruit. By adding ethanol or methanol to the pulp, flavonoid extraction may be increased tenfold. Ascorbic acid, anthocyanins and phenolic acids are significantly higher in fresh juice than in juice after pasteurisation and concentration. Changes in antioxidant activity during juice concentration are relatively small (Arena *et al.* 2001). For example, during thermal concentration of orange juice, there is a change in carotenoid distribution between the pulp and serum, and the pulp structure is modified (Arena *et al.* 2000). During refrigerated storage of fresh juice, about half the soluble flavanones precipitate and integrate into the solid (cloud) fraction (Gil-Izquierdo *et al.* 2001). This occurs in commercial orange juice, and these precipitated flavones are not available for absorption in the intestine.

Theoretically, membrane filtration is an alternative to hydrothermal processing of liquid products. It allows removal of microorganisms and separation of harmful fractions. It also causes minimal damage to antioxidants (Lindley 1998; Jiao *et al.* 2004).

9.3.2 Microwave cooking

In the literature, many comparative publications may be found that assess the influence of different home processing methods (cooking), among them studies of microwave treatment on antioxidants (Vadivambal & Jayas 2007). Comparative research on the influence of conventional and microwave cooking on antioxidants in broccoli showed that antioxidant losses were high and antioxidant levels declined continuously with cooking time. After 5 minutes of cooking, total phenolics were reduced to 28%, ascorbic acid to 34% and total carotenoids to 77% (Zhang & Hamauzu 2004). As an example, during boiling of spinach, flavonoid losses reached over 50% (Gil *et al.* 1999). These losses should be taken into account when calculating the dietary intake of these antioxidants. There were contrary results obtained by Turkmen *et al.* (2005), who processed pepper, squash, green beans, peas, leek, broccoli and spinach by boiling, steaming and microwaving. Total antioxidant capacity of pepper, green beans, broccoli and spinach significantly increased during cooking compared to the fresh products. In other plant products, antioxidant activity decreased. In most cases, the changes in antioxidant capacity in the products were not significantly dependent on this processing method. The authors state that moderate heat treatment may be assumed to be a means of enhancement of the health properties of some vegetables. In other comparative studies, losses of carotenoids during cooking of green beans in a covered pot, steaming in a covered pot, pressure cooking and microwave cooking were investigated (De la Cruz-Garcia *et al.* 1997). In all heating methods, an increase in carotenoids in green beans was observed. The highest quantity of available lutein was observed in pot cooking, available β-carotene in steamed beans, and the lowest quantity of both carotenoids in microwave cooking.

9.3.3 Ohmic heating

Some electromagnetic heating methods may also be used for thermal processing of food products. These methods include ohmic heating, infrared radiation and microwave radiation. In ohmic heating, the biological material is placed between two electrodes, between which a faradic current is applied. In biological materials, compounds that have specific electrical charges move in the direction of the oppositely charged electrode. When the charge on the electrode is changed, the compounds change their direction of movement. Material is heated due to friction and water is vapourised. In the infrared radiation method, the biological material is placed under radiators. Heat accumulates in the material and its temperature rises. Microwave heating is the best-known technique of electromagnetic heating. It is used on an industrial-scale as well as in home food processing in microwave ovens. Temperature control inside heated material is easiest in ohmic heating and infrared heating, while the fastest heating occurs in microwave heating. Comparative research on orange juice heating with all three methods showed that antioxidant losses depend strongly on the final temperature and time of exposure to this temperature. Vikram *et al.* (2005) showed that the biggest losses of vitamin C were observed during microwave heating due to a temperature rise up to 100–125°C. The degradation of vitamin C, for all compared methods of heating, followed first-order kinetics. In infrared heating, the temperature of the juice reached 40–80°C. The smallest losses of vitamin C were observed in the ohmic heating system. The activation energy for vitamin C destruction was within the range of 7.54–125.6 kJ/mol. Degradation kinetics of ascorbic acid were also studied at different temperatures (40–80°C), powers (0–300 W) and electrical conductivity (0.25–1.0% NaCl) (Assiry *et al.* 2006). The results of this investigation confirmed that ascorbic acid degradation can be described successfully by a first-order model during both conventional and ohmic heating. The Arrhenius equation showed negative values for the temperature coefficient during most ohmic treatments due to a combination of factors that may alter the reaction mechanisms.

9.3.4 Pulsed electric fields

The pulsed electric field (PEF) is a relatively new food processing technique (Knorr 1999; Ode-Omowaye *et al.* 2001). This treatment involves applying very short (10–1000 μs) electric pulses at high electric intensities (20–80 kV/cm). It is an alternative to thermal food sterilisation and results in the inactivation of microorganisms and enzymes. This technique is very useful for treatment of citrus juice, especially orange juice. Microbial inactivation in orange juice with high intensity PEF reaches levels as high as those achieved with heat pasteurisation. This method also inactivates peroxidase, which is involved in the oxidation of a wide range of natural compounds. General assessment of the technique is positive because it destroys antioxidants at a very low level. The use of PEF method to stabilise freshly squeezed orange juice decreased ascorbic acid content by only 5.3% (Sanchez-Moreno *et al.* 2004a). When compared to heat processing, PEF resulted in higher retention of vitamin C in orange juice-based drinks. Elez-Martinez and Soliva-Fortuny (2006) showed that vitamin C retention after PEF treatment was 91.2%, whereas it was only 82.8% after pasteurisation. After 56 days of storage, the vitamin C content of the juice processed with PEF was 49.2% of the initial value. In another study, the shelf-life of orange–carrot juice treated by two pulses at 25 kV/cm was compared with a juice that was heat-treated at 98°C for 21 seconds and kept in refrigerated storage. The residual concentration of ascorbic acid in the PEF-treated juice was 90%, compared to only 83% in the pasteurised juice (Torregosa *et al.* 2006).

9.3.5 Extrusion

Extrusion is a kind of hydrothermal processing. It involves specific conditions of treatment, namely, high pressure combined with high temperature. It is well-known that thermal processing causes chemical changes in organic materials. Extrusion is often performed at temperatures reaching 160–220°C, which destroys vitamins C, E and A, and these losses can reach over 50% (Harper 1979). Therefore, this processing method is drastic with regards to impact on active compounds. High losses of bioactive compounds during extrusion have been reported by Zieliński *et al.* (2001). The authors showed that extrusion of wheat, barley, rye and oat caused a significant decrease in tocopherols and tocotrienols (63–94%) and in glutathione (20–50%), depending on the cereal treated, but simultaneously a significant increase in the content of phenolic acids was observed. Especially high losses in tocopherols are observed in extruded plant materials containing lipids. The treatment of raw materials by heating can also result in limited changes. For example, the extrusion of buckwheat flour at 170°C for 10 minutes did not cause any change in antioxidant activity (tested by DPPH) but an increase of polar compounds was observed (Sensoy *et al.* 2006).

9.3.6 Roasting and frying

Significant changes in antioxidants are also observed during roasting and frying. Deep-fat frying has been used ubiquitously at home from the sixth century BC onwards, and on a commercial scale since the late twentieth century. For this reason, one can propose that frying is one of the principal cooking methods (Saguy & Dana 2003). Unfortunately, it is especially harmful from a nutritional point of view. It is characterised by high oil temperature (160–180°C). When the food product contains suitable moisture, its temperature does not exceed 100°C. However, during frying a significant amount of water is evaporated, which enhances heat transfer. A large amount of hot oil results in an interaction between oil and food, and enhances rapid oxidation and production of lipid radicals. These destroy fragile compounds such as tocopherols, which are decomposed quickly compared with non-saturated fatty acids. A decrease in tocopherols leads to oxidation of other antioxidants such as vitamins A, C and catechin. By this means, the overall antioxidant potential of the product is radically lowered. During frying, the formation of non-volatile decomposition products and polymerisation of unsaturated fatty acids are observed.

Comparative research on the influence of cooking, frying and roasting of tomatoes on their antioxidant potential showed that these pretreatments resulted in a significant reduction in the ascorbic acid, total phenolic and lycopene contents when compared to fresh fruits. Frying caused the largest loss of antioxidants and nutrients. The reason was the high temperature of the process. Losses in ascorbic acid reached over 60% (Sahlin *et al.* 2004). The effects of boiling and baking were similar, but losses of both antioxidants and nutrients were significantly lower.

Research on frying of carrot chips showed that this treatment allows good preservation of carotenoids. Slices from peeled carrots, fermented with a *Lactobacillus* strain and deep-fried in palm oil at 170°C, maintained 88% of the initial carotene capacity and about 20% of it was identified in the *cis*-form (Skrede *et al.* 1997).

To retard undesirable changes in oil and food products, in some countries, oil manufacturers add antioxidants to the fat products. This helps to protect the fat when exposed to high temperature (Du & Li 2008). Phenolic antioxidants react with lipid radicals and form stable products, which interrupt the oxidative chain reaction. The most commonly used antioxidants are butylated hydroxyanisole and butylated hydroxytoluene. However, there is a growing tendency to use natural antioxidants contained in spices. High resistance against the destructively high temperature of 180°C was observed in spice oils used to prevent lipid oxidation during frying (Che Man & Jaswir 2000; Tomaino *et al.* 2005). The introduction of primary antioxidants to oils and fats may reduce losses of tocopherols. For example, rosemary extract and ascorbyl palmitate can markedly retard losses of tocopherols during deep-fat frying (Gordon & Kourimska 1995). However, the frying process still significantly decreases the antioxidant capacity of some oils. In comparative studies, it was shown that sunflower oil underwent more chemical changes during frying than did olive and virgin olive oils (Quiles *et al.* 2002). Antioxidant capacity of the mentioned oils was correlated with polar components and ultraviolet indices but not with peroxide index or acidity value. Interestingly, the virgin olive and sunflower oils, with a very different fatty acid composition, had the same response to frying time. Oils with the highest content of vitamin E and phenolic compounds demonstrated the highest resistance to oxidation. Another comparative study on deterioration of plant oils during deep-frying was reported by Naz *et al.* (2005). The authors compared refined olive, corn and soybean oils exposed to air and air/light for 30 days, and then used for deep-frying. They ascertained that the most detrimental conditions were in the following order: deep-frying > air/light exposure > air exposure. The peroxide, *p*-anisidine and iodine values of the oils increased in magnitude according to the order soybean > corn > olive. Deterioration rate was not only due to oxidation reactions but also oil hydrolysis, and depended on fat content, oil moisture and frying time.

Roasting, carried out in an oven at high temperature (180–220°C), also causes some changes in the antioxidant potential of foods. Some of changes are undesired and cause a decrease of nutritional value of food products. This is especially true in relation to the oxidation of polyunsaturated lipids. On the other hand, MRPs which develop during the roasting process improve the oxidative stability of many foods and play an important role as health-promoting factors. The formation of MRPs leads to an increase in the content of enediol structure in reductones, which slow the oxidation of lipids and sugars by donating a hydrogen and therefore breaking the radical chain reaction. The compounds are effective metal-chelating agents, as well as oxygen scavengers. There are some reports describing the development of antioxidant MRPs in real food products. An example is an investigation on the roasting process of coffee beans (Nicoli *et al.* 1997a). All the roasted coffee brews studied showed a greater oxygen-scavenging capacity than a crude coffee brew. The main oxygen scavengers in the crude coffee brew were caffeic acid and chlorogenic acid, naturally present in the coffee beans. During the roasting process, most of the phenolic compounds are destroyed and incorporated into browning products. These compounds can act both as primary and secondary antioxidants. Oxygen consumption reached a maximum after samples were roasted for 10 minutes, when intermediate products of the Maillard reaction were formed. The MRP concentration in roasted coffee ranges from 15 to 25% on a dry matter basis. There are also hypotheses that polyphenolic compounds are partially lost during thermal treatment. Results obtained with buckwheat flour roasting showed that processing did not cause any changes in total phenolic content, but the antioxidant activity, measured by DPPH test, slightly decreased (Sensoy *et al.* 2006).

Roasted seeds are used for production of different plant oils. During the roasting process, heat treatment of seeds causes formation of a pleasant (nut-like) aroma or taste, which improves the sensory quality of the oils concerned. The conventional processing of condiment oils involves seed cleaning, roasting and mechanical pressing by expeller. A key step of seed processing is the roasting process because of the formation of the colour, flavour and sensory quality of the product. Such technology is applied in the production of sesame, red pepper and perillar oils. Another interesting product is safflower oil, which is produced using the seed roasting temperature of 140–180°C. In a study by Lee *et al.* (2004), the fatty acid composition of safflower oils did not change with treatment temperature, and the major component was linoleic acid (~80%). The major tocopherol in the oil is α-tocopherol, and its content in the oil gradually increased by approximately 20% as temperature increased from 140 to 180°C. Heat treatment significantly affected antioxidant activity of the oil. When the roasting temperature increased, the oxidative stability of safflower oil increased.

Roasting and frying processes can also be conducted by microwave technique. Yoshida *et al.* (2006) exposed pumpkin seeds to microwaves at a frequency of 2450 MHz and found that more than 85% of tocopherols remained after 20 minutes of roasting. Microwave heating of olive oil showed an α-tocopherol retention of 51%, while the frying pan heated samples showed 38% retention (Ruiz-Lopez *et al.* 1995).

9.3.7 Membrane filtration

Recently, membrane filtration has been commonly used in the food industry for fractionation of food compounds, cold sterilisation, clarification and concentration of liquid foods. Fractionation and concentration of some compounds on the basis of their molecular weight can strongly influence the antioxidant capacity of treated foods. In consequence, ultrafiltration can be proposed as an efficient method to enrich the food products with compounds showing very high antioxidant activity.

In the literature, there are many reports on the effect of ultrafiltration on the antioxidant activity of food products. The best-investigated materials are proteins and peptides. A number of studies have been devoted to antioxidant potential of enzymatic hydrolysates of soybean, which contained hydrophobic, 5–16 amino acid residues. The antioxidant activity of soybean hydrolysates depends on the type of protease used and degree of hydrolysis. Moure *et al.* (2006) studied the antioxidant activity of raw soybean proteins and their hydrolysates. They found that the molecular weight fraction below 10 kDa presented the highest antioxidant activity. Hydrolysates from the fraction between 30 and 50 kDa showed the highest hydroxyl radical scavenging capacity. However, a higher trolox equivalent antioxidant capacity value was determined in the fraction with molecular weight over 50 kDa. The authors found that emulsions prepared by mixing the hydrolysates with linoleic acid and Tween 40 demonstrated higher reducing power and higher antioxidant potency than the original fractions. Soy protein hydrolysates can also be produced from continuous membrane reactor systems. Chiang *et al.* (1999) used continuous hollow-fibre bioreactor and a combination of Alcalase and Flavourzyme for hydrolysis of isolated soy protein. The authors concluded that the antioxidant activity of isolated soy protein was remarkably enhanced by enzyme hydrolysis. The hydrolysates from the fraction between 30 and 3 kDa membrane (retentate) had a much higher antioxidant activity than that from the 3 kDa (permeate). This result suggests that peptides with higher molecular weight possess higher antioxidant activity.

Ultrafiltration has been proposed to improve the stability of cold tea produced industrially. In the final product, due to the tendency of water-soluble tannins to form complexes with proteins, sometimes a precipitation residue is observed. Using ultrafiltration, it is possible to separate high-molecular-weight proteins and, therefore, to prevent tannin–protein complex precipitation. The results of investigations by Todisco *et al.* (2002) showed that using a 40 kDa membrane it is possible to obtain a final product with polyphenolic concentration and colour parameters (CIE, *L*, *a*, *b*) that remain stable for 2 months. Losses of polyphenols were below 1.2% at the end of storage in dark bottles at $-4°C$.

Ultrafiltration is often used for clarification and concentration of citrus and vegetable juices as an alternative to traditional techniques. Fruit juices present in the market constitute two types of products: fresh juice obtained by fruit squeezing and submitted to mild pasteurisation; and, more frequently, juices reconstituted from concentrates. Replacement of heat evaporation by membrane filtration has many advantages. It allows a reduction in losses of antioxidants and volatile aroma substances and simultaneously eliminates all microorganisms, colloids and large solids. Processing usually consists of two steps: ultrafiltration for juice clarification; and reverse osmosis, membrane distillation or osmotic distillation to increase juice concentration. A comparative study by Cassano *et al.* (2003) on the concentration of blood, orange juice demonstrated that the total antioxidant activity of juice concentrated by evaporation was lower than that of the fresh juice. During ultrafiltration, the total antioxidant activity was maintained in both permeate and retentate. When reverse osmosis was applied, a small decrease of the total antioxidant activity was observed. Osmotic distillation, applied as subsequent concentration step after reverse osmosis, did not cause any significant loss in antioxidant activity of the juice. Integrated membrane processing can be postulated as an efficient method of processing fruit and vegetable juices. As an example, an integrated membrane process was proposed for the production of kiwifruit juice (Cassano *et al.* 2006). Losses of total antioxidant activity after ultrafiltration and osmotic distillation relative to the fresh juice were 4.4 and 11.1%, respectively. The reduction of vitamin content in the final concentrate was also very limited.

In the literature, there are other reports on the effect of membrane filtration on the chemical composition and antioxidant activity of fruit juices (Wang *et al.* 2005), extraction of polyphenols from grape seeds (Nawaz *et al.* 2006), treatment of olive mill wastewater (Turano *et al.* 2002), extraction of the bioactive components of green tea in ethanol solvent (Nwuha 2000), extraction of carnosine from chicken muscle extracts (Maikhunthod & Intarapichet 2005), recovery of isoflavones from a waste water (Xu *et al.* 2004) and clarification of vinegar (Lopez *et al.* 2005).

9.3.8 High hydrostatic pressure processing

High hydrostatic pressure processing (HHPP) is a relatively new non-thermal technology applied in the production of fruit juices such as orange juice (Torres & Velazquez 2005). It allows inactivation of pathogenic and spoilage microorganisms in foods with fewer changes in texture, colour and flavour as compared to conventional technologies. It has been introduced in Japan, France, the United States and Spain and is mainly used to obtain fresher products with reduced microbiological levels. High pressures, reaching 100–1000 MPa, lead to the disruption of the food matrix and cell membranes, and release of antioxidants from the food matrix and from its complexes with plant polymers. Research by Sanchez-Moreno *et al.* (2003a,b) showed that this technology significantly enhanced the extrability of carotenoids and flavanones that migrate from orange fruit to juice and increased the antioxidant potential of the juice. These authors (Sanchez-Moreno 2004b) also showed the positive effect of HHPP on carotene extractability and the resulting increase of the antioxidant potential of tomato juice. The higher the pressure applied, the higher the total amount of carotenoids extracted. The authors conclude that pressure treatment at 400 MPa led to the highest vitamin A value among different treatments assayed, and that this is best for maintaining the potential health benefits. These observations are in agreement with the research by Butz & Tauscner (2002) who showed that high pressure, around 600 MPa, in most cases did not induce the loss of beneficial substances. The authors processed carrots, tomatoes and broccoli. Even after 60 minutes of high-pressure treatment, no changes in the total concentration of carotenoids were observed. Also, the relatively high temperature of the process, which reached 95°C for 60 minutes, did not destroy carotenoids. Lycopene is especially resistant to high temperature. During drastic thermal processing, covalent bonds in lycopene may be cleaved. But high pressure does not lead to disruption of covalent bonds (Tauscher 1995). The stability and isomerisation of lycopene by HHPP were evaluated by Qui *et al.* (2006). It was found that high pressure affected the content of total lycopene and the percentage of the presumptive 13-*cis*-isomer. The higher the storage temperature, the greater was the loss of total lycopene and the higher the percentage of 13-*cis*-isomer. The highest stability of lycopene was found when tomato puree was pressurised at 500 MPa and stored at 4°C for 6 months. The mechanism of lycopene loss appeared to be isomerisation.

The use of the HHPP (600 MPa, 40°C, 4 minutes) compared to thermal pasteurisation (80°C, 60s) also led to a better retention of antioxidant capacity of orange juice, mainly due to lower ascorbic acid degradation rates (Polydera *et al.* 2005).

9.3.9 Gamma-irradiation

Food irradiation is used mainly to control foodborne pathogens and reduce the number of spoilage microorganisms. It is used in processing of dried food ingredients, particularly herbs and spices. Allowed doses of radiation reach 30 kGy (http://www.iaea.org/icgfi). A sterilisation effect is reached at doses over 10 kGy and this dose is rarely surpassed in industrial practice (Crawford & Ruff 1996). The range of ionising radiation usage is increasing, especially because of the ban on fumigation with ethylene oxide. According to widely accepted data, irradiation does not make food products radioactive and radiolytic products formed during irradiation pose no danger to humans (Smith & Pillai 2004).

Results of many studies show that gamma-irradiation of spices and vegetables leads to relatively low losses in antioxidants. Kitazuru *et al.* (2004) investigated the effect of irradiation of cinnamon at doses of 5–25 kGy. After irradiation, extraction with ether, ethanol and distilled water was performed. It was stated that a small decrease in antioxidant capacity was observed at doses of about 20–25 kGy. These losses may be limited by removing oxygen from the product. A decrease of antioxidant activity after gamma-irradiation of lupin extracts was reported by Lampart-Szczapa *et al.* (2003). On the other hand, the literature also reports that gamma-irradiation can enhance the antioxidant activity in some foods. Gamma-irradiation of fresh vegetable juice with a 3 kGy dose followed by 3-day cold storage increased significantly the total phenolic content, while that of the non-irradiated control decreased (Song *et al.* 2006). Data obtained in the experiments indicated that immediately after irradiation the antioxidant activity decreased, but between 0 and 1 day of storage it significantly increased. This observation is confirmed by similar results from many authors (Moussaid *et al.* 2000; Beaulieu *et al.* 2002; Fan *et al.* 2003). This phenomenon is probably due to the oxygenation of phenolic compounds. Gamma-irradiation for prevention of contamination was also applied to the mushroom *Agaricus blazei* with a dose from 2.5 to 20 kGy. The analysis of methanolic extracts from this material showed that antioxidant activities were significantly higher than those extracts from non-irradiated control, and reached the highest value in a sample irradiated with 20 kGy (Huang & Mau 2006).

Gamma-irradiation used to treat fresh vegetables has a positive effect on antioxidant extractability. Research on red beetroot showed that at doses of 2.5–10 kGy, the diffusion coefficient of betanin increases by 25%. However, it should be noted that radiation at close range can also destroy betanins (Nayak *et al.* 2006).

A model study on non-enzymatic browning reaction in amino acid–sugar solutions showed that the degree of browning of an irradiated solution increased with increasing irradiation dose and was dependent on the type of sugar (Oh *et al.* 2005). The non-reducing sugars did not react with lysine by heating alone, but did react by irradiation. The highest degree of reaction was achieved with sucrose, followed by fructose, arabinose, xylose and glucose.

Gamma-irradiation can also be successfully used as a method of protection of minimally processed food against microbial spoilage. It was found that Chinese cabbage packed with air, CO_2 and CO_2/N_2 and irradiated at doses up to 2 kGy could be stored at low temperature for 3 weeks without microorganism development (Ahn *et al.* 2005). Antiradical and antioxidant activities, and the phenolic content, were slightly increased by irradiation at 0.5 kGy. However, the phenolic content was reduced by irradiation over 1 kGy. This dose also ensured microbial safety of the vegetable.

9.3.10 Fermentation and enzymatic processing

Research to date has shown that fermentation processes are very good for antioxidant preservation. They run in moderate temperature and under limited oxygen access, which assures a low level of lipid oxidation. Due to the hydrolytic action of enzymes, antioxidant activity is increased. This mainly involves compounds present in conjugated form, with sugar residues linked to hydroxyl groups and with other compounds such as esters of organic acids, lipids and amines. After the decomposition of these compounds, their acid or aglycone forms are released. Examples of such antioxidants are quercetin and myricetin. Hydrolysis of glycoside linkages can be catalysed by β-glucosidase, produced by many fungi and bacteria. Enzymatic cleavage of proteins also leads to an increase in antioxidant activity. Amino acids released to the environment act synergistically with antioxidants and are an additional protection for the product. It has been shown that soy protein, egg albumen and casein hydrolysates act synergistically with tocopherols (Kim *et al.* 1989).

Soybean is traditionally the most important legume and the major source of protein in East Asian countries. It contains many valuable nutritional components and after fermentation it is considered as a healthy food (Minamiyama *et al.* 2003; Kataoka 2005). Fermented soybean is a rich source of antioxidants. Fermentation of soybeans increases the antioxidant potential of the product by precursor modifications. This may be observed in fermented products such as tempeh. The presence of enzymes that decompose hydrogen peroxide and bind oxygen in carbohydrate oxidation to organic acids should also be mentioned. Additionally, fermented soy is able to complex with ferrous and cupric ions. Fermented soybean showed excellent scavenging activity on DPPH radicals at concentrations from 20 to 100%, whereas non-fermented soybeans showed 100% scavenging activity only at 100% concentration (Yang *et al.* 2000). Soybean may react with free radicals, particularly of peroxy radicals, thereby terminating the chain reaction of lipid oxidation. Fermented soy sauce (Shoyu) contains an antibacterial and antihypertensive component (Kataoka 2005). The author demonstrated that soy sauce possesses antimicrobial activity against pathogenic bacteria such as *Staphylococcus aureus*, *Shigella flaxneri*, *Vibrio cholera*, *Salmonella enteritidis* and enteropathogenic *Escherichia coli* (*E. coli* O157:H7).

There have also been studies on the fermentation of soybean with probiotic cultures of lactic acid bacteria and bifidobacteria with a view to developing a probiotic soymilk diet adjunct (Wang *et al.* 2003; Wang *et al.* 2006). The authors used probiotic strains of *Lactobacillus acidophilus*, *Streptococcus thermophilus*, *Bifidobacterium infantis* and *Bifidobacterium longum* individually, or in combination. In fermented soymilk, both the inhibition of ascorbate autoxidation and the reducing activity and scavenging effect of superoxide anion radicals varied with the starter cultures utilised. The antioxidant activity of fermented soymilk was significantly higher than that found in unfermented soymilk. The use of mixed cultures of lactic acid bacteria was more efficient than fermentation with individual strains. Antioxidant activity increased as the fermentation period was extended. Additionally, it was found that freeze-drying caused a significantly lower reduction in the antioxidant activity of soymilk than did spray drying. However, spray-dried fermented soymilk did demonstrate a higher antioxidant activity than that of dried unfermented soymilk.

Fermentation of grass pea with *Rhizopus oligosporus* significantly increased the antiradical properties of seeds. Methanol and buffer extracts from fermented seeds inhibited oxidation of linoleic acid. The highest antioxidant activity showed in buffer extracts and this activity was correlated with presence of phenols (Starzyńska-Janiszewska *et al.* 2008).

By-products obtained from the juice and wine industry can be a source of valuable products such as phenolic antioxidant supplements for food production. Microbiological processing leads usually to better release of antioxidants from the plant tissue matrix. This was confirmed in the processing of cranberry pomace by solid-state fermentation using filamentous fungus *R. oligosporus*, which is known to be used in tempeh fermentation (Vattem *et al.* 2004a). It was shown that during fermentation the quantity of total extractable phenols and capacity of ellagic acid increased, and antioxidant potential of the product rose. It was also shown that in the presence of a water-soluble phenolic fraction extracted from the product, the growth of pathogenic bacteria *Listeria monocytogenes*, *Vibrio parahaemolyticus* and *E.* O157:H7 was restrained. Antibacterial action of phenols is probably due to hyperacidification of the plasma membrane of microorganisms and disruption of the H^+-ATPase activity. These results showed that there is a possibility to join the antioxidant and antibacterial action of phenols. Similar effects were obtained with the use of the food-grade fungus *Lentinus edodes* (Vattem *et al.* 2004b). This shows the opportunity to use phenol-rich by-products as antibacterial food additives. Some fungi produce hydrolytic enzymes able to cleave complex polyphenols such as tannins. These substances also inhibit microorganisms by sequestering metal ions, and inhibiting ion channels and proteolytic enzymes. The idea of fruit waste enrichment in phenolic antioxidants by solid-state fermentation is also of interest for pineapple wastes. Here, there is also a possibility to use fungi from the species *R. oligosporus*, which are able to produce β-glucosidase and to hydrolyse phenolic glycosides (Correia *et al.* 2004). Due to their action, the level of free phenolics in pomace increases. The release of free phenolics from the solid matrix rises when soy flour is added to the pomace. High antioxidant activity is observed during the first stage of fermentation due to the release of insoluble polymeric phenolics. However, a decrease in antioxidant activity also occurs due to the further hydrolysis of polymeric forms in the late stages of fermentation. The main factor that regulates decomposition of these polymeric forms is the level of nitrogen in pomace. To produce enhanced levels of free phenolics, a high nitrogen capacity to the end of fermentation is recommended (Correia *et al.* 2004). Microbiological processing has also been proposed to increase the antioxidant activity of by-products of rabbiteye blueberry (Su & Silva 2006) and grapes (Louli *et al.* 2004). Different processing techniques of rabbiteye blueberry lead to different products such as juice pomace, wine pomace and vinegar pomace. It was found that wine pomace of blueberry had the highest total phenolic content, antioxidant activities and antiradical activities (as measured by DPPH). The lowest antioxidant potential was observed in vinegar pomace, which was due to the aerobic conditions of the process. However, vinegar pomace still maintained a significant antioxidant capacity. Louli *et al.* (2004) showed that ethyl acetate extraction of wine industry wastes, followed by supercritical carbon dioxide extraction, results in very active phenolic preparations with high antioxidant potential.

There are also data that show the positive influence of fermentation using lactic acid bacteria on olive oil enrichment in polyphenols. The method depends on oil incubation with olive wastewater previously fermented by *Lactobacillus plantarum*. During fermentation, reductive depolymerisation of polyphenols which are more soluble in olive oil than in wastewater was observed (Kachouri & Hamdi 2004). The release of simple phenolic compounds from olive oil by-products has been achieved by treatment with enzymes from *Aspergillus niger* showing cinnamoyl esterase activity (Bouzid *et al.* 2005). The enzyme cut the bond between phenolic compounds and polysaccharides. One of the main phenolic compounds released from the matrix in this reaction was hydroxytyrosol.

Wine, especially red, is a food product with high antioxidant capacity. This potential depends mostly on raw material quality and fermentation conditions. The critical roles in the release of antioxidants from mash are played by pectinolytic and cellulitic enzymes, the fermentation of skins, mash heating and oxygen access elimination. Research by Netzel *et al.* (2003) showed that very high concentrations of flavonoids, stilbenes and antioxidant capacity were found in red wines produced from mash heated up to 65°C and fermented with the skins. Efficient extraction of anthocyanins, flavan-3-ols, flavonols and resveratrols was achieved. In contrast, the phenolic acids in wine did not increase. Most flavonoids in grape juice are present in a glycoside form. Comparative research on grape juice, wine and non-alcoholic wine composition showed that glycosides are also present in fermented products, at a lower level compared to the juice (Bub *et al.* 2001). The total amount of anthocyanidin glycosides present in wine is only 50.3% of those present in fresh juice, and in non-alcoholic wine this amount is reduced to 42.8%. It means that glycosides are being lost during the ethanol fermentation process and further during dealcoholisation by a vacuum rectification process. It can be expected that some of these glycosides are hydrolysed during processing.

Fermentation for tea manufacturing can significantly affect the oxidation of tea polyphenols. This is especially true for black tea, which is produced from fermented leaves and is exposed to enzymatic oxidation during processing. Tea flavonoids are chemically and enzymatically very reactive. Research by Manzocco *et al.* (1998) showed that the antioxidant potential of green tea, produced from fresh leaves, is significantly higher than that of black tea. The enzymatic browning during fermentation and drying caused a marked decrease in the radical scavenging properties of tea catechins. It was shown, though, that during tea extract pasteurisation there is an unexpected increase in the chain-breaking activity. This is probably due to repolymerisation of phenolic compounds to form brown-coloured macromolecular polyphenolics.

Catechins undergo partial depolymerisation in the presence of oxidative enzymes. This is followed by a significant increase in antioxidant potential. This effect is, however, partially lost during processing because shorter polyphenol polymers may bind to other chemical compounds and produce large brown macropolymers (Cheigh *et al.* 1995).

Fermented vegetables represent high antioxidant activity. Fermented cabbage is an example of such a product. Fermented and fresh cabbage juices are able to induce cellular detoxication mechanisms (glutathione and glutathione-*S*-transferase). Additionally, fermented cabbage juice has been shown to demonstrate antimutagenic activity.

Chocolate is another product rich in antioxidants. In dark chocolate, the polyphenol content varies from 3.6 to 8 mg per 1 gram of the product. Cocoa beans that are to be processed into chocolate undergo 5–6 days of long fermentation (Wollgast & Anklam 2002). The pods are cut from the trees and the beans, with the adhering pulp, are removed and transferred to boxes for spontaneous anaerobic fermentation. During the first day of fermentation, the adhering pulp turns to liquid and drains away. In the third day of fermentation, the temperature reaches 45–50°C, and the mass is occasionally stirred to improve aeration. During fermentation of cocoa beans, polyphenols diffuse with cell liquids from their storage cells and undergo oxidation to condensed high molecular mostly insoluble tannins. This is associated with a decrease in epicatechin content by 10% between the second and third days of fermentation. These reactions are both non-enzymatic and catalysed by the enzyme polyphenol oxidase. During the fermentation process, anthocyanins are hydrolysed to anthocyanidins, which further polymerise with simple catechins to form complex tannins. Anthocyanins are usually reduced during the fermentation process by 93% after 4 days.

Fermentation of cereals by lactic acid bacteria usually leads to high polyphenol losses. Towo *et al.* (2006) showed that fermentation of sorgo gruels reduced the amount of total phenols, namely, catechol groups, by 32% as compared to untreated sorgo due to the acidic environment and rearrangement of the phenolic structures.

9.3.11 Storage

During long-time storage, polyphenolic compounds are gradually oxygenated chemically and enzymatically. In enzymatic oxidation, after a short-term increase in antioxidant potential, a rapid decrease is observed. In chemical oxidation changes are slower and strongly depend on temperature. Research by Sharma and Le Maguer (1996) showed that lycopene loss was 76% in dried tomato pulp solids after 4 months of storage at room temperature. Zanoni *et al.* (1999) showed that lycopene losses after storage of tomato halves in air and darkness at 37°C reached up to 74–81% after 3 months, depending on the drying temperature.

A wide study on phenolic compound changes in fruit and vegetables, depending on processing and storage conditions, was introduced by Tomas-Barberan and Espin (2001). The authors showed the significant role of enzymes from the phenol oxidase and peroxidase groups in phenolic compound cleavage in damaged fruits. Low-temperature storage of fruit variably influences the antioxidant behaviour. In red fruits that contain anthocyanins, the capacity of antioxidants increases. This was observed in strawberries, blueberries, grapes and pomegranates. At low temperatures of storage many fruits turn brown, which is due to an increase in phenolic compounds. The influence of storage conditions on antioxidant potential of fruit was also investigated by Kalt *et al.* (1999). Low-temperature storage of cured and non-cured roots of sweet potatoes increased the antioxidant activity, especially in periderm tissue (Padda & Picha 2008).

Hussein *et al.* (2000) investigated the influence of the packaging system on vitamin C and β-carotene losses during the storage of broccoli and green pepper after 10 days at 4°C. Vegetables were air packed or 'pillow packed' by sealing the vegetables in a polyethylene bag, squeeze packed with removal of excess air and vacuum packed with total removal of air. There was a significant loss of vitamin C over a 10-day period in all packaging systems used. However, the overall loss of β-carotene was not statistically significant. In most cases, there was no difference in loss of the two vitamins between the packaging systems. Also, cutting vegetables into pieces did not lead to any differences in the decrease of antioxidant levels. The immersion of carrots in ascorbic and glucose solutions protected the antioxidant properties of freeze- and hot-air-dried carrots during long-term storage (Yen *et al.* 2008).

High losses in antioxidants were also observed by Daood *et al.* (1996) during ground paprika storage. After 3 months of storage, all the α-tocopherol and most of the ascorbic acid were lost. The carotenoid content stayed at a high level until the end of the second month, but after this period dramatically shifted to lower levels.

Prolonged storage is not always responsible for depletion in the antioxidant properties of foods. In some food products, compounds with novel antioxidant properties can be formed, which can enhance or maintain overall antioxidant potential. A good example of this may be red wine, which showed a progressive increase in chain-reaction breaking activity with prolonged air exposure (Tybaro & Ursini 1995).

In the presence of ethylene and at low temperature, in some vegetables, lettuce for instance, negative changes in quality are observed. Leaves become brown at the edges, which is due to the formation of peroxidase isomers through stimulation by ethylene. Storage in freezers in a low-oxygen atmosphere and at increased CO_2 level may lead to a decrease in anthocyanins and fruit browning.

Packaging system and storage conditions also have an influence on antioxidant losses. One of the major methods of food protection against oxidation is vacuum packing or packing in a modified gas atmosphere reduced or even devoid of oxygen. The use of metallicised polyester/polyester polypacking systems to store dried plant materials is an example of such a system (Hymavathi & Khader 2005).

9.4 Conclusions

Research on antioxidant changes in the food production chain is incomplete. To date, most information has focused on changes of vitamin C, tocopherol and carotenoid content. Studies on polyphenols are even less extensive, but during the last few years this field has had more attention. Most research has been directed towards examination of lipid oxidation and model experiments with chemically pure antioxidants. Consumers show decreasing trust in chemical additives and strongly prefer foods that contain natural antioxidants. Maintaining these compounds in a complicated chemical environment such as food requires in-depth investigation into the interactions of these compounds with other food constituents, and detailed analyses into oxidation and enzymatic cleavage. The oxidative stability of foods is a very complicated matter because it demands the control of pro-oxidative and antioxidant compounds. It is often difficult to choose the best processing technique. The use of thermal processing leads to pro-oxidative enzyme inactivation, but on the other hand denaturation of antioxidant enzymes and proteins that chelate transient metals. Unless all factors that influence oxidation processes of specific food compounds are known, it is impossible to engineer a technological process that would efficiently protect the product. From the nutritional point of view, methods of food processing should be promoted that allow maintaining the maximum level of natural food antioxidants. There is also an important role for synergy in the process of food product protection against oxidation processes. Among methods that are less harmful to antioxidants are new physical methods such as high-pressure and pulsed electric field processing, as well as low-temperature vacuum drying, puffing, short cooking in small water amounts and fermentation processes. To assess the protective role of antioxidants in decreasing the risk of cancer and cardiovascular disease, there is a fundamental need to study the role of interactions between chemical changes and biological activity of antioxidants in human beings. Expanding knowledge in this area should lead to improvements in technological processing methods in such a way as to produce high-quality food with effective biological activity. In this respect, genetic engineering has much to offer because it could allow the engineering of plant materials that would supply high antioxidant capacity, while reducing losses due to oxidation. These methods could also lead to higher functional activity of food products. There is also a widening field in enzymatic processing of food matrices in the direction of increasing antioxidant extractability, and also structure modification to increase natural product diversity and produce more active and easily available antioxidants (Willits *et al.* 2004). In the growing long-term consciousness of consumers, technological development should lead to an increase in the role of the functional food segment.

References

Abonyi, B.I., Feng, H., Tang, J., Edwards, C.G., Mattinson, D.S. & Fellman, J.K. (2002). Quality retention in strawberry and carrot purees dried with Refractance Window System. *Journal of Food Science*, **67**, 1051–1056.

Ahn, H.-J., Kim, J.-H., Kim, J.-K., Kim, D.-H., Yook, H.-S. & Byun, M.-W. (2005). Combined effects of irradiation and modified atmosphere parking on minimally processed Chinese cabbage (*Brassica rapa* L.). *Food Chemistry*, **89**, 589–597.

Amin, I., Norazaidah, Y. & Hainida, K.I.E. (2006). Antioxidant activity and phenolic content of raw and blanched *Amaranthus* species. *Food Chemistry*, **94**, 47–52.

Arena, E., Fallico, B. & Maccarone, E. (2000). Influence of carotenoids and pulps on the color modification of blood orange juice. *Journal of Food Science*, **65**, 458–460.

Arena, E., Fallico, B. & Macarone, E. (2001). Evaluation of antioxidant capacity of blood orange juices as influenced by constituents, concentration process and storage. *Food Chemistry*, **74**, 423–427.

Assiry, A.M., Sastry, S.K. & Samaranayake, C.P. (2006). Influence of temperature, electrical conductivity, power and pH on ascorbic acid degradation kinetics during ohmic heating using stainless steel electrodes. *Bioelectrochemistry*, **68**, 7–13.

Beaulieu, M., D'Aprano, G. & Lacroix, M. (2002). Effect of dose rate of gamma irradiation on biochemical quality and browning of mushrooms *Agaricus bisporus*. *Radiation Physics and Chemistry*, **63**, 311–315.

Boekel, M.A.J.S. & Jongen, W.M.F. (1997). Product quality and food processing: how to quantify the healthiness of a product. *Cancer Letters*, **114**, 65–69.

Bouzid, O., Navarro, D., Roche, M., Asther, M., Haon, M., Delattre, M., Lorquin, J., Labat, M., Asther, M. & Lasage-Meesen, L. (2005). Fungal enzymes as a powerful tool to release simple phenolic compounds from olive oil by-product. *Process Biochemistry*, **40**, 1855–1862.

Bub, A., Watzl, B., Heeb, D., Rechkemmer, G. & Briviba, K. (2001). Malvidin-3-glucoside bioavailability in humans after ingestion of red wine, dealcoholized red wine and red grape juice. *European Journal of Nutrition*, **40**, 113–120.

Butz, P. & Tauscner, B. (2002). Emerging technologies: chemical aspects. *Food Research International*, **35**, 279–284.

Capecka, E., Mareczek, A. & Leja, M. (2005). Antioxidant activity of fresh and dry herb of some *Lamiaceae* species. *Food Chemistry*, **93**, 223–226.

Carbonell, J.V., Pinaga, F., Jasa, V. & Pena, J.L. (1986). The dehydration of paprika with ambient heated air and the kinetics of color degradation during storage. *Journal of Food Engineering*, **5**, 179–193.

Cassano, A., Drioli, E., Galaverna, G., Marchelli, R., Di Sivestro, G. & Cagnasso, P. (2003). Clarification and concentration of citrus and carrot juices by integrated membrane process. *Journal of Food Engineering*, **57**, 153–163.

Cassano, A., Figoli, A., Tagarelli, A., Sindona, G. & Drioli, E. (2006). Integrated membrane process for the production of highly nutritional kiwifruit juice. *Desalination*, **189**, 21–30.

Chang, C.-H., Lin, H.-Y., Chang, C.-Y. & Liu, Y.-C. (2006). Comparison on the antioxidant properties of fresh, freeze-dried and hot-air-dried tomatoes. *Journal of Food Engineering*, **77**, 478–485.

Che Man, Y.B. & Jaswir, I. (2000). Effect of rosemary and sage extracts on frying performance of rafined, bleached and deodorized (RBD) palm olein during deep-fat frying. *Food Chemistry*, **69**, 301–307.

Cheigh, H.S., Um, S.H. & Lee, C.Y. (1995). Antioxidant characteristics of melanin-related products from enzymatic browning reaction of catechin in a model system. In: *American Chemical Society Symposium*, Series *no. 600*. Korea, pp. 200–208.

Chiang, W.-D., Shih, C.-J. & Chu, Y.-H. (1999). Functional properties of soy protein hydrolysate produced from a continuous membrane reactor system. *Food Chemistry*, **65**, 189–194.

Correia, R.T.P., McCue, P., Magalhaes, M.M.A., Macedo, G.R. & Shetty, K. (2004). Production of phenolic antioxidant by the solid-state bioconversion of pineapple waste mixed with soy flour using *Rhizopus oligosporus*. *Process Biochemistry*, **39**, 2167–2172.

Crawford, L.M. & Ruff, E.H. (1996). A review of the safety of cold pasteurization through irradiation. *Food Control*, **7**, 87–97.

Daood, H.G., Vinkler, M., Markus, F., Hebshi, E.A. & Biacs, P.A. (1996). Antioxidant vitamin content of spice red pepper (paprika) as affected by technological and varietal factors. *Food Chemistry*, **55**, 365–372.

De la Cruz-Garcia, C., Gonzales-Castro, M.J., Oruna-Soncha, M.J., Lopez-Hernandez, J., Simal-Lozano, J.A. & Simal-Gandara, J. (1997). The effect of various culinary treatments on the pigment content of green beans (*Phaseolus vulgaris*, L.). *Food Research International*, **30**, 787–791.

Dekker, M., Verkerk, R., Van Der Sluis, A.A., Khokhar, S. & Jongen, W.M.F. (1999). Analysing the antioxidant activity of food products: processing and matrix effects. *Toxicology in Vitro*, **13**, 797–799.

Del Caro, A., Piga, A., Vacca, V. & Agabbio, M. (2004). Changes of flavonoids, vitamin C and antioxidant capacity in minimally processed citrus segments and juices during storage. *Food Chemistry*, **84**, 99–105.

Dewanto, V., Wu, X.Z., Adom, K.K. & Liu, R.H. (2002). Thermal processing enhances the nutritional value of tomatoes by increasing total antioxidant activity. *Journal of Agricultural and Food Chemistry*, **50**, 3010–3014.

Du, H. & Li, H. (2008). Antioxidant effect of *Cassia* essentials oil on deep-fried blef during the frying process. *Meat Science*, **78**, 461–468.

Elez-Martinez, P. & Soliva-Fortuny, R. (2006). Comparative study on shelf life of orange juice processed by high intensity pulsed electric fields or heat treatment. *European Food Research and Technology*, **222**, 321–329.

Ewald, C., Fjelkner-Moding, S., Johansson, K., Sjoholm, I. & Akesson, B. (1999). Effect of processing on major flavonoids in processed onions, green beans, and peas. *Food Chemistry*, **64**, 231–235.

Fan, X., Toivonen, P.M.A., Rajkowski, K.T. & Sokorai, K.J.B. (2003). Warm water treatment in combination with modified atmosphere packing reduces undesirable effects of irradiation on the quality of fresh-cut iceberg lettuce. *Journal of Agricultural and Food Chemistry*, **51**, 1231–1236.

Feng, H., Tang, J., Mattison, D.S. & Fellman, J.K. (1999). Microwave and spouted bed drying of frozen blueberries: the effect of drying and pretreatment methods on physical properties and retention of flavor volatiles. *Journal of Food Processing and Preservation*, **23**, 463–479.

Gayathri, G.N., Platel, K., Prakash, J. & Srinivasan, K. (2004). Influence of antioxidant spices on the retention of β-carotene in vegetables during domestic cooking process. *Food Chemistry*, **84**, 35–43.
Gil, M.I., Ferreres, F. & Tomas-Barberan, F.A. (1999). Effect of postaharvest storage and processing on the antioxidant constituents (flavonoids and vitamin C) of fresh-cut spinach. *Journal of Agricultural and Food Chemistry*, **47**, 2213–2217.
Gil-Izquierdo, A., Gil, M.I., Ferreres, F. & Tomás-Barberán, F.A. (2001). In vitro availability of flavonoids and other phenolics in orange. *Journal of Agricultural and Food Chemistry*, **49**, 1035–1041.
Giovanelli, G., Zanoni, B., Lavelli, V. & Nani, R. (2002). Water sorption, drying and antioxidant properties of dried tomato products. *Journal of Food Engineering*, **52**, 135–141.
Gordon, M.H. & Kourimska, L. (1995). Effect of antioxidants on losses of tocopherols during deep-fat frying. *Food Chemistry*, **52**, 175–177.
Goula, A.M., Konstantinos, G. & Adamopoulos, G. (2005). Stability of lycopene during stray drying of tomato pulp. *Lebensmittel-Wissenschaft und-Technologie*, **38**, 479–487.
Harper, J.M. (1979). Food extrusion. *CRC Critical Reviews in Food Science and Nutrition*, **11**, 1550–1215.
Hiranvarachat, B., Suvarnakuta, P. & Devahastin, S. (2008). Isomerisation kinetics and antioxidant activities of β-carotene in carrots undergoing different drying techniques and conditions. *Food Chemistry*, **107**, 1538–1546.
Huang, S.-J. & Mau, J.-L. (2006). Antioxidant properties of methanolic extracts from *Agaricus blazei* with various doses of gamma irradiation. *LWT – Food Science and Technology*, **39**, 707–716.
Hunter, K.J. & Fletcher, J.M. (2002). The antioxidant activity and composition of fresh, frozen, jarred and canned vegetables. *Innovative Food Science and Emerging Technologies*, **3**, 399–406.
Hussein, A., Odumeru, J.A., Ayanbadejo, T., Faulkner H., McNab W.B., Hager H. & Szijarto L. (2000). Effects of processing and packing on vitamin C and β-carotene content of ready-to-use (RTU) vegetables. *Food Research International*, **33**, 131–136.
Hymavathi, T.V. & Khader, V. (2005). Carotene, ascorbic acid and sugar content of vacuum dehydrated ripe mango powders stored in flexible packing material. *Journal of Food Compositions and Analysis*, **18**, 181–192.
Jägerstad, M. & Skog, K. (2005). Genotoxicity of heated-processed foods. *Mutation Research*, **574**, 156–172.
Jiao, B., Cassano, A. & Drioli, E. (2004). Recent advances on membrane processes for the concentration of fruit juices: a review. *Journal of Food Engineering*, **63**, 303–324.
Kachouri, F. & Hamdi, M. (2004). Enhancement of polyphenols in olive oil by contact with fermented olive mill wastewater by *Lactobacillus plantarum*. *Process Biochemistry*, **39**, 841–845.
Kalt, W., Forney, C.F., Martin, A. & Prior, R.L. (1999). Antioxidant capacity, vitamin C, phenolics, and anthocyanins after fresh storage of small fruits. *Journal of Agricultural and Food Chemistry*, **47**, 4638–4644.
Kataoka, S. (2005). Functional effects of Japanese fermented soy sauce (shoyu) and its components. *Journal of Bioscience and Bioengineering*, **100**, 227–234.
Kim, S.B., Yeun, D.M., Yeo, S.G., Ji, C.I., Lee, Y.M. & Park, Y.H. (1989). Antioxidant effects of food protein hydrolysates by protease. *Korean Journal of Food Science and Technology*, **21**, 492–497.
Kim, S.-Y., Jeong, S.-M., Park, W.-P., Nam, K.C., Ahn, D.U. & Lee, S.-C. (2006). Effect of heating conditions of grape seeds on the antioxidant activity of grape seed extracts. *Food Chemistry*, **97**, 472–479.
Kitazuru, E.R., Moreira, A.V.B., Mancini-Filho, J., Delincee, H. & Villavicencio, A.L.C.H. (2004). Effects of irradiation on natural antioxidants of cinnamon (*Cinnamomum zeylanicum* N). *Radiation Physics and Chemistry*, **71**, 37–39.
Knorr, D. (1999). Novel approaches in food-processing technology: new technologies for preserving foods and modifying function. *Current Opinion in Biotechnology*, **10**, 485–491.
Lampart Szczapa, E., Korczak, J., Nogala-Kałucka, M. & Zawirska-Wojtasiak, R. (2003). Antioxidant properties of lupin seed products. *Food Chemistry*, **83**, 279–285.
Lana, M.M. & Tijskens, L.M.M. (2006). Effects of cutting and maturity on antioxidant activity of fresh-cut tomatoes. *Food Chemistry*, **97**, 203–211.
Lee, S.C., Jeong, S.M., Kim, S.Y., Park, H.R., Nam, K.C. & Ahn, D.U. (2006). Effect of far-infrared radiation and heat treatment on the antioxidant activity of water extracts from peanut hulls. *Food Chemistry*, **94**, 489–493.
Lee, Y.C., Oh, S.W., Chang, J. & Kim, I.H. (2004). Chemical composition and oxidative stability of safflower oil prepared from safflower seed roasted with different temperatures. *Food Chemistry*, **84**, 1–6.
Lehtinen, P., Kiiliainenet, K., Lehtomaki, I. & Laakso, S. (2003). Effect of heat treatment on lipid stability in processed oats. *Journal of Cereal Science*, **37**, 215–221.

Lin, C.-H. & Chang, C.-Y. (2005). Textural change and antioxidant properties of broccoli under different cooking treatments. *Food Chemistry*, **90**, 9–15.

Lin, T.M., Durance, T.D. & Scaman, C.H. (1998). Characterization of vacuum microwave, air and freeze dried carrot slices. *Food Research International*, **31**, 111–117.

Lindley, M.G. (1998). The impact of food processing on antioxidants in vegetable oils, fruits and vegetables. *Trends in Food Science and Technology*, **9**, 336–340.

Lombard, K., Peffley, E., Geoffriau, E., Thompson, L. & Herring, A. (2005). Quercetin in onion (*Allium cepa* L.) after heat-treatment simulating home preparation. *Journal of Food Compositions and Analysis*, **18**, 571–581.

Lopez, F., Pescador, P., Güell, C., Morales, M.L., Garcia-Parrilla, M.C. & Troncoso, A.M. (2005). Industrial vinegar clarification by cross-flow microfiltration: effect on colour and polyphenol content. *Journal of Food Engineering*, **68**, 133–136.

Louli, V., Ragoussis, N. & Magoulas, K. (2004). Recovery of phenolic antioxidants from wine industry by-products. *Biosource Technology*, **92**, 201–208.

Maikhunthod, B. & Intarapichet, K.O. (2005). Heat and ultrafiltration extraction of broiler meat carnosine and its antioxidant activity. *Meat Science*, **71**, 364–374.

Manzocco, L., Anese, M. & Nicoli, M.C. (1998). Antioxidant properties of tea extracts as affected by processing. *Lebensmittel-Wissenschaft und-Technologie*, **31**, 694–698.

Manzocco, L., Calligaris, S., Mastrocola, D., Nicoli, M.C. & Lerici, C.R. (2001). Review of non-enzymatic browning and antioxidant capacity in processed foods. *Food Sciences and Technology*, **11**, 340–346.

Manzocco, L., Mastrocola, D. & Nicoli, M.C. (1999). Chain-breaking and oxygen scavering properties of wine as affected by some technological procedures. *Food Research International*, **31**, 673–678.

Marx, M., Stuparic, M., Schieber, A. & Carle, R. (2003). Effects of thermal processing on *trans-cis*-isomerization of β-carotene in carrot juices and carotene-containing preparations. *Food Chemistry*, **83**, 609–617.

Mayer-Miebach, E., Bessnilian, D., Regier, M. & Schumann, H.P. (2005). Thermal processing of carrots: lycopene stability and isomerisation with regard to antioxidant potential. *Food Research International*, **38**, 1103–1108.

Methakhup, S., Chiewchan, N. & Devahastin, S. (2005). Effects of drying methods and conditions on drying kinetics and quality of Indian gooseberry flake. *Lebensmittel-Wissenschaft und-Technologie*, **38**, 579–587.

Minamiyama, Y., Takemura, S., Yoshikawa, T. & Okada, S. (2003). Fermented grain products, production, properties and benefits to heath. *Pathophysiology*, **9**, 221–227.

Moure, A., Dominiquez, H. & Parajo, J.C. (2006). Antioxidant properties of ultrafiltration-recovered soy protein fraction from industrial effluents and their hydrolysates. *Process Biochemistry*, **41**, 447–456.

Moussaid, M., Lacroix, M., Nketsia-Tabiri, J. & Boubekri, C. (2000). Phenolic compounds and the colour of oranges subjected to a combination treatment of waxing and irradiation. *Radiation Physics and Chemistry*, **57**, 273–275.

Mujumdar, A.S. (2006). *Handbook of Industrial Drying*, 3rd edn. Taylor & Francis Group, Philadelphia, PA.

Nawaz, H., Shi, J., Mittal, G.S. & Kakuda, Y. (2006). Extraction of polyphenols from grape seeds and concentration by ultrafiltration. *Separation and Purification Technology*, **48**, 176–181.

Nayak, C.A., Chethana, S., Rastogi, N.K. & Raghavarao, K.S.M.S. (2006). Enhanced mass transfer during solid-liquid extraction of gamma-irradiated red beetroot. *Radiation Physics and Chemistry*, **75**, 173–178.

Naz, S., Siddiqi, R., Sheikh, H. & Sayeed, S.A. (2005). Deterioration of olive, corn and soybean oils due to air, light, heat and deeo-frying. *Food Research International*, **38**, 127–134.

Netzel, M., Strass, G., Birtsch, I., Koenitz, R., Christmann, M. & Birtsch, R. (2003). Effect of grape processing on selected antioxidant phenolics in red wine. *Journal of Food Engineering*, **56**, 223–228.

Nicoli, M.C., Anese, M., Manzocco, L. & Lerici, C.R. (1997a). Antioxidant properties of coffee brews in relation to the roasting degree. *Lebensmittel-Wissenschaft und-Technologie*, **30**, 292–297.

Nicoli, M.C., Anese, M. & Parpinel, M. (1999). Influence of processing on the antioxidant properties of fruit and vegetables. *Trends in Food Science and Technology*, **10**, 94–100.

Nicoli, M.C., Anese, M., Parpinel, M.T., Franceschi, S. & Lerici, C.R. (1997b). Loss and/or formation of antioxidants during food processing and storage. *Cancer Letters*, **114**, 71–74.

Nicoli, M.C., Calligaris, S. & Manzocco, L. (2000). Effect of enzymatic and chemical oxidation on the antioxidant capacity of catechin model systems and apple derivatives. *Journal of Agricultural and Food Chemistry*, **48**, 4576–4580.

Nilsson, J., Stegmark, R. & Akesson, B. (2004). Total antioxidant capacity in different pea (*Pisum sativum*) varieties after blanching and freezing. *Food Chemistry*, **86**, 501–507.
Nindo, C.I., Sun, T., Wang, S.W., Tang, J. & Powers, J.R. (2003). Evaluation of drying technologies for retention of physical quality and antioxidants in asparagus (*Asparagus officinalis* L.). *Lebensmittel-Wissenschaft und-Technologie*, **36**, 507–516.
Nwuha, V. (2000). Novel studies on membrane extraction of bioactive components of green tea in organic solvents: part I. *Journal of Food Engineering*, **44**, 233–238.
Ode-Omowaye, B.I.O., Angersbach, A., Taiwo, K.A. & Knorr, D. (2001). Use of pulsed electric field pre-treatment to improve dehydration characteristics of plant based foods. *Trends in Food Science and Technology*, **12**, 285–295.
Oh, S.-H., Lee, Y.-S., Lee, J.-W., Kim, M.R., Yook, H.-S. & Byun, M.-W. (2005). The effect of γ-irradiation on the non-enzymatic browning reaction in the aqueous model solutions. *Food Chemistry*, **92**, 357–363.
Padda, M.S. & Picha, D.H. (2008). Effect of low temperature storage on phenolic composition and antioxidant activity of sweetpoptatoes. *Postharvest Biology and Technology*, **47**, 176–180.
Piga, A., Agabbio, M., Gambella, F. & Nicoli, M.C. (2002). Retention of antioxidant activity in minimally processed mandarin and satsuma fruits. *Lebensmittel-Wissenschaft und-Technologie*, **35**, 344–347.
Pokorny, J. & Schmidt, S. (2001). Natural antioxidant functionality during food processing. In: *Antioxidants in food. Practical Applications*. Pokorny, J., Yanishlieva, N. & Gordon, M. (eds), CRC Press Woodhead Publishing Ltd, Cambridge, pp. 331–354.
Polydera, A.C., Stoforos, N.G. & Taoukis, P.S. (2005). Effect of high hydrostatic pressure treatment on post processing antioxidant activity of fresh Navel orange juice. *Food Chemistry*, **91**, 495–503.
Qui, W., Jiang, H., Wang, H. & Gao, Y. (2006). Effect of high hydrostatic pressure on lycopene stability. *Food Chemistry*, **97**, 516–523.
Quiles, J.L., Ramirez-Tortosa, M.C., Gomez, J.A., Huertas, J.R. & Mataix, J. (2002). Role of vitamin E and phenolic compounds in the antioxidant capacity, measured by ESR, of virgin olive, olive and sunflower oils after frying. *Food Chemistry*, **76**, 461–468.
Ramesh, M.N., Wolf, W., Tevini, D. & Jung, G. (1999). Studies on inert gas processing of vegetables. *Journal of Food Engineering*, **40**, 199–205.
Rocha-Guzman, N.E., Gonzales-Laredo, R.F., Ibarra-Perez, F.J., Nava-Berumen, C.A. & Gallegos-Infante, J.A. (2007). Effect of pressure cooking on the antioxidant activity of extracts from three common bean (*Phaseolus vulgaris* L.) cultivars. *Food Chemistry*, **100**, 31–35.
Ruiz-Lopez, M.D., Artacho, R., Pineda, M.A.F., de la Serrana, H.L.G. & Martinez, M.C.L. (1995). Stability of α-tocopherol in virgin olive oil during microwave heating. *Lebensmittel-Wissenschaft und-Technologie*, **28**, 644–646.
Saguy, I.S. & Dana, D. (2003). Intefrated approach to deep frying: engineering, nutrition, health and consumer aspects. *Journal of Food Engineering*, **56**, 143–152.
Sahlin, E., Savage, G.P. & Lister, C.E. (2004). Investigation of the antioxidant properties of tomatoes after processing. *Journal of Food Compositions and Analysis*, **17**, 635–647.
Sanchez-Moreno, C., Cano, M.P., de Ancos, B., Martin-Belloso, O. & Martin, A. (2004a). Pulsed electric fields-processed orange juice consumption increases plasma vitamin C and decreases F2-isoprostanes in healthy humans. *Journal of Nutritional Biochemistry*, **15**, 601–607.
Sanchez-Moreno, C., Plaza, L., de Ancos, B. & Cano, M.P. (2003a). Effects of high-pressure processing on health-promoting attributes of freshly squeezed orange juice (*Citrus sinensis* L.) during chilled storage. *European Food Research and Technology*, **216**, 18–22.
Sanchez-Moreno, C., Plaza, L., de Ancos, B. & Cano, M.P. (2003b). Vitamin C, provitamin A, carotenoids and other carotenoids in high-pressurized orange juice during refrigerated storage. *Journal of Agricultural and Food Chemistry*, **51**, 647–653.
Sanchez-Moreno, C., Plaza, L., de Ancos, B. & Cano, M.P. (2004b). Effect of combined treatments of high-pressure and natural additives on carotenoid extractability and antioxidant activity of tomato puree (*Lycopersicum esculentum* Mill.). *European Food Research and Technology*, **219**, 151–160.
Sant'Ana, P.H.M., Stringheta, P.C., Brandao, S.C.C. & de Azered, R.M.C. (1998). Carotenoid retention and vitamin A value in carrot (*Daucus carota* L.) prepared by food service. *Food Chemistry*, **61**, 145–151.
Schieber, A. & Carle, R. (2005). Occurrence of carotenoid *cis*-isomers in food: technological, analytical, and nutritional implications. *Trends in Food Science and Technology*, **16**, 416–422.
Schreiner, M., Huyskens-Keil, S., Krumbein, A., Prono-Widayat, H. & Luedders, P. (2003). Effect of film packing and surface coating on primary and secondary plant compounds in fruit and vegetable products. *Journal of Food Engineering*, **56**, 237–240.

Schweiggert, U., Schieber, A. & Carle, R. (2005). Inactivation of peroxidase, polyphenoloxidase, and lipoxygenase in paprika and chili powder after immediate thermal treatment of the plant material. *Innovative Food Science and Emerging Technologies*, **6**, 403–411.

Sensoy, I., Rosen, R.T., Ho, C.-T. & Karwe, M.V. (2006). Effect of processing on buckwheat phenolics and antioxidant activity. *Food Chemistry*, **99**, 388–393.

Sharma, S.K. & Le Maguer, M. (1996). Kinetics of lycopene degradation in tomato pulp solids under different processing and storage conditions. *Food Research International*, **29**, 309–315.

Shi, J., Le Maguer, M., Kakuba, Y., Liptay, A. & Niekamp, F. (1999). Lycopene degradation and isomerization in tomato dehydration. *Food Research International*, **32**, 15–21.

Skrede, G., Nilsson, A., Baardseth, P., Rosenfeld, H.J., Enersen, G. & Slinde, E. (1997). Evaluation of carrot varieties for production of deep fried carrot chips – III. Carotenoids. *Food Research International*, **30**, 73–81.

Smith, J.S. & Pillai, S. (2004). Irradiation and food safety. *Food Technology*, **7**(2), 48–55.

Song, H.P., Kim, D.H., Jo, C., Lee, C.H., Kim, K.S. & Byun, M.W. (2006). Effect of gamma irradiation on the microbiological quality and antioxidant activity of fresh vegetable juice. *Food Microbiology*, **23**, 372–378.

Soysal, Y. (2004). Microwave drying characteristics of parsley. *Biosystems Engineering*, **89**, 167–173.

Soysal, Y., Öztekin, S. & Eren, Ö. (2006). Microwave drying of parsley: modelling, kinetics and energy aspects. *Biosystems Engineering*, **93**, 403–413.

Starzyńska-Janiszewska, A., Stodolak, B. & Jamróz, M. (2008). Antioxidant properties of extracts from fermentum and cooked seeds of Polish cultivars of *Lathyrus dativus*. *Food Chemistry*, **109**, 285–292.

Su, M.-S. & Silva, J.L. (2006). Antioxidant activity, anthocyanins, and phenolics of rabbiteye blue berry (*Vaccinium ashei*) by-products as affected by fermentation. *Food Chemistry*, **97**, 447–451.

Sumnu, G., Turabi, E. & Oztop, M. (2005). Drying of carrots in microwave and halogen lamp-microwave combination ovens. *Lebensmittel-Wissenschaft und-Technologie*, **38**, 549–553.

Tauscher, B. (1995). Pausterisation of food by hydrostatic high pressure: chemical aspects. *Zeitschrift für Lebensmitteluntersuchung und – Forschung*, **200**, 3–13.

Todisco, S., Tallarico, P. & Gupta, B.B. (2002). Mass transfer and polyphenols retention in the clarification of black tea with ceramic membranes. *Innovative Food Science and Emerging Technologies*, **3**, 255–262.

Tomaino, A., Cimino, F., Zimbalatti, V., Venuti, V., Sulfaro, V., De Pasquale, A. & Saija, A. (2005). Influence of heating on antioxidant activity and the chemical composition of some spice essential oils. *Food Chemistry*, **89**, 549–554.

Tomas-Barberan, F.A. & Espin, J.C. (2001). Phenolic compounds and related enzymes as determinants of quality in fruits and vegetables. *Journal of the Science of Food and Agriculture*, **81**, 853–876.

Toor, R.K. & Savage, G.P. (2006). Effect of semi-drying on the antioxidant components of tomatoes. *Food Chemistry*, **94**, 90–97.

Torregosa, F., Esteve, M.J., Frigola, A. & Cortes, C. (2006). Ascorbic acid stability during refrigerated storage of orange-carrot juice treated by high pulsed electric field as compared with pasteurized juice. *Journal of Food Engineering*, **73**, 339–345.

Torres, J.A. & Velazquez, G. (2005). Commercial opportunities and research challenges in the high pressure processing of foods. *Journal of Food Engineering*, **67**, 95–112.

Towo, E., Matuschek, E. & Svanberg, U. (2006). Fermentation and enzyme treatment of tannin sorghum gruels: effect on phenolic compounds, phytate and in vitro accessible iron. *Food Chemistry*, **94**, 369–376.

Turano, E., Curcio, S., de Paola, M.G., Calabro, V. & Iorio, G. (2002). An integrated centrifugation-ultrafiltration system in the treatment of olive mill wastewater. *Journal of Membrane Science*, **209**, 519–531.

Turkmen, N., Sari, F. & Velioglu, S. (2005). The effect of cooking methods on total phenolics and antioxidant activity of selected green vegetables. *Food Chemistry*, **93**, 713–718.

Tybaro, F. & Ursini, F. (1995). Misura di attivita antiossidante tramite l'ananlisi di una cinetica di competizione. In: *Proceeding of the Workshop Antiossidanti Naturali Negli Alimenti. Aspetti Tecnologici e Qualitativi*. Conte, L., Rosa, D. & Zamorani, A. (eds), CLEUP, Padova, pp. 45–53.

Vadivambal, R. & Jayas, D.S. (2007). Changes in quality of microwave-treated agricultural products – a review. *Biosystems Engineering*, **98**, 1–16.

Vattem, D.A., Lin, Y.-T., Labbe, R.G. & Shett, K. (2004a). Antimicrobial activity against select food-born pathogens by phenolic antioxidants enriched in cranberry pomace by solid-state bioprocessing using the food grade fungus *Rhizopus oligosporus*. *Process Biochemistry*, **39**, 1939–1946.

Vattem, D.A., Lin, Y.-T., Labbe, R.G. & Shetty, K. (2004b). Phenolic antioxidant mobilization in cranberry pomace by solid-state bioprocessing using food grade fungus *Lentinus edodes* and effect on antimicrobial activity against select food borne pathogens. *Innovative Food Science and Emerging Technologies*, **5**, 81–91.

Vikram, V.B., Ramesh, M.N. & Prapulla, S.G. (2005). Thermal degradation kinetics of nutrients in orange juice heated by electromagnetic and conventional methods. *Journal of Food Engineering*, **69**, 31–40.

Wachtel-Galor, S., Wong, K.W. & Benzie, I.F.F. (2008). The effect of cooking on *Brassica* vegetables. *Food Chemistry*, **110**, 706–710.

Wang, B.J., Wei, T.C. & Yu, Z.R. (2005). Effect of operating temperature on component distribution of West Indian cherry juice in a microflitration system. *Lebensmittel-Wissenschaft und-Technologie*, **38**, 683–689.

Wang, J. & Xi, Y.S. (2005). Drying characteristics and drying quality of carrot using a two-stage microwave process. *Journal of Food Engineering*, **68**, 505–511.

Wang, Y.C., Yu, R.C. & Chou, C.C. (2006). Antioxidant activities of soymilk fermented with lactic acid bacteria and bifidobacteria. *Food Microbiology*, **23**, 128–135.

Wang, Y.C., Yu, R.C., Yang, H.Y. & Chou, C.C. (2003). Sugar, acid and B-vitamin contents in soymilk fermented with lactic acid bacteria alone or simultaneously with bifidobacteria. *Food Microbiology*, **20**, 333–338.

Willits, M.G., Giovanni, M., Prata, R.T.N., Kramer C.M., De Luca V., Steffens J.C. & Graser G. (2004). Bio-fermentation of modified flavonoids: an example of in vivo diversification of secondary metabolites. *Phytochemistry*, **65**, 31–41.

Wollgast, J. & Anklam, E. (2002). Review on polyphenols in *Theobroma cacao*: changes in composition during the manufacture of chocolate and methodology for identification and quantification. *Food Research International*, **33**, 423–447.

Wu, X. Gu, L., Holden, J., Haytowitz, D.B., Gebhardt, S.E., Beecher, G. & Prior R.L. (2004). Development of a database for total antioxidant capacity in foods: a preliminary study. *Journal of Food Composition Analysis*, **17**, 407–422.

Xu, L., Lamb, K., Layton, L. & Kumar, A. (2004). A membrane-based process for recovering isoflavones from a waste stream of soy processing. *Food Research International*, **37**, 867–874.

Yamauchi, R., Yamamoto, N. & Kato, K. (1995). Iron-catalyzed reaction products of alpha-tocopherol with methyl 13(S)-hydroperoxy-9(Z), 11(E)-octadecadienoate. *Lipids*, **30**, 395–404.

Yang, J.-H., Mau, J.-L., Ko, P.-T. & Huang, L.-C. (2000). Antioxidant properties of fermented soybean broth. *Food Chemistry*, **71**, 249–254.

Yen, Y.-H., Shih, C.-H. & Chang, C.-H. (2008). Effect of adding ascorbic acid and glucose on the antioxidant properties during storage of dried carrot. *Food Chemistry*, **107**, 265–272.

Yilmaz, Y. & Toledo, R. (2005). Antioxidant activity of water-soluble Maillard reaction products. *Food Chemistry*, **93**, 273–278.

Yoshida, H., Tomiyama, Y., Hirakawa, Y. & Mizushina, Y. (2006). Microwave roasting effcts on the oxidative stability of oils and molecular species of triacylglycerols in the kernels of pumpkin (*Cucurbita* spp.) seeds. *Journal of Food Composition Analysis*, **19**, 330–339.

Zanoni, B., Peri, C., Nani, R. & Lavelli, V. (1999). Oxidative heat damages of tomato halves as affected by drying. *Food Research International*, **31**, 395–401.

Zhang, D. & Hamauzu, Y. (2004). Phenolics, ascorbic acid, carotenoids and antioxidant activity of broccoli and their changes during conventional and microwave cooking. *Food Chemistry*, **88**, 503–509.

Zieliński, H., Kozłowska, H. & Lewczuk, B. (2001). Bioactive compounds in the cereal grains before and after hydrothermal processing. *Innovative Food Science and Emerging Technologies*, **2**, 159–169.

10 Development and commercialization of microalgae-based functional lipids

Jaouad Fichtali and S.P.J. Namal Senanayake

10.1 Introduction

Microalgae are a very diverse group of organisms that consist of both prokaryotic and eukaryotic forms. *Crypthecodinium cohnii* is a unicellular, non-photosynthetic, marine dinoflagellate and is found naturally in association with decaying seaweed. It is evident that *C. cohnii* is probably the major source of a triacylglycerol (TAG) oil rich in docosahexaenoic acid (DHA; 22:6*n*-3) and essentially free of all other polyunsaturated fatty acids (PUFAs), making it not only unique in such a fatty acid profile, but also highly desirable commercially. Because it grows easily in bioreactors, it is an ideal source of DHA. For these reasons, *C. cohnii* represents a promising microalga for the commercial production of DHA. *Schizochytrium* sp. is a thraustochytrid and is a member of the Chromista kingdom. Thraustochytrids are single-cell organisms that produce a high oil and long-chain PUFA content. There are no reports of toxicity or pathogenicity associated with *Schizochytrium* in the literature (Hammond *et al.* 2002), and it contains no algal toxins as determined by analytical methods. *C. cohnii* and *Schizochytrium* sp. have been specifically selected for commercial production because they have been in culture for many years and in all the numerous studies using these species there have never been any indications of pathogenicity or toxigenicity. Successful cultivation of *C. cohnii* and *Schizochytrium* sp. to produce commercial algal lipids containing DHA has been achieved only by Martek Biosciences Corporation in the United States.

10.2 Industrial production of microalgal lipids

10.2.1 Fermentation

C. cohnii and *Schizochytrium* sp. are grown by heterotrophic fermentation. The heterotrophic fermentation is independent of light and occurs in bioreactors, which can be operated axenically (pure culture of one species only) and under controlled optimum conditions. Further benefits of heterotrophic fermentation include higher biomass concentrations, increased reproducibility and straightforward scale-up of the fermentation processes. The industrial DHA production potential of microalgae mainly depends on the primary strain selection for the fatty acid composition, yield and adaptability to the fermentor. As microalgae are highly environmentally dependent and the synthesis of fatty acids, especially PUFAs, is influenced by many parameters, such as culture age, salinity, medium composition, temperature and

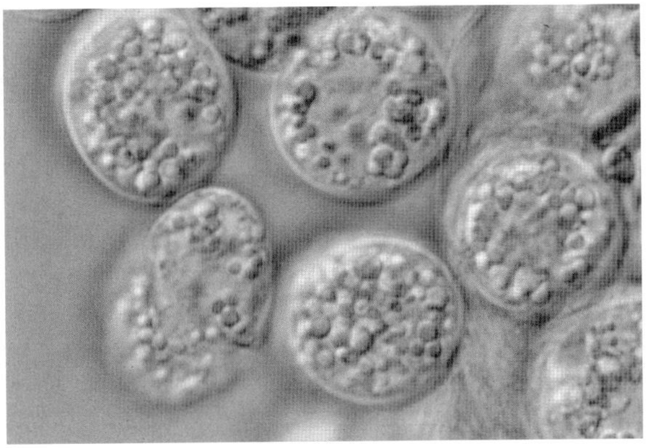

Fig. 10.1 Cells of *C. cohnii* showing high concentration of DHA-rich oil bodies. For a colour version of this figure, please see the plate section.

aeration, a cost-effective fermentation process should be established through systematic investigations into the effects of various nutrient and environmental conditions and the use of various high-cell-density strategies for the growth and DHA production of the selected microalgae. The production strains can be selected for rapid growth and high levels of production of the specific oils. Master seed banks of all strains are maintained under liquid nitrogen conditions and working seed stocks, prepared from this master seed bank, are also maintained cryogenically. On initiation of a production run, an individual ampoule from a working seed is used to inoculate a shake flask. The medium used to grow *C. cohnii* and *Schizochytrium* sp. from shake flask to production-scale contains a carbon source, nitrogen source, salts and a number of other micronutrients. The cultures are transferred successively to large fermentors based on specific growth parameters. Throughout the process, the concentration of carbon substrate, pH, temperature, pressure, airflow, agitation and dissolved oxygen are regularly monitored and controlled. All fermentations involved in such high-value products require axenic culture. Therefore, at each transfer stage in the inoculum sequence, broth samples are plated to establish the microbial purity. In addition, the purity of the cultures is also monitored every 24 hours by manual observation of a sample under a microscope and the plating of culture broth samples under several conditions to confirm the presence or absence of any microbial contamination. In the final fermentation vessel, the cultures are allowed to go into nitrogen limitation, when they begin producing their storage products, TAGs, which are high in DHA. During this lipid accumulation phase, *C. cohnii* cells lose their flagella and become 'cyst-like' cells packed with DHA-rich lipid bodies (Figure 10.1). The lipid portion at this stage constitutes over one-half the cell dry weight and the cells are ready for harvesting.

10.2.2 Harvesting

Downstream processing of DHA-rich microalgae involves one or more solid–liquid separation techniques. At the conclusion of the fermentation, the biomass may need to be separated from the culture medium via centrifugation, filtration, sedimentation or flocculation. Table 10.1 provides information on some centrifuges and filters that have been used for the recovery of microalgae.

Table 10.1 Performance of centrifuges and filters used for harvesting of microalgae

Harvest method	Equipment	Manufacturer(s)	Operation mode	Type of concentration procedure
Centrifugation	Disc-stack centrifuge	Westfalia, Alfa Laval	Manual, intermittent or continuous	One step
	Decanter centrifuge	Westfalia, Alfa Laval	Continuous	For final concentration
	Nozzle bowl centrifuge	Westfalia, Alfa Laval	Continuous	For preconcentration or final concentration
Pressure Filtration	Filter basket	Seitz, Dinglinger	Discontinuous	For preconcentration
	Chamber filter press	Netzsch	Discontinuous	One step
	Suction filter	Seitz	Discontinuous	For preconcentration
Vacuum Filtration	Vacuum drum filter	Dorr Olliver	Continuous	One step
	Belt filter press	Dinglinger, Tefsa-USA	Continuous	For preconcentration

10.2.2.1 *Centrifugation*

Centrifugation is often considered as the preferred harvesting method of microalgae. A centrifuge is a piece of equipment, generally driven by a motor, which puts an object in rotation around a fixed axis, applying force perpendicular to the axis. Centrifugation is a process that involves the use of the centripetal force for the separation of mixtures, used in industry and in laboratory settings. More dense components of the mixture migrate away from the axis of the centrifuge, while less dense components of the mixture migrate toward the axis. The rate of centrifugation is specified by the acceleration applied to the sample, typically measured in revolutions per minute (rpm) or the equivalent multiple of earth's gravitational force (*g*'s). The particles' settling velocity in centrifugation is a function of their size and shape, centrifugal acceleration, the volume fraction of solids present, the density difference between the particle and the liquid, and the viscosity. The selection of a centrifuge for harvesting microalgae may involve expensive pilot-scale evaluations. A few simple laboratory-scale centrifugation and gravimetric sedimentation tests would provide indications about the efficiency of separation via centrifugation.

There are several different basic types of centrifuges that can be used: disc-stack, bowl, tubular bowl and scroll discharge decanter centrifuge, depending on the particle size of the wet biomass. Disc-stack centrifuges come in several types depending on whether the solids are discharged or retained and the mechanism of discharge of solids. Disc-stack centrifuges are suitable for a wide range of separation tasks that involve lower solid concentrations and smaller particle and droplet sizes. Disc-stack centrifuges normally feature four main sections: inlet zone, disc-stack area, liquid discharge section and the solid discharge section. A disc-stack centrifuge separates solids from the liquid phase in one single continuous process using extremely high centrifugal forces. When the denser solids are subjected to such forces, they are forced outward against the rotating bowl wall, while the less dense liquid phases form inner concentric layers. The area where these two different liquid phases meet is referred to as the interface position. This can be easily varied in order to ensure that the separation takes place with maximum efficiency. Inserting special plates (disc stack) provides additional surface settling area, which contributes to speeding up the separation process considerably.

It is the particular configuration, shape and design of these disc stacks that make it possible for this type of centrifuge to undertake the continuous separation of a wide range of different solids from the liquid phase. The concentrated solids so formed can be removed manually, continuously or intermittently, depending on the centrifuge type and the amount of solids involved in the specific application.

The tubular bowl centrifuge has been used longer than most other designs of centrifuge. It is generally operated vertically, the tubular rotor providing a long flow path, enabling clarification. The tubular bowl centrifuge is based on a very simple geometry: it is formed by a tube, of length several times its diameter, rotating between bearings at each end. The process stream enters at the bottom of the centrifuge and high centrifugal forces act to separate out the solids. The bulk of the solids will adhere on the walls of the bowl, while the liquid phase exits at the top of the centrifuge. As this type of system lacks a provision of solid rejection, the solids can only be removed by stopping the machine, dismantling it and scraping or flushing the solids out manually. The tubular bowl attains high g-force that permits good solid dewatering. However, the solid capacity of the tubular bowl centrifuge is very limited. Foaming can be a problem unless the system includes special skimming or centripetal pumps. This type of centrifuge may be suitable for diluted small volumes of fermentation broth.

The decanter centrifuge is based on the simple concept of a settling tank, in which solids gradually fall to the bottom due to the force of gravity. However, the decanter centrifuge is designed to provide the continuous mechanical separation of solids from the liquid phase, to keep pace with modern industrial demands. The scroll discharge decanter consists of a solid bowl, tapered at one end, and a close-fitting helical screw, which rotates at a slightly different speed to the bowl. The g-forces are smaller (4000–10 000 g), and the clarity of the fermentation broth is not as good as with disc-stack centrifuges. The scroll discharge decanter is suitable for slurries with high solid content (up to 80% v/v solids). Depending on the particular configuration and equipment, a decanter centrifuge can be used to separate a wide range of different solids from the liquid phase on a continuous basis.

10.2.2.2 *Filtration*

Filtration is a mechanical/physical operation, which is used for the separation of solids from fluids by interposing a medium in the fluid flow through which the fluid can pass, but in which the solids in the fluid are retained. The degree of separation will depend on the pore size and the thickness of the medium, as well as the mechanisms that occur during filtration. It is possible to perform filtration using different conditions, but a number of factors will influence the choice of suitable type of equipment to meet the specified requirements at minimum cost, including (a) the physical properties of the filtrate (i.e. viscosity and density), (b) the nature of solid particles (i.e. size, shape and packing characteristics), (c) the solid to liquid ratio, (d) the scale of operation, (e) batch or continuous operation, (f) pressure or vacuum operation and (g) the need for aseptic conditions. The simplest method of filtration is to pass a solution of a solid and fluid through a porous interface so that the solid is trapped, while the fluid passes through. This principle relies on the size difference between the particles making up the fluid, and the particles making up the solid. In the laboratory, a Büchner funnel is often used, with a paper filter serving as the porous barrier.

Rotary vacuum drum filters and chamber filter presses appear to be commonly employed types of filters for recovery of biomass from mass cultures of microalgae (Mohn 1980). These filters have the advantage of continuous operation and are useful when sterility requirements are not rigorous. Rotary filters are available for vacuum or pressure filtration. The rotary vacuum drum filter is one of the oldest filters applied to industrial liquid filtration. A rotary drum filter consists of a drum frame covered with filter cloth. Half the drum is submerged in the algal slurry, with the other half above it. As the drum rotates, the algal slurry is sucked into the cloth. As the drum rotates out of the slurry, the cake is dried. This drying is caused by the vacuum continuously being drawn through the cake in the exposed section of the drum. At the end of the rotation cycle, the filter cake is discharged and the process repeats itself. The filters may incorporate a drum cloth that is caulked onto the drum itself, or they may utilize an endless belt, which tracks off and discharges away from the drum.

Membrane microfiltration and ultrafiltration are possible alternatives to conventional filtration for recovery of microalgal biomass. Microfiltration may be suitable for recovery of fragile cells (Petrusevski *et al.* 1995). However, for the large-scale production of algal biomass, membrane filtration technology may not be a feasible process. Small aquaculture farms generally use membrane filtration for harvesting algal cells for feeding shellfish larvae (Borowitzka 1997). The recovery of two different algal cultures (varying in size) using filtration under pressure or vacuum conditions was evaluated by Mohn (1980). The belt press and the chamber filter press were found to be the most adequate devices operating under pressure or vacuum conditions (Table 10.1).

Spray-drying

Spray-drying has become the most important method for dehydration of fluid foods in the Western world. Spray-drying involves the atomization of a liquid feedstock into a spray of droplets and contacting the droplets with hot air in a drying chamber. The drying medium is typically air. The drying proceeds until the desired moisture content is reached in the sprayed particles and the product is then separated from the air. The sprays are produced by either rotary or nozzle atomizers. Evaporation of moisture from the droplets and formation of dry particles proceed under controlled temperature and airflow conditions. As soon as droplets of the spray come into contact with the drying air, evaporation takes place from the saturated vapor film, which is quickly established at the droplet surface. Due to the high specific surface area and the existing temperature and moisture gradients, an intense heat and mass transfer result in an efficient drying. The evaporation leads to a cooling of the droplet and thus to a small thermal load. Powder is discharged continuously from the drying chamber. Operating conditions and dryer design are selected according to the drying characteristics of the product and powder specification.

Every spray-dryer consists of feed pump, atomizer, air heater, air disperser, drying chamber, and systems for exhaust air cleaning and powder recovery. Widely varying drying characteristics and quality requirements of the various spray-dried products determine the selection of the atomizer, the most suitable airflow pattern, and the drying chamber design. For operation of a spray-dryer, it is usual practice to preconcentrate the liquid as much as possible for several reasons: (a) economy of operation, (b) increased capacity, (c) increase of particle size, (d) increase of particle density, (e) more efficient powder separation and (f) improved dispersibility of the product.

Spray-dryers can dry a product very quickly compared to other methods of drying. They also turn a solution, or slurry, into a dried powder in a single step, which can be advantageous for profit maximization and process simplification. Relatively high temperatures are needed for spray-drying operations. However, heat damage to products is generally only slight, because of an evaporative cooling effect during the critical drying period, and because the subsequent time of exposure to high temperatures of the dry material may be very short. The typical surface temperature of a particle during the constant drying phase is relatively low (45–50°C). For this reason, it is possible to spray-dry some algal suspensions without destruction of the organisms. The physical properties of the products are intimately associated with the powder structure, which is generated during spray-drying. It is possible to control many of the factors which influence powder structure in order to obtain the desired properties.

The most important responsibility for an operator of a spray-drier is to maintain a constant moisture content of the powder. This is required to meet legal standards and for maintaining a uniform quality. It is important to understand how the final moisture content can be controlled by changing the operating conditions. But first, it should be noted that the final moisture content is controlled by the relative humidity of the outlet air. If that value is too high, then the powder particles will absorb moisture rather than give moisture away. The primary conditions which may be controlled directly by the operator are (a) inlet temperature, (b) flow rate of liquid feed (pump speed and pump pressure), (c) airflow rate (fan speed and position of baffles) and (d) the particle size (adjustment of atomizer). Among other operating conditions, outlet temperature and relative humidity of the outlet air are particularly important and need careful attention. However, these can only be indirectly controlled by adjusting the primary conditions. For outlet temperature, the condition is dependent on liquid feed intake. If the feed intake is increased, the outlet temperature will drop. If the intake is reduced, the outlet temperature will increase and approach the inlet temperature. The outlet temperature will also be affected by the airflow rate. For a constant inlet temperature and constant feed intake, an increase in the airflow will raise the outlet temperature.

Lyophilization

Lyophilization is a process more commonly known as freeze-drying. Lyophilization is the process of removing water from a product by sublimation and desorption. This process is performed in lyophilization equipment, which consists of a drying chamber with temperature controlled shelves, a condenser to trap water removed from the product, a cooling system to supply refrigerant to the shelves and condenser, and a vacuum system to reduce the pressure in the chamber and condenser to facilitate the drying process.

Lyophilization cycles consist of three phases: freezing, primary drying and secondary drying. Conditions in the dryer are varied through the cycle to insure that the resulting product has the desired physical and chemical properties, and that the required stability is achieved.

During the freezing phase, the goal is to freeze the mobile water of the product. Significant supercooling may be encountered, so the product temperature may have to be much lower than the actual freezing point of the solution before freezing occurs. The rate of cooling will influence the structure of the frozen matrix. If the water freezes quickly, the ice crystals will be small. This may cause a finer pore structure in the product with higher resistance to flow of water vapor and longer primary drying time. If freezing is slower, ice crystals will grow from the cooling surface and may be larger. The resultant product may have coarser pore structure and perhaps a shorter primary drying time.

In the primary drying phase, the chamber pressure is reduced, and heat is applied to the product to cause the frozen mobile water to sublime. The water vapor is collected on the surface of a condenser. The condenser must have sufficient surface area and cooling capacity to hold all the sublimed water from the batch at a temperature lower than the product temperature. If the temperature of the ice on the condenser is warmer than the product, water vapor will tend to move toward the product, and drying will stop.

Table 10.2 Pilot- and industrial-scale techniques for disruption of microalgal cells

Method	Technique	Principle
Physical	High-pressure homogenization	Cells forced through small orifice are broken by high shear
	Microfluidization	By orienting the liquid stream containing cells in precisely defined microchannels and impinging them against one another
	Bead milling	Cells are ruptured by high mechanical forces
	Ultrasonication	Cells are broken with ultrasonic cavitation
	Crushing	Cells are crushed between glass or steel balls
	Grinding	Cells are ruptured by grinding with abrasives
	Osmotic shock	Osmotic rupture of cell membranes
Chemical	Alkali	Solubilize cell membranes and cause saponification of lipids
	Acids	Solubilize cell membranes
	Detergents	Solubilize cell membranes
	Organic solvents	Dissolve cell walls
Biological	Enzymes	Cell walls are digested, providing disintegration

It is important to control the drying rate and the heating rate during this phase. If the drying proceeds too rapidly, the dried product can be displaced out of the container by escaping water vapor. If the product is heated too rapidly, it will melt or collapse. This will cause degradation of the product, and will change the physical characteristics of the dried material, making it harder to reconstitute and visually unappealing. While frozen mobile water is present, the product must be held below the eutectic temperature or glass transition temperature.

10.2.3 Cell disruption

Disruption of microalgal cells focuses on obtaining the desired products from within the cell, and it is the cell wall that must be disrupted to allow the contents of the cell out. The choice of disruption methods will vary depending on the type of cell and its particular cell wall structure. The objectives of cell disruption are to (a) solubilize the maximum amount of the product present in the cell while maintaining optimum biological activity, (b) avoid secondary alteration of the product and (c) limit the detrimental effects of disruption on the subsequent processing steps. A wide range of disintegration methods have been developed in trying to achieve these objectives. These can be grouped into physical, chemical and biological methods, depending on the basic effects which cause disintegration (Table 10.2). The method selected for large-scale cell disruption will be different in every case, but will depend on (a) susceptibility of cells to disruption, (b) product stability, (c) ease of extraction from cell debris, (d) speed of the method and (e) cost of the method. Regardless of the method, disruption must be effective and the product must remain in an active form.

Currently, intracellular products are released from microalgal cells, mainly by mechanical disruption of the cells. In this process, the cell envelope is physically broken, releasing all intracellular components into the surrounding medium. Mechanical cell disruption methods suffer from several drawbacks. Because cells are broken completely, all intracellular materials are released. Therefore, the product of interest must be separated from a complex mixture of proteins, nucleic acids and cell wall fragments. Released nucleic acids may increase the viscosity of the solution and may complicate subsequent processing steps. The cell debris, produced by mechanical means, often consists of small cell fragments, making the solution difficult to clarify. Complete product release often requires more than one pass through the disruption device, exacerbating the problem by further reducing the size of the fragments. These are difficult to remove by continuous centrifugation because the throughput of the device is inversely related to the particle size. Filtration is complicated by the gelatinous nature of the homogenate and by its tendency to foul membranes. Furthermore, mechanical methods expose the cells, and hence the extracted product, to very harsh conditions. Equipment for cell disruption includes (a) bead mills, (b) high-pressure homogenizers and (c) ultrasonics, among others.

10.2.3.1 *Bead mills*

Bead mills consist of either a horizontal grinding chamber containing impellers or rotating discs mounted concentrically or off-centered, on a motor-driven shaft. The grinding action is due to glass or plastic beads typically occupying 80–85% of the working volume of the chamber. Bead mills are usually operated in batch or continuous mode. The beads are retained in the grinding chamber either by a sieve plate or a similar device to retain the beads. The units require high-capacity cooling systems. Horizontal units are preferred for cell disruption to reduce the fluidizing effect in the vertical units. The parameters involved in the process are numerous and include bead type and size, configuration of disc/impeller, speed of disc/impeller, loading and density, cell properties, cell concentration, process feed flow rate and residence time, among others. Given the large number of variables that need to be optimized for disruption of cells in a bead mill, it is not surprising that this technique lags behind high-pressure homogenization. Although the cell disruption can be batchwise, continuous operation is more practical with large fluid volumes. The bead mill has been used in the disruption of yeast cells, bacteria, algae and filamentous organisms. The performance of a bead mill is related to a number of factors such as selecting the proper grinding bead, operating at the correct disc peripheral speed and finding the right feed pump rate.

10.2.3.2 *High-pressure homogenizers*

High-pressure homogenizers lend themselves to a relatively user-friendly operation. Cell disruption is essentially achieved by passing the cells at high pressures through a small valve or orifice. Several types of equipment including APV high-pressure homogenizers (APV Gaulin, Germany), Rannie hyper homogenizers (APV Rannie AS, Denmark), ultra high-pressure cell disrupters (Constant Systems Ltd, United Kingdom) and microfluidizers (Microfluidics, USA) are currently available in the market. Some of the high-pressure homogenizers can achieve pressures of up to 1500 bar. Relatively new valve design and material have improved the cell breakage efficiency and reduced wear and tear. Microfluidics design, which is based on two impinging process fluid jets, can operate at high pressures even though requiring a supply of pressurized air to activate the equipment. Ultra high-pressure cell disrupters are relatively new to the field. The equipment has been tested for containment and can reach very high pressures of up to 2700 bar, depending on the prototype.

The Niro VHP homogenizer consists of a horizontal single-acting reciprocating multi-plunger pump (single plunger for table top lab units), with a homogenizing valve installed on the high-pressure outlet manifold. The homogenizer pump operates at a near constant flow rate regardless of the homogenizing valve set pressure (back pressure to the pump). The homogenizing valve consists of three main components: the impact head, impact ring and passage head. Homogenizing pressure is increased when pressure is applied by the pneumatic actuator to the valve shaft, closing the adjustable gap (flow area) between the impact head and the passage head. The homogenizing effect is caused by the product entering the valve inlet at pressure, and as it passes through the minute gap, the velocity rapidly increases while the pressure rapidly decreases to atmospheric pressure. The product is subject to shear/impingement as it is forced through the valve, then impacts as it hits the impact ring at high velocity. The intense energy change also causes turbulence and cavitation. The homogenizing efficiency is due to a combination of the pressure applied and the geometry of the valve. There are different valve geometries available depending on the application, whether you are trying to obtain a stable emulsion, particle size reduction of a suspension or cell rupture. The Niro R type sharp profile valve design is a result of fluid dynamic calculations, R&D testing and experience gained in the field. The Niro Soavi R type valve requires substantially less pressure than flat profile valves or alternative homogenizing technologies to achieve the same degree of particle size reduction or cell rupture.

10.2.3.3 *Ultrasonication*

The treatment of microalgal cells in suspension with inaudible ultrasound (typically 20–50 kHz) results in their inactivation and disruption. Ultrasonication utilizes the rapid sinusoidal movement of a probe within the liquid. It is characterized by high frequency (18 kHz to 1 MHz), small displacements (less than about 50 mm), moderate velocities (a few m/second), steep transverse velocity gradients (up to 4000/second) and very high acceleration (up to about 80 000 *g*). Ultrasonication produces cavitation phenomena when acoustic power inputs are sufficiently high to allow the multiple production of microbubbles at nucleation sites in the fluid. The bubbles grow during the rarefying phase of the sound wave, and then collapse during the compression phase. On collapse, a violent shock wave passes through the medium. The whole process of gas bubble nucleation, growth and collapse due to the action of intense sound waves is called cavitation. The collapse of the bubbles converts sonic energy into mechanical energy in the form of shock waves equivalent to several thousand atmospheric pressures (300 MPa). This energy imparts motions to parts of cells which disintegrate when their kinetic energy content exceeds the wall strength. Equipment for the large-scale use of ultrasonics has been available for many years and is widely used by the chemical industry, but has not yet found extensive use in microalgal cell disruption. As with most cell disintegration methods, very fine cell debris particles may be produced, which can hinder further processing. Disadvantages of this method include (a) heat generated by the ultrasound process must be dissipated, (b) high noise levels (most systems require hearing protection and sonic enclosures), (c) variability in yield and (d) generation of free radicals that can react with other molecules. Ultrasonication remains, however, a popular, useful and simple small-scale method for microbial cell disruption.

10.2.4 Extraction and refining

The extraction methods should be rapid, efficient and gentle in order to reduce degradation of the microalgal lipids. The extraction solvents should be inexpensive, volatile, free from toxic compounds, able to form a two-phase system with water, and be poor extractors of undesirable components (Grima *et al.* 1999). The oil is extracted from the dried algal biomass and processed using procedures that have been well established in the edible oil industry. The oil can also be winterized to remove a higher melting oil fraction. Winterized oil is then refined to remove free fatty acids and other impurities such as phospholipids, proteinaceous and mucilaginous substances. The refined oil is then bleached, a process in which citric acid, activated silica and bleaching clay are added to absorb any remaining polar materials and heavy metals, and to break down lipid oxidation products. Next, the oil is deodorized. The deodorized oil is diluted with high-oleic sunflower oil to bring the DHA level to an industry standard of 40%. The oil is then stabilized by adding antioxidants, mainly ascorbyl palmitate and tocopherols. The oil is finally packaged in nitrogen-purged containers and stored frozen until shipment. Oil samples are collected at each step of the process for analysis of peroxide value, free fatty acids, residual soaps, phosphorus content and fatty acid composition. All unit operations are carried out according to current good manufacturing practice regulations as required by the US Food and Drug Administration (FDA).

10.2.5 The integrated process

Successful development and commercialization of microalgae-based functional lipids rely on satisfying both operation and innovation need. The usual challenge is how to operate efficiently in the present while also innovating effectively for the future. An example of an integrated process is given in Figure 10.2, which shows the main unit operations to produce algae oil. Different variations to this process have shown to be feasible and the choice is generally based on cost (both capital and operating), safety, quality and speed. Oxidation is a major issue with PUFA oils and special attention should be paid to the five 'enemies' of oxidation – temperature, oxygen, transition metals, light and time – when making decisions about equipment design and manufacturing parameters. Establishing final manufacturing parameters for doing the same things better and faster with less waste and higher quality takes major time and effort. A successful transition from development to manufacturing should be based on risk minimization and needs a collaborative effort between development engineers/scientists and manufacturing staff. Manufacturing should take into consideration that processes developed at pilot-scale are not final, and issues neglected in the development phase (such as impurities and change in feed characteristics) could have major impact on the process and should be addressed in production. The goal is to achieve a production environment where (1) work is defined, repetitive and clear with shared goals; (2) information is readily available; (3) rules and procedures are applicable; (4) operational specs are complete; (5) efficiency, technical perfection and measurable productivity are achieved; (6) environment is stable with predefined outcomes; and (7) downtime is reduced.

10.3 Composition of algal biomass

Results of proximate analysis conducted on dried microalgae from *C. cohnii* and *Schizochytrium* sp. are reported in Table 10.3. Crude fat (40–66%) and protein (20–23%) were the predominant components of dried microalgae, with minor amounts of ash (~4.0%), crude fiber (2.0–5.8%) and moisture (~3.0%) accounting for the balance. DHA content, calculated on a dry weight basis, in the whole cell microalgae from *C. cohnii* and *Schizochytrium* sp. comprised 20.4–21.0% of the cells. Lipid class composition of the dried microalgae from *Schizochytrium* sp. was determined by extracting crude lipid from the cells and fractionating into various lipid class components. TAGs represented the major lipid class fraction in the crude lipids, accounting for 90–92% of the total crude lipid. Minor amounts of fatty acid

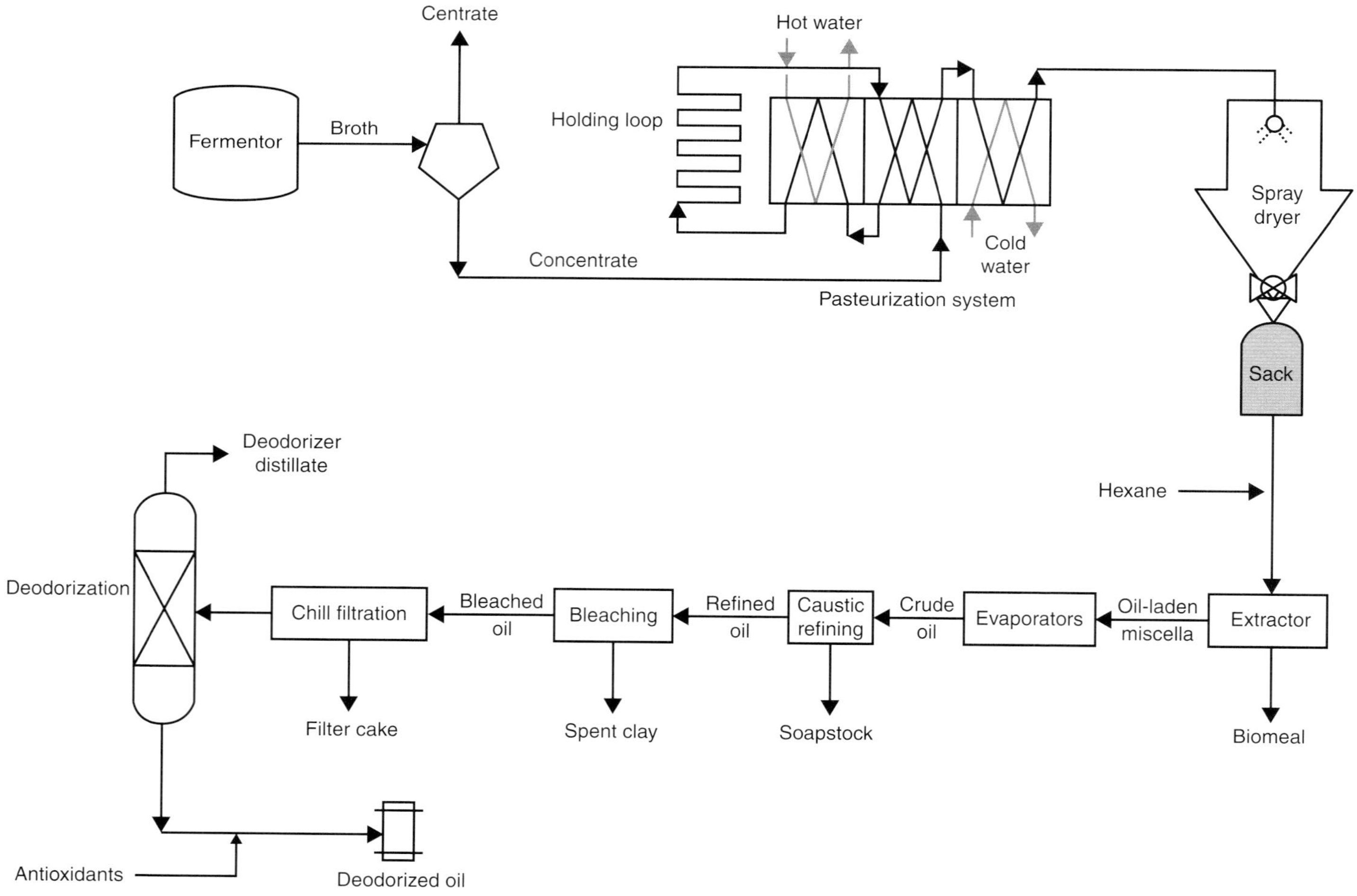

Fig. 10.2 Illustration of a microalgae process for the production of DHA-rich oils. For a colour version of this figure, please see the plate section.

Table 10.3 Proximate compositions of *Schizochytrium* sp. and *C. cohnii* whole cell microalgae produced via fermentation process

Parameter (%)	*Schizochytrium* sp.	*C. cohnii*
Moisture	3.0	2.7
Ash	4.0	3.9
Crude protein	20.0	23.1
Total lipids	66.0	40.0
Crude fiber content	2.0	5.8
DHA (dry weight basis)	21.0	20.4

sterol esters (0.4%), diacylglycerols (1%), free sterols (1%) and free fatty acids (0.1%) were present in the crude lipid fraction isolated from *Schizochytrium* sp. β-Carotene was identified as the primary carotenoid present in crude lipid. Fatty acid profiles of crude lipids have been determined. In *Schizochytrium* sp., DHA (C22:6*n*-3) and docosapentaenoic acid (DPA) (C22:5*n*-6) were shown as the major PUFAs in the crude lipid fraction, accounting for 42 and 12.5%, respectively. In addition, myristic (9%) and palmitic (26%) were the other major fatty acids. Cholesterol, brassicasterol and stigmasterol were identified as the major sterol components in the fatty acid sterol ester and free sterol fraction in the crude lipid isolated from *Schizochytrium* sp.

10.4 Characteristics of algal lipids

DHASCO® oil, produced by Martek Biosciences Corporation, is extracted from the marine microalgal species *C. cohnii*. The final product contains approximately 40% (w/w) DHA. It is a free-flowing liquid oil, which is yellow–orange in color due to the co-extraction of carotene pigments. The final product contains about 95% TAGs, with some diacylglycerols and unsaponifiable material, as is typical for all food-grade vegetable oils. Because of the controlled manufacturing process of a single-cell oil, as discussed earlier, the potential for contamination with environmental pollutants (i.e. pesticide residues, dioxin, etc.,) or heavy metals (i.e. Pb, As and Hg) is eliminated. The fatty acid composition of DHASCO oil is given in Table 10.4. The fatty acid profile of this algal oil is unique in that it contains no PUFAs other than DHA, except a small quantity of linoleic acid (~0.5%) from high-oleic sunflower oil diluent (Kyle 1996). The DHASCO oil has been used for the supplementation of infant formulas.

The unsaponifiable matter of DHASCO oil is generally about 1.5% and is made up of mainly sterols (Kyle 2001). The main sterol has been identified as the 4-methylsterol, dinosterol. The principal components of the sterol fraction in DHASCO (i.e. dinosterol) are found in the normal metabolic pathway of cholesterol biosynthesis and have been identified in several common food sources including fish and shellfish. A study providing large amounts of the isolated unsaponifiable fraction of crude DHASCO to rats concluded that these sterols had no adverse effects on growth or lipid metabolism (Kritchevsky *et al.* 1999).

About 45% of the DHA found in algal oil is located at the *sn*-2 position of TAG molecules (Myher *et al.* 1996). The TAG structure of algal oil is nearly identical to that of human milk, with respect to the positional distribution of DHA in TAGs. Martin *et al.* (1993) reported that in human milk about 50–60% of the DHA is preferentially esterified at the *sn*-2 position of TAGs. Thus, digestion and absorption of DHA in algal oil are expected to be similar to that of DHA in human milk fat.

Table 10.4 Typical analyses of DHASCO and DHASCO-S oils

Parameter	DHASCO oil[a]	DHASCO-S oil[b]
Docosahexaenoic acid (g/kg)	400 (minimum)	350 (minimum)
Docosahexaenoic acid (%)	40–45	34–39
Arachidonic acid (g/kg)	—	—
Arachidonic acid (%)	—	—
Peroxide value (meq/kg)	0–0.5	5.0 (maximum)
Free fatty acids (%)	0.03–0.1	0.5 (maximum)
Moisture and volatiles (%)	0.0–0.02	0.05 (maximum)
Unsaponifiable matter (%)	1–2	4.5 (maximum)
Insoluble impurities (%)	Below detection	Below detection
Trans fatty acids (%)	Below detection	Below detection
Heavy metals (ppm)	Below detection	Below detection
Major fatty acids (%)		
10:0	0–0.5	—
12:0	2–5	0–0.5
14:0	10–15	9–15
16:0	10–14	24–28
16:1	1–3	0.2–0.5
18:0	0–2	0.5–0.7
18:1*n*-9	10–30	0.5–3.0
18:2*n*-6	0.4	0.5–1.3
20:0	<0.1	0.2–0.3
20:3*n*-6	—	0–0.5
20:4*n*-6	<0.1	0.5–0.8
22:0	0.1	0.1–0.2
22:5*n*-3	0.25	12–16
22:6*n*-3	40–45	36–41

[a]DHASCO oil from *C. cohnii*.
[b]DHASCO-S oil from *Schizochytrium* sp.

DHASCO®-S oil is TAG oil, extracted from the *Schizochytrium* sp., which is enriched to about 40% (w/w) in DHA. It is described as a yellow to light orange-colored oil and contains greater than 90% (w/w) of TAGs, with some diacylglycerols, free fatty acids, carotenoids, squalene and phytosterols. β-Carotene has been identified as the primary carotenoid component of the lipid fraction. The oil contains a range of fatty acids, including EPA (eicosapentaenoic acid) and DPA, as well as DHA. However, DHA is the most abundant PUFA component of the oil. Compositional analyses of other components of the oils compare favorably with typical commercial edible oils. In general, the residual extraction solvent is undetectable, and there are no detectable *trans* fatty acids, pesticide residues, or heavy metals such as arsenic, mercury or lead.

The non-saponifiable fraction of DHASCO-S oil is generally about 1.5% by weight and made up primarily of squalene, sterols and carotenoids. These components are all present in the food supply. Cholesterol, brassicasterol and stigmasterol have been identified as the major sterol components of the oil.

10.5 Safety studies of algal lipids

Many safety studies have been conducted with DHA-rich algal oils from *C. cohnii* and *Schizochytrium* sp. These studies were done according to FDA's safety guidelines and generally made in compliance with FDA good laboratory practice regulations. DHASCO

Table 10.5 Milestone advisories related to use of microalgal oils in infant formula

Milestone	Year
Regulatory Office of the Ministry of Health in The Netherlands independently evaluated and approved *life'sDHA* (DHASCO) and *life'sARA*™ (ARASCO) as safe for use in preterm and term infant formula	1995
An independent panel of preeminent toxicologists and nutritionists in the United States concluded that *life'sDHA* (DHASCO) and *life'sARA* (ARASCO) are GRAS for use in preterm and term infant formula	1995
The same panel concluded that *life'sDHA* is GRAS for use by adults, including pregnant and lactating women	1996
The Committee on Toxicology of the United Kingdom independently evaluated and approved *life'sDHA* (DHASCO) and *life'sARA* (ARASCO) as safe for use in preterm and term infant formula	1996
The Ministry of Health in France also approved *life'sDHA* (DHASCO) and Martek's ARA (ARASCO) for use in infant formula	1996
The US Food and Drug Administration completed a favorable review of Martek's GRAS notification regarding the use of its *life'sDHA* (DHASCO) and *life'sARA* (ARASCO) oil blend in infant formula	2001
Health Canada completed a favorable review of Martek's submission supporting the use of its proprietary *life'sDHA* (DHASCO) and *life'sARA* (ARASCO) oils in infant formulas in Canada	2002
Food Standards Australia New Zealand concluded that *life'sDHA* (DHASCO) and *life'sARA*(ARASCO) are safe for use in infant formula	2002

oil from *C. cohnii* has undergone extensive safety studies in animals and has been shown to have no acute toxicity when given at the maximum possible dose (20 g/kg body weight) to rats (Boswell *et al.* 1996). The safety of DHA™-S from *Schizochytrium* sp. is based on the safety of the source organism, the safety of the oil-soluble components of the source organism, i.e. the fatty acid and sterol components, and a battery of classic toxicity studies conducted on the algae. Results of toxicology studies have been published and were conducted by dietary administration or gavage of the source algae in laboratory animals and target species of food-producing animals (Hammond *et al.* 2002). Safety was further supported by the historical safe use of the algae as a commercial dietary ingredient in several commercial animal species. The milestone advisories related to the use of microalgal oils in infant formula are summarized in Table 10.5. In March 2004, FDA completed a favorable review of Martek's generally recognized as safe (GRAS) notification for use of the DHA-rich oil derived from *Schizochytrium* sp. in food applications. The agency has issued a letter informing the company that the FDA has no questions regarding the company's notification that the Martek DHA under review is safe for use in food products.

10.6 Applications

10.6.1 Infant formula

Martek's oils have been used in infant formula since 1994 and to date it has been estimated that over 24 million babies, including more than 500 000 preterm infants, have been fed infant formulas containing *life'sDHA*™. Currently, *life'sDHA* oils are licensed and sold to 24 infant formula manufacturers, which represent more than 70% of the worldwide wholesale

Table 10.6 List of companies that produce DHA-supplemented infant formula

Abbott Laboratories
Aspen Pharmacare
Heinz Wattie's Limited – a subsidiary of the H.J. Heinz Company
Royal Numico
Laboratorios Ordesa
Materna Ltd
Mead Johnson Nutritionals – Bristol-Myers Squibb
Medici Medical
Nestle
Nutrition and Sante Iberia, S.L.
Pasteur Milk
PT Sanghiang Perkasa
Semper AB
Synutra, Inc.
Wyeth Ayerst
PBM Products, LLC
Arla Foods
Murray Goulburn
Namyang Dairy Products Co., Ltd
Parmalat Colombia
Hain Celestial Group
Alter Farmacia

infant formula market (Table 10.6). DHA is important for healthy visual and mental development throughout infancy. Studies with both preterm and term infants suggest that adequate DHA nutrition, provided through either breast milk or DHA-fortified formula, is associated with optimal mental and visual development and function (Makrides *et al.* 1995; Birch *et al.* 1998, 2000, 2002; Hoffman *et al.* 2003).

Numerous studies have examined the effect of DHA supplementation on visual function in term and preterm infants. Birch *et al.* (2002) found that infants who were breastfed and then weaned to formula supplemented with DHA and ARA (arachidonic acid) demonstrated more mature visual acuity than those breastfed infants weaned to non-supplemented formula. In a similar study, Hoffman *et al.* (2003) concluded that infants who were breastfed for 4–6 months and then weaned to DHA and ARA-supplemented infant formula demonstrated more mature visual acuity than those infants who were weaned to non-supplemented formula. Studies have shown that DHA-supplemented infants exhibited better visual acuity than that of non-supplemented infants (equivalent to one line on the eye chart), and similar to that of breastfed infants (Makrides *et al.* 1995; Birch *et al.* 1998). Some studies have failed to report similar effects of DHA and ARA supplementation on infant visual and mental development, but these studies typically involved supplemented infant formulas with lower levels of DHA and ARA (Lucas *et al.* 1999; Makrides *et al.* 2000; Auestad *et al.* 2001).

As with vision, dietary sources of ARA and DHA benefit neurological development in term infants. Birch *et al.* (2000) concluded that infants who were fed formula supplemented with DHA and ARA at recommended levels demonstrated improved mental development and scored 7 points higher on the Bayley Mental Development Index (MDI). Both the cognitive and motor subscales of the MDI showed a significant developmental age advantage for DHA- and DHA+ARA-supplemented groups over the control group. The formula-supplemented infants also exhibited better visual acuity (equivalent to one line on the eye chart) than the non-supplemented infants and similar visual acuity to that of the breastfed infants.

In addition to the vision and cognitive development mentioned above, long-term vascular benefits are demonstrated as the result of DHA and ARA supplementation. Infants fed formula supplemented with DHA and ARA had significantly lower blood pressure compared to infants fed non-supplemented formula, and similar to that of breastfed infants. Since blood pressure during childhood is reflective of that later in life, it has been suggested that early intake of DHA and ARA may support cardiovascular health later in life (Forsyth *et al.* 2003).

Although a developing fetus requires DHA and ARA throughout pregnancy, these vital nutrients are especially important during the third trimester. These last few months are when the most significant neurological, visual and nervous system development occurs, making the fatty acids transferred to the fetus through the placenta particularly critical.

The tissue accretion of long-chain PUFAs is compromised in infants born prematurely. Preterm babies have decreased stores of DHA and ARA due to their shortened time *in utero* (Vanderhoof *et al.* 1997). Clandinin *et al.* (1980) have shown that the accretion of DHA and ARA by the fetal brain during the last trimester of gestation is essential; therefore, infants born prematurely are at an increased risk of having a decreased level of these two fatty acids. Several studies have shown that preterm infants fed formula supplemented with DHA and ARA achieved normal growth in terms of weight, length and head circumference (Leaf *et al.* 1992a, b). Carlson *et al.* (1993) found that premature infants fed DHA had better visual acuity at 2 and 4 months of age. Werkman and Carlson (1996) also showed a positive difference in premature infants fed DHA and ALA at ages up to 12 months after term. In another study, Uauy *et al.* (1990) documented a positive effect of DHA on retinal function and visual acuity in premature infants. Due to the rapid development of their brains and eyes, and rapid physical growth rates, preterm babies have heightened demands for DHA and ARA. In addition, premature infants have been shown to be at greater risk for neurological deficits, such as learning disabilities, social/behavioral problems and lower intelligence scores. And because they have less developed enzyme systems, their ability to efficiently produce DHA and ARA from precursor fatty acids is limited.

There have also been several studies that have compared regular infant formula to formula supplemented with DHA and ARA as to their effect on mental development. One study showed that preterm infants fed DHA- and ARA-supplemented formula demonstrated improved visual and mental development when compared to the infants fed non-supplemented formula (O'Connor *et al.* 2001). Another study found an IQ advantage of preterm infants fed breast milk *by tube* compared to a non-supplemented formula-fed group (Lucas *et al.* 1992), indicating an effect beyond the actual act of breastfeeding.

10.6.2 Nutritional supplements and functional foods

As consumer awareness of DHA grows, so does its marketability. Consumers have a strong preference for healthier foods. This preference translates into increased demand for foods containing healthy ingredients. In fact, 93% of US consumers surveyed believe some foods have health benefits that go beyond good nutrition because they contain ingredients that may help prevent or cure disease (International Food Information Council 2002). Thus, omega-3 DHA continues to gain recognition as a healthy ingredient.

In 2002, the Institute of Medicine at the National Academy of Sciences issued adequate intake levels for linolenic acid, the initial building block for all omega-3 fatty acids found in the body. For male teenagers and adult men, 1.6 g/day was recommended. For female teenagers and adult women, the recommended amount was 1.1 g/day. These guidelines do not seem as well matched to the existing health research on omega-3 fatty acids as guidelines issued by the Workshop on the Essentiality of and Recommended Dietary Intakes for Omega-6 and Omega-3 Fatty Acids in 1999, sponsored by the National Institutes of Health. This panel of experts recommended that people consume at least 2% of their total daily calories as omega-3 fats. To meet this recommendation, a person consuming 2000 calories/day should eat sufficient omega-3-rich foods to provide at least 4 g of omega-3 fatty acids.

Food processors and producers can help consumers increase their *n*-3 fatty acid intake by developing DHA-fortified and other *n*-3 fatty acid-enriched foods. Nutritional supplements and food ingredient products with high levels of DHA can be produced from dried microalgae, refined algal oil, DHA concentrates, algal phospholipids and DHA-enriched eggs. The market for DHA-rich oil for human consumption can be divided into four categories: food ingredients, nutraceutical/functional foods, health foods and pharmaceuticals. One of the applications of DHA-rich oil in food products can make use of microencapsulated oil. Microencapsulation provides protection against oil oxidation and imparts oxidative stability. Hence, algal oils enriched in DHA can be microencapsulated into a powdered product that is relatively stable for storage at ambient temperatures. Algal oil may be concentrated in the form of TAGs, as free fatty acids, or as the simple alkyl esters. Most of the algal oil products sold is in the TAG form. Potential food candidates for incorporation of DHA-rich algal oils include yogurt, cheese, nutrition bars, baked goods and several other applications.

Martek's DHA oil, derived from *C. cohnii*, has been approved for sale in the United States, where it is widely sold through health food stores under the brand name Neuromins®. Martek also supplies the bulk TAG oil from *C. cohnii* to infant formula manufacturers in a number of countries. Martek's DHA oil derived from *Schizochytrium* sp. is designed specifically for use as a food ingredient. It is manufactured using technology that was obtained by Martek in its 2002 acquisition of OmegaTech Inc. (Boulder, CO).

10.6.3 Animal feed

Microalgae can be incorporated into the feed of a wide variety of animals ranging from fish to pets and farm animals. In fact, 30% of the current world algal production is sold for animal feed applications. Many safety studies have proved the suitability of algal biomass as a feed supplement. Microalgae are also used in other animal feeds. For example, microalgae such as *Spirulina*, *Chlorella*, *Haematococcus* and *Phaeodactylum* not only provide nutrition to animals such as chickens, but their carotenoids are excellent sources of pigmentation for egg yolks. In poultry feed, *Haematococcus* has been shown to be useful in pigmentation of both egg yolks and muscle tissues (Elwinger *et al.* 1997; Inborr 1998). Research indicates that omega-3 fatty acids play an important role in animal health. Studies have shown that omega-3 fatty acids play a role in reducing pulmonary inflammation (Khol-Parisini 2007), improving stallion performance (Brinsko *et al.* 2005) and increasing immunity (Hall *et al.* 2004) in horses.

Currently, numerous pet food products in the market are supplemented with omega-3 fatty acids. From complete and balanced diet foods to veterinary supplements, new products are emerging all the time, and many others have been available since the mid-1990s (Hillyer 2007). In 2006, the National Academy of Science's National Research Council updated the 'Nutrient Requirements of Dogs and Cats,' which now recommends amounts of omega-3 fatty acids as being conditionally essential.

10.6.4 Aquaculture feed

Microalgae play a vital role in the rearing of aquatic animals such as shrimp, fish and mollusks (e.g. oysters, scallops, clams and mussels) and are of major interest for aquaculture. They can be utilized in aquaculture as live feeds for all growth stages of bivalve mollusks, for the larval/early juvenile stages of abalone, crustaceans and some fish species, and for zooplankton used in aquaculture food chains. The main role of microalgae for aquaculture is related to nutrition, being used as a sole component or as a food additive. In addition to providing protein (essential amino acids) and energy, they provide other key nutrients such as vitamins, essential PUFAs, pigments and sterols, which are transferred through the food chain. The PUFAs, derived from microalgae, have been identified as important nutrients that contribute significantly to larval growth and survival (Sargent *et al.* 1997). Larvae ultimately acquire these fatty acids from algae, either by directly feeding on microalgae or by feeding on rotifers and Artemia that have been reared on microalgae high in PUFAs.

The use of carotenoids as pigments in aquaculture species is well documented. It appears that their primary functions include a role as an antioxidant, enhancing immune response, reproduction, growth, maturation and photoprotection. Microalgae such as *Dunaliella salina*, *Haematococcus pluvialis* and *Spirulina* are also used as a source of natural pigments for the culture of prawns, salmonid fish and ornamental fish. The largest potential use of natural astaxanthin produced by *Haematococcus* is for salmonid aquaculture feeds. In aquaculture, these feed applications include salmon, sea bream, shrimp culture and ornamental fish. The continued growth of salmonid has created a vast demand for pigments. The flesh color of salmonids is the result of the absorption and deposition of dietary astaxanthin. Salmonids are unable to synthesize astaxanthin *de novo*; therefore, carotenoid pigments must be supplied in their aquaculture diet.

Microalgae are an important food source and feed additive in the commercial rearing of many aquatic animals, especially the larvae and spat of bivalve molluscs, penaeid prawn larvae and live food organisms such as rotifers. The importance of algae in aquaculture is not surprising as algae are the natural food source of these animals. Although several alternatives for algae exist, such as yeasts and microencapsulated feeds, live algae are still the best and the preferred food source.

The major constraint on microalgal production for aquaculture is the cost. New technologies are being developed to reduce these costs. Large-scale photobioreactors and continuous culture systems are now being developed, and these allow better control of culture conditions to optimize the nutritional value of the algae cultured and reduce production costs. New methods of concentrating and preserving the algae are required to permit larger volumes of algae to be grown, concentrated and shipped to the hatcheries, thus reducing production costs and providing a more reliable supply of algae. Furthermore, a better understanding of algal physiology and the effect of culture conditions on the chemical composition and nutritional value of the algae is of great importance to the aquaculture industry.

References

Auestad, N., Halter, R., Hall, R.T., Blatter, M., Bogle, M.L., Burks, W., Erickson, J.R., Fitzgerald, K.M., Dobson, V., Innis, S.M., Singer, L.T., Montalto, M.B., Jacobs, J.R., Qiu, W. & Bornstein, M.H. (2001). Growth and development in term infants fed long-chain polyunsaturated fatty acids: a double-masked, randomized, parallel, prospective, multivariate study. *Pediatrics*, **108**, 372–381.

Birch, E.E., Garfield, S., Hoffman, D.R., Uauy, R. & Birch, D.G. (2000). A randomized controlled trial of early dietary supply of long-chain polyunsaturated fatty acids and mental development in term infants. *Developmental Medicine and Child Neurology*, **42**, 174–181.

Birch, E.E., Hoffman, D.R., Castaneda, Y.S., Fawcett, S.L., Birch, D.G. & Uauy, R. (2002). A randomized controlled trial of long-chain polyunsaturated fatty acid supplementation of formula in term infants after weaning at 6 wk of age. *American Journal of Clinical Nutrition*, **75**, 570–580.

Birch, E.E., Hoffman, D.R., Uauy, R., Birch, D.G. & Prestidge, C. (1998). Visual acuity and the essentiality of docosahexaenoic acid and arachidonic acid in the diet of term infants. *Pediatric Research*, **44**(2), 201–209.

Borowitzka, M.A. (1997). Microalgae for aquaculture: opportunities and constraints. *Journal of Applied Phycology*, **9**, 393–401.

Boswell, K., Kosketo, E.K., Carl, L., Glaza, S., Hensen, D.J., Williams, K.D. & Kyle, D.J. (1996) Preclinical evaluation of single-cell oils that are highly enriched with arachidonic acid and docosahexaenoic acid. *Food and Chemical Toxicology*, **34**, 585–593.

Brinsko, S., Varner, D. & Love, C. (2005). Effect of feeding a DHA-enriched nutriceutical on the quality of fresh, cooled and frozen stallion semen. *Theriogenology*, **63**(5), 1519–1527.

Carlson, S.E., Werkman, S.H., Rhodes, P.G. & Tolley, E.A. (1993). Visual-acuity developments in healthy preterm infants: effect of marine-oil supplementation. *American Journal of Clinical Nutrition*, **58**, 35–42.

Clandinin, M.T., Chappell, J.E., Leong, S., Heim, T., Swyer, P.R. & Chance, G.W. (1980). Intrauterine fatty acid accretion rates in human brain: implications for fatty acid requirements. *Early Human Development*, **4**, 121–129.

Elwinger, K., Lignell, A. & Wilhelmson, M. (1997). Astaxanthin rich algal meal (*Haematococcus pluvialis*) as carotenoid source in feed for laying hens. In: Proceedings of the VII European Symposium on the Quality of Eggs and Egg Products. Poznan, Poland, pp. 52–59.

Forsyth, J.S., Willatts, P., Agostoni, C., Bissenden, J., Casaer, P. & Boehm, G. (2003). Long chain polyunsaturated fatty acid supplementation in infant formula and blood pressure in later childhood: follow up of a randomised controlled trial. *British Medical Journal*, **326**, 953.

Grima, E.M., Medina, A.R. & Gimenez, A.G. (1999). Recovery of algal PUFAs. In: *Chemicals from Microalgae*. Cohen, Z. (ed.), Taylor & Francis, Boca Raton, FL, pp. 108–144.

Hall, J.A., Van Saun, R.J., Tornquist, S.J., Gradin, J.L., Pearson, E.G. & Wander, R.C (2004). Effect of type of dietary polyunsaturated fatty acid supplement (corn oil or fish oil) on immune responses in healthy horses. *Journal of Veterinary Internal Medicine*, 2004 Nov-Dec; **18**(6), 880–886.

Hammond, B.G., Mayhew, D.A., Kier, L.D., Mast, R.W. & Sander, W.J. (2002). Safety assessment of DHA-rich microalgae from *Schizochytrium* sp. *Regulatory Toxicology and Pharmacology*, **35**, 255–265.

Hillyer, C.D. (2007). Omega-3 fatty acids in pet food. *Inform*, **18**, 373–375.

Hoffman, D.R., Birch, E.E., Castaneda, Y.S., Fawcett, S.L., Wheaton, D.H., Birch, D.G. & Uauy, R. (2003). Visual function in breast-fed term infants weaned to formula with or without long-chain polyunsaturates at 4 to 6 months: a randomized clinical trial. *Journal of Pediatrics*, **142**(6), 669–677.

Inborr, J. (1998). *Haematococcus*, the poultry pigmentor. *Feed Mix*, **6**(2), 31–34.

International Food Information Council (2002). *Functional Foods: Attitudinal Research Study*. International Food Information Council Foundation.

Khol-Parisini, A. (2007). Effects of feeding sunflower oil or seal blubber oil to horses with recurrent airway obstruction. *Canadian Journal of Veterinary Research*, **71**(1), 59–65.

Kritchevsky, D., Tepper, S.A., Czarnecki, S.K. & Kyle, D.J. (1999). Effects of 4-methylsterols from algae and of β-sitosterol on cholesterol metabolism in rats. *Nutrition Research*, **19**(11), 1649–1654.

Kyle, D.J. (1996). Production and use of a single cell oil which is highly enriched in docosahexaenoic acid. *Lipid Technology*, **9**, 107–110.

Kyle, D.J. (2001). The large scale production and use of a single cell oil highly enriched in docosahexaenoic acid. In: *Omega-3 Fatty Acids. Chemistry, Nutrition and Health Effects*. Shahidi, F. & Finely, W.J. (eds), ACS Symposium Series 788, American Chemical Society, Washington, DC, pp 92–107.

Leaf, A.A., Leighfield, M.J., Costeloe, K.L. & Crawford, M.A. (1992a). Factors affecting long-chain polyunsaturated fatty acid composition of plasma choline phosphoglycerides in preterm infants. *Journal of Pediatric Gastroenterology and Nutrition*, **14**, 300–308.

Leaf, A.A., Leighfield, M.J., Costeloe, K.L. & Crawford, M.A. (1992b). Long chain polyunsaturated fatty acids in fetal growth. *Early Human Development*, **30**, 183–191.

Lucas, A., Morley, R., Cole, T.J., Lister, G. & Leeson-Payne, C. (1992). Breast milk and subsequent intelligence quotient in children born preterm. *Lancet*, **339**(8788), 261–264.

Lucas, A., Stafford, M., Morley, R., Abbott, R., Stephenson, T., Macfadyen, U., Elias-Jones, A. & Clements, H. (1999). Efficacy and safety of long-chain polyunsaturated fatty acid supplementation of infant-formula milk: a randomised trial. *Lancet*, **354**, 1948–1954.

Makrides, M., Neumann, M., Simmer, K., Pater, J. & Gibson, R. (1995). Are long-chain polyunsaturated fatty acids essential nutrients in infancy? *Lancet* **345**(8963), 1463–1468.

Makrides, M., Neumann, M.A., Simmer, K. & Gibson, R.A. (2000). A critical appraisal of the role of dietary long-chain polyunsaturated fatty acids on neural indices of term infants: a randomized, controlled trial. *Pediatrics*, **105**, 32–38.

Martin, J.C., Bougnoux, P., Antoine, J.M., Lanson, M. & Couet, C. (1993). Triacylglycerol structure of human colostrum and mature milk. *Lipids*, **28**, 637–643.

Mohn, F.H. (1980). Experiences and strategies in the recovery of biomass from mass cultures of microalgae. In: *Algae Biomass*. Shelef, G. & Soeder, C.J. (eds), Elsevier/North Holland Biomedical Press, Amesterdam, pp. 547–571.

Myher, J.J., Kuksis, A., Geher, K., Park, P.W. & Diersen-Schade, D.A. (1996) Stereospecific analysis of triacylglycerols rich in long-chain polyunsaturated fatty acids. *Lipids*, **31**, 207–215.

O'Connor, D.L., Hall, R., Adamkin, D., Auestad, N., Castillo, M., Connor, W.E., Connor, S.L., Fitzgerald, K., Groh-Wargo, S., Hartmann, E.E., Jacobs, J., Janowsky, J., Lucas, A., Margeson, D., Mena, P., Neuringer, M., Nesin, M., Singer, L., Stephenson, T., Szabo, J. & Zemon, V. (2001). Growth and development in preterm infants fed long-chain polyunsaturated fatty acids: a prospective, randomized controlled trial. *Pediatrics*, **108**, 359–371.

Petrusevski, B., Bolier, G., Van Bremen, A.N. & Alaerts, G.J. (1995). Tangential flow filtration: a method to concentrate freshwater algae. *Water Research*, **29**, 1419–1424.

Sargent, J.R., McEvoy, L.A. & Bell, J.G. (1997). Requirements, presentation and sources of polyunsaturated fatty acids in marine fish larval feeds. *Aquaculture*, **155**, 117–127.

Uauy, R.D., Birch, D.G., Birch, E.E., Tyson, J.E. & Hoffamn, D.R. (1990). Effect of dietary omega-3 fatty acids on retinal function of very-low-birth-weight neonates. *Pediatric Research*, **28**, 485–492.

Vanderhoof, J., Gross, S., Hegvi, T., Clandinin, T., Porcelli, P. & DeCristofaro, J. (1997). A new arachidonic acid (ARA) and docosahexanoic acid (DHA) supplemented preterm formula: growth and safety assessment. *Pediatric Research*, **41**, 242A.

Werkman, S.H. & Carlson, S.E. (1996). A randomized trial of visual attention of preterm infants fed docosahexaenoic acid until nine months. *Lipids*, **31**, 91–97.

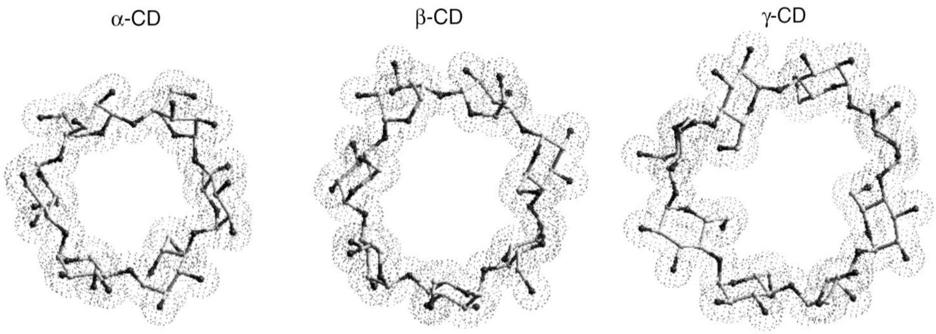

Plate 1 Chemical structure of α-, β- and γ-cyclodextrins. All molecular models were obtained with *Hyperchem Professional, Release 7.5*.

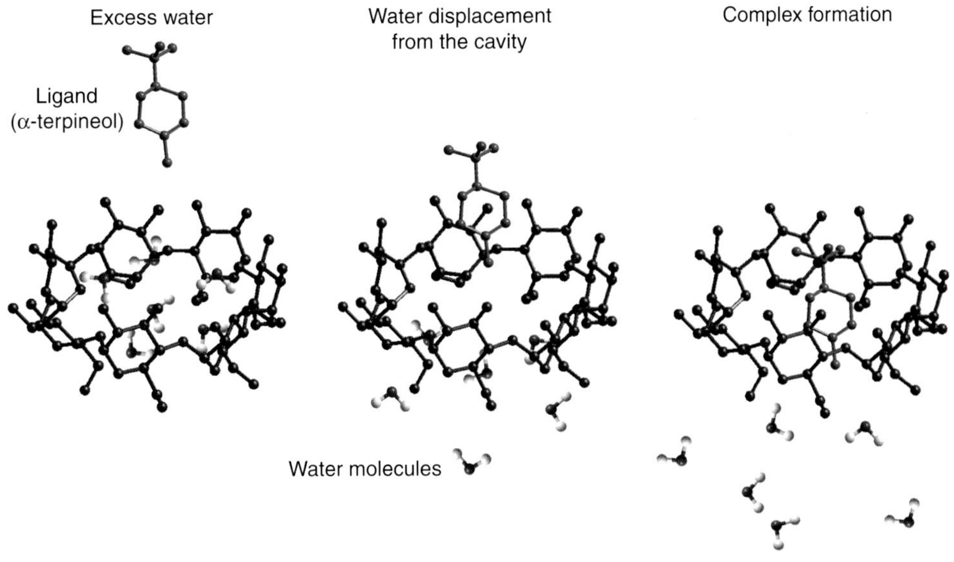

Plate 2 Scheme showing the displacement of water during complex formation and of ligand displacement in excess water. Exemplified for α-terpineol as guest ligands and β-CD as a host.

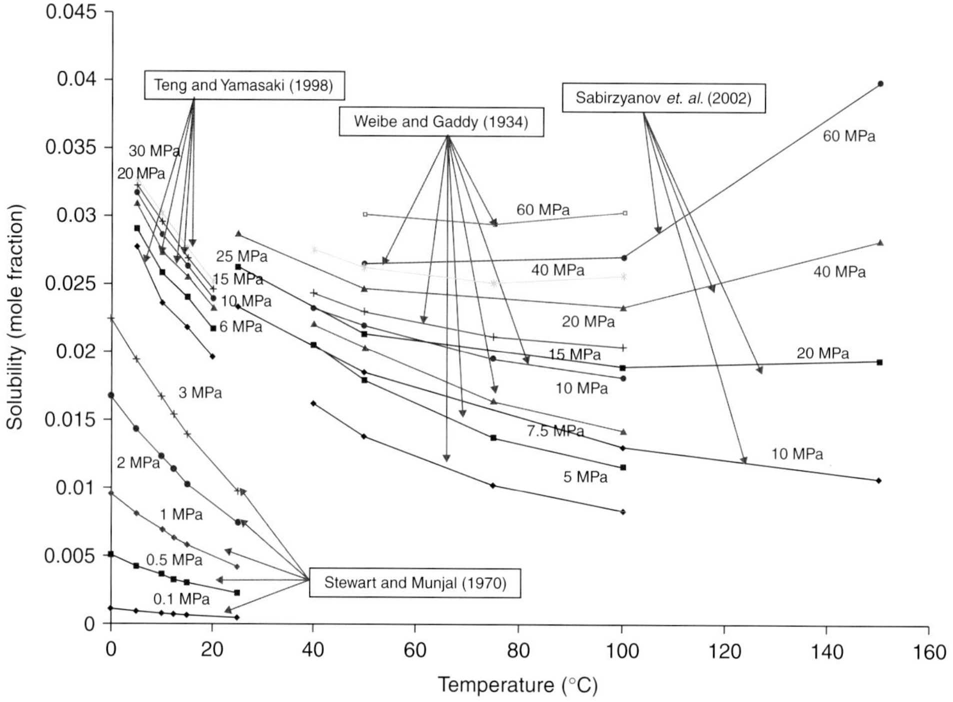

Plate 3 Mole fraction solubility of CO_2 in water as a function of temperature and pressure.

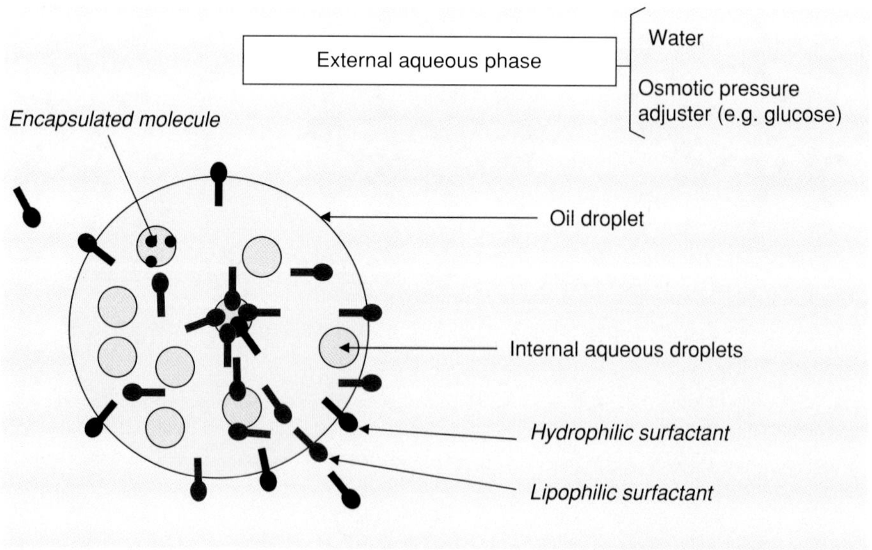

Plate 4 Schematic representation of multiple emulsion.

Plate 5 Carnation® Breakfast Anytime® products produced by Nestlé© contain Prebio™. Prebio is Nestlé's proprietary blend of inulin and other oligofructoses, which are added to food products as a prebiotic for the promotion gastrointestinal health. Reproduced with permission of Nestlé Canada.

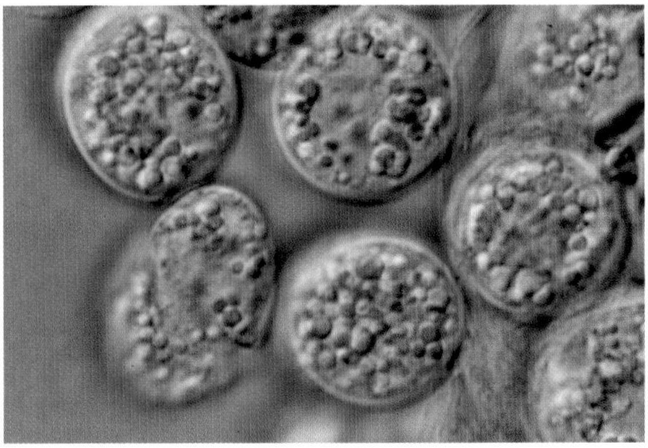

Plate 6 Cells of *C. cohnii* showing high concentration of DHA-rich oil bodies.

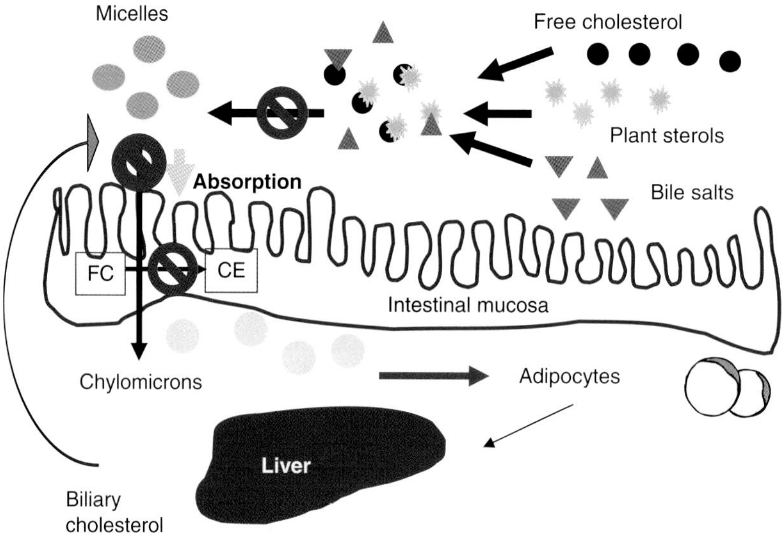

Plate 10 Phytosterols mode of action. FC, free cholesterol; CE, cholesterol ester.

Part III

Product design and regulation

11 New trends for food product design

Juan-Carlos Arboleya, Daniel Lasa, Idoia Olabarrieta and Iñigo Martínez de Marañón

11.1 Introduction

11.1.1 Functional food

Consumers' global demand for healthier diets is increasing. New products must fit not only the needs of consumers but also their lifestyle and income. Health aspects will have to be considered for all foods, but there is a growing market for specific functional foods. For instance, every year the Netherlands spends €165 million in this market and this figure is expected to double over the next 5 years. In particular, the global market for dietetic food products is increasing considerably from about $US32 billion in 2000 to $US55 billion in 2006 (CIAA 2005). It seems clear then that market will demonstrate a larger range of innovative health food products in the near future.

Functional food development implies the incorporation of specific bioactive compounds with positive physiological effects (Hamer *et al.* 2005). There are different technological strategies that modify food composition to accomplish health improvement by adding those bioactive compounds (Duchateau & Klaffke 2008). Another approach to develop healthy food is based on the appropriate selection of food components and their relative amounts, such as incorporating air as small dispersed bubbles, use of non-caloric ingredients and immobilizing large quantities of water (Zuñiga & Aguilera 2008). In addition to the importance of the health effects of food, taste, new textures and convenience remain a crucial factor for the consumer.

11.1.2 Structuring food for health and wellness

Research in food microstructure related to the production of healthy foods could be a key factor in the design of effective functional products. Improved knowledge of the molecular, physicochemical and physiological processes that occur during food ingestion, digestion and absorption will facilitate the rational design and fabrication of functional foods, improving health and wellness (McClements *et al.* 2008). Research is therefore crucial for understanding the relationship between the bioavailability of functional ingredients and food microstructure. For instance, one way to obtain good bioavailability is the use of the intrinsic self-assembly abilities of food ingredients for delivering active ingredients from food (Sanguansri & Augustin 2006). Additionally, despite the need to produce functional food for health concerns, it will be necessary to achieve other consumer demands. Consumers want to eat healthily without missing the enjoyment and pleasure of eating (Wansink

2007). Consumers want to keep the desired food texture but without the calories and with nutritional added value. Thus, the food industry must have a strong focus on delivering innovation to meet market and consumer trends in health, texture, nutrition and targeted delivery solutions (Berry 2008). Food developers must be reminded of the following consumer demands: Is it good for me? Is it good for the world? Does it taste good? Does it consider gastronomic culture? Does it provide extra value? At present, the most important food companies such as Nestlé and Unilever and prestigious international research teams (Institute of Food Research, UK; University of Wageningen, Netherlands; University of Massachusetts, USA) have already paid attention to these demands and considerable research has been developed focusing on the understanding of food microstructure in order to produce healthy but tasty food. It seems clear that a close relationship between food researchers, food technologists, nutritionists and food designers is needed for the good design and development of functional food products, but at the same time maintaining optimal organoleptic properties.

The last requirement, but of no less importance, is the consideration of 'structuring food' not only to give nutritional and health benefits but also for creating an added value in the sense of culture or knowledge, which could be referred to as wellness. Food can contribute to other values rather than health or a way of eating. Consumers, nowadays more than ever, demand food that makes them feel something more. This will be a continuing challenge for the food industry.

11.1.3 New food for thought

New food product design has been traditionally driven by three different approaches to launch new product ideas: surveying competitors, identifying needs through market research and developing new technologies (Stern *et al.* 2007). Moreover, it has been reported that there is a clear need for a new approach, which could lead to exclusive discoveries that can take the marketplace by surprise (Goldenber & Mazursky 2002). The emotional position of the consumer should be borne in mind when designing a new food product. After all, the sense of taste responds to the joy of eating. For this approach, creativity is no doubt a crucial skill for success. Chefs involved in developing innovative foods are usually well-known for their creativity (van der Linden *et al.* 2008). Top-level cooks are continually looking for innovation to make stimulating dishes where generation of new ideas is always needed. These chefs have already adopted a cooking style, which incorporates a basic understanding of science and technology, creating dishes that defy our preconceptions of what food should be and making us pay more attention to what we eat (Vega & Ubbink 2008). Some of the most world-famous restaurants (Mugaritz, Spain; Fat Duck, UK; El Bulli, Spain) also show a real concern about creating healthier, innovative and tastier food with modern gastronomy. Therefore, it is clear to see that chefs could play a key role in inspiring the food industry to look for interesting new developments on high-quality functional food. However, for business innovation, creativity is not enough. Developing new functional food would involve, as mentioned above, food researchers, food technologists, nutritionists and food designers working together as a multidisciplinary team to generate innovation (Linnemann & van Boekel 2007). The challenge lies in integrating chefs in this team to discover ways to mass-produce a chef's idea or recipe in a consistent, high-quality food for the general public, at a reasonable cost together with a focus on originality, health and wellness. Highly palatable food, through culinary manipulation, maximises the natural pleasure and perceptual properties of food. Thus, an appreciation of, and attention to, the sensory and

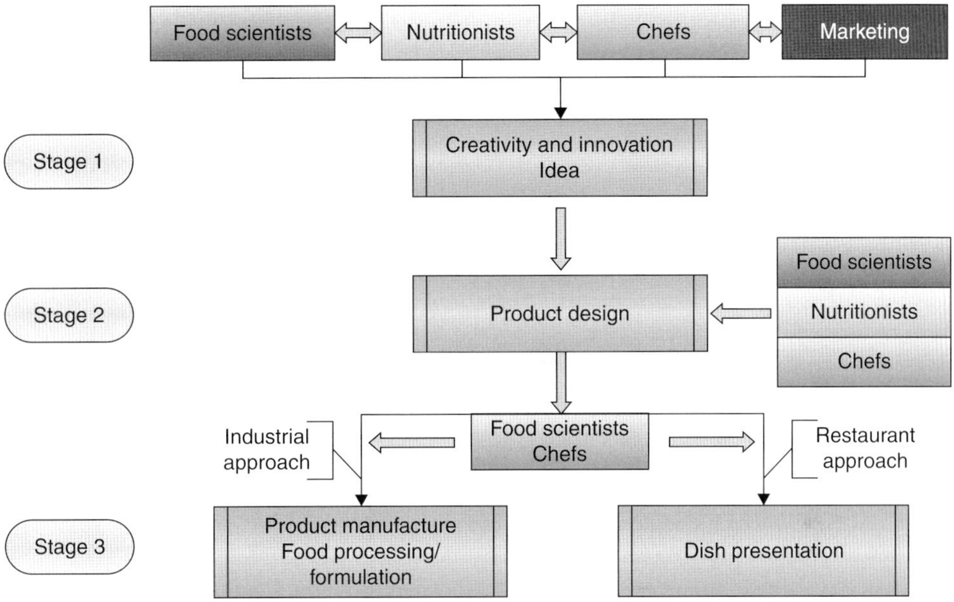

Fig. 11.1 Process of functional food product design. For a colour version of this figure, please see the plate section.

nutrient qualities of food enables the food scientist, nutritionist or marketing researcher to construct or strengthen food products with intention and foresight (Witherly 1987).

11.1.4 A new approach for food product design

In our view, chefs should thoroughly contribute to the trend of producing healthy and tasty products into manufactured foods. From this initial need, an alternative method for new product development involves a systematic process for creating new ideas. Figure 11.1 illustrates how functional product design is a combined effort of a multidisciplinary working team. The way in which functional food product design process is organised depends largely on the input provided by the specialists in marketing, food science, nutrition and gastronomy. Led by chefs' creativity alongside food scientists and nutritionists, they outline the needs and requirements of the consumer and draw new ideas (stage 1) to design new functional products that would meet consumer expectations (stage 2). At this point, this new idea of a functional product can be applied by two different approaches: (a) a 'restaurant approach' that will result in the presentation of a dish in the restaurant and (b) an 'industrial approach' that will allow the manufacture of a new functional product. Again, chefs may play an important role by helping in the formulation and use of new technologies and controlling final product appearance.

This work shows a practical way that optimises the high level of chef's creativity to produce rational approaches to functional food product design from the initial chef's idea, followed by the development of scientific work, the consequent knowledge transfer, first into a dish for the restaurant, and second into a functional product for the food industry. This chapter reports on interesting work at the interface between science and creative cooking, and it is particularly focused on specific 'case studies' in order to come to new ways of producing innovative products.

Our approach is based on proper formulation of functional components, study of the food microstructure and how this affects health, for instance, by (i) replacing butter by using gels, (ii) use of edible films to incorporate functional ingredients, (iii) lowering fat content by using new technologies and (iv) incorporating air as small dispersed bubbles to alter sensation of fullness and satiety.

11.2 Functional food product design: Case studies

11.2.1 Designing functional food by fat replacement

11.2.1.1 *Stage 1: Innovation and creativity*

The sauce is a creative outlet in the gastronomic world as it gives a dish properties such as flavour, moisture, texture, and appearance and emphasises an original style of cooking (Miller 1999). A sauce, in the most basic terms, is a thickened, flavoured liquid designed to enhance flavour in food. The most important part of a sauce is the base, often formed by a significant amount of either animal or vegetable fats (emulsion sauces). The presence of fat in sauce allows an increase in viscosity, and it will have a positive, strong influence on the sauce's overall sensory characteristics including its flavour, colour, visual appeal and sensation in the mouth. The major concern for the chefs in our team was to replace a base sauce made with butter into a healthier one, keeping the consistency and improving emulsion stability, brightness and greasiness but adding a new sensation. They also looked for brighter and cleaner flavours with the aim of avoiding the distinctive and intense flavour that fat gives in emulsion sauces. Replacement of fat itself in this sauce means a considerable improvement in the diet.

11.2.1.2 *Stage 2: Product design*

Regarding technical requirements, fat should be replaced by an ingredient, which improves the elasticity, consistency and stability of sauces. On a basic level, many starches and hydrocolloids add viscosity and a creamy mouthfeel to lower-fat sauces, contributing to the final texture of a sauce. Pulp pectin, for example, can be used in certain foods such as ketchup tomato sauce as a thickener or as an agent, increasing viscosity (Mesbahi *et al.* 2005). Addition of some hydrocolloids can also alter the sensation of fullness and satiety, thus reducing the energy intake (Juvonen *et al.* 2007). Gelatin has several functions for foods (Pranoto *et al.* 2007) and it is widely used by cooks. It is one of the most versatile and utilised gelling agents in food application, and it shows a wide spectrum of sensory and physicochemical properties of food gels, for example, appearance, taste and stability that make them valuable as food matrices (Renard *et al.* 2006). Fish skins represent a considerable amount of waste in the restaurant. This waste is an excellent raw material for the preparation of high-protein foods. Fish gelatin was extracted from desalted, cleaned cod skins in water for 2 hours at a temperature of 80°C. The broth thereafter was filtered and extensive characterisation of extracted gelatin was carried out to determine physicochemical properties (protein content, molecular weight, colour measurements) showing excellent gelling properties (Figure 11.2). A squid-derived solution was mixed with butter and cod sauce to make two different sauces (Figure 11.3). Viscosity of a sauce made with butter showed similar values than cod sauce at 40°C (right temperature for serving a sauce). In addition, cod sauce did show great stability as temperature decreased. On the contrary, butter sauce showed creaming below 30°C.

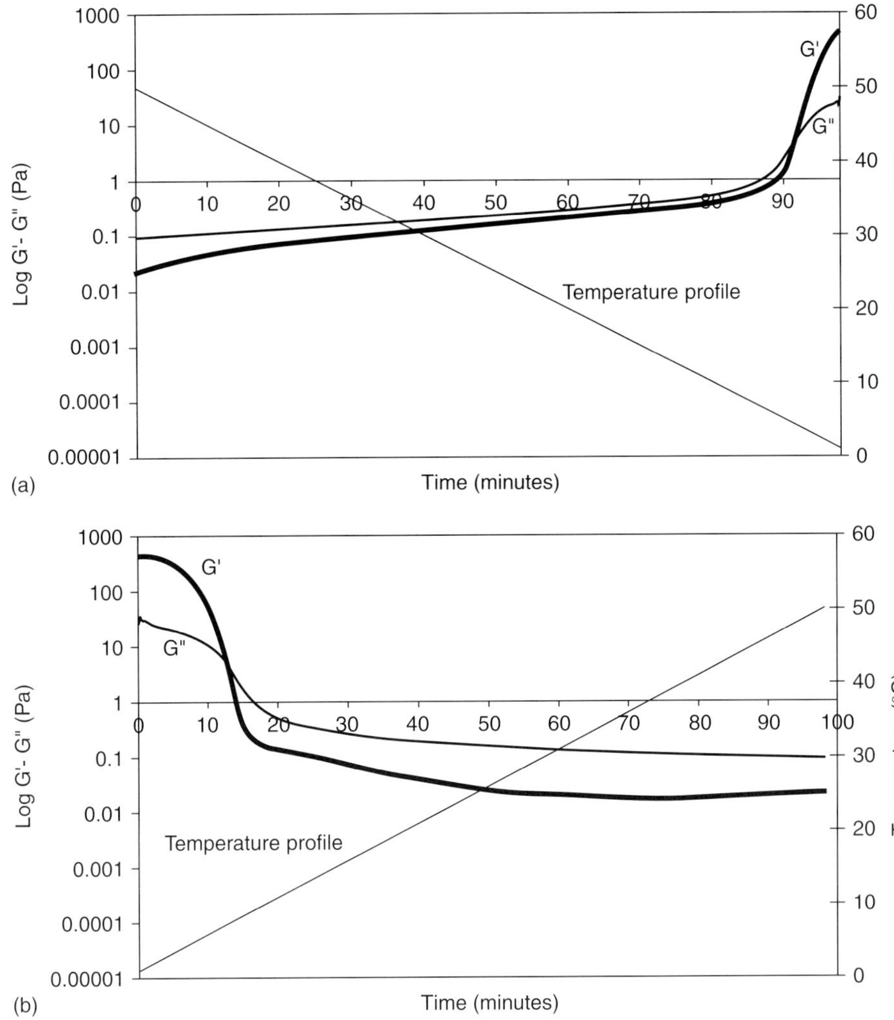

Fig. 11.2 Rheological properties of gelatin extracted from cod skin (a) cooling process (50–1°C), gelling point = 4.2°C ; (b) heating process (1–50°C), melting point = 7.4°C; 0.5 Hz frequency, 0.06 shear stress.

Regarding organoleptic requirements, the most important attributes for a sauce are flavour, texture and brightness. In traditional cuisine, butter has been used extensively as a sauce base in order to get more dense, intense and brighter textures. This characteristic plays an important role in the palatability and intensity of flavours since fatty mouthfeel is more durable and invasive. Nevertheless, the use of butter has monopolised almost all kinds of sauces. Replacing butter by cod sauce introduces interesting possibilities. Once the food scientist tackled the issues of sauce stability and texture, the importance of maintaining a consistent flavour and an acceptable colour was considered. At this stage, chefs were crucial to achieve optimal organoleptic properties for the sauce due to their highly developed sensory skills. Gelatin-made sauce showed a more enhanced and persistent flavour than the sauce made with butter. Once fish skins are thoroughly washed, fish flavour, colour and aroma are almost non-existent; therefore, it produces a longer mouthfeel and retronasal aroma release

Fig. 11.3 Sauce elaboration: extracted gelatin and squid-made solution.

from the rest of ingredients. In addition, presence of gelatin fostered colour intensity and brightness whereas fat-made sauce showed a greyish appearance with reduced tonality.

The knowledge acquired during the product design stage was used to elaborate several dishes in the kitchen (Figure 11.4; *restaurant approach*). This sauce, after reducing its original volume to half, was also used for the casting of edible films. Covering the white fish with

Fig. 11.4 Presentation of final dish based on gelatin-based sauce.

Fig. 11.5 Roast monk fish loin 'with skin'.

the film provided an enhancement of brightness, greasiness, taste and texture (Figure 11.5; Arboleya *et al.* 2008). In this film, it is possible to add different aromas (e.g. tuna, garlic, onion and anise) providing taste enhancement. Likewise, it would be even possible to add encapsulated flavours or bioactive compounds.

11.2.1.3 *Industrial approach*

In this specific case, the cod sauce produced in the kitchen itself could be a perfectly suitable healthy commercial product for the food industry. It corresponds to healthy food guidelines and can be used for various purposes. Industrial manufacturers are keen on creating authentic, high-quality products. Like chef-run restaurants, industrial manufacturers have to produce consistent, quality sauces. In order to accomplish this achievement, developers need to add functionality to the sauce to provide healthy products. Besides organoleptic and nutritional considerations, cod sauce mass production would also involve the use of cod skins which are considered as waste. The management of this waste could consequently result in high economic benefits for the fish industry.

It has been reported that replacing butter in sauces can be achieved by using other starches and hydrocolloids such as chitosan, so the oil content of a fish sauce could be halved (Agullo *et al.* 1998). Organoleptic properties were improved in this type of sauces in the presence of acetic, lactic and glutamic acids to dissolve the chitosan and therefore shelf-life stability. A simple and low-cost process has recently been developed by researchers from University College Dublin to produce a selection of agarose microparticles that have a range of functionalities, for example, as fat replacers or as microencapsulation vehicles for functional ingredient protection and delivery (Ellis & Jacquier 2008).

This type of healthy sauce could also be improved in terms of nutritional value by adding bioactive compounds with positive physiological effects. For instance, lycopene (Djuric & Powell 2001) or water-soluble xylan (Hayashi & Ibuki 2006), which are both associated with decreased cancer risk, have been incorporated in sauces with satisfactory results. Another way to improve the nutritional properties of foods may be modifying the starch component, often included to stabilise the food product (Raben *et al.* 1997).

As explained above with the *restaurant approach*, these types of sauces could be used as a film or coating by covering a food product. Such films or coatings, aside from the improvement of the organoleptic properties, show considerable potential in terms of accurate and functional food development potentially useful in controlling satiety and improving health (Norton *et al.* 2006). There are potentially many applications for edible films in the food industry, ranging from the protection of sensitive ingredients to increasing the efficacy of food additives. They may be used either for the entrapment and controlled release of nutraceuticals and other bioactive molecules (Siro *et al.* 2006; Thompson *et al.* 2007; Wang *et al.* 2007) or for their capacity to protect encapsulated flavours (Hambleton *et al.* 2008). Functional peptides, for instance, can bind to receptors in the gastrointestinal tract and influence the feeling of satiety. One possibility would be to encapsulate these materials entrapped by an edible film or gel (Norton *et al.* 2007). Hence, tailoring the structure of an edible film or a gel would give more opportunities for protection or release of a physiologically bioactive component in the gut.

Another way for replacing fat in foodstuffs would be the application of new technologies which allow getting adequate organoleptic properties with reduced content in fat or without oil or fat. To achieve those purposes, it is very important either to develop the process and the recipe together or to adapt specific products for each technology. That is the case of snacks, which are all produced from pellets specifically formulated for frying.

In the traditional process for producing snacks, water distributed on the food surface is rapidly converted into steam when foodstuff contacts with hot frying oil. After water evaporation, spaces generated in fried snacks are filled with frying oil. Since fried foods usually contain high amounts of fat, it is desirable to reduce the absorption of frying oil or to remove it. Frying technology could be modified in order to reduce the absorption of frying oil by foodstuffs. Fried foods can hence be defatted to produce fat-free or low-fat products using the existing frying systems. Defatting processes can be performed by injecting and forcing steam through the product to remove the excess of fat on the product surface. Instead of adding steam, some treatment units can be equipped with centrifuges to remove fat products after frying.

Nowadays, snack pellets suppliers usually develop pellets for expansion either by frying or by hot air. Oil-free puffing processes, such as hot air or steam impingement and microwaves–vacuum puffing, allow getting healthier snacks. As described before, the water evaporation causes an instantaneous opening or expansion of the inside cell structure. This increase in porosity is sometimes related to an expansion of the product volume. The puffing degree depends on salt content; some types of pellets are difficult to expand when salt content goes down to 0.5%. Besides fat reduction, by increasing surface area those puffing processes could make the expanded product a better substrate for digestive enzymes. Furthermore, the higher porosity enables the development of crispy snacks.

Contrary to fried snacks, oil-free snacks do not require strong flavours to mask oil-fried taste. More delicate flavours could then be used, which should help to create new products. Likewise, for savoury products, small amounts of oil can be sprayed onto the product surface to improve the taste and the texture. Oil or a more complex coating could also incorporate flavours or functional ingredients.

11.2.2 Designing dietetic products by using aerated foodstuffs

11.2.2.1 *Stage 1: Innovation and creativity*

Bubbles are desirable elements in gastronomic creations, offering novel structures and textures, attractive appearance and improved volumes in culinary art. Aerated systems are however thermodynamically unstable and in a fluid system will break down eventually (Dickinson & Wasan 1996). Controlling the size distribution of air bubbles and the spatial dispersion of the gaseous phase is crucial to control the quality of the product (Lau & Dickinson 2005). Aerated foods, such as mousses and soufflés, are classic examples in which the incorporation and retention of bubbles are a critical factor in the success of the dish (Zuñiga & Aguilera 2008). An aerated structure can also facilitate mastication and enhance flavour delivery (Campbell & Mougeot 1999).

The original idea from the chef was to achieve the following design for a dish: 'Trap the food within a bubble, so trapping the aroma which may be either from the food itself or vapor. That will initiate a sudden explosion of senses for the consumer'. From this initial idea, the aim was to create a hemispheric bubble by making a protein solution to lower the interfacial tension which promotes bubble formation, create a viscoelastic interface to stabilise the foam against coalescence, and increase bulk viscosity to slow down drainage (Wilde 2000). Another important issue for the chef was to create the sensation of lightness in a dish. Meanwhile, the aerated product could also potentially achieve a reduction in the caloric density and it could induce satiety through novel gastronomic structures. Common strategies for promoting healthy eating habits are controlling portion size and reducing the energy density of the meal (Hazen 2007). Satiety may not be reached in portion size alone. Given the same volume of food, it may be advantageous to replace more energy-dense ingredients with equally satisfying flavour components. It has been reported that satiety signals differ as the meal moves through the gut but include oral (taste and texture), gastric (distension and emptying) and intestinal (distension and nutrient absorption) factors (Read *et al.* 1994). However, our prestigious chefs would include another factor in the process of satiety: the visual factor, as they firmly believe that feeling of fullness starts before food is eaten, at the point when the food is just being viewed by the consumer.

11.2.2.2 *Stage 2: Product design*

Protein–polysaccharide mixtures are widely used by cooks as they play an essential role in the microstructure of many foodstuffs. Egg white proteins, for instance, are optimal ingredients for foam formation. By controlling the rheology of the aqueous phase, food polysaccharides, such as xanthan, reduce the thinning rate of aqueous films between bubbles and hence increase the stability of the foam (Dickinson & Izgi 1996). Formation of protein–polysaccharide complexes is sometimes related to an enhancement in functionality (Arboleya & Wilde 2005).

Our research was therefore focused to study how xanthan concentration and pH affected the surface properties of an egg white protein solution (Figure 11.6; Arboleya *et al.* 2008). The solution at pH 4.6 showed much higher values in shear stress than the one at pH 7, suggesting that at pH 4.6 there was a substantial increase in the firmness of the foam. These results were then communicated to the chef by transferring them into recipes (*restaurant approach*): egg white protein and xanthan solutions at pH 4.6 (by adding a beetroot solution) would give more foam stability and would produce smaller and more compact foams (Figure 11.7), whereas solutions at pH 7 (by adding cocoa powder to a water solution) would make less stable foam and produce bigger bubbles (Figure 11.8).

By using the knowledge acquired from the studies described above, the next step in the process design was to create an aerated product that induces a sense of fullness, based on the principle that satiety begins when you first look at the product. Additionally, we strove to lower sugar content in some products such as meringue. The recipe was formed by ovalbumin, methylcellulose, maltodextrin and amorphous silica, which helped for the reinforcement of

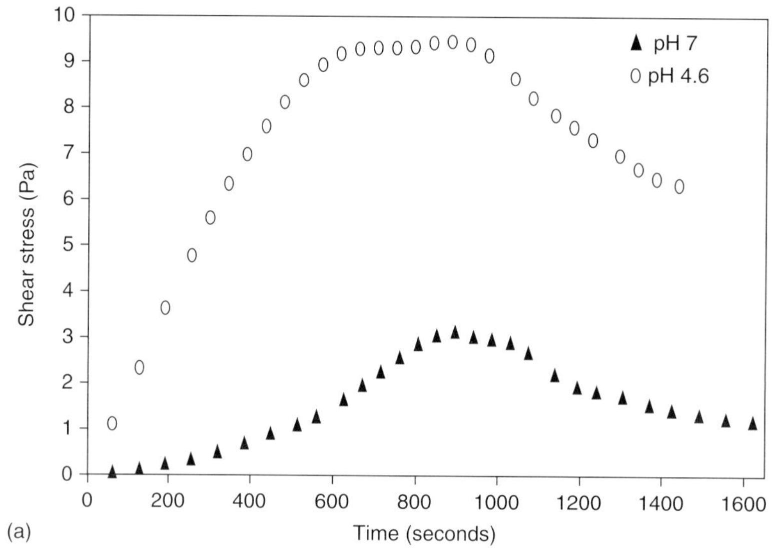

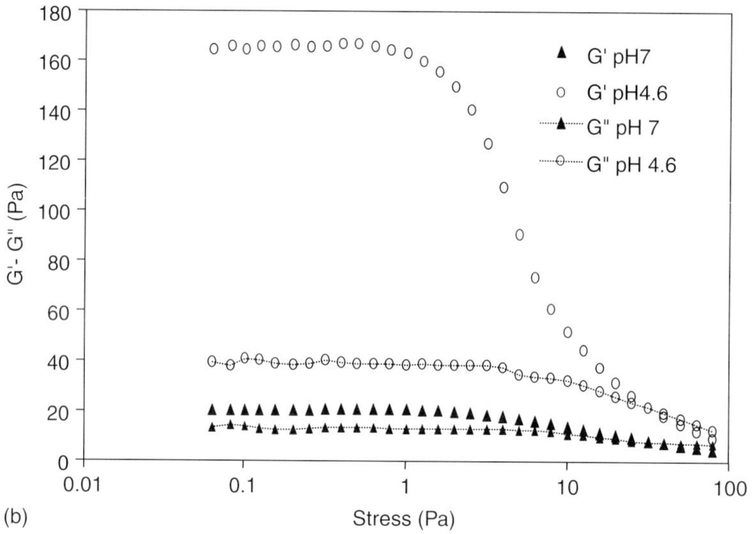

Fig. 11.6 Rheological properties of foamed solutions (10 mg/mL protein solution and 1 mg/mL xanthan). (a) Steady-state flow experiments. Shear range from 0.00083 to 3 s^{-1}. (b) Oscillatory stress sweep experiments. 0.008–100 Pa; 1 Hz frequency; pH 4.6 (○) and pH 7 (▲).

foam structure. Apart from the considerable interest in the effects of silica on human health (Martin 2007), these partially hydrophobic particles are also known to stabilise interfaces, even in the absence of any added surfactant (Aveyard *et al.* 2003), and extend storage stability (Castro *et al.* 2006). Remarkably, stable foams are generated even from polymers that are liquid at room temperature, and hence are otherwise unfoamable (Thareja *et al.* 2008). These results were then transformed into a final dish (Figure 11.9).

Regarding organoleptic requirements, the hugely aerated dish that was created seems to initially induce a feeling of fullness despite the reduction in product intake. There is some

Fig. 11.7 Sun ripened berry fruits lightly covered in virgin olive oil and lime. Cold beetroot bubbles.

evidence that food intake is influenced by both the weight and volume of foods. It has been reported that increasing the air content may be an effective strategy to reduce energy intake from energy-dense products (Osterholt *et al.* 2007).

Aroma is another key factor in the acceptance of foods by consumers. Aromas kept in aerated products will be rapidly released into the oral cavity during the mastication process (Bruin 1999). The addition of aromas in aerated food structures would provide new opportunities in culinary ideas and product design. For instance, it is suggested that the brain response of a food odour sensed retronasally is related to satiation. The extent of retronasal aroma release during consumption depends on the physical structure of a food such that solid foods produce a longer retronasal aroma release than liquid foods. This fact implies that perceived satiation is increased by altering the extent of retronasal aroma release (Ruijschop *et al.* 2008). A step forward for this product might be the inclusion of different flavours by encapsulating aromas on each cavity.

Fig. 11.8 Vanity. Mouth-watering chocolate cake, garnished with chilled cream from real milk, complemented with traces of gold and smoky bubbles of cocoa.

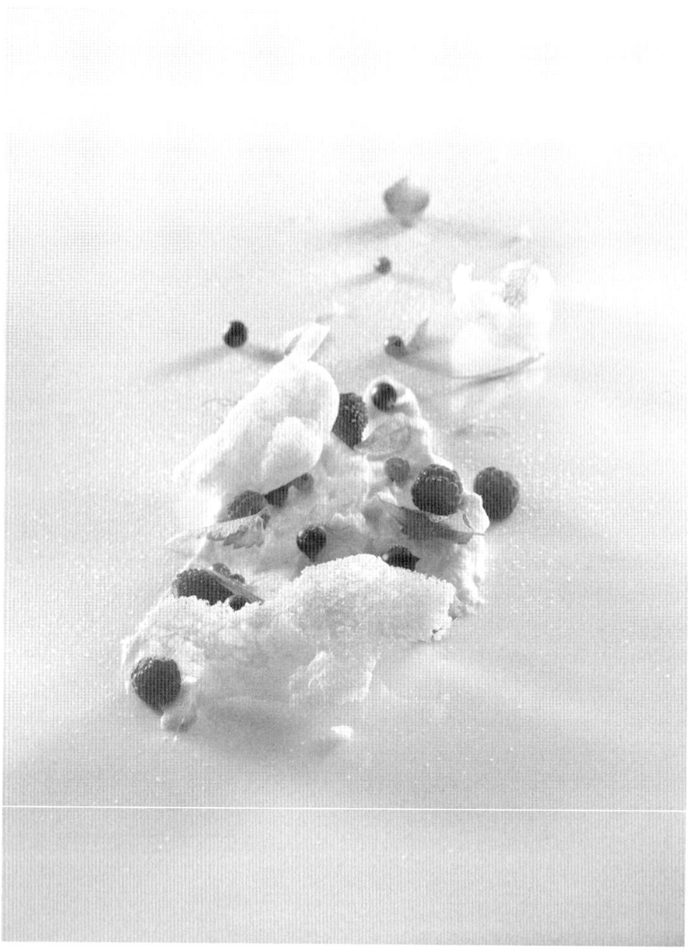

Fig. 11.9 Evoking a spring morning. Milk ice cream, red sun ripened berry fruits, aniseed herb buds and insipid textures.

The microstructure of these aerated dishes influences the rate of uptake into the body, and thus satiety. Controlling the rate of uptake of nutrients such as sugars can also have a significant influence on the health of the consumer. In the case of *evoking a spring morning* (Figure 11.9), sugar was completely replaced by maltodextrin. By using this compound, chefs could obtain similar texture without having an intense taste of sweetness.

11.2.2.3 *Industrial approach*

The same concepts from the gastronomic products mentioned above could easily be used as a suitable commercial health product for the food industry. The challenges of bubbles in food products are measuring them, understanding their behaviour and translating this understanding into commercial advantage. Designed aerated foodstuffs with customised texture, calorie reduction and flavour properties will surely contribute towards developing new dietetic foods for the treatment of obesity. A strategy in food design may then be to maintain taste perception of demanded energy-dense foods while imperceptibly adding small bubbles to the system in order to lower caloric content per portion.

Food microstructure has a significant influence on the feeling of fullness and various aspects of health by lowering breakdown in the gastrointestinal tract (Norton *et al.* 2006). Some studies suggest that it is possible to increase the sense of satiety by altering the physical properties of meals, for instance, by increasing viscosity in the presence of some polysaccharides or fibres (Hoad *et al.* 2004). Certain types of gels, therefore, show an important potential in terms of the development of functional foods that are useful in controlling satiety and improving health. A combination of these gels with air can create an edible foam product, which can be used in the treatment or prevention of obesity. The intake of the edible foam product can produce a more prominent feeling of fullness than a similar product that does not have the in-mouth stability of the edible foam product (Cox *et al.* 2008). In this sense, strong aerated gels could increase the initial sense of fullness, but the effect of an aerated structure on the rate of breakdown in the gastrointestinal tract needs to be determined (Zuñiga & Aguilera 2008).

11.3 Conclusions

The food industry could use the knowledge gained from chefs to discover ways to mass-produce a chef's idea or recipe in a consistent, high-quality food for the general public. Another key requirement for functional food product design would be a better understanding of the mechanisms underlying the functional behaviour of a given food structure and how this affects consumer health. Integrating chefs within a food product development process would have great benefits for the design of functional food with a focus on originality and health, while retaining cultural, social and anthropological aspects which hugely affect consumer feelings of well-being. This could encourage the food industry to seek new developments of functional food. Chefs are invaluable for the food industry, not only because they can be extremely useful for covering all the stages of a functional food product design process, but also because of their ability to connect easily with the public. They could well be an interesting bridge to communicate the use of functional food and balanced diet for healthier consumers. The food industry would do well to seek out such an integration that may be the key to unlock the next big revolutions in the food industry and consumer market.

References

Agullo, E., Ramos, V. & Varillas, M.A. (1998). Chitosan: its use in a low-fat food product. *Anales de la Asociacion Quimica Argentina*, **86**(1–2), 1–4.

Arboleya, J.C., Olabarrieta, I., Luis-Aduriz, A., Lasa, D., Vergara, J., Sanmartin, E., Iturriaga, L., Duch, A. & de Maranon, I.M. (2008). From the chef's mind to the dish: how scientific approaches facilitate the creative process. *Food Biophysics*, **3**(2), 261–268.

Arboleya, J.C. & Wilde, P.J. (2005). Competitive adsorption of proteins with methylcellulose and hydroxypropyl methylcellulose. *Food Hydrocolloids*, **19**(3), 485–491.

Aveyard, R., Binks, B.P. & Clint, J.H. (2003). Emulsions stabilised solely by colloidal particles. *Advances in Colloid and Interface Science*, **100**, 503–546.

Berry, D. (2008). Texture without the calories. Available online http://www.foodproductdesign.com (accessed 5 May 2008).

Bruin, S. (1999). Phase equilibria for food product and process design. *Fluid Phase Equilibria*, **160**, 657–671.

Campbell, G.M. & Mougeot, E. (1999). Creation and characterisation of aerated food products. *Trends in Food Science and Technology*, **10**(9), 283–296.

Castro, I.A., Motizuki, M., Murai, H., Chiu, M.C. & Silva, R.S.S.F. (2006). Effect of anticaking agent addition and headspace reduction in the powdered-drink mix sensory stability. *Journal of Food Quality*, **29**(3), 203–215.

CIAA (2005). *European Technology Platform on Food for Life. The Vision for 2020 and Beyond.* Confédération of the Food and Drink Industries in the EU, Brussels, Belgium.

Cox, A.R., De Groot, P.W.N., Melnikov, S.M., Stoyanov, S.D. & Anon (2008). Edible foam product useful in manufacturing foodstuff or nutritional product for use in treating or preventing overweight or obesity, contains water, protein, and carbohydrates. Patent Number: WO2008046729-A1; EP2081451-A1; AU2007312442-A1; CA2666111-A1; IN200900616-P3; CN101528065-A.

Dickinson, E. & Izgi, E. (1996). Foam stabilization by protein-polysaccharide complexes. *Colloids and Surfaces A-Physicochemical and Engineering Aspects*, **113**(1–2), 191–201.

Dickinson, E. & Wasan, D.T. (1996). Food colloids, emulsions, gels and foams – editorial overview. *Current Opinion in Colloid and Interface Science*, **1**(6), 709–711.

Djuric, Z. & Powell, L.C. (2001). Antioxidant capacity of lycopene-containing foods. *International Journal of Food Sciences and Nutrition*, **52**(2), 143–149.

Duchateau, G.S.M.J. & Klaffke, W. (2008). Product composition, structure, and bioavailability. *Food Biophysics*, **3**(2), 207–212.

Ellis, A. & Jacquier, J.C. (2008). Manufacture and characterisation of agarose microparticles. *Journal of Food Engineering*, **90**(2), 141–145.

Goldenber, J. & Mazursky, D. (2002). *Creativity in Product Innovation.* Cambridge University Press, Cambridge, UK.

Hambleton, A., Debeaufort, F., Beney, L., Karbowiak, T. & Voilley, A. (2008). Protection of active aroma, compound against moisture and oxygen by encapsulation in biopolymeric emulsion-based edible films. *Biomacromolecules*, **9**(3), 1058–1063.

Hamer, M., Owen, G. & Kloek, J. (2005). The role of functional foods in the psychobiology of health and disease. *Nutrition Research Reviews*, **18**(1), 77–88.

Hayashi, Y., Ibuki, I. & Anon (2006). Colon-cancer preventive agent used in foodstuff such as jelly, pudding, yoghurt, mayonnaise, curry, meat sauce, ham, fried fish paste and cow milk, contains water-soluble xylan containing polysaccharide as active ingredient. Patent Number JP2008127370-A.

Hazen, C. (2007). Serving up calorie-controlled entrées. Available online http://www.foodproductdesign.com/articles/463/7ah3110372951050.html (accessed 31 October 2007).

Hoad, C.L., Rayment, P., Spiller, R.C., Marciani, L., Alonso, B.D., Traynor, C., Mela, D.J., Peters, H.P.F. & Gowland, P.A. (2004). In vivo imaging of intragastric gelation and its effect on satiety in humans. *Journal of Nutrition*, **134**(9), 2293–2300.

Juvonen, K.R., Flander, S.M. & Karhunen, L.J. (2007). Effect of dietary fibre and protein on postprandial satiety and satiety peptides. *Agro Food Industry Hi-Tech*, **18**(1), 49–51.

Lau, C.K. & Dickinson, E. (2005). Instability and structural change in an aerated system containing egg albumen and invert sugar. *Food Hydrocolloids*, **19**(1), 111–121.

Linnemann, A. & van Boekel, M. (2007). *Food Product Design: An integrated approach.* Wageningen Academic Publishers, Wageningen, The Netherlands.

Martin, K.R. (2007). The chemistry of silica and its potential health benefits. *Journal of Nutrition Health and Aging*, **11**(2), 94–98.

McClements, D.J., Decker, E.A., Park, Y. & Weiss, J. (2008). Designing food structure to control stability, digestion, release and absorption of lipophilic food components. *Food Biophysics*, **3**(2), 219–228.

Mesbahi, G., Jamalian, J. & Farahnaky, A. (2005). A comparative study on functional properties of beet and citrus pectins in food systems. *Food Hydrocolloids*, **19**(4), 731–738.

Miller, M. (1999). Gold-standard sauces – culinary connection. Available online http://www.foodproductdesign.com/articles/463/463_0999cc.html (accessed 1 September 1999).

Norton, I., Moore, S. & Fryer, P. (2007). Understanding food structuring and breakdown: engineering approaches to obesity. *Obesity Reviews*, **8**, 83–88.

Norton, I.T., Frith, W.J. & Ablett, S. (2006). Fluid gels, mixed fluid gels and satiety. *Food Hydrocolloids*, **20**(2–3), 229–239.

Osterholt, K.M., Roe, D.S. & Rolls, B.J. (2007). Incorporation of air into a snack food reduces energy intake. *Appetite*, **48**(3), 351–358.

Pranoto, Y., Lee, C.M. & Park, H.J. (2007). Characterizations of fish gelatin films added with gellan and kappa-carrageenan. *Lwt-Food Science and Technology*, **40**(5), 766–774.

Raben, A., Andersen, K., Karberg, M.A., Holst, J.J. & Astrup, A. (1997). Acetylation of or beta-cyclodextrin addition to potato starch: beneficial effect on glucose metabolism and appetite sensations. *American Journal of Clinical Nutrition*, **66**(2), 304–314.
Read, N., French, S. & Cunningham, K. (1994). The role of the gut in regulating food-intake in man. *Nutrition Reviews*, **52**(1), 1–10.
Renard, D., van de Velde, F. & Visschers, R.W. (2006). The gap between food gel structure, texture and perception. *Food Hydrocolloids*, **20**(4), 423–431.
Ruijschop, R.M.A.J., Boelrijk, A.E.M., De Ru, J.A., de Graaf, C. & Westerterp-Plantenga, M.S. (2008). Effects of retro-nasal aroma release on satiation. *British Journal of Nutrition*, **99**(5), 1140–1148.
Sanguansri, P. & Augustin, M.A. (2006). Nanoscale materials development – a food industry perspective. *Trends in Food Science and Technology*, **17**(10), 547–556.
Siro, I., Fenyvesi, E., Szente, L., De Meulenaer, B., Devlieghere, F., Orgovanyi, J., Senyi, J. & Barta, J. (2006). Release of alpha-tocopherol from antioxidative low-density polyethylene film into fatty food simulant: influence of complexation in beta-cyclodextrin. *Food Additives and Contaminants*, **23**(8), 845–853.
Stern, B., Taraflin, E. & Larry, S. (2007). New thought for food. *Food Technology*, **61**(10), 34.
Thareja, P., Ising, B.P., Kingston, S.J. & Velankar, S.S. (2008). Polymer foams stabilized by particles adsorbed at the air/polymer interface. *Macromolecular Rapid Communications*, **29**(15), 1329–1334.
Thompson, A.K., Mozafari, M.R. & Singh, H. (2007). The properties of liposomes produced from milk fat globule membrane material using different techniques. *Lait*, **87**(4–5), 349–360.
Van Der Linden, E., McClements, D.J. & Ubbink, J. (2008). Molecular gastronomy: a food fad or an interface for science-based cooking? *Food Biophysics*, **3**(2), 246–254.
Vega, C. & Ubbink, J. (2008). Molecular gastronomy: a food fad or science supporting innovative cuisine? *Trends in Food Science and Technology*, **19**(7), 372–382.
Wang, X.Y., Du, Y.M., Luo, J.W., Lin, B.F. & Kennedy, J.F. (2007). Chitosan/organic rectorite nanocomposite films: structure, characteristic and drug delivery behaviour. *Carbohydrate Polymers*, **69**(1), 41–49.
Wansink, B. (2007). Helping consumers eat less. *Food Technology*, **61**(5), 34–38.
Wilde, P.J. (2000). Interfaces: their role in foam and emulsion behaviour. *Current Opinion in Colloid and Interface Science*, **5**(3–4), 176–181.
Witherly, S. (1987). Physiological and nutritional influences on cuisine and product development. In: *Food Acceptance and Nutrition*. Solms, J. & Pangborn, R. (eds), Academic Press, San Diego, CA, pp. 403–415.
Zuñiga, R.N. & Aguilera, J.M. (2008). Aerated food gels: fabrication and potential applications. *Trends in Food Science and Technology*, **19**(4), 176–187.

12 Reverse pharmacology for developing functional foods/herbal supplements: Approaches, framework and case studies

Anantha Narayana D.B.

12.1 What is reverse pharmacology?

The search for actives for maintenance of health or for treating health problems is always going on. The latter is primarily in the domain of pharmaceutical industry. The search for non-medicinal treatment approach and for maintenance of health through food and dietary modalities as well as lifestyle approaches is on the anvil and increasing in the last decades.

Pharmacological approaches to modern medicines have long been based on classical pharmacological screening either in isolated organs or through receptor binding assays.

Newer approaches have emerged with combinatorial chemistry coming up, which was soon followed by high-throughput screening, ultra-high-throughput screening and high-content screening. These technological breakthroughs have brought faster methods and greater speed in increasing capacities to screen large number of compounds (potential actives) for any specific therapeutic activity or more than one therapeutic activity concurrently. Cell-based assays or directional methods are routinely used for the first step of screening in the drug discovery process (Vogel & Gerhard 2002).

The drug discovery process:

- Target identification
- Screening of the actives
- Prioritized actives
- Toxicity studies
- Preclinical safety/toxicological and pharmacological studies
- Phase 1 trial
- Phase 2a and b trials
- Phase 3 trial–marketing authorization
- Launch product
- Phase 4 (post-marketing surveillance)
- Long-term safety monitoring

However, well-documented traditional knowledge such as Ayurveda, traditional Chinese medicines (TCM), Bhutanese medicines and other such holistic health care systems provides recipes and actives that have been found effective and safe for the past many centuries

(Williamson 2006). The safety and efficacy of recipes in traditional systems is well established and documented in texts. 'Ayurveda' is a Sanskrit word derived from two root words, *Ayu*, meaning Life and, *Veda*, meaning knowledge or science. It is understood to be the oldest health care system in the world, the origin of which can be traced back to 4500 BC (Williamson 2006) in the Indian subcontinent. The basic objective of Ayurveda is 'to maintain the health of the healthy and to mitigate the disease in the diseased ones'. *Charaka Samhita* (Sharma 1981–1994) distinctly categorizes two kinds of herbs – one of which promotes and maintains health (*Swasthasyarojaskaram*) and the other that alleviates the disease process (*Arthasya Roganut*).

Ayurveda is understood to have been systematically documented in 1000 BC (*Introduction to Ayurveda* – http://indianmedicine.nic.in/ayurveda.asp), in the form of *Charaka Samhita* (Sharma 1981–1994) and *Sushruta Samhita* (Sharma 2001), and later many other texts were added to it. While some of the texts have been lost, a number of them are available even today. The Drugs and Cosmetics Act of India (Deshpande & Gandhi 2004) recognizes 57 such books, which collectively contain about 35 000 recipes and processes to make them. These recipes are said to have been documented only after conducting numerous trials on humans from different corners of India (Valiathan 2006). The recipes and therapies mentioned in Ayurveda are in use in the Indian subcontinent since the inception of Ayurveda, and 70% of the population in this region still relies on Ayurveda for its primary health care (Mukherrjee & Wahile 2006). The use of Ayurveda, however, is not restricted to India alone. In the United States, 751 000 people have received Ayurvedic treatment according to the National Center for Health Statistics, 2004. Besides the United States, Ayurvedic products are also exported from India to Canada, Germany, Japan, Malaysia, Australia, New Zealand, Middle East, France, Switzerland, South Africa and Russia (DGCIS 2002).

In India, Ayurvedic practitioners receive state-licensed, institutionalized medical training regulated by the government (Indian Medicine Central Council Act of India 1970 – http://www.ccimindia.org/1.htm). As of 1 April 2005, there were 438 721 registered Ayurvedic Practitioners, 753 hospitals, 15 193 government-owned dispensaries and 7997 manufacturers of Ayurvedic products in India (*Ayurveda: Infrastructure* – http://indianmedicine.nic.in/summary-of-infrastructure.asp).

Ayurvedic companies in India make two categories of Ayurvedic products, which are duly recognized by the Drugs and Cosmetics Act of India (Deshpande & Gandhi 2004): classical preparations – i.e. recipes, which are mentioned in the textbooks, and proprietary preparations – new rational combinations of Ayurvedic ingredients on the basis of newer knowledge generated through usage and clinical studies or experience of the Ayurvedic experts.

A large numbers of these recipes are used even today and their safety is ensured. In such cases, WHO's assessment of traditional medicines recommends the need for a confirmatory human study where safety and efficacy of the recipe or active is documented using modern scientific methods. The actives used in such recipes can be then screened through high-throughput screening to identify promising actives.

This approach can help identify new leads for specific health benefits much quicker than the conventional drug discovery route.

In such cases, drawing information from these documents, listing potential recipes/actives, performing their confirmatory human studies for the intended health benefit form the first step in identifying 'actives' and their safety and efficacy. Such studies can be conducted adopting today's tools of conventional diagnosis and medicines, adopting appropriate methods, end points, biomarker testing to predefined protocols.

This approach involves:

1. Identifying the potentially effective recipes/actives from a list of potential recipes that have been or are already in use in traditional documents
2. Basing therapeutic activities of the recipes/actives from the documentations and their interpretations
3. Screening the potential list of recipes/actives adopting appropriate screening technique(s), and short-listing a selected few of them
4. Developing necessary quality control specifications, deciphering the levels of usage, working out the dosages, documenting the history of usages and safety profiles known from the traditional literature as well as current published literatures, if any
5. Developing appropriate protocols for human studies
6. Confirming safety and efficacy of one among them through 'human pharmacology studies' (human clinical studies adopting good clinical practices guidelines) leading to 'actives' being identified
7. Following up the identified active through more studies of randomized, placebo-controlled, double-blind designs (RPCDB)

Such a process would lead to:

- Identification of intended health benefit
- Literature screening for listing of potential recipes/actives from traditional documentations for the identified health benefit
- Prioritized actives by screening through appropriate method(s)
- Decide dose levels
- Quality control specifications
- Develop appropriate protocols for human studies
- Confirmatory human studies (Phase 2 trial)
- RPCDB human studies (Phase 3 trial)
- Marketing authorization
- Launch product
- Post-marketing surveillance

Since this process, in essence, starts from evaluation of those recipes/actives already known and may be in current use, it is referred to as 'reverse pharmacology', a term that has been gaining popularity as a strategy for drug development. It came into use in India in the 1990s among scientists. This term has now been used even in official guidelines of the Indian Council of Medical Research (Ethical Guidelines for Biomedical Research on Human Participants 2006). Patwardhan *et al.* (2004) provide an extensive review of reverse pharmacology.

12.2 Ayurveda's strength for functional foods

The basic objective of Ayurveda is categorically 'to maintain the health of the healthy and to mitigate the disease in the diseased ones'. *Charaka Samhita* (Sharma 1981–1994) distinctly categorizes two kinds of herbs – one those promote and maintain health (*Swasthasyarojaskaram*) and other those alleviate disease process (*Arthasya Roganut*). As against popular misconceptions, Ayurveda is not 'medicines' alone but is primarily a health care system for preventive, protective and prophylactic care of human beings. It achieves this by prescribing a lifestyle of healthy living through cleansing and rejuvenation of both mind and body. For the mind, it prescribes *yoga*, *prananyam* (breathing exercises), meditation and several behavioral modalities. For the body, in addition to these measures, it prescribes certain forms of therapies such as *Abhyangam* and *Panchakarma* periodically, proper diets and rejuvenators. Ayurveda's strength lies in the fact that it states clearly '*anna* ⟫ *anna-aushadhi* ⟫ *aushadhi*' continuum, which means 'diet ⟫ diet as medicine ⟫ medicine' continuum. Such wisdom is important when one wishes to tap traditional knowledge for developing new actives for health maintenance, the most common aim of food supplements.

One of the most important parts of Ayurvedic system is the part on *rasayana therapy*, loosely translated as immunomodulators and rejuvenators. *Rasayana* (*rasa* – the essence of nutrients derived from digestion of food; *ayana* – the body channels) means circulation of the *rasa*, the nutrients in the body to all the parts of the body. According to *Charaka Samhita* (Sharma 1981–1994), benefits of *rasayana* are to attain longevity, enhancement of memory, intelligence, freedom from disease, youthful age, luster, complexion and voice, optimum strength of physique and sense organs, respectability and brilliance (Katiyar *et al.* 1997). These are exactly the benefits that are being sought to be delivered through food supplements or health supplements or dietary supplements. Dhawan (2006) describes the benefits of *rasayana* and elaborates how tea can be considered as a *rasayana* due to its positive health aspects, improvement of resistance to diseases of varied etiology, and promote longevity.

Ayurvedic wisdom also have provided for many types of *rasyaynas* depending on the need – namely, '*nitya rasyaynas*' (*rasyaynas* for daily consumption), 'naimittika *rasyaynas*' (specific *rasyaynas*) to be taken at times of specified needs, say, e.g. during changing seasonal times, when some symptoms start appearing, '*samanya rasyaynas*' (general all-purpose *rasyaynas*) recipes – that can be taken any time or all throughout the year. *Rasayanas* are generally either single herb or mineral-based preparations or a recipe processed in a specific way as given in the documents made of polyherbal/herbomineral-based formulations. Some of the formulations may have more than 25 herbs or mixture of 25 herbs and minerals processed together. Ayurvedic texts recognized by the Drugs and Cosmetics Act provide detailed formulae, processing methods, usage levels, indications, contraindications (if any) and directions for usages. Such information is available in oriental language of Sanskrit, and would need help of a qualified Ayurvedic *Vaidya* (doctor) or an Ayurvedic researcher to understand them. In many countries, many of such ingredients or recipes have culturally become part of food or items to be taken along with or after food in small proportions and it has become a routine for people to do so. In fact, this is the easiest and sure method of staying healthy and age healthily. *Rasayana* chapter of Ayurveda hence provides a great source of information for identifying and developing food supplements for many benefits.

So would be the case with other traditional systems of health care such as TCM. A number of publications have provided information on Ayurvedic ingredients (Katiyar 1997; Thatte & Dhanukar 1997; Anand 2003; Mishra 2004).

No concrete estimates have been made of the official texts available in Ayurveda providing recipes and herbs for various health benefits including *rasayanas*. Rough estimates would be more than 59 000 recipes as per a brochure of the Traditional Knowledge Digital Library prepared by the National Institute of Science Communication and Information Resources, which has documented in an electronically searchable form, all such recipes (Traditional Knowledge Digital Library – http://www.tkdl.res.in/tkdl/langdefault/common/Abouttkdl.asp?GL=Eng).

More than 2000 herbs and minerals form part of these recipes. These cover only about eight of the official books recognized by the Drugs and Cosmetics Act in India. Such is the veritable database that Ayurveda offers with their history of use, and many of these are in current use.

12.3 Framework for functional food development

There can be many approaches for functional food development, but the reverse pharmacology approach is unique. A conceptual framework has been proposed based on observations of the work that has happened in India in the last 15 years, both in academic institutions, national laboratories and corporate firms (Figure 12.1). The explanatory note to be read along with this framework is also given in Figure 12.1.

The framework is the result of experience across India over two decades and in many projects undertaken both by academic and national laboratories and few industries. The work suggested at specific stages in the framework is not sacrosanct and needs to be addressed depending on the information available and can be varied on a case-to-case basis.

12.4 Case studies

12.4.1 Case of development of a digestive and antiflatulent

The search for some products that can help improve digestion and remove gas formed in the stomach revealed that Ayurveda offers many recipes, out of which a powder formulation, namely 'Hingwastak *churna*', was a potential candidate (*churna* is an Ayurvedic dosage form where powder of several herbs and ingredients are blended as a solid dosage form and meant for oral administration). In this recipe, apart from cumin, pepper, long pepper and ginger, 'asafetida' was an ingredient, which in both traditional medicines (TM) and modern literature reported for activity as a good carminative and could be expected to help expel gas formed in stomach due to indigestion. Animal models for testing such activity were not very useful and the *churna* was also not very consumer-friendly for taking as it had an off-taste, stuck to the tongue and buccal cavity when taken. The recipe was modified to improve the taste and converted to small-sized tablets that could be chewed slowly in the mouth. The tablets could be easily packed in glass bottles for better preservation and improved stability. The composition of Hingwastak *churna* and the modified product currently sold as tablets Hingoli^{TM} (trademark of M/S Dabur India Ltd, a leading Ayurvedic product company) are given in Tables 12.1a and 12.1b. Consumer studies have proved its efficacy and safety. It has been well accepted and is a big seller (http://www.dabur.com).

12.4.2 Case of single-ingredient herbal products

The emergence of DSHEA law in the United States (Centre for Food Safety and Applied Nutrition, US Food and Drug Administration, Dietary Supplement and Health Education Act 1994 – http://www.cfsan.fda.gov/~dms/dietsupp.html) triggered development of dietary supplements and Ayurvedic herbs. Adopting the reverse pharmacology approach, many institutes developed single herbs selected from Ayurvedic wisdom. Development of phytochemical standardization approaches for testing the herbs or their extracts for either

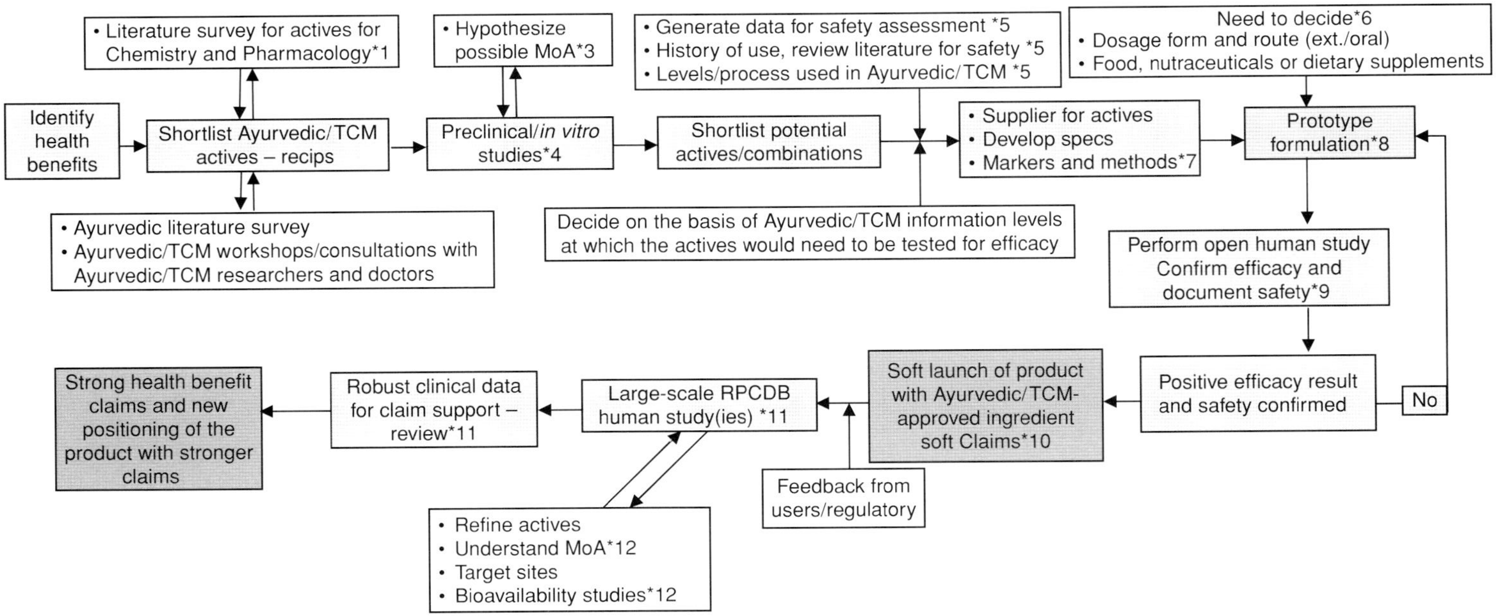

Fig. 12.1 Proposed framework for reverse pharmacology approach for developing food supplements based on leads from Ayurveda/TCM and other traditional medicines.

1. At this stage, collection and collation of all published literature of anything known about the short-listed ingredients covering habitat, description, chemistry, pharmacology, any published human and animal study data, patents, availability in various geographies in the world are required. Such searches would be required to not limit to information available in searchable web-based databases but also go to cover from books, treatises, journals and Masters and PhD dissertation and thesis, which have valuable information.

Fig. 12.1 *Continued*

2. This is recommended as an important step to list out possible actives for health benefits based on actual knowledge and experience of practicing TM doctors and researches who work on TM and, if possible, practitioners who are manufacturing products. Such literature survey would need to be done by Ayurveda/TCM or TM researchers who understand the language and terminologies of these TMs. Such researchers would also need to be able to interpret the language of TM to today's biological and pharmacological sciences.
3. At this stage, it would be good to have an appreciation or hypothesize the possible mechanism of action (MoA). While such MoA can be thought about based on published scientific work, it is always not possible to do so due to lack of such scientific work in the published domain about all or many of the leads from TM. Discussions with experts of TM would reveal the way a particular product/intervention works as per the traditional health science (say, *Ayurveda* – pacifying aggravated *vata* or improving the low level of *Agni* etc. or TCM – pacifying the aggravated *Yin* or strengthening *yang* etc.). A pharmacologist who has also exposure to such TM knowledge will be able to transliterate or draw up possible MoAs using current human pharmacological understandings. At this stage it may be possible also to shortlist the systems/organs in the body on which the TM recipe may be working. Such an understanding at this stage would be adequate to guide possible preclinical screens/evaluation models to be decided. Information on the exact MoA may really be not required in all cases, but explanation based on traditional knowledge has been seen to be quite acceptable to users except discerning opinion-making doctors and scientists. Strategic decision on whether to invest detailed studies to understand the exact MoA at this stage needs to be taken.
4. Performing studies using laboratory animals is one route, while adoption of molecular screening techniques, organ culture techniques and such other techniques which look at gene expression, action on specific enzymes or targets are the other routes to be adopted. Since naturals may not act necessarily on a single target, attempts at this stage of the preclinical screening to involve a battery of such screenings techniques are recommended with a view to increase selection of larger numbers of leads and also not to loose a good active because of the limitation of screening techniques/targets. The approach of this stage is to maximize the testing protocols and get the best possible hits.
5. Generation of maximum information at this stage would help in taking data-based decision regarding safety of the natural as well as limit animal-based toxicological data generation to answer specific concerns only on safety. A useful dialogue with toxicologists at this stage is encouraged. Specific data on quantum of the actives, cultivated or produced, sold and used by consumers, possible extrapolated data on extend of human exposure and related safety aspects known or reported as also additional information on safety from TM experts at this stage would be very beneficial. Trade and industry production-based data would be useful source of information for the same. Over all clear guidelines for toxicological studies specific to naturals are not available, but most regulators expect studies done adopting current toxicological models applicable to synthetic compounds. Other aspect currently being discussed is the need for genotoxicity data at this stage.

6. This is perhaps one of the important but difficult stage to decide on the levels at which the natural actives are to be used in the product. The factors that dictate such a decision are the following:
   - The intended uses of the product, namely, food, food supplement, or quasi-medicine or an over-the-counter product.
   - Health benefit being aimed – preventive, maintenance, adjuvant to current medications, deliver specific health benefits, say, improve immunity, reduce cholesterol, etc.
   - Is it possible to use functional levels of the natural actives to deliver the above two aims, in the health supplement being formulated?
   - Technological capabilities to incorporate such functional levels and manage texture, taste and sensorial and other aspects.
   - Cost and economies.
   - Regional/global regulations related to above.

   Mapping of all known information about the natural actives in the traditional recipes and a level at which they are used, dose, duration of entire treatment, quantum of use per day, extractive values of natural actives, process by which the actives used in the traditional medicine at this stage forms the first step in determining dose/levels of usage in the product. In many cases it would be possible to divide the levels decided into multiple usage portions of the product per day, thereby making possibility to improve texture, taste and sensorial per each serving as the amount per serving would be reducible manageable levels. An experienced researcher in collaboration with a TM expert can take a call and decide the levels of usage, which would be an expert assessment-based decision. However, the scientific approach which is expensive and time-consuming would be to undertake phase 1 or phase 2 or dose-finding studies. In many cases relationship between therapeutic and prophylactic levels may be possible to be drawn.
7. It is advisable to broad base the specifications generated at this stage in collaboration with the ingredient supplier(s), working with them to adopt processes, adopt more than one marker compound testing or develop quality profiles for consistency testing, adoption of DNA mapping for conformation of identity of plants, and adoption of at least one biological testing model is recommended to reduce the possibility of failures in the finished product(s).
8. The prototype development at this stage is to aim at making a reasonably good product in the format decided, which meets minimum requirements as regards sensorial. Such prototypes should be capable of not adversely affecting compliance during open human studies, while at the same time incorporation of desired levels of naturals and arrive at a stable product format are required. Concurrently, development work needs continue to improve texture, mouthfeel, product appeal and other such characteristics including sensorial. Formats such as capsules and tablets are easy to formulate, but the final usage in such formats may expose them to be 'treated as medicinal and not amenable to dietary supplement formats'. However, this is a call that is to be taken carefully as it will have impact on development costs, timelines, marketing strategy, marketability and success in market. It is also important that the ingredients being tested and/or the prototype products comply with the quality specifications adopted, for each batch made for the human studies. These prototypes are also confirmed for shelf-life for reasonable period of storage required for the completion of the study.

Continued

Fig. 12.1 *Continued*

9. It is to be noted that the framework recommends an open study to check the safety and efficacy of the product. At this stage it is important to generate data that the product is effective and safe or the safety concerns are known for finding ways to communicate them to users. The efficacy parameters decided for the human studies or expected from the lead/prototype need to be carefully and practically decided. Expecting or targeting 'effectiveness that are normally obtained from drugs', from prototypes designed as a dietary supplement, often causes problems. (By drug-like effects, it refers to effectiveness in short duration and with doses that are not necessarily determined for such activity.) Food supplements normally work slowly and need some time before the health benefits are experienced by the user. Human clinical studies need to be carefully done with appropriate protocols that specifically take into consideration, aspects applicable to traditional medicine. Factors such as volunteer/patient profiles based on traditional knowledge such as *Prakriti* (one's psychosomatic constitution) or similar assessment as per TCM such as *Yin* and *Yang* levels (roughly analogous to genotypes/phenotypes) need to be carefully incorporated in the inclusion and exclusion criteria. Otherwise, possibility of recruiting of large numbers of non-responders, who are likely to show specific side effects, would give false indications and results. Since the food supplements are designed to work slowly, the duration of such human studies is generally longer than most drug trials, many extending to several months with study material/prototype interventions. In such cases it is important to find ways to ensure and monitor as well as document 'proper compliance' to treatments by each group in the study. In addition to adoption of modern diagnostic tools for biochemical markers or other measurements, assessment of parameters as per traditional knowledge by TM expert concurrently would add great value to this study and generate highly useful information. It is also recommended to consider induction of one or two knowledgeable TM experts in the research ethics committees (RECs) and make them intersystem RECs. Such recommendations form part of the Indian Government's ICMR guidelines too (Ethical Guidelines for Biomedical Research on Human Participants 2006). At this stage of taking a decision about the product to be introduced in the market as a soft launch or as a test launch conduction of a RPCDB, human studies are not recommended due to its cost, time-consuming and logistical problems.
10. If the product in the above open study provides efficacy and safety data, the product may be reviewed for a soft launch or a test launch for specific period in a predecided territory after developing the proper packaging and positioning statements. Due regulatory approvals need to be taken before such soft launch. At this point, broad claims based on TM knowledge and published information would be possible and would be adequate. Claims may be ingredient claims linked to the ingredient being known in tradition for health benefits or specific consumer benefits. This would actually help getting highly useful feedback on all aspects including sensorial of the product, while giving ability to be a first-mover advantage or faster to market access. This would also help getting additional data on safety basis actual usage.
11. Based on experience during the test marketing, if the acceptance of the product is found to be good, it is suggested to invest heavily on the product at this stage, such investment to improve product format, texture and sensorial as well as generate highly credible claim support data. At this stage well-planned multicentric RPCDB human studies with common protocols will help generate strong claim support data. Based on results of such studies the pack claims, specific literature and strategy as well as positioning of the product can be strengthened and a regional/global launch undertaken.
12. Since heavy investment will now be made on this product at this stage, one of the approaches would be to inform and sensitize key opinion formers, influencers of product usage such as physicians, dentists, cosmeticologists, dieticians and other professionals; it would be required to address one of their questions regarding MoA. Scientific studies have to be undertaken to confirm or disprove the hypothesis of MoA that had been built while making selection of possible actives. Credible scientific information at this stage would be required for dissemination to the key opinion formers and influencers. In the same direction, scientific studies to generate data on pharmacokinetics and/or pharmacodynamics (bioavailability) should be undertaken and data so generated be shared with key opinion formers and influencers.

Table 12.1(a) Composition of *hingwastak churna*

Sl. No.	Name of the herb (Ayurvedic name)	Botanical name, and part used
1.	Shunti	*Zingiber officinale*, dried rhizomes
2.	Maricha	*Piper nigrum*, dried fruits
3.	Pippali	*Piper longum*, dried fruits
4.	Saindava	Rock salt
5.	Jeera	*Cuminum cyminum*, dried fruits
6.	Kala jeera	*Carum carvi*, dried fruits
7.	Heeng	Asafetida, gum resin

chemical/analytical marker compounds or biochemical markers gave way to food supplements. These food supplements were easily formulated by filling them into capsules to deliver specified dosages, as DSHEA allowed capsules as a dosage for supplements. A number of firms market such single-ingredient herbs/herbal extract-based products for specific health benefits. Table 12.2 provides a non-exhaustive list of such supplements, their composition, health benefits listed. It is not intended to provide details of all the information on each of these products as they are available in the public domain, provided by the firms that manufacture them. Himalaya Healthcare (http://www.himalayaheatlhcare.com) is one such source of information and is a leading Indian Ayurvedic firm.

12.5 Factors to make reverse pharmacology work

The proposed framework, while it looks simple, depends on a number of factors for its effective implementation. Some of the most important ones are as follows:

1. Clear definition of health benefit.
2. Clear articulation of the benefit in the terminologies used in TM and proper search of the literature.

Table 12.1(b) Composition of Hingoli™ tablets

Sl. No.	Name of the herb (Ayurvedic name)	Botanical name, and part used	Composition per tablet
1.	Shunti	*Zingiber officinale*, dried rhizomes	5.70 mg
2.	Maricha	*Piper nigrum*, dried fruits	8.55 mg
3.	Pippali	*Piper longum*, dried fruits	8.55 mg
4.	Saindava	Rock salt	114.0 mg
5.	Jeera	*Cuminum cyminum*, dried fruits	11.0 mg
6.	Kala jeera	*Carum carvi*, dried fruits	—
7.	Heeng	Asafetida, gum resin	5.70 mg
8.	Kala lavan	Unaqua Sodium chloride	14.25 mg
9.	Sarakara	Crystal sugar	59.85 mg
10.	Nimbu sar	Citrus limonum	25.85 mg
11.	Suddha swarna gairic	Purified silica material	8.55 mg

Note: Hingwastak *churan* and Hingoli™ Trade Marks of M/S Dabur India Ltd, India.

Table 12.2 Selected list of single herb-based (with Ayurvedic wisdom base) dietary supplements, marketed by Himalaya™

Sl No.	Name of the herb [Ayurvedic name]	Western/English name, if any known	Botanical name	Part of plant used	Health benefit[a]
1	Amalaki	Gooseberry	*Emblica officinalis*	Fruit, dried fruits	Antioxidant, immunity enhancer
2	Arjuna	—	*Terminalia arjuna*	Bark	Heart health
3	Ashwagnadha	Indian ginseng	*Withania somnifera*	Roots	Calms the nerves, revives mind and body, immunomodulator
4	Bael	—		Fruit, bark, leaves	Gut health and intestinal infections
5	Brahmi	—	*Baccopa monneri*	Aerial parts	Cognition and nervine tonic
6	Gokshura	—	*Tribulus terrestris*	Fruits	Energizer, performance enhancer
7	Guduchi	—	*Tinospora cardifolia*	Dried stems	Immunomodulator, builds resistance to infections
8	Hardira	Turmeric	*Curcuma longa*	Rhizomes	Cytoprotective, sugar metabolism, cognition, builds resistance to infections
9	Haritaki	Myrobalan	*Terminalia chebula*	Dried fruits	Gentle laxative, immunomodulator and antioxidant
10	Karela	Bitter gourd	*Mimordica charantia*	Fruits and seeds	Sugar control, skin health and bitter tonic
11	Lasuna	Garlic bulbs	*Allium sativum*	Bulbs, fresh or dried	Blood circulation, lipid profile modulation, digestive
12	Mandukaparni	—	*Centella asiatica*	Aerial parts	Cognition, enhance concentration and attention, nervine tonic
13	Shallaki	—	*Boswellia serrata*	Gum resin	Joint health, anti-inflammatory
14	Shatavari	—	*Asparagus racemosus*	Dried roots	Women's health, immunomodulator, builds resistance to infections
15	Guggulu	—	*Cammiphora mukul*	Gum resin	Lipid regulator, weight management
16	Shunti	Ginger	*Zingiber officinale*	Fresh or dried rhizome	Digestive, gut health, joint health and antinausea
17	Tagara	—	*Valerian officinalis*	Roots	Calming the mind, sleep enhancer
18	Tulasi	Basil	*Ocimum basilicum* and other species	Leaves and aerial parts	Immunomodulator, builds resistance to infections, respiratory health
19	Vasaka	—	*Adhataodha vasaka*	Leaves	Respiratory health
20	Vriksahamla	—	*Garcinia indica* and other species	Fresh or dried fruits	Weight management, metabolism promoter
21	Yasthimadhu	Liquorice	*Glycyrrhiza glabra*	Rhizomes	Gut health, antacids, immunomodulator, sweetener

[a]Modified by the author.

Note: Himalaya™ Drug Company (trademarks owner: MMI corporation, India, and licensed user: Himalaya Drug Company, India).

3. Availability and adoption of proper screening methods, correct interpretation of results of such screening, testing right herbs/minerals, extracts and samples processed as in TM or otherwise.
4. Availability or ability to arrive at the processing methods of the lead and its dose.
5. Development of proper dosage form and quality profiles for each batch prepared, and a stable prototype.
6. Development of an appropriate and statistically worked out human study protocol with appropriate end points and biomarkers.
7. Meticulous working to agreed protocols/methods, periodical review of the results and changes in studies decided and documentation of all work.
8. And most importantly, a well-coordinated 'cross-functional team' of scientists who bring their individual and collective expertise. A group of traditional doctors/researchers, botanists, biologists, natural product chemists, pharmacologists, pharmacists, conventional medicine doctors, safety experts and clinical researchers would be required. Constant consultation with a regulatory expert is a must. Constant hand holding and building trust and confidence in the team especially with the TM experts are important. Most importantly, a business champion would drive the development work.

Acknowledgments

It is not intended nor is it possible to acknowledge individually the large numbers of scientists from cross-functional disciplines of sciences who have contributed to the development of reverse pharmacology approach. Many have toiled to produce success stories. The author acknowledges the scientists at national laboratories, universities and other academic institutions (both publicly and privately funded), Ayurvedic and allopathic hospitals and doctors working there, and scientists from privately run businesses and their research and development wings. Continuing governmental funding by the Government of India has largely promoted this scientific culture and endeavors. Business leaders need special acknowledgment as without their support none of such work could have happened.

References

Anand, N. (2003). *Ayurveda: Chemical Constituents of Some CNS Active Plants in Ayurveda & Allopathic Medicines and Mental Health*. Bharatiya Vidya Bhavan, India, New Delhi, pp. 152–163.

Deshpande, S.W. & Gandhi, N. (2004). Drugs and Cosmetics Act of India, 1940 and Rules, 1945, 3rd edn. *The First Schedule*. Susmit Publishers, India, Mumbai, p. 169.

DGCIS. (2002). Ministry of Commerce and Industry, Govt. of India.

Dhawan, B.N. (2006). Tea as Rasayana. In: *Protective Effects of Tea on Human Health*. Jain, N.K., Siddiqui, M.A. & Weisburger, J.H. (eds), CABI, UK, pp. 6–15.

Ethical Guidelines for Biomedical Research on Human Participants, Indian Council of Medical Research, India. (2006). Clinical Evaluation of traditional *Ayurveda, Siddha Unani*, remedies and medicinal plants. Available online http:/www.icmr.nic.in (29 August 2008).

Katiyar, C.K., Brindavanam, N.B., Tiwari, P. & Narayana, D.B.A. (1997). *Immunomodulatory Products from Ayurveda: Current Status and Future Perspectives, Immunomodulation*. Upadhyay, S.N. (ed.), Narosa Publications, Delhi, pp. 163–187.

Mishra, L.C. (ed.) (2004). *Scientific Basis for Ayurvedic Therapies*. CRC Press, Washinton, DC.

Mukherrjee, P.K. & Wahile, A. (2006). Integrated approaches towards drug development from Ayurveda and other Indian system of medicines. *Journal of Ethnopharmacology*, **103** (1), 25–35.

Patwardhan, B., Vaidya, A.D.B. & Chorghade, M. (2004). Ayurveda and natural products drug discovery. *Current Science*, **86** (6), 789–799.

Sharma, P.V. (ed.) (2001). *Susruta-Samhita: With English Translation of Text and Dalhana's Commentary Along with Critical Notes*. Chaukhambha Publications, Visvabharati, India.

Sharma, P.V. (ed.) (1981–1994) *Charaka Samhita*, Vols 4. Chaukhambha Sanskrit Series, Varanasi, India.

Thatte, U. & Dhanukar, S. (1997). Rasayana concept: clues from immunomodulatory therapy. In: *Immunomodulation*. Upadhyay, S.N. (ed.), Narosa Publications, Delhi.

Valiathan, M.S. (2006). *Towards Ayurvedic Biology, A Decadal Vision Document*. Indian Academy of Science, Bangalore.

Vogel, G.H. (ed.) (2002). *Drug Discovery and Evaluation, Pharmacological Assays*, 2nd edn. SpringerVerlag, Berlin.

Williamson, E. (2006). Ayurveda: introduction to pharmacists. *The Pharmaceutical Journal*, **276**, 108–110.

13 An overview of functional food regulation in North America, European Union, Japan and Australia

Paula N. Brown and Michael Chan

13.1 Introduction

Most jurisdictions regulate consumer products using distinct product categories, such as food and drugs. Within these regulatory hierarchies, there exist specific divisions that are tasked with the evaluation and regulation of each distinct product category. This highly departmentalised framework has been the classic method by which regulation of consumer products is accomplished. For example, foods fall under the purview of departments and sections specific for foods, while drugs are the responsibility of departments specialising in drugs. This framework is sufficient for the majority of products that can be clearly classified and defined.

In a regulatory sense, functional foods and nutraceuticals are not clearly classified and defined. Other than in Japan, the term 'functional food' is neither defined nor recognised as a distinct product category by any regulatory authority. Even the Japanese definition fails to encompass all products that could be considered functional foods or nutraceuticals. The lack of a formal definition essentially requires regulatory authorities to place functional food products into one of their already established and regulated categories.

Functional foods and nutraceuticals have properties inherent to both foods and drugs; they straddle the border between conventional food and drugs. In general, a discussion on the regulation of functional food and nutraceuticals in a particular jurisdiction involves discussion of that jurisdiction's framework to allow claims for products that are or resemble foods. Although these products need not make an explicit claim, the definitions for functional foods and nutraceuticals declare these products as exerting some sort of effect on health and are viewed by regulatory authorities to be within the realm of drugs, medicines and/or therapeutic products.

This chapter discusses how functional foods and nutraceuticals are addressed within the risk-based regulatory frameworks adopted by regulatory authorities in five specific jurisdictions; Canada, the United States, the European Union (EU), Japan and Australia. The mechanisms that each of these jurisdictions have developed to regulate these products are still in development and still evolving. This chapter provides a basic overview of the frameworks used and a general sense of how functional foods and nutraceuticals are treated in each jurisdiction.

13.2 The Canadian regulatory framework

The regulation of food and drug products in Canada is the responsibility of Health Canada (Health Canada 2008). As with other jurisdictions, various directorates and branches within

Health Canada oversee specific categories of products. No formal definitions for functional foods or nutraceuticals exist in Canada; these products can be considered foods, drugs or natural health products (NHPs) under the Canadian regulatory system. Ultimately, the regulatory fate of the products depends on the category to which they are assigned.

Canada does not possess a formal body to make these determinations. Manufacturers typically submit their product for review to whichever branch of Health Canada they believe to be responsible. If it is determined that the product would be better regulated in a separate branch, the applicant is directed there. The issue of proper classification in Canada is most evident in products bordering the food and NHP interface. In Canada, products making claims are usually classified as drugs and subject to drug law, although, there are exceptions for products meeting certain conditions to be regulated in a different way.

13.2.1 Foods

Foods are subject to relevant sections in the Food and Drug Act (Queens Printer 1985) and Regulations (Queens Printer n.d.). The responsibility for regulating foods in Canada is shared between Health Canada's Food Directorate (Health Canada 2005) and the Canadian Food Inspection Agency (CFIA) (Canadian Food Inspection Agency 2008). The Food Directorate is responsible for establishing policies and standards that govern nutrition and health. This would include food safety and food labelling as they relate to nutrition and health. The CFIA conducts the enforcement and inspection duties related to the food supply, as well as the development of standards and guidelines for packaging, labelling and advertising of foods.

In general, conventional foods do not require premarket approval provided there is evidence of traditional use and no claims are made. It is possible, however, that some functional foods be considered novel foods or genetically modified foods as described in the Food and Drug Regulations (Queens Printer n.d.). These foods require a premarket safety assessment prior to going to market. With the exception of plants with novel traits, which are assessed by the CFIA, most functional foods that are considered novel have their safety assessment performed by the Food Directorate. Guidelines to assist manufacturers in making submissions can be found on the Health Canada website (Health Canada 2006b).

The guidelines established and the amount of evidence required in establishing the safety of these foods by Health Canada are very rigorous (Health Canada 2006b). The safety assessment, however, is flexible and allows different types of evidence to be considered in the review. Each application is reviewed on a case-by-case basis. When making decisions, the safety review also considers the effects the introduction of the food into the Canadian marketplace would have on Canadian nutrition. Novel foods in Canada need to be safe for consumption and should refrain from jeopardising the nutritional status of Canadians.

13.2.2 Nutrient content claims

Certain claims are permitted on foods in Canada. The type of permitted claims falls under two general categories: nutrient content claims and health claims. Section B.01.500 of the Canadian Food and Drug Regulations outlines the requirements for a food to be allowed to have nutrient content claims (Queens Printer n.d.). These claims allow a food to have a characterisation of its nutrient content on its label. The label can thus state the food is 'low', 'high', 'a good source' or other descriptor of a particular nutrient. In order to make these claims, the food must meet the specifications set out in the table in Section B.01.513 of the Food and Drug Regulations (Queens Printer n.d.).

13.2.3 Health claims in food

The definition of a drug under the Food and Drugs Act includes any product 'represented for use in the diagnosis, treatment, mitigation or prevention of a disease, disorder or abnormal physical state or its symptoms, in human beings or animals or restoring, correcting or modifying organic functions in human beings or animals' (Queens Printer 1985). Under this definition, any products that make such therapeutic claims are considered drugs and are subject to drug law. To allow for foods to have claims without being subject to requirements for drugs, the Food and Drug Regulations were amended in 2002 (Queens Printer n.d.). The amendment exempted qualifying foods making specific diet-related health claims from many of the provisions specified for drugs (Queens Printer n.d.).

A diet-related health claim describes the characteristics of a diet that may reduce the risk of developing a diet-related disease or condition. There are currently five disease–diet relationship claims that are allowed in Canada summarised as follows:

1. A diet low in sodium and high in potassium, and the reduction of risk of hypertension
2. A diet adequate in calcium and vitamin D, and the reduction of risk of osteoporosis
3. A diet low in saturated and trans fats, and the reduction of risk of heart disease
4. A diet rich in vegetables and fruits, and the reduction of risk of some types of cancer
5. Minimal fermentable carbohydrates in gum, hard candy or breath-freshening products, and the reduction of risk of dental carries

There are strict criteria that manufacturers must follow if they want to use these dietary health claims. These criteria are set out in Section B.01.603 of the Food and Drug Regulations (Queens Printer n.d.). The table specifies precisely what claim can appear on the label, as well as what requirements must be met to be able to use the claim. In addition to the claim-specific requirements in the table, any food making any diet-related health claim must not be intended solely for children under 2 years of age or represented for use in a very low-energy diet. Any food can use these claims provided the food meets the specified compositions.

Each of these diet-related health claims required individual amendments to the Food and Drug Regulations to become valid. Health Canada is currently reviewing other claims that have been approved for use in the United States for consideration of approval in Canada. Interested parties can also make submissions to Health Canada for the approval of drug-like claims that have not yet been reviewed. For any claim that is approved, a regulatory amendment would be made and the claim would then be added to Section B.01.603 of the Food and Drugs Regulation and be permitted for use by any food that meets the requirements set out in the section.

In March of 2009, Health Canada published an updated interim guidance document that described the principles and criteria they would use to evaluate health claims as well as the information required to obtain regulatory approval for the claim (Health Canada 2009b). The safety of the food must be established for the claim to be considered. Unless the product is unmodified and already has an established history of safe use, an additional safety evaluation as per the novel food requirements would need to be performed prior or concurrent to the submission (Health Canada 2009b). The claim must be supported by high-quality human studies. The type and number of studies required to demonstrate sufficient evidence depends on a variety of factors. For instance, health claims associated with food constituents would require intervention studies, whereas health claims associated with a food category are better served with well-designed observational studies.

Only health claims that are considered to be 'drug-like' need undergo the submission process and subsequent regulatory amendment to receive approval. Function claims are not considered 'drug-like' and are not required to undergo the above approval process, provided they are truthful and not misleading (Health Canada 2009d). Function claims are claims that describe the link between the characteristics of a diet, a food or food constituents associated with health or performance (Health Canada 2009d). Nutrient function claims must meet requirements set out in the regulations; these include the quantity of the nutrient in the food and the permitted wording of the claim. Those that are making non-nutrient function claims in their foods are required to have evidence that supports the truthfulness and validity of the claim. Although premarket approval for these claims is voluntary, Health Canada encourages the practice.

13.2.4 Natural health products

The Natural Health Products Regulations came into effect on 1 January 2004 and officially introduced the term 'natural health product' into the Canadian regulatory framework (Queens Printer 2003). NHPs include vitamins and minerals, herbal remedies, homeopathic medicines, traditional medicines, probiotics and other products, such as amino acids and essential fatty acids (Queens Printer 2003). Under the NHP Regulations, these products are regulated as a subset of drugs under the Food and Drug Act (Queens Printer 2003). Similar to foods making the health claims described earlier, NHPs are exempt from many of the drug provisions set out in the Food and Drug Regulations. These products must meet the requirements set out in the Natural Health Products Regulations. The regulating authority for NHPs in Canada is the Natural Health Products Directorate (NHPD) (Health Canada 2009c). To be legally sold in Canada, all NHPs must undergo a premarket review process and obtain a product license and natural product number.

The definition of an NHP includes a substance component and a function component. The substance component refers to the medicinal ingredient or ingredients in the NHP, and includes 'herbal remedies, traditional and homeopathic medicines and materials derived from plants, algae, bacteria, fungi, or non-human animal material, amino acids, essential fatty acids, probiotics, minerals, several vitamins and synthetic versions of the natural ingredients' (Queens Printer 2003).

The function component of the definition defines an NHP as follows:

- a 'substance manufactured, sold or represented for use in the diagnosis, treatment, mitigation or prevention of a disease, disorder or abnormal physical state or its symptoms in humans;
- restoring or correcting organic functions in humans; or
- modifying organic functions in humans, such as modifying those functions in a manner that maintains or promotes health' (Queens Printer 2003).

An NHP can take on any form, provided it complies with the two components of the definition. Thus, NHPs can exist in typical supplement forms, such as capsules, pills, tablets, powders and liquids, as well as in more conventional food forms, such as bars, gums, wafers or beverages. Because NHPs are considered a subset of drugs, they are permitted to make health claims. With the combination of limited restrictions on product form and the ability to make claims, manufacturers have seen the NHP Regulations as a mechanism through which they can add claims to their food products. Although this was not the original intent of the regulations, the NHPD has received numerous submissions of this type (Health Canada 2009c). These so-called food-like NHPs or NHPs in food formats have proved to be a significant regulatory challenge for the NHPD (Health Canada 2009c). Health Canada has stated that proper classification of these products is required in order to ensure that they are evaluated under the appropriate regulations. These products possess characteristics of both foods and NHPs so it is not always simple to classify them as either one or the other.

As such, new guidelines have been developed to facilitate the classification of these products (Health Canada 2009a). These guidelines base their decisions on four criteria: product composition, product representation, product format and public perception and history of use (Health Canada 2009a). A separate committee made up of representatives from both the Food and NHP Directorates will be responsible for making these decisions (Health Canada 2009a). Although it is likely under the new guidelines the majority of functional foods will be classified as foods and subject to food law and the health claims framework described above, some products may be classified as NHPs. Additionally, nutraceuticals and functional ingredients may be more likely classified as NHPs, due to their presentation format.

NHP submissions are reviewed by the NHPD. Applications must include evidence detailing the product's safety, efficacy and quality. Also required in the product license application are the proposed label and the claims that are being made with the product. The standards of evidence required to demonstrate safety, quality and efficacy will vary between products. The amount and type of evidence will depend on the nature of the product and the type of claim being used. Factors such as the intended use, dosage, duration of use and target population are all important considerations in the safety assessment of the product. This is in contrast to the review of novel foods and health claims in foods which are both considered to be consumed *ad libitum* and thus may require more stringent review.

NHP claims can fall under two categories: traditional use claims and non-traditional use claims. Traditional use claims are for products that have been used within a cultural belief system or healing paradigm for at least 50 consecutive years. The demonstration of the traditional use of the product in another culture or country other than Canada is acceptable. Non-traditional use claims are for those claims that do not meet the requirements of traditional use claims. These claims can use evidence demonstrating traditional use, but they will also require other sources of scientific evidence.

The NHPD has prepared monographs for certain ingredients and formulations (Health Canada 2006a). These monographs can be used as evidence for safety and efficacy in an application. The monographs provide specific details including, but not limited to, the medicinal ingredient's proper name, the medicinal ingredient's common name, the source of the medicinal ingredient, the route of administration, the dose for the product, quality specifications for the product, the duration of use for the product and the targeted sub-population group (Health Canada 2006a). The monographs also describe a recommended use or purpose of the ingredient and risk information associated with the ingredient. An applicant may make alterations to the wording of these elements, but cannot alter their meaning.

If the product matches the required criteria, as set out in the monograph, a special type of NHP application called 'a compendial application' can be made. The monograph can be used as the sole source of evidence for the safety and efficacy of the product. Otherwise, additional evidence is usually required to supplement the monograph. It is unlikely that functional foods would be eligible for compendial applications; however, the published monographs could be used as a source of evidence for some applications.

The NHPD has published a list of non-medicinal ingredients that can be used in the manufacture of NHPs (Health Canada 2003). The list also includes limits and specifications for the product. Manufacturers may use these ingredients provided they meet the specifications outlined in the list. If the applicants wish to use an ingredient not on this list or use an ingredient on the list outside the specified purpose or at a level not specified in the list, they would need to ensure the safety of the ingredient in the product.

All manufacturers, packagers, labellers and importers of NHPs are required to hold a site license issued by the NHPD. The site license will specify exactly which of the previously listed activities the applicant is allowed to do with respect to NHPs. In order to obtain a site license, the applicants must demonstrate that they are in compliance with the Good Manufacturing Practices (GMP) requirements set out in the Natural Health Product Regulations (Queens Printer 2003).

13.2.5 Summary of the Canadian regulatory framework

The Canadian regulatory system for functional foods is still in its infancy. In Canadian law, any food product that makes a therapeutic claim is by definition a drug; this is not changed by introduction of food claims and NHP regulations. Products that are under these regulations are still drugs; the products, however, can be considered special subsets of drugs having to comply with their own special provisions. It is not clear how effective this system will be at regulating the industry as the number and complexity of products being developed for the Canadian market increases. Already Health Canada has had to face significant regulatory challenges in the classifying of food-like NHPs or NHPs in food format (Queens Printer 2003). It is unknown how effective the new guidelines for classifying products would be in addressing these challenges.

13.3 The United States regulatory framework

The United States Food and Drug Administration (FDA) controls food and drug regulations in the United States (US Food and Drug Administration 2008). Functional foods and nutraceuticals are not defined in the United States regulatory framework. As with other jurisdictions, such products are regulated as either foods or drugs. In 1994, the US Congress passed the Dietary Supplements Health and Education Act (DSHEA) (Government Printing Office 1994). The DSHEA was designed to regulate products that supplemented the diet. This act amended the US Food Drug and Cosmetic Act to add dietary supplements as a subcategory of food (Government Printing Office 2004). As a consequence, those generally considered to be functional food and nutraceutical products are likely to fall within food law in the United States.

13.3.1 Dietary Supplements and the Dietary Supplements Health and Education Act

Dietary supplements are a subset of foods and subject to US food law (Government Printing Office 2004). DSHEA also states that the Secretary of Health and Human Services may, by regulation, prescribe specific GMPs for dietary supplements (Government Printing Office 1994). In June 2007, the FDA issued a final rule establishing regulations to require current good manufacturing practices (cGMP) specific for dietary supplements (Office of Federal Regulations 2008). The cGMPs in place have strict requirements to ensure that dietary supplements are safe and not adulterated; they put particular emphasis on identity testing of raw materials and testing of finished products to ensure specifications are met (Office of Federal Regulations 2008).

Dietary supplements do not require premarket approval from the FDA (FDA Center for Food Safety and Applied Nutrition 2008a). FDA reviews neither the safety nor the efficacy of dietary supplements prior to their introduction to the market (FDA Center for Food Safety and Applied Nutrition 2008a). The FDA may take enforcement action on products in the marketplace if the products pose a health risk or do not meet the requirements set out in the legislation. The manufacturers of the products are responsible for ensuring that the requirements are met.

The FDA may only remove dietary supplements from the market if the products 'present a significant or unreasonable risk of illness or injury' (Government Printing Office 2004) or poses an 'imminent hazard to public health or safety' (Government Printing Office 2004). The burden of proof is on the FDA to show the aforementioned conditions prior to removing products from market. This can become a long and expensive process for the FDA as was prominently illustrated by the ban on ephedrine-alkaloid dietary supplements (EDS). FDA's formal investigation into EDS began in 1997 and a ban on EDS was issued 7 years later (Nutraceutical Corp 2006). A further 2 years of court proceedings (a lower court's initial decision to overturn the ban was later reversed upon appeal) settled the matter in the FDA's favour (Nutraceutical Corp 2006); a further appeal to the US Supreme Court was denied in May 2007 (FDA Center for Food Safety and Applied Nutrition 2006b). The FDA's decision to take enforcement action can be very resource and time-intensive. Thus, a more common course of action taken by the FDA is the issuance of warning letters to manufacturers and public advisories on dangerous products.

A dietary supplement is defined as a product (other than tobacco) intended to supplement the diet that bears or contains one or more of the following dietary ingredients:

- a vitamin;
- a mineral;
- a herb or other botanical;
- an amino acid;
- a dietary substance for use by man to supplement the diet by increasing the total dietary intake; or
- a concentrate, metabolite, constituent, extract, or combination of any ingredient described in clause (A), (B), (C), (D), or (E), (Government Printing Office 1994).

Products falling within this definition must clearly state on their labels that they are dietary supplements. Additionally, a dietary supplement must be 'intended for ingestion in tablet, capsule, powder, softgel, gelcap or liquid form or if not intended for ingestion in such a form not be represented for use as a conventional food or as a sole item of a meal or the diet' (Government Printing Office 1994). While some drink products could meet this requirement and are considered dietary supplements, most products that resemble conventional foods would not.

In theory, this should prevent functional foods that appear as conventional foods from being marketed and sold. However, the lack of premarket approval allows manufacturers to bring products to market without initial objections from the FDA. Only after the FDA becomes aware of a product's characteristics are they able to take enforcement action.

Although considered foods, dietary supplements cannot be added freely to conventional foods. Under the Food Drug and Cosmetic Act, only ingredients that are generally recognised as safe (GRAS) or are approved food additives can be added to conventional foods (Government Printing Office 2004). Of the two, GRAS is the simpler and more common approach used to obtain permission to add functional ingredients into foods.

13.3.2 Generally recognised as safe

To obtain GRAS status for an ingredient, the applicant can use a self-affirmation process (FDA Center for Food Safety and Applied Nutrition 2004b). The process requires a panel of experts to review the data and evidence demonstrating the safe use of the ingredient in the product. The review panel need not be selected by the FDA but must consist of experts qualified by scientific training and experience to evaluate the safety of the product. The evidence needed for a GRAS self-affirmation includes toxicology studies, exposure or consumption data, manufacturing information and product specifications, such as intended use and dosage levels.

Once the expert panel reviews the data and comes up with a favourable GRAS determination, the manufacturer may begin using the ingredient. Manufacturers can choose to notify the FDA of their GRAS self-affirmation findings (FDA Center for Food Safety and Applied Nutrition 2001). This is the usual course of action, as a manufacturer can then be reasonably sure that, should the FDA accept their findings, the GRAS status of the ingredient will not be challenged by the FDA at a later date.

Upon receipt of the notification, the FDA will publish the information on their website for comment (FDA Center for Food Safety and Applied Nutrition 2008d). Within 90 days of receiving a completed notice, the FDA will inform the applicant in writing whether they have any objections to the findings. Even if the FDA disagrees with the manufacturer's GRAS assessment, the manufacturer may still continue using the GRAS product. However, the manufacturer will be risking future regulatory enforcement action from the FDA. The GRAS status is tied to the ingredient's use, not the ingredient itself. Thus, the use of the ingredient in other food products or at levels other than those applied for is not necessarily allowed. To be certain, another GRAS self-affirmation study is needed.

13.3.3 Dietary ingredients

The establishment of the DSHEA also saw the enactment requiring premarket safety notification for new dietary ingredients (NDIs) (Government Printing Office 1994). In the early years of DSHEA, little attention was paid to this part of the Act. In recent years, the FDA has stepped up enforcement of this provision and has begun to use it as a form of premarket approval for dietary supplements (Noonan & Noonan 2006).

An NDI is defined in DSHEA as 'a dietary ingredient that was not marketed in the United States before October 15, 1994' (Government Printing Office 2004). All NDIs require a premarket review of its safety by the FDA. Dietary ingredients used prior to 15 October 1994 are presumed to be safe and do not require the premarket safety evaluation. Although dietary supplements do not specifically require premarket approval, a large number of dietary supplements are being rejected or removed because they contain NDIs (Noonan & Noonan 2006). The reasons cited by FDA include mismatches between the marketed and tested ingredient, the use of therapeutic language, the use of food language and clinical studies that have been done via non-oral routes (Noonan & Noonan 2006). However, the most frequently cited reason for rejection is the lack of sufficient safety data or that the substance fails to meet the safety standard (Noonan & Noonan 2006).

If a dietary supplement contains an NDI, the manufacturer or distributor is required to notify and provide safety information to the FDA at least 75 days prior to the product going to market (Government Printing Office 1994). This notification is not required for dietary ingredients that were legally marketed prior to 15 October 1994 (Government Printing Office 1994). Such ingredients were grandfathered into the legislation as 'old' dietary ingredients (Government Printing Office 1994).

The FDA has stated that simple presence of a dietary ingredient in the marketplace is not sufficient for it to be considered an 'old' dietary ingredient (Noonan & Noonan 2006). The ingredient must have been lawfully marketed as a dietary ingredient prior to 15 October 1994 to qualify (Government Printing Office 1994). The FDA stated that for an ingredient to be considered lawfully marketed, the manufacturer or distributor must have written evidence that the ingredient is chemically identical to a dietary ingredient legally marketed in the United States before this date (Noonan & Noonan 2006). The evidence required to prove that an ingredient was lawfully marked must include a product invoice, bill of lading, product label or a catalogue with a date showing evidence of marketing before 15 October 1994 (Noonan & Noonan 2006). The FDA also made it clear that the ingredient itself must have been marketed, not just a formulation containing the ingredient (Noonan & Noonan 2006). The mere presence of the ingredient in food is not sufficient evidence to prove prior marketing; the ingredient must be clearly marketed for its own properties (Noonan & Noonan 2006).

The safety evidence required in an NDI application is usually scientific (FDA Center for Food Safety and Applied Nutrition 2005) because showing safe history of use for NDIs to the FDA is difficult (Noonan & Noonan 2006). FDA will only accept data that show the NDI in the exact form and at the dosage level described in the NDI submission (FDA Center for Food Safety and Applied Nutrition 2005). Furthermore, since the NDI will be in a dietary supplement, and a dietary supplement is a food, the safety evidence provided must support the NDI's use for food purposes or as an article of food (FDA Center for Food Safety and Applied Nutrition 2005). Evidence demonstrating the use of the dietary ingredient for therapeutic purposes, such as traditional herbal medicinal uses, will not be considered sufficient (FDA Center for Food Safety and Applied Nutrition 2005). The safety evidence must clearly show that the ingredient in the reference or study is identical to the ingredient in the NDI submission and at the dosage levels stated in the NDI submission (FDA Center for Food Safety and Applied Nutrition 2005).

13.3.4 New drugs

In the United States a substance cannot be classified as both a drug and a food; the Food Drug and Cosmetic Act states that 'any article that has been approved as a new drug or an article authorized for investigation as a new drug, antibiotic or biological for which substantial clinical investigations have been instituted and for which the existence of such investigations has been made public, cannot be considered a dietary supplement unless it was marketed as a dietary supplement or as a food prior to obtaining the license' (Government Printing Office 2004). A single study published in scientific literature can be adequate for classifying a substance as a drug under this definition (Noonan & Noonan 2006).

This is especially important as drug companies are increasingly investigating the use of constituents in foods as sources for new drugs (Noonan & Noonan 2006). Although these compounds may naturally occur in foods, the actual compounds may not have been marketed in the United States as a food or dietary supplement. The addition of an ingredient that is considered a new drug into a product would make the product a drug and subject to drug law. FDA's definition of marketed for new drugs is the same as for NDIs (Noonan & Noonan 2006). Therefore, it is important for manufacturers interested in manufacturing dietary supplements ensure that the ingredients in their product have not had drug applications associated with them. In addition, manufacturers should ensure that the ingredient in question has been previously marketed as a dietary supplement. Otherwise, the ingredient can only be marketed as a drug and not a dietary supplement or ingredient in food.

13.3.5 Structure function claims

Under DSHEA, dietary supplements are allowed to have structure function claims (Government Printing Office 1994). These claims are not nutrient claims or drug claims. A dietary supplement cannot claim 'to diagnose, mitigate, treat, cure or prevent a specific disease or class of diseases' (Government Printing Office 2004). Structure function claims are statements of nutritional support. The statements allowed can be made 'if the statement claims a benefit related to a classical nutrient deficiency disease and discloses the prevalence of such disease in the United States, describes the role of a nutrient or dietary ingredient intended to affect the structure or function in humans, characterizes the documented mechanism by which a nutrient or dietary ingredient acts to maintain such structure or function, or describes general well-being from consumption of a nutrient or dietary ingredient' (Government Printing Office 1994). The claims, or statements, are only allowed if 'the manufacturer of the dietary supplement has substantiation that such statement is truthful and not misleading, and the statement contains, prominently displayed and in boldface type, the following statement "This statement has not been evaluated by the Food and Drug Administration. This product is not intended to diagnose, treat, cure or prevent any disease"' (Government Printing Office 1994).

No premarket approval is required for the claim, but the manufacturer of the dietary supplement must notify the Secretary of Health and Human Services within 30 days after the first marketing of the dietary supplement and indicate the statement being used (Government Printing Office 1994). The FDA has published guidance papers to assist industry in understanding the requirements for using structure function claims (FDA Center for Food Safety and Applied Nutrition 2002) and the amount of evidence needed to support them (FDA Center for Food Safety and Applied Nutrition 2004c).

Conventional foods can also have structure function claims (FDA Center for Food Safety and Applied Nutrition 2004a); however, the type of structure function claims used for conventional foods is much more limited than those for dietary supplements (FDA Center for Food Safety and Applied Nutrition 2004a). The claims can only be based on the nutrients the food contains for which a daily requirement, such as a recommended daily intake, has been established (FDA Center for Food Safety and Applied Nutrition 2004a). An example of such a claim would be 'Calcium builds strong bones' (FDA Center for Food Safety and Applied Nutrition 2004a). Manufacturers making these types of claims do not have to notify FDA and no disclaimer is necessary (FDA Center for Food Safety and Applied Nutrition 2008b).

13.3.6 Nutrient claims

Conventional foods are more likely to have nutrient content claims than structure function claims. Nutrient content claims were permitted for foods following the passage of the Nutrition Labeling and Education Act of 1990 (NLEA) (Government Printing Office 1990). These claims allow for the characterisation of particular nutrients in a food product. With nutrient content claims, manufacturers can characterise their products as being 'low', 'high', 'reduced', or other descriptive words with respect to the amount of a particular nutrient (Government Printing Office 1990). The manufacturers may also claim their product to be 'healthy' provided the product contains a certain level of total fat, saturated fat, cholesterol and sodium (Government Printing Office 1990). The requirements for these types of claims can be found in Appendix A and Appendix B of the FDA's Food Labeling Guide (FDA Center for Food Safety and Applied Nutrition 2008b). Only authorised nutrient claims listed in the two appendixes are allowed (FDA Center for Food Safety and Applied Nutrition 2008b).

For claims not currently authorised, the manufacturer may notify the FDA of its intent to use the claim provided the claim has been 'authorized and published by a scientific body of the United States Government with official responsibility for public health protection or human nutrition research' (Government Printing Office 2004). Examples of such bodies include the National Institutes of Health and the Centers of Disease Control and Prevention (Government Printing Office 2004). There are also restrictions in place to prevent the use of certain nutrient claims if other nutrients in the product are at specific levels set by FDA in the interest of maintaining public health (Government Printing Office 2004).

13.3.7 Health claims in food

Health claims were first permitted for foods with the passage of NLEA in 1990 (Government Printing Office 1990). Significant updates to health claims regulation occurred with the passage of the Food and Drug Administration Modernization Act in 1997 (Government Printing Office 1997) and the establishment of the Consumer Health Information for Better Nutrition Initiative of 2003 (FDA Center for Food Safety and Applied Nutrition 2008c). Two types of health claims are permissible under these rules: authorised health claims and qualified health claims (Government Printing Office 1997; FDA Center for Food Safety and Applied Nutrition 2008c). In each case, only claims approved by the FDA can be used. Furthermore, health claims are only permitted on conventional foods that meet specific nutrient content requirements (FDA Center for Food Safety and Applied Nutrition 2008b). They must contain 10% or more of the daily value for one or more of the six nutrients – vitamin A, vitamin C, iron, calcium, protein or fibre – without fortification (FDA Center for Food Safety and Applied Nutrition 2008b); the food cannot contain more than the allowable amounts of the disqualifying nutrients – total fat, saturated fat, cholesterol and sodium (FDA Center for Food Safety and Applied Nutrition 2008b). Dietary supplements are exempt from the nutrient requirements (FDA Center for Food Safety and Applied Nutrition 2008b).

13.3.8 Authorised health claims

Authorised health claims were the first type of claims permissible under the NLEA (Government Printing Office 1997). In order for a health claim to be authorised, there must be 'sufficient scientific acceptance among qualified scientific experts that the claim is supported by the totality of publicly available scientific evidence derived from well-designed studies conducted in a manner which is consistent with generally recognized scientific procedures and principles' (Government Printing Office 1990). A guidance document has been published by the FDA to further explain what they consider to be 'adequate significant scientific agreement' (FDA Center for Food Safety and Applied Nutrition 1999).

With the passage of NLEA, the US Congress directed the FDA to evaluate ten specific nutrient–disease relationships which they considered possible health claim topics (Rowlands & Hoadley 2006). The FDA authorised five of these claims and rejected two (Rowlands & Hoadley 2006). For the remaining three claims, the FDA concluded that the evidence of these claims could be accepted with alternative wording (Rowlands & Hoadley 2006). Under the NLEA, interested parties could also submit a petition to request that the FDA authorise additional health claims (Government Printing Office 1990). Four more claims were authorised through such submissions, but the process to accomplish this was considered very slow (Rowlands & Hoadley 2006).

The passage of the Food and Drug Administration Modernization Act allowed for an expedited health claim authorisation process (Government Printing Office 1997). A petitioner could now submit a claim to the FDA based on 'current published authoritative statements from scientific bodies of the United States Government with official responsibility for public health protection or human nutrition research or from the National Academy of Sciences' (Government Printing Office 1997). The petitioner may then, provided there are no objections from the FDA, use the claim 120 days following submission and notification of the use of the claim to the FDA (Government Printing Office 1997).

There have been few claims authorised through the use of this system (Rowlands & Hoadley 2006). The requirements for sufficient scientific acceptance were so high that few health claims could meet them (Rowlands & Hoadley 2006). Generating the evidence necessary was an expensive and time-consuming process (Rowlands & Hoadley 2006). In response, manufacturers began to launch lawsuits challenging the FDA's blanket prohibition on dietary supplement claims on conventional foods (FDA Center for Food Safety and Applied Nutrition 2008c). Eventually, the FDA was ordered by the US Court of Appeals for the District of Columbia to develop a framework to allow for claims without requiring significant scientific agreement (*Pearson v. Shalala* 1999). In response, the FDA developed the framework that allows for qualified health claims.

13.3.9 Qualified health claims

In 2003, the FDA launched the Consumer Health Information for Better Nutrition Initiative for foods and supplements (FDA Center for Food Safety and Applied Nutrition 2008c). The result of the initiative allowed the use of qualified health claims in foods. These claims did not have sufficient scientific agreement in support of them and would therefore be required to have a qualifying statement detailing the limitations of the evidence supporting the claim.

The exact procedures for submitting a qualified health claim have not been set. The FDA has released guidance on their current thinking on the topic and has stated a final rule may or may not be forthcoming (FDA Center for Food Safety and Applied Nutrition 2006c).

Petitioners are required to submit a petition to the FDA to make a qualified health claim (FDA Center for Food Safety and Applied Nutrition 2006c). The procedures to be followed and the information to be included in petitions are detailed in guidance documents released by the FDA (FDA Center for Food Safety and Applied Nutrition 2003a,b). These are interim policies that the FDA may change at a later date (FDA Center for Food Safety and Applied Nutrition 2006c). Upon completion of the review, the FDA will announce their decision on the matter through a letter to the applicant. Similar to the FDA's policy for GRAS self-affirmation, the FDA will not officially approve any qualified health claims; instead, the FDA will determine whether they will exercise enforcement discretion or deny the claim (FDA Center for Food Safety and Applied Nutrition 2006c).

The actual qualified health claim that can be used is specified in the letter. The wording of the claim is selected by the FDA based on the FDA's estimate of the strength of the evidence supporting the claim. The claims are quite verbose and very conservative in nature. For example, the most recent qualified health claim to be issued a letter of enforcement discretion is as follows:

> Limited and not conclusive scientific evidence suggests that eating about 1½ tablespoons (19 grams) of canola oil daily may reduce the risk of coronary heart disease due to the unsaturated fat content in canola oil. To achieve this possible benefit, canola oil is to replace a similar amount of saturated fat and not increase the total number of calories you eat in a day. One serving of this product contains [x] grams of canola oil. (FDA Center for Food Safety and Applied Nutrition 2006a)

Surveys found that the way these claims were presented did not necessarily have the intended effect on the consumers (International Food Information Council 2005). Surveyed consumers considered qualified health claims too 'wordy' and preferred the unqualified structure function claims (International Food Information Council 2005). They were unable to differentiate between the levels of evidence presented and felt that the words used amounted to marketing trickery, rather than a scientific review (International Food Information Council 2005). A proposed graphical grading system using letter grades to show the level of evidence behind the claim was also viewed negatively by consumers (International Food Information Council 2005). Consumers viewed lower letter grades negatively, which would deter them from purchasing a product with such a claim (International Food Information Council 2005).

13.3.10 Summary of the United States regulatory framework

The FDA has been reluctant to allow any sort of health claims to appear on conventional foods. This reluctance is reflected in the historic development of the current frameworks that allow for claims in foods. Permission to use health claims in food, whether they be authorised or qualified, is very difficult to obtain, and manufacturers need to be prepared to invest significant resources in order to accomplish this. In the case of qualified health claims, it may be considered a waste of effort given the negative response these claims have garnered from consumers.

Dietary supplements are a different matter. The category of dietary supplements in the United States is unique. The broad freedom that manufacturers have in making claims on their products without any premarket authority is not seen in other jurisdictions. Although this freedom is limited to products meant to supplement the diet and not on food products per se, this distinction has not prevented products appearing as conventional foods from being sold as dietary supplements. The lack of premarket approval means the FDA cannot perform enforcement action on these products until after they have reached the market. Even then, the burden of proof is upon the FDA to demonstrate the illegality of a product; ultimately, the level of enforcement is dependent on where the FDA chooses to direct its limited resources. Recent actions by the FDA, such as the increased enforcement of NDIs and the introduction of cGMPs, show that the FDA is willing to devote resources to further regulation of these products.

13.4 The European Union's regulatory framework

There are 27 member states in the EU, and each member state is responsible for passing and enforcing their own laws within their borders. Thus, each member state has their own legislation that deals with food and drugs. The laws that the member states pass, however, have to meet the terms of the various treaties, directives and regulations passed and agreed upon by the EU as a whole. The EU strives for greater harmonisation within its borders and passes legislation to achieve this. Although the individual laws passed by each member state may not be identical, they are all shaped and driven by the overall EU legislative framework. Disagreements and discrepancies still arise, but, in general, a discussion of any one member state's policies towards food and drug regulation can be accomplished through a discussion of the overall EU policies towards food and drug regulation.

As with other jurisdictions, the EU has no legal definitions and no formal regulatory framework for functional foods, dietary supplements or nutraceuticals. The exact rules and regulations that a product is subject to would depend on the product's nature and use. The General Food Law Regulations are applicable to all foods. Legislation on dietetic foods, on food supplements or on novel foods may also be applicable for certain products. Medicines are subject to the EU Medicinal Product's Directives.

The main issue for these products is simply determining if they are foodstuffs or medicinal products. There is no overarching guideline for the EU to address classification of borderline products; it is each member state's responsibility to make these determinations. The amount of resources devoted to classification of these types of products varies from state to state. Some states, such as the United Kingdom, have set up review panels to make such determinations, while others have not (Medicines and Healthcare Products Regulatory Agency 2008). Decisions can vary and will depend on the nature and appearance of the product and which member state the application is submitted to. These products can fall under any number of directives or regulations governing food and drug law.

The following summary describes the more pertinent regulations and directives that could potentially apply to functional foods and nutraceuticals. The laws that apply are very much dependent on the individual product itself and in which member state the product and application is first introduced.

13.4.1 Regulation (EC) No. 178/2002 (General Food Law Regulation)

Regulation (EC) No. 178/2002 of the European Parliament and of the Council of 28 January 2002, commonly referred to as the General Food Law Regulation, describes all the general principles and requirements of food law (European Union Publications Office 2002b). Since this regulation applies to all foodstuffs, functional foods and nutraceuticals that are considered foods in the EU are subject to its provisions.

The regulation defines 'food' as 'any substance or product, whether processed, partially processed or unprocessed, intended to be, or reasonably expected to be ingested by humans' (European Union Publications Office 2002b). It includes drinks, chewing gum and any substance including water, intentionally incorporated into the food during its manufacture, preparation or treatment (European Union Publications Office 2002b). The dividing line between food and medicinal products is not specified in the food law. Rather, it is stated in the Medicinal Product's Directives, which is discussed in more detail later in this summary.

The general food law is rooted in risk management, and it firmly establishes that scientific risk analysis is the basis through which all decisions are to be made in matters of food (Coppens *et al.* 2006). Among the provisions of the general food law was the creation of an independent food authority, the European Food Safety Agency (EFSA), which was given the task of providing scientific advice based on scientific risk assessment with clearly separated responsibilities for risk assessment, risk management and risk communication (European Food Safety Commission 2008c).

Among the more significant provisions of the general food law that may apply to a functional food is the addition of the precautionary principle into regulations. The precautionary principle, a major part of EU law originally used to address environmental issues, was aimed at protecting the environment and public health from uncertain risks resulting from an action. There is a large focus on scientific evidence in the evaluation of risk and harm. In the case of the general food law, the principle can be evoked whenever an analysis of the available scientific information suggests that there is a possibility of harm but uncertainty is still present as there is a lack of scientific consensus. Under such a situation, the precautionary principle allows the European Community to adopt provisional risk management measures to ensure health protections, pending more scientific information and a more comprehensive risk assessment (European Union Publications Office 2002b).

The issuing of such measures requires a lot of considerations in order to balance the freedom and rights of individuals, industry and organisations, with the need to minimise risk to health (Coppens *et al.* 2006). There are many conditions that are attached to application of the precautionary principle, including the requirement that any measures must be proportionate and no more restrictive of trade than is required to achieve health protection. Other factors, such as technical and economic feasibility, are also considered. Measures need to be reviewed within a reasonable period, depending on the nature of the risk identified and the type of scientific information needed to clarify the scientific uncertainty.

13.4.2 Regulation EC No. 258/97 (regulation concerning novel foods and novel food ingredients)

In January of 1997, Regulation EC No. 258/97, more commonly known as the Novel Foods Regulation, came into effect (European Union Publications Office 1997). The regulation sought to protect public health by ensuring that new or novel foods or food ingredients in Europe would be subjected to a safety assessment before being placed in the market.

The deciding factor on whether a food would be considered novel is whether it has already been sold under food law in the EU market prior to 1997. To show that a food or food ingredient is not novel, the manufacturer needs to prove significant sale of the product in the EU as food prior to May 1997. Otherwise, the manufacturer is required to submit the product for authorisation prior to introducing it to the market.

There are two different procedures for authorisation: a simplified procedure for substantially equivalent food or food ingredients and a full authorisation procedure for all other foods and food ingredients. In order to use the simplified procedure, the applicant needs to provide evidence showing the novel food or food ingredient is substantially equivalent to an existing food ingredient in regards to composition, nutritional value, metabolism, intended use and level of undesirable substances (European Union Publications Office 1997). Foods can show substantial equivalence through generally recognised scientific evidence or from an opinion of a competent food assessment body in the member states. The regulation does not clearly define what is meant by generally recognised scientific evidence, so all the applications submitted thus far have relied on opinions of food assessment bodies (Coppens *et al.* 2006).

With the simplified procedure, the applicant must provide a dossier with the evidence to the European Commission. The Commission will then forward a copy of the dossier to all member states whereupon the evidence is reviewed. If the Commission considers the evidence of substantial equivalence acceptable and none of the member states raise any objections, the novel food or food ingredient is approved. If the evidence is deemed insufficient or a member state raises objections or doubts, the Commission will ask a competent member state's food assessment body of its own choosing to provide an opinion of the product. If the opinion is positive, the novel food or food ingredient is approved; otherwise, it must go through the full authorisation procedure.

For those foods or food ingredients where no substantial equivalence can be shown, a full authorisation procedure is required. The applicant must make a request for marketing approval and prepare a dossier for one of the EU member state authorities for initial assessment. The member state authority will review the dossier and then provide its opinion on the application in an assessment report. If the report indicates the application is acceptable, the report is provided to the European Commission whereupon it is forwarded to the other member states. The member states then have 60 days to object to the opinion. If there are no objections, the product is approved, but as of yet there has been no application where at least one member state has not objected on the grounds of public health or safety concerns.

Once the objection is made, a relevant EFSA scientific panel will be established to review the application. The panel will provide their opinion to a committee composed of member state experts. It is this standing committee that will determine if the novel food or food ingredient can be placed in the EU market or not.

13.4.3 Directive 2002/46/EC (Food Supplements Directive)

Directive 2002/46/EC, commonly known as the Food Supplements Directive, was established to harmonise food supplements in the EU (European Union Publications Office 2002a). Prior to its incorporation in 2002, products falling under this category were subject to national law and free movement of these products across member borders was restricted.

This directive defines food supplements as 'foodstuffs the purpose of which is to supplement the normal diet and which are concentrated sources of nutrients or other substances with a nutritional or physiological effect, alone or in combination, marketed in dose form, namely forms such as capsules, pastilles, tablets, pills and other similar forms, sachets of powder, ampoules of liquids, drop dispensing bottles, and other similar forms of liquids and powders designed to be taken in measured small unit quantities' (European Union Publications Office 2002a).

The directive states that the food supplements are foodstuffs and that they must meet all provisions that are described in the food regulation. Thus, safety and quality need to be demonstrated through scientific evidence. The directive essentially recognises products possessing physiological effects, but non-medicinal properties, and allows them to be regulated outside of medicine law.

Articles in the directive currently limit food supplements to only vitamins and minerals. The directive further limits the vitamins and minerals that can be considered food supplements through the attachment of a positive list (European Union Publications Office 2002a). The use of a positive list was upheld by the European Court (Cases C-154 2005), and this form of authorisation has been commonly used by some member states to regulate food supplements not explicitly described in this directive (Coppens *et al.* 2006).

13.4.4 Regulation 1925/2006 (Addition of Vitamins and Minerals and Certain Other Substances to Foods)

The intended purpose of the Regulation 1925/2006 is similar to that of the Food Supplements Directive in that it allows for the free movement of these products across member states while maintaining a high level of safety for human health (European Union Publications Office 2007b). Under this regulation, scientific testing is required to assure the quality and safety of the products. Prior to the passage of this regulation in 2006, each member state had their own frameworks to regulate vitamins and minerals.

Much like the food additives directive, this regulation uses positive lists to establish which vitamins and minerals can be added to food. There are provisions in the regulation to allow for submissions to add other vitamins and minerals not on the list. The addition of these substances is dependent on the evaluation of an appropriate scientific dossier concerning the safety and bioavailability of the individual substance by the EFSA. The regulation allows maximal levels of these substances to be added.

Of significance for functional food manufacturers is that the regulation also allows for the possibility of placing substances on a negative list. Substances that appear on Annex III of the regulation are prohibited from use in food, restricted or under Community scrutiny (European Union Publications Office 2007b). The EFSA can add to the list any substances that could represent a potential risk to consumers. The EFSA can also set limits on the amount of a particular substance that can be added to food to protect the health of consumers. If a manufacturer wants to use a substance on the list, they must first establish the safety of the substance through scientific evidence and seek permission from the regulatory authority. This provision can have considerable consequences for those substances in use today, especially those that are used at higher levels.

13.4.5 Regulation EC No 1924/2006 (Regulation on Health and Nutrition Claims)

EC 1924/2006, also known as the Health and Nutrition Claims Regulation, was published in the official journal of the EU at the end of December 2006 (European Union Publications Office 2007a). Prior to passage, each state had their own framework allowing, or in some cases preventing, health and nutrition claims on foods. Like the previously mentioned pieces of legislation, the Health and Nutrition Claims Regulation strives to harmonise the rules and procedures and allow for consistent application of them throughout all the member states. The goals of free movement of goods across all member states and the assurance of safety and reliance of scientific evidence are very prominent themes in this piece of legislation. All previous frameworks used by individual states were required to conform to this framework. Upon passage of the regulation, claims previously approved by the individual states were no longer valid and had to be reassessed under the new principles set out by this regulation.

Under this regulation, nutrition and health claims are only permitted if the following conditions are fulfilled:

- 'the presence, absence or reduced content in a food or category of food of a nutrient or other substance in respect of which the claim is made has been shown to have a beneficial nutritional or physiological effect, as established by generally accepted scientific evidence;
- the nutrient or other for which the claim is made:
  is contained in the final product in a significant quantity as defined in Community legislation or, where such rules do not exist, in a quantity that will produce the nutritional or physiological effect claimed as established by generally accepted scientific data; or
  is not present or is present in a reduced quantity that will produce the nutritional or physiological effect claimed as established by generally accepted scientific data;
- where applicable, the nutrient or other substance for which the claim is made is in a form that is available to be used by the body;
- the quantity of the product that can reasonably be expected to be consumed provides a significant quantity of the nutrient or other substance to which the claim relates, as defined in Community legislation or, where such rules do not exist, a significant quantity that will produce the nutritional or physiological effect claimed as established by generally accepted scientific evidence' (European Union Publications Office 2007a).

The regulation differentiates between two categories of claims: Article 13 claims and Article 14 claims (European Union Publications Office 2007a). Article 14 claims include claims relating a foodstuff with reduction of disease risk or to children's development or health. Article 13 claims are for any other claims and essentially equivalent to what are described in the other jurisdictions as structure–function claims.

Under the regulation, all authorised health and nutrition claims are put on a list detailing the claim and conditions for the claim's use (European Union Publications Office 2007a). For those claims not on the lists, authorization applications need to be submitted for approval and placement on the list.

A guidance document detailing the requirements of the submission, with particular evidence to the scientific data required, was published by the EFSA to assist in the submission process (European Food Safety Authority 2008a).

Applications are first submitted to the EFSA for review of the data and evidence. Once their evaluation is complete, the EFSA presents its scientific opinion and recommendations for the claim to the European Commission and the member states. These opinions are published on the EFSA website and applicants and members of the public may make comments regarding these opinions to the Commission. These opinions are non-binding and ultimately the decision of whether to proceed is up to the European Commission. If they wish to proceed, the Commission consults with the member states on whether to authorise the use of the claim and its subsequent addition to the register.

At the end of 2007, a number of submissions to the EFSA were made by member states. Submissions included claims that had been previously approved by the states themselves and new claims submitted by manufacturers to their member state representatives. In October 2009, the EFSA released their first series of claim opinions (European Food Safety Authority, 2009a). The series included 94 opinions covering 523 claims. A second series of opinions is expected to be announced in 2010 (European Food Safety Authority, 2009b).

Approximately one-third of the opinions were favourable. These favourable opinions mainly dealt with claims related to the function of vitamins and minerals. Opinions on claims relating dietary fibres and fatty acids with maintenance of cholesterol levels and sugar-free chewing gum with the maintenance of dental health were also favourable. Approximately half of the unfavourable opinions were due to a lack of information on the substance on which the claim was based (European Food Safety Authority, 2009a).

The EFSA is also responsible for providing scientific advice to the European Commission concerning the development of nutrient profiles to be used with health claims. Under the regulation, health and nutrition claims are limited to foods that meet certain nutrient profiles. In January 2008, the EFSA released an opinion providing scientific advice on the major considerations and issues that need to be addressed for the development of the profile(s) (European Food Safety Authority 2008b). Throughout the year, comments, views and opinions on these profiles were also submitted to the Commission and individual member states' food regulatory agencies.

Although the nutrient profiles were to be developed and published by the European Commission with the EFSA's assistance on 19 January 2009, the latest draft proposal, released on 13 February 2009, suggested a final draft was expected to be available to the Commission standing committee on the food chain and animal health for their 27 March 2009 meeting (European Commission 2009b). As the draft proposal was not well received and there is a lack of consensus between the member states on a variety of issues, the Commission decided against presenting the proposal at the 27 March 2009 meeting (European Commission 2009a). At the time of writing this chapter it is unknown when the committee will review the proposal.

13.4.6 Commission Directive 2001/83/EC (Directive on Medicinal Products for Human Use)

Directive 2001/83/EC is the EU's medicines law (European Union Publications Office 2001). Like food legislation, medicine legislation relies on scientific evidence and methods to ensure safety, quality and efficacy. Applications are reviewed and evaluated by the European Medicines Agency (EMEA) (European Medicines Agency 2008). The EMEA then submits their determination to the Commission for final approval. The EMEA also monitors the medicines that are already in the market to ensure their safety and quality.

The directive defines a medicinal product as:

- 'Any substance or combination of substances presented as having properties for treating or preventing disease in human beings; or
- Any substance or combination of substances which may be used in or administered to human beings either with a view to restoring, correcting or modifying physiological functions by exerting a pharmacological, immunological or metabolic action or to making a medical diagnosis' (European Union Publications Office 2001).

The second part of this definition has caused many of the inconsistencies concerning functional food products in the different member states. This part of the definition states that if any substance in a product exerts a pharmacological, immunological or metabolic action, the product can be considered a medicinal product. This is the case even if no statements are made in its labelling or presentation.

The directive provides the definition, but little guidance, so ultimately each individual member state determines what is a medicinal product. A given product may be considered medicinal in one member state but not another which results in certain products being sold as foods or added as food supplements in one member state, while in another they are sold as medicines. With the passing of the food supplements regulation and the promise of further harmonization of foodstuffs that would come along with it, it is anticipated that some of these issues will be resolved. However, in its current state the food supplements regulations only apply to vitamins and minerals. Everything else is still subject to the determination of individual member state.

The usual determining factors used by member states to differentiate foodstuffs and medicinal products are dosage, presentation and claims. There are no strict rules that are followed in this regard and each product is considered on a case-by-case basis.

If a medicinal claim is made, the product will automatically be considered a medicinal product. However, member states can still consider a product medicinal even if no claims are made. Products presented in unit dose form, such as tablets or pills, are usually considered medicinal unless they are vitamins or minerals, which fall under the food supplements regulations. Many member states use positive and negative lists to state which botanicals are permitted in foods and which are considered medicines. None of the member states use the same lists.

13.4.7 Summary of the EU's regulatory framework

There is no single framework for functional food regulation in the EU. How a product treated is very much dependent on how it is viewed in the community. Prior to the passage of the directives and regulations detailed above, how a product was viewed in the community would very much depend on which part of the community the product was currently in. The efforts at greater harmonisation have improved this. The establishment of agencies, such as the EFSA and EMEA, has provided a place whereby universal opinions and evaluations can be made on products. Decisions within the EU rely heavily on scientific evidence. The safety, efficacy and quality of products are all accessed through review of scientific evidence. The current level of evidence required is high; some argue the level is set too high. Many of the regulations and directives are still very new. The procedures to meet the provisions are in some cases still being developed and in other cases being tested out. The framework is still evolving and it remains to be seen exactly what it will turn into.

13.5 The Japanese regulatory framework

The regulation of foods and drugs in Japan is the responsibility of the Ministry of Health Labour and Welfare (MHLW) (Ministry of Health, Labour and Welfare 2006). However, it was the Japanese Ministry of Education, Culture, Sports, Science, and Technology (MEXT) that paved the way for functional food development in Japan (Arai 1996). In 1984, the MEXT conducted a series of studies focusing on the 'tertiary functions of foods' (Ohama *et al.* 2006). The studies looked at the use of food in the prevention of disease during the semi-healthy state of the body (Ohama *et al.* 2006). The results of these studies led the MHLW to introduce a system to regulate such foods (Ohama *et al.* 2006). Introduced in 1991, the Foods for Specified Health Uses (FOSHU) system was the first piece of legislation designed to deal with functional foods and nutraceuticals in the world (Arai 1996).

The fact that Japan explicitly includes functional foods and nutraceuticals in their regulatory framework does not mean that their framework is any less complicated than the other frameworks reviewed here. As with the other jurisdictions, the major issue that needs to be resolved with functional foods and nutraceuticals regulation in Japan is product classification. The varying definitions used for functional foods and nutraceuticals are not collectively recognised in the Japanese regulatory framework. As such, a product that may be considered a functional food by a manufacturer could fall into a myriad of different categories under the Japanese system.

Functional food and nutraceutical products are classified as either drugs or non-drugs. Non-drugs are further classified as foods with health claims (FHCs) or general foods. FHCs are further classified based on the type of claim used, either foods with nutrient function claims (FNFC) or foods for specified health uses (FOSHU). There are even publicly recognised health foods that do not include any claims and are thus regulated simply as general foods.

13.5.1 The Pharmaceuticals Affairs Law

The distinction between food and drug is set out in the Pharmaceutical Affairs Law (Ohama *et al.* 2006). The Pharmaceutical Affairs Law specifies drugs as:

- 'items recognized in the Japanese Pharmacopoeia,
- items (other than quasi-drugs) that are intended for use in the diagnosis, cure or prevention of disease in man and animal, and which are not equipment or instruments, and
- items (other than quasi-drugs and cosmetics) that are intended to affect the structure or functions of the body of man or animal' (Ohama *et al.* 2006).

The Pharmaceutical Affairs Law also includes two lists that help differentiate between drug and non-drug substances (Ohama *et al.* 2006). The first list identifies ingredients used exclusively as drugs and the second list identifies ingredients used as materials not judged as drugs as long as no drug efficacy claim is made. Each of these lists is further broken down into three subcategories:

- 'Substances originating from plants,
- Substances originating from animals and
- Other substances such as chemicals, minerals and others, which are synthesized or highly purified substances obtained from living organisms' (Ohama *et al.* 2006).

The first two subcategories, for non-drugs, in the Pharmaceutical Affairs Law's categorisation of foods, plant and animal substances, list ingredients that can be used as foods or food additives. The third subcategory lists ingredients that are usually only approved food additives.

13.5.2 Food additives

Food additives are a subset of food and are defined as:

- 'substances used in or on food in the process of manufacturing food, or
- substances used for the purpose of processing or preserving food' (Ministry of Health, Labour and Welfare n.d.).

Thus, food additives include substances remaining in the finished food product and substances not remaining in the finished product, such as extraction solvents and other processing aids. Only food additives on the above non-drug list are allowed to be used in food (Ministry of Health, Labour and Welfare n.d.). Any new ingredient or raw material not yet on the lists requires an evaluation by the MHLW (Ministry of Health, Labour and Welfare n.d.). Guidelines for submissions are provided by the MHLW. Submissions must include evidence of the safety and efficacy of the food additive (Ministry of Health, Labour and Welfare n.d.). Inclusion of the food additive must be for a technical purpose (Ministry of Health, Labour and Welfare n.d.).

The MHLW does perform evaluations of certain food additives that are authorised for use in other jurisdictions without applications from manufacturers (Ministry of Health, Labour and Welfare n.d.). This is one of the efforts at deregulation Japan has implemented to allow for greater global harmonisation. Another measure taken in the same vein was the allowance of substances appearing in dosage form, such as tablets or capsules, to be considered as foods. Prior to the change, substances in dosage form were automatically considered drugs.

13.5.3 Food vs drugs

Under the Food Sanitation Law, any substance ingested orally is defined either as a drug or non-drug (Japan External Trade Organization 2006). FHCs are considered 'non-drugs' and are subject to food regulations. They are the only type of 'non-drugs' allowed to make claims and only ingredients or raw materials that have been classified as 'non-drug' can make claims. For certain raw materials and ingredients that have yet to be classified, manufacturers or importers need to make a request for evaluation to the MHLW (Ohama *et al.* 2006). Submission to the MHLW includes data and documents detailing the scientific names of the ingredients, the site of use, the pharmacological and/or physiological actions, any narcotic, psychotropic drug or stimulant drug-like action, any previous examples of approval as a drug in Japan or elsewhere and the eating customs in Japan and elsewhere (Ohama *et al.* 2006). Historical uses and eating customs play a large role in the consideration of the ingredient's classification (Ohama *et al.* 2006). For substances in the non-food category that are further processed to obtain specific ingredients, the ingredients obtained also need to be reinvestigated to determine whether they are drugs (Ohama *et al.* 2006). The exception is when the substance is extracted using either water or ethanol (Ohama *et al.* 2006). The use of these solvents is permitted in manufacturing and processing without a review of its classification (Ohama *et al.* 2006). It is also possible to have a product previously classified as a drug reclassified as a non-drug by subjecting a petition to review by pharmaceutical and medical authorities followed by submission to the MHLW (Ohama *et al.* 2006). There have been three successful reclassifications of this type to date; Coenzyme Q_{10} in March 2001, L-carnitine in November 2002 and α-lipoic acid (thioctic acid) in March 2004 (Ohama *et al.* 2006) were all reclassified as non-drugs.

Like other jurisdictions, Japan has incorporated articles in their legislation to deal with new, novel products (Japan External Trade Organization 2006). For foods, the Japanese MHLW can prohibit the sale of any substance that has not commonly been served for human consumption, if the Pharmaceutical Affairs and Food Sanitation Council are of the opinion that the substance could be harmful to human health (Japan External Trade Organization 2006). Furthermore, even substances that have commonly been served for human consumption can also be prohibited by the MHLW if the substance is processed or prepared in a new way that raises concerns about its effects on human health (Japan External Trade Organization 2006).

13.5.4 Nutritional content claims in food

Manufacturers can include statements detailing the nutritional content of their products (Ministry of Health, Labour and Welfare n.d.). The MHLW allows for the declaration of energy value and nutrients in a product provided that manufacturers follow the standards set out in the 'Nutrition Labeling Standards' (Ministry of Health, Labour and Welfare n.d.). Along with nutrient listing and amounts, the standards allow for nutrient content claims such as 'high in', 'rich in', 'source of', 'low', 'zero', 'light' and other descriptors for relative nutrient amounts (Ministry of Health, Labour and Welfare n.d.).

Foods that do make claims must comply with and are regulated by the FHC system. There are two categories under the FHC system: 'FOSHU (Ministry of Health, Labour and Welfare n.d.) and FNFC (Ministry of Health, Labour and Welfare n.d.). Only claims that maintain or promote health conditions in non-sick individuals are acceptable.

13.5.5 Foods for specified health uses

Under the FOSHU system, foods containing a functional ingredient can claim its physiological effects on the human body upon its label and bear the FOSHU seal (Ministry of Health, Labour and Welfare n.d.). So far, only single functional ingredients are allowed, so products containing combinations of functional ingredients are not accepted. To gain FOSHU approval, the safety and efficacy of the product must be demonstrated and the claim must be approved by the MHLW. The requirements listed by the MHLW are as follows:

- 'Effectiveness on the human body is clearly proven;
- absence of any safety issues (animal toxicity tests, confirmation of effects in cases of excess intake, etc.);
- use of nutritionally appropriate ingredients (e.g. no excessive use of salt, etc.);
- guarantee of compatibility with product specifications by the time of consumption;
- established quality control methods, such as specifications of products and ingredients, processes and methods of analysis' (Ministry of Health, Labour and Welfare n.d.).

The requirements for FOSHU approval are relatively steep, and applicants must be willing to invest significant time and resources to achieve approval. Demonstration of efficacy requires evidence from both animal and human studies. The size and number of studies varies with the complexity of the ingredient and product in which the ingredient is formulated. The evaluation of the efficacy evidence is performed by the Council on Pharmaceutical Affairs and Food Sanitation division of the MHLW (Ministry of Health, Labour and Welfare n.d.).

The safety of the product is evaluated by the Food Safety Commission (Ministry of Health, Labour and Welfare n.d.). Safety is typically demonstrated through toxicity studies and presentation of historical dietary use of the product and ingredient in Japan and foreign countries. The dosage level must be set at an appropriate level to ensure efficacy and safety (Ohama *et al.* 2006). Clinical toxicity studies must also show that an overdose level of three to five times the established level is safe (Ohama *et al.* 2006). Carcinogenicity, teratogenicity and mutagenicity test may also be required. Stability studies for the functional ingredient and shelf-life studies for the product are required.

Clinical studies are expected to be randomised, placebo-controlled, double-blind trials conducted on healthy test subjects, with results showing statistical significance against the control at a *p*-value less than 0.05. The application must also include the manufacturing procedure of the product, include proposed labelling and must clearly identify the functional ingredient.

There are currently nine categories of approved FOSHU products (Ministry of Health, Labour and Welfare n.d.). They are as follows:

- 'Foods to modify gastrointestinal conditions
- Foods related to blood cholesterol level
- Foods related to blood pressure
- Foods related to dental hygiene
- Cholesterol plus gastrointestinal conditions, triacylglycerol plus cholesterol
- Foods related to mineral absorption
- Foods related to osteogenesis
- Foods related to triacyglycerol' (Ministry of Health, Labour and Welfare n.d.).

The FOSHU system has gone through a number of revisions since its inception (Ohama *et al.* 2006). In an effort to make it easier to obtain FOSHU approvals, the MHLW introduced the 'qualified FOSHU', 'standardised FOSHU' and 'reduction of disease risk FOSHU' systems.

The qualified FOSHU system is a slightly relaxed FOSHU approval process. In a qualified FOSHU application, the exact functional ingredient in the product need not be identified. A qualified FOSHU application also relaxes the standard of clinical study required. If a randomised controlled trial is used, the significance levels need only be less than 10% (Ohama *et al.* 2006). A non-randomised controlled trial can be used instead, but in this case the significance level must be less than 5% (Ohama *et al.* 2006).

All other documents and evidence required in the normal FOSHU application are required to be submitted for the qualified FOSHU application. The other requirement for a qualified health claim must be a declaration that the scientific basis for the health claim has not yet been established.

The standardised FOSHU system was developed to allow for those products that contain a functional ingredient that has already obtained FOSHU status in other products to go through a more simplified application process. In order for the product to qualify for the simplified process, it must meet the FOSHU standards and specifications set up by the MHLW and include an ingredient that has already received FOSHU approval. Provided these conditions are met, a simplified application process can be used and clinical studies are not necessarily required to prove safety and efficacy.

The approved ingredients are selected by the MHLW based on past FOSHU applications. These ingredients must have gone through a sufficient number of FOSHU applications and have a high level of scientific evidence to ensure safety and efficacy. The dosage of these ingredients in the product must follow the specifications set out in the MHLW. The list of MHLW approved functional ingredients includes:

- 'indigestible dextrin (3–8 g/day)
- polydextrose (7–8 g/day)

- xylooligosaccharide (1–3 g/day)
- fructooligosaccharide (3–8 g/day)
- soybean oligosaccharide (2–6 g/day)
- isomaltooligosaccharide (10 g/day)
- lactofructooligosaccharide (2–8 g/day)
- galactooligosaccharide (2–5 g/day)
- partially hydrolyzed guar gum (5–12 g/day)' (Ohama *et al.* 2006).

Generally, a new regular FOSHU application process can take from 2 to 3 years (Ohama *et al.* 2006). The standardised FOSHU application takes about 3 months (Ohama *et al.* 2006).

The third modified FOSHU system is the reduction of disease risk FOSHU system. This system allows for the use of FOSHU claims establishing a link between disease reduction and the functional ingredient. So far, only calcium and folic acid have been approved for claims under the reduction of disease risk FOSHU system (Ohama *et al.* 2006).

13.5.6 Food with nutrient function claims

Whereas FOSHU claims are tied to specific products, FNFCs are tied to specific nutrients. Introduced in April 2001, the FNFC category of foods allows functional claims to be made on a product if it contains an approved nutrient in amounts within the MHLW established guidelines. To be valid, the label of the product must include the nutrient claim, the daily intake guidelines, the storage methods and other precautionary statements (Ministry of Health, Labour and Welfare n.d.).

The list of permitted nutrients was developed by the MHLW and is based on scientific evaluation and evidence. Currently, there are 12 vitamin and 5 mineral claims permitted (Ministry of Health, Labour and Welfare n.d.). The FNFC is a completely voluntary system. No premarket approval from or notification to the MHLW is necessary so long as the food meets the established standards and specifications.

13.5.7 Health foods without claims

All foods, regardless of claims or not, cannot have advertising or labels that hint at drug efficacy through promotional statements or phrases, descriptions of ingredients, descriptions of the manufacturing process, descriptions of the origin or history of the product or ingredient or reference to articles from newspapers and magazines or quotes from medical doctors or scientists. Food labels cannot have instructions for use in a format similar to those on drugs, such as 'take before meals', 'take two times daily', 'take two tablets a day', 'adults take three to six tablets a day' or similar regimens. Also, food products cannot be in the form of ampoules, sublingual tablets or sprays. These forms and statements are all limited to drugs.

There are a large number of products not regulated under the FHC system (Ohama *et al.* 2006). These products avoid the requirements of the FHC system by simply not making any claim. Yet, consumers can still recognize these products as foods that impart health benefits or as 'health foods'. As these products need only comply with Japan's Basic Food Law, their efficacy may be questionable.

13.5.8 Summary of Japanese regulatory system

Getting a product with a health claim into the Japanese market is not a simple thing. Safety and efficacy needs to be demonstrated regardless of what category the product is in. However, the Japanese market is very large and well-informed. The Japanese consumer already associates certain ingredients with healthy choices even in the absence of a claim. With the amount of resources that are needed to get a claim approved, it is clear why many manufacturers choose to forego claims.

The MHLW has recognised this trend and is looking at ways to make claims easier to get. The recent changes to the FHC system are steps in this direction. There is also funding available to assist manufacturers to perform the necessary trials and investigations. The MHLW hopes these things will encourage more companies to use the FHC system rather than be unregulated health foods. In this way, they can ensure legitimacy to the market without stifling its growth.

13.6 The Australian regulatory framework

The regulation of functional foods and nutraceuticals in Australia is handled in a number of ways. Again, the type of regulatory framework that a particular functional food or nutraceutical product is subject to is very much dependent on how it is classified.

Australia regulates products as either therapeutic or non-therapeutic. Therapeutic products are regulated at a national level and are the responsibility of the Australian Therapeutic Goods Administration (TGA) (Australian Government Department of Health and Ageing 2007c), while non-therapeutic products are handled at the local territorial or state levels. Although state and territorial governments have the ability to make their own regulations and standards for foods, most states and territories in Australia choose to follow standards developed by the Food Standards Australia New Zealand (FSANZ) (Food Standards Australia New Zealand 2008b).

Under Section 7 of the Australian Therapeutic Goods Act 1989, products can be declared as therapeutic goods or not (Australian Office of Legislative Drafting and Publishing 2008). Such Section 7 declarations can be used to categorise products on the food and therapeutic product interface. The TGA refers questionable products to a joint TGA/FSANZ committee, the External Reference Panel on Interface Matters, which recommends whether the product should be regulated as a food or a therapeutic good (Australian Government Department of Health and Ageing 2004a). The considerations for regulators when making these decisions include the composition or nature of the product, the presentation of the product and the intended use of the product (Australian Government Department of Health and Ageing 2004b).

Typically, products that make a therapeutic claim are considered therapeutic regardless of presentation. There are, however, some structure function claims permitted in foods, and the proposed standard for health claims in foods, when it is adopted, would allow foods to carry approved claims.

13.6.1 General good regulations

Each state or territorial parliament in Australia has the power to pass legislation on any subject matter as long as it is not inconsistent with Commonwealth powers. For locally manufactured products, enforcement and administration of the laws is the responsibility of the state health departments and local municipal councils. The regulatory system varies between each state and territory. Some states and territories have set up separate authorities to specifically deal with food law, whereas others have one department that handles all aspects of public health.

Overall, state and territorial governments within Australia have adopted the Australia New Zealand Food Standards Code through enacting legislation. The piece of legislation is usually the state or territories' food Act or equivalent and will state a legal obligation to enforce the Australia New Zealand Food Standards Code, with or without reservations.

Imported products are subject to Commonwealth authority and are examined by the Australian Quarantine and Inspection Service and the FSANZ. These agencies ensure that all imported food complies with the Food Standards Code regardless of which state or territory the product will ultimately be sold in.

FSANZ has developed a standard to address the regulation of novel foods in Australia (Food Standards Australia New Zealand 2008e). This standard is used to evaluate ingredients that have not been traditionally used in foods. Novel food ingredients require a premarket safety assessment that is performed by FSANZ. Applications require significant evidence to demonstrate their safety. Once approved, the novel food ingredient will be placed on a positive list and can then be used or sold within Australia. The standard allows applicants to request a period of exclusivity in their application. This allows applicants a period of 15 months whereby they have exclusive permission to use or sell the product.

FSANZ will not review an ingredient until an applicant submits a formal application. FSANZ, however, has set up an Advisory Committee to perform preliminary reviews of ingredients to determine if the ingredient requires a novel food application. Before making a formal novel food application, sponsors can seek a recommendation from the Committee to see if the application is truly necessary. The Committee also performs reviews of ingredients that have appeared in the global market and makes recommendations for them. The Committee publishes these recommendations on the FSANZ website (Food Standards Australia New Zealand 2008c).

13.6.2 Food claims

Australia and New Zealand currently allow nutrient content claims and some function maintenance claims (Food Standards Australia New Zealand 2008d). Beginning in 1998, FSANZ began running a pilot study to allow approved foods to display a health claim stating the relationship of pregnant women consuming folate and a reduced risk of neural tube defects in unborn babies. This study was used to evaluate a framework to allow health claims in food (Food Standards Australia New Zealand 2008).

In 2003, the Australia and New Zealand Food Regulation Ministerial Council released a Policy Guideline on Nutrition, Health and Related Claims (Food Standards Australia New Zealand 2005). FSANZ used the guideline to develop a proposed standard to more effectively regulate health claims (Food Standards Australia New Zealand 2005) allowing for three types of claims: nutrient content claims, general level claims and high-level claims (Food Standards Australia New Zealand 2008a). Nutrient content claims would allow for statements regarding the amount of specific nutrients or other biologically active ingredients to be made. In essence, these claims are the same type of nutrient content claims described in the other jurisdictions above. The proposed standard would contain a table detailing the permitted substances, the content claims that can be used with them, and the content levels required for the claim to be valid. Under this proposed standard, manufacturers need only provide evidence that the substance is present at the required levels in order to make the claim. FSANZ does not consider these as health effect claims but rather as the presence or absence of particular nutritional properties and as such these products do not have to meet any specific nutritional profile other than for the substance that is the subject of the claim (Food Standards Australia New Zealand 2008a).

General level health claims are equivalent to the structure function claims previously described for the other jurisdictions. These claims can refer to the maintenance of good health, a component and its function in the body, specific benefits for performance and well-being in relation to foods, as well as how a diet, food or component can modify a function beyond its role in normal growth and development, or potential for a food or component to assist in reducing the risk of or helping to control a non-serious disease or condition (Food Standards Australia New Zealand 2008a). These claims must be substantiated by authoritative, current and generally accepted information or on a structured review of the totality of evidence supporting the claim (Food Standards Australia New Zealand 2008a). Evidence must be provided to demonstrate the product contains the component that is the subject of the claim in sufficient quantities to merit the claim. FSANZ intends to develop a list of claims that manufacturers may use.

Food products that make general level claims must not exceed a specific nutrient profile score. The maximal nutrient profile score allowed for the product is determined based on the category of food the product falls under. The scoring system is present to prevent 'unhealthy' foods containing higher levels of 'risk-increasing' substances such as sugar, saturated fats or sodium, from making 'healthy' claims. A food would be assigned a baseline score that is based on the amount of 'risk-increasing' substances in a 100 g or 100 mL serving. This baseline score can be reduced if the food has certain desired or healthy substances, such as protein and/or fibre.

The standard will include a model list of preapproved statements that can be used (Food Standards Australia New Zealand 2008a) and other claims would be accepted provided they may meet the criteria stated above. General claims would not require premarket approval; however, it is up to the manufacturer to have the scientifically substantiated evidence ready to be produced if requested by enforcement officials.

High-level health claims can refer to the presence of a nutrient or substance in a food and its relationship to a serious disease or condition or to an indicator of a serious disease. The claims need to be based on a diet–disease relationship. These relationships require premarket approval from FSANZ and are to be evaluated on a case-by-case basis. Applicants need to provide sufficient evidence from human trials establishing the relationship.

Once a diet–disease relationship has been substantiated by FSANZ, it will be added to the standard and a high-level health claim based on the relationship can then be used (Food Standards Australia New Zealand 2008a). Applications for approval of diet–disease relationships can be made without public notification. Applicants can choose to only consult with the Food Regulation Standing Committee and the Expert Advisory Committee to seek review and approval. This allows the applicant to have a slight first-to-market advantage, although, once published in the standard, any food that met the criteria could use a claim based on the diet–disease relationship. Like foods making general level health claims, foods making high-level health claims cannot exceed their categorical nutrient profile score to be considered eligible.

Eight diet–disease relationships were pre-approved for high level claims as part of the proposed standard:

1. Calcium, vitamin D status and osteoporosis
2. Calcium and enhanced bone mineral density
3. Folic acid and neural tube defects
4. Saturated fatty acids and LDL cholesterol
5. Saturated and *trans* fatty acids and LDL cholesterol
6. Sodium and blood pressure
7. Increased intake of vegetables and fruit and coronary heart disease
8. A high intake of vegetables and fruits and coronary heart disease (Food Standards Australia New Zealand 2008a)

FSANZ submitted their final assessment report on the proposed standard to the Ministerial Council in April 2008 (Food Standards Australia New Zealand 2008a). This report recommended that the draft standard be used (Food Standards Australia New Zealand 2008a). After consideration of the proposed standard and recommendations, the Ministerial Council determined there were a number of issues that required further review (Australia New Zealand Therapeutic Products Authority 2003).

Chief among the Ministerial Council's concerns was the apparent difficulty in enforcing compliance with the standard (Australia New Zealand Therapeutic Products Authority 2003). The Council also did not favour allowing nutrient content claims to be exempt from the nutrient profiling scoring criteria. Although FSANZ consumer surveys indicated that consumers based their purchase on the nutrient offered and not on the assumption of any further health benefits (Food Standards Australia New Zealand 2008a), the Council felt that any claims implied healthy food choices and unhealthy foods should not be eligible for claims (Australia New Zealand Therapeutic Products Authority 2003).

The Council stated that the standard was highly complex, would be extremely difficult and resource-intensive for industry to comply with and would be extremely difficult for regulators to monitor and enforce (Australia New Zealand Therapeutic Products Authority 2003). The council was particularly concerned with the procedures outlined for general health claims. As these claims did not require premarket assessment, there was a potential for foods with untruthful or misleading claims to enter marketplace without regulatory knowledge. Since enforcement would be predominantly handled by individual provincial and territorial health agency, although the Australian Competition and Consumer Commission does have some authority to prevent and remedy misleading and deceptive representations in consumer products, the Council is concerned that the burden on regulators monitoring and enforcing the standard may be too much to handle. The subjectivity in the weight of evidence required to substantiate claims would make it difficult for regulators to adequately assess claims quickly and unequivocally. Foods with untruthful and misleading claims could thus appear and remain in the market.

The Council asked FSANZ to review the proposal and provide a report to address the issues identified. During 2009, an independent ministerial review of labelling law and policy is to be conducted in Australia and New Zealand (Australian Government Department of Health and Aging 2008d). The Ministerial Council has requested the FSANZ to present their review of the health claims proposal in March 2010 so that it can be considered alongside the ministerial review on food labelling law and policy (Australian Government Department of Health and Aging 2008d).

13.6.3 The Australia New Zealand Therapeutic Products Authority

In 2003, the Australian and New Zealand governments signed a treaty to establish a single binational agency to regulate therapeutic products (Food Standards Australia New Zealand 2009). The proposed Australia New Zealand Therapeutic Products Authority is currently in limbo as New Zealand's government did not proceed with legislation required to enable the Authority's establishment (New Zealand State Services 2007). Although the government has stated the bill is not abandoned and will remain on the Order Paper, sufficient parliamentary support is required for the bill to be revisited. It is unknown what will become of the proposed legislation but it is believed that the majority of the proposed amendments will be incorporated into Australian legislation. Stakeholders have noted that a transition to the proposed framework would not be as difficult for Australia as it would be for New Zealand since the TGA's current regulatory system already operates in a similar manner to the proposed system. For now, all therapeutic products in Australia are the responsibility of the TGA.

13.6.4 The Australian regulatory framework for therapeutic products

Under the current Australian framework, functional foods and nutraceuticals considered to be therapeutic are classified as complementary medicines (Australian Government Department of Health and Ageing 2006). Products containing vitamins, minerals or nutritional supplements can fall under this classification and most complementary medicines are classified as 'Listed' medicines under the Australian system (Australian Government Department of Health and Ageing 2006). These Listed medicines cannot carry high-level indications or claims and are considered to be of lower risk than registered medicines. Under the proposed new framework, the classification terms 'listed' and 'registered' would be replaced with the terms 'Class 1' and 'Class 2' (Australia New Zealand Therapeutic Products Authority 2006). Regardless of their classification, all therapeutic products in Australia require a product license. Once approved for sale by the TGA, a therapeutic product will be granted a license and placed on the Australian Register of Therapeutic Goods (Australian Government Department of Health and Ageing 2008a). All therapeutic products must also be manufactured in TGA-licensed facilities. Manufacturers must follow GMP requirements set out by the TGA (Australian Government Department of Health and Ageing 2008b).

Listed medicines have a simplified approval process. The TGA has made available an electronic application and validation process for listed medicines (Australian Government Department of Health and Ageing 2008c). The application process requires the sponsor to enter product information including specifications, claims and warnings, into the system. The sponsor is required to make a statutory declaration that all information entered into the system is correct and the product complies with all standards and guidelines for a therapeutic product. Following the payment of an application fee, the TGA will process the application and then provide a registration number. Postmarket confirmation on the product will sometimes be done as the TGA will randomly select sponsors for postmarket review. The TGA will request samples and evidence to support the claims and information made in the original application. If there are discrepancies, the TGA will work with the sponsor and take appropriate action to ensure the product meets the requirements of the law.

Listed medicines are only allowed to contain ingredients listed in a Permitted Ingredients List (Australian Government Department of Health and Ageing 2007b). New substances can be added to the list through a new substance application to the TGA (Australian Government Department of Health and Ageing 1999). Applications are reviewed by the TGA with the assistance of advisory committees. The primary concern for the addition of new ingredients onto the list is safety. Efficacy of the product will be dependent on the claim, and these items are typically evaluated more thoroughly during the electronic application for the 'listed' medicines. The types of claims permitted in listed medicines include 'symptomatic relief of conditions (other than serious disease, disorders, or conditions), health maintenance, health enhancement and risk reduction' (Australian Government Department of Health and Ageing 2006). The type of evidence required to support the claim is dependent on the type of claim made. This evidence can include evidence of traditional use.

The Complementary Medicines Evaluation Committee is available to provide advice and make recommendations to the TGA concerning the evaluation and regulation of complementary medicines (Australian Government Department of Health and Ageing 2007a). The committee also assists in the evaluation of ingredients to be added to the Permitted Ingredients List and the setting of standards that products and ingredients must meet.

13.6.5 Summary of the Australian regulatory framework

In Australia, for a functional food or nutraceuticals to make a claim, the product would have to be classified as a therapeutic product. These 'low-risk' complementary medicines go through a simplified process, whereby products that meet the requirements for a 'listed' medicine, i.e. the ingredients are on the permitted list, can achieve a low level claim through a simple application process. If ingredients are not on the permitted list, the process becomes more difficult and complicated as the sponsor must either apply for the ingredient's inclusion on the permitted list or make an application as a 'registered' medicine. Regardless of the application process taken, these products are considered therapeutic products and not food. This limits the number of functional food products able to obtain claims, as many foods would not be eligible for inclusion to the list of Listed Medicines.

With the new proposed standard for health claims for foods, non-therapeutic products would be allowed to bear claims. It is unknown and when if this proposal may obtain ministerial approval. As it stands, the Ministerial Council has identified some major issues with the proposed standards and how FSANZ will respond to these issues is unknown. The major concern with the proposal revolve around general level health claims and their lack of a premarket approval process. It may be that the most viable solution would be to require general level claims to follow the same requirements as high-level claims. The ministerial council, however, has also stated concerns about potential cost to industry and regulators that could be associated with the implementation and use of this proposed standard. Forcing premarket approval of general level claims would add an increased burden to an already resource-intensive standard.

13.7 Conclusions on food regulation

The regulation of functional foods and nutraceuticals presents a number of challenges with risk-based regulatory frameworks. Many jurisdictions have incorporated different methodologies to handle these products, but while specific policies and approaches may differ, the fundamental principles remain the same.

Each jurisdiction reviewed above utilises some form of product categorisation in their regulations. The categories themselves may vary, but all the jurisdictions place a border between therapeutic and non-therapeutic products. Historically, this interface was quite clear; medicinal products are for treatment and prevention of health conditions whereas food is for nutrition. Functional foods are much harder to place.

For the most part their regulators have two major considerations when placing these products, safety and efficacy. Mechanisms to address safety are relatively straightforward; novel food legislation ensures that non-traditional food products will be safe for consumption *ad libitum*. There is some concern that these standards are set arbitrarily high and the review processes are onerous and expensive; however, the fact remains that mechanisms do exist.

The evaluation of efficacy is an entirely different matter. In the past those responsible for food regulation did not have to worry about efficacy in their product evaluations. Jurisdictions have now cautiously allowed claims to be made. With the possible exception of the EU, it is not individual products that are approved for claims per se; rather, jurisdictions approve relationships between substances and effects. Foods meeting the proper criteria are then free to make a claim based on the relationship. The framework the EU utilises does prescribe the use lists with set criteria; however, the recent opinions from the EFSA suggest permitted claims would have to be for very specific products or product groups.

Mechanisms to allow claims in foods are still not fully implemented. Delays in implementation have been common themes in all the jurisdictions and are indicative of the complexity of this subject. It is difficult to predict the exact effect proposed regulations will ultimately have on the food, functional food and nutracutical marketplace. Of course, these new pieces of legislation are not immutable and will undoubtedly develop and improve over time as new information and experience is obtained.

References

Arai, S. (1996). Studies on functional foods in Japan – state of the art. *Bioscience, biotechnology, and biochemistry*, **60**(1), 9–15.

Australia New Zealand Therapeutic Products Authority (2003). Australia and New Zealand sign treaty to regulate medicines and therapeutic products. Available online http://www.anztpa.org/media/031210j.htm (accessed 23 July 2008).

Australia New Zealand Therapeutic Products Authority (2006). Australia New Zealand Therapeutic Products Regulatory Scheme (Medicines) Rule 2006. Available online http://www.anztpa.org/consult/dr-medrule.pdf (accessed 23 July 2008).

Australian Government Department of Health and Ageing (1999). Application for an evaluation of a new complementary medicine substance. Available online http://www.tga.gov.au/docs/pdf/newcmapp.pdf (accessed 23 July 2008).

Australian Government Department of Health and Ageing (2004a). Medicines regulation and the TGA. Available online http://www.tga.gov.au/docs/html/medregs.htm (accessed 23 July 2008).

Australian Government Department of Health and Ageing (2004b). Section 7 declarations – food or therapeutic good? Available online http://www.tga.gov.au/docs/html/cmec/section7.htm (accessed 23 July 2008).

Australian Government Department of Health and Ageing (2006). The regulation of complementary medicines in Australia – an overview. Available online http://www.tga.gov.au/cm/cmreg-aust.htm (accessed 23 July 2008).

Australian Government Department of Health and Ageing (2007a). Complementary Medicines Evaluation Committee (CMEC). Available online http://www.tga.gov.au/docs/html/cmec/cmec.htm (accessed 23 July 2008).

Australian Government Department of Health and Ageing (2007b). Substances that may be used in listed medicines in Australia. Available online http://www.tga.gov.au/cm/listsubs.htm (accessed 23 July 2008).

Australian Government Department of Health and Ageing (2007c). What the TGA does. Available online http://www.tga.gov.au/about/tga.htm (accessed 23 July 2008).

Australian Government Department of Health and Ageing (2008a). ARTG. Available online https://www.ebs.tga.gov.au/ebs/ANZTPAR/PublicWeb.nsf/cuMedicines?OpenView (accessed 4 January 2010).

Australian Government Department of Health and Ageing (2008b). Australian Code of Good Manufacturing Practice for Medicinal Products. Available online http://www.tga.gov.au/docs/pdf/gmpcodau.pdf (accessed 23 July 2002).

Australian Government Department of Health and Ageing (2008c). Electronic Listing Facility (ELF). Available online http://www.tga.gov.au/online/elf.htm (accessed 23 July 2008).

Australian Government Department of Health and Aging (2008d). Notice of Publication of Request for Review of P293–20 June 2008. Available online http://www.health.gov.au/internet/main/publishing.nsf/Content/request-review-of-P293 (accessed 29 April 2009).

Australian Office of Legislative Drafting and Publishing (2008). *Therapeutic Goods Act 1989*. Australian Office of Legislative Drafting and Publishing, Canberra.

Canadian Food Inspection Agency (2008). Canadian Food Inspection Agency – table of contents. Available online http://www.inspection.gc.ca/english/toce.shtml (accessed 4 June 2008).

Cases C-154/04 and C-155/04 (2005). *The Queen, on the application of Alliance for Natural Health and Nutri-Link Ltd v Secretary of State for Health (C-154/04) and The Queen, on the application of National Association of Health Stores and Health Food Manufacturers Ltd v Secretary of State for Health and National Assembly for Wales (C-155/04) ECR I-06451*.

Coppens, P., Fernandes da Silva, M. & Pettman, S. (2006). European regulations on nutraceuticals, dietary supplements and functional foods: a framework based on safety. *Toxicology*, **221**, 59–74.

European Commission (2009a). Standing Committee on the Food Chain and Animal Health Section on General Food Law, Summary Record of the meeting held on 27 March 2009. Available online http://ec.europa.eu/food/committees/regulatory/scfcah/general_food/sum_27032009_en.pdf (accessed 4 January 2010).

European Commission (2009b). Working Document on the Setting of Nutrient Profiles. Available online http:/www.senat.fr/europa/subsidarite/11_03_2009/Texte.pdf (accessed 21 April 2009).

European Food Safety Authority (2008a). Final scientific and technical guidance for applicants for preparation and presentation of the application for authorisation of a health claim. Available online http://www.efsa.europa.eu/EFSA/efsa_locale-1178620753812_1178623592471.htm (accessed 23 June 2008).

European Food Safety Authority (2008b). The setting of nutrient profiles for foods bearing nutrition and health claims pursuant to Article 4 of the regulation (EC) No 1924/2006. *The EFSA Journal*, **644**, 1–44. Available online http://www.efsa.europa.eu/EFSA/efsa_locale-1178620753812_118689506673.htm (accessed 21 April 2009).

European Food Safety Commission (2008c). EFSA: Home. Available online http://www.efsa.eu.int/EFSA/efsa_locale-1178620753812_home.htm (accessed 23 June 2008).

European Food Safety Authority (2009a) EFSA delivers its first series of opinions on 'general function' health claims. Available online http://www.efsa.europa.eu/en/press/news/nda091001.htm (accessed 4 January 2010).

European Food Safety Authority (2009b) Technical Report of EFSA. Briefing document for member states and European Commission on the evaluation of Article 13.1 Health Claims. *The EFSA Journal*, **7**(11), 1386. Available online http://www.efsa.europa.eu/en/efsajournal/doc/1386,1.pdf (accessed 4 January 2010).

European Medicines Agency (2008). European Medicines Agency. Available online http://www.emea.europa.eu/ (accessed 23 June 2008).

European Union Publications Office (1997). Regulation (EC) No 258/97 of 27 January 1997 concerning novel foods and novel food ingredients. European Union Publications Office, Gare, Luxembourg.

European Union Publications Office (2001). Directive (EC) 2001/83/EC of 6 November 2001 on the Community code relating to medicinal products for human use. European Union Publications Office, Gare, Luxembourg.

European Union Publications Office (2002a). Directive (EC) 2002/46/EC of 10 June 2002 on the approximation of the laws of the member states relating to food supplements. European Union Publications Office, Gare, Luxembourg.

European Union Publications Office (2002b). Regulation (EC) No 178/2002 of the European Parliament and of the Council of 28 January 2002 laying down the general principles and requirements of food law, establishing the European Food Safety Authority and laying down procedures in matters of food safety. European Union Publications Office, Gare, Luxembourg.

European Union Publications Office (2007a). Regulation (EC) 1924/2006 of 20 December 2006 on nutrition and health claims made on foods. European Union Publications Office, Gare, Luxembourg.

European Union Publications Office (2007b). Regulation (EC) 1925/2006 of 20 December 2006 on the addition of vitamins and minerals and of certain other substances to foods. European Union Publications Office, Gare, Luxembourg.

FDA Center for Food Safety and Applied Nutrition (1999). Guidance for industry significant scientific agreement in the review of health claims for conventional foods and dietary supplements. Available online http://www.cfsan.fda.gov/~dms/ssaguide.html (accessed 11 June 2008).

FDA Center for Food Safety and Applied Nutrition (2001). How to submit a GRAS notice. Available online http://www.cfsan.fda.gov/~dms/opa-frgr.html (accessed 11 June 2008).

FDA Center for Food Safety and Applied Nutrition (2002). Structure/function claims small entity compliance guide. Available online http://www.cfsan.fda.gov/~dms/sclmguid.html (accessed 11 June 2008).

FDA Center for Food Safety and Applied Nutrition (2003a). Interim procedures for qualified health claims in the labeling of conventional human food and human dietary supplements. Available online http://www.cfsan.fda.gov/~dms/hclmgui3.html (accessed 11 June 2008).

FDA Center for Food Safety and Applied Nutrition (2003b). Interim evidence-based ranking system for scientific data. Available online http://www.fda.gov/ohrms/dockets/dailys/03/Aug03/080103/03n-0069-rpt0001-04-Attachment-b-vol4.pdf (accessed 11 June 2008).

FDA Center for Food Safety and Applied Nutrition (2004a). Food labeling – structure/function claims. Available online http://www.cfsan.fda.gov/~dms/labstruc.html (accessed 11 June 2008).

FDA Center for Food Safety and Applied Nutrition (2004b). Guidance for industry frequently asked questions about GRAS. Available online http://www.cfsan.fda.gov/~dms/grasguid.html (accessed 11 June 2008).

FDA Center for Food Safety and Applied Nutrition (2004c). Substantiation for dietary supplement claims made under Section 403® (6) of the Federal Food, Drug, and Cosmetic Act. Available online http://www.cfsan.fda.gov/~dms/dsclmgui.html (accessed 11 June 2008).

FDA Center for Food Safety and Applied Nutrition (2005). A dietary supplement labeling guide: Chapter VII. Premarket notification of new dietary ingredients. Available online http://www.cfsan.fda.gov/~dms/dslg-7.html (accessed 11 June 2008).

FDA Center for Food Safety and Applied Nutrition (2006a). Qualified health claims: letter of enforcement discretion – unsaturated fatty acids from canola oil and reduced risk of coronary heart disease. Available online http://www.cfsan.fda.gov/~dms/qhccanol.html (accessed 11 June 2008).

FDA Center for Food Safety and Applied Nutrition (2006b). Sales of supplements containing ephedrine alkaloids (Ephedra) prohibited. Available online http://www.fda.gov/oc/initiatives/ephedra/february2004/ (accessed 11 June 2008).

FDA Center for Food Safety and Applied Nutrition (2006c). FDA's implementation of 'qualified health claims': questions and answers. Available online http://www.cfsan.fda.gov/~dms/qhcqagui.html (accessed 11 June 2008).

FDA Center for Food Safety and Applied Nutrition (2008a). US FDA/CFSAN – dietary supplements: overview. Available online http://www.cfsan.fda.gov/~dms/supplmnt.html (accessed 11 June 2008).

FDA Center for Food Safety and Applied Nutrition (2008b). A food labeling guide. Available online http://www.cfsan.fda.gov/~dms/2lg-toc.html (accessed 11 June 2008).

FDA Center for Food Safety and Applied Nutrition (2008c). Consumer health infromation for better nutrition initiative task force final report. Available online http://www.cfsan.fda.gov/~dms/nuttftoc.html (accessed 11 June 2008).

FDA Center for Food Safety and Applied Nutrition (2008d). Numerical listing of GRAS notices. Available online http://www.cfsan.fda.gov/~rdb/opa-gras.html (accessed 11 June 2008).

Food Standards Australia New Zealand (2005). Nutrition, health and related claims a guide to the development of a food standard for Australia and New Zealand. Available online http://www.foodstandards.gov.au/_srcfiles/Health%20Claims%20Short%20Guide%20with%20Summary1.pdf#search=%22health%20claims%22 (accessed 23 July 2008).

Food Standards Australia New Zealand (2008a). Final assessment report proposal P293 nutrition, health and related claims. Available online http://www.foodstandards.gov.au/_srcfiles/P293%20Health%20Claims%20FAR%20and%20Att%201%20&%202%20FINAL.pdf (accessed 23 July 2008).

Food Standards Australia New Zealand (2008b). Food Standards Australia New Zealand. Available online http://www.foodstandards.gov.au/ (accessed 23 July 2008).

Food Standards Australia New Zealand (2008c). Record of views formed in response to inquiries. Available online http://www.foodstandards.gov.au/foodmatters/novelfoods/novelfoodrecordofvie3934.cfm (accessed 23 July 2008).

Food Standards Australia New Zealand (2008d). Standard 1.2.8 nutrition information requirements. Available online http://www.foodstandards.gov.au/_srcfiles/ACF2A90.pdf (accessed 23 July 2008).

Food Standards Australia New Zealand (2008e). Standard 1.5.1 novel foods. Available online http://www.foodstandards.gov.au/_srcfiles/Standard_1_5_1_Novel_Foods_v95.pdf (accessed 23 July 2008).

Food Standards Australia New Zealand (2009). Nutrition, health and related claims. Available online http://www.foodstandards.gov.au/foodmatters/healthnutritionandrelatedclaims/index.cfm (accessed 29 April 2009).

Food Standards Australia New Zealand (n.d.). A pilot for a health claims system using folate/neural tube defects health claims. Available online http://www.foodstandards.gov.au/newsroom/publications/evaluatingthefolateneuraltubedefecthealthclaimpilot/anzfaapilotforahealt1038.cfm (accessed 23 July 2008).

Government Printing Office (1990). Nutrition Labeling and Education Act of 1990, PL 101–535 (Nov. 8, 1990). Government Printing Office, Washington, DC.

Government Printing Office (1994). Dietary Supplement Health and Education Act of 1994, PL 103–417 (Oct. 25, 1994). Government Printing Office, Washington, DC.

Government Printing Office (1997). Food and Drug Administration Moderization Act of 1997, PL 105–115 (Nov. 21, 1997). Government Printing Office, Washington, DC.

Government Printing Office (2004). Federal Food, Drug and Cosmetic Act, 21 U.S.C. §301 (2004). Government Printing Office, Washington, DC.

Health Canada (2003). List of acceptable non-medicinal ingredients. Available online http://www.hc-sc.gc.ca/dhp-mps/prodnatur/legislation/docs/nmi-imn_list1-eng.php (accessed 4 June 2008).

Health Canada (2005). Food directorate. Available online http://www.hc-sc.gc.ca/ahc-asc/branch-dirgen/hpfb-dgpsa/fd-da/index-eng.php (accessed 4 June 2008).

Health Canada (2006a). Compendium of monographs – natural health products. Available online http://www.hc-sc.gc.ca/dhp-mps/prodnatur/applications/licen-prod/monograph/index-eng.php (accessed 4 June 2008).

Health Canada (2006b). Guidelines for the safety assessment of novel foods. Available online http://www.hc-sc.gc.ca/fn-an/legislation/guide-ld/nf-an/guidelines-lignesdirectrices-eng.php (accessed 4 June 2008).

Health Canada (2008). Health Canada – homepage. Available online http://www.hc-sc.gc.ca/index-eng.php (accessed 4 June 2008).

Health Canada (2009a). Classification of products at the food-natural health product interface: products in food fomats. Available online http://www.hc-sc.gc.ca/dhp-mps/prodnatur/legislation/docs/food-nhp-aliments-psn-guide-eng.php (accessed 29 April 2009).

Health Canada (2009b). Guidance document for preparing a submission for food health claims. Available online http://www.hc-sc.gc.ca/fn-an/legislation/guide-ld/health-claims_guidance-orientation_allegations-sante-eng.php (accessed 29 April 2009).

Health Canada (2009c). New guidance document – classification of products at the food-natural health product interface: products in food format. Available online http://www.hc-sc.gc.ca/dhp-mps/prodnatur/bulletins/food_nhp_aliments_psn-2009-eng.php (accessed 29 April 2009).

Health Canada (2009d). Questions and answers on health claims. Available online http://www.hc-sc.gc.ca/fn-an/label-etiquet/claims-reclam/qa-qr_claims-allegations-eng.php (accessed 29 April 2009).

International Food Information Council (2005). Qualified health claims consumer research project executive summary. Available online http://www.ific.org/research/qualhealthclaimsres.cfm (accessed 11 June 2008).

Japan External Trade Organization (2006). Food sanitation law in Japan. Available online http://www.jetro.go.jp/en/market/regulations/pdf/food-e.pdf (accessed 16 July 2008).

Medicines and Healthcare Products Regulatory Agency (2008). Independent Review Panel on the classification of borderline products. Available online http://www.mhra.gov.uk/Committees/Medicinesadvisorybodies/IndependentReviewPanelontheClassificationofBorderlineProducts/index.htm (accessed 23 June 2008).

Ministry of Health, Labour and Welfare (2006). Welcome to the Ministry of Health, Labour and Welfare. Available online http://www.mhlw.go.jp/english/index.html (accessed 16 July 2008).

Ministry of Health, Labour and Welfare (n.d.). Ministry of Health, Labour and Welfare: food additives. Available online http://www.mhlw.go.jp/english/topics/foodsafety/foodadditives/index.html (accessed 16 July 2008).

Ministry of Health, Labour and Welfare (n.d.). Ministry of Health, Labour and Welfare: labeling system for nutrient. Available online http://www.mhlw.go.jp/english/topics/foodsafety/fhc/04.html (accessed 16 July 2008).

Ministry of Health, Labour and Welfare (n.d.). Ministry of Health, Labour and Welfare: food with health claims, food for special dietary uses, and nutrition labeling. Available online http://www.mhlw.go.jp/english/topics/foodsafety/fhc/02.html (accessed 16 July 2008).

Ministry of Health, Labour and Welfare (n.d.). Ministry of Health, Labour and Welfare: food with nutrient function claims. Available online http://www.mhlw.go.jp/english/topics/foodsafety/fhc/01.html (accessed 16 July 2008).

Ministry of Health, Labour and Welfare (n.d.). The guidelines for designation of food additives, and for revision of standards for use of food additives (Excerpt). Available online http://www.ffcr.or.jp/zaidan/FFCRHOME.nsf/pages/PDF/$FILE/Guideline.pdf (accessed 16 July 2008).

New Zealand State Services (2007). Therapeutics products and medicines bill on hold. Available online http://www.beehive.govt.nz/release/therapeutics+products+and+medicines+bill+hold (accessed 23 July 2008).

Noonan, C. & Noonan, P.W. (2006). Marketing dietary supplements in the United States: a review of the requirements for new dietary ingredients. *Toxicology*, **221**, 4–8.

Nutraceutical Corp (2006). *Nutraceutical Corp. v. Von Eschenbach*, 459 F.3d 1033 (10th Cir. 2006).

Office of Federal Regulations (2008). Current Good Manufacturing Practice in Manufacturing, Packaging, Labeling, or Holding Operations for Dietary Supplements, 21 C.F.R. § 111 (2008). Office of Federal Regulations, Washington, DC.

Ohama, H., Ikeda, H. & Moriyama, K. (2006). Health foods and foods with health claims in Japan. *Toxicology*, **221**, 59–74.

Pearson v. Shalala (1999). *Pearson v. Shalala*, 164F.3d 650 (D.C. Cir. 1999).

Queen's Printer (1985). Food and Drugs Act, R.S., 1985, c. F-27. Queen's Printer, Ottawa.

Queen's Printer (n.d.). Food and Drug Regulations, C.R.C., c. 870. Queen's Printer, Ottawa.

Queen's Printer (2003). Natural Health Products Regulations, S.O.R. 2003–196. Queen's Printer, Ottawa.

Rowlands, C.J. & Hoadley, J.E. (2006). FDA perspectives on health claims for food labels. *Toxicology*, **221**, 35–43.

US Food and Drug Administration (2008). U.S. Food and Drug Administration. Available online http://www.fda.gov/ (accessed 11 June 2008).

Part IV

Functional foods and health

14 Functional foods that boost the immune system

Calvin London

14.1 The rise of immune-boosting functional foods

Perhaps of all the functional foods, those that have an effect on the immune system of humans, birds, animals and fish, may have one of the brightest and as yet largely untapped futures.

As consumers, we are becoming more conscious of the role of food in our lives. This is in response to the global paradox that shows that in the next 20 years we will struggle to satisfy the food needs of the growing population on one hand, as we wrestle with unprecedented rates of obesity and associated diseases on the other. We have also established as a way of life, the indiscriminate use of antibiotics as an 'easy, one-step fix' for human diseases and the indiscriminate supplementation of animal and fish feeds with antibiotics to enhance growth. This has led to an increase in the genetic pool of resistant microorganisms that can effectively transfer resistance to many human antibiotics of therapeutic importance to such an extent that some antibiotics are now almost useless (Philips *et al.* 2004; Turnidge 2004).

Consumers are becoming more aware of these issues and growing numbers are choosing to return to the origins of all medications and fighting disease by choosing foods that can improve on the function of the immune system.

The concept of eating foods that naturally, or through supplementation, provide immune benefits is not new. More than 4000 years after Egyptians began applying honey called *Leptospermum* honey (manuka honey) to wounds, bees that collect nectar from manuka and jelly bushes in Australia and New Zealand are now farmed as a source of this honey that is claimed to have the ability to reduce infection and relieve the symptoms of flu and sore throats (Buontanotte 2009).

It is estimated that by 2010 the North American functional food market is going to grow to about US$167 billion, driven by consumers having an increased awareness of the role that functional food can play in supporting their health, longevity and well-being. Health issues are becoming increasingly important as a result of an aging population, and lifestyle diseases are becoming more common. Society, concerned about the potential side effects from modern-day medications provided by the pharmaceutical industry, is starting to seek more natural forms of prevention and cure. It has also been shown that a high proportion of patients in hospitals are malnourished and that malnourishment impairs immune function (McWhirter & Pennington 1994; Girodon 1999). In addition, it is generally accepted that there is a major burden of ill health in the population from malfunctions in the immune system, e.g. rheumatoid arthritis, inflammatory bowel disease and asthma (Yang & Xia 1995; Navarro-Alarcon & Lopez-Martinez 2000; Grimble 2001).

It is commonly recognised that at least 90 nutrients are needed to maintain life. This includes 63–74 minerals and trace elements, all 16 vitamins, 12 amino acids and 3 essential fatty acids. It is important to note that without minerals none of the other nutrients can be utilised. On the other hand, mineral molecules in their pure form are too large to be absorbed (DiSilvestro 2005; Groff *et al.* 2005).

'Immunonutrition' is a term that has been used to describe 'modulation of the activities of the immune system, and the consequences on the patient of immune activation, by nutrients or specific food items fed in amounts above those normally encountered in the diet' (Clancy 2003). These nutrients have been initially identified in studies on animal models, but are now widely used in clinical practice. While the animal studies have indicated the mechanisms by which immunonutrition may work, the evidence of clinical efficacy is controversial (Grimble 1998; O'Flaherty & Bouchier-Hayes 1999).

Like conventional drugs, immunonutrients should produce biological effects in a dose-dependent manner. However, as they do not produce immediately measurable changes in physiological or pathological conditions, they should be considered nutrients, not drugs. In a multicentre trial of 326 critically ill intensive care patients fed an immune-enhancing formula or a standard high-protein formula, only patients that received adequate doses of the immunonutrient formula (on average 821 mL/day during the early feeding period) demonstrated reduction in length of hospital stay. Conversely, a large study of well-nourished, relatively young (median age 64–66 years) surgical patients who received immunonutrients at a subtherapeutic level (e.g. <30% of nutritional needs, <500 mL/day) did not demonstrate differences in outcome compared with patients who received intravenous fluid (Schloerb 2001).

Immunonutrients occupy a space somewhere between essential nutrients (those nutrients critical to normal health, such as vitamins) and drugs with defined impacts on specific diseases. This group of bioactive chemicals derived from foods but taken as supplements at much higher concentration than diet alone could provide is the foundation of functional foods that affect the immune system. They include antioxidants from fruits and berries, fatty acids found in cold-water fish, and potentially disease-fighting compounds from common spices such as cinnamon and turmeric. Claims have been made for their role in everything from fighting cancer and cardiovascular disease to maddeningly vague notions about 'supporting healthy living' (Coppens *et al.* 2006; Espín *et al.* 2007).

The functional food revolution began developing at least 10 years ago; customers have made their intentions clear. By 2000, consumers had delivered a clear message – 'we would rather get our supplementation via food than through pills or powders'. For many, however, they do not understand how or why these foods work and in many cases there is little scientific data to support claims. For others, there is a developing body of well-structured, scientific understanding about the mechanisms of action.

Today, the success of functional foods is one largely enjoyed by the players in the 'inherently functional' category. These include foods that are by nature functional, for example, soy, green tea and yoghurt. Modified or 'fortified foods' are more difficult to classify, especially in a market flooded by products with nebulous health claims, clever and sometimes misleading marketing and a lack of regulatory classification.

14.2 Review of the immune system

The immune system functions to protect the body against intruding microbes and environmental agents. The skin surface and the gut lining of animals form the primary defence mechanism of any immune system whether invertebrate or vertebrate, cold-blooded or warm-blooded. The great majority of infections in normal individuals are of limited duration and leave very little permanent damage. This is due to the individual's immune system, which combats infectious agents.

There are two functional divisions to the immune system, namely, the innate immune system and the adaptive immune system. Innate immunity acts as a first line of defence against infectious agents. Most potential pathogens are checked before they establish an overt infection. The adaptive (or acquired immunity) system produces a specific reaction to each infectious agent, which normally eradicates that agent. Furthermore, the adaptive immune system remembers that particular infectious agent and can prevent it causing disease later (Roitt *et al.* 1985).

Most readers will have a basic understanding of the concepts of immunity. Indeed, many may have a much greater degree of understanding. For those interested there are several good references that can provide a more detailed discussion of the immune system, which is clearly beyond the scope of this chapter (Roitt *et al.* 1985; Janeway *et al.* 2001; Peakman & Vergani 2003; Clayton 2008a, b).

The skin and the gut are principal components of the innate immune system; they are responsible for the first response to invasion by foreign particles. The skin (the largest organ in the body) serves as the primary barrier to outside antigens. The intestine provides a protective interface between the internal environment and the constant challenge of food-derived antigens and microorganisms from the external environment. As observed later in this chapter, many functional foods have been recognised for their ability to protect and enhance the interactions at the gut barrier.

Unlike the acquired immune system, the innate immune system is generally in a state of high alert. It springs into action the moment it recognises the presence of a pathogen. The innate immune system is supported by a second layer of very specific innate immune support agents. They include vitamin D, the trace element selenium, the plant extract β-sitosterol and the 1-3,1-6-β-glucans derived from yeast, or, more expensively, from mushrooms such as the Shiitake. Not surprisingly, such compounds have also become favoured functional food components (Clayton 2008b).

It will be appreciated that the innate and adaptive immune systems do not act in isolation. Antibodies produced by lymphocytes help phagocytes to recognise their targets. Following clonal activation by antigen, T lymphocytes produce lymphokines, which stimulate phagocytes to destroy infectious agents more effectively (Roitt *et al.* 1985).

The ability of functional foods to affect the immune system is mediated through the gastrointestinal tract. The intestine, the largest immune organ of the body, provides a protective interface between the internal environment and the constant challenge of food-derived antigens and microorganisms from the external environment. There exists a multiplicity of opportunities through which to boost immunity by inherently functional food. In order for the immune system to remain charged and ready, there are a number of essential nutrients and building blocks that are also required: vitamin D, the trace elements selenium, calcium, to name a few. Essential nucleotides, building blocks for protein, can also be incorporated into functional foods.

14.3 Immune-enhancing nutrients

What exactly is an 'immune-enhancing nutrient'? When compared with conventional dietary nutrients, an immune-enhancing nutrient is a substance that provides identifiable salutary effects upon the immune system. Some nutrients such as glutamine, arginine, omega-3 fatty acids, nucleotides and probiotics have been shown to have a considerable influence on immune function (delayed hypersensitivity). For this reason, they are called 'immunonutrients' or 'immunity regulators' (Schloerb 2001).

Over the past decade, for example, glutamine has been shown to be beneficial in the prevention of infectious morbidity and mortality in seriously ill patients. Glutamine is also considered a conditionally essential amino acid. One of glutamine's major roles is as an oxidative fuel for rapidly replicating cells including gastrointestinal mucosal cells – enterocytes and colonocytes, and immune cells – lymphocytes and macrophages. During stressed states the demand for glutamine by the liver, kidney and intestinal tract exceeds the supply and the body requires an exogenous source of glutamine. Without an exogenous source of glutamine, plasma levels drop dramatically and the body begins breaking down muscle tissue as a means of extracting glutamine and producing glutamine through transamination or α-ketoglutarate (first to glutamate and then to glutamine) generated in the Krebs cycle (Lacey & Wilmore 1990; Calder & Yaqoob 1999; Huang *et al.* 2003).

Not all functional foods that affect immunity are specifically immune-enhancing nutrients. Indeed, the origins of many functional foods that impart immunity are examples of complex foods many of which lack decisive scientific explanation but are instead based on anecdotal information. These functional foods – sometimes referred to as 'smart foods' in human nutrition – can play a major role in improving all-round performance (Varley 2007).

Other groups such as grape seed and skin extracts (presumably due to the resveratrol content) inhibit the growth of tumours and stop cancer-causing chemicals, but the exact mechanisms are unclear. Japanese green tea has been shown to reduce the weight of tumours and works well with chemotherapy drugs, enhancing their effects and reducing the incidence of cancer tumours (Diamond & Cowden 1997; Denehey 2006).

Functional foods affecting the immune system can be divided into three main categories (Table 14.1). The first of these are referred to as inherent functional foods. These tend to be characterised by the more complex foods such as whole vegetables, fruits, berries, herbs and spice or nuts (see Tables 14.1 and 14.2).

The second group is characterised more by chemical components that include building blocks for the immune system such as nucleotides, vitamins and minerals for immune system maintenance (Tables 14.1 and 14.3). These constitute fortified or modified foods. These terms are slightly misleading because the representative components are not so much foods themselves, but rather compounds that can be added to basic foods to enhance their activity. In this sense, the food acts as a vector, or carrier for the compound, that has the effect in addition to any inherent activity that might be found in the food group. Whole grain bread, which for example has an inherent activity due to fibre, vitamins and minerals from the grain components, can be fortified with omega-3, recognised as a potent stimulant for phagocytic cells of the immune system.

The third group is those foods or compounds that have a more ancillary effect on the immune system, rather than a direct effect (Table 14.1). Probiotics, for example, create an environment through their own activity in the gut that reduces the challenge on the immune system and also reinforces some of the primary defence mechanisms in the body. There are many studies that discuss the extremely positive effect that probiotics can have on the immune system and host's general health and well-being.

14.4 Inherent functional foods

Some of the actions of the inherent food examples shown in Table 14.1 have been described in more detail in Table 14.2. Certain food groups have similar components, many of which can

Table 14.1 Examples of representative immune-boosting foods or compounds in each of three designated subcategories

Sub-category		Example
Inherent	Fruit	Apples
		Apricots
		Avocados
		Grapefruit
		Guava
		Kiwi fruit
		Mangoes
		Oranges
		Pineapples
		Red grapes
		Satsumas
		Tomatoes
	Vegetables	Beetroot
		Broccoli
		Cabbage
		Carrot
		Cauliflower
		Frozen peas
		Lettuce and salad greens
		Onion
		Peas
		Peppers
		Potatoes
		Red peppers
		Spinach
		Squash
		Sweet potatoes
	Seeds	Rapeseed
	Berries	Blackberries
		Blueberries
		Cranberries
		Strawberries
	Herbs and spices	Astragalus
		Coriander
		Echinacea
		Garlic
		Ginger
		Ginkgo biloba
		Ginseng
		Turmeric
		Watercress
	Beans	Lentils
		Soy plant
	Miscellaneous	Bee pollen
		Brazil nuts
		Brown rice
		Eggs
		Fish
		Honey
		Meat
		Milk
		Mushrooms

(Continued)

Table 14.1 *(Continued)*

Sub-category		Example
		Raw almonds
		Seaweed
		Tea
		Whole grain bread
		Yeast and yeast extract
Modified/fortified	Nucleotides	Omega-3
		Arginine
		Glutamine
		Glycine
		Cystine
	Amino acids	Alanine
		Calcium D-glucarate
		S-amino acids
	Vitamins	Biotin
		Vitamin C
		Bioflavonoids
		Quercetin
		Coenzyme Q-10
		Vitamin E
	Minerals	Selenium
		Iron
	Other	3-Hydroxy-3-methylobutyrate
		β-Glucan
		Isoprinosine
		Mannan oligosaccharides
		Polyunsaturated fatty acids
		Thymostimuline
Ancillary	Prebiotics	Fructo-oligosaccharides
		Galacto-oligosaccharides
		Inulin
		Lactose
	Probiotics	*Lactobacillus*
		Streptococcus
	Other	Asparagus
		β-Carotene

also be used to fortify or modify other foods (see Table 14.3). This is perhaps not surprising. Food after all is the 'best medicine' for the body. A stable and balanced diet has long been established as essential for normal body functioning and sustained health.

Single deficiencies of pantothenic acid, thiamine, pyridoxine, riboflavin, folic acid and vitamin B_{12} have been associated with reduced antibody responses and impairment of the immune system (Beisel 1982). More commonly, the activity associated with boosting the immune system may be associated to particular components of the food, but the overall response may be due to a combination of different elements. Many fruits and vegetables are rich in vitamins (especially vitamin C, vitamin E and folate) and iron. Proteins such as lean meat, fish, eggs and milk are recognised as excellent sources of these vitamins and iron (Beisel 1982; Klausner 2000; Schloerb 2001; Guy 2007; Turner 2007; Rosenberg 2008).

Table 14.2 Immune-boosting actions of selected 'inherent' functional foods

Food	**Claimed activity**
Apples	Rich in antioxidants, quercetin and polyphenoles
Bee pollen	Good source of essential fatty acids, enzymes and helps eliminate many food allergies
Beetroot	High in minerals, anticancer, anti-inflammatory, antioxidant, immune-boosting and detoxifying properties
Blackberries	Rich in virus-fighting bioflavonoids and anthocyanins (types of antioxidants)
Blueberries	Rich in antioxidants
Brazil nuts	Rich in selenium; an anticancer nutrient
Broccoli	Guards against cancer and rich in a wide range of antioxidants, vitamins C and E, folate, iron and sulphaparazines
Carrots	High in carotenes
Dried shredded coconut	Excellent source of fibre, a factor in cancer prevention
Echinacea	Highly effective immune-system tonic; other North American herbs that boost immunity are chaparral (*Larrea divaricate*), yerba mansa (*Anemopsis californica*) and osha (*Ligusticum porteri*)
Elderberry	Boosts resistance to infection and illness, flu remedy
Garlic	Like onions, rich in sulfur compounds that stimulate the immune system by boosting the activity of natural killer and T-helper cells; garlic is a potent anti-inflammatory agent that blocks carcinogens, boosts production of anticancer enzymes and inhibits cancerous cells from spreading
Ginger	Fresh ginger root acts as an anti-inflammatory by inhibiting COX-2 enzymes, part of the chemical pathway that produces inflammatory chemicals
Ginseng	Ginseng boosts immune function; claimed to be particularly useful for colds and flu
Guava	Rich in cancer-fighting vitamin C and β-carotene (contains four times more vitamin C than oranges)
Kiwi fruit	Enhances the immune response by promoting the production of antibodies
Lettuce and salad greens	Contain valuable amounts of vitamins, minerals and antioxidants; varieties, such as chicory and endive, stimulate the liver, making them great detoxifiers
Mangoes and papaya	Rich in β-carotene to boost the immune system and possibly protect against cancer
Mushrooms	Boost the immune system, help fight viruses and cancer and could even help fight HIV and AIDS
Onions	High in quercetin (a type of antioxidant that inhibits enzymes that trigger inflammation) and sulfur compounds (used to manage the body's immune system)
Pineapple	Bromelain, found in the pineapple stem, decreases inflammation and is immune-enhancing; an excellent source of the antioxidant vitamin C
Propolis	Propolis (created by bees when resins from plants are mixed with wax), stimulates the body's immune system against colds and other respiratory infections
Raw almonds	One of the best natural sources of essential amino acids and essential fatty acids, required for optimum immunity

(*Continued*)

Table 14.2 (*Continued*)

Food	Claimed activity
Red grapes	Contain a compound called resveratrol which has anticancer effects and is anti-inflammatory, quercetin and other antioxidants
Red peppers	One of the very best sources of immune-boosting β-carotene, and rich in vitamin C (they contain twice the vitamin C of oranges)
Seaweed	Stimulates the immune system and guards against dermatitis, obesity, heavy metal poisoning, depression, congestion and anaemia
Spinach	Rich in inflammation-fighting carotenoids, as well as immune-boosting vitamin E; other green leafy vegetables that are great for immunity are kale, chard, turnip greens and mustard greens
Strawberries	Rich in vitamin C and the soluble fibre pectin, which helps rid the body of toxins and cholesterol
Sweet potatoes	Rich in carotenoids and antioxidants that boost immunity and minimise inflammation
Tea	Both black and green teas contain powerful antioxidants such as quercetin and other polyphenols that may inhibit tumours
Tomatoes	Contain lycopene (the carotenoid in tomatoes that makes them red), which reduces the risk of some cancers, including prostate, lung and colon
Turmeric	The key component in curry, turmeric contains curcumin, a compound that has anti-inflammatory effects as a COX-2 inhibitor
Watercress	It is another member of the cancer-fighting Brassica family and a source of iron, calcium and folic acid
Wheat grass	A blood cleanser and detoxifier attributed to both the 'natural plant enzymes' and the chlorophyll content claimed to 'detoxify the body'; wheat grass contains the antioxidant, apigenin

14.4.1 Fruits and vegetables

The value of fruits, vegetables and protein sources (whole foods) as modulators of the immune system is well established (Table 14.2). Most whole foods are recognised as sources of vitamins and minerals but worthy of note is some that have additional properties unique to that food type. Cabbage and other brassica vegetables (such as cauliflower and brussel sprouts) contain active compounds called isothiocyanates, which prompt precancerous cells in the gut to self-destruct. Cranberries are known for helping to prevent and treat urinary tract infections (especially cystitis in women), and have both antifungal and antiviral properties. Ginger stimulates the immune system and circulation (Rose 2007).

Seaweed has also become revered as a nutrient powerhouse with claims that it not only stimulates the immune system, but also boosts brainpower, guards against dermatitis, obesity, heavy metal poisoning, depression, congestion and anaemia. While the body of information to support these claims grows, the reality is that almost all nutrients in seaweeds are found in all green vegetables. In addition, seaweed produces natural toxins called microcystins that can cause liver damage in humans and are thought to be potentially carcinogenic (Hill 2007).

14.4.2 Mushrooms

Mushrooms may be 90% water, but some varieties appear to possess anticancer powers. Research shows that some can boost your immune system, help fight viruses and cancer and could even help fight HIV and AIDS (Reid 1989; Dowden 2000a, b; Miller 2006).

Table 14.3 Examples of immune-boosting components of selected modified functional food ingredients

Food component	Claimed activity
Alanine	Stimulation of lymphocyte proliferation, enhancement of antibody production
Antioxidants	Specific vitamins, plant and animal compounds that help to maintain and protect the integrity of the immune system
Arginine	Stimulates growth hormone synthesis, improves helper T-cell number, stimulates lymphocyte molecules, and cytokine production
β-Glucan	One of the most successful components at stimulating the activity of the immune system, especially macrophage, T and B cells and natural killer cells
Calcium D-glucarate	Natural detoxification of carcinogens
Colostrum	Exerts immunostimulating effects on non-specific immunological mechanisms
Conjugated linoleic acid	Healthy immune function
Diallyl sulfide, allyl methyl trisulfide	Healthy immune function
Dithiolethiones	Healthy immune function
Glutamine	Nutrient for immune cells, improves gut barrier function, acts as a precursor for glutathione
Iron	Assists in the formation and regeneration of immune cells such as leukocytes and eosinophils
Isoflavones – daidzein, genistein	Healthy brain and immune function
Isoprinosine	Enhances immune responses
Lignans	Healthy immune function
Magnesium	Healthy immune function
n-3 Polyunsaturated fatty acid	Acts as anti-inflammatory agents, reverses immunosuppression
Nucleotides	Reverse malnutrition and starvation-induced immunosuppression, enhance T-cell maturation and function, enhance natural killer cell activity, improve delayed cutaneous hypersensitivity, and aid in resistance to infection; precursors for new DNA and RNA
Omega-3	Omega-3 fatty acids are found in oily fish (mackerel, herring, sardines, tuna, trout, salmon), flaxseed oil, canola, soy and walnut oils, dark green vegetables, parsley, seaweeds, nuts, seeds (pumpkin and sesame seeds, tahini), legumes (hummus) and wholegrain cereals; omega-3 polyunsaturated fatty acids prevent the development of some tumours
S-amino acids	Enhance antioxidant status via glutathione synthesis
Selenium	An antioxidant which has been shown to help prevent cancer by increasing killer cells
Vitamins	Vitamins A, B, C and E act as antioxidants and aid in the production of immune cells and antibodies
Zinc	A mineral that is effective in boosting the production of specific white cells such as killer T cells and antibodies
Miscellaneous	Thymostimuline, astaxanthine, 3-hydroxy-3-methylobutyrate

The best-known example of immune-active mushrooms is the Shiitake mushroom that contains an immune-boosting ingredient called lentinan. Pure lentinan given as an injection has been shown to help hepatitis B patients. It is licensed as an anticancer drug by the Japanese Food and Drug Association and has been shown to reduce the rate at which tumours spread.

Polysaccharides in maitake mushrooms also have the ability to act as immunomodulators and, as such, they may be helpful for HIV sufferers. AHCC (active hexose-correlated compound) is also made from mushrooms. This supplement, first developed in Japan in the late 1980s, has been used for its effect on the immune system. Research has shown that AHCC can boost the activity of natural killer cells, which destroy the cells that have become infected with a virus (Dowden 2000a, b; Miller 2008).

14.4.3 Seeds and nuts

Seeds and nuts contain a variety of vitamins and minerals that have an effect on the immune system. Walnuts are recognised as a source of omega-3 (see later section), and other healthful compounds, including vitamin E, which is a powerful immune booster. Brazil nuts contain a combination of vitamin E, selenium and vitamin B for boosting the immune system, and raw almonds are one of the best natural sources of essential amino acids and essential fatty acids, both of which are required for optimum immunity. Pumpkin seeds are also a great source of zinc and vitamin A, which have positive effects on the immune system (Klausner 2000; Rose 2007; Turner 2007).

14.4.4 Herbs

Chinese herbal polysaccharides have aroused great interest because of their natural origin, lack of drug residue and low side effects. In Chinese medicines the most widely used herbs for enhancing immunity are ginseng, ginkgo, astragalus, gotu kola, ligustrum and codonopsis. Astragalus is one of the finest immune-boosting herbs and is better suited as a functional food ingredient, as it is not as bitter as other herbs such as *Echinacea*. It supports white blood cells and tends to increase their number and efficiency (Chen *et al.* 2003; Rose 2007).

The North American herb *Echinacea* (purple Kansas cornflower) is also a highly effective immune-system tonic and was widely used by the Plains Indians to cure and prevent many ailments. Many studies as far back as 1970 have shown that *Echinacea* boosts the immune system by increasing protection against infectious invaders. Some studies claim that *Echinacea* can reduce the incidence of developing a cold by 58% and shortens the length to 1.4 days (Miller 2008), but there is probably a greater amount of scientific evidence mounting that shows this is still to be substantiated (Giles 2000; Barrett *et al.* 2002; Dalby-Brown *et al.* 2005; Michaud *et al.* 2007).

There is some evidence that *Echinacea* stimulates the body to produce more infection-fighting white blood cells, such as T lymphocytes and killer white blood cells and stimulates the release of interferons, one of the body's most potent infection-fighting weapons. *Echinacea* has also been reported to produce more phagocytic macrophages by preventing bacteria from secreting an enzyme called hyaluronidase, which enables them to break through protective membranes, such as the lining of the intestines and respiratory tract, and invade tissues. *Echinacea* also seems to search out and destroy some viruses, such as the common cold and flu viruses (Collins 1999; Sears 2007).

Preparations of *Echinacea* are widely used as alternative remedies to prevent the common cold and infections in the upper respiratory tract, and it has become one of the most successful functional food components. The action of *Echinacea* is not due to a single component, but rather the interaction of no less than ten different active extracts. Other North American herbs that boost immunity are chaparral (*Larrea divaricate*), yerba mansa (*Anemopsis californica*) and osha (*Ligusticum porteri*). Cats fed 5 mg bixin generally showed the highest immune stimulatory and antioxidative action (Park *et al.* 2007). In this treatment, bixin enhanced lymphoblastogenic response, populations of T helper and T cytotoxic cells, natural killer cytotoxicity and IgG production. Bixin also inhibited DNA damage. Other herbs have also established credibility for their effect on the immune system (Reid 1989; Dowden 2000a; Messonnier 2001; Messonnier 2007; Rose 2007; Turner 2007; Miller 2008). Turmeric (the key component in curry) contains curcumin, which like ginger works as a COX-2 inhibitor and has anti-inflammatory effects. Watercress, another member of the cancer-fighting Brassica family, is a useful provider of iron, calcium and folic acid. A study found that adults who were taking American ginseng daily during the winter months caught fewer colds and needed less sick days, compared to those taking a placebo. American ginseng as well as Siberian and Asian ginseng can also build defences against winter germs.

Andrographis has long been used in Ayurvedic and traditional Chinese medicine to boost the immune system. Actually, andrographis was first used during the Indian flu epidemic of 1919, where it was credited with stalling the spread of the disease. According to research, andrographis works better than a placebo for reducing the symptoms of respiratory infections and it may even prevent the infection in the first place (Miller 2008).

Elderberry was considered in Roman times to be a flu remedy. Recently, elderberry extract has been researched for its role in treating influenza infections, especially when it is taken within the first 24 hours of developing symptoms. One recent study proved that individuals who were taking elderberry recovered 4 days sooner from influenza than those taking the placebo. Additionally, the use of other medications was less for those who used elderberry (Miller 2008).

14.4.5 Onions and garlic

Onions are high in quercetin, a type of antioxidant that inhibits enzymes that trigger inflammation. Onions and garlic also contain sulfur compounds that are used to manage the body's immune system by boosting the activity of natural killer and T helper cells, which manage the immune system (Svoboda *et al.* 2007; Turner 2007).

Garlic has antibacterial, antifungal and antiviral properties, and is thought to help lower blood pressure. Garlic contains allicin, which has cancer-fighting potential and antioxidants and organosulfur compounds that may block carcinogens from forming, boost production of anticancer enzymes and inhibit cancerous cells from spreading (Klausner 2000).

A number of garlic components have found favour as additives or ingredients for functional foods thereby creating modified or fortified functional foods. Some examples are shown in Table 14.3. Several of these are regularly added to foods for human consumption. They can also be added to feeds for animals, both livestock or production animals and companion animals such as dogs and cats.

14.5 Fortified and modified food components

14.5.1 Vitamin C

There has been more research about the immune-boosting effects of vitamin C than perhaps any other nutrients. Vitamin C increases the production of infection-fighting white blood

Table 14.4 Immunological effects of β-glucan

Immunological effects[a]
Enhances macrophage phagocytic functions
Enhancement of macrophage number
Initiates phagocytic activity
Enhanced macrophage activity
Stimulates the innate immune system
Non-specific immunostimulation
Polysaccharide immunomodulators
Immunomodulation
Immunostimulatory
Tumour inhibitory and antibacterial activity
Radiation injury to skin
Metastasis
Genotoxicity
Immune activator
Against oxidative organ injury
Activation of leukocytes
Immune potentiation
Prophylaxis against surgical infection

[a]Adapted from Beta Glucan Research Organization (2006).

cells and antibodies and increases levels of interferon that coats cell surfaces, preventing the entry of viruses. Around 200 mg/day can be automatically obtained by eating at least six servings of fruits and vegetables a day.

Fruit juices such as orange juice have long been recognised as an excellent source of vitamin C, but other fruits such as guava and kiwi contain four times the amount of vitamin C (Daniells 2008). Vegetables such as red capsicums, parsley, broccoli and cabbage, coloured berries (strawberries, blueberries and boysenberries), tomatoes and avocados are also rich sources of vitamin C (Klausner 2000; Rose 2007).

While freshly squeezed orange juice is a good source of vitamin C, ripe fruits have higher vitamin C content than 'green' or preripe fruits. As a result, many fruit juices are fortified with vitamin C (as an additional ingredient indicated in Table 14.4) (Landen 2006).

Red peppers contain twice the vitamin C content of oranges and are also a valuable source of carotene (Dowden 2000a; Rose 2007).

14.5.2 Vitamin E

Vitamin E is also recognised as an important antioxidant and immune booster but has not received the attention given to vitamin C. Vitamin E stimulates the production of natural killer cells (that seek out and destroy germs and cancer cells) and enhances the production of B cells (the immune cells that produce antibodies that destroy bacteria).

It is not difficult to get 30–60 mg every day but difficult for most people to consume more than 60 mg/day consistently. Vitamin E-rich foods include wheat germ, whole oats, cold pressed olive oil, fruits, dark green leafy vegetables, avocado, fish, poultry, meat, eggs and raw nuts and seeds (Table 14.2). Supplements may be necessary to get enough vitamin E to boost your immune system, which requires 100–400 mg/day (see Table 14.3) (Guy 2007; Turner 2007).

14.5.3 Zinc

This valuable mineral increases the production of white blood cells that fight infection and helps them fight more aggressively. It also increases the number of infection-fighting T cells, killer cells that fight against cancer, and helps white cells release more antibodies. Zinc is found in a wide variety of foods including lean meat, chicken, fish, milk and other dairy foods (cheese), brewers yeast, egg yolks, legumes (soy beans, lima beans, lentils, peas), wholegrain (bread), sunflower seeds, pumpkin seeds and pecans. A moderate amount of zinc is found in vegetables (Guy 2007).

Zinc is particularly important in aged people, as the immune system often weakens with age. While some studies claim that zinc supplements in the form of lozenges can lower the incidence and severity of infections, other studies have failed to show this correlation. Zinc is still recognised as a fortifying agent that has an effect on the immune system (see Table 14.3), and one of the best sources of zinc for infants and young children is zinc-fortified cereals, which provide 0–15 mg/ounce (Sears 2007).

14.5.4 Amino acids and nucleotides

A number of amino acids are currently used to fortify foods because of their beneficial effect on the immune system. Arginine, when taken with synergistic cofactors such as vitamin B, stimulates the pituitary to secrete growth hormone, a vital immune regulator. Arginine also enlarges the thymus gland (which produces T cells), greatly enhances the body's healing powers and helps prevent cancer (Reid 1989; Ogilvie 2002).

It makes sense that during times of extraordinary demand (such as growth, reproduction, environmental change or challenge, combating disease and recovery from injury) trillions of additional nucleotides must be readily available for cell proliferation. A strong immune system is equally important in an organism's ability to respond to other stress factors, such as injury, sudden environmental changes, physical exertion and growth. Such pressures tax the immune system and, consequently, the ability of an animal to survive and adequately react to drastic changes during its life. Moreover, challenges of the immune system reduce the ability to maintain or improve performance and productivity in livestock. Supplementary nucleotides have been demonstrated to affect a number of immune functions, including reversing malnutrition and starvation-induced immunosuppression, enhancing T-cell maturation and function, enhancing natural killer cell activity, improving delayed cutaneous hypersensitivity, and aiding in resistance to such infectious agents as *Staphylococcus aureus* and *Candida albicans* (Hoffman 2007a, b, c). In numerous trial experiments in different species, a positive effect of dietary RNA and nucleotides has been observed. The response of the immune system to vaccinations is accelerated and, at the same time, shows an increase in the antibody titres (Hoffman 2007a, b).

Dietary nucleotides are typically consumed in amounts of 1-2 g/day from sources such as animal protein, peas, yeast, beans and milk. Human breast milk contains approximately 70 mg of nucleotides per litre, and consequently, many infant formulae are now supplemented with dietary nucleotides. Several immune-enhancing enteral formulae also contain nucleotides (Schloerb 2001).

Nucleotides may also be particularly important as a supplement in newborn or young babies and animals. In general, newborn animals are not provided with a functional immune system. Innate (non-specific) and/or acquired (specific) immunity have to be developed in the early stages of life to enable the animal to react to environmental disturbances and health challenges. The gradual decrease of maternal antibodies and the simultaneous increase of the internal immune defence are antidromic processes associated with, for example, the development of the alimentary system in young animals.

As an example, glutamine can also be synthesised in many cells and tissues of the body. However, only certain tissues are able to release significant amounts of glutamine. These include the lung, brain and skeletal muscle. Skeletal muscle is considered to be the most important glutamine producer in the body (~60%).

Animal studies indicate that intramuscular and plasma glutamine concentrations are decreased in stress situations such as sepsis (Ardawi & Majzoub 1991), following surgery and athletic training (Jensen *et al.* 1996; Hack *et al.* 1997). The lowered plasma glutamine concentration which occurs is most likely the result of demand for glutamine (by the liver, kidney, gut and immune system) exceeding the supply, and it is proposed that glutamine be considered a conditionally essential amino acid during stress (Ardawi & Newsholme 1991]; Lacey & Wilmore 1990; Jensen *et al.* 1996; Hack *et al.* 1997).

Other amino acids, including glutamate, aspartate and arginine, cannot substitute for glutamine to support lymphocyte proliferation (Ardawi & Newsholme 1983; Calder 1995).

14.5.5 Selenium

This mineral increases natural killer cells, mobilises cancer-fighting cells and acts as an antioxidant. Best food sources of selenium are tuna, red snapper, lobster, shrimp, whole grains, vegetables (depending on the selenium content of the soil they are grown in), brown rice, egg yolks, cottage cheese, chicken (white meat), sunflower seeds, garlic, Brazil nuts and lamb chops (Table 14.2) (Rose 2007; Sears 2007).

People living in parts of China, New Zealand and Europe historically have had diets low in selenium. The 'Spanish flu' epidemic of 1918 originated in China, and new, more virulent strains of flu have recently surfaced in Asia and may be attributed to mutant viruses abetted by deficient selenium status in the host. Studies show that a non-virulent virus, in a selenium-deficient host, mutates to become 'super virulent' – a 'killer' virus (Nelson *et al.* 2001).

Selenium is also an important functional food supplement for some production animals. Organic selenium yeast has been added to swine grower diet. Because it is organic selenium yeast, the pigs actually incorporate that selenium into the muscle, not into the fat. Independent testing conducted at the University of Prince Edward Island has determined that the selenium content can be raised by 63.6% compared to regular pork when supplemented in the diet (D. Hurnick, personal communication).

14.5.6 Antioxidants

Antioxidants are among the most commonly used supplements that assist with a variety of medical disorders and maintain general health and well-being. Antioxidants are specific vitamins (such as vitamin C and vitamin E) or plant and animal compounds (bioflavonoids such as grape seed extract, quercetin, coenzyme Q-10, *Ginkgo biloba*) that are responsible for maintaining health. Technically, they do not boost the immune system (although proanthocyanidins appear to potentiate the immune system via enhancement of T-lymphocyte activity and modulation of neutrophil and macrophage responses) but rather clean up the products of cell damage (oxidising chemicals) (Bendich 1993; Knight 2000; Messonnier 2001).

14.5.7 Calcium d-glucanate

Calcium D-glucanate speeds up your body's natural detoxification of carcinogens. Our bodies remove harmful substances through the liver in a process called glucoronidation. Many are known carcinogens. It protects against breast, lung, colon, liver, bladder, skin and prostate cancers. Some research indicates that many people who develop cancer have reduced glucoronidation.

14.5.8 Colostrum

Colostrum is a rich source of immunoglobulines, IgA and IgG in particular but also IgD, IgE and IgM. Many new and exciting discoveries are being made about the healing properties of this very old cure. Bovine colostrum has been proved not only to support the body's immune system, but also contains valuable growth factors that promote tissue repair (Cara 2008).

Colostrum powder of which there are several types available as nutraceuticals or functional food supplements for humans and animals (Krakowski *et al.* 1999, 2002; Villadsen 2005) contains a combination of bioactive components to support a healthy immune system. Lactoferrin (an iron-binding whey protein found in naturally occurring body fluids, i.e. tears, saliva and breast milk) has the highest concentration in colostrum. It is a major constituent of leukocytes and owing to its iron-binding properties is able to bind pathogens invading the body, thus having an antibacterial, antiviral, antifungal, anti-inflammatory, antioxidant and immunomodulatory activities (Lonnerdal & Lyer 1995; Huang *et al.* 2003).

Colostrum also contains lysozyme and lactoperoxidase. Lysozyme (sometimes referred to as the body's own antibiotic), which has the ability to destroy cell walls of certain bacteria and lactoperoxidase, also an enzyme, possesses antimicrobial properties (Cara 2008).

14.5.9 Omega-3

Omega-3 oils from oily fish, such as salmon, sardines, mackerel and tuna, and flaxseed oil, nuts and seeds, block chemicals called cytokines that can cause low mood. Studies of polyunsaturated fatty acids of the *n*-3 series, especially eicosapentaenoic and docosahexaenoic acid, indicate that these fatty acids may prevent the development of carcinogen-induced tumours, the growth of solid tumours, as well as the occurrence of cachexia and metastatic disease in experimental tumour models. Fatty acids of the *n*-3 series have been shown to normalise elevated blood lactic acid and insulin levels in non-malignant conditions. In contrast, polyunsaturated fatty acids of the *n*-6 series appear to enhance tumour development and metastases (Kolakowski *et al.* 2006).

Some evidence suggests that the omega-3 fatty acids in fish might slow or prevent tumour growth. In a large Italian study, people who ate fish just twice a week had 30–50% less risk of cancers of the mouth, oesophagus, stomach, pancreas, colon and rectum than those who ate fewer than one serving of fish a week (Klausner 2000).

Readers are referred to more specific references on omega-3 and its potential to boost the immune system such as Kolakowski *et al.* or the review by the University of Maryland Medical Centre (Kolakowski *et al.* 2006; University of Maryland Medical Center 2008).

14.5.10 β-Glucan

Of all the natural compounds known to stimulate the innate immune system, the most documented and most effective are the 1-3,1-6-β-glucans, generally derived from brewer's

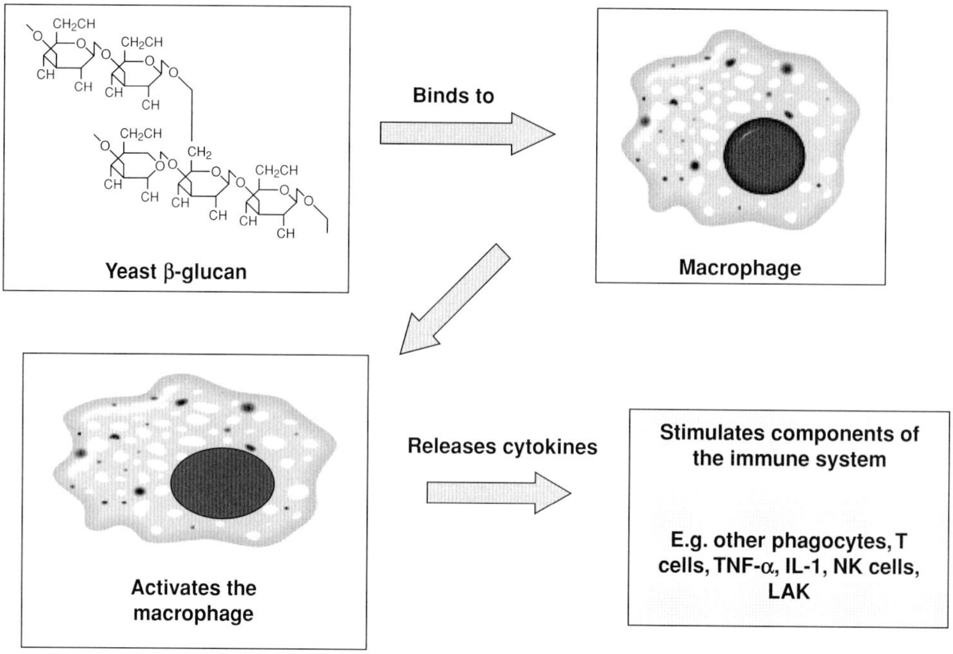

Fig. 14.1 Response of the immune system to β-glucan stimulation. For a colour version of this figure, please see the plate section.

yeast. These molecules activate the innate immune system in humans and other mammals, in birds, fish and even crustacea (Hahn & Albersheim 1978; Robertsen *et al.* 1994; Song & Hsieh 1994; Zentec *et al.* 2002).

Macrophages have receptors which specifically recognise 1-3,1-6-β-glucans as they occur naturally in the cell walls of many bacteria and fungi. The innate immune system recognises these as a challenge and responds accordingly by mounting an immune response, as shown in Figure 14.1. The use of so-called immune modulators has become a noteworthy development in recent years. Research on β-1,3/1,6-glucans is today deeply rooted in modern medical science, while a positive track record from practical use of immune-enhancing products in this category can be dated back to ancient medicine. It was discovered more than 40 years ago that a cell wall preparation (zymosan) of baker's yeast had a remarkable ability to activate white blood cells and enhance disease resistance in animals and humans. Some years later it was found that the active component in zymosan was a polysaccharide, consisting of a branched chain of glucose molecules (glucan) connected by β-1,3 and β-1,6 linkages (Manners *et al.* 1973a; b; Williams *et al.* 1996).

Intact cells of baker's yeast do not activate the immune system because the biologically active component, β-1,3/1,6-glucan, is enclosed within the cell wall and covered with mannoproteins. These surface components have to be removed in a process which simultaneously releases and exposes the β-1,3/1,6-glucan in its active form.

The cell wall of brewer's yeast also contain β-1,3-glucan, but the biological activity of this molecule is lower, even in pure form, because it has few of the 1,6 branches needed. Brewer's yeast has a slightly different structure from baker's yeast, but has almost the same mode of action as found in extracts of the Shiitake mushroom (*Lentinus edodes*) and the mushroom known as Lingzhi (*Ganoderma lucidum*) in China. These have a recognised position in traditional oriental medicine and are still regarded as a 'spiritual essence', which strengthens the body and cures cancer, urinary disorders, fever diseases and arthritis/rheumatism (Rosenhaugh 2002).

β-Glucans are known to act as adjuvants and immunostimulants, and a number of effects on the components of the immune system have been identified (see Table 14.3). β-Glucans enhance the activities of macrophages, neutrophils, natural killer cells, T cells and B cells. As vaccine adjuvants, they potentiate humoral and cellular immune responses to microbial and tumour antigens and non-specifically enhance host resistance to disease. This broad spectrum of activities may be primarily attributed to macrophage activation. *In vivo* and *in vitro* studies have shown that β-glucans increase macrophage phagocytosis and macrophage cytokine production. (Table 14.4) β-Glucan, extracted from *Saccharomyces cerevisiae* cell wall (β-1,3-glucan with long β-1,6-glucan branches), is used as an immunomodulator to improve animal health status, primarily because of its ability to enhance the resistance of the host to viral, bacterial, fungal and parasitic infections (Diluzio 1983; Bohn & Bemiller 1995).

β-Glucan has an opportunity to become a primary immune-stimulating functional component for the addition to many food stocks for commercial meat-producing animals. As consumer desire for more natural productivity enhancement grows, particularly with the reduction of antibiotics, the opportunity grows stronger and more detailed research confirms its efficacy. For example, trials in early weaning pigs have shown that β-glucan can help to overcome the lack of a mature gut-associated immune defence system against viral infection and bacterial infections. Similar results have also been demonstrated for other animals such as broiler hens (Johnson 1997; Kim *et al.* 2000; Hiss & Sauerwein 2003; Burkey *et al.* 2004; Jamroz *et al.* 2004; London 2008).

Beneficial effects for β-glucans have also been reported for companion animals such as dogs, cats and horses, where the stimulation of the innate or non-specific immune system provides additional protection against opportunistic infections and generally improves the well-being of the animals (Krakowski *et al.* 1999; Rosenhaugh 2002; Hayek *et al.* 2004). A detailed bibliography of references relating to the effects of β-glucans can be found in the Beta Glucan Research Review (2006).

14.6 Ancillary functional food components

The final group of functional food components referred to in Table 14.1 is the ancillary components. These are typified by prebiotics and the probiotics.

14.6.1 Pre- and probiotics

The prebiotic market is growing directly in accordance with the increased awareness of functional foods for health in humans. Examples of prebiotics include inulin, fructo-oligosaccharides, galacto-oligosaccharides and lactulose. These occur naturally in many foods but can also be incorporated into beverages, confectionery and dairy products.

A new term that has entered the literature is 'symbiotic', which is the mixture of prebiotics and probiotics that beneficially affect the host by improving the survival and multiplication of live microbial dietary supplements in the gastrointestinal tract.

Interest in probiotic-enhanced immunity dominates contemporary literature. Recent reviews of immune effects of probiotics in humans demonstrate that these microbes stimulate cell-mediated immune effector functions, with enhanced secretion of INF-γ by blood cells, enhanced phagocytosis and an increase in expression of complement receptors on phagocytes leading to additional local mechanisms such as an alternation of the gut milieu, a downregulation of the mucosal secretary response to pathogens and an activation of a local immune response, which may contribute a clinical benefit (Clancy 2003; O'Sullivan *et al.* 2005).

Probiotics have three possible modes of action: generating immunostimulants, producing antimicrobial compounds and competitively excluding other bacteria. Probiotics are living organisms; as such, their stability through processing is an issue that needs to be addressed, as extrusion/expansion processing inactivates these biological products (Clancy 2003).

There are also certain criteria that have been developed to evaluate the potential of microorganisms to function as probiotics. Adequate numbers of viable organisms must reach the intestinal tract. Probiotic organisms must be able to survive transit through the acidic environment of the stomach and resist bile digestion. Organisms must have the ability to adhere to intestinal epithelial cells, colonise the intestinal tract, produce an antimicrobial factor and inhibit enteric pathogens. Other properties such as immunomodulation, modulation of metabolic activities and inactivation of procarcinogens are also desirable.

The use of probiotics such as lactobacilli animals seems sensible because these bacteria are present in large numbers in the gastrointestinal tracts of many animals. The indiscriminate use of antibiotics and the dramatic increase in antibiotic-resistant microbial pathogens have renewed interest in the possibility of deliberately feeding 'beneficial' microorganisms to humans as an alternative to antibiotic therapy (Gatesoupe 1999; Guillot 2000; Hansen 2000; Newman 2007; Simon *et al.* 2007; Timmerman 2007).

A group of less readily recognised immune modulators and stimulants have also potential for fortification in foods although the data to support these indications are not fully developed. These may well be the new-age functional food components, for example the following.

14.6.2 Astaxanthin

Astaxanthin is common in fish and shrimp that enhance immune functions and the central nervous system. For more than 10 years, astaxanthin's role in enhancing the immune system and preventing oxidative stress has been the subject of international research and there are more than 300 publications and patents regarding astaxanthin (Adams 2005).

14.6.3 Isoprinosine

A synthetic immunostimulant (1-dimetyloamino-2-propanolo-4-acetoamidobenzate with inosine 3:1) and thymostimuline (TFX) – an extract from the calf thymus – are strong immunostimulants, and 3-hydroxy-3-methylobutyrate (HMB) – a metabolite of leucine used as a dietetic fodder supplementation – induces the immune system to stimulate the metabolic activity of phagocytes and enhances proliferation of B and T lymphocytes.

TFX, isoprinosine and HMB when used in pregnant sows had a positive influence on the immunologic quality of the sow's colostrum, providing enhanced immunological support for weaning piglets (Krakowski *et al.* 2002).

In the last 20 years, it has become much clearer that psychological and neurological factors influence immunity. A healthy diet, not deficient in energy, nutrients and micronutrients, is essential to prevent and fight off diseases. Stress influences immunity through the hypothalamic pituitary adrenal and autonomic mechanisms, and it has become clear that immune mediators influence behaviour. Furthermore, immune mediators have a profound influence on feed intake and on the energy and nitrogen balance (Niewold 2008).

An important distinction needs to be made between malfunction of the immune system because of (simple) deficiencies, and the purported enhancement and modulation of the immune system above the mere nutritional requirements by certain compounds, the so-called functional foods. Micronutrient requirements can differ considerably depending on the physiological state, i.e. during disease, lactation, pregnancy, etc. Similar variability in results is seen with pro- and prebiotics, and even adverse reactions have been described. The beneficial effect of antibiotics on growth has been amply demonstrated, which is most likely due to the known direct anti-inflammatory effect of antibiotics on immune cells.

There is a fine balance between beneficial and detrimental effects on the body's immune system. These problems may be linked by a common thread – the overreaction of the immune system to a dietary component, most probably a protein. True allergies to foods are the result of an antigen eliciting an IgE-mediated response. In this stimulation of the immune system, an offending (antigenic) protein is thought to bridge, or cross-link, primed IgE receptors on the mast cell, a specialised 'battleship' cell of the immune system. The formation of this cross-link causes the mast cell to degranulate, releasing chemicals such as histamines, cytokines, prostaglandins and enzymes into the circulation. Their job is to destroy the offending antigen and shore up the defences at the tissue level to repel and expel the invaders via vomiting, sneezing, swelling, etc.

14.7 Functional immune-boosting animal feeds

To date, there has been little discussion on the use of functional foods for immune-boosting effects in animals. This is in no way a reflection of the lack of effect. On the contrary, immune-boosting functional foods are used extensively in animal applications. Although the intended outcomes may differ from those of human applications, for the main part, most functional food components already discussed have a similar effect in animals. Studies are now confirming the advantages of adding functional components to foods for both companion and production animals. The use of functional foods for agricultural benefit is an exciting new dimension (Bartelme 2003; Hoffman 2007a, b, c; Koeleman 2008; Niewold 2008; Phillips 2008).

Novel forms of functional food components have also shown promise in animal husbandry. Modified animal cold viruses (adenoviruses) that ferry natural immune-boosting molecules into cells have also the potential to improve the health and growth of livestock raised under intensive conditions, and cytokines such as γ-interferon could provide an effective alternative to the use of antibiotics for disease prophylaxis and growth promotion (Pyper 2000). Organic mixtures of functional food components have also shown potential. For example, three components of the naturally occurring compounds – capsaicin, cinnamaldehyde and eugenol – found in clove, cinnamon and capsicum, produced positive effects in cattle (Vikari 2008) and other herbs and spices especially for production animals (Zentec & Mader 2002).

14.8 The future of immune-boosting functional foods

Having reviewed the significant opportunities that seem to exist for the use of foods (complete and modified) that have a positive influence on the immune system, it would seem appropriate to close with some discussion of the limiting factors, the 'puddles that need to be jumped'. These obstacles fall into three categories – (i) scientific-based evidence, (ii) regulation and (iii) public acceptance – although not necessarily in that order, and accepting that none of these are mutually exclusive.

It is possible to envisage a future where 'designer foods' become an accepted part of our diet and existence. We are now starting to focus attention on more exacting scientific clinical studies to validate the efficacy of functional foods and/or components. We know that nuts are rich in vitamin E whereas bananas and beef meat are not. A range of different foods – Brazil nuts, eggs or fish, for example – are better choices as a source of selenium.

In Canada, to be approved as a source of omega-3, food must contain a minimum of 0.3 mg/100 g of meat. Enriching the diet of pigs through the choices of oilseeds, grains, vitamins and minerals has the ability to increase the omega-3 content in pork between 100 and 1000 times greater than you would find in typical pork (D. Hurnick, personal communication).

On a broad scale, it can be demonstrated that beef is an excellent immune-boosting food and that antioxidants take care of the free radicals. It can also be shown that tea has more antioxidants per part than fruits and vegetables. However, the exact mechanisms of action of each of these examples are complex. Are they due to single ingredients, or is a composite of several ingredients (the whole food) required? For the large part, we know that some work for some people, others do not (Muller 2008).

Interpreting the phrase 'significant scientific agreement' is crucial in influencing the development of functional foods (as well as dietary supplements). There are a few functional foods where every single clinical trial has shown positive results. Whether this would still hold true, if significantly more than one or two trials were conducted, is difficult to predict given the multitude of variables in clinical research of this kind (Hunter 2002).

As few as one or two successful clinical trials are considered adequate by some regulatory authorities governing functional foods. It is generally accepted, however, that in order for a compound to claim efficacy, there needs to be greater than five clinical trials and in order to purport that data show a trend for a product, there needs to be three to ten clinical trials. Many functional foods or functional food components have not been tested in a single, controlled and regulated clinical study, which leads to scepticism by some regulators and scientists.

Other issues for immune-boosting functional foods are concerned with the nature of the actual ingredient. Not all β-glucans, for example, have the same activity. Glucan is primarily obtained from bacteria (curdlan), baker's and brewer's yeast, fungi (scleroglucan), and edible mushrooms (lentinan, shizophyllan). Mannan, often named MOS meaning 'mannan oligosaccharides', is also derived from common yeast and is used in the feed industry to prevent the colonisation of pathogens in the intestinal tract of farm animals.

Most products are derived from yeast and are intended for health food applications or as feed ingredients. Glucan contents range from 2 to 92%, whereas commercial MOS products usually have mannan contents below 30%. It is now becoming clear that the actual level of β-glucan (i.e. purity) has a directly proportional relationship to the immune-boosting ability. However, of even greater importance is the type of β-glucan and whether it is soluble. Insoluble yeast β-glucans produce the greatest immune effect (Simon *et al.* 2007).

Closely related is the quality of probiotics. Unfortunately, probiotic products in health food stores and in feed supplements labelled as containing *Lactobacillus acidophilus* often do not contain any viable *Lactobacillus* species, let alone the advertised *L. acidophilus*. The benefits of probiotics are often demonstrated under defined experimental laboratory conditions, but these beneficial effects fail to materialise in clinical trials, often because the trials are not properly controlled or consist of too few subjects. Large, double-blind, clinical trials are essential for establishing the practical and scientific logic of the probiotic concept (Berg 1998).

One of the key issues for functional foods is to have a strong regulatory backbone. This has to be based on good science, and a lack of it will eventually cause problems in the industry. The issue with functional foods and regulations is not the lack of functional food regulations; it is the enforcement of the regulations.

Clinically relevant biomarkers are the ones that are keenly interpreted by consumers; 'I cough less', 'I have more energy'. These can also be defendable. If the label says that it contains something it should be there. If it says that it does something it should do it. There is no place for feeding the public with data and not meeting the claims. Many current functional food components are deliberately and ambiguously labelled to avoid regulations in different countries because of the lack of supporting, controlled scientific data and the complacency of many regulatory systems.

Nutrient content claims should indicate the presence of a specific nutrient at a certain level. Structure and function claims should describe the effect of dietary components on the normal structure or function of the body. Dietary guidance claims should describe the health benefits of broad categories of foods. Health claims should confirm a relationship between components in the diet and risk of disease or health condition, and be supported by significant scientific agreement. This is, however, not always the case.

The two main federal agencies in the United States responsible for regulating the accuracy of health-related claims for functional foods are the Food and Drug Administration (FDA) and the Federal Trade Commission (FTC). The FDA regulates the claims on labels; the FTC regulates the claims in advertisements, including print media, radio, television and the internet. The U.S. Department of Agriculture is involved to a lesser extent; it regulates product label claims for functional foods that contain a certain amount of meat or poultry.

The manufacturer may deem the ingredient 'generally recognised as safe' (GRAS). This approach is called 'self-affirmation'. The FDA accepts this decision on faith, and may not challenge it unless questions arise after the product is marketed. In functional foods, GRAS is relevant, especially for new dietary ingredients (Hunter 2002).

In Japan, functional foods carry a special designation 'foods for special health use'. They are defined as 'foods which are, based on the knowledge concerning the relationship between foods or food components and health, expected to have certain health benefits, and have been licensed to bear a label claiming that a person using them for specific health use may expect to obtain the health use through the consumption thereof' (Hunter 2002). In Europe, most foods are regulated under the laws of the European Union (EU) although the EU has no official definition of functional foods (Hunter 2002).

It would appear that legislation and regulation are in place in most Western countries. What lacking is the enforcement of the regulation. This lax enforcement contributes to the production of poor-quality products, such as a product with low levels compared to the stated label claim. Another problem with quality is the lack of adequate recommendations for use, leading consumers to establish their own dosage regime, which may or may not be effective, or even worse may be harmful rather than beneficial. A third problem with quality is that some products can be contaminated with harmful materials such as lead, pesticides, and DMSI (dimethyl sulfoxide, an anti-inflammatory) (Mendell 2007).

Regulators do not seem to provide guidance to companies on the type of safety-related information on labels for functional foods (and dietary supplements). The absence of such information poses a significant safety risk to some consumers. Although the labels of functional foods often declare the amounts of active ingredients, there are still many unknowns.

Consumers have no idea how much of an ingredient to take; what the recommended dietary allowance is, if it exists at all; whether there are any long-term risks or allergy problems associated with its regular use; or whether there is a risk of toxicity at a certain usage level (International Food Information Council Foundation and Institute of Food Technologists 2005).

As consumers increase their consumption of functional foods in the hopes of improving health, the regulators must determine acceptable health claims, safety standards and quality control of manufacturing. The new challenge for the industry is creating the business opportunities for the next generation of functional foods. This will require different concepts, technologies, imagination and flair (Heasman 2003; Mendell 2007).

One of the confusing aspects for consumers is that currently there is no unified definition of functional foods. Functional foods – nutraceuticals, pharmafoods, designer foods – all are marketing terms. Attempts have been made to distinguish between a functional food, a fortified food and an enriched food. A functional food offers health benefits beyond basic nutrition. A fortified food has basic nutrients, such as vitamins or minerals, added to increase the food's nutritional value or has nutrients added that much of the population lacks.

The term 'functional' was originally applied to 'a food that was specifically created to have health-promoting benefits. It quickly became any food that could be helpful to your health. Fruits and vegetables in their natural state would qualify today as a functional food, so consumers are confused about what really is a functional food' (Hunter 2002).

Public acceptance of functional foods is growing at a fast rate. This is due in part to functional foods enjoying a boost in advertising awareness as natural methods of enjoying healthy life or medicinal benefits, combined with a dwindling confidence in man-made therapeutics. Key factors in the acceptance of functional foods, especially those that have an effect on the immune system, will be the level of understanding and resulting confidence in products above and beyond the more obvious factors such as acceptable taste and presentation (London 2008). As evidence is accumulated to support long-standing anecdotal claims that foods or components of foods can and do influence the immune system, a growing proportion of the population will opt for a more 'organic' or natural way to prevent disease, rather than treat disease.

Whether this progresses totally to prescription diets to suit the changing immune system requirement of the young, the aged, the immune compromised, or merely a healthier, balanced maintenance program, remains to be seen. It is more likely that a transition will occur from the current status quo through dietary changes supported by specific nutritional or immune-boosting supplementation.

As a final example, the tremendous potential that immune-boosting foods offer can be seen from the incidence of flu pandemics. Many predict there will be another pandemic soon. In the best case we expect billions to fall ill, with 2–7 million deaths – but it could be far worse. Our ability to deal with such a catastrophe has been put into perspective by Clayton (Clayton 2008a). This is described as follows:

> History shows that flu pandemics occur every 30 years or so. After this time the genetic makeup of a flu virus has changed so much that immunity built up from previous strains becomes irrelevant; so that overall population immunity, our main defence against pandemics, has become negligible.
>
> There were three pandemics in the 20th century, and all spread worldwide within a year of being detected. The Spanish flu in 1918–19 killed up to 50 million people. In the 50's the Asian flu pandemic killed a mere million, and in 1968 Hong Kong flu killed another million or so. Prime candidate is the bird flu now gathering momentum in Asia, and which has already shown human-to-human transmission.

As governments stock up on anti-viral treatments, there is doubt over the validity of the underlying two assumptions: firstly, that the emergency could be managed, and secondly that the anti-viral drugs will be reasonably effective. Both of these assumptions are very questionable. Our ability to deal with the fall-out of a contagious and highly lethal viral epidemic is, realistically, inadequate. The efficacy of anti-virals (which was never very high) is being seriously undermined by some governments to give the anti-viral drugs (for example, amantadine) to infected flocks of poultry. Nothing is more likely to breed new drug-resistant viruses.

If one in four people deemed sufficiently important (army, police, medical personnel and the political classes) will be protected, what should the rest of us, the expendable folk, do?

The best defence against infection is to prepare and maintain your innate immune system. 'Prevention is better than cure', which is probably the most relevant phrase to summarise the significant potential of functional foods to boost the immune system.

References

Adams, M. (2005). Food companies begin to include immune-boosting ingredients. Available online http://www.newstarget.com/z012560.html (accessed 29 November 2007).

Ardawi, M.S.M. & Majzoub, M.F. (1991). Glutamine metabolism in skeletal muscle of septic rats. *Metabolism*, **40**, 155–164.

Ardawi, M.S.M. & Newsholme, E.A. (1983). Glutamine metabolism in lymphocytes of the rat. *Biochemical Journal*, **212**, 835–842.

Barrett, B.P., Locken, K., Maberry, R., Bobula, J.A. & D'Alessio, D. (2002). Treatment of the common cold with unrefined *Echinacea*. A randomized, double-blind, placebo controlled trial. *Annals Internal Medicine*, **37**, 939–946.

Bartelme, T. (2003). Beta glucan as a biological defense modulator: helping fish to help themselves. *Advanced Aquarist's Online Magazine*. Available online http://www.advancedaquarist.com/issues/sept2003/feature.htm (accessed 11 November 2005).

Beisel, W.R. (1982). Single nutrients and immunity. *American Journal of Clinical Nutrition*, **35**(2), 417–468.

Bendich, A. (1993). Physiological role of antioxidants in the immune system. *Journal of Dairy Science*, **76**, 2789–2794.

Berg, R.D. (1998). Probiotics, prebiotics or 'conbiotics'? *Trends in Microbiology*, **6**(3), 89–92.

Beta Glucan Research Organization (2006). Beta glucan research – *Saccharomyces cerevisiae*. Condition, function and disease. Indexed References. Available online http://www.betaglucan.org (accessed 3 October 2008).

Bohn, J.A. & Bemiller, J.N. (1995). (1-3)-β-D-glucans as biological response modifiers: a review of structure–functional activity relationships. *Carbohydrate Polymers*, **28**, 3–14.

Buontanotte, F. (2009). Honey the new antibiotic. Available online http://www.articlebiz.com/article/142876-1-honey-the-new-antibiotic (accessed 6 January 2010).

Burkey, T.E., Dritz, S.S., Nietfeld, J.C., Johnson, B.J. & Minton, J. E. (2004). Effect of dietary mannanoligosaccharide and sodium chlorate on the growth performance, acute-phase response, and bacterial shedding of weaned swine challenged with Salmonella enterica serotype typhimurium. *Journal of Animal Science*, **82**, 397–404.

Calder, P.C. (1995). Requirement for both glutamine and arginine by proliferating lymphocytes. *In Proceedings of Nutrition Society*, **54**, 123A.

Calder, P.C. & Yaqoob, P. (1999). Glutamine and the immune system. *Amino Acids*, **17**, 227–241.

Cara, E. (2008). Colostrum – a natural boost for the immune system. Available online http://www.articlesbase.com/supplements-and-vitamins-articles/colostrum-a-natural-boost-for-the-immune-system-481166.html (accessed 6 January 2010).

Chen, H.L., Li, D.F., Chang, B.Y., Gong, L.M., Dai, J.G. & Yi, G.F. (2003). Effects of Chinese herbal polysaccharides on the immunity and growth performance of young broilers. *Poultry Science*, **82**, 364–370.

Clancy, R. (2003). Immunobiotics and the probiotic evolution. *FEMS Immunology and Medical Microbiology*, **38**, 9–12.

Clayton, P. (2008a). Bird flu. Available online http://www.beta-glucan.co.uk/bird_flu.htm (accessed 25 January 2008).
Clayton, P. (2008b). Diet. Available online http://www.beta-glucan.co.uk/diet.htm (accessed 25 January 2008).
Collins, E. (1999). *Your Complete Guide to Echinacea and Immunity*. Prima Publications, Toronto, Ontario.
Coppens, P., da Silva, M.F. & Pettman, S. (2006). European regulations on nutraceuticals, dietary supplements and functional foods: a framework based on safety. *Toxicology*, **221**(1), 59–74.
Dalby-Brown, L., Barsett, H., Lanbo, A.R., Meyer, A.S. & Mølgaard, P. (2005). Synergistic antioxidative effects of alkamides, caffeic acid derivatives, and polysaccharide fractions from *Echinacea purpurea* on *in vitro* oxidation of human low-density lipoproteins. *Journal of Agriculture and Food Chemistry*, **53**, 9413–9423.
Daniells, S. (2008). Yellow kiwifruit may boost immune health: study. Available online http://www.nutraingredients.com/news/ng.asp?n=84761&c=nYKpq%2FDPy1K1d9Mk (accessed 28 April 2008).
Denehey, C. (2006). The use of herbs and medicines in gynecological surgery: an evidence-based review. *Journal of Midwifery and Women's Health*, **51**(6), 402–409.
Diamond, W.J. & Cowden, W.L. (1997). *Definitive Guide to Cancer*. Future Medicine Publishing, California.
Diluzio, N.R. (1983). Immunopharmacology of glucan – a broad spectrum enhancer of host defense mechanisms. *Trends in Pharmacological Science*, **4**, 344–347.
DiSilvestro, R. (2005). *Handbook of Minerals as Nutritional Supplements*. Thompson Wadsworth, Belmont, CA.
Dowden, A. (2000a). Health zone: the greengrocer's top 30 foods to help keep you fighting fit – fruit and veg. *The Mirror*, London, England (13 July 2000).
Dowden, A. (2000b). Health zone: medicinal foods – I suppose you think that's fungi. *The Daily Mirror*, London, England (21 September 2000).
Espín, J.C., García-Conesa, M.T. & Tomás-Barberán, F.A. (2007). Nutraceuticals: facts and fiction. *Phytochemistry*, **68**(22–24), 2986–3008.
Gatesoupe, F.J. (1999). The use of probiotics in aquaculture. *Aquaculture*, **180**, 147–165.
Giles, J.T. (2000). Evaluation of *Echinacea* for treatment of the common cold. *Pharmacotherapy*, **20**(6), 690–697.
Girodon, F. (1999). Impact of trace elements and vitamin supplementation on immunity and infections in institutionalized elderly patients. *Archives of Internal Medicine*, **159**, 748–754.
Grimble, R.F. (1998). Nutritional modulation of cytokine biology. *Nutrition*, **14**, 634–640.
Grimble, R.F. (2001). Nutritional modulation of immune function. *Proceedings of the Nutrition Society*, **60**, 389–397.
Groff, J.L., Gropper, S.A.S. & Hunt, S.M. (2005). *Advanced Nutrition and Human Metabolism*, 4th edn. West Publishing Co., St Paul, MI.
Guillot, J.F. (2000). Make probiotics work for poultry. *Feed Mix* (Special 2000), 28–30.
Guy, L. (2007). Immune-boosting foods for kids. Available online http://www.artofhealing.net.au/articles/immune-boosting-for-kids (accessed 6 January 2010).
Hack, V., Weiss, C., Friedmann, B., Suttner, S., Schykowski, M., Batsch, P. & Dodge, W. (1997). Decreased plasma glutamine level and $CD4^+$ T cell number in response to 8 wk of anaerobic training. *American Journal of Physiology*, **272**, E788–E795.
Hahn, M.G. & Albersheim, P. (1978). Host-pathogen interactions. XIV. Isolation and partial characterization of an elicitor from yeast extract. *Plant Physiology*, **62**, 107.
Hansen, G.H. (2000). Use of probiotics in marine aquaculture. *Feed Mix*, **8**(4), 32–34.
Hayek, M.G., Massimino, S.P., Ceddia, M.A. (2004). Modulation of immune response through nutraceutical interventions: implications for canine and feline health. *Veterinary Clinic Small Animals*, **34**, 229–247.
Heasman, M. (2003). Addressing the functional foods paradox: the first generation of functional foods has passed and it is time for companies to make room for what's next. Nutraceuticals World. Available online http://www.nutraceuticalsworld.com/articles/2003/11/addressing-the-functional-foods-paradox (accessed 11 December 2007).
Hill, A. (2007). Forget superfoods, you can't beat an apple a day. *The Observer* (13 May 2007).
Hiss, S. & Sauerwein, H. (2003). Influence of dietary β-glucan on growth performance, lymphocyte proliferation, specific immune response and haptoglobin plasma concentrations in pigs. *Journal of Animal Physiology and Animal Nutrition*, **87**, 2–11.

Hoffman, K. (2007a). The use of nucleotides in animal feed (Part 1). *Feed Mix*, **15**(4), 25–27.
Hoffman, K. (2007b). The use of nucleotides in animal feed (Part 2). *Feed Mix*, **15**(5), 14–16.
Hoffmann, K. (2007c). The use of nucleotides in animal feed (Part 3). *Feed Mix*, **15**(6), 36–38.
Huang, Y., Shao, X.M. & Neu, J. (2003). Immunonutrients and neonates. *European Journal of Pediatrics*, **162**, 122–128.
Hunter, B.T. (2002). Functional foods are poorly regulated (food or supplement?). *Consumers' Research Magazine* (1 February 2002).
International Food Information Council Foundation and Institute of Food Technologists (2005). Guidelines for communicating the emerging science of dietary components for health functional foods. Available online http://internal.ific.org/nutrition/functional/guidelines/guidelinesfulldoc.cfm (accessed 6 January 2010).
Jamroz, D., Wiliczkiewicz, A., Orda, J., Wertelecki, T. & Skorupinska, J. (2004). Response of broiler chickens to the diets supplemented with feeding antibiotic or mannanoligosaccharides. *Electronic Journal of Polish Agricultural Universities*, **7**, 250–254.
Janeway, C.A., Travers, P., Walport, M. & Shlomchik, M. (2001). *Immunobiology – The Immune System in Health and Disease*, 5th edn. Garland Publishing, New York.
Jensen, G.L., Miller, R.H., Talabiska, E.D.G., Fish, J. & Gianferante, L. (1996). A double blind, prospective, randomized study of glutamine-enriched compared with standard peptide-based feeding in critically ill patients. *American Journal of Clinical Nutrition*, **64**, 615–621.
Johnson, R.W. (1997). Inhibition of growth by pro-inflammatory cytokines: an integrated review. *Journal of Animal Science*, **75**, 1244–1255.
Kim, J.D., Hyum, Y., Sohn, K.S., Kim, T.J., Woo., H.J. & Han, I.K. (2000). Effects of mannan oligosaccharide and protein levels on growth performance and immune status in pigs weaned at 21 days of age. *Journal of Animal Science and Technology*, **42**, 489–498.
Klausner, A. (2000). From A to tea: EN picks 10 top cancer-fighting foods. *Environmental Nutrition*. Available online http://www.highbeam.com/doc/1G1–61825165.html (accessed 6 January 2010).
Knight, J.A. (2000). Review: free radicals, antioxidants, and the immune system. *Annals of Clinical and Laboratory Science*, **30**(2), 145–158.
Koeleman, E. (2008). Animal friendly feeding. Available online http://www.allaboutfeed.net/weblog/animal-nutrition/animal-friendly-feeding-id2375.html (accessed 7 January 2010).
Kolakowski, A., Domiszewski, Z. & Bienkiewicz, G. (2006). Effect of biological and technological factors on the utility of fish as a source of n-3 PUFA. In: *Omega 3 Fatty Acid Research*. Teale, P.C. (ed.), Nova Science Publishers, Hauppauge, NY, pp. 83–107.
Krakowski, L., Krzyzanowski, J., Wrona, Z. & Siwicki, A.K. (1999). The effect of non-specific immunostimulation of pregnant mares with 1,3/1,6 glucan and levamisole on the immunoglobulins levels in colostrum, selected indices of non-specific cellular and humoral immunity in foals in neonatal and postnatal period. *Veterinary Immunology and Immunopathology*, **68**, 1–11.
Krakowski, L., Krzyzanowski, J., Wrona, Z., Kostro, K. & Siwicki, A.K. (2002). The influence of non-specific immunostimulation of pregnant sows on the immunological value of colostrum. *Veterinary Immunology and Immunopathology*, **87**, 89–95.
Lacey, J.M. & Wilmore, D.W. (1990). Is glutamine a conditionally essential amino acid? *Nutrition Reviews*, **48**, 297–309.
Landen, S. (2006). Fruit juice, nutrition and health: a review. Position Paper, The Australian Fruit Juice Association. Available online http://www.afja.com.au/publications/db/afja_position_paper2.pdf (accessed 6 January 2010).
London, C.J. (2008). Are neutraceuticals the next big thing in animal biotechnology? Available online http://www.engormix.com/e_articles_view.asp?art=997 (accessed 17 July 2008).
Lonnerdal, B. & Lyer, S. (1995). Symposium on 'evidence-based nutrition', nutrition modulation of immune function. *Proceedings of Nutrition Society*, **60**, 389–397.
Manners, D.J., Masson, A.J. & Patterson, J.C. (1973a). Structure of a β-(13)-D-glucan from yeast cell walls. *Biochemical Journal*, **135**(1), 19–30.
Manners, D.J., Masson, A.J., Patterson, J.C., Bjorndal, H. & Lindberg, B. (1973b). Structure of a β-(16)-D-glucan from yeast cell walls. *Biochemical Journal*, **135**(1), 31–36.
McWhirter, J.P. & Pennington, C.R. (1994). Incidence and recognition of malnutrition in hospital. *British Medical Journal*, **308**, 945–948.
Mendell, C. (2007). Oral joint supplements: do they work? Available online http://www.thehorse.com/PrintArticle.aspx?ID=10073 (3 January 2008).

Messonnier, S. (2001). *Natural Health Bible for Dogs & Cats: Your A-Z Guide to Over 200 Conditions, Herbs, Vitamins, and Supplements*. Random House, New York.
Messonnier, S. (2007). Immune system support in pets. Available online http://www.petcarenaturally.com/articles/immune-system-support.php (accessed 17 July 2008).
Michaud, L.B., Karpinski, J.P., Jones, K.L. & Espirito, J. (2007). Dietary supplements in patients with cancer: risks and key concepts, Part 2. *American Journal Health-System Pharmacy*, **64**(5), 467–480.
Miller, D. (2006). Medicinal mushrooms can change your life. Available online http://www.articlesbase.com/health-articles/medicinal-mushrooms-can-change-your-life-87061.html (accessed 28 September 2008).
Miller, D. (2008). 10 top winter cold and flu supplements. Available online http://www.articlesbase.com/health-articles/10-top-winter-cold-and-flu-supplements-312894.html.
Muller, J. (2008). Which foods best boost your immune system? Available online http://www.mindbodyhealth.com/boost_immune.htm.
Navarro-Alarcon, M. & Lopez-Martinez, M.C. (2000). Essentiality of selenium in the human body: relationship with different diseases. *Science and Total Environment*, **249**, 347–371.
Nelson, H.K. *et al.* (2001). Host nutritional selenium status as a driving force for influenza virus mutations. *The FASEB Journal*, **15**, 1846–1848.
Newman, K. (2007). Probiotics gain more interest. *Feed Mix*, **15**(1), 12–14.
Niewold, T. (2008). Effect nutrition on stress and immunity. *Feed Mix*, **16**(3), 22–226.
O'Flaherty, L. & Bouchier-Hayes, D.J. (1999). Immunonutrition and surgical practice. *Proceedings of the Nutrition Society*, **58**, 831–837.
O'Sullivan, G.C., Kelly, P., O'Halloran, S., Collins, C., Collins, J.K., Dunne, C. & Shanahan, F. (2005). Probiotics: an emerging therapy. *Current Pharmaceutical Design*, **11**, 3–10.
Ogilvie, G.K. (2002). Nutrition and cancer: exciting advances for 2002. Proceedings of 27th WSAVA Congress, Granada, Spain.
Park, J.S., Hayek, M.G., Ceddia, M.A., Reinhart, G.G. & Chew, B.P. (2007). Bixin stimulates feline immunity. *The Journal of the Federation of American Societies for Experimental Biology*, **21**, 546.
Peakman, M. & Vergani, D. (2003). *Basic and Clinical Immunology*. Churchill Livingstone (Elsevier), London.
Philips, I., Casewell, M., Cox, T., De Groot, B., Friis, C., Jones, R., Nightingale, C., Preston, R. & Waddell, J. (2004). Does the use of antibiotics in food animals pose a risk to human health? A critical review of published data. *Journal of Antimicrobial Chemotherapy*, **53**(1), 28–52.
Phillips, T. (2008). Functional fixes. *Petfood Industry*, **50**(8), 24–25.
Pyper, W. (2000). A therapeutic role for viruses (Brief Article). *Ecos*, **102**, 36.
Reid, D. (1989). *The Tao of Health, Sex and Longevity: A Modern Practical Guide to the Ancient Way*. Fireside Books, Florida.
Robertsen, B., Engstad, R.E. & Jorgensen, J.B. (1994). Beta-glucans as immunostimulants in fish. In: *Modulators of Fish Immune Responses*. Stolen, J. & Fletcher, T.C. (eds), Fair Haven, NJ, Vol **1**, pp. 83–99.
Roitt, I.M., Brostoff, J. & Male, D.K. (1985). *Immunology*. Gower Medical Publishing, London.
Rose, T. (2007). *Vitamins and Minerals (How to Get the Nutrients Your Body Needs)*. Bounty Books, London.
Rosenberg, M. (2008). The top anti-cancer foods. Available online http://ezinearticles.com/?The-Top-Anti-Cancer-Foods&id=1593537 (accessed 8 January 2010).
Rosenhaugh, H.V. (2002). Beta-1,3/1,6-glucans. A promising way to improve health in companion animals. *Feed Mix*, **10**(3), 26–28.
Schloerb, P.R. (2001). Immune enhancing diets: products, components, and their rationales. *Journal of Parenteral and Enteral Nutrition*, **25**(2), S3–S7.
Sears, J.M. (2007). Feeding your immune system. Available online http://www.askdrsears.com/html/4/t042500.asp (accessed 8 January 2010).
Simon, O., Vahjen, W. & Taras, D. (2007). Potentials of probiotics in pig nutrition. *Feed Mix*, **15**(1), 25–27.
Song, Y.L. & Hsieh, Y.T. (1994). Immunostimulation of tiger shrimp hemocytes for generation of microbicidal substances: analysis of reactive oxygen species. *Developmental and Comparative Immunology*, **1**(3), 201–209.
Svoboda, M., Nechvatalova, K., Krejci, J., Drabek, J. & Ficek, R. (2007). The absence of iron deficiency effect on the humoral immune response of piglets to tetanus toxoid. *Veterinarni Medicina*, **52**(5), 179–185.
Timmerman, H. (2007). Probiotics reduce costly problems in calves. *Feed Mix*, **15**(1), 15–17.
Turner, L. (2007). Putting out the fire: 10 foods to fight inflammation. *Better Nutrition Magazine* (1 April 2007).

Turnidge, J. (2004). Antibiotics in animals – much ado about something. *Australian Prescriber*, **24**, 26–27.
University of Maryland Medical Center (2008). Omega-3 review. Available online http://www.umm.edu/altmed/articles/omega-3-000316.htm (accessed 8 January 2010).
Varley, M. (2007). Smart feed for pigs. *Pig Progress*, **23**(8), 26–27.
Vikari, A. (2008). Positive results of plant extracts for beef cattle. *Feed Mix*, **16**(2), 34–35.
Villadsen, J.K. (2005). Bovine colostrum – a lucrative solution in piglet rearing. *Feed Mix*, **13**(1), 25–27.
Williams, D.L., Mueller, A. & Browder, W. (1996). Glucan-based macrophage stimulators. *Clinical Immunotherapy*, **5**(5), 392–399.
Yang, G.Q. & Xia, Y.M. (1995). Studies on human dietary requirements and safe range of dietary intakes of selenium in China and their application in the prevention of related endemic diseases. *Biomedical Environmental Science*, **8**(3), 187–201.
Zentec, J. & Mader, A. (2006). The impact of plant extracts on the immune system. *Engormix*. Available online http://www.engormix.com/e_articles_view.asp?art=284 (accessed 8 January 2010).
Zentec, J., Marquart, B. & Pietrzak, T. (2002). Intestinal effects of mannanoligosaccharides, transgalactooligosaccharides, lactose and lactulose in dogs. *Journal of Nutrition*, **132**, 1682S–1684S.

15 The Mediterranean diets: Nutrition and gastronomy

Federico Leighton Puga and Inés Urquiaga

15.1 Mediterranean diet definition

In the first half of twentieth century, it was observed that coronary heart disease mortality in southern Europe was lower than in northern Europe, with the population of Crete having the lowest coronary heart disease mortality and the longest life expectancy. In this perspective, the observation was attributed to the dietary habits of this region.

The Seven Countries Study constitutes the first nutritional epidemiological investigation that provided solid data for cardiovascular disease rates in different populations, a study designed to investigate the relationship between diet and cardiovascular disease (Keys 1970). In this study the results for all-cause death rates in Greece, Japan and Italy were quite favorable compared with the United States, Finland, the Netherlands, and the former Yugoslavia; the results also showed a lower incidence of coronary disease after a 5-year follow-up for the same countries that exhibited low mortality. The diet consumed by the Mediterranean cohorts was associated with a very low incidence of coronary heart disease, and was called the Mediterranean diet by Keys (1995). Following the Seven Countries Study, the Mediterranean diet has been singled out as a healthy diet. Scientific evidence increasingly supports this epidemiological theory, which defines the relationship between a certain variety of foods and chronic diseases. Additional studies later confirmed the association of Mediterranean diet with decreased incidence and prevalence of chronic diseases, mainly cardiovascular disease, in countries where it was consumed. There are food peculiarities for the different populations in the Mediterranean basin; however, beyond the apparent differences, there are nutritional characteristics common to all or most of the diets in the Mediterranean region. The Mediterranean diet is characteristically low in saturated and high in monounsaturated fats (olive oil), low in animal protein, rich in carbohydrates, and rich in vegetables and leguminous fiber. People with Mediterranean diet eat a relatively large amount of fish and white meat, abundant fruits and vegetables, and a low amount of red meat; they also drink moderate amounts of red wine (Simopoulos 2001). The health benefits of Mediterranean diets have been attributed, at least in part, to the high consumption of antioxidants provided by fruits, vegetables and wine and to the type of fat, rich in monounsaturated and omega-3 fatty acids from vegetables and fish, and especially to a balanced omega-6/omega-3 fatty acid ratio, as is found in the traditional diet of Greece prior to 1960 (Simopoulos 2001).

Several studies have used dietary scores as instruments to measure adherence to Mediterranean diet and have reported an inverse association with overall mortality (Tabengwa *et al.* 2003; Knoops *et al.* 2004; Martinez-Gonzalez *et al.* 2004; Trichopoulou & Critselis 2004;

Trichopoulou *et al.* 2005; Bach *et al.* 2006; Panagiotakos *et al.* 2006). These studies are mostly made on Mediterranean populations. One of these studies, published in 1995, included 182 elderly (>70 years old) residents of three rural Greek villages, followed up for approximately 5 years (Trichopoulou *et al.* 1995). Diet was assessed with a validated extensive semiquantitative questionnaire on food intake. The authors used the food groups recommended by Davidson and Passmore (1979) to design the score; however, they combined starchy roots with cereals and did not consider sugars and syrups for which, so far, no systemic health implications had been documented over and beyond their contribution to net energy intake. The traditional Mediterranean diet was also defined in terms of these food groups with the addition of moderate intake of ethanol, and the score was built in terms of eight characteristic components: high monounsaturated/saturated fat ratio, moderate ethanol consumption, high consumption of legumes, high consumption of cereals (including bread and potatoes), high consumption of fruits, high consumption of vegetables, low consumption of meat and meat products, and low consumption of milk and dairy products. They used as cut-off points the corresponding median values specific for each sex. The hypothesis was that an individual diet with more of these components was beneficial, whereas a diet with fewer of these components would be less healthy. In the sample studied, 34 subjects (19% of the total) had two or fewer of the eight desirable dietary components, whereas 104 subjects (57%) had four or more of the eight desirable components. The results showed that one unit increase in diet score was associated with a significant 17% reduction in overall mortality.

Another prospective investigation by the same group of authors analyzed a population of 22 043 Greek adults, 20–86 years old, during a median of 44 months of follow-up (Trichopoulou *et al.* 2003). In this case, adherence to the traditional Mediterranean diet was assessed by the same scale developed by Trichopoulou *et al.* (1995), but revised to include fish intake. A value of either 0 or 1 was assigned to each of the nine indicated components with the use of the sex-specific median as the cut-off. For each beneficial component (vegetables, legumes, fruits and nuts, cereal and fish), persons whose consumption was below the median were assigned a value of 0, and persons whose consumption was at or above the median were assigned a value of 1. For each component presumed to be detrimental (meat, poultry, and dairy products which are rarely non-fat or low fat in Greece), consumption below the median received a value of 1, and for consumption at or above the median, a value of 0. For ethanol, a value of 1 was assigned to men who consumed between 10 and 50 g/day and to women who consumed between 5 and 25 g/day. For fat intake, the ratio of monounsaturated lipids to saturated lipids was used. Results showed that a higher degree of adherence to the Mediterranean diet was associated with a reduction in total mortality, deaths due to coronary heart disease and deaths due to cancer. Associations between individual food groups contributing to the Mediterranean diet score and total mortality were generally not significant. To apply the score to non-Mediterranean populations, Trichopoulou *et al.* (2005) substituted monounsaturated lipid with the sum of monounsaturated and polyunsaturated lipids in the numerator of lipid ratio. These authors investigated the relation of the modified score with overall mortality in a large sample of elderly Europeans participating in EPIC (the European Prospective Investigation into Cancer and Nutrition Study), 74 607 men and women aged 60 or more, from nine European countries (Denmark, France, Germany, Greece, Italy, the Netherlands, Spain, Sweden and UK). The conclusion was that the Mediterranean diet, measured with a modified score applicable across Europe, was associated with increased survival among older people. In another investigation in elderly Europeans, the HALE Project, that includes 1507 apparently healthy men and 832 women, aged 70–90 years in 11 European countries, the conclusion was that adherence to a Mediterranean diet and healthful lifestyle is associated with more than 50% lower rate of all-causes and cause-specific mortality (Knoops *et al.* 2004). Studies carried out in other parts of the world to evaluate Mediterranean diet adherence, for example in Denmark (Osler & Schroll 2007) and Australia (Kouris-Blazos *et al.* 1999; Harriss *et al.* 2007), also showed that a greater adherence to the traditional Mediterranean diet is associated with a significant reduction in total mortality. Japanese diet also appears to be very healthy in the Seven Countries Study. Japanese are known for their longevity or healthy life expectancy; they consume large amounts of cereals (rice), vegetables and fruits, and fish, but characteristically they have a much lower intake of energy and oils (Tokudome *et al.* 2000).

15.2 Food components in the Mediterranean diet

Clearly, interventional and epidemiological studies show that certain dietary habits and light-to-moderate alcohol consumption are associated with cardiovascular protection. These effects were summarized indicating that foods of animal origin were directly correlated with coronary heart disease mortality, whereas vegetable food groups, fish and alcohol were inversely correlated (Kouris-Blazos *et al.* 1999; Menotti *et al.* 1999). Populations consuming a diet rich in vegetables, olive oil, fish and wine (Mediterranean diet) had the lowest mortality among the seven countries included in that study. The benefits of this diet can apparently be transferred to populations of other ethnic origins (Kouris-Blazos *et al.* 1999) and have also been reproduced in secondary prevention trials (de Lorgeril *et al.* 1994).

Components and characteristic dietary pattern of Mediterranean diets are as follows:

- Fruits and vegetables in large quantities
- Fresh food minimally processed
- Olive oil as main source of fat
- Cereals, mainly whole meal bread and pasta, daily
- Leguminous, seeds, nuts and dry fruits, daily
- Dairy products, mostly fermented, yoghurt and cheese, moderate daily consumption
- Eggs: one to four per week
- Fish – moderate consumption (two to four times per week)
- White meat – moderate consumption (two to four times per week)
- Red meat – consumption in small amount, or sometimes per month
- Wine – moderate consumption, regularly, one to two glasses/day, mostly with meals
- Common use of species and various condiments, lemon, vinegar, garlic, aromatic herbs, pepper, etc.

15.2.1 Mediterranean ecosystems and agriculture

The Mediterranean ecosystems in the world are associated to characteristic agricultural products that have marked repercussions on the health of the populations consuming them. The Mediterranean basin, northern California, central Chile, the Western Cape of South Africa and southwestern Australia – all share a Mediterranean climate characterized by cool, wet winters and hot, dry summer. These five regions have similarities in natural vegetation that are being enhanced by the increasing level of biotic exchange between the regions as time passes since European settlement in each region (Groves 1992). These are the five main terrestrial Mediterranean ecosystems described in the world and constitute the Mediterranean biome.

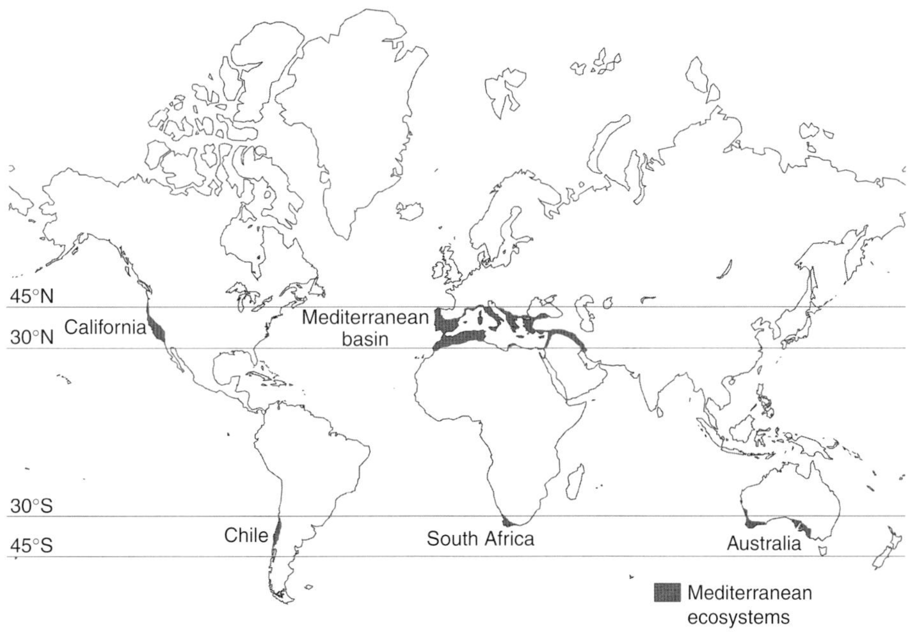

Fig. 15.1 World Mediterranean ecosystems. Five main ecosystems recognized following Köppen (Köppen 1936) climate definitions.

The Mediterranean ecosystems are climatically controlled and have distinctive vegetation. They represent less than 2% of the emerged land, yet they support 20% of the plant species on earth. Mediterranean climate is a dry-summer subtropical type climate as defined by Köppen (1936) who recognized the five regions. The Mediterranean ecosystems are localized on the western coasts of continental landmasses, at latitudes comprised between 30 and 45°, north or south of the equator (Figure 15.1).

Mediterranean climate and Mediterranean ecosystems have a restricted distribution and they are also limited in time since the Mediterranean climate developed for the first time in the Pleistocene. The Mediterranean climate depends to a considerable extent on the existence of cold ocean currents; therefore, it is dependent on their preservation. If world climatic changes eliminate them, the Mediterranean climate will disappear as a transient episode on a geological scale (di Castri 1981).

In these five areas, the Mediterranean biome is characterized by evergreen or drought deciduous shrublands. The Mediterranean maquis in the old world is the same as the chaparral of southern California, the Chilean matorral, South African fynbos and the Australian mallee scrub communities. Also, there is frequently a high degree of endemism in the flora and fauna, due to the limited extent and isolation (almost island-like) of these areas of Mediterranean biome. This isolated condition has advantages for the agriculture, especially in terms of plague control. The proportional distribution of the Mediterranean indigenous flora among the five ecosystems is shown in Figure 15.2.

The conversion to agricultural production, also recognized as the European invasion, has had great impact in the local flora and fauna of Mediterranean ecosystems.

Agricultural production resorts to a tiny fraction of the existing plant and animal species. Pre-European mammal fauna and native vegetation, in all Mediterranean ecosystems, have been profoundly impacted by agriculture with great loss of diversity.

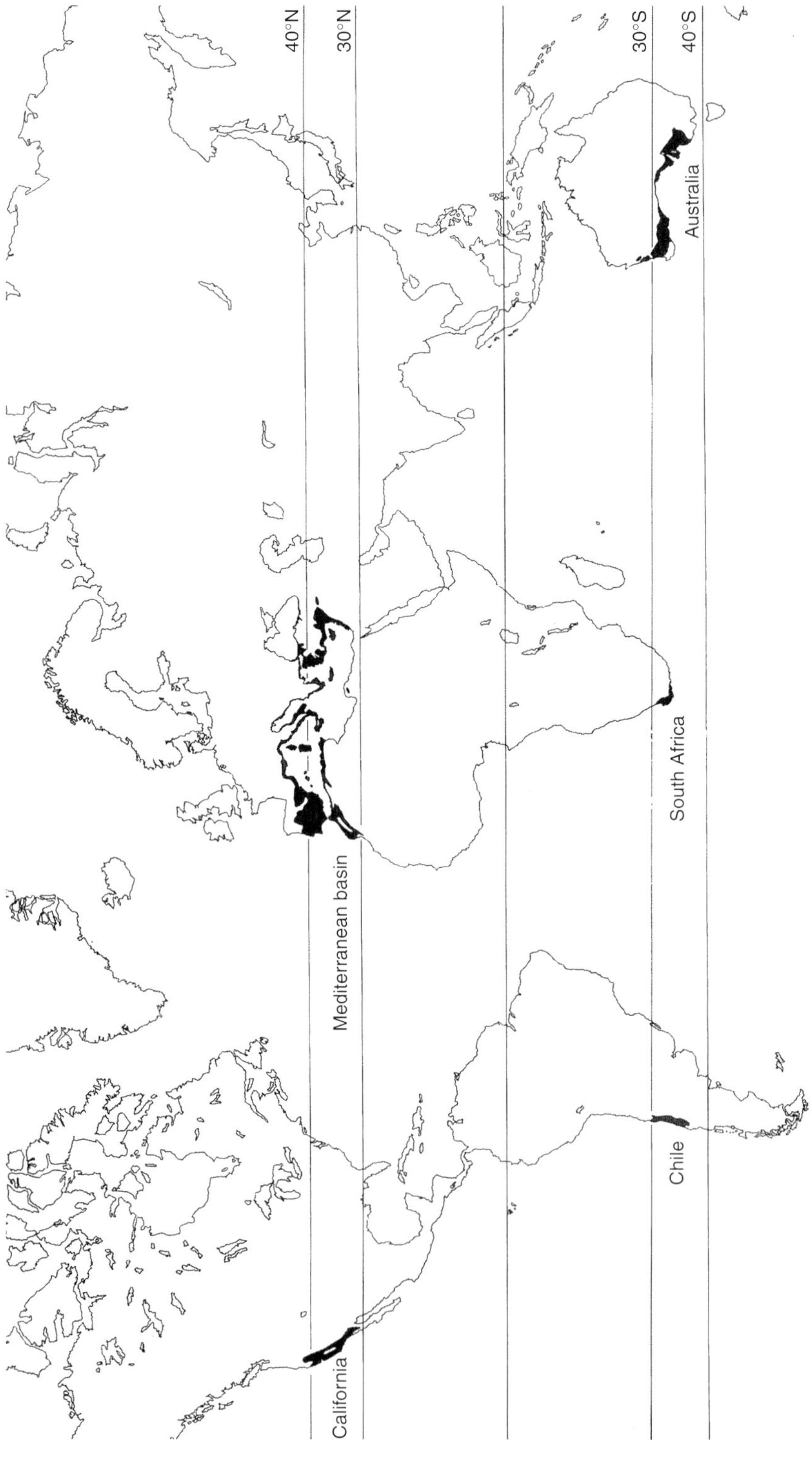

Fig. 15.2 World Mediterranean-type shrublands distribution.

15.2.2 Mediterranean agriculture

15.2.2.1 *Origins of Mediterranean agriculture*

Humans created agricultural settlements around 8500 years BC, in a process that characteristically took place in the Fertile Crescent, the land between the Tigris and Euphrates rivers, in Mesopotamia, that also extends to the Mediterranean Sea, in what corresponds to the eastern part of the Mediterranean basin ecosystem. There is evidence that this process also occurred, approximately at the same time or later, in China, in the Highlands of New Guinea, Mexico, the Andes, the south eastern United States, Ethiopia, Africa's, Sahel zone and tropical West Africa. But none of these original agricultural settlements developed inside Mediterranean ecosystems. These, in fact, would later receive most of its domesticated species from the Mediterranean basin ecosystem.

The process originated in the Fertile Crescent gave birth to Middle Eastern civilization; plant and animal domestication, irrigation and new tools led to the agricultural revolution and the transformation of hunter–gatherers into a more sedentary and socially complex permanent society. Agriculture developed in the narrow swath of hills, extending only from southeast Turkey to western Iran. Food production slowly extended westward, reaching the Mediterranean basin in 2000–3000 years. In contrast to native flora and fauna, species domestication concentrated in very few species. There are approximately 200 000 species of higher plants and some 4000 species of wild mammals around the world. Some 150 mammals could be considered candidates for domestication, actually only 14 became valuable domestic animals; of those 14, 13 were species of the Eurasian continent and one, the llama, the only big mammal animal domesticated in the New World. Around 850 BC, hunter–gatherers of the Fertile Crescent assembled a package of 8 valuable crops, founder crops, providing them with carbohydrate and protein, fat and fiber, and they also assembled a package of 4 valuable domestic animals (cows, sheep, goat, pigs and later horses) that provided them with meat and hides and leather and milk and power (Diamond 1997).

The west–east axis of the Eurasian land mass, as well as of the Fertile Crescent, permitted crops, livestock and people to expand at the same latitude without having to adapt to new day lengths, climates and even diseases. In contrast, the north–south orientation in the Americas, Africa and the Indian subcontinent probably slowed the diffusion of agricultural innovations, accounting for the headstart some societies had on others in the march of human history. For example, wheels were invented in Mesoamerica and llamas were domesticated in the central Andes by 3000 BC, yet 5000 years later the Americas' only beast of burden and only wheels had still not encountered each other. In a similar way, iron metallurgy and pottery reached the Sahel zone, north of the equator, at least as early as they reached Western Europe. However, pottery reached the southern tip of Africa around AD 1 and metallurgy arrived there later, from Europe, on ships (Diamond 1997).

Mediterranean agriculture is defined by its products such as cereals, vegetables, fruit, olive oil and wine. These products are crucial for the Mediterranean diets, together with marine food (Table 15.1). The top ten vegetable foodstuffs in Mediterranean basin are olive, olive oil, garlic, lemon, orange, tomato, chickpeas, eggplant, peppers and fennel. Mediterranean agriculture and geography also determined the type of farm animals, principally sheep and goats, leading to a diet low in red meat and butter, but rich in cheese and yogurt.

Olive oil is perhaps the most representative product of Mediterranean agriculture. The world distribution of olive tree plantations shows excellent correlation with the Mediterranean ecosystems. Only 0.5% of the world area of olive trees occurs in countries without Mediterranean climate (Grigg 2001). Countries from the Mediterranean basin produce 98%

Table 15.1 Main products from Mediterranean and continental agricultural activities

Mediterranean products
- Fruits and fresh vegetables
- Cereals
- Leguminous, nuts and dried fruits
- Wine
- Olive oil
- Fish and seafood

In contrast, *continental agriculture* is defined by the following products:
- Oil seeds
- Bovine meat and milk
- Sugar beet

Source: J. Lama de Espinosa en 'La agricultura mediterránea en el siglo XXI', 2000.

of the olive oil in the world, and consume 90%. Spain, Italy, Greece, Tunisia and Turkey are the main Mediterranean olive oil producers (Table 15.2). The production from the other four Mediterranean ecosystems accounts for approximately 0.5% of world production. However, Australia and Chile have increased tenfold their production in the last decade. In Chile, olive plantations are growing even faster in the last 2–3 years.

The olive was first domesticated somewhere in what is now Syria, Lebanon and Israel, more than 6000 years ago (Zohary 1993). From there it traveled to Egypt, Asia Minor and finally Greece; it was found in Crete and the Greek mainland in the fourth millennium BC. It appears that the Phoenicians introduced it to South Italy and then the entire Mediterranean. By the end of the Roman Empire, olive tree grew, and olive oil was made, throughout the Mediterranean region. The Spanish introduced the olive to America in 1540. The tree was taken northward by Jesuit and Franciscan missionaries in the eighteenth century, and grown in what is now the Baja California region of Mexico and in California in the United States

Table 15.2 World olive oil production by country (1000 tonnes/year)

Country/region	1996–2000	2006
Mediterranean basin	2319.8	2532.7
Spain	891.8	1108.7
Italy	527.5	603.0
Greece	417.6	370.0
Tunisia	183.6	170.0
Turkey	131.0	160.0
Others in the Mediterranean basin[a]	168.3	121.0
California, United States	0.9	1.0
Australia	0.7	8.0
Chile	0.2	2.4
South Africa	1.0	
Canada	0.0	0.0
Mediterranean ecosystem total	2321.5	2545.1
World total	2480.6	2573.5
Mediterranean Ecosystem/total	93.6%	98.%

From the International Olive Oil Council.
[a]Others in the Mediterranean basin: Morocco, Argelia, Portugal, Palestine, Lebanon, Croatia, France, Israel, Serbia and Montenegro, Slovenia and Cyprus.

(Fernandez 1997). The tree was recorded in desert Peru in 1560 and later became established in northern Chile and Argentina, sustained by irrigation in Peru and Argentina (Gade 2000). In Chile, olive production extended from the Limarí Valley in the North to the Bio-Bio River in the South. In 1952, there was some olive plantations, but only in 1990 commercialization strategies started. Now Chile produces only extra virgin olive oil and its present production is approximately 2400 tonnes per year.

Curiously, olive oil has not become part of the diet in the Latin America in spite of the presence of Spaniards and Portuguese. Possibly, olive oil was not an important part of the diet of the early Iberian settlers in their home countries. There is some evidence to suggest that consumption in the past was lower than at present (Grigg 2001). In Spain, the olive industry was neglected especially after the expulsion of the Moriscos in 1609, and experimented expansion only since the 1880s (Bull 1936).

Although there is abundant evidence of the production and consumption of olive oil in ancient Greece and Rome, there is no evidence on the quantities produced and consumed, and what proportion was used for non-food purposes. Also, in the sixteenth century Crete estimated that production would provide per capita consumption much below the present level (Hamilakis 1999).

Wine is also an emblematic product from Mediterranean agriculture. A world map illustrating the principal wine-producing regions in the world looks remarkably similar to Figure 15.1, the map of the Mediterranean ecosystems, with the addition of some other wine-producing regions, mainly Argentina, China and eastern United States. Mediterranean agricultural products come principally, but not exclusively, from Mediterranean ecosystems.

Almost 75% of the world wine production occurs in the Mediterranean ecosystems. Of this production, three-quarters occur in the Mediterranean basin ecosystem and the rest in Australia, South Africa and Chile. As shown in Table 15.3, the contribution from Chile and

Table 15.3 World wine production by country (1000 hL/year)

Country/region	1996–2000	2005
Mediterranean basin	165 165	156 440
France	56 271	52 004
Italy	54 386	50 556
Spain	34 162	34 750
Portugal	6828	7254
Greece	3832	4093
Hungary (southwest)	4126	2900
Croatia	2096	1900
Others in the Mediterranean basin[a]	3464	2983
California, United States	19 457	26 110
Australia	7380	14 000
South Africa	7837	8410
Chile	5066	7890
Canada	403	425
Mediterranean ecosystems total	204 905	212 850
World total	270 826	286 175
Mediterranean ecosystems/total	75.7%	74.4%

From the Wine Institute.
[a]Others in the Mediterranean basin: Macedonia, Slovenia, Morocco, Tunisia, Turkey, Cyprus, Lebanon, Bosnia-Herzegovina and Israel.

Australia is rapidly increasing, while South Africa and Europe maintain or decrease production. Wine shows, just as olives, that from an agricultural point of view, the Mediterranean ecosystems outside of the Mediterranean basin, experience an active process of growth, increasing production of the Mediterranean agriculture products. Among the four, for the time being, South Africa appears to be the slowest to adopt this growth strategy.

Some investigators place the discovery of winemaking, or at least its development, in the southern Caucasus. It is also thought that the domestication of wine grapes (*Vitis vinifera*) initially occurred within this area. The earliest firm evidence for winemaking to date is from the Neolithic period site called Hajji Firuz Tepe, Iran, where a deposit of sediment preserved in the bottom of an amphora proved to be a mix of tannin and tartrate crystals, familiar to anyone who drinks wine from corked bottles today. Hajji Firuz Tepe has been dated to 5400–5000 BC (McGovern *et al.* 1996).

Grape growing and winemaking traveled southward from the Caucasus to Palestine, Syria, Egypt and Mesopotamia. Wine consumption and its socioreligious context then helped to expand winemaking around the Mediterranean.

European grape vines were introduced to America during Columbus' second voyage, but the first attempts to grow grapevines in the Caribbean region failed, mainly due to climatic problems. In 1519, the higher altitude of Mexico allowed the first successful planting of grapevines in America. In the middle of the sixteenth century, grapevine reached Cuzco from Mexico or directly from Spain. And then the vines spread to Chile. Chilean wine exports began at the end of colonial period; the first data correspond to the 5-year period 1784–1789 and the destination was New Spain, as Mexico was known during that period. In 1851, there was a spectacular transformation of the Chilean winemaking industry. Grape growers decided to import from France cuttings of winemaking varieties, Cabernet Sauvignon, Cot, Merlot, Pinot Noir, Sauvignon, Semillon and Riesling. These varieties adapted very well to the Chilean Mediterranean climate and are the only pre-Philoxera clones that exist in the world. Today, Chile is the tenth world wine producer and the fifth world wine exporter (Castellucci 2007).

Wine consumption dates to 5000 years BC. Hippocrates, considered the 'father of medicine', used wine as medicine 2500 years ago. Wine was common in classical Greece and Rome. In medieval Europe, the Christian Church was a staunch supporter of wine, which was necessary for the celebration of the Catholic mass, whereas wine consumption was forbidden in the Islamic civilization. Strikingly, some Muslim chemists pioneered the distillation of wine, which was used for other purposes, including cosmetic and medical uses. In Greece, the earliest evidence of winemaking is a stone foot press at Vathipetro, a Minoan villa on Crete, dated to 1600 BC.

Today, the highest average consumption of wine, per capita and per year, is traditionally seen in the producer countries; however, these countries, in general, are undergoing a decline in their individual levels of yearly consumption; in 2004, France 55 liters per capita; Italy 49 liters per capita; Portugal 48 liters per capita; Argentina 29 liters per capita; Spain 34 liters per capita; Chile 16 liters per capita; Greece 30 liters per capita; and Romania 26 liters per capita. In other countries traditionally producers and consumers of wine, a stabilization or growth in their individual consumption is observed. It is the case for Germany and Austria, stabilized at levels close to 24 and 30 liters per capita; Switzerland close to 42 liters per capita; Hungary with 31 liters per capita; and Croatia with 42 liters per capita. In North America, consumption levels have increased more significantly since 2002, to reach 8.2 liters per capita in the United States and 11.4 liters per capita in Canada. In Australia, consumption reached 21.9 liters per capita in 2004, and in South Africa stabilization was observed in 2004

Table 15.4 World fruits and vegetables production by country (1000 tonnes/year)

Country/region	1989–1991	2004
Mediterranean basin	139 295	163 834
Turkey	27 080	36 046
Italy	32 004	34 276
Spain	24 529	29 401
France	18 189	19 843
Morocco	5223	7810
Greece	8076	7782
Algeria	2900	5151
Others in the Mediterranean basin[a]	21 294	23 525
California, United States	35 387	43 572
South Africa	5801	7769
Chile	4539	7407
Australia	3850	5486
Canada	2776	3228
Mediterranean ecosystem total	188 871	228 069
World total	812 733	1 383 649
Mediterranean ecosystem/total	23.2%	16.5%

Source: FAO Stat, California Agricultural Resource Directory 2006.
[a]Others in the Mediterranean basin: Portugal, Hungary, Tunisia, Israel, Lebanon, Yugoslavia, Serbia and Montenegro, Bosnia and Herzegovina, Croatia, Macedonia, Slovenia, Palestine and Cyprus.

with 7.8 liters per capita. In contrast, countries that are not wine producers show, in general, a growing trend of consumption, as in the United Kingdom (18 liters per capita in 2004), Norway, Finland and Iceland (OIV 2004).

On the whole, vegetables and fruits are central to Mediterranean agriculture and Mediterranean diet, yet in contrast to olives and wine, vegetables and fruits are produced in various agricultural ecosystems. The Mediterranean ecosystems accounted for 23% of the world production in 1989–1991 and only 16% in 2004. Production increases more rapidly outside the Mediterranean ecosystems with some remarkable exceptions, such as Chile, Australia, Algeria and Morocco (Table 15.4). Again, as for olives and wine, the Mediterranean ecosystems away from the Mediterranean basin experiment an accelerated growth, clearly exemplified by the growth of production in Chile. The productivity of various countries, expressed as food production per capita, is growing mostly outside the Mediterranean basin, with the exception of Bosnia and Herzegovina and Macedonia, which also exhibit marked growth (Table 15.5). Again, the dynamic development of the four extra Mediterranean basin ecosystems can be appreciated (Figure 15.3).

The Mediterranean diet is consequence of a global exchange of domesticated animal and plant species and the gastronomic art of Mediterranean people. They devised agricultural, cooking and gastronomic practices that rendered them attractive and healthy, and a key element in their culture.

The top ten foods Mediterranean rim countries have in common are olives, olive oil, garlic, lemons, oranges, tomatoes, chickpeas, eggplants, fennel and sweet peppers. The Mediterranean food basket is the result of domesticated food species exchanged mainly between America, Europe, north of Africa and Asia.

Some emblematic foods such as potato and tomato are native to America, specifically from the Andes (Ecuador, Peru and Chile). Common beans and green beans, sweet and hot peppers are native from Mexico, Central America, Bolivia and Peru. Corn is from Mexico and north of Central America.

Table 15.5 Per capita food production

Country	1996	2005
Bosnia and Herzegovina	95.77	116.61
Chile	99.96	113.20
Canada	97.23	107.54
Macedonia	97.94	104.29
Hungary	99.51	105.24
Australia	91.64	96.85
Israel	99.88	105.49
South Africa	102.12	105.79
United States	97.92	100.38
Turkey	103.33	103.94
Croatia	90.98	90.39
Italy	98.88	97.80
Slovenia	104.04	102.42
Spain	93.56	90.70
Greece	99.01	95.00
Morocco	121.41	115.37
France	102.03	92.80
Portugal	100.69	90.95
Tunisia	124.07	104.18
Lebanon	117.39	97.96
Cyprus	107.59	86.04

From FAO Stat, FAO indices of agricultural production relative to the base period 1999–2001. Disposable production for any use except as seeds and feed.

From north of Africa are broad beans and artichoke. Chicken and rice are from China; citrus from Southeast Asia; garlic and onion from Central Asia. Eggplants are from India, lot of spices from Asia and other places.

Also, other typical vegetables in the Mediterranean diet as cabbage, fennel, parsley, oregano, cilantro, asparagus and borage are originating in the Mediterranean region.

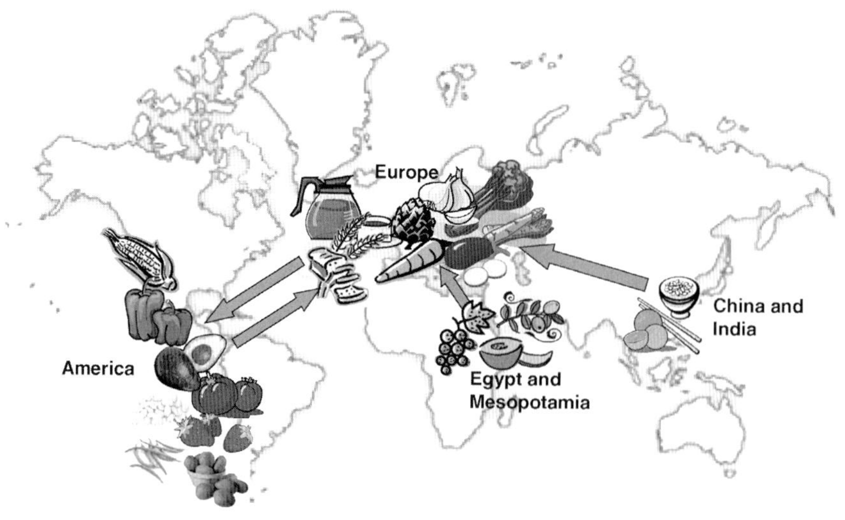

Fig. 15.3 Mediterranean diet: integrated by animal and plant food species which originated in different regions of the earth.

15.3 Some health mechanisms of the Mediterranean diet

The Mediterranean diet or diets, as originally defined, are associated with lower rate of all-causes mortality, coronary heart disease, cardiovascular disease and cancer. The benefits of Mediterranean diets have been attributed principally to a large consumption of antioxidants, especially polyphenols, provided by vegetables, fruits, extra virgin olive oil and wine; and to the type of fat, especially monounsaturated fat from virgin olive oil, and omega-3 polyunsaturated fatty acids (PUFA) from vegetables and fish (Trichopoulou *et al.* 1999; Simopoulos 2001). In the case of wine, both ethanol and polyphenols are considered the bioactive components with regard to health effects especially to cardiovascular risk factors. Important aspects in the understanding of the biological effects of Mediterranean diets are endothelial function, antioxidants, monounsaturated fatty acids (MUFA)/PUFA ratio, omega-6/omega-3 ratio and postprandial control.

15.3.1 Endothelial function

The vascular endothelium is the thin layer of cells that cover the interior surface of blood vessels, forming the first barrier between circulating blood in the lumen and the rest of the vessel wall. Endothelial cells line the entire circulatory system, from the heart to the smallest capillary. It is not a simple limit between circulating blood in the lumen and the smooth muscle cells of the vessel wall; it modulates the structure and vascular function. In normal conditions it works in favor of vasodilatation and platelet antiaggregation. Hypertension, diabetes, hypercholesterolemia, oxidized low-density lipoprotein (LDL), hyperhomocysteinemia, estrogen deficit and smoking damage endothelial cells and drive to endothelial dysfunction, initiating atherosclerosis.

Endothelial dysfunction characterized by loss of endothelium-dependent vasodilatation occurs early in vascular disease, before any morphologically evident alteration can be detected. The initial cause is decreased production, increased degradation or decreased sensitivity to nitric oxide.

Nitric oxide is a gas produced in the endothelial cells by an enzyme named endothelial nitric oxide synthase (eNOS). It diffuses into medial smooth muscle cells causing vasodilatation, and into the blood causing inhibition of platelet aggregation and monocyte adhesion. Reactive oxygen species, generated for example from oxidized LDL or smoking, oxidize nitric oxide in the vessel wall. So decreased nitric oxide bioavailability occurs, explaining oxidative stress as a direct determinant of endothelial dysfunction.

Endothelial dysfunction can be detected by the dilatory response of a peripheral vessel, as the brachial artery, measured by an ultrasonic probe, after the flow of blood that accompanies the release of a sphygmomanometric cuff temporal blockade of circulation, a flow-mediated vascular reactivity response.

The eNOS enzyme that produces nitric oxide in endothelial cells is a principal mediator of the positive health effects of Mediterranean diet and wine (Leighton *et al.* 2006). Other factors which are capable of enhancing eNOS function are antioxidants, omega-3 fatty acids, exercise, high-density lipoprotein (HDL) and estrogens. Some depressing factors are oxidative stress, smoking, homocysteine, oxidized LDL, high-fat diets, triglycerides, free fatty acids, abdominal fat, hypertension, insulin resistance and diabetes. Factors which are capable of enhancing or depressing the function of eNOS produce their effect via gene expression, enzyme regulation, substrate availability, product stability or others. Depressed eNOS would result in metabolic syndrome, atherosclerosis, thrombosis, inflammation and vasoconstriction.

15.3.2 Antioxidants

A good example of the harmful effect of smoking and the beneficial effect of antioxidants on endothelial function was published by Papamichael *et al.* (2004). They measured the flow-mediated dilation (FMD) in the brachial artery of volunteers before and 15, 30, 60 and 90 minutes after they smoked a single cigarette, or smoked and drank 250 mL of red wine, or smoked and drank 250 mL of dealcoholized red wine. They found that acute smoking of one cigarette caused a reduction in FMD ($P < 0.001$), which was statistically significant 15, 30 and 60 minutes after the inhalation of smoke compared to baseline levels. However, simultaneous ingestion of either red wine or dealcoholized red wine while smoking almost completely prevented the change in FMD. So smoking caused a significant impairment in endothelial function, and consumption of red wine or dealcoholized red wine with smoking decreased smoke's harmful effect on endothelium. Vitamin C and a diet rich in antioxidant and wine as Mediterranean diet protect endothelial function (Cuevas *et al.* 2000).

Fruits, vegetables, tea, chocolate and wine are good sources of antioxidants as vitamin C, vitamin E, carotenoids, glutathione, lycopene and polyphenols. Antioxidants diminish oxidative stress in the human body and protect from all the chronic diseases and conditions that are associated with it such as atherosclerosis, cancer, hypertension, metabolic syndrome, Alzheimer disease and the natural aging progression.

15.3.3 MUFA/PUFA ratio, omega-6/omega-3 ratio

It is now clear that there is not one Mediterranean diet. There are several Mediterranean nations with varied cultures, traditions and dietary habits. Their diets differ in amount of total fat, amount of olive oil, kind of meat and wine intake, milk versus cheese intake, fruits and vegetables. However, beyond the apparent differences, some nutritional characteristics are common to all or most of the diets in Mediterranean region; that are, low in saturated fat and high in monounsaturated fatty acids (olive oil), balanced in omega-6/omega-3 fatty acids ratio, low in animal protein and rich in antioxidants and fiber.

Nowadays, it is well established that the principal dietetic cause of high blood cholesterol is the consumption of saturated fat. People that eat diets rich in butter and red meat have elevated blood LDL cholesterol and increased risk of coronary heart disease. On the other hand, people who eat diets rich in MUFA (virgin olive oil) have lower LDL cholesterol level and less risk of coronary heart disease. Also, virgin olive oil has abundant phytochemicals with antioxidants, antithrombotic, anti-inflammatory and antiatherogenic properties that protect the organism from oxidative damage and chronic diseases.

The balance in omega-6/omega-3 fatty acids ratio especially from vegetable origin is an important point to consider. Extensive studies on the traditional diet of Greece show that because of consumption of wild plants and eggs and meat from grazing poultry and animals, the omega-3 fatty acids were found throughout the food chain, giving an omega-6/omega-3 fatty acids ratio, similar to the ratio of the Paleolithic diet. On the basis of estimates from studies on Paleolithic nutrition and modern-day hunter–gatherer populations, it appears that human beings evolved consuming a diet that contained small and roughly equal amounts of omega-6 and omega-3 polyunsaturated fatty acids. The current Western diet is very high in omega-6 fatty acids because of agricultural practices and the recommendation to substitute saturated fats by omega-6 fatty acids to lower serum cholesterol concentrations. The vegetable oils such as corn, sunflower, safflower, cottonseed oil are all very rich in linoleic acid – vegetable omega-6 fatty acid and margarines made from them contain trans fatty acids that behave like saturated fat in terms of raising blood cholesterol while lowering HDL. It is now recognized that dietary linoleic acid favors oxidative modification of LDL cholesterol and increases platelet response to aggregation. In contrast, α-linolenic fatty acid intake (vegetable omega-3 fatty acid from canola oil, flax and nuts) is associated with inhibitory effects on the clotting activity of platelet and on the regulation of arachidonic acid metabolism. In clinical studies, α-linolenic fatty acid contributes to lower blood pressure, and a prospective epidemiological study (the Lyon Diet Heart study) showed that α-linolenic fatty acid is inversely related to the risk of coronary heart disease in men (de Lorgeril *et al.* 1994).

The health advantage of the consumption of long-chain omega-3 fatty acids – eicosapentaenoic acid and docosahexaenoic acid – from fish is well documented. Demonstrated effects are reduction in susceptibility to cardiac arrhythmia, diminished blood clotting, enhanced endothelial function, lower blood pressure and decreased triglycerides levels.

15.3.4 Control of postprandial risk factors

Accumulating data from multiple lines of evidence suggest that exaggerated elevation in blood glucose and triglycerides after a meal, postprandial hyperglycemia and postprandial hyperlipemia, respectively, named postprandial dysmetabolism, is an important and fundamental disturbance involved in the genesis of inflammation, endothelial dysfunction, hypercoagulability and sympathetic hyperactivity. Postprandial dysmetabolism is an independent predictor of future cardiovascular events in non-diabetic individuals. Highly processed, calorie-dense and easily digestible foods cause abnormal surges in blood glucose and triglycerides level that increase free radical and oxidative damage. Diets high in fiber, vegetables and fruits, whole grains, legumes and nuts, as Mediterranean diets, moderate the post-meal increase in glucose, triglycerides and inflammatory mediators.

The amount and type of carbohydrate consumed with a meal is a determinant of the magnitude of postprandial glucose excursion. Equally important is the additional food; for instance, recent studies show that almonds, pistachios or peanuts, when eaten with white bread or mashed potatoes, will reduce the postprandial glucose excursion by approximately 30–50%. Also, nuts lower post-meal oxidative stress and additionally provide antioxidants.

Another observation is that a mixture of vinegar and olive oil, the traditional salad dressing used in the Mediterranean diet, reduces post-meal glycemia, probably because acetic acid slows gastric emptying and thus delays carbohydrate absorption and improves satiety.

Red wine consumption with a meal mitigates plasma postprandial lipoperoxides. Volunteers ate test meal consisting of 'Milanese' meat and fried potatoes, with 400 mL of red wine or with an isocaloric hydroalcoholic solution. Those who drank the isocaloric hydroalcoholic solution presented postprandial LDL more susceptible to oxidation than red wine drinkers. So red wine counteracts the oxidative stress induced by a meal. Also ethanol, in general, and especially red wine reduced post-meal glucose by 38% when consumed with white bread.

Physical activity enhances insulin sensitivity and lowers postprandial glucose and triglyceride levels in a dose-dependent fashion. Exercise improves inflammation directly by lowering postprandial glucose, and indirectly by reducing excess abdominal fat.

15.4 Mediterranean diet and gastronomy

15.4.1 The way of cooking and the art of food combination

The Mediterranean diet is a varied one, and it includes lots of foods. No foods are prohibited and no ones are obligatory – the clue is the frequency of consumption of those foods. Some foods are preferred, such as fruits and vegetables, cereals and legume, fish and white meat. Others are eaten in moderation, such as dairy products principally yogurt and cheese. Red meat, butter and sugar are consumed in low quantity. Olive oil is their principal source of fat, and wine accompanies the meals. People in Mediterranean culture eat and drink in moderation and enjoy the meal.

But the special characteristic of the Mediterranean diet is that it is healthy, savory and delicious. Dishes are attractive and appetizing and that make eating a pleasure. The approach to cooking consists in using different products that complement nutritional and gastronomic requirements. Additionally, the meal makes you feel pleasure and satiety, the reward after work.

Diets are not a list of products to eat; more important are the way of cooking and the combination of those products.

The stomach is a prime location for interaction of compounds as high-fat foods that contain lipid oxidation products, with other food constituents. The stomach acts as a bioreactor and the gastric fluid as a medium for further dietary lipid peroxidation and/or antioxidation (Kerem *et al.* 2006). The gastrointestinal tract is constantly exposed to dietary oxidized food compounds produced during the processing and storage of foods or during their digestion in the stomach (Halliwell *et al.* 2000). High-fat, high-cholesterol foods containing oxidized products affect endogenous lipoprotein production and catabolism, and lead to transient exposure of arteries to lipid oxidation products. Humans have been shown to excrete increased amounts of malondialdehyde (MDA, a marker of lipid oxidation) in their urine after ingestion of oxidized fats. So the stomach and the gastrointestinal tract may be a main biological site of action for polyphenolic antioxidants in general.

In a randomized, crossover study, the effect of red wine polyphenols on postprandial levels of plasma and urine MDA was investigated, when red wine polyphenols were used in the preparation of meals of turkey cutlets. The control was red turkey meat cutlets cooked on an electric grill for 6 minutes until well done, and volunteers ate them supplemented by water. Three meals of 250 g turkey cutlets were tested: control, prepared without wine poliphenols and accompanied by water as the only drink; another consisting in cooking the meat, afterwards soaked in red wine polyphenols plus 200 mL of red wine as drink; and a third meal consisting in turkey soaked in red wine polyphenols prior to cooking plus 200 mL of red wine as drink with the meal. Results showed that the meal of turkey meat cutlets soaked in red wine polyphenols after cooking reduced 75% the level of plasma MDA compared with control. However, when the meat was cooked with red wine polyphenols, the elevation of plasma MDA produced by the control meal was completely prevented. This inhibition of MDA absorption by polyphenols could be due to the prevention of lipid peroxidation during cooking and in the stomach medium, and it could also be due to the formation of an adduct between polyphenols and aldehydes, which may prevent the absorption of these compounds in humans. This result is an example of the importance of the way of using food ingredients. Cooking meat with polyphenols can preserve lipids from oxidation. Drinking wine can preserve lipids from oxidation in the digestive system, before absorption.

15.4.2 The spices and herbs

Another characteristic of the Mediterranean diet is the use of spices and herbs for preparing meals. Species are powerful antimicrobial agents, and this was a reason why they were used in the past. A demonstration of this is the positive correlation between the number of spices used in different countries and annual temperature, an indicator of relative spoilage rates of unrefrigerated foods (Billing & Sherman 1998).

Vegetables, including herbs and spices, are the most important sources of phenolics with antioxidant capacity in the Mediterranean diet. Between herbs, garden thyme, garden sage, rosemary and marjoram have the highest concentration of phenolics and the highest antioxidant capacity value in an Italian study (Ninfali *et al.* 2005). The values are many times higher than those of the vegetables reported in that study, such as broccoli, garlic, leek and fennel. Cumin and fresh ginger are the top in the list of spices in this study, and in other study, ground clove, ground ginger, ground cinnamon, turmeric power and ground yellow mustard seeds are the first one (Ninfali *et al.* 2005).

So a main benefit of consuming plant polyphenols in the human diet as an integral part of the meal may arise from their ability to prevent generation and absorption of cytotoxic lipid oxidation products, such as reactive carbonyls or other reactive compounds commonly found in our foods. The use of spices and herbs in foods preparations, as seasoning to impart flavor, protects from oxidation. This protection can begin during the storage of food, continuing during the preparation and cooking depending on the culinary procedure and ingredients used. Polyphenols in the gastrointestinal tract, in blood and in the cells act as antioxidant, antiplatelet, anti-inflammatory, anticancer, antiaging, and so on.

Other important point is the portion size. Spices and herbs are used in small quantity, but their antioxidant capacity is so high that can contribute significantly to a meal. Ninfali *et al.* quantify the contribution of 3 g of marjoram to the total antioxidant capacity of a salad composed of 75 g of lettuce and 122 g of tomatoes (Ninfali *et al.* 2005). In the study they reported a contribution of 33% to the total antioxidant capacity and a 27% to total polyphenol from marjoram to the salad portion.

15.4.3 Cooking practices

The effect of common cooking practices – boiling, steaming and frying – on phytochemical contents (polyphenols, carotenoids, glucosinolates and ascorbic acid) and total antioxidant capacities has been investigated. Results suggest that for each vegetable a cooking method would be preferred to preserve the nutritional and physicochemical qualities. Water-cooking treatments better preserved carotenoids in vegetables, and ascorbic acid in carrots and courgettes (Miglio *et al.* 2008). But for phenolic compounds, steaming and frying are better than boiling to preserve them into the vegetal products. Also, the extent of heat treated is another factor to be considered – extended treatment destroys the compounds (Furniss *et al.* 2008). There has been observed an overall increase of total antioxidant capacity values in cooked tomatoes, carrots, courgettes and broccoli by boiling, steaming or frying, probably because of matrix softening and increased extractability of compounds, which could be partially converted into more antioxidant chemical species.

Tomatoes and tomato products are typical components of the Mediterranean diet and these are a major dietary source of lycopene. Several studies have reported that lycopene from fresh tomatoes or tomato juice is poorly absorbed in comparison with lycopene from processed tomatoes as tomato paste. Other studies have reported that boiling tomato juice in corn oil increased the uptake of lycopene compared with straight tomato juice (Fielding *et al.* 2005). Finally, the consumption of diced tomatoes cooked with olive oil as Mediterraneans do resulted in higher plasma lycopene concentrations than consumption of diced tomatoes cooked without olive oil. So the addition of olive oil to diced tomatoes during cooking greatly increases the absorption of lycopene. The results highlight the importance of how a food is prepared and consumed in determining the bioavailability of dietary carotenoids such as lycopene (Fielding *et al.* 2005).

Traditionally, Mediterranean start with olive oil to sauté onions, garlic, parsley, oregano, and fresh tomato or tomato paste is used. This recipe has differences in different countries. To this zucchini, potatoes, rice, eggplant, meat, fish or beans may be added. This way of cooking has the characteristic of using a variety of food all at once that provides lot of antioxidants to preserve lipids, protein and carbohydrate from oxidation. Some antioxidants from vegetables dissolve into the water–lipid interface, contributing to the protection of lipid meat oxidation during cooking. Others may get more available to the gastrointestinal digestion, protecting oxidation during this process. And the rest get into the human body protecting from chronic disease.

How much the Mediterranean approach of cooking contributes to the health is not resolved. So Mediterranean cuisine continues to be a mystery, but there is no doubt that it is good for your well-being.

15.5 Mediterranean diet 'food at work' intervention

In Chile, one-fifth of the workers, approximately 1.2 million, receive food at the workplace, predominantly lunch. With the purpose of evaluating the feasibility of Mediterranean diet food at work interventions in Chile and in general, we performed a pilot study (Leighton *et al.* 2009).

A small industry was selected (a heavy machinery maintenance and repair workshop) with approximately 180 workers, 88% of whom were men. The industry owners and administrators, the local workers union and the canteen contractor accepted to collaborate for the intervention. The canteen contractor, who received US$3.75 for each lunch, accepted to modify the food offered, with emphasis on diet Mediterranisation, meanwhile keeping the option of the previous diet for those not convinced to make the change. The participants were informed about Mediterranean diets and health, and were informed of the relevance of the metabolic syndrome as a risk condition for chronic diseases. The metabolic syndrome is a cluster of metabolic components, associated with a high risk of cardiovascular disease (Grundy *et al.* 2004). The National Cholesterol Education Program's Adult Treatment Panel III clinical definition of the metabolic syndrome requires the presence of at least three abnormal risk factors, among a set of five: abdominal obesity, high plasma triacylglycerol, low plasma HDL, high blood pressure and high fasting plasma glucose. All the employees or workers were invited to participate, and 90% of them accepted and 68% of these completed all the controls programmed for the 12-month intervention period. At months 0, 4, 8 and 12, clinical, biochemical, anthropometric and nutritional evaluations were performed. A specially designed Mediterranean diet score that evaluates 14 items was applied by nutritionists, giving an overall measure of the evolution of dietary habits, since it covered food at work as well as food during the rest of the day and during holidays. As shown in Figure 15.4, there is an apparent continuous improvement in the degree of Mediterranisation of the workers' diets during the intervention; median score values for months 0, 4, 8 and 12 were 4.8, 5.5,

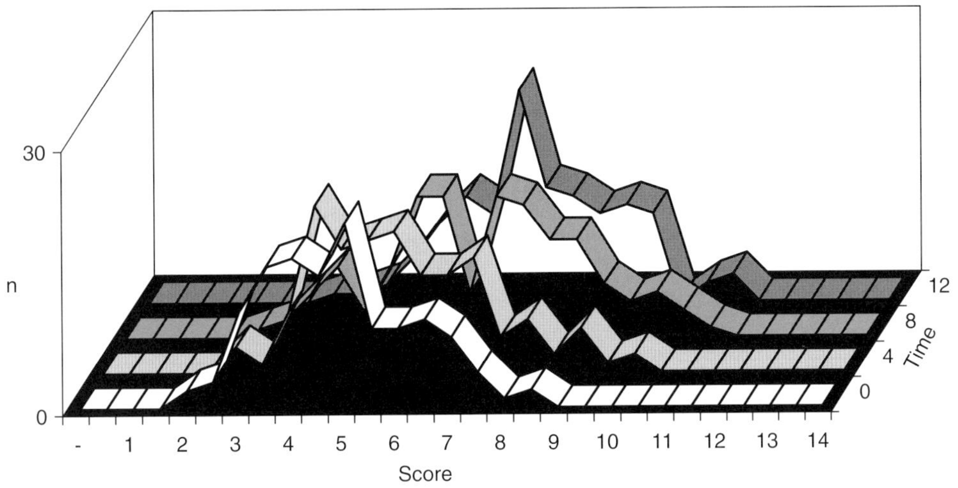

Fig. 15.4 Evolution of the Mediterranean diet score in a food at work intervention study.

7.0, and 7.0, respectively. Mean score values ± SD for these time points were 4.8 ± 1.4, 5.6 ± 1.6, 6.7 ± 1.9 and 7.4 ± 1.5. From 3-day food records, macronutrient intake and other parameters were measured. As shown in Table 15.6, after 12 months there was a slight decrease in caloric intake and significant increases in fiber consumption and in the ratios of monounsaturated/saturated fatty acids, but a decrease in the ratio of omega-6/omega-3 fatty acids. This is consistent with the Lyon Heart Study based on a modified diet of Crete and a decrease in the omega-6/omega-3 fatty acid ratio (Renaud *et al.* 1995). Although the various scores to evaluate the Mediterranean dietary pattern have not included the omega-6/omega-3 fatty acid ratio and its importance in reducing the risk of chronic diseases, especially coronary heart disease, the study by Simopoulos (2001), Renaud *et al.* (1995) and Ambring *et al.* (2006), as well as the work of Guebre-Egziabher *et al.* (2008), indicates that the dietary omega-6/omega-3 fatty acid ratio should be evaluated in all studies on Mediterranean diets.

Lunch at the canteen was the opportunity to make objective changes in the food offered to the workers, and it was also the opportunity to illustrate the theoretical concepts of healthy diet that were repeatedly explained to them. As shown in Table 15.7, the intervention was successful in modifying habits; fruit and vegetable consumption markedly increased while red meat consumption decreased almost by two-thirds; whole meal bread was offered and

Table 15.6 Macronutrients consumption, from 3-day food records, in male volunteers

	Month 0	Month 12	*P* value for difference
Calories	2251 ± 656	2091 ± 550	0.046
Carbohydrates (%)	54.2	54.3	
Proteins (%)	13.8	14.4	
Total fats (%)	31.9	31.3	
Fiber (g)	19.5 ± 6.2	21.5 ± 7.7	0.011
Monounsaturated/saturated fat ratio	1.1 ± 0.3	1.5 ± 0.5	0.0001
Omgega-6/omega-3 fat ratio	23.9 ± 12.0	9.8 ± 6.2	0.0001

Table 15.7 Main changes in average daily food consumption (lunch time) in the workers' canteen in 12 months of intervention

Food	Month 0[a] (g/person per day)	Month 12[a] (g/person per day)	Change
Vegetables (without potatoes)	174.5	264.2	+51%
Fruits	47.2	106.2	+125%
Cereals	44.9	61.7	+37%
White bread	80.1	34.0	−57%
Whole bread	0	32.8	+
Red meat	103.8	38.5	−63%
White meat	53.1	68.8	+29%
Fish	2.8	21.0	+750%
Canola oil	0	21.3	+
Olive oil	0	6.5	+

[a] Average daily consumption per person, recorded daily for 4 weeks, during the first and the last month of intervention.

reached 49% of bread consumption; fish consumption showed a large relative increase, starting from a very small value, and canola and olive oil were introduced in the diet. The evidence for a change in the alimentary pattern of the workers, a Mediterranisation of their diet, was accompanied by a reduction in the risk factors that define metabolic syndrome. As shown in Figure 15.5, after 12 months of intervention, the proportion of workers without any risk factor doubled and those with metabolic syndrome decreased by at least one-third or more if those under treatment for hypertension are not considered.

The food at work intervention, made with the purpose of obtaining a Mediterranisation of the worker's diet, was successful. The Mediterranisation was done not only in the type of food but also in the way of cooking. The menus were designed every week according to the frequency of consumption of food in the Mediterranean diet and choosing recipes from the Mediterranean approach of cooking adapted to the Chilean reality. The diet improved

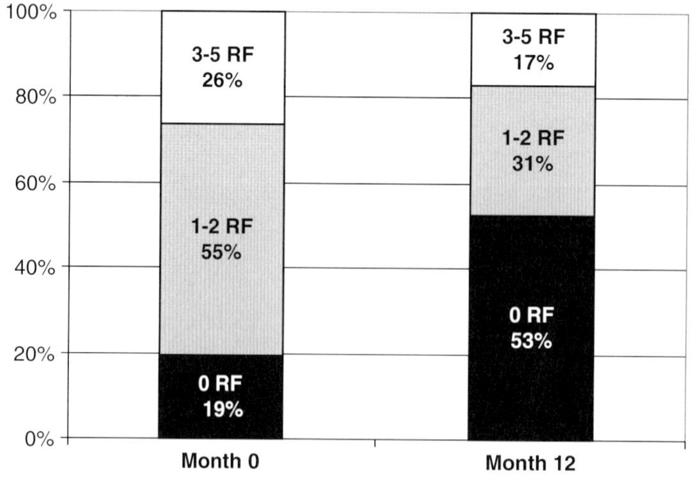

Fig. 15.5 Metabolic syndrome components present in volunteers at the beginning and after completion of the 12 months intervention study.

at the canteen, and also the global diet of the workers improved and the change in diet was accompanied by a marked decrease in the prevalence of cardiovascular risk factors, those that define the metabolic syndrome. Workers were grateful by the appetizing meals and their enhanced health. Clearly, Mediterranisation of the diet is a feasible objective, and the observed reduction in risk factors supports the hypothesis that it is a powerful tool to reduce chronic diseases in the population. Another very relevant conclusion is that adults are valid targets for chronic disease preventive measures.

References

Ambring, A., Johansson, M., Axelsen, M., Gan, L., Strandvik, B. & Friberg, P. (2006). Mediterranean-inspired diet lowers the ratio of serum phospholipid n-6 to n-3 fatty acids, the number of leukocytes and platelets, and vascular endothelial growth factor in healthy subjects. *American Journal of Clinical Nutrition*, **83**, 575–581.

Bach, A., Serra-Majem, L., Carrasco, J.L., Roman, B., Ngo, J., Bertomeu, I. & Obrador, B. (2006). The use of indexes evaluating the adherence to the Mediterranean diet in epidemiological studies: a review. *Public Health Nutrition*, **9**, 132–146.

Billing, J. & Sherman, P.W. (1998). Antimicrobial functions of spices: why some like it hot. *The Quarterly Review of Biology*, **73**, 3–49.

Bull, W.E. (1936). The olive industry of Spain. *Journal of Economic Geography*, **12**, 136–154.

Castellucci, F. (2007). *World Vitiviniculture in 2006*. 5th General Assambly of OIV, Budapest.

Cuevas, A.M., Guasch, V., Castillo, O., Irribarra, V., Mizon, C., San Martin, A., Strobel, P., Perez, D., Germain, A.M. & Leighton, F. (2000). A high-fat diet induces and red wine counteracts endothelial dysfunction in human volunteers. *Lipids*, **35**, 143–148.

Davidson, S.S. & Passmore, R. (1979). *Human Nutrition and Dietetics*. Churchill Livingstone, Edinburgh.

de Lorgeril, M., Renaud, S., Mamelle, N., Salen, P., Martin, J.L., Monjaud, I., Guidollet, J., Touboul, P. & Delaye, J. (1994). Mediterranean alpha-linolenic acid-rich diet in secondary prevention of coronary heart disease. *Lancet*, **343**, 1454–1459.

di Castri, F. (1981). *Mediterranean-Type Shrublands*. Elsevier, Amsterdam.

Diamond, J. (1997). *Guns, Germs and Steel: The Fates of Human Societies*. Norton, New York.

Fernandez, A.G. (1997). Table olives. In: *Production and Processing*. Chapman and Hall, London.

Fielding, J.M., Rowley, K.G., Cooper, P. & O'Dea, K. (2005). Increases in plasma lycopene concentration after consumption of tomatoes cooked with olive oil. *Asia Pacific Journal of Clinical Nutrition*, **14**, 131–136.

Furniss, C.S., Bennett, R.N., Bacon, J.R., LeGall, G. & Mithen, R.F. (2008). Polyamine metabolism and transforming growth factor-beta signaling are affected in Caco-2 cells by differentially cooked broccoli extracts. *Journal of Nutrition*, **138**, 1840–1845.

Gade, D.W. (2000). *South America*. Cambridge University Press, Cambridge.

Grigg, D. (2001). Olive oil, the Mediterranean and the world. *Geojournal*, **53**, 163–172.

Groves, R.H. (1992). *Biogeography of Mediterranean Invasions*. Cambridge University Press, Cambridge.

Grundy, S.M., Brewer, H.B. Jr., Cleeman, J.I., Smith, S.C., Jr & Lenfant, C. (2004). Definition of metabolic syndrome: report of the National Heart, Lung, and Blood Institute/American Heart Association conference on scientific issues related to definition. *Circulation*, **109**, 433–438.

Guebre-Egziabher, F., Rabasa-Lhoret, R., Bonnet, F., Bastard, J.P., Desage, M., Skilton, M.R., Vidal, H. & Laville, M. (2008). Nutritional intervention to reduce the n-6/n-3 fatty acid ratio increases adiponectin concentration and fatty acid oxidation in healthy subjects. *European Journal of Clinical Nutrition*, **62** (11), 1287–1293.

Halliwell, B., Zhao, K. & Whiteman, M. (2000). The gastrointestinal tract: a major site of antioxidant action? *Free Radical Research*, **33**, 819–830.

Hamilakis, Y. (1999). Food technologies of the body: the social context of wine and oil production and consumption in Bronze Age Crete. *World Archaeol*, **31**, 38–54.

Harriss, L.R., English, D.R., Powles, J., Giles, G.G., Tonkin, A.M., Hodge, A.M., Brazionis, L. & O'Dea, K. (2007). Dietary patterns and cardiovascular mortality in the Melbourne Collaborative Cohort Study. *American Journal of Clinical Nutrition*, **86**, 221–229.

Kerem, Z., Chetrit, D., Shoseyov, O. & Regev-Shoshani, G. (2006). Protection of lipids from oxidation by epicatechin, trans-resveratrol, and gallic and caffeic acids in intestinal model systems. *Journal of Agricultural and Food Chemistry*, **54**, 10288–10293.

Keys, A. (1970). Coronary heart disease in seven countries. *Circulation*, **41**, 1–211.

Keys, A. (1995). Mediterranean diet and public health: personal reflections. *American Journal of Clinical Nutrition*, **61**, 1321S–1323S.

Knoops, K.T., de Groot, L.C., Kromhout, D., Perrin, A.E., Moreiras-Varela, O., Menotti, A. & van Staveren, W.A. (2004). Mediterranean diet, lifestyle factors, and 10-year mortality in elderly European men and women: the HALE project. *The Journal of the American Medical Association*, **292**, 1433–1439.

Köppen, V. (1936). *Das Geographische System der Klimate*. Gebrüder Borntraeger, Berlin.

Kouris-Blazos, A., Gnardellis, C., Wahlqvist, M.L., Trichopoulos, D., Lukito, W. & Trichopoulou, A. (1999). Are the advantages of the Mediterranean diet transferable to other populations? A cohort study in Melbourne, Australia. *British Journal of Nutrition*, **82**, 57–61.

Leighton, F., Miranda-Rottmann, S. & Urquiaga, I. (2006). A central role of eNOS in the protective effect of wine against metabolic syndrome. *Cell Biochemistry and Function*, **24**, 291–298.

Leighton, F., Polic, G., Strobel, P., Pérez, D., Martínez, C., Vásquez, L., Castillo, O., Villarroel, L., Echeverría, G., Urquiaga, I., Mezzano, D. & Rozowski, J. (2009). Health impact of Mediterranean diets in food at work. *Public Health Nutrition*, **12**(9A), 1635–1643.

Martinez-Gonzalez, M.A., Fernandez-Jarne, E., Serrano-Martinez, M., Wright, M. & Gomez-Gracia, E. (2004). Development of a short dietary intake questionnaire for the quantitative estimation of adherence to a cardioprotective Mediterranean diet. *European Journal of Clinical Nutrition*, **58**, 1550–1552.

McGovern, P.E., Glusker, D.L., Exner, L.J. & Voigt, M.M. (1996). Neolitic resinated wine. *Nature*, **381**, 480–481.

Menotti, A., Kromhout, D., Blackburn, H., Fidanza, F., Buzina, R. & Nissinen, A. (1999). Food intake patterns and 25-year mortality from coronary heart disease: cross-cultural correlations in the Seven Countries Study. The Seven Countries Study Research Group. *European Journal of Epidemiology*, **15**, 507–515.

Miglio, C., Chiavaro, E., Visconti, A., Fogliano, V. & Pellegrini, N. (2008). Effects of different cooking methods on nutritional and physicochemical characteristics of selected vegetables. *Journal of Agricultural and Food Chemistry*, **56**, 139–147.

Ninfali, P., Mea, G., Giorgini, S., Rocchi, M. & Bacchiocca, M. (2005). Antioxidant capacity of vegetables, spices and dressings relevant to nutrition. *British Journal of Nutrition*, **93**, 257–266.

OIV (2004). *Situation Report for the World Vitivinicultural Sector*. Organisation Internationale de la Vigne et du Vin, Madeleine, Paris.

Osler, M. & Schroll, M. (1997). Diet and mortality in a cohort of elderly people in a north European community. *International Journal of Epidemiology*, **26**, 155–159.

Panagiotakos, D.B., Chrysohoou, C., Pitsavos, C. & Stefanadis, C. (2006). Association between the prevalence of obesity and adherence to the Mediterranean diet: the ATTICA study. *Nutrition*, **22**, 449–456.

Papamichael, C., Karatzis, E., Karatzi, K., Aznaouridis, K., Papaioannou, T., Protogerou, A., Stamatelopoulos, K., Zampelas, A., Lekakis, J. & Mavrikakis, M. (2004). Red wine's antioxidants counteract acute endothelial dysfunction caused by cigarette smoking in healthy nonsmokers. *American Heart Journal*, **147**, E5.

Renaud, S., de Lorgeril, M., Delaye, J., Guidollet, J., Jacquard, F., Mamelle, N., Martin, J.L., Monjaud, I., Salen, P. & Toubol, P. (1995). Cretan Mediterranean diet for prevention of coronary heart disease. *American Journal of Clinical Nutrition*, **61**, 1360S–1367S.

Simopoulos, A.P. (2001). The Mediterranean diets: what is so special about the diet of Greece? The scientific evidence. *Journal of Nutrition*, **131**, 3065S–3073S.

Tabengwa, E.M., Grenett, H.E., Parks, D.A. & Booyse, F.M. (2003). *Moderate Alcohol Increases Clot Lysis in Vivo in a Mouse Model by Increasing t-PA and u-PA and Decreasing PAI-1 Expression*. American Heart Association Scientific Sessions. Orlando, FL, November 9–12.

Tokudome, S., Nagaya, T., Okuyama, H., Tokudome, Y., Imaeda, N., Kitagawa, I., Fujiwara, N., Ikeda, M., Goto, C., Ichikawa, H., Kuriki, K., Takekuma, K., Shimoda, A., Hirose, K. & Usui, T. (2000). Japanese versus Mediterranean diets and cancer. *Asian Pacific Journal of Cancer Prevention*, **1**, 61–66.

Trichopoulou, A., Costacou, T., Bamia, C. & Trichopoulos, D. (2003). Adherence to a Mediterranean diet and survival in a Greek population. *New England Journal of Medicine*, **348**, 2599–2608.

Trichopoulou, A. & Critselis, E. (2004). Mediterranean diet and longevity. *European Journal of Cancer Prevention*, **13**, 453–456.

Trichopoulou, A., Kouris-Blazos, A., Wahlqvist, M.L., Gnardellis, C., Lagiou, P., Polychronopoulos, E., Vassilakou, T., Lipworth, L. & Trichopoulos, D. (1995). Diet and overall survival in elderly people. *British Medical Journal*, **311**, 1457–1460.
Trichopoulou, A., Orfanos, P., Norat, T., Bueno-de-Mesquita, B., Ocke, M.C., Peeters, P.H., Van Der Schouw, Y.T., Boeing, H., Hoffmann, K., Boffetta, P., Nagel, G., Masala, G., Krogh, V., Panico, S., Tumino, R., Vineis, P., Bamia, C., Naska, A., Benetou, V., Ferrari, P., Slimani, N., Pera, G., Martinez-Garcia, C., Navarro, C., Rodriguez-Barranco, M., Dorronsoro, M., Spencer, E.A., Key, T.J., Bingham, S., Khaw, K.T., Kesse, E., Clavel-Chapelon, F., Boutron-Ruault, M.C., Berglund, G., Wirfalt, E., Hallmans, G., Johansson, I., Tjonneland, A., Olsen, A., Overvad, K., Hundborg, H.H., Riboli, E. & Trichopoulos, D. (2005). Modified Mediterranean diet and survival: EPIC-elderly prospective cohort study. *British Medical Journal*, **330**, 991.
Trichopoulou, A., Vasilopoulou, E. & Lagiou, A. (1999). Mediterranean diet and coronary heart disease: are antioxidants critical? *Nutrition Reviews*, **57**, 253–255.
Zohary, D. (1993). Domestication of plants in the Old World. In: *The Origin and Spread of Cultivated Plants in West Asia, Europe and the Nile Valley*. Clarendon Press, Oxford.

16 Functional foods for the brain

Ans Eilander, Saskia Osendarp and Jyoti Kumar Tiwari

16.1 Introduction

In the last few years the importance and role of certain nutrients for children's brain and mental development has been gaining scientific interest. Advanced measurement techniques provided new insights on the role and function of nutrients in the brain, and intervention trials assessed the impact of nutrition on certain age-related cognitive functions. As science advanced, new products and food supplements came in the market, targeting children's behaviour, attention and mental development. This chapter aims to summarise the current scientific evidence on the role of nutrition in children's development by defining for each nutrient what is known and what is yet not known with regard to its physiological function in the brain and the impact on children's attention, and cognitive development. Finally, we describe how nutrients can be added responsibly to foods targeted to children.

Historically, the prime focus of nutrition research was on physical health, leading to a wealth of recommendations on nutritional requirements for normal physical growth. However, in recent decades, longer life expectancies and changes in societies have posed more challenges on cognitive health, and as such, the area of neurobiology and brain research has gained significant importance. As a result, numerous studies have been performed trying to describe the causal relationship of nutrition and mental development into a defined relation that is linking nutrients to building blocks of brain cells and brain function.

Specific nutrients are essential for promoting healthy brain development of infants and children and preserving cognitive function of seniors. Many studies have been reviewed demonstrating that nutrition can influence the development of the brain in children (Gordon 1997; Bryan *et al.* 2004; Bourre 2006a,b; Georgieff 2007) and protect the brain against degeneration during aging (Smorgon *et al.* 2004; Bourre 2006a,b; Kotani *et al.* 2006; Scott *et al.* 2006). These discoveries had been able to link the effects of specific nutrients and its manifestation on cognitive function. In this chapter, we focus on research in children only. The nutritional need and impact of the growing and aging brain are likely to be rather different, and describing both areas properly would require another chapter.

The term 'cognition' is broad and covers various high-level psychological processes, such as memory, learning, reasoning, attention, language and coordination of motor outputs. Cognitive development refers to the changes of the cognitive processes observed over longer periods (months or years) and is usually assessed in children by batteries of performance tests assessing specific cognitive abilities. Carroll developed a hierarchical model of intelligence consisting of three strata (see Figure 16.1) (Carroll 1993). In this model, overall intelligence

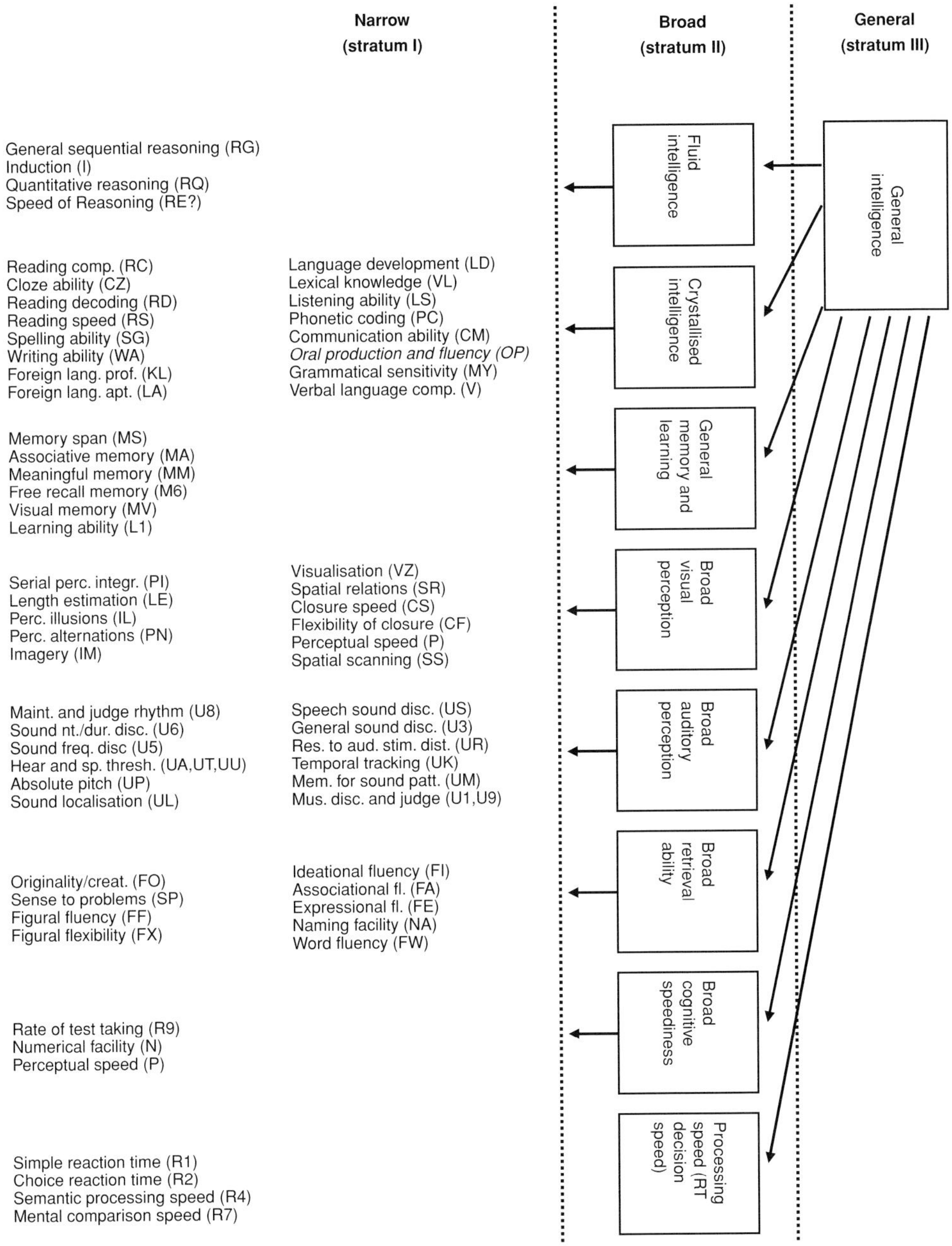

Fig. 16.1 The three-stratum structure of cognitive abilities by Carroll (Carroll 1993).

(stratum III) is composed by eight intellectual abilities (stratum II), which all correlate with overall intelligence. Stratum I consists of a fairly large number of more basic and specific cognitive processes, such as general sequential reasoning, reading comprehension, language development or memory span, underlying each of the intellectual abilities.

The development of infants and children is particularly vulnerable to nutrient deficiencies due to the high growth rate, especially of the brain. The brain achieves 90% of its full size

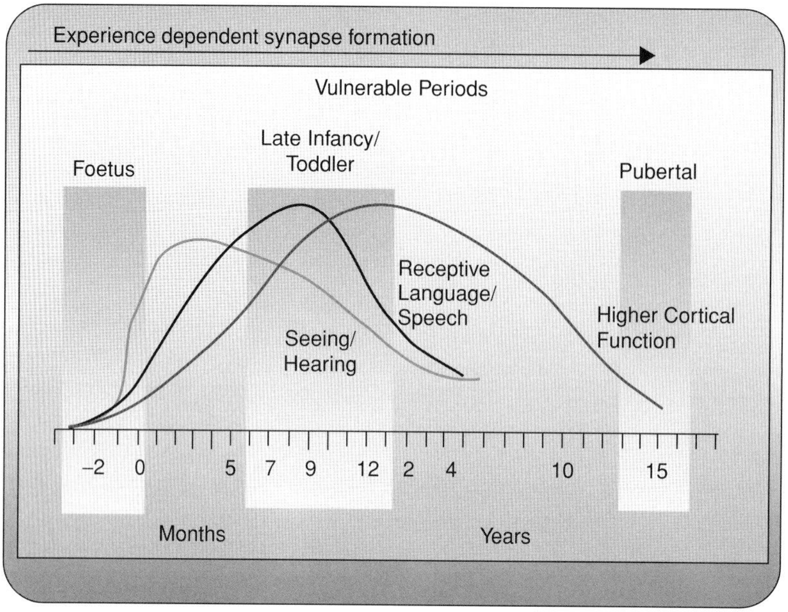

Fig. 16.2 Development of cognitive functions (adapted) (Thompson & Nelson 2001).

by the age of 5, with most rapid brain growth during the first 2 years of life. When rapid cell growth stops, the brain progresses to maturity into adulthood by further organisation of brain cells as well as generation of new connections between them (Paus 2005). The various parts of the brain develop and mature at different rates (see Figure 16.2). In the first 2 years the most basic functions such as the senses (taste, audition, smell, vision, touch) and mobility apparatus (movement, walking) are developed. Areas involved in advanced spatial orientation and language follow later. The frontal lobes mature last, which is the area of the brain where more complex functions are positioned, such as planning ahead, complex problem solving and memory.

Nutritional interventions can have an impact on cognitive functioning directly and indirectly. Micronutrients may have a direct effect on cognitive development, via the modification of brain structure and brain functions (Fernstrom 2000). It has been suggested that nutrition can affect the brain's macrostructure (e.g. development of brain areas such as the hippocampus), microstructure (e.g. myelination of neurons), and level and operation of neurotransmitters (e.g. dopamine levels or receptor numbers), all of which can have an impact on mental development.

Indirectly, nutrition may improve cognitive development via a general improvement of health and energy metabolism (Beard 2001). Overall improvement in health status of a young child will result in more explorative behaviour as the child will be more active and as a result the caregiver will interact more with the child, thereby improving social stimulation, and indirectly this will impact cognitive development of the child. In older children, improved health status will have a positive impact on school attendance, and also improve activity and social interactions with other children, leading to improve cognitive development and performance. When general aspects of psychological development, such as well-being, motivation and concentration, improve, it is likely that intelligence will also be positively affected.

The timing of the nutrition intervention is also very important because some brain functions that develop during early life can be irreversibly impaired due to poor brain development owing to nutritional deficits and lead to poor cognitive functioning in later life. One such study showed that anaemic children younger than 5 years performed less well at school and at 10 years of age even after anaemia had been corrected, compared to children who were not anaemic during early childhood (Hurtado *et al.* 1999).

The effect of nutritional interventions on cognitive performance of children can be measured through a set of psychological assessment using a cognitive test battery. In infants and young children, mental development is multidimensional and nonlinear. It is a result of physical growth, neurological maturation, interactions with the environment and the integration of stimuli provided by immediate caregivers and broader social and economic context (Flanagan & Harrison 2005). Therefore, the assessment of mental development of infants requires the examination of multiple domains and multiple sources of information; the comparison of developmental milestones with standardised population is essential. The 'Bayley Scales of Infant Development' is used to measure general development of infants and young children. More specific cognitive functions, such as gross motor skills, parent–child interaction, language comprehension, sleep–wake behaviour and attention, can be assessed in infants and young children with more precise methods. In children older than three years, intelligence is usually assessed by sets of short tests measuring various cognitive abilities, all correlated with a general factor of intelligence, such as the Kaufman Assessment Battery for Children and Wechsler tests.

16.2 Evidence from intervention trials

16.2.1 Iron

Iron deficiency is the most common deficiency in the world affecting 1.2 billion people worldwide, mostly infants, children and women. Iron requirements are especially high during pregnancy and therefore many pregnant women become iron deficient. Iron is important for normal development of the fetus. The iron content of the human brain increases continuously during the development and up throughout the teenage period by active transferrin receptor-mediated transport of iron into the brain. Approximately 10% of brain iron is present at birth, and at age 10, the human brain reaches only one-half of its normal iron content; optimal amount is reached at the age of 20–30 years (Beard 2001). The iron content is in particular high in dopaminergic brain regions (Beard & Connor 2003). Brain iron is required for myelination, neurotransmitter neurochemistry and neuronal energetics (Beard & Connor 2003). There is substantial evidence from animal research associating iron deficiency and/or anaemia with changes in the structure and function of the central nervous system. Studies have shown that in iron deficiency with or without anaemia, neurotransmitters and the related receptors are altered. This is most likely due to reduced oxygen availability (Agarwal 2001). There may also be indirect relations between anaemia and poor development. Anaemic children tend to explore and move around their environment less than non-anaemic children, and they induce less stimulating behaviour in their caretakers (Grantham-McGregor & Ani 2001). Further to this, iron deficiency may increase susceptibility to infections (Baynes & Bothwell 1990), which in turn affects school attendance and achievement. Finally, iron deficiency may inhibit the synthesis of thyroid hormones that are essential for growth and development of the brain and central nervous system (Zemmermann *et al.* 2000; Delange 2000).

Longitudinal studies consistently indicate that children with early childhood anaemia continue to have poor cognitive and motor development and school achievement into middle childhood (Grantham-McGregor & Ani 2001). Supplementing anaemic children (Soemantri *et al.* 1985; Lynn & Harland 1998) and even non-anaemic, iron-deficient children (Bruner *et al.* 1996) with iron has been shown to have positive effects on cognitive performance. The dosages used in these studies were 50–60 mg/day. The effects were mainly on certain measures of cognition, e.g. language and verbal learning skills and not on school achievements or IQ (Grantham-McGregor & Ani 2001). In anaemic children of younger than 2 years, short-term trials of iron treatment at dosages of 3–5 mg iron per kg bodyweight have generally failed to benefit development and only limited data from preventive studies support a causal relationship (Grantham-McGregor & Ani 2001). It has been suggested that the lack of effect in younger children might be due to problems with the sensitivity of tests used to measure cognitive performance in this age group (Pollitt 2001). It is also possible that, especially in the younger children, other factors apart from nutritional status have greater impact on the child's development. There is considerable evidence that anaemia is primarily associated with lower socioeconomic background and biomedical disadvantages that can themselves affect children's development (Grantham-McGregor & Ani 2001). The evidence for a beneficial effect of iron treatment on cognition in anaemic older children (>2 years) is reasonably convincing, although not conclusive (Grantham-McGregor & Ani 2001). A meta-analysis of 17 studies in infants and children on iron supplementation and effect on mental and motor development scores concluded that there was no effect in children younger than 2 years, while there was a positive effect of iron supplementation on mental performance in children older than 2 years (Sachdev *et al.* 2005).

In summary, iron deficiency is the most prevalent micronutrient deficiency worldwide and may have detrimental effects on cognitive development and performance of infants and children. There is convincing evidence that iron supplementation has positive effects on mental performance of children that are older than 2 years.

16.2.2 Iodine

Iodine is an essential micronutrient that forms a vital component of the thyroid hormones, thyroxine and triiodothyronine, which are crucial regulators of the metabolic rate and physical and mental development in humans. Thyroid hormones play a major role in the growth and development, function and maintenance of the central and peripheral nervous system (Hetzel 1983; Delange 2000).

In utero iodine deficiency leads to structural alterations in the brain of the fetus with long-term effects on mental development and performance. Severe iodine deficiency during pregnancy can cause cretinism, characterised by severe intellectual disability (Hetzel 1983). Studies in pregnant women provided conclusive evidence that cretinism can be prevented by iodine supplementation before and during pregnancy (Pharoah *et al.* 1971; Fierro-Benitez *et al.* 1972). Moreover, the beneficial effects of iodine supplementation during pregnancy on cognitive development are still apparent in childhood (Connolly *et al.* 1979; Fierro-Benitez *et al.* 1988; Pharoah & Connolly 1991).

Observational studies comparing populations from iodine-deficient areas with those in iodine-replete areas widely support the view that iodine deficiency reduces mental performance and leads to poorer achievement later in life. A meta-analysis of 18 studies completed in humans aged 2 months to 45 years indicated a general loss of 13.5 IQ points in chronically iodine-deficient populations, compared to non-iodine-deficient groups (Bleichrodt & Born 1994).

Eight randomised, placebo-controlled intervention trials have addressed the question of reversibility of the consequences of iodine deficiency later in life. Of these studies, three found that iodine supplementation can improve mental performance in deficient school children (Dodge *et al.* 1969; Shrestha *et al.* 1994; Zimmermann *et al.* 2006) and two showed that improvement in iodine status following iodine supplementation improved cognitive performance in school children (Bautista *et al.* 1982; Van Den Briel *et al.* 2000). Three studies did not find an improvement on cognition, which may have been because the iodine supplementation was not sufficient to improve the iodine status or other underlying factors in the study design (Bleichrodt *et al.* 1989; Isa *et al.* 2000; Huda *et al.* 2001).

In conclusion, iodine is essential for brain development of the fetus and cognitive performance of infants and children. Studies have shown that iodine supplementation either during pregnancy or during childhood is effective in improving brain development and cognitive performance of children.

16.2.3 Zinc

Zinc is an important mineral that is present in all body tissues and fluids. Zinc is an essential component of over 300 enzymes participating in the synthesis and degradation of carbohydrates, lipids, proteins and nucleic acids as well as in the metabolism of other micronutrients. Zinc plays a central role in the immune system, affecting a number of aspects of cellular and humoral immunity (Hotz & Brown 2004). Zinc contributes to normal brain function. Zinc is essential for neurogenesis, neuronal migration and synaptogenesis (important processes in the brain), and its deficiency could interfere with neurotransmission and subsequent neuropsychological behaviour (Bhatnagar & Taneja 2001).

Zinc deficiency may affect cognitive development by alterations in attention, activity, neuropsychological behaviour and motor development. The exact mechanisms that are impaired by zinc deficiency leading to poor neuropsychological behaviour are not clear. Research in animals showed that severe zinc deficiency particularly during periods of rapid growth, such as gestation or adolescence, is associated with alterations in brain development, increased emotional response to stress, reduced motor activity, and less accurate performance on measures of attention and short-term memory (Golub *et al.* 1995).

Relatively few studies have examined the effect of zinc supplementation at levels of 10–20 mg of zinc on behaviour and development in children and results are inconsistent. Low maternal intakes of zinc during pregnancy and lactation were found to be associated with less focused attention in neonates and decreased motor functions at 6 months of age in Egypt (Kirksey *et al.* 1994). Although zinc supplementation trials in infants and preschool children providing 10–20 mg of zinc have found positive effects on various aspects of cognitive development (Sazawal *et al.* 1996; Bentley *et al.* 1997), one study has found no effect (Ashworth *et al.* 1998) and one trial even observed lower scores on a mental development index (Hamadani *et al.* 2001). In older children, there is some evidence that zinc supplementation benefited neuropsychological performance in Chinese school-aged children (Sandstead *et al.* 1998). However, two studies in Guatemala and Canada did not observe any beneficial effects on cognition after zinc supplementation in stunted children (Gibson *et al.* 1989; Cavan *et al.* 1993). There is presently no data available on any possible long-term effects of zinc supplementation or zinc deprivation on cognitive performance. A recent review concluded that, at present, there is no clear consensus on the beneficial impact of

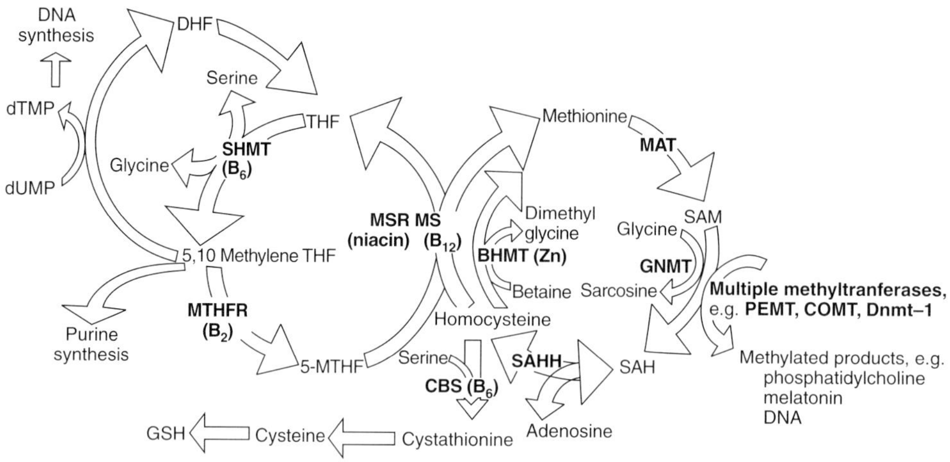

Fig. 16.3 The interaction between folate metabolism, the methionine cycle and methyl groups metabolism (Sugden 2006). DHF, dihydrofolate; THF, tetrahydrofolate; SHMT, serine hydroxymethyltransferase; MAT, methionine adenosyltransferase; MSR, methionine synthase reductase; MS, methionine synthase; SAM, *S*-adenosylmethionine; BHMT, betaine–homocysteine methyl transferase; GNMT, glycine *N*-methyltransferase; MTHFR, methylenetetrahydrofolate reductase; PEMT, phosphatidylethanolamine; COMT, catechol-O-methyltransferase; Dnmt-1, DNA methyltransferase; 5-MTHF, 5-methyltetrahydrofolate; CBS, cystathionine β-synthase; SAH, *S*-adenosylhomocysteine; SAHH, SAH hydrolase.

zinc supplementation on mental performance in infants, young children and school children (Black 2003).

Zinc plays an important role as a coenzyme in many physiological processes and may therefore have an impact on brain function; however, current evidence is inconclusive for effects of zinc on cognitive development and performance in infants and children.

16.2.4 B-vitamins

The principal role of vitamins from the B complex is to function as coenzyme in numerous metabolic processes in the human body. Thiamine, riboflavin, pyridoxine, niacin and pantothenic acid are needed for metabolism of carbohydrate, protein or fat substrates needed for energy utilisation and important for synthesis of amino acids, fatty acids, cholesterol, steroids and glucose. Folate, cobalamin, pyridoxine and riboflavin are needed for methyl group transfer (homocysteine–methionine–SAM pathway) (Figure 16.3). Folate also has an important role as methyl donor required for RNA and DNA synthesis. Via their coenzyme role in metabolic processes, B-vitamins contribute to optimal functioning of the nervous system. The evidence linking thiamine, riboflavin, pyridoxine, folate and cobalamin to cognitive development and performance is discussed in more detail.

16.2.4.1 *Thiamine*

The predominant form of thiamine in the brain is thiamine diphosphate, which is an essential cofactor for the three key enzymes involved in brain energy metabolism: α-ketoglutarate dehydrogenase complex, pyruvate dehydrogenase complex and transketolase (Westermarck & Antila 2000). Moreover, thiamine has several other roles in brain metabolism such as involvement in synthesis of glutamate, γ-aminobutyric acid and acetylcholine, activation of large conductance chloride channels and catecholamine synthesis (Bettendorff 1994). There is some evidence from rat studies that thiamine deficiency has adverse affects on learning and memory (Mair *et al.* 1985, 1991).

Disturbances of thiamine metabolism have been associated with several human diseases, including Wernicke's encephalopathy, infantile subacute necrotizing encephalopathy and certain forms of inherited ataxias. The Wernicke–Korsakov syndrome is by far the most common consequence of thiamine deficiency, resulting in neurological disorders associated with impaired cognitive function. The pathology is restricted to the central nervous system. A more severe form of thiamine deficiency may result in beriberi, a cardiac and peripheral nervous system disease, and it has been implicated in the pathogenesis of cerebellar degeneration and peripheral neuropathy (Kril 1996).

Evidence for beneficial effects of thiamine supplementation in infants and children is limited to one double-blind study of 9- to 19-year-olds. Children supplemented with 2 mg of thiamine daily for 1 year scored better on tests of memory and intelligence and had quicker reaction times than their matched pair controls (Harrell 1946). This improvement was seen despite the adequate dietary thiamine intake of all children in the study.

16.2.4.2 *Riboflavin*

Riboflavin (vitamin B2) is as coenzyme involved in redox reactions. Riboflavin in the form of the coenzyme flavin adinine dinucleotide interferes with the metabolism of other B-vitamins, including folate and cobalamin (homocysteine–methionine–SAM pathway), and the formation of the vitamin B6 coenzyme pyridoxal phosphate. Furthermore, poor riboflavin status interferes with the iron metabolism by mobilisation of ferritin and increasing iron absorption as shown in both animal and human studies (Powers 2003). Through these mechanisms, riboflavin may be related to cognition. In fact, animal studies have shown that riboflavin deficiency may lead to neurodegeneration and peripheral neuropathy (Norton *et al.* 1976; Jortner *et al.* 1987; Johnson & Storts 1988; Wada *et al.* 1996).

Only one human case study linked riboflavin to cognitive functioning by finding a reversible effect on neurologic and visual abnormalities, and anaemia in a 30-month-old riboflavin-deficient girl after supplementation of a high dose of riboflavin (Leshner 1981).

16.2.4.3 *Pyridoxin*

Pyridoxin (vitamin B6) in the form of pyridoxal phosphate is a coenzyme involved in the synthesis of neurotransmitters. Experiments with vitamin B6-deficient animals have shown that levels of neurotransmitters such as serotonin, dopamine and γ-aminobutyric acid were reduced (Guilarte 1993). Pyridoxal phosphate is also needed for the formation of haemoglobin and therefore deficiency of vitamin B6 may cause microcytic anaemia. Studies in adults with extremely low intakes of vitamin B6 showed abnormal electroencephalogram tracings (Canham *et al.* 1964; Kretsch *et al.* 1991); however, no supplementation studies have been conducted on the role of vitamin B6 on cognitive performance in children.

16.2.4.4 *Folate*

Folate functions in multiple coenzyme forms in the acceptance and donation of one-carbon units (methyl groups) in key metabolic pathways, including the metabolism of nucleotides and certain amino acids. Folate is required for the synthesis of methionine from homocysteine, and methionine is required for the synthesis of *S*-adenosylmethionine (SAM). SAM is a methyl group donor used in many biological methylation reactions, including the methylation of a number of sites within DNA and RNA. Riboflavin, pyridoxin and cobalamin are needed to convert folate conjugates, and deficiencies of these B-vitamins will therefore affect the function of folate. In the brain, folate is needed for synthesis of the neurotransmitters – dopamine, norepinephrine and serotonin. Also, folate may be required for methylation of phospholipids in neuronal membranes and have an impact on membrane-bound receptors, second messenger systems and ion channels. Folate deficiency may cause pernicious anaemia with similar effects on cognitive development and functioning as anaemia caused by iron deficiency (Hutto 1997; Sugden 2006).

Most of the research on the effects of folate on the brain in children has centred on the role of folate in the development of the neural tube. Nevertheless, there is still little understanding of the exact mechanisms by which supplementation of women with folate (400 μg/day) from 2 months prior to conception to 2 months after conception reduces the prevalence of neural tube defects in newborns (Fleming 2001). No studies have been performed to investigate the link between folic acid and cognitive performance in infants and children.

16.2.4.5 *Cobalamin*

Cobalamin (vitamin B12) in the form of methylcobalamin is also involved as a cofactor in methylation reactions (homocysteine–methionine–SAM pathway) that are also needed to metabolise folate. Because of the interactions with folate metabolism, cobalamin is indirectly involved in neurotransmitter synthesis and deficiency of cobalamin will also lead to pernicious anaemia. Furthermore, the vitamin B12 cofactors adenosylcobalamin and methylcobalamin are involved in myelination of the spinal cord and brain. Possibly, vitamin B12 may be required for maintaining an optimal balance of neurotrophic and neurotoxic cytokines, preventing neurodegeneration (Dror & Allen 2008).

Similar to folate, vitamin B12 deficiency during pregnancy has been associated with an increased risk of neural tube defects (Refsum 2001). The evidence for linking vitamin B12 status to cognitive performance in children is limited to two observational studies. The first study, conducted in the Netherlands, showed that infants of 15 months of age of macrobiotic mothers had a lower vitamin B12 status and delayed motor and language development compared to control children (Dagnelie *et al.* 1989; Dagnelie & van Staveren 1994). When these children were 10–16 years of age, those consuming a macrobiotic diet early in life performed less well on cognitive tests, measuring fluid intelligence, spatial ability and short-term memory compared to omnivore children (Louwman *et al.* 2000). Similarly, a second study conducted in Guatemala demonstrated that 8- to 12-year-old school children with vitamin B12 deficiency had poorer scores on cognitive tests than children with an adequate vitamin B12 status. However, these findings were not controlled for socioeconomic status, haemoglobin, iron status and blood lead levels (Allen *et al.* 1999).

16.2.4.6 *B-vitamins in combination*

While there are few studies assessing the effect of single B-vitamins on cognition in children, there are two studies that have investigated the effect of a combination of B-vitamins. The first study conducted in Indian school children aged 5–11 years showed that supplementation of thiamine, riboflavin, niacin, pyridoxine, folate and cobalamin for 12 months resulted in higher scores on psychomotor performance compared to controls (Bamji *et al.* 1982). The second study in 9- to 11-year-old Chinese children demonstrated that children receiving a supplement containing thiamine, riboflavin, niacin and folate for 1 month performed better on a test of visual attention compared to children in the placebo group (Jiang 2006). Unfortunately, the designs of these studies were not optimal and therefore more studies are needed to conclude whether supplementation with B-vitamins would benefit children's cognition.

Although there are plausible mechanisms for the role of B-vitamins in functioning of the central nervous system, more research is needed to conclude whether supplementation with B-vitamins could improve cognitive performance during childhood.

16.2.5 Fatty acids

Fatty acids are important for children's growth and development. Approximately 25–30% of the fatty acids in the human brain consist of polyunsaturated fatty acids of which the *n*-3 fatty acid, docosahexaenoic acid (DHA), and the *n*-6 fatty acid, arachidonic acid (AA), are major components. DHA and AA are rapidly incorporated in the nervous tissue of retina and brain during the brain's growth spurt, which mainly takes place from the last trimester of pregnancy up to 2 years of age (Dobbing & Sands 1973; Clandinin *et al.* 1980a,b; Martinez 1992). Beyond development of the central nervous system, *n*-3 and *n*-6 fatty acids may influence brain function throughout the life by modifications of neuronal membrane fluidity, membrane activity-bound enzymes, number and affinity of receptors, function of neuronal membrane ionic channels, and production of neurotransmitters and brain peptides (Yehuda 2003).

The human body can synthesize DHA and AA from α-linolenic acid (ALA) and linoleic acid (LA), respectively. ALA and LA must be supplied by the diet and are therefore called essential fatty acids. However, the conversion of ALA to eicosapentaenoic acid (EPA) and DHA is only marginal (Emken *et al.* 1994; Salem *et al.* 1999; Vermunt *et al.* 1999; Pawlosky *et al.* 2001), with an overall conversion rate of at most some 4% (Brenna 2002), and it is presently unclear whether conversion of dietary essential fatty acids is under all circumstances sufficient to meet optimal DHA and AA requirements. Moreover, intakes of polyunsaturated fatty acids and the fish fatty acids, EPA and DHA, are especially low in children (Elmadfa & Weichselbaum 2005; Kris-Etherton *et al.* 2007).

Observational studies have shown positive effect of fish intake during pregnancy on cognitive development of infants (Cohen *et al.* 2005). Based on these observations, it has been recommended for pregnant and lactating women to consume at least 200 mg DHA per day (Koletzko *et al.* 2007). Four randomised, controlled trials have studied the effect of EPA and DHA intake during week 15–20 of pregnancy to delivery. Two of these trials showed a positive effect on problem solving at 9 months of age and eye–hand coordination at 2.5 years of age (Judge *et al.* 2007; Dunstan *et al.* 2008). Another study showed that 4-year-old children of mothers who were supplemented with DHA from week 17–19 of pregnancy to 3 months postpartum scored 4 IQ points more than children of mothers who received a control oil (Helland *et al.* 2003). Similarly, supplementing lactating women with DHA seems beneficial for cognitive development of infants, as two of three studies found positive effects on problem solving at 9 months and psychomotor development at 30 months (Jensen *et al.* 2005; Lauritzen *et al.* 2005).

Evidence for supplementing infants with AA and DHA on mental development seems less obvious. The Cochrane review of 11 well-conducted studies in term infants concluded that two of them showed beneficial effects on mental development of infants born at term (Simmer *et al.* 2008a). Similarly, Cochrane concluded that there is no evidence for a beneficial effect of AA and DHA supplementation of preterm infants on cognitive development based on a meta-analysis of seven studies. However, three of the seven studies showed beneficial effects on mental development (Simmer *et al.* 2008b).

Research in healthy children older than 2 years is limited to an observational study and a randomised, controlled trial. The prospective cohort study conducted in 7-year-old children in the Netherlands did not find a relationship between DHA status measured at 7 years of age and cognitive performance as measured by Kaufman Assessment Battery for Children (Bakker *et al.* 2003). Also, the study in 7- to 10-year-old children in Indonesia and Australia failed to demonstrate a significant effect of 110 mg EPA and DHA supplementation on cognitive performance (Osendarp *et al.* 2007). The evidence of beneficial effects of fish oil on attention, learning and visual and motor function is limited to studies in children with neurodevelopmental disorders (Richardson & Puri 2002; Stevens *et al.* 2003; Richardson & Montgomery 2005; Vaisman *et al.* 2008) and childhood phenylketonuria who consume diets restricted in animal foods and therefore have a low intake of AA and DHA (Agostoni *et al.* 2000; Beblo *et al.* 2001, 2007).

It can be concluded that there is growing evidence that EPA and DHA are needed for development of the fetal brain. However, evidence for beneficial effects of AA and DHA supplementation in infants is inconclusive, and evidence is lacking for effects of DHA on cognitive performance in healthy children older than 2 years. More research, such as investigating dose and type of fatty acids and domains of attention and behaviour, is needed to conclude whether supplementing healthy older children with EPA and DHA would be beneficial.

16.2.6 Multiple micronutrients

The single micronutrient deficiencies are helpful in identifying its possible role and effect on cognitive performance. Consumption of monotonous diets lacking important micronutrients will result in coexisting micronutrient deficiencies, so multiple micronutrient interventions may be more effective in improving cognitive functioning than single micronutrient interventions (Mason *et al.* 2001). Six studies assessed the effect of multiple micronutrient-fortified foods on cognitive performance in school-aged children (van Stuijvenberg *et al.* 1999; Jinabhai *et al.* 2001; Solon *et al.* 2003; Vazir *et al.* 2006; Osendarp *et al.* 2007; Manger *et al.* 2008). Most of the studies were conducted in developing countries in malnourished children. Five of the six studies have shown significantly beneficial effects on at least one of the cognitive performance indicators measured in children (van Stuijvenberg *et al.* 1999; Vazir *et al.* 2006; Osendarp *et al.* 2007; Manger *et al.* 2008).

In summary, there is evidence for a positive effect of multiple micronutrient fortification on intelligence scores in school-aged children.

16.3 Challenges in fortification of foods for children

Food fortification is one of the strategies used by the World Health Organization (WHO) and the Food and Agricultural Organization of the United Nations to increase the intake of micronutrients in order to lower the prevalence of micronutrient deficiencies in population

(Allen *et al.* 2006). The addition of micronutrients to processed foods may lead to relatively rapid improvements in the micronutrient status of a population at reasonable costs, especially when existing technology and local distribution networks are used. To make food fortification a successful cost-effective method, it is important that the food chosen to be fortified is consumed in adequate amounts by a large proportion of the target group in which the micronutrient status needs to be improved. The foods should be consumed on a regular and frequent basis to ensure a continuous supply of micronutrients and maintain body stores, which is in particular essential for optimal growth and development of children. Furthermore, the fortificants should have a good bioavailability and should not affect the sensory properties of foods. It is preferred to use food vehicles that are centrally processed, and to have the support of the food industry. Monitoring of the effectiveness of the fortification programs is needed to evaluate whether the target population is reached and micronutrient status improves. Monitoring needs to be conducted at the level of processing of foods, i.e. regulatory monitoring, as well as at household or individual level.

While food fortification is an effective strategy to combat micronutrient deficiencies in children older than 2 years, it is more difficult to target infants and young children. Older children usually consume adequate amounts of foods prepared in the household. Despite the requirements for micronutrients for children are different from those for adults, it is assumed that food fortification for the general population can cover the needs for older children. However, younger children consume much smaller amounts and may also not be able to eat the same foods as other household members. In addition, rapid growth during the first few years of life necessitates higher requirements for micronutrients per kilogram body weight in infants and young children compared to older children and adults. Therefore, infants and young children may not receive adequate amounts of micronutrients through fortified foods targeted at the general population. Therefore, fortification of complementary foods especially designed to meet the nutritional requirements of infants and young children may be an effective way to supply them with appropriate levels of micronutrients.

Finally, the marketing of fortified foods to children, especially foods targeting such a sensitive area as brain development, requires special care. Principles for marketing to children have been proposed by the WHO (WHO 2006) and, voluntarily, by some international food producers who agreed to voluntarily ban all paid marketing communications directed primarily at children younger than 6 or in some cases 12 years.

Children younger than 12 are known to have difficulties differentiating between fact and fiction. Marketing to children should therefore not convey misleading messages or undermine parental influence. Marketing practices to children should not suggest time pressure or price pressure, encourage unhealthy dietary habits or blur the boundary between promotion and content.

16.4 Conclusions

In examining the challenges for public policy with regard to cognitive enhancement, it is important to consider the full range of possibilities that are available, and their individual characteristics. The role of nutrients in influencing physiology in general and cognitive benefit has been highlighted in this chapter. Some of the key challenges in food for the brain are as follows:

- Interaction/compatibility of various nutrients (e.g. Fe and DHA oxidation)
- Product taste and colour stability

- Reduced shelf-life of the product
- Influence on bioavailability of the nutrients

Further, the role of psychosocial intervention is widely recognised in mental development. The scope of findings must be viewed with caution, as environmental influences can be paramount.

We discussed the effect of nutrition on brain and cognitive performance where substantial intervention studies were available for justification. This chapter provides direction for fortification of key nutrients and possible benefit or improvement in the performance of some cognitive subsystem without reference to identifiable pathology or dysfunction. We recognise that psychosocial interventions, i.e. education, training and enriched positive environment, are other important aspects that influence cognition. Methods of enhancing cognition through functional foods/nutrition need firm validation, taking into account safety, efficacy and social consequences.

References

Agarwal, K.N. (2001). Iron and the brain: neurotransmitter receptors and magnetic resonance spectroscopy. *British Journal of Nutrition*, **85**, S147–S150.

Agostoni, C., Massetto, N., Biasucci, G., Rottoli, A., Bonvisutto, M., Bruzzese, M.G., Giovannini, M. & Riva, E. (2000). Effects of long-chain polyunsaturated fatty acid supplementation on fatty acid status and visual function in treated children with hyperphenylalaninemia. *Journal of Pediatrics*, **137**, 504–509.

Allen, L., de Benoist, B., Hurrell, R.F. & Dary, O. (2006). *Guidelines on Food Fortification with Micronutrients*. World Health Organization and Food and Agriculture Organization of the United Nations, Geneva.

Allen, L.H., Penland, J.G., Boy, E., DeBaessa, Y. & Rogers, L.M. (1999). Cognitive and neuromotor performance of Guatemalan schoolers with deficient, marginal, and normal plasma vitamin B-12. *Faseb Journal*, **13**(4), A544.

Ashworth, A., Morris, S.S., Lira, P.I.C. & Grantham-McGregor, S.M. (1998). Zinc supplementation, mental development and behaviour in low birth weight term infants in northeast Brazil. *European Journal of Clinical Nutrition*, **52**(3), 223–227.

Bakker, E.C., Ghys, A.J.A., Kester, A.D.M., Vles, J.S.H., Dubas, J.S., Blanco, C.E. & Hornstra, G. (2003). Long-chain polyunsaturated fatty acids at birth and cognitive function at 7y of age. *European Journal of Clinical Nutrition*, **57**(1), 89–95.

Bamji, M.S., Arya, S., Sarma, K.V.R. & Radhaiah, G. (1982). Impact of long-term, low-dose B-complex vitamin supplements on vitamin status and psychomotor performance of rural school boys. *Nutrition Research*, **2**(2), 147–153.

Bautista, A., Barker, P.A., Dunn, J.T., Sanchez, M. & Kaiser, D.L. (1982). The effects of oral iodized oil on intelligence, thyroid status, and somatic growth in school-age children from an area of endemic goiter. *American Journal of Clinical Nutrition*, **35**(1), 127–134.

Baynes, R.D. & Bothwell, T.H. (1990). Iron-deficiency. *Annual Review of Nutrition*, **10**, 133–148.

Beard, J.L. (2001). Iron biology in immune function, muscle metabolism and neuronal functioning. *Journal of Nutrition*, **131**(2S-2), 568S–579S.

Beard, J.L. & Connor, J.R. (2003). Iron status and neural functioning. *Annual Review of Nutrition*, **23**, 41–58.

Beblo, S., Reinhardt, H., Demmelmair, H., Muntau, A.C. & Koletzko, B. (2007). Effect of fish oil supplementation on fatty acid status, coordination, and fine motor skills in children with phenylketonuria. *Journal of Pediatrics*, **150**(5), 479–484.

Beblo, S., Reinhardt, H., Muntau, A.C., Mueller-Felber, W., Roscher, A.A. & Koletzko, B. (2001). Fish oil supplementation improves visual evoked potentials in children with phenylketonuria. *Neurology*, **57**(8), 1488–1491.

Bentley, M.E., Caulfield, L.E., Ram, M., Santizo, M.C., Hurtado, E., Rivera, J.A., Ruel, M.T. & Brown, K.H. (1997). Zinc supplementation affects the activity patterns of rural Guatemalan infants. *Journal of Nutrition*, **127**(7), 1333–1338.

Bettendorff, L. (1994). Thiamine in excitable tissues – reflections on a non-cofactor role. *Metabolic Brain Disease*, **9**(3), 183–209.
Bhatnagar, S. & Taneja, S. (2001). Zinc and cognitive development. *British Journal of Nutrition*, **85**, S139–S145.
Black, M.M. (2003). The evidence linking zinc deficiency with children's cognitive and motor functioning. *Journal of Nutrition*, **133**(5 Suppl 1), 1473S–1476S.
Bleichrodt, N. & Born, M.P. (1994). A meta-analysis of research on iodine and its relationship to cognitive development. In: *The Damaged Brain of Iodine Deficiency*. Stanbury, J.B. (ed.), Cognizant Communication Corporation, New York, pp. 195–200.
Bleichrodt, N., Escobar del Rey, F., Morreale de Escobar, G., Garcia, I. & Rubio, C. (1989). Iodine deficiency, implications for mental and psychomotor development in children. In: *Iodine and the Brain*. Delong, G.R., Robbins, J. & Connolly, K. (eds), Plenum Press, New York, pp. 269–288.
Bourre, J.M. (2006a). Effects of nutrients (in food) on the structure and function of the nervous system: update on dietary requirements for brain. Part 1: micronutrients. *Journal of Nutrition, Health and Aging*, **10**(5), 377–385.
Bourre, J.M. (2006b). Effects of nutrients (in food) on the structure and function of the nervous system: update on dietary requirements for brain. Part 2 : macronutrients. *Journal of Nutrition, Health and Aging*, **10**(5), 386–399.
Brenna, J.T. (2002). Efficiency of conversion of alpha-linolenic acid to long chain n-3 fatty acids in man. *Current Opinion in Clinical Nutrition and Metabolic Care*, **5**(2), 127–132.
Bruner, A.B., Joffe, A., Duggan, A.K., Casella, J.F. & Brandt, J. (1996). Randomised study of cognitive effects of iron supplementation in non-anaemic iron-deficient adolescent girls. *Lancet*, **348**(9033), 992–996.
Bryan, J., Osendarp, S., Hughes, D., Calvaresi, E., Baghurst, K. & van Klinken, J.W. (2004). Nutrients for cognitive development in school-aged children. *Nutrition Reviews*, **62**(8), 295–306.
Canham, J.E., Nunes, W.T. & Eberlin, E.W. (1964). Electroencephalographic and central nervous system manifestations of vitamin B-6 deficiency and induced vitamin B-6 dependency in normal human adults. In: *Proceedings of the Sixth International Congress on Nutrition*. E&S Livingstone, Edinburgh.
Carroll, J.B. (1993). *Human Cognitive Abilities: A Survey of Factor-Analytic Studies*. Cambridge University Press, New York.
Cavan, K.R., Gibson, R.S., Grazioso, C.F., Isalgue, A.M., Ruz, M. & Solomons, N.W. (1993). Growth and body composition of periurban Guatemalan children in relation to zinc status: a longitudinal zinc intervention trial. *American Journal of Clinical Nutrition*, **57**(3), 344–352.
Clandinin, M.T., Chappell, J.E., Leong, S., Heim, T., Swyer, P.R. & Chance, G.W. (1980a). Extrauterine fatty acid accretion in infant brain: implications for fatty acid requirements. *Early Human Development*, **4**(2), 131–138.
Clandinin, M.T., Chappell, J.E., Leong, S., Heim, T., Swyer, P.R. & Chance, G.W. (1980b). Intrauterine fatty acid accretion rates in human brain: implications for fatty acid requirements. *Early Human Development*, **4**(2), 121–129.
Cohen, J.T., Bellinger, D.C., Connor, W.E. & Shaywitz, B.A. (2005). A quantitative analysis of prenatal intake of n-3 polyunsaturated fatty acids and cognitive development. *American Journal of Preventive Medicine*, **29**(4), 366–374.
Connolly, K.J., Pharoah, P.O. & Hetzel, B.S. (1979). Fetal iodine deficiency and motor performance during childhood. *Lancet*, **2**(8153), 1149–1151.
Dagnelie, P.C. & van Staveren, W.A. (1994). Macrobiotic nutrition and child health: results of a population-based, mixed-longitudinal cohort study in the Netherlands. *American Journal of Clinical Nutrition*, **59**(5 Suppl), 1187S–1196S.
Dagnelie, P.C., van Staveren, W.A., Vergote, F.J., Dingjan, P.G., Van Den, B.H. & Hautvast, J.G. (1989). Increased risk of vitamin B-12 and iron deficiency in infants on macrobiotic diets. *American Journal of Clinical Nutrition*, **50**(4), 818–824.
Delange, F. (2000). The role of iodine in brain development. *Proceedings of the Nutrition Society*, **59**(1), 75–79.
Dobbing, J. & Sands, J. (1973). Quantitative growth and development of human brain. *Archives of Disease in Childhood*, **48**(10), 757–767.
Dodge, P.R., Palkes, H., Fierro-Benitez, R. & Ramirez, I. (1969). Effect on intelligence of iodine in oil administered to young Andean children – a preliminary report. In: *Endemic Goiter*. Stanbury, J.B. (ed.), PAHO, Washington, D.C., pp. 378–380.

Dror, D.K. & Allen, L.H. (2008). Effect of vitamin B12 deficiency on neurodevelopment in infants: current knowledge and possible mechanisms. *Nutrition Reviews*, **66**(5), 250–255.

Dunstan, J.A., Simmer, K., Dixon, G. & Prescott, S.L. (2008). Cognitive assessment of children at age 2(1/2) years after maternal fish oil supplementation in pregnancy: a randomised controlled trial. *Archives of Disease in Childhood. Fetal and Neonatal Edition*, **93**(1), F45–F50.

Elmadfa, I. & Weichselbaum, E. (2005). *European Nutrition and Health Report 2004*. Forum of Nutrition edn, Karger, Basel.

Emken, E.A., Adlof, R.O. & Gulley, R.M. (1994). Dietary linoleic-acid influences desaturation and acylation of deuterium-labeled linoleic and linolenic acids in young-adult males. *Biochimica et Biophysica Acta-Lipids and Lipid Metabolism*, **1213**(3), 277–288.

Fernstrom, J.D. (2000). Can nutrient supplements modify brain function?. *American Journal of Clinical Nutrition*, **71**(6 Suppl), 1669S–1675S.

Fierro-Benitez, R., Cazar, R., Stanbury, J.B., Rodriguez, P., Garces, F., Fierrorenoy, F. & Estrella, E. (1988). Effects on school-children of prophylaxis of mothers with iodized oil in an area of iodine deficiency. *Journal of Endocrinological Investigation*, **11**(5), 327–335.

Fierro-Benitez, R., Ramirez, I. & Suarez, J. (1972). Effect of iodine correction early in fetal life on intelligence quotient. A preliminary report. *Advances in Experimental Medicine and Biology*, **30**, 239–247.

Flanagan, D.P. & Harrison, P.L. (2005). *Contemporary Intellectual Assessment, Theories, Tests and Issues*, 2nd edn. The Guilford Press, New York.

Fleming, A. (2001). The role of folate in the prevention of neural tube defects: human and animal studies. *Nutrition Reviews*, **59**(8), S13–S20.

Georgieff, M.K. (2007). Nutrition and the developing brain: nutrient priorities and measurement. *American Journal of Clinical Nutrition*, **85**(2), 614S–620S.

Gibson, R.S., Vanderkooy, P.D., Macdonald, A.C., Goldman, A., Ryan, B.A. & Berry, M. (1989). A growth-limiting, mild zinc-deficiency syndrome in some southern Ontario boys with low height percentiles. *American Journal of Clinical Nutrition*, **49**(6), 1266–1273.

Golub, M.S., Keen, C.L., Gershwin, M.E. & Hendrickx, A.G. (1995). Developmental zinc deficiency and behavior. *Journal of Nutrition*, **125**(8 Suppl), 2263S–2271S.

Gordon, N. (1997). Nutrition and cognitive function. *Brain and Development*, **19**(3), 165–170.

Grantham-McGregor, S. & Ani, C. (2001). A review of studies on the effect of iron deficiency on cognitive development in children. *Journal of Nutrition*, **131**(2S-2), 649S–666S.

Guilarte, T.R. (1993). Vitamin B6 and cognitive development: recent research findings from human and animal studies. *Nutrition Reviews*, **51**(7), 193–198.

Hamadani, J.D., Fuchs, G.J., Osendarp, S.J.M., Khatun, F., Huda, S.N. & Grantham-McGregor, S.M. (2001). Randomized controlled trial of the effect of zinc supplementation on the mental development of Bangladeshi infants. *American Journal of Clinical Nutrition*, **74**(3), 381–386.

Harrell, R.F. (1946). Mental response to added thiamine. *Journal of Nutrition*, **31**, 283–298.

Helland, I.B., Smith, L., Saarem, K., Saugstad, O.D. & Drevon, C.A. (2003). Maternal supplementation with very-long-chain n-3 fatty acids during pregnancy and lactation augments children's IQ at 4 years of age. *Pediatrics*, **111**(1), e39.

Hetzel, B.S. (1983). Iodine deficiency disorders (IDD) and their eradication. *Lancet*, **2**(8359), 1126–1129.

Hotz, C. & Brown, K.H. (2004). Assessment of the risk of zinc deficiency in populations and options for its control. International Zinc Nutrition Consultative Group (IZiNCG) technical document #1. *Food and Nutrition Bulletin*, **25**(Suppl. 2), S94–S204.

Huda, S.N., Grantham-McGregor, S.M. & Tomkins, A. (2001). Cognitive and motor functions of iodine-deficient but euthyroid children in Bangladesh do not benefit from iodized poppy seed oil (Lipiodol). *Journal of Nutrition*, **131**(1), 72–77.

Hurtado, E.K., Claussen, A.H. & Scott, K.G. (1999). Early childhood anemia and mild or moderate mental retardation. *American Journal of Clinical Nutrition*, **69**(1), 115–119.

Hutto, B.R. (1997). Folate and cobalamin in psychiatric illness. *Comprehensive Psychiatry*, **38**(6), 305–314.

Isa, Z.M., Alias, I.Z., Kadir, K.A. & Ali, O. (2000). Effect of iodized oil supplementation on thyroid hormone levels and mental performance among Orang Asli schoolchildren and pregnant mothers in an endemic goitre area in Peninsular Malaysia. *Asia Pacific Journal of Clinical Nutrition*, **9**(4), 274–281.

Jensen, C.L., Voigt, R.G., Prager, T.C., Zou, Y.L., Fraley, J.K., Rozelle, J.C., Turcich, M.R., Llorente, A.M., Anderson, R.E. & Heird, W.C. (2005). Effects of maternal docosahexaenoic acid intake on visual function and neurodevelopment in breastfed term infants. *American Journal of Clinical Nutrition*, **82**(1), 125–132.

Jiang, Y.Y. (2006). Effect of B vitamins-fortified foods on primary school children in Beijing. *Asia-Pacific Journal of Public Health*, **18**(2), 21–25.
Jinabhai, C.C., Taylor, M., Coutsoudis, A., Coovadia, H.M., Tomkins, A.M. & Sullivan, K.R. (2001). A randomized controlled trial of the effect of antihelminthic treatment and micronutrient fortification on health status and school performance of rural primary school children. *Annals of Tropical Paediatrics*, **21**(4), 319–333.
Johnson, W.D. & Storts, R.W. (1988). Peripheral neuropathy associated with dietary riboflavin deficiency in the chicken. I. Light microscopic study. *Veterinary Pathology*, **25**(1), 9–16.
Jortner, B.S., Cherry, J., Lidsky, T.I., Manetto, C. & Shell, L. (1987). Peripheral neuropathy of dietary riboflavin deficiency in chickens. *Journal of Neuropathology and Experimental Neurology*, **46**(5), 544–555.
Judge, M.P., Harel, O. & Lammi-Keefe, C.J. (2007). Maternal consumption of a docosahexaenoic acid-containing functional food during pregnancy: benefit for infant performance on problem-solving but not on recognition memory tasks at age 9 mo. *American Journal of Clinical Nutrition*, **85**(6), 1572–1577.
Kirksey, A., Wachs, T.D., Yunis, F., Srinath, U., Rahmanifar, A., Mccabe, G.P., Galal, O.M., Harrison, G.G. & Jerome, N.W. (1994). Relation of maternal zinc nutriture to pregnancy outcome and infant development in an Egyptian village. *American Journal of Clinical Nutrition*, **60**(5), 782–792.
Koletzko, B., Cetin, I. & Brenna, J.T. (2007). Dietary fat intakes for pregnant and lactating women. *British Journal of Nutrition*, **98**(5), 873–877.
Kotani, S., Sakaguchi, E., Warashina, S., Matsukawa, N., Ishikura, Y., Kiso, Y., Sakakibara, M., Yoshimoto, T., Guo, J. & Yamashima, T. (2006). Dietary supplementation of arachidonic and docosahexaenoic acids improves cognitive dysfunction. *Neuroscience Research*, **56**(2), 159–164.
Kretsch, M.J., Sauberlich, H.E. & Newbrun, E. (1991). Electroencephalographic changes and periodontal status during short-term vitamin B-6 depletion of young, nonpregnant women. *American Journal of Clinical Nutrition*, **53**(5), 1266–1274.
Kril, J.J. (1996). Neuropathology of thiamine deficiency disorders. *Metabolic Brain Disease*, **11**(1), 9–17.
Kris-Etherton, P.M., Innis, S., Ammerican, D.A. & Dietitians of Canada (2007). Position of the American Dietetic Association and Dietitians of Canada: dietary fatty acids. *Journal of the American Dietetic Association*, **107**(9), 1599–1611.
Lauritzen, L., Jorgensen, M.H., Olsen, S.F., Straarup, E.M. & Michaelsen, K.F. (2005). Maternal fish oil supplementation in lactation: effect on developmental outcome in breast-fed infants. *Reproduction Nutrition Development*, **45**(5), 535–547.
Leshner, R.T. (1981). Riboflavin deficiency – a reversible neurodegenerative disease. *Annals of Neurology*, **10**, 294–295.
Louwman, M.W.J., van Dusseldorp, M., van de Vijver, F.J.R., Thomas, C.M.G., Schneede, J., Ueland, P.M., Refsum, H. & van Staveren, W.A. (2000). Signs of impaired cognitive function in adolescents with marginal cobalamin status. *American Journal of Clinical Nutrition*, **72**(3), 762–769.
Lynn, R. & Harland, E.P. (1998). A positive effect of iron supplementation on the IQs of iron deficient children. *Personality and Individual Differences*, **24**(6), 883–885.
Mair, R.G., Anderson, C.D., Langlais, P.J. & Mcentee, W.J. (1985). Thiamine deficiency depletes cortical norepinephrine and impairs learning processes in the rat. *Brain Research*, **360**(1–2), 273–284.
Mair, R.G., Otto, T.A., Knoth, R.L., Rabchenuk, S.A. & Langlais, P.J. (1991). Analysis of aversively conditioned learning and memory in rats recovered from pyrithiamine-induced thiamine deficiency. *Behavioral Neuroscience*, **105**(3), 351–359.
Manger, M.S., McKenzie, J.E., Winichagoon, P., Gray, A., Chavasit, V., Pongcharoen, T., Gowachirapant, S., Ryan, B., Wasantwisut, E. & Gibson, R.S. (2008). A micronutrient-fortified seasoning powder reduces morbidity and improves short-term cognitive function, but has no effect on anthropometric measures in primary school children in northeast Thailand: a randomized controlled trial. *American Journal of Clinical Nutrition*, **87**(6), 1715–1722.
Martinez, M. (1992). Tissue levels of polyunsaturated fatty acids during early human development. *Journal of Pediatrics*, **120**(4 Pt 2), S129–S138.
Mason, J.B., Lotfi, M., Dalmiya, N., Sethuraman, K. & Deitchler, M. (2001). *Current Progress and Trend in the Control of Vitamin A, Iodine, and Iron Deficiencies*. The Micronutrient Initiative, Ottawa.
Norton, W.N., Daskal, I., Savage, H.E., Seibert, R.A. & Lane, M. (1976). Effects of riboflavin deficiency on the ultrastructure of rat sciatic nerve fibers. *American Journal of Pathology*, **85**(3), 651–660.
Osendarp, S.J., Baghurst, K.I., Bryan, J., Calvaresi, E., Hughes, D., Hussaini, M., Karyadi, S.J., van Klinken, B.J., Van Der Knaap, H.C., Lukito, W., Mikarsa, W., Transler, C. & Wilson, C. (2007). Effect of a 12-mo micronutrient intervention on learning and memory in well-nourished and marginally nourished

school-aged children: 2 parallel, randomized, placebo-controlled studies in Australia and Indonesia. *American Journal of Clinical Nutrition*, **86**(4), 1082–1093.
Paus, T. (2005). Mapping brain maturation and cognitive development during adolescence. *Trends in Cognitive Sciences*, **9**(2), 60–68.
Pawlosky, R.J., Hibbeln, J.R., Novotny, J.A. & Salem, N. (2001). Physiological compartmental analysis of alpha-linolenic acid metabolism in adult humans. *Journal of Lipid Research*, **42**(8), 1257–1265.
Pharoah, P.O., Buttfield, I.H. & Hetzel, B.S. (1971). Neurological damage to the fetus resulting from severe iodine deficiency during pregnancy. *Lancet*, **1**(7694), 308–310.
Pharoah, P.O. & Connolly, K.J. (1991). Effects of maternal iodine supplementation during pregnancy. *Archives of Disease in Childhood*, **66**(1), 145–147.
Pollitt, E. (2001). Statistical and psychobiological significance in developmental research. *American Journal of Clinical Nutrition*, **74**(3), 281–282.
Powers, H.J. (2003). Riboflavin (vitamin B-2) and health. *American Journal of Clinical Nutrition*, **77**(6), 1352–1360.
Refsum, H. (2001). Folate, vitamin B12 and homocysteine in relation to birth defects and pregnancy outcome. *British Journal of Nutrition*, **85**, S109–S113.
Richardson, A.J. & Montgomery, P. (2005). The Oxford-Durham study: a randomized, controlled trial of dietary supplementation with fatty acids in children with developmental coordination disorder. *Pediatrics*, **115**(5), 1360–1366.
Richardson, A.J. & Puri, B.K. (2002). A randomized double-blind, placebo-controlled study of the effects of supplementation with highly unsaturated fatty acids on ADHD-related symptoms in children with specific learning difficulties. *Progress in Neuropsychopharmacology and Biological Psychiatry*, **26**(2), 233–239.
Sachdev, H.P.S., Gera, T. & Nestel, P. (2005). Effect of iron supplementation on mental and motor development in children: systematic review of randomised controlled trials. *Public Health Nutrition*, **8**(2), 117–132.
Salem, N., Pawlosky, R., Wegher, B. & Hibbeln, J. (1999). In vivo conversion of linoleic acid to arachidonic acid in human adults. *Prostaglandins Leukotrienes and Essential Fatty Acids*, **60**(5–6), 407–410.
Sandstead, H.H., Penland, J.G., Alcock, N.W., Dayal, H.H., Chen, X.C., Li, J.S., Zhao, F. & Yang, J.J. (1998). Effects of repletion with zinc and other micronutrients on neuropsychologic performance and growth of Chinese children. *American Journal of Clinical Nutrition*, **68**(2 Suppl), 470S–475S.
Sazawal, S., Bentley, M., Black, R.E., Dhingra, P., George, S. & Bhan, M.K. (1996). Effect of zinc supplementation on observed activity in low socioeconomic Indian preschool children. *Pediatrics*, **98**(6), 1132–1137.
Scott, T.M., Peter, I., Tucker, K.L., Arsenault, L., Bergethon, P., Bhadelia, R., Buell, J., Collins, L., Dashe, J.F., Griffith, J., Hibberd, P., Leins, D., Liu, T., Ordovas, J.M., Patz, S., Price, L.L., Qiu, W.Q., Sarnak, M., Selhub, J., Smaldone, L., Wagner, C., Wang, L., Weiner, D., Yee, J., Rosenberg, I. & Folstein, M. (2006). The Nutrition, Aging, and Memory in Elders (NAME) study: design and methods for a study of micronutrients and cognitive function in a homebound elderly population. *International Journal of Geriatric Psychiatry*, **21**(6), 519–528.
Shrestha, S.R., West, C.E., Bleichrodt, N., van de Vijver, F. & Hautvast, J.G.A.J. (1994). Supplementation with iodine and iron improves mental development in Malawian children. In: *Effect of Iodine and Iron Supplementation on Physical, Psychomotor and Mental Development in Primary School Children in Malawi*. Thesis Wageningen, Grafisch Service Centrum, Wageningen, pp. 41–58.
Simmer, K., Patole, S.K. & Rao, S.C. (2008a). Longchain polyunsaturated fatty acid supplementation in infants born at term. *The Cochrane Database of Systematic Reviews*, 1, CD000376.
Simmer, K., Schulzke, S.M. & Patole, S. (2008b). Longchain polyunsaturated fatty acid supplementation in preterm infants. *The Cochrane Database of Systematic Reviews*, 1, CD000375.
Smorgon, C., Mari, E., Atti, A.R., Dalla, N.E., Zamboni, P.F., Calzoni, F., Passaro, A. & Fellin, R. (2004). Trace elements and cognitive impairment: an elderly cohort study. *Archives of Gerontology and Geriatrics. Supplement*, **38**, 393–402.
Soemantri, A.G., Pollitt, E. & Kim, I. (1985). Iron deficiency anemia and educational achievement. *American Journal of Clinical Nutrition*, **42**(6), 1221–1228.
Solon, F.S., Sarol, J.N., Bernardo, A.B.I., Solon, J.A., Mehansho, H., Sanchez-Fermin, L.E., Wambangco, L.S. & Juhlin, K.D. (2003). Effect of a multiple-micronutrient-fortified fruit powder beverage on the nutrition status, physical fitness, and cognitive performance of schoolchildren in the Philippines. *Food and Nutrition Bulletin*, **24**(4 Suppl), S129–S140.

Stevens, L., Zhang, W., Peck, L., Kuczek, T., Grevstad, N., Mahon, A., Zentall, S.S., Arnold, I.E. & Burgess, J.R. (2003). EFA supplementation in children with inattention, hyperactivity, and other disruptive behaviors. *Lipids*, **38**(10), 1007–1021.

Sugden, C. (2006). One-carbon metabolism in psychiatric illness. *Nutrition Research Reviews*, **19**, 117–136.

Thompson, R.A. & Nelson, C.A. (2001). Developmental science and the media. Early brain development. *American Psychologist*, **56**(1), 5–15.

Vaisman, N., Kaysar, N., Zaruk-Adasha, Y., Pelled, D., Brichon, G., Zwingelstein, G. & Bodennec, J. (2008). Correlation between changes in blood fatty acid composition and visual sustained attention performance in children with inattention: effect of dietary n-3 fatty acids containing phospholipids. *American Journal of Clinical Nutrition*, **87**(5), 1170–1180.

Van Den Briel, T., West, C.E., Bleichrodt, N., van de Vijver, F.J.R., Ategbo, E.A. & Hautvast, J.G.A.J. (2000). Improved iodine status is associated with improved mental performance of schoolchildren in Benin. *American Journal of Clinical Nutrition*, **72**(5), 1179–1185.

van Stuijvenberg, M.E., Kvalsvig, J.D., Faber, M., Kruger, M., Kenoyer, D.G. & Benade, A.J.S. (1999). Effect of iron-, iodine-, and beta-carotene-fortified biscuits on the micronutrient status of primary school children: a randomized controlled trial. *American Journal of Clinical Nutrition*, **69**(3), 497–503.

Vazir, S., Nagalla, B., Thangiah, V., Kamasamudram, V. & Bhattiprolu, S. (2006). Effect of micronutrient supplement on health and nutritional status of schoolchildren: mental function. *Nutrition*, **22**(1), S26–S32.

Vermunt, S.H.F., Mensink, R.P., Simonis, A.M.G. & Hornstra, G. (1999). Effects of age and dietary n-3 fatty acids on the metabolism of [C-13]-alpha-linolenic acid. *Lipids*, **34**, S127.

Wada, Y., Kondo, H. & Itakura, C. (1996). Peripheral neuropathy of dietary riboflavin deficiency in racing pigeons. *Journal of Veterinary Medical Science*, **58**(2), 161–163.

Westermarck, T. & Antila, E. (2000). Diet in relation to the nervous system. In: *Human Nutrition and Dietetics*, 10th edn. Garrow, J.S., James, W.P.T. & Ralph, A. (eds), Churchill Livingstone, London, pp. 715–730.

World Health Organization (2006). Forum on the Marketing of Food and Non-alcoholic Beverages to Children. *Marketing of Food and Non-Alcoholic Beverages to Children: Report of a WHO Forum and Technical Meeting*. Oslo, Norway, 2–5 May.

Yehuda, S. (2003). Omega-6/omega-3 ratio and brain-related functions. *World Review of Nutrition and Dietetics*, **92**, 37–56.

Zimmermann, M., Adou, P., Torresani, T., Zeder, C. & Hurrell, R. (2000). Persistence of goiter despite oval iodine supplementation in goitrous children with iron deficiency anemia in Cote d'Ivoire. *American Journal of Clinical Nutrition*, **71**(1), 88–93.

Zimmermann, M.B., Connolly, K., Bozo, M., Bridson, J., Rohner, F. & Grimci, L. (2006). Iodine supplementation improves cognition in iodine-deficient schoolchildren in Albania: a randomized, controlled, double-blind study. *American Journal of Clinical Nutrition*, **83**(1), 108–114.

17 Tangible health benefits of phytosterol functional foods

Jerzy Zawistowski

17.1 Introduction

Functional foods and nutraceuticals are the fastest-growing category in the food market today. The main driving force for the market sustainability of these foods is the consumer himself, who is more and more aware of the benefits of healthy eating. An informed consumer is actively looking for foods that may provide risk reduction for certain chronic diseases such as coronary heart disease, in addition to their nutritional value. In the last decade, the food industry has been engaged in both product and process innovation research to yield new functional foods to satisfy customers' needs.

The world market for nutraceutical products was estimated to gross an approximate US$56 billion in 2001 alone, which represented 37% of the global nutrition industry. There were three major markets – the United States (33.2%), Europe (32.3%) and Japan (25.1%) – accounted for over 90% of all functional food sales in 2001. The estimated growth of functional food sales in 2001 and 2002 was 7.3 and 9%, respectively, with growth as large as 35% for products such as fish oil and omega-3 fatty acids in 2002 (The Natural Marketing Institute 2003). In 2006, the US sales of functional foods and beverages reached about US$24.8 billion, with sales projected to leap by 56% to US$38.8 billion by 2011 (Lukovitz 2007).

Phytosterols are well-established functional ingredients in the global market. They represent one of the most intensely and actively researched groups of nutraceuticals in the area of cardiovascular diseases. This group of compounds has been clinically proved to reduce blood cholesterol. Based on numerous clinical studies in which the low-density lipoprotein (LDL) level decreased by about 10% after phytosterol consumption, the US National Cholesterol Education Program has advised that phytosterol consumption in the range of 2 g/day will have a contributing effect to the primary prevention of coronary heart disease (Expert Panel on Detection, Evaluation, and Treatment of High Blood Cholesterol in Adults 2001). With nearly 100 million Americans and another 100 million Europeans experiencing elevated cholesterol levels, phytosterol-containing functional foods may offer preventive solutions that are safe and effective alternatives or complementary to medical strategies to lower blood cholesterol.

This chapter provides a review of phytosterol properties, their efficacy and safety, as well as the challenges regarding formulation, regulation and commercialisation of food products enriched with phytosterols. In this review, the term 'phytosterols' encompasses both sterols and stanols.

Table 17.1 Sterol content in selected foods (Moreau *et al.* 1999)

Food source	mg/kg
Rice bran oil	11 900
Corn oil	9 680
Sesame seeds	7 140
Soybean oil	2 500
Olive oil	2 210
Garbanzo beans	350
Bananas	160
Carrots	120
Tomatoes	70

17.2 Phytosterol properties

17.2.1 Chemistry and occurrence

Phytosterols are lipid-like compounds found in vegetable oil, seeds, nuts and coniferous trees and are essential for stabilising and regulating the fluidity and the permeability of cell membranes in plants. A similar role is shared by cholesterol in animals. Phytosterols are present in some foods as minor constituents. Vegetable oils are major sources of dietary phytosterols and may contain 0.1–1.0% of sterols (Verleyen *et al.* 2002). Comprehensive compositional analysis of edible oils and fats is provided by Phillips *et al.* (2002). Some of the specialised oils, such as that extracted from corn fibre, may contain sterols up to 10 g/100 g of oil (Moreau 2005). Cereals, nuts, vegetables, berries and other fruits contain sterols in the range of 0.05–0.22 g/100 g (Piironen *et al.* 2000). Table 17.1 shows phytosterol content in selected foods (Moreau *et al.* 1999). In plants, phytosterols occur in various forms: as free alcohols, as esters (esterified with long-chain fatty acids or with ferulic acid as in rice bran – γ-oryzanol), as steryl glucosides and as acetylated steryl glucosides (Moreau *et al.* 2002).

There are approximately 200 different phytosterols and related compounds in plant and marine materials (Piironen *et al.* 2000), with the most common ones being β-sitosterol, campesterol and sitostanol (Figure 17.1). Minor phytosterols are avenasterol, oryzanol, fucosterol and ergosterol (Ling & Jones 1995). β-Sitosterol, campesterol and stigmasterol are predominant in sterol fractions of the soy and corn oils. Δ-5-Avenasterol is found in oat bran and γ-oryzanol is a natural component of rice, while brassicasterol is a characteristic sterol isolated from cruciferous plants of the Brassicaceae family, such as canola. Vegetable sources provide mainly unsaturated sterols with a low concentration (about 2%) of saturated counterparts (stanols). Table 17.2 shows the phytosterol composition of selected cereals, fruits and vegetables (Piironen & Lampi 2004). On the other hand, phytosterols from coniferous trees, such as pine and spruce, contain a significant (about 10% and more) amount of stanols.

Chemically, phytosterols are triterpenes, composed of a steroid skeleton and an alkyl side chain. They are closely related to cholesterol, but they are also structurally distinct, with the presence of a side chain substitution of a methyl (campesterol) or an ethyl (sitosterol) group on carbon 24. Moreover, the additional double bond at position 22 is unique for stigmasterol (Moreau 2005). Hydrogenation of plant sterols results in the formation of plant stanols; thus, saturation of the 5α position of campesterol produces campestanol, and a similar positional saturation of β-sitosterol will yield sitostanol (Figure 17.1).

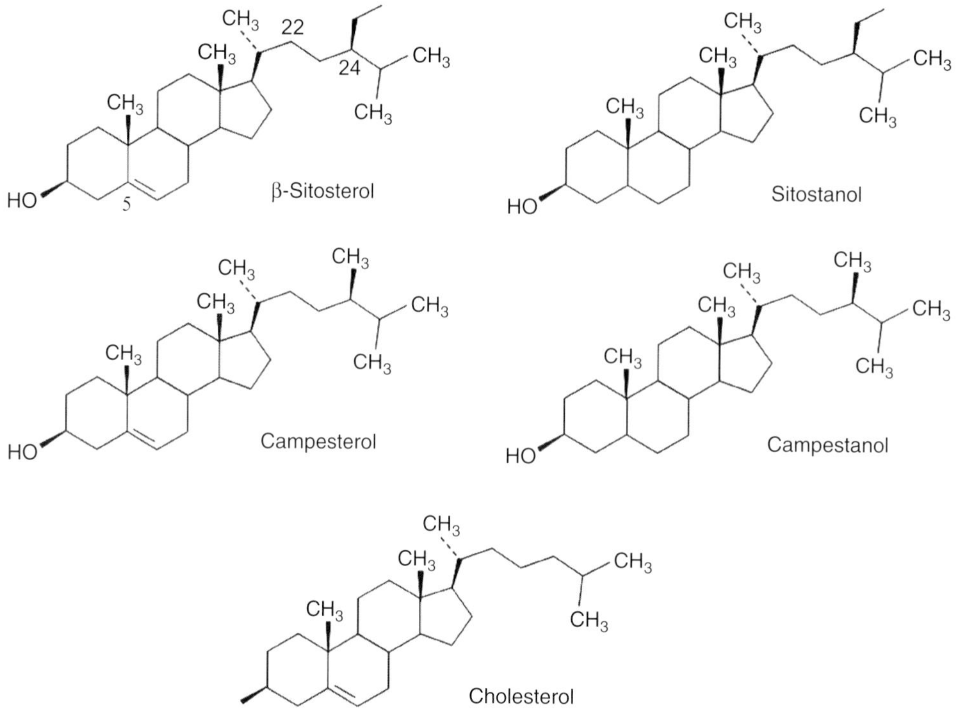

Fig. 17.1 Chemical structure of the main phytosterols and cholesterol.

Stanols are a less abundant class of sterols that are found in oil seeds and wood pulp alternatives. They contain a fully saturated ring structure, as evidenced by the lack of the carbon–carbon double bond found in both cholesterol and phytosterols.

The presence of phytosterols in an average Western diet varies from about 100 to 300 mg/day, including 20–50 mg/day of stanols. In contrast, vegetarian and Japanese diets produce a daily consumption of phytosterols that approximates 450 mg/day (Nair *et al.* 1984; Ling & Jones 1995; Morton *et al.* 1995). The amount of phytosterols in the typical Western diet is not sufficient to exert a lowering effect on blood cholesterol in the human body.

Table 17.2 Phytosterol composition of several cereals, vegetables and fruits (mg/kg)[a]

Sample	Campesterol	Sitosterol	Stigmasterol	Avenasterol	Stanols	Total
Barley	150–192	437–484	24–36	56–69	17–19	720–801
Millet	112	371	18	87	—	770
Corn	—	—	—	—	—	662–1205
Rice	146	375	104	20	32	723
Broccoli	67–69	285–310	8–11	2	18	367–390
Carrot	10	70	30	—	—	120
Lettuce	10	50	40	—	—	100
Tomato	10	30	30–35	—	—	70
Apple	3.6–9.0	130–157	1	7	8	130–185
Banana	20	110	30	—	—	160
Raspberry	9	233	—	10	2	274

[a]Adapted from Piironen and Lampi (2004).

Therefore, phytosterol-enriched functional foods are gaining popularity among health-conscious consumers.

17.2.2 Solubility

Phytosterols are lipophilic compounds and by their nature are practically insoluble in aqueous solutions. Their solubility in fats and oils is limited and dependent on the type of oil, as well as the temperature. It is possible to incorporate 10–20% of phytosterols into vegetable oil at 50–80°C; however, phytosterols tend to crystallise on cooling (Engel & Schubert 2005). It has been shown that the solubility of phytosterols in corn oil at lower temperatures (below 50°C) is limited to 2–6%. A low solubility of sterols and stanols was also shown in sunflower and olives oils, at low temperatures, by Melnikov *et al.* (2004). Yet, sterols and mixtures of sterols/stanols have higher solubility than stanols alone. Among these, wood stanols were slightly better solubilised in one study than their soya counterparts (Vaikousi *et al.* 2007).

The addition of water to oil tends to decrease phytosterol solubility by twofold (Christiansen *et al.* 2002). About 3.5–4% of β-sitosterol was dissolved in pure medium-chain triglyceride oil (MCT), in contrast to 1.5–2% of the same sterol in MCT oil that contained 10% water (Christiansen *et al.* 2002). The lower solubility of phytosterols in oil/water mixtures can be explained by the formation of monohydrate molecules (Jondacek 1997). Generally, phytosterols are better solubilised in more polar oils such as diacetyl glyceride oil (e.g. Enova oil marketed by ADM). The partially polar nature of phytosterols, due to the presence of hydroxyl group, may account for this (Vaikousi *et al.* 2007).

The high melting point of phytosterols, which is in the range of 138–145°C (Vaikousi *et al.* 2007), affects their solubility. An inverse relationship between log solubility and melting points for many aromatic compounds has been shown before by Yalkowsky (1981).

Esterification of phytosterols with long-chain polyunsaturated fatty acids was introduced in the late 1980s in order to increase fat solubility and to enable their incorporation into fatty foods such as vegetable fat spreads (Mattson *et al.* 1977). This technique increases phytosterol average fat solubility by tenfold (Ostlund 2002). It is worthwhile to note that the solubility of phytosterol esters increases with the number of double bonds, as well as with the length of the fatty acid chain attached via a hydroxyl moiety (Noakes *et al.* 2005). This is also associated with a decrease in the melting point. It is evident that esterification of phytosterols with even mid-chain fatty acids reduces their melting point. Vaikousi *et al.* (2007) have shown that the esterification of stanols with C8–C12 fatty acids caused a substantial reduction in melting point of about 40°C. The melting points of β-sitosterol esters with C6–C14 fatty acids were within the range 74–87°C, which was lower than for stanols esters (Leeson & Floter 2002). Commercial sterol esters with long-chain polyunsaturated fatty acids have melting points in the range of 25–45°C (ADM 2007; Cognis 2007).

Furthermore, sterols and stanols are soluble in organic solvents such as hexane, isooctane and acetate, while their esters are soluble in isooctane, hexane and 2-propanol (FAO 2008).

17.2.3 Antioxidant and antimicrobial properties

A number of scientific reports suggest that phytosterols possess antioxidant activity. This property is linked to the chemical structure of phytosterol molecules. Sterols with an ethyldiene side chain, such as avenasterol and fucosterol, have been found to be effective antioxidants (Boskou 1998; Guillen & Manzanos 1998). In plants, it is believed that phytosterols also act as antioxidants by intercalating and stabilising membranes (Yoshida & Niki 2003). Gordon and Magos (1983) have shown that these phytosterols inhibit thermal oxidation of olive oil at 180°C. It is most likely that free lipid radicals react with phytosterols at allylic carbon atoms, forming stable isomers and slowing down further oxidation by interrupting the chain of autoxidation (Yoshida & Niki 2003). White and Armstrong (1986) also showed that avenasterol, but not β-sitosterol, prevents deterioration of soybean oil when added and heated at 180°C. The latter sterol lacks a double bond on a side chain. In current studies, mixed sterols containing β-sterol, campesterol, and smaller amounts of stigmasterol and avenasterol exerted antioxidant and antipolymerisation properties when heated in oil at 180°C (Winkler & Warner 2008). Both stigmasterol and avenasterol have unsaturated bonds in their side chains. Interestingly, the antipolymerisation effect was concentration-dependent and favourable for phytosterols in soybean oil, in the range of 0.25–2.5%. However, when phytosterols were added to purified sunflower oil at 2.5%, an increased polymerisation was observed (Winkler & Warner 2008). It is possible that some of the phytosterols present in the mixture may have a pro-oxidant effect, as has been shown for stigmasterol (Yoshida & Niki 2003).

Wang *et al.* (2002) showed that the antioxidant and antipolymerisation properties of oryzanol and other steryl/stanyl ferulates are concentration-dependent. In addition, the oxidative stability of oil with added rice phytosterols at frying temperature (180°C) was varied and dependent on the type of sterols present. Phytosterol mixtures from rice bran oil exhibited the highest oxidative stability value, followed by oryzanol and sitostanol ferulate, indicating an additive effect of the various phytosterols present.

Our results revealed that a wood sterol/stanol mixture provided oxidative stability to MCT heated at 180°C (Du & Zawistowski 2002). The oxidation of essential polyunsaturated fatty acids, which had been used to enrich MCT, was significantly inhibited. In addition, the presence of phytosterols in oil showed a positive effect on the thermal stability (60°C) of the MCT oil-in-water emulsion (Du & Zawistowski 2002). The above findings are similar to results by Yoshida and Niki (2003), who report that a mixture of β-sitosterol, stigmasterol and campesterol inhibited oxidation of methyl linoleate in oil during heating. Wood phytosterols in free form were more effective in protecting against thermal oxidation of canola oil when compared to phytosterol esters. Moreover, reduced polymerisation attributed to thermal degradation of oils that had undergone heat stress and oxidation during frying was also greater for non-esterified phytosterol mixtures (Zawistowski *et al.* 2005).

The antioxidant properties of phytosterols may be related to the antimicrobial activities as has been confirmed in other phytochemicals (Dykes *et al.* 2003). Although the information is too limited to make a conclusive judgement, some preliminary research has shown that β-sitosterol obtained from several plants possesses antimicrobial activity against *Bacillus subtilis* (Beltrame *et al.* 2002) and *Candida albicans* (Moshi *et al.* 2004). Our own studies indicate that a mixture of free sterols and stanols, when emulsified, delays onset of microbiological spoilage of dairy products (Monu *et al.* 2008).

17.2.4 Thermal behaviour

Phytosterols are natural constituents of edible oils and other food products as well as essential bioactive compounds of cholesterol-lowering functional foods. Because foods may be exposed to frying, or even deep-frying, it is important to understand the thermal behaviour of these components.

Phytosterols, by virtue of their antioxidant properties, undergo oxidation to spare more thermally labile compounds such as polyunsaturated fatty acids. Studies have shown that phytosterols, when heated, may form polar and non-polar compounds (Dutta & Savage 2002). Heating may cause reduction of phytosterol content, and lower or even defunct efficacy. In addition, oxidised phytosterols may have harmful properties similar to oxidised cholesterol (Dutta 2002).

Phytosterols are relatively thermal resistant, as has been shown by a number of studies. Continuous frying in various oils for 9 hours resulted in phytosterol losses in the range of 4–6% (Winkler *et al.* 2007). Similar studies, where phytosterols in oils were submitted to commercial frying for 2 days, showed only slight changes in their content in the range of 6–8% (Dutta & Appelqvist 1996). Changes that occurred during extreme thermal treatment (180°C, and prolonged time) were dependent on the type of oils. For example, when various oils were continuously fried for 35 hours, phytosterol content in corn, soybean and expeller-pressed soybean oils dropped by 15, 9.4 and 4%, respectively (Winkler *et al.* 2007). Frying phytosterol-fortified (5% wood phytosterols) oils at 180°C for 6 hours resulted in a significant reduction of phytosterol concentration by 5–20%. Of particular interest was the finding that the overall rate of loss of major phytosterol content was lower in thermally processed soybean oil compared to sesame oil over an identical heating period at 180°C (Zawistowski *et al.* 2003). Some studies (Ghavami & Morton 1984; Soupas *et al.* 2004) have shown that the thermal stability of phytosterols added to oils depends on the degree of oil saturation. In studies conducted using phytosterol-enriched (2.5–4%) rapeseed oil, losses of sterols were higher in hydrogenated oil than in regular, deodorised oil (Ghavami & Morton 1984). This may suggest that unsaturated fatty acids are oxidised faster, providing some thermal protection to phytosterols.

Thermal stability of phytosterols also depends on their chemical structure. Stanols are more resistant than their unsaturated counterparts. Lampi *et al.* (1999) showed that β-sitostanol was much more stable than ergosterol in paraffin and rapeseed oils heated for 24 hours at 180°C. Significantly lower losses of sitostanol as compared with other sterols (campesterol, stigmasterol, sitosterol and avenasterol) in corn oil, during deep-frying, were also reported by others (Winkler *et al.* 2007).

A few studies were focused on oxidative products arising from thermal degradation of phytosterols. The majority of them are related to phytosterols that are inherently occurring and present in low concentration in edible oils and other foods. Home cooking vegetables and fruits showed no significant differences in phytosterols in raw and cooked products (Normén *et al.* 1999). In commercial deep-fat frying, these components in oil partition into the foods fried in these oils. Heating oil (conventional and microwave heating) at 180°C for 2 hours revealed no significant sterol differences of untreated and treated oils (Albi *et al.* 1997). No marked differences in phytosterol composition were detected when French fries and potato chips were deep-fried at 250°C for 2 days in a rapeseed/palm oil blend, sunflower or high-oleic sunflower oils (Dutta & Appelqvist 1997). However, a slight increase of the sterol oxide level has been detected in oils and in fried products. In a rapeseed/palm oil blend, sterol oxide increased from 41 to 60 ppm, and in sunflower oil it increased from 40 to 57 ppm, while in high-oleic sunflower oil it increased from 46 to 56 ppm (Dutta 1997). The sterol oxidation products identified in this study were hydroxyl, dihydroxy, keto, and epoxy derivatives of sitosterol and campesterol (Dutta 1997). Figure 17.2 shows some of these products.

Przybylski *et al.* (1999) evaluated the changes in the phytosterol fraction of canola oils with modified fatty acid composition after 72 hours of heating at 190°C. The oils used were regular canola oil, low-linolenic canola oil, high-oleic canola oil and hydrogenated canola oil. The study showed that cholesterol is more resistant to deterioration than phytosterols. Cholesterol

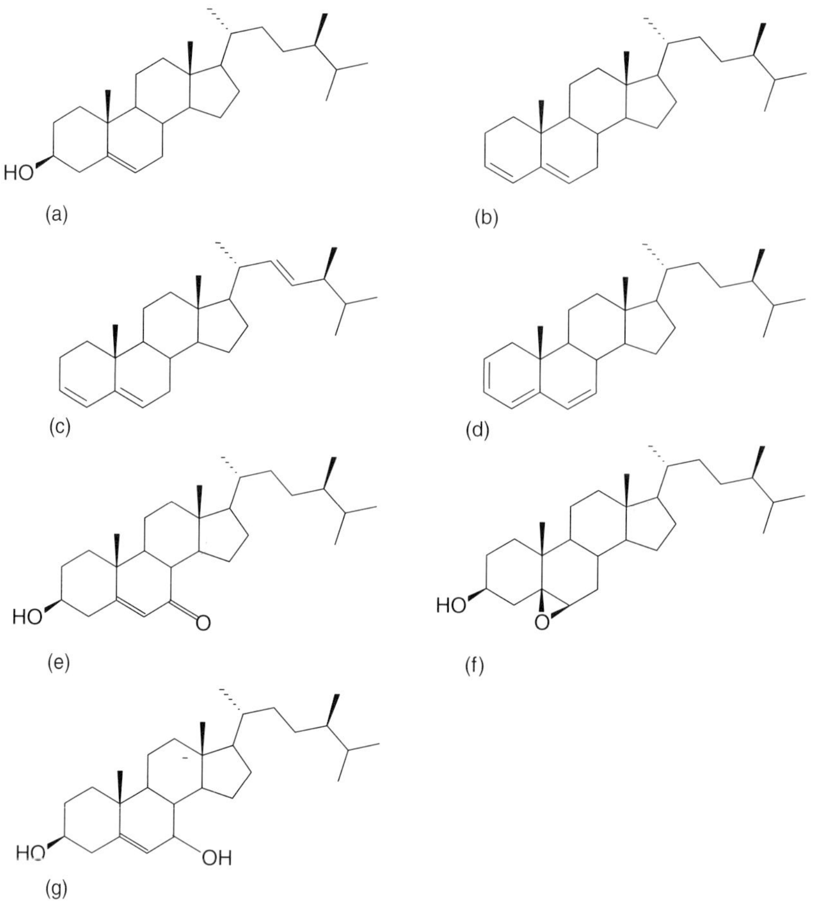

Fig. 17.2 Structure of the main phytosterol oxides and dehydrates as exemplified by campesterol. Analogue oxides and dehydrates are generated by the other phytosterols. (a) Campesterol, (b) campesta-3,5-diene, (c) campesta-3,5,22-triene, (d) campesta-2,4,6-triene, (e)7- keto-campesterol, (f) 5,6-epoxy-campesterol, (g) 7-hydroxy-campesterol.

deteriorates faster at temperatures above 120°C, whereas β-sitosterol and campesterol deteriorate faster at 95°C, which indicates that phytosterols can oxidise faster than cholesterol. This may impart that phytosterols may provide antioxidant protection to cholesterol. The study also concluded that in the four types of oil evaluated, brassicasterol was the least resistant to deterioration, followed by β-sitosterol, and campesterol, respectively.

Oxidative stability is very important for phytosterol-containing functional foods (Katan *et al.* 2003), since these foods are widely available for consumers in many countries. Generally, commercially available phytosterols are stable during processing such as spray-drying, pasteurisation, ultra-high temperature (UHT, 135°C, 1–2 seconds) treatment and long-term storage (Soupas 2006). In addition, phytosterols were proved to be very stable during esterification at 90 and 120°C for an extended period (Azadmard-Damirchi & Dutta 2008). However, a high-temperature treatment, such as deep-frying or even pan-frying, may induce phytosterol oxidation. The effect of frying for 6 hours at 180°C on commercial wood phytosterols formulated (5%) into soybean and sesame oils was investigated. Upon heating, phytosterols were converted into sterol dienes and trienes, yielding 20 and 28% dehydrates in sesame oil, and soybean oil, respectively (Figure 17.2). It was observed that campesterol was more thermally labile than β-sitosterol (Zawistowski *et al.* 2003). Dehydrates may be very initial products of phytosterol conversion caused by heat. It has been suggested that the types of oxidative products depend on the phase of oxidation. In the initial stage of oxidation, keto-phytosterols are accumulated. Once the oxidation reaches the dynamic stage, these compounds are converted to epoxyphytosterols and hydroxyphytosterols (Figure 17.2) (Soupas 2006). Phytosterol esters, due to the presence of polyunsaturated fatty acids, are the most susceptible to thermal oxidation, followed by free sterols and stanols. Generally, phytosterols are oxidised in the same way as other lipids, and the rate of oxidation increases with an increasing number of double bonds (Porter *et al.* 1995).

17.3 Efficacy of phytosterols

The therapeutic value of phytosterols has been recognised for over 50 years. The drug product Cytellin™ derived from tall oil phytosterols was developed and marketed by Eli Lilly between 1950 and 1980. A number of clinical studies conducted on this product with doses between 9 and 50 g/day reported efficacy and no side effects (Pollak & Kritchevsky 1981). However, the product was ultimately withdrawn from the market since prohibitively large daily doses were required to achieve therapeutic potency.

In many of the clinical research studies conducted using stanol esters, researchers have concluded that stanols are more potent at lowering plasma total and LDL cholesterol than sterol esters, yet stanols are relatively poorly absorbed or not absorbed at all (Heinemann *et al.* 1986). However, the last 5 years of research brought about clinical results indicating that sterol esters are equally efficacious to stanol esters, when administered to individuals as a component of food. Reduction in LDL cholesterol was in the range of 10–15% when daily doses of 1–3 g (based on free phytosterols) of stanol or sterol esters were used (Heinemann *et al.* 1986; Vanhanen & Mietinen 1992; Hallikainen *et al.* 2000; Vuorio *et al.* 2000; Judd *et al.* 2002). An interesting point of view regarding the differences between sterol and stanol effectiveness is presented by some researchers, who claim that stanol esters maintain their efficacy longer than sterol esters (Miettinen & Gylling 2004; O'Neill *et al.* 2005). This observation is based on the greater plasma absorbability of sterols as opposed to stanols. It has been suggested that absorbed sterols downregulate bile acid synthesis, which in turn attenuates their long-term efficacy. In contrast, consumption of stanols lowers the sterol level in plasma and does not affect bile acid synthesis (O'Neill *et al.* 2005).

Research conducted over the past two decades indicates that phytosterols in free forms are as effective as their esterified counterparts. Nestel and co-workers conducted a head-to-head clinical study with free phytosterols versus phytosterol esters and showed that consumption of 2.4 g/day of phytosterols enriched in bread, cereal and margarine led to comparable efficacy (Nestel *et al.* 2001). Some researchers found free phytosterols to be even more effective than phytosterol esters in lowering cholesterol absorption in human (Mattson *et al.* 1982). It has been shown that consumption of as little as 0.7–0.8 g/day of free phytosterols yielded a LDL cholesterol reduction in the range of 6–15% (Pelletier *et al.* 1995). Doses of 1.5 g/day of free stanols and sterols were effective in lowering LDL cholesterol levels in hypercholesterolemic subjects (Heinemann *et al.* 1986; Jones *et al.* 1999).

Numerous clinical studies conducted with phytosterols utilised various food matrices such as vegetable oils, margarine, mayonnaise, dressings, ground beef, chocolate as well as low-fat foods such as cereal bars, ready-to-eat cereals, juices, yogurt, milk drink and soy drink (Law 2000; Katan *et al.* 2003; Clifton *et al.* 2004; Deveray *et al.* 2004; Moreau 2004; Normén *et al.* 2004; Weidner *et al.* 2008). For example, free wood phytosterol-enriched chocolate (6% sterols in chocolate) was the subject of a randomised, double-blind, placebo-controlled trial conducted in the Netherlands. Over a 4-week period, 70 participants consumed three servings per day of either phytosterol-enriched chocolate (1.8 g of sterols/day) or placebo chocolate. Total LDL cholesterol levels were significantly ($P < 0.05$) reduced without affecting high-density lipoprotein (HDL) cholesterol and triglyceride levels (de Graaf *et al.* 2002). Recently, a multicentre, double-blind, placebo-control, parallel clinical study was conducted using 83 hypercholesterolemic patients. The treatment group received 100 mL/day of low-fat yogurt drink enriched with 1.6 g of phytosterols for 42 days. Consumption of phytosterol-containing yogurt significantly reduced LDL cholesterol by over 12% after 3 weeks, and almost 11% after the next 6 weeks. Interestingly, triglyceride levels were also significantly reduced by 14% ($P < 0.018$) (Plana *et al.* 2008).

The overwhelming majority of human clinical studies on the effect of phytosterols have monitored established markers for coronary heart disease such as total and LDL cholesterol. However, a new predictor of cardiovascular disease, C-reactive protein (CRP), is emerging as supported by scientific data. Growing evidence suggests that CRP plays a more direct role in the pathogenesis of atherosclerosis and may contribute to the progression of atherosclerotic plaques (Libby & Ridker 2004; Nissen *et al.* 2005). Some newer clinical studies have investigated the effect of phytosterols on inflammation as assessed by CRP. In an 8-week, placebo-controlled, double-blind, randomised study, 72 healthy volunteers with mildly elevated cholesterol levels were fed twice a day a reduced-calorie orange juice drink enriched with phytosterols (1 g/250 mL). The results showed a significant, simultaneous reduction of LDL cholesterol (9.4%; $P < 0.001$) and CRP (12%; $P < 0.005$) (Devaraj *et al.* 2006). This indicates that phytosterols may have anti-inflammatory effects in addition to their cholesterol-lowering capabilities. The decrease in CRP was also reported in another study, where subjects were treated with phytosterol-enriched, low-fat fermented milk. However, the decrease was not significant, most likely due to the large variability in CRP at baseline (Hansel *et al.* 2007). Clifton *et al.* (2008) also reported a tendency for a modest, but statistically non-significant, decrease of CRP on consumption of 3 g of phytosterols in the cohort study.

The cholesterol-lowering results of clinical trials using food products containing free phytosterols are consistent with the findings obtained from trials conducted with stanol and sterol esters, independent of the phytosterol sources (Clifton *et al.* 2008). As a general rule, free sterol, stanol or sterol ester consumption results in an 8–12% reduction in the plasma LDL cholesterol levels (Katan *et al.* 2003). It has been generally accepted among scientists that there are no differences in efficacy to lower plasma cholesterol among various physical forms of phytosterols (Katan *et al.* 2003). Notwithstanding this, however, it is important to recognise that the method used to incorporate phytosterols into the food matrix will result in differences in the magnitude of change in circulating total plasma and lipoprotein cholesterol.

17.4 Mechanism of action of phytosterols

The presence of cholesterol in the intestine originates from three sources: diet (200–400 mg/day), bile (up to 1 g/day) and exfoliated enterocytes (300–400 mg/day) (Northfield & Hoffman 1975; Vuoristo & Mittinen 1985). Of this, about 50% of cholesterol is absorbed in the intestinal lumen, re-esterified by intestinal cells, incorporated into chylomicrons and secreted into the circulation (Turley & Dietschy 2003). The amount of

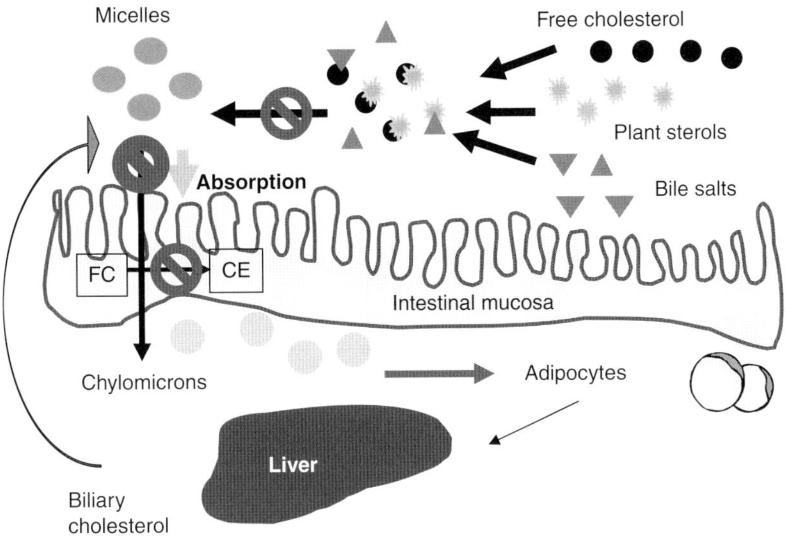

Fig. 17.3 Phytosterols mode of action. FC, free cholesterol; CE, cholesterol ester. For a colour version of this figure, please see the plate section.

cholesterol absorbed into the circulation depends on genetic variability and dietary factors (Sehayek 2004). Consumption of phytosterols plays a significant role in lowering cholesterol absorption.

There are a number of proposed mechanisms underlying the hypocholesterolemic effect of phytosterols (Figure 17.3). The main one involves the competition between phytosterols and cholesterol for micellar solubilisation based on similarities in their chemical structures (Gylling *et al.* 1999; Hendriks *et al.* 1999). It has been suggested that phytosterols have higher affinity for micelles than has cholesterol. Subsequently, phytosterols lower solubility of cholesterol, displace it from the mixed micelle and inhibit its absorption (Ostlund 2002).

The coprecipitation of cholesterol and phytosterols is considered to be another mechanism in the inhibition of cholesterol absorption. Both compounds have poor solubility in fat and mutually limit their incorporation into micelles. Once phytosterol consumption reaches a critical level in the intestinal lumen, the formation of mixed cholesterol/phytosterol crystals occurs. This leads to precipitation of cholesterol in the gastrointestinal tract and the reduction of its intestinal intake, since cholesterol in the crystalline form is not absorbable (Trautwein *et al.* 2003).

Phytosterols can also compete with cholesterol for uptake by the intestinal mucosal cell or by interfering with the esterification in enterocytes (Gylling *et al.* 1999). This mechanism may include the competition with cholesterol for the sterol transporter, Niemann-Pick C1-like 1 protein, to obtain entry into enterocytes. There could also be some competition for the ATP proteins, ABCG5 and ABCG8, which are responsible for efflux of both cholesterol and phytosterols between the intestinal mucosa and lumen (von Bergmann *et al.* 2005). Transport of phytosterols into the circulation is inefficient, since only 5% or less of phytosterols are absorbed from the intestinal lumen into the circulation as compared with up to 60% for cholesterol (Heineman *et al.* 1993; Yonkah 2006). However, this interference may also lower intestinal absorption of cholesterol.

The competition with cholesterol for esterification may account for the additional hypocholesterolemic effect of phytosterols. However, this effect is not significant since the rate of phytosterol esterification is much lower than that of cholesterol. Campesterol, β-sitosterol and stigmasterol are esterified at 89, 79 and 34% of the rate of cholesterol, respectively (Moghadasian 2000).

Studies have shown, as well, that the individual's cholesterol balance is a factor in the degree of reduced cholesterol bioavailability. The finding that the higher the baseline cholesterol absorption efficiency, the lower the cholesterol absorption when phytosterols are introduced, explains the positive correlation between high initial cholesterol absorption efficiency and increased magnitude of serum total and LDL cholesterol reductions (Gylling *et al.* 1999).

Although the reduction in cholesterol absorption is partially compensated by the increase of cholesterol synthesis, this increase is not large enough, and consequently total and low-density lipoprotein cholesterol (LDL-C) circulating in plasma are reduced. Levels of HDL cholesterol and triacylglycerides are typically not affected (Gylling *et al.* 1999; Nguyen *et al.* 1999).

It is worthwhile to mention that not all phytosterols are confined only to the inhibition of cholesterol absorption. Δ^{22}-Unsaturated phytosterols, primarily stigmasterol, but also ergosterol and brassicasterol are inhibitory to cholesterol synthesis. It has been shown that these types of phytosterols inhibit Δ^{24}-reductase, an enzyme that is involved in the conversion of lanosterol into cholesterol (Fernandaz *et al.* 2002).

17.5 Safety of phytosterols

Numerous clinical studies, safety evaluation studies, as well as the history of use of phytosterol-enriched foods indicate that sterols and stanols are generally recognised as safe (Hendriks *et al.* 2003; Katan *et al.* 2003; Korpela *et al.* 2006). However, an important concern is their potential interference with the absorption of fat-soluble nutrients, such as vitamins and antioxidants. Clinical trials conducted with sterol and stanol esters have shown that α- and β-carotene (provitamins) levels were reduced to 10–25%, with no effect on serum retinol levels (Weststrate & Meijer 1998; Gylling & Miettinen 1999; Gylling *et al.* 1999; Hendriks *et al.* 1999). A similar phytosterol-induced reduction of plasma lycopene levels has been shown, but less uniform for all products tested (Weststrate & Meijer 1998). Both α- and β-carotene, precursors of vitamin A, are transported mainly in LDL particles, which are reduced in phytosterol ester treatment. The clinical importance of the reduction in plasma β-carotene is unknown because the fat-soluble vitamin retinal, for which β-carotene is the precursor, is seemingly unaffected by phytosterol ester intake (Nguyen *et al.* 1999). Finally, with some exceptions (Gylling *et al.* 1999), many of the clinical trials that have reported cholesterol-lowering properties of phytosterols, as well as reductions in β-carotene, are relatively short-term studies lasting up to 3.5 weeks. Longer, placebo-controlled clinical studies are required to confirm the potential delayed side effects of dietary fortification with phytosterols. This is particularly important for those subjects who may suffer from phytosterolemia (Lutjohann & von Bergmann 1997). The European Food Safety Authority (EFSA) has evaluated the effect of phytosterols on carotenoids. The EFSA concluded that the reduction was in the normal range of variation for these compounds in the blood of adults (EFSA 2002), and generally of no concern.

Decreases in plasma vitamin D (25-OH-vitamin D), vitamin K and α-tocopherol have not been observed (Hendriks *et al.* 1999). However, a recent, long-term (15 weeks) safety study conducted in Finland reported a decrease in serum α-tocopherol levels in the treatment groups receiving foods (bread, yogurt and meat products) enriched with phytosterols in the range of 1.25–5.00 g/day as compared with the placebo group. The differences were not significant when α-tocopherol values were adjusted for reduced serum total cholesterol concentration (Tuomilehto *et al.* 2009). Similar results were reported earlier from a clinical study that administered low-fat dairy products enriched with phytosterols (Korpela *et al.* 2006).

Absorbability of phytosterol into the blood stream is one of the controversies around the safety of phytosterols that has been discussed recently in the literature. Some authors suggested that in spite of low absorbability of phytosterols, increased levels of sterols in blood may be associated with an increased risk of coronary heart disease (Weingartner *et al.* 2009). This is probably true in the case of people who are suffering from sitosterolemia, a rare autosomal, recessively inherited lipid metabolic disorder. Only about 40 cases of this disease had been identified worldwide as of the year 2000 (Lee *et al.* 2001). However, in non-sitosterolemics, there was no clear association between circulating levels of phytosterols and coronary heart disease risk (Wilund *et al.* 2004; Chan *et al.* 2006).

17.6 Manufacturing of phytosterols

Commercial phytosterols are primarily produced from soybean and rapeseed oils, as well as from tall oil. Sterols derived from soybean and other oilseeds are obtained during oil processing as by-products of vitamin E manufacturing. Coniferous phytosterols are derived from tall and pitch oil, products of the alkaline pulping of wood, using extraction, distillation and crystallisation processes (Quilez *et al.* 2003). It has been estimated that to produce 1 tonne of phytosterols approximately 2500 tonnes of vegetable oil or the same volume of coniferous trees is required. Commercial phytosterols are also obtained during production of biodiesel fuel. This technology utilises a variety of vegetable oils. Phytosterols are obtained primarily from rapeseed oil residues in the oil deacidification process, or from deodoriser distillates arising in the production of fatty acid methyl esters, known as a biodiesel fuel. Approximately 54 kg of phytosterols may be produced from 1 tonne of distillate (Verhe *et al.* 2007).

Currently, Wong (2008) has shown potential economic benefits for the commercial production of phytosterols from grain processing residues in breweries and distilleries. The proposed process can yield 100 tonnes of sterols from 18 000 to 65 000 tonnes of grain residues.

17.7 Challenges in formulation, regulatory approval and commercialisation of phytosterol-containing foods

17.7.1 Formulation of functional foods

Formulation of food products with phytosterols is difficult from a technological and food quality point of view, since phytosterols are insoluble in water and poorly soluble in fats and oils. At the same time, because of their waxiness, phytosterols may have an adverse effect on sensorial properties of foods (Salo & Wester 2005). There is also a risk of creating oxidative products if phytosterol-containing foods are exposed to extreme thermal treatments, as has been discussed in the previous section.

Commercial phytosterols are available in two forms: phytosterol esters (semiliquid paste) and free phytosterols in crystalline form (see Table 17.3 for examples of commercial food products enriched with phytosterols as approved under specific regulatory venues in various countries). The latter phytosterols can be purchased in various particle sizes, typically 30–300 μm. Phytosterol esters have significantly higher fat solubility and lower melting point, making them much easier to formulate in fatty-based foods than free phytosterols. However, esters are less stable than free phytosterols because of their polyunsaturated fatty acid moiety. Smith (1981) has reported that β-sitosterol was resistant to oxidation during 8-week storage at 20°C, while its esterified counterpart, β-sitosterol linoleate, was extensively autoxidised in the linoleate moiety and sterol nucleus, presumably at the C-7 position. Oxidative lability of phytosterol esters may also have adverse effects on the stability of the foods that contain them. We have shown that canola oil enriched with phytosterol esters is oxidised much faster than the control oil when heated at 100°C and 190°C for 2 and 6 hours. On the contrary, the same phytosterols present in the free form exhibit protection against thermal oxidation of canola oil (Zawistowski *et al.* 2005).

Another challenge to the formulator of food products is posed by the inherent 'powderiness' of free phytosterols due to their crystalline form. The particle size of solid ingredients has a significant impact on sensorial attributes of foods, such as taste and texture. Insoluble particles can contribute to grittiness, and coarseness, which define in part the overall 'mouthfeel' of foods (Ringel 1970; Birch *et al.* 1977). For example, chocolate is perceived as having a gritty texture when the cocoa powder used in the formulation has particles larger than 35 μm. If at least 20% of the cocoa powder particles are larger than 22 μm, the chocolate texture is defined as being coarse (Rostagno 1969). When formulating phytosterol-containing foods, it is important to use as small particle sizes as possible, preferably below 20 μm, so that foods can be perceived by consumers as smooth in texture and acceptable in 'mouthfeel'.

Historically, free phytosterols have been used to formulate foods with little success. The breakthrough in food formulation came with the introduction of technology, which enabled the conversion of crystalline phytosterols into fat-soluble esters. The process of stanols esterification was patented by the Finish food manufacturer, Raisio Oy, in 1989 (Thompson & Grundy 2005). Consequently, margarines and vegetable fat spreads were the very first stanol-containing functional foods introduced in the market. In 1995, Raisio developed margarine under the name Benecol, for the EU market. The introduction of this product marked the beginning of functional foods in Europe and subsequently in the United States. Since then, sterol and stanol esters have been used to formulate low-fat vegetable spreads. Some of these spreads contain less than 30% fat. In addition to their health benefits, phytosterols provide texturising effect and minimise the use of hard stock (Wester 1998). Vegetable fat spreads are ideal vehicles for phytosterol esters, although cheese, salad dressings and yogurts are also used as delivery systems. It is worthwhile to note that the inclusion of phytosterol esters into the food matrix adds extra fat and calories. Therefore, phytosterol esters are not suitable for enrichment of fat-free food products. Also, the use of phytosterol esters in the formulation of low-fat foods, including vegetable spreads, may present challenges. For example, cholesterol-lowering fat spreads typically contain 12–13% phytosterol esters. This adds about 5% fat and requires the formulator to keep the oil concentration in the fat phase below 20%.

In spite of these challenges, a number of phytosterol ester-containing food products including extra-light vegetable fat spreads, such as a 23% low-fat product called Flora Pro-Active, can be found in the market.

Although formulation of vegetable fat spreads with an efficacious amount of free phytosterols is fraught with difficulties, some manufactures prefer to use this form of sterols instead of their esterified form for a number of reasons. Firstly, the incorporation of sterol and stanol esters into fat spreads is generally protected by patents and patent applications. Secondly, phytosterols are less expensive than their esters. Thirdly, as it has been already discussed, free phytosterols do not have any caloric value from fat that is carried into food

Table 17.3 Examples of commercial food products enriched with phytosterols as approved under specific regulatory venues in various countries

Country	Regulatory venue	Type of health claim	Example of health claim wording	Example of commercial foods enriched with phytosterols[a]
The United Stats	GRAS notification Self-GRAS	Disease risk reduction (the Nutrition Labelling and Education Act – sterol/stanols and coronary heart disease claim)	'May reduce risk of coronary heart disease'	*Fat spreads* Take Control (Unilever Bestfoods) Benecol (McNeil) Natucol (GFA) *Orange juice* Heart Wise (Minute Maid) *Yogurt and yogurt drink* Benecol (McNeil) Promise Active (Unilever) Healthy Heart (Yoplait) *Breakfast cereals* Heart Wise (Healthy Snacks) Corazonas Tortilla Chips (Corazonas Foods)
European Union	Novel food regulations	Disease risk reduction (Regulation EC 1924/2006) on nutrition and health claims made on foods, Article 14)	'Plant sterols have been shown to lower/reduce blood cholesterol. Blood cholesterol lowering may reduce the risk of coronary heart disease'	*Fat spreads* Benecol (McNeil) Benecol (Minerva Greece) Becel Pro-activ (Unilever Bestfoods) Dairygold Heart (Dairygold Ireland) Deli Reform Active (Walter Rau Germany) *Yogurt drink* Emmi Benecol (Emmi AG) Benecol (McNeil) Flora Pro-activ (Unilever Bestfoods) Danacol (Danone) Tesco Reducol (Tesco) Salutare Reducol (Pingo Doce Portugal) Healthy Heart (Yoplait) Optifit Cholest Drink (Aldi) *Yogurt* Benecol (Valio) Flora Pro-activ (Unilever Bestfoods) Tesco Reducol (Tesco) Pirkka Reducol (Ruokakesko)

(*Continued*)

Table 17.3 *(Continued)*

Country	Regulatory venue	Type of health claim	Example of health claim wording	Example of commercial foods enriched with phytosterols[a]
				Milk Flora Pro-activ (Unilever Bestfoods) Mimosa Benvital (Lactogal, Portugal) *Cheese* Heartfelt Reducol (Fayrefield Foods, UK) Heartily Healthy (The Cheese Company, UK) *Bread* Pirkka Reducol (Kesko Oy. Finland)
Australia and New Zealand	Novel food standard, Standard 1.5.1	Proposal P293 – high-level health claims. Requires preapproval by FSANZ	'With natural plant sterols which reduce cholesterol uptake'	*Fat spreads* Logicol (MeadowLea Foods) Lochol (Coles) Flora Pro-activ (Van den Bergh Unilever) *Yogurt* Logicol (MeadowLea Foods) *Milk* Logicol (MeadowLea Foods)
Japan	Food of specified health uses	Structure function claim	'Good for those concerned about serum cholesterol' 'Good for those having relatively high serum cholesterol and triglycerides with mild obesity'	*Edible oils* Econa DAG/sterol oil (Kao Co.) Oillio corn fibre oil (Nisshin Oillio Group Ltd) *Fat spreads* Rama Pro-activ (Unilever)
South Korea	Health Functional Food Act	Disease risk reduction Structure function claim	'Phytosterols may reduce risk of coronary heart disease'	Cholzero green tea, coffee and milk drinks (Eugene Science Inc.) Cholzero cooking oil (Eugene Science Inc.)
Taiwan	Health Food Control Act	Yes, approved by DOH	'Consumption of the product may help lower blood total cholesterol'	Plant sterol milk (Uni-President)
Malaysia	Food Act and Regulation, Food Supplement	Nutrient function claims	'Plant sterol or stanol helps lower or reduce cholesterol'	ActiCol omega plus milk (Nestle) ActiCol omega plus soy drink (Nestle)

Indonesia	Drug and Food Control Regulations	Disease risk reduction	'May reduce risk of coronary heart disease'	NesVita Proheart milk (Nestle) Flora Pro-activ fat spread (Unilever) Benecol smoothes (Kalbe Nutritionals)
Singapore	Food Regulation Act, approval on case-by-case bases by Food Control Department of Singapore	No health claims permitted	'For people who want to lower their blood cholesterol level'	ActiCol omega plus milk (Nestle)
Thailand	The notification for foods for special dietary uses	Nutrient function claims	'Plant sterol may help lower cholesterol'	Benecol yogurt drink (Saha Pathanapibul)
The Philippines	Food Fortification Act	Structure/function claim	'This product contains natural plant sterols that help lower cholesterol'	NesVita Proheart milk (Nestle) NesVita Proheart yogurt drink (Nestle)
Brazil	Technical regulation on procedures for registration of foods and or new ingredients	Structure/function claims	'Phytosterols help in reducing the absorption of cholesterol'	Benecol fat spread (BBN)

[a]The term 'phytosterols' includes free sterols and stanols, as well as sterol and stanols esters.

products. Theoretically, phytosterols can be incorporated into a fat phase by dissolving them at 80°C or higher. However, on cooling, sterols tend to recrystallise, making the emulsion very viscous, and therefore impairing processing. The recrystallisation of phytosterols, which is enhanced by the high water content of low-fat spreads, also contributes to the graininess and waxiness of the final product, sensorial attributes that are not acceptable to consumers. To create acceptable low-fat spreads, Auriou and Ferreres (2002) have proposed to disperse phytosterol powder in an aqueous phase in the presence of an emulsifier with a higher HLB (hydrophilicity–lipophilicity balance) value than phytosterols, such as lysolecithin. Zawistowski (2007) has incorporated phytosterols into the aqueous phase of vegetable fat spread, with the stabilising aid of modified starch and protein.

Formulation of phytosterols into aqueous-based food matrices requires stabilised liquid dispersions. Once suspended into a liquid medium, phytosterol particles may cream or sediment, which can adversely affect sensory attributes such as taste and appearance. In some products the sterol particles may aggregate, making it difficult or even impossible to redisperse uniformly on shaking. Reducing the particle size of phytosterols may help to stabilise dispersion, since smaller particles have more surface area per unit volume, and therefore they can be mixed more efficiently with food ingredients. Size reduction can be achieved by mechanical grinding (e.g. dry milling and micronisation) as well as wet milling (e.g. colloidal/jet milling and microfluidisation). The latter technique allows for processing phytosterols in high-concentration slurries and creates uniform dispersions, comprising particles of less than 10 μm. Microfluidisation, or particle collision technology, was successfully used to formulate yogurt, soya drink, cereal bars and chocolate enriched with phytosterols (Zawistowski 2003). Stabilisation of aqueous dispersions can also be achieved by admixing phytosterols with lyophilic colloids (carrageenin, alginate, guar gum, xanthan gum, etc.), stabilisers such as microcrystalline cellulose, and emulsifiers such as polysorbates, sucrose fatty acid esters, and lecithin (Zawistowski 2003), as well as protein (Moreau 2004). Lecithin was also used to formulate water-dispersible stanols as emulsified micelles with potential applications in fat-free beverages (Ostlund *et al.* 1999). However, the high cost of lecithin has so far precluded commercial applications. A less expensive approach was proposed recently by Japanese researchers, who developed a method to produce a phytosterol–egg yolk lipoprotein complex (PSY). Apparently, PSY is readily dispersible in water due to the emulsifying properties of egg phospholipids, and suitable for use in processed foods. The cholesterol-lowering properties of PSY are comparable with those of free phytosterols and esters (Matsuoka *et al.* 2008).

Another novel approach that might provide solutions to the aforementioned problems is to utilise nanotechnology. This refers to the use of free phytosterols with very small particle sizes (e.g. 300 nm), in forms of nanoparticles and/or nanoemulsions. Nanotechnology is increasingly being used in food applications. It has been suggested that this technology may improve food processing, packaging and stability, as well as result in enhanced flavour and nutrition (Weiss 2007). Dolhaine *et al.* (2007) used well-established methods including rapid expansion of supercritical solutions to produce nanoscale phytosterols and phytosterol esters, with particle sizes in the range of 10–300 nm. This invention provides phytosterols that are readily dispersible in a number of food matrices. In addition, nanosterols are claimed to be more absorbable into the blood serum upon consumption in comparison with their conventional counterparts (Dolhaine *et al.* 2007). The latter property, however, may present a safety problem. Therefore, until proven safe, this form of phytosterols cannot be used in food products.

Free phytosterols are the preferred sterol type to use in low-fat or fat-free matrices, the food categories that are in demand among consumers who are consciously avoiding fatty foods. Phytosterols were successfully incorporated into low-fat and fat-free yogurt and yogurt drink, milk, orange juice, cereal bars and bread. In 1993, the Swiss dairy company Emmi introduced an Emmi Benecol yogurt drink in Europe. The low-fat drink is a single daily shot and delivers about 2 g of stanols in a 100 mL volume. The marketing success of this yogurt drink encouraged many other producers to introduce yogurt shots in the European market, containing either free or esterified phytosterols. In 2006, the sales of yogurt drinks approached 65% of those generated by well-established fat spreads (AC Nielsen 2006). Yogurt drink was also a subject of clinical studies. A 9.5% decrease in serum cholesterol concentration was achieved when a single-dose drink was consumed with a meal (Noakes *et al.* 2005). The European success of a single-shot yogurt drink was so big that Unilever introduced a three-ounce 'Promise Activ SuperShot' yogurt drink in the US market in 2007. The Promise drink, similarly to European yogurt drinks, was formulated to deliver an efficacious dose of phytosterol (2 g) per single serving size. However, in contrast to European regulations, the US Food and Drug Administration (FDA) authorised the risk reduction health claim only with consumption of phytosterols twice a day (FDA 2005). To encourage consumers to drink yogurt once a day, yet be in line with the regulations, the producer of Promise SuperShot made the following label statement: 'Drink Promise Activ SuperShots at least once a day with meals. Foods containing plant sterols should be eaten twice a day with meals. At least 0.4 g plant sterols per serving for a total daily intake of at least 0.8 g, as part of a diet low in saturated fat and cholesterol, may reduce the risk of heart disease. A serving of this product contains 2 g plant sterols. Contains 2 grams of natural plant sterols per serving'. A drawback of this approach is unnecessary overconsumption of phytosterols (two serving sizes contain 4 g of sterols) by some consumers, which is not allowed in Europe.

17.7.2 Current regulatory framework

One of the challenges of food formulators is to follow current regulations and law specific for the target market. The regulations have a significant impact on the choice of food matrix, number of serving sizes that are used for delivering phytosterols, ingredients and nutritional composition of food products, as well as allowable maximum daily doses. Manufacturers and marketers are responsible for ensuring that a food label and associated advertising material is truthful, not misleading and substantiated. In order to allow marketing of a new phytosterol-containing food product, the large majority of international jurisdictions require a cumbersome dossier containing information regarding food composition, formulation and processing, as well as clinical and safety substantiation. In Japan, clinical data are very crucial for approval of a food product. For cholesterol-lowering foods, the authorities required substantiation by results of clinical trials conducted for more than 12 weeks using a specific food product (Ohama *et al.* 2008).

In the United States, phytosterols have been approved, since 1998, for use in a number of foods including fat-based products such as vegetable fat spreads and salad dressing, dairy products and fat-free drinks such as orange juice. The approval was granted under the GRAS (generally recognised as safe) notification and self-GRAS rulings. In September 2000, the FDA issued an interim ruling that allows a health claim that links the consumption of plant sterol and stanol esters, and more recently free phytosterols, to the reduction of risk of coronary heart disease. The stated health claim – 'may reduce the risk of heart disease'- is typical of what is allowed to be on a label of food products that contain phytosterols. To make such a claim, the food product must contain 'at least 0.65 grams per serving of plant sterols (or 1.7 grams per serving of plant stanols), and be eaten twice a day with meals for a daily total intake of at least 1.3 grams of sterols (or at least 3.4 grams of stanols), as a part of a diet low in saturated fat and cholesterol' (FDA 2005).

The Novel Food Regulations (258/57) that were legislated in the European Community (EU) in 1997 are currently used for the approval of functional foods in 27 European countries. These regulations apply to novel foods and novel food ingredients, which have not been 'previously used to a significant degree for human consumption' and which have properties that vary from their conventional counterparts, consequently having no history of safe food use. In July 2000, the Commission approved phytosterol esters as novel food ingredients, and authorised their use in yellow fat spreads for lowering the cholesterol level in humans (EC 2000). In 2004, the EU Commission approved phytosterol-containing foods such as milk, yogurt and other fermented milk-type drinks, soy drinks, fat spreads, salad dressings, spicy sauces and cheese-type products with fat content not exceeding 12% (EC 2004a, b, c, d). Currently, phytosterols have been approved for use in rye bread as long as bread contains at least 50% rye flour and no added fat (EC 2006). It is noteworthy to emphasise that all food products must be made in such a manner that they can easily be divided into portions that contain either a maximum of 3 g (in the case of one portion per day) or a maximum of 1 g (in the case of three portions per day) of added phytosterols. The amount of phytosterols added to a container of beverages should not exceed 3 g. Although several phytosterol-containing foods have already borne health claims governed by the regulations of various European countries, only recently has the phytosterol claim been authorised by the EFSA. In 2008, a health claim that links consumption of phytosterols with lowering blood cholesterol and the risk of coronary heart disease was approved under Article 14 of Regulation (EC) No 1924/2006 (reduction of disease risk claims) based on convincing scientific substantiation (EFSA 2008).

In 2001, with the introduction of the Novel Food Standard (FSANZ 2008), the act that is similar to the European Novel Food Regulations, the Food Standards Australia New Zealand (FSANZ), formerly Australia New Zealand Food Authority, has approved the use of phytosterol esters, and tall oil free phytosterols in edible oil spreads. Both forms of sterols can be added to spreads up to 13.7 g/100 g (sterol esters) and 8 g/100 g (free sterols), as long as table spreads contain less than 28% total saturated and trans fatty acid combined (FSANZ 2002). As of today, phytosterol esters and tall oil phytosterols have also been approved in Australia and New Zealand, for enrichment of low-fat milk and yogurt, which must contain not more than 1.5% fat. In addition, phytosterol esters may be added to breakfast cereals if the total fibre content of the breakfast cereal is no less than 3 g/50 g, while sugar content is no more than 30 g/100 g of total sugars (FSANZ 2006).

The use of phytosterols in functional foods is also approved in some Asian countries. In Japan, phytosterol-containing foods are approved under the Food of Specified Health Uses regulations by the Ministry of Health, Labor and Welfare. Approved food products enriched with phytosterols may bear the health claim 'good for those concerned about serum cholesterol', while food products that contain a combination of diacylglycerol and phytosterols may claim 'good for those having relatively high serum cholesterol and triglycerides with mild obesity' (Ohama *et al.* 2008). In Taiwan, phytosterol-containing foods are approved under the Health Food Control Act by the Executive Yuan, Department of Health. In 2007, milk enriched with tall oil phytosterols (Uni President) was granted approval to be sold on the Taiwanese market (Zawistowski 2008). Phytosterol-containing foods are also approved by other Asian countries including South Korea, Malaysia, Indonesia, Thailand, the Philippines and Singapore. The leading commercial food product of this type in the Pacific Rim countries is milk for heart health by Nestle (Zawistowski 2008).

In Brazil, free phytosterols and their esters are approved, as functional ingredients for use in food products, by the National Health Surveillance Agency. In this jurisdiction, food products must guarantee to deliver a daily intake of 1–3 g of sterols. In contrast to the EU and US regulations, the daily intake of phytosterols can be incorporated into one, two or three serving sizes of foods (De Figueiredo & Lajolo 2008).

17.8 Conclusion

The efficacy and safety profile of phytosterols is proved by an overwhelming number of clinical studies that have been conducted over the last 50 years. This has been acknowledged by many international regulatory jurisdictions by granting an approval for using phytosterols in food products. Phytosterol-containing functional foods are surging onto the world market, as they represent a good alternative for risk reduction of coronary heart disease for people with borderline and higher cholesterol levels.

Acknowledgement

The author thanks Mira Laza of Omnia Foods Ltd for stimulating discussion and proofreading of the manuscript.

References

AC Nielsen (2006). Top 100 grocery brands. *AC Nielsen Retail News*, March 2006.

ADM (2007). CardioAid-S Plant Sterol Esters. Product Code 040087 – Technical Data Sheet. *Food Ingredient Catalog 2007–2008*. ADM Natural Health and Nutrition, Decatur, Illinois. Available online http://www.admworld.com (accessed 10 February 2009).

Albi, T., Lanzón, A., Guinda, A., Pérez-Camino, M.C. & León, M. (1997). Microwave and conventional heating effects on some physical and chemical parameters of edible fats. *Journal of Agricultural and Food Chemistry*, **45**, 3000–3003.

Auriou, N. & Ferreres, V. (2002). Emulsions and aqueous dispersions of phytosterols. *PCT Patent Application*, WO 02/065859 A1, 29 August 2002.

Azadmard-Damirchi, S. & Dutta, P.C. (2008). Stability of minor lipid components with emphasis on phytosterols during chemical interesterification of a blend of refined olive oil and palm stearin. *Journal of the American Oil Chemists Society*, **85**, 13–21.

Beltrame, F.L., Pessini, G.L., Doro, D.L., Dias Filho, B.D., Bazotte, R.B. & Cortez, D.A.G. (2002). Evaluation of the antidiabetic and antibacterial activity of *Cissus secyoides*. *The Brazilian Archives of Biology and Technology*, **45**, 21–25.

Birch, G.G., Brennan, J.G. & Parker, K.J. (1977). *Sensory Properties of Foods*. Applied Science Publishers, London, p. 326.

Boskou, D. (1998). Frying temperatures and minor constituents of oils and fats. *Grasas y Aceites*, **49**, 326–330.

Chan, Y.M., Varady, K.A., Lin, Y., Trautwein, E., Mensink, R.P., Plat, J. & Jones, P.J. (2006). Plasma concentrations of plant sterols: physiology and relationship with coronary heart disease. *Nutrition Reviews*, **64**, 385–402.

Clifton, P.M., Mano, M., Duchateau, G.S.M.J.E., Van Der Knaap, H.C.M. & Trautwein, E.A. (2008). Dose-response effects of different plant sterol sources in fat spreads on serum lipids and C-reactive protein and on the kinetic behavior of serum plant sterols. *European Journal of Clinical Nutrition*, **62**, 968–977.

Clifton, P.M., Noakes, M., Sullivan, D., Erichsen, N., Ross, D., Annison, G., Fassoulakis, A., Cehun, M. & Neste, P. (2004). Cholesterol-lowering effects of plant sterol esters differ in milk, yoghurt, bread and cereal. *European Journal of Clinical Nutrition*, **58**, 503–509.

Cognis (2007). Vegapure. The product Range, Product Specification Sheet, Cognis Nutrition and Health. Available online http://www.cognic.com (accessed 18 February 2009).

Christiansen, L.I., Rantanen, J.T., von Bonsdorff, A.K., Karjalainen, M.A. & Yliruusi, J.K. (2002). A novel method of producing a microcrystalline β-sitosterol suspension in oil. *European Journal of Pharmaceutical Sciences*, **15**, 261–269.

De Figueiredo, M.C. & Lajolo, F.M. (2008). Supplements and functional foods legislation in Brazil. In: *Nutraceutical and Functional Food Regulations in the United States and Around the World*. Bagchi, D. (ed.), Academic Press, Amsterdam, New York, pp. 349–364.

de Graaf, J., de Sauvage-Nolting, P.R.W., van Dam, M., Belsey, E.M., Kastelein, J.J.P., Pritchard, P.H. & Stalenhoef, A.F.H. (2002). Consumption of tall oil-derived phytosterols in a chocolate matrix significantly decreases plasma total and low-density lipoprotein-cholesterol levels. *British Journal of Nutrition*, **88**, 479–488.

Devaraj, S., Autret, B.C. & Jialal, I. (2006). Reduced-calorie orange juice beverage with plant sterols lowers C-reactive protein concentrations and improves the lipid profile in human volunteers. *American Journal of Clinical Nutrition*, **84**, 756–761.

Deveray, S., Jialal, I. & Vega-Lopez, S. (2004). Plant sterol-fortified orange juice effectively lowers cholesterol levels in mildly hypercholesterolemic healthy individuals. *Arteriosclerosis, Thrombosis, and Vascular Biology*, **24**, 25–28.

Dolhaine, H.G., Kropf, C., Christophliemk, P., Biermann, M. & Schroeder, C. (2007). Use of nanoscale sterols and sterol esters. US Patent, US 6,352,737 B1, 5 March 2002.

Du, K. & Zawistowski, J. (2002). Oxidative stability of designer oil. *Ann. Conf. Inst. Food Tech.* Abstract No. 13665, Anaheim, CA, 15–19 June.

Dutta, P.C. (1997). Studies on phytosterol oxides. II: content in some vegetable oils and in French fries prepared in these oils. *Journal of the American Oil Chemists Society*, **74**, 659–666.

Dutta, P.C. (2002). Determination of phytosterol oxidation products in foods and biological samples. In: *Cholesterol and Phytosterol Oxidation Products: Analysis, Occurrence, and Biological Effects*. Guardiola, F., Dutta, P.C., Codony, R. & Savage, G.P. (eds), AOCS Press, Champaign, IL, pp. 335–374.

Dutta, P.C. & Appelqvist, L.A. (1996). Sterols and sterol oxides in the potato products, and sterols in the vegetable oils used for industrial frying operations. *Grasas y Aceites*, **47**, 38–47.

Dutta, P.C. & Appelqvist, L.A. (1997). Studies on phytosterol oxides. I: effect of storage on the content in potato chips prepared in different vegetable oil. *Journal of the American Oil Chemists Society*, **74**, 647–657.

Dutta, P.C. & Savage, G.P. (2002). Formation and content of phytosterol oxidation products in foods. In: *Cholesterol and Phytosterol Oxidation Products: Analysis, Occurrence, and Biological Effects*. Guardiola, F., Dutta, P.C., Codony, R. & Savage, G.P. (eds), AOCS Press, Champaign, IL, pp. 319–334.

Dykes, G.A., Amarowicz, R. & Pegg, R.B. (2003). Enhancement of nisin antibacterial activity by a bearberry (*Arctostaphylos uva-ursi*) leaf extract. *Food Microbiology*, **20**, 211–216.

EC (European Commission) (2000). Commission Decision 2000/500/EC. *Official Journal European Union*, L 200/59, p. 1, 8 August 2000.

EC (European Commission) (2004a). Commission Decision 2004/333/EC. *Official Journal European Union*, L 105/40, p. 1, 14 April 2004.

EC (European Commission) (2004b). Commission Decision 2004/334/EC. *Official Journal European Union*, L 105/43, p. 1, 14 April 2004.

EC (European Commission) (2004c). Commission Decision 2004/335/EC. *Official Journal European Union*, L 105/46, p. 1, 14 April 2004.

EC (European Commission) (2004d). Commission Decision 2004/336/EC. *Official Journal European Union*, L 105/49, p. 1, 14 April 2004.

EC (European Commission) (2006). Commission Decision 2006/845/EC. *Official Journal European Union*, L 31/18, p. 1, 3 February 2006.

EFSA (2002). General view of the Scientific Committee on Food on the long-term effects of the intake of elevated levels of phytosterols from multiple dietary sources, with particular attention to the effects on β-carotene. European Food Safety Authority, Scientific Committee on Foods, 3 October 2002.

EFSA, European Food Safety Authority (2008). Plant sterols and blood cholesterol. Scientific substantiation of a health claim related to plant sterols and lower/reduced blood cholesterol and reduced risk of (coronary) heart disease pursuant to Article 14 of Regulation (EC) No 1924/2006. *The EFSA Journal*, **781**, 1–12.

Engel, R. & Schubert, H. (2005). Formulation of phytosterols in emulsions for increased dose response in functional foods. *Innovative Food Science and Emerging Technologies*, **6**, 233–237.

Expert Panel on Detection, Evaluation, and Treatment of High Blood Cholesterol in Adults (2001). Executive summary of the Third Report of the National Cholesterol Education Program (NCEP) Expert Panel on detection, evaluation, and treatment of high blood cholesterol in adults (Adult Treatment Panel III). *The Journal of the American Medical Association*, **285**, 2486–2497.

FAO (2008). Phytosterols, phytostanols and their esters – Monographs 5. *FAO 69the Meeting of Joint Expert Committee on Food Additives*, 17–26 June 2008, Rome, Italy.

FDA (US Food and Drug Administration) (2005). Code of federal regulations, 'plant sterol/stanol esters and risk of coronary heart disease (CHD). *Food and Drugs*, 21, Vol. 2, Chap. I, Part 101 Food Labeling, Subpart E, Specific Requirements for Health Claims, July 21, 2005.

Fernandaz, C., Suarez, Y., Ferruelo, A.J., Gomez-Coronado, D. & Lasuncion, M.A. (2002). Inhibition of cholesterol biosynthesis by Δ^{22}-unsaturated phytosterols via competitive inhibition of sterol Δ^{24}-reductase in mammalian cells. *Biochemical Journal*, **366**, 109–119.

FSANZ, Food Standards Australia New Zealand (2002). Food Standards Code. Commonwealth of Australia Gazette No. FSC 2, 20 June 2002, pp. 3–40.

FSANZ, Food Standards Australia New Zealand (2006). Food Standards Code. Commonwealth of Australia Gazette No. FSC 31, 9 November 2006, pp. 3–9.

FSANZ, Food Standards Australia New Zealand (2008). Novel food standard. In: *Australia New Zealand Food Standards Code*. Ansat Pty Ltd CAN.

Ghavami, M. & Morton, I.D. (1984). Effect of heating at deep-fat frying temperature on the sterol content of soya bean oil. *Journal of the Science of Food and Agriculture*, **35**, 569–572.

Gordon, M.H. & Magos, P. (1983). The effect of sterols on the oxidation of edible oils. *Food Chemistry*, **10** (2), 141–147.

Guillen, M.D. & Manzanos, M.J. (1998). Study of the composition of the different parts of a Spanish Thrus vulgaris L. plant. *Food Chemistry*, **63**(3), 373–383.

Gylling, H. & Miettinen, T.A. (1999). Cholesterol reduction by different plant stanol mixtures and with variable fat intake. *Metabolism*, **48**, 575–580.

Gylling, H., Puska, P., Vartiainen, E. & Miettinen, T.A. (1999). Retinol, vitamin D, carotenes and alpha-tocopherol in serum of a moderately hypercholesterolemic population consuming sitostanol ester margarine. *Atherosclerosis*, **145**, 279–285.

Hallikainen, M.A., Sarkkinen, E.S., Gylling, H. & Uusitupa, M.I. (2000). Comparison of the effects of plant sterol ester and plant stanol ester-enriched margarines in lowering serum cholesterol concentrations in hypercholesterolemic subjects fed a low-fat diet. *European Journal of Clinical Nutrition*, **54**, 715–725.

Hansel, B., Nicolle, C., Lalanne, F., Tondu, F., Lassel, T., Donazzolo, Y., Ferrières, J., Krempf, M., Schlienger, J.-L., Verges, B., Chapman, M.J. & Bruckert, E. (2007). Effect of low-fat, fermented milk enriched with plant sterols on serum lipid profile and oxidative stress in moderate hypercholesterolemia. *American Journal of Clinical Nutrition*, **86**, 790–796.

Heineman, T., Axtmann, G. & von Bergmann, K. (1993). Comparison of intestinal absorption of cholesterol with different plant sterols in man. *European Journal of Clinical Investigation*, **23**, 827–831.

Heinemann, T., Leiss, O. & von Bergmann, K. (1986). Effect of low-dose sitostanol on serum cholesterol in patients with hypercholesterolemia. *Atherosclerosis*, **61**, 219–223.

Hendriks, H.F.J., Brink, E.J., Meijer, G.W., Princen, H.M.G. & Ntanios, F.Y. (2003). Safety of long-term consumption of plant sterol esters-enriched spread. *European Journal of Clinical Nutrition*, **57**, 681–692.

Hendriks, H.F.J., Weststrate, J.A., van Vliet, T. & Meijer, G.W. (1999). Spread enriched with three different levels of vegetable oil sterols and the degree of cholesterol lowering in normocholesterolaemic and mildly hypercholesterolaemic subjects. *European Journal of Clinical Nutrition*, **53**, 319–327.

Jones, P.J.H., Ntanios, F.Y., Raeini-Sarjaz, M. & Vanstone, C.A. (1999). Cholesterol-lowering efficacy of a sitostanol-containing phytosterol mixture with a prudent diet in hyperlipidemic men. *American Journal of Clinical Nutrition*, **69**, 1140–1150.

Judd, T.J., Baer, D.J., Chen, S.C., Clevidence, B.A., Muesing, R.A., Kramer, M. & Meijer, G.W. (2002). Plant sterol esters lower plasma lipids and most carotenoids in mildly hypercholesterolemic adults. *Lipids*, **37**, 33–42.

Katan, M.B., Grundy, S.M., Jones, P., Law, M., Miettinen, T. & Paoletti, R. (2003). Efficacy and safety of plant stanols and sterols in the management of blood cholesterol levels. *Mayo Clinic Proceedings*, **78**, 965–978.

Korpela, R., Tuomilehto, J., Hogstrom, P., Seppo, L., Piironen, V., Salo-Vaananen, P., Toivo, J., Lamberg-Allardt, C., Karkkainen, M., Outila, T., Sundvall, J., Vilkkila, S. & Tikkanen, M.J. (2006). Safety aspects and cholesterol-lowering efficacy of low fat dairy products containing plant sterols. *European Journal of Clinical Nutrition*, **60**, 633–642.

Lampi, A.M., Dimber, L.H. & Kamal-Eldin, A. (1999). A study on the influence of fucosterol on thermal polymerization of purified high oleic sunflower triacylglycerols. *Journal of the Science of Food and Agriculture*, **79**, 573–579.

Law, M. (2000). Plant sterol and stanol margarines and health. *British Medical Journal*, **320**, 861–864.

Lee, M., Lu, K. & Patel, S.B. (2001). Genetic basis of sitosterolemia. *Current Opinion in Lipidology*, **12**, 141–149.

Leeson, P. & Floter, E. (2002). Solidification behavior of binary sitosteryl ester mixtures. *Food Research Institute*, **35**, 983–991.

Libby, P. & Ridker, P.M. (2004). Inflammation and atherosclerosis: role of C-reactive protein in risk assessment. *American Journal of Medicine*, **116**, 9–16.

Ling, W.H. & Jones, P.J.H. (1995). Dietary phytosterols: a review of metabolism', benefits and side effects. *Life Sciences*, **57**, 195–206.

Lukovitz, K. (2007). Functional food sale hit $ 25 billion in US in 2006. *Marketing Daily, Media Post Communications*, 21 February, p. 3.

Lutjohann, D. & von Bergmann, K. (1997). Phytosterolaemia: diagnosis, characterization and therapeutical approaches. *Annals of Medicine*, **29**, 181–184.

Matsuoka, R., Muto, A., Kimura, M., Hoshina, R., Wakamatsu, T. & Masuda, Y. (2008). Cholesterol-lowering activity of plant sterol-egg yolk lipoprotein complex in rats. *Journal of Oleo Science*, **57**, 309–314.

Mattson, F.H., Grundy, S.M. & Crouse, J.R. (1982). Optimizing the effect of plant sterols on cholesterol absorption in man. *American Journal of Clinical Nutrition*, **35**, 697–700.

Mattson, F.H., Volpenhein, R.A. & Erickson, B.A. (1977). Effect of plant sterol esters on the absorption of dietary cholesterol. *Journal of Nutrition*, **107**, 1139–1146.

Melnikov, S., Seijen ven Hoorn, J. & Bertrand, B. (2004). Can cholesterol absorption be reduced by phytosterols and phytostanols via a co-crystallization mechanism? *Chemistry and Physics of Lipids*, **127**, 15–33.

Miettinen, T.A. & Gylling, H. (2004). Plant stanol and sterol esters in prevention of cardiovascular diseases. *Annals of Medicine*, **36**, 126–134.

Moghadasian, M.H. (2000). Pharmacological properties of plant sterol *in vivo* and *in vitro* observations. *Life Sciences*, **67**, 605–615.

Monu, E., Blank, G., Holley, R. & Zawistowski, J. (2008). Phytosterol effects on milk and yogurt microflora. *Journal of Food Science*, **73** (3), 121–126.

Moreau, R.A. (2004). Plant sterols in functional foods. In: *Phytosterols as Functional Food Components and Nutraceuticals*. Dutta, P.C. (ed.), Marcel Dekker, New York, pp. 317–345.

Moreau, R.A. (2005). Phytosterols and phytosterol esters. In: *Healthful Lipids*. Akoh, C.A. & Lai, O.M. (eds), AOCS Press, Champaign, Illinois, pp. 335–360.

Moreau, R.A., Norton, R.A. & Hicks, K.B. (1999). Phytosterols and phytostanols lower cholesterol. *Inform*, **10**, 572–577.

Moreau, R.A., Whitaker, B.D. & Hicks, K.B. (2002). Phytosterols, phytostanols, and their conjugates in foods: structural diversity, quantitative analysis, and health-promoting uses. *Progress in Lipid Research*, **41**, 457–500.

Morton, G.M., Lee, S.M., Buss, H., David, H. & Lawrence, P. (1995). Intakes and major dietary sources of cholesterol and phytosterols in the British diet. *Journal of Human Nutrition and Dietetics*, **50**, 429–440.

Moshi, M.J., Joseph, C.C., Innocent, E. & Nkunya, M.H.H. (2004). In vitro antibacterial and antifungal activities of extracts and compounds from *Uvaria scheffleri*. *Pharmaceutical Biology*, **42**, 269–273.

Nair, P.P., Turjman, N., Kessie, G., Calkins, B., Goodman, G.T., Davidovitz, H. & Nimmagadda, G. (1984). Diet, nutrition intake, and metabolism in populations at high and low risk for colon cancer. Dietary cholesterol, beta-sitosterol, and stigmasterol. *American Journal of Clinical Nutrition*, **40**(Suppl. 4), 927–930.

Nestel, P., Cehun, M., Pomeroy, S., Abbey, M. & Weldon, G. (2001). Cholesterol-lowering effects of plant sterol esters and non-esterified stanols in margarine, butter and low-fat foods. *European Journal of Clinical Nutrition*, **12**, 1084–1090.

Nguyen, T.T., Dale, L.C., von Bergmann, K. & Croghan, I.T. (1999). Cholesterol-lowering effect of stanol ester in a US population of mildly hypercholesterolemic men and women: a randomized controlled trial. *Mayo Clinic Proceedings*, **74**, 1198–1206.

Nissen, S.E., Tuzcu, E.M., Schoenhagen, P., Crowe, T., Sasiela, W.J., Tsai, J., Orazem, J., Magorien, R.D., O'Shaughnessy, C. & Ganz, P. (2005). Statin therapy, LDL cholesterol, C-reactive protein, and coronary artery disease. *The New England Journal of Medicine*, **352**, 29–38.

Noakes, M., Clifton, P.M., Doornbos, A.M.E. & Trautwein, E.A. (2005). Plant sterol ester enriched milk and yoghurt effectively reduce serum cholesterol in modestly hypercholerolemic subjects. *European Journal of Nutrition*, **44**, 214–222.

Normén, L., Frohlich, J. & Trautwein, E. (2004). Role of plant sterols in cholesterol lowering. In: *Phytosterols as Functional Food Components and Nutraceuticals*. Dutta, P.C. (ed.), Marcel Dekker, New York, pp. 243–315.

Normén, L., Johnsson, M., Andersson, H., van Gameren, Y. & Dutta, P. (1999). Plant sterols in vegetables and fruits commonly consumed in Sweden. *European Journal of Nutrition*, **38**, 84–89.

Northfield, T.C. & Hoffman, A.F. (1975). Biliary lipid output during three meals and an overnight fast. I. Relationship to bile acid pool size and cholesterol saturation in gallstone and control subjects. *Gut*, **16**, 1–11.

O'Neill, F.H., Sanders, T.A.B. & Thompson, G.R. (2005). Comparison of efficacy of plant stanols ester and sterol ester: short-term and longer-term studies. *American Journal of Cardiology*, **96**(Suppl.) 29D–36D.

Ohama, H., Ikeda, H. & Moriyama, H. (2008). Health foods and foods with health claims in Japan. In: *Nutraceutical and Functional Food Regulations in the United States and Around the World*. Bagchi, D. (ed.), Academic Press, Amsterdam, New York, pp. 249–280.

Ostlund, R.E. (2002). Phytosterols in human nutrition. *Annual Review of Nutrition*, **22**, 533–549.

Ostlund, R.E., Jr, Spilburg, C.A. & Stenson, W.F. (1999). Sitostanol administration in lecithin micelles potently reduces cholesterol absorption in humans. *American Journal of Clinical Nutrition*, **79**, 826–831.

Pelletier, X., Belbraouet, S., Mirabel, D., Mordret, F., Perrin, J.L., Pages, X. & Debry, G. (1995). A diet moderately enriched in phytosterols lowers plasma cholesterol concentrations in normocholesterolemic humans. *Annals of Nutrition and Metabolism*, **39**, 291–295.

Phillips, K.M., Ruggio, D.M., Toivo, J.I., Swank, M.A. & Simpkins, A.H. (2002). Free and esterified sterol composition of edible oils and fats. *The Journal of Food Composition and Analysis*, **15**, 123–142.

Piironen, V. & Lampi, A.-M. (2004). Occurrence and levels of phytosterols in foods. In: *Phytosterols as Functional Food Components and Nutraceuticals*. Dutta, P.C. (ed.), Marcel Dekker, New York, pp. 1–32.

Piironen, V., Lindsay, D.G., Miettinen, T.A., Toivo, J. & Lampi, A.M. (2000). Review. Plant sterols: biosynthesis, biological function and their importance to human nutrition. *Journal of the Science of Food and Agriculture*, **80**, 939–966.

Plana, N., Nicolle, C., Ferre, R., Camps, J., Cos, R., Villoria, J. & Masana, L. (2008). Plant sterol-enriched fermented milk enhances the attainment of LDL-cholesterol goal in hypercholesterolemic subjects. *European Journal of Nutrition*, **77**, 32–39.

Pollak, O.J. & Kritchevsky, D. (1981). *Monographs on Atherosclerosis*. Clarkson, T.B., Kritchevsky, D. & Pollak, O.J. (eds), Kargel, Basel.

Porter, N.A., Caldwell, S.E. & Mills, K.A. (1995). Mechanisms of free radical oxidation of unsaturated lipids. *Lipids*, **30**, 277–290.

Przybylski, R., Zambiazi, R. & Li, W. (1999). Kinetics of sterols changes during storage and frying of canola oils. *Journal of the American Oil Chemists Society*, **64**, 388–392.

Quilez, J., Garcia-Lorda, P. & Slas-Salvado, J. (2003). Potential uses and benefits of phytosterols in diet: present situation and future directions. *Clinical Nutrition*, **22**(4), 343–351.

Ringel, R.L. (1970). Studies of oral region texture perception. In: *Oral Sensation and Perception*. Bosma, J.F. (ed.), Charles C. Thomas Publisher, Springfield, IL, pp. 323–331.

Rostagno, W. (1969). Chocolate particle size and its organoleptic influence. *Manufacturing Conference*, **49**, 81785.

Salo, P. & Wester, I. (2005). Low-fat formulations of plant stanols and sterols. *American Journal of Cardiology*, **96**, 51–54.

Sehayek, E. (2004). Genetic regulation of cholesterol absorption and plasma plant sterol levels: commonalities and differences. *Journal of Lipid Research*, **44**, pp. 2030–2038.

Smith, L.L. (1981). Other oxidations: cholecalciferol, sitosterol, ergocalciferol. In: *Cholesterol Autoxidation*. Smith, L.L. (ed.), Published by Springer, Plenum Press, New York and London, p. 246.

Soupas, L. (2006). *Oxidative Stability of Phytosterols in Food Models and Foods*. Ph.D. Dissertation, EKT-series 1370, University of Helsinki. Department of Applied Chemistry and Microbiology, Helsinki, Finland.

Soupas, L., Juntunen, L., Lampi, A.M. & Piironen, V. (2004). Effects of sterol structure, temperature, and lipid medium on phytosterol oxidation. *Journal of Agricultural and Food Chemistry*, **52**, 6845–6491.
The Natural Marketing Institute (2003). *The 2003 Health and Wellness Trends Report*. The Natural Marketing Institute, Harleysville, PA, 15 January, p. 4.
Thompson, G.R. & Grundy, S.M. (2005). History and development of plant sterol and stanols esters for cholesterol-lowering purposes. *American Journal of Cardiology*, **96** (Suppl.), 3D–9D.
Trautwein, E.A., Duchateau, G.S., Lin, Y., Melnikov, S.M., Molhuizen, H.O. & Ntanois, F.Y. (2003) Proposed mechanisms of cholesterol-lowering action of plant sterols. *European Journal of Lipid Science and Technology*, **105**, 171–185.
Tuomilehto, J., Tikkanen, M.J., Högström, P., Keinänen-Kiukaanniemi, S., Piironen, V., Toivo, J., Salonen, J.T., Nyyssönen, K., Stenman, U.-H., Alfthan, H. & Karppanen, H. (2009). Safety assessment of common foods enriched with natural nonesterified plant sterols. *European Journal of Clinical Nutrition*, **63**(5), 684–691.
Turley, S.D. & Dietschy, J.M. (2003). Sterol absorption by the small intestine. *Current Opinion in Lipidology*, **14**, 233–240.
Vaikousi, H., Lazaridou, A., Biliaderis, C.G. & Zawistowski, J. (2007). Phase transitions, solubility, and crystallization kinetics of phytosterols and phytosterol-oil blends. *Journal of Agricultural Food Chemistry*, **55**, 1790–1798.
Vanhanen, H.T. & Mietinen, T.A. (1992). Effects of unsaturated and saturated dietary plant sterols on their serum contents. *Clinica Chimica Acta*, **205**, 97–107.
Verhe, R., Echim, C. & De Greyt, W. (2007). Biodiesel production and utilisation. *Third International Conference on Renewable Resources and Biorefineries, Session 9: Bio-Energy I*, 4–6 June 2007, Ghent, Belgium.
Verleyen, T., Forcades, M., Verhe, R., Dewettinck, K., Huyghebaert, A. & De Greyt, W. (2002). Analysis of free and esterified sterols in vegetable oils. *Journal of the American Oil Chemists Society*, **79** (2), 117–122.
von Bergmann, K., Sudhop, T. & Lutjohann, D. (2005). Cholesterol and plant sterol absorption: recent insights. *American Journal of Cardiology*, **96** (Suppl.), 10D–14D.
Vuorio, A.F., Gylling, H., Turtola, H., Kontula, K., Ketonen, P. & Miettinen, T.A. (2000). Stanol ester margarine alone and with simvastatin lowers serum cholesterol in families with familial hypercholesterolemia caused by the FH-North Karelia Mutation. *Arteriosclerosis, Thrombosis, and Vascular Biology*, **20**, 500–506.
Vuoristo, M. & Mittinen, T.A. (1985). Increased biliary lipid secretion in coeliac disease. *Gastroenterology*, **88**, 134–142.
Wang, T., Hicks, K.B. & Moreau, R. (2002). Antioxidant activity of phytosterols, oryzanol, and other phytosterol conjugates. *Journal of the American Oil Chemists Society*, **79**, 1201–1206.
Weidner, C., Krempf, M., Bard, J.-M., Cazaubiel, M. & Bell, D. (2008). Cholesterol lowering effect of a soy drink enriched with plant sterols in a French population with moderate hypercholesterolemia. *Lipids in Health and Disease*, **7**, 35–43.
Weingartner, O., Bohm, M. & Laufs, U. (2009). Controversial role of plant sterol esters in management of hypercholesterolaemia. *European Heart Journal*, **30**, 404–409.
Weiss, J. (2007). Nanotechnological food applications – with a focus on novel delivery system for bioactives. In: *Nanotechnology Short Course*, IFT, Chicago, 27 July 2007.
Wester, I. (1998). Texturizing compositions for use in fat blends in food. International Patent WO/1998/019556. 14 May 1998.
Weststrate, J.A. & Meijer, G.W. (1998). Plant sterol-enriched margarines and reduction of plasma total- and LDL-cholesterol concentrations in normocholesterolemic and mildly hypercholesterolemic subjects. *European Journal of Clinical Nutrition*, **52**, 334–343.
White, P.J. & Armstrong, L.S. (1986). Effect of selected oat sterols on the deterioration of heated soybean oil. *Journal of the American Oil Chemists Society*, **63**(4), 525–529.
Wilund, K.R., Liqing Yu, L., Xu, F., Vega, G., Grundy, S., Cohen, J.C. & Hobbs, H.H. (2004). Plant sterol levels are not associated with atherosclerosis in mice and men. *Arteriosclerosis, Thrombosis, and Vascular Biology*, **24**, 1–7.
Winkler, J.K. & Warner, K. (2008). The effect of phytosterol concentration on oxidative stability and thermal polymerization of heated oils. *European Journal of Lipid Science and Technology*, **110**(5), 455–464.
Winkler, J.K., Warner, K. & Glynn, M.T. (2007). Effect of deep-fat frying on phytosterol content in oils with differing fatty acid composition. *Journal of the American Oil Chemists Society*, **84**, 1023–1030.

Wong, A. (2008). Phytosterols in selected grain processing residues. *Electronic Journal of Environmental, Agricultural and Food Chemistry*, **7**(6), 2948–2958.
Yalkowsky, S.H. (1981). *Techniques of Solubilization of Drugs*. Marcel Dekker, New York, pp. 1–14.
Yonkah, V.V. (2006). Phytosterol in human health. In: *Handbook of Functional Lipids*. Akoh, C.C. (ed.), CRC Press, Boca Raton, FL, pp. 408–418.
Yoshida, Y. & Niki, E. (2003). Antioxidant effects of phytosterol and its components. *Journal of Nutritional Science and Vitaminology*, **49** (4), 277–280.
Zawistowski, J. (2003). Method of preparing microparticles of phytosterols or phytostanols. European Patent, EP 1 148 793 B1, 13 August 2003.
Zawistowski, J. (2007). Emulsions comprising non-esterified phytosterols in the aqueous phase. US Patent Application Publication, US 2007/0141123 A1, 21 June 2007.
Zawistowski, J. (2008). Regulations of functional foods in selected Asian countries in the Pacific Rim. In: *Nutraceutical and Functional Food Regulations in the United States and Around the World*. Bagchi, D. (ed.), Academic Press, Amsterdam, New York, pp. 365–401.
Zawistowski, J., Du, W. & Kitts, D.D. (2005). Presence of free sterols but not sterol esters enhances thermal stability of oils. *26th World Congress and Exhibition of the International Society for Fat Research, 25–28 September*, Prague, Czech Republic, p. 34.
Zawistowski, J., Kitts, D.D., Stoynov, N. & Du, K. (2003). Thermal oxidation behavior of a phytosterol mixture in different oil systems. *94th American Oil Chemist' Society Annual Meeting, 4–7 May*, Kansas City, MI.

18 Obesity and related disorders

Yanwen Wang

18.1 Definition of obesity and commonly used measures

Obesity is the condition of excessive body fat that contributes to weight gain. Although anabolic steroids can increase lean mass and thus body mass (Ferreira 1998), accumulation of fat mass or triacylglycerides is generally the cause of excessive increase of body weight as other forms of energy storage do not share the same potential as adipose tissue to exceed the limits of requirement (Sikaris 2004). However, the definition of obesity cannot be simply based on body weight. There is a need to normalize the body weight against body height. The most commonly used measure is body mass index (BMI), which is the ratio of body weight in kilograms to the square of height in meters. The World Health Organization (WHO) has established a weight classification system based on the BMI to evaluate overweight and obesity as well as associated health risks among adults (Table 18.1). This system provides a simple reference for individual assessment, national and international comparisons, clinical diagnosis, as well as interstudy comparisons in scientific research. BMI is a population-level measure of overweight and obesity as it is the same for both sexes and for all ages of adults. However, it does not account for the body composition or fat distribution in the body, and thus, BMI may not be an accurate predictor of health risk for certain groups, including youths who have not reached their full height, adults who are very lean or muscular, pregnant women, and the elderly. Accordingly, waist circumference or the ratio of waist circumference is used as a supplementary measure for identifying additional health risk. The normal range is below 0.95 in men and 0.85 in women. The ratio of waist to hip circumference is a better prognostic disease indicator than BMI, particularly among those with BMI measurements below 35 (Health Canada 2003; Sikaris 2004). Skinfold is another measure that is determined by the amount of subcutaneous fat. Therefore, the ratio of waist to hip circumference and skinfold are more informative on body fat mass and fat distribution, while BMI is used as a guide for the management of body weight. The combinatorial use of these three measures is able to assess collectively the weight range and corresponding contributory components of overweight or obesity. They are also used to determine the type of obesity, where the risk of developing metabolic syndrome and other obesity-related diseases is higher in subjects with central obesity than those with general obesity.

It should be noted that the weight classification system described above is for adults only. The new WHO Child Growth Standards, launched in April 2006, include BMI charts for infants and young children up to age 5. For children and teens, BMI is age- and sex-specific and is often referred to as BMI-for-age. In this regard, BMI ranges for children and teens

Table 18.1 Body weight classification for adults

Body mass index (kg/m^2)	Classification	Risk of developing health problems
<18.5	Underweight	Increased risk
18.5–24.9	Normal weight	Least risk
25.0–29.9	Overweight	Increased risk
≥30.0	Obese	
30.0–34.9	Obese class I	High risk
35.0–39.9	Obese class II	Very high risk
≥40.0	Obese class III	Extremely high risk

Source: Health Canada (2003).

have recently been defined by the WHO International Obesity Task Force. These ranges take into account normal differences in body fat between boys and girls and differences in body fat at various ages. Percentiles are the most commonly used indicator to assess the size and growth patterns of individual children. The percentile indicates the relative position of a child's BMI number among other children of the same sex and age. BMI is used to screen for overweight, at risk of overweight and underweight, but it is not a diagnostic tool. A child may have a high BMI for age and sex, but to determine if excess fat is a problem, a health care provider needs to perform further assessments. These assessments may include skinfold thickness measurements, evaluations of diet, physical activity, family history and other appropriate health screenings.

Although the BMI number is calculated the same way for children and adults, the criteria used to interpret the meaning of the BMI number for children and teens are different from those used for adults. BMI-for-age growth charts take into account these differences and allow translation of a BMI number into a percentile for a child's sex and age. Conversely for adults, BMI is interpreted through categories that do not take into account sex or age. Healthy weight ranges cannot be provided for children and teens because healthy weight ranges change with increasing height and age.

18.2 Prevalence of overweight and obesity

The prevalence of obesity and overweight has been increasing at an alarming rate since the 1970s and has now become a worldwide epidemic. In Canada, the overall age-standardized prevalence of obesity in adults at the age of 20 years or more has increased from 10% in 1970 to 23% in 2004. The annual deaths attributed to obesity were 8414 in 2004. If obesity trends remain unchanged, the prevalence of obesity in adults is predicted to reach 27% in men and 24% in women by the year 2010. The situation is worse in the United States. Since the mid-1970s, the prevalence of overweight and obesity has increased sharply for both adults and children. Data from two National Health and Nutrition Examination Survey surveys show that among adults aged 20–74 years the prevalence of obesity increased from 15.0% (in the 1976–1980 survey) (Flegal *et al.* 2002) to 32.9% (in the 2003–2004 survey) (Ogden 2006). Since the 1990s, obesity-caused deaths in the United States have accounted for over 300 000 cases per year. The morbidity and poor health-related quality of life associated with obesity is now greater than smoking, alcoholism and poverty (Sturm & Wells 2001). In Europe, the prevalence values for obesity translate similarly to those for North America. In 1995, obesity in Europe was estimated to be 10–20% for men and 15–25% for women (Seidell & Flegal

1997). European obesity has been estimated to be as high as 26% in men and 22% in women (Wilborn 2005).

The prevalence of obesity is increasing throughout the world population. However, the distribution varies greatly between and within countries. Pacific populations have some of the world's highest prevalence rates of obesity. The proportion of men and women with BMI above 30 in Nauru was 77% in 1994, and people living in New Zealand in the early 1990 had an obesity rate of approximately 65–70% (Swinburn *et al.* 2004). In Asian countries, the absolute obesity rates are low; however, the relative rate of increase has become higher. Over the past two decades, China has gone through an impressive pace of economic development, and parallel with this change, the Chinese population has experienced many dramatic changes in lifestyle which are linked to increases in obesity and chronic diseases. From 1992 to 2002, prevalence of overweight and obesity increased in all genders and age groups as well as in all geographic regions. According to the WHO BMI-based classification system, the combined prevalence of overweight and obesity increased from 15 to 22%, whereas the Chinese standard shows an increase from 20% in 1992 to 30% in 2002 (Wang *et al.* 2007). Unfortunately, this situation is worsening worldwide and has raised an urgent warning to the general public, health professionals, governments and industry. Moreover, whereas obesity previously occurred in adults, it has now expanded toward children and adolescents (Popkin *et al.* 2006). In the United States, two surveys conducted in 1976–1980 (Flegal *et al.* 2002) and 2003–2004 (Ogden 2006) have shown quick increases in overweight among children and teens. For children aged 2–5 years, the prevalence of overweight increased from 5.0 to 13.9%; for those aged 6–11 years, prevalence increased from 6.5 to 18.8%; and for those aged 12–19 years, prevalence increased from 5.0 to 17.4%. This pandemic is also penetrating the poorest nations of the world, initially affecting urban middle-aged adults, but is now increasingly affecting semiurban and rural areas (Prentice 2006). According to the WHO's projections in 2005, approximately 1.6 billion people aged 15 years and over were overweight and at least 400 million adults were obese. It is predicted that by 2015, approximately 2.3 billion adults will be overweight and more than 700 million will be obese.

18.3 Health costs related to obesity

In previous generations, overweight and obesity were viewed more as an aesthetic rather than a health problem by the general population and even health care professionals. The societal impact of overweight and obesity had not garnered their attention until the last decade when health threats and the economic burden of obesity increased to a significant level. Obesity causes a number of health problems including premature mortality and morbidity, type 2 diabetes and insulin resistance, dyslipidemia, gallbladder disease, obstructive sleep apnea and respiratory problems, cardiovascular disease (CVD), hypertension, osteoarthritis, some types of cancer (breast, endometrial, colon, prostate and kidney), psychosocial problems, functional limitations and impaired fertility (Pi-Sunyer 1991; Willett *et al.* 1999; Shari & Bassuk 2006). As for economic consequences, the costs associated with obesity in Canada increased from 2.4% in 1997 to 2.6% of in 2001 (Birmingham *et al.* 1999; Katzmarzyk & Janssen 2004). As for the United States, the figures of direct health care costs are even higher, and as recent as 2003, it had been estimated to be 5–7% of the total health care costs (Finkelstein *et al.* 2003). Comparable figures are somewhat lower for other Western countries. Moreover, it should be noted that these figures are in fact underestimations since the costs were calculated only for the population with a BMI above 30 and omitted any burdens brought on by overweight. The direct costs of obesity result predominantly from diabetes, CVD and hypertension. Indirect costs, which significantly surpass direct ones, include workdays lost, physician visits, disability pensions, and premature mortality, which all increase as BMI increases. The costs derived from impaired quality of life have not been included, but given the social and psychological consequences of obesity, they are likely to be of a substantial amount (Swinburn *et al.* 2004). Looking toward the future, the WHO warns that the burden of obesity and diabetes will spread its effects to developing countries and the projected numbers of new cases of diabetes will run into the hundreds of millions within next two decades.

18.4 Etiology of obesity

The etiology of obesity is complicated, and the mechanisms of obesity have not been well understood. However, understanding the causes of overweight and obesity is crucial in developing effective prevention and intervention procedures and products. To date, it is known that a number of factors are involved in the development of overweight and obesity, including diet evolution, change of lifestyle, physical activity, genetic factors, food behavior, socioeconomic status, environment (home, school and workplace), medium, food and drink supply (prepared foods and drinks at restaurants and processed foods and drinks from the food industry), and high technology (television, video games, computers). Details are given for some of these factors in the following sections.

18.4.1 Evolution of diet

A huge diversity in dietary habits and genetic traits has developed among humans during the evolutionary process (Cordain 2000). Data from the Stone Age (35 000–10 000 BC) suggest a low fat intake (<22% of total energy) and a high protein and carbohydrate consumption (about 40% of energy). The hunter–gatherer lifestyle influenced the animal and plant food availability for the subsequent 10 000 years and markedly affected the patterns of macronutrient intake, with the energy provided by fat increasing to 28–58% while protein energy declined to 19–35%. During the last century, with advances in agriculture and livestock breeding, the composition of diet intake had changed again to consist 10–17% of energy from protein, 40–60% from carbohydrates and 20–59% from fat. The advances of agricultural and animal farming technology resulted in a sharp decrease in the cost of vegetable oils and sugar, making them more competitive than cereals as food ingredients in the world (Drewnowski 2000). This change has led to a marked reduction of foods derived from grain products and greatly increased energy consumption as a consequence (Popkin 2001). As populations become more urban and incomes rise, diets high in fat, sugar and animal products replace more traditional diets that were high in complex carbohydrates and fibers (Drewnowski 2000; Popkin 2001). Unique traditional food habits are being replaced by Westernized fast foods, soft drinks and increased meat consumption (Drewnowski 2000). As a result, the population in most industrialized countries eat more fat (40–50% of total energy), especially saturated fats, and less carbohydrates. The reduction of carbohydrates is mainly due to decreased starch consumption. This global Westernization of diet has increased dietary energy density and passive energy overconsumption (Popkin 2001), and is linked to a growing prevalence of obesity and other chronic diseases in both high- and low-income countries (Marti & Martinez 2006).

18.4.1.1 *High-fat diet and obesity*

The amount and composition of the macronutrient intake appear to strongly influence the regulation of food intake, energy metabolism and expenditure (Buchholz & Schoeller 2004; Heymsfield 2004). The ratio of fat to carbohydrate is the main factor that easily causes overconsumption of energy relative to energy expenditure, leading to weight gain (Saris 2003). A large body of epidemiological data have supported the relationship between high fat intake and obesity development. Although there is not a conclusive answer to the question whether high-fat diet versus low-fat diet is associated with weight gain when the energy consumption is controlled (Martinez 2000; Bray 2004), at *ad libitum*, the changes in food composition with increased fat content are associated with overall energy overconsumption and the development of overweight and obesity. Therefore, high fat consumption likely does not directly cause obesity, but instead indirectly leads to it since eating foods containing high fat results in overintake of total energy or of high energy-dense foods. This overconsumption is a result of a diet high in fat, having a high energy density but weak influence on satiety. In addition, the high palatability of high-fat foods and the relatively weak metabolic autoregulation of a high-fat diet are also contributors. However, the degree of the influence of dietary fat on energy consumption and body weight depends on diet composition. While most high-fat diets tend to be energy-dense diets and are thus weight promoting, important exceptions must be noted. For example, many processed low-fat foods are quite energy-dense, and conversely vegetable-based foods are quite energy dilute and in fact appear to protect against weight gain despite having significant added fat (Swinburn *et al.* 2004). Increased intake of vegetables and fruits can combat or balance the unwanted influence of fat on passive energy overconsumption and weight gain. This effect may not be simply a result of reducing energy or food intake but a consequence of decreased energy intake and increased energy metabolism and expenditure.

18.4.1.2 *Carbohydrate intake and obesity*

The definition of carbohydrates can often be confusing; however, it is very important in understanding their influences on body weight. Sugars are predominantly monosaccharides and disaccharides, which are referred to as 'free sugars', if they are present naturally in foods such as fruit juice, honey and syrups or added by the manufacturer, cooker or consumer. The value of simple sugars is in their ability to improve the palatability of many foods, but sweetening leads to energy overconsumption in the same way that fat does. Research has demonstrated that similar amounts of energy are consumed on high-fat and high-sucrose diets, while there is a lower energy intake and weight loss associated with high-starch diets (Raben *et al.* 1997). Processed foods with high fat and/or sugar content may lead to weight gain (Drewnowski *et al.* 1992; Anderson 1995).

Polysaccharides include starch and non-starch, the latter having considerable commonality with the term 'dietary fiber' (Swinburn *et al.* 2004). There is a reciprocal relationship between the fat percentage and the carbohydrate percentage in the diet, and these two components contribute over 80% of total energy. The structure of starch is complex in terms of chain length and number of side chains. As such, starches increase blood glucose and serum insulin to varying extents when the same amount but different types are consumed. Thus, the relationship between the consumption of starch and body weight may be better explained by their glycemic index (GI). Lower GI starches produce greater satiety, reduce food intake and promote weight loss (Ludwig 2000; Brand-Miller *et al.* 2002). High-GI diets result in rapid absorption of glucose, thereby altering hormonal and metabolic functions and promoting excessive food intake (Ludwig 1999). Moreover, high-glycemic-load diets can lower the metabolic rate and produce greater voluntary food intake (Agus *et al.* 2000). Low-GI diets may influence fuel storage by promoting fat oxidation instead of carbohydrate oxidation, whereas raised insulin levels in response to high-GI diets inhibit lipolysis and encourage fat storage, limiting fuels available for oxidation and encouraging overeating (Ludwig 1999; Brand-Miller *et al.* 2002).

It is well-known that dietary fiber is beneficial to weight control, in addition to many other health benefits. Evidence from the majority of studies has shown that a high-fiber diet results in lower energy intake and lower body weight. In general, there are no differences between soluble fiber, insoluble fiber, mixed fiber or between fiber as a supplement and within foods (Swinburn *et al.* 2004). Fiber affects energy balance through its intrinsic effects (i.e. energy density and palatability), hormonal effects (i.e. gastric empting and postprandial glycemia and insulinemia) and colonic effects (i.e. fermentation to short-chain fatty acids and effects on satiety). Overall, evidence convincingly supports the benefits of a high fiber intake in managing body weight (Howarth *et al.* 2001; Swinburn *et al.* 2004).

Consumption of drinks is another important contributor, confounded with carbohydrate intake, to obesity. However, this factor is generally not well recognized by the public since the major component of drinks is water, meaning the energy density of drinks is not comparable with foods. However, drinks with high sugar do promote energy overconsumption and weight gain, and is of most relevance in populations with a high intake of drinks (Swinburn *et al.* 2004). From a US national survey, children ingesting nine or more ounces of soft drinks per day consumed nearly 200 kcal more than those who did not drink soft drinks (Harnack *et al.* 1999). Another study in the United States showed that high intake of sugar drinks predicted the development of obesity over 19 months in 12-year-old children (Ludwig *et al.* 2001). It is estimated in this study that an increase of one can of soda per day increased the risk of obesity 1.6 times, but this association was not seen with diet soda drinks. Some fruit drinks and cordial drinks can also be high in sugar and may promote energy intake and weight gain if drunk in large quantities (Swinburn *et al.* 2004).

18.4.1.3 *Dietary protein and obesity*

Contribution of protein to overall energy intake is relatively small, ranging from 10 to 15%, across all populations and periods (Gurr 1991). This may limit the scope for considering protein intake as a population measure to combat obesity. Nevertheless, protein is generally considered a most satiating macronutrient, especially among people with low habitual protein intake. Increasing protein intake may be helpful for individuals who are on a weight control program. However, at a population level, protein intake is likely not an important determinant of obesity prevalence (Swinburn *et al.* 2004).

18.4.2 Changes in food behavior

18.4.2.1 *Increased snack consumption*

The industrial revolution has brought remarkable changes to food production, processing and packaging. These changes have been linked directly to the dramatic increase in the variety of packed and processed foods. Snacks that were previously unavailable have recently become very common. In the United States, the frequency of eating snacks and the energy density of snack foods have continuously increased, and consequently the contribution of energy from snacks to total energy intake has significantly increased as well (Zizza *et al.* 2001). Currently, snacks contribute up to 20–25% of total energy intake in the United States and the UK. Similar situations have happened in Europe and other countries. Although there is insufficient evidence to determine the effect of high frequency of eating on weight gain, the intake of high-energy-density snack foods is likely to promote it. A typical example is that a binge eating disorder and night eating syndrome have been associated with a high risk of weight gain (Schenck & Mahowald 1994; Yanovski & Sebring 1994).

18.4.2.2 *Increased frequency of eating outside the home*

Another important contributory factor of food behavior related to weight gain is the increased frequency of eating food outside the home. In Western countries, the frequency of eating food prepared outside the home is increasing, and this is most apparent and best documented in the United States (Swinburn *et al.* 2004). Generally, food prepared away from the home is higher in total energy, total fat, saturated fat and cholesterol than homemade food. The fat content prepared at home has fallen to 32%, but food prepared outside the home is still 38% (Swinburn *et al.* 2004). Moreover, it is possible that foods consumed in or bought from a restaurant will have fewer amounts of vegetables and fruits. Evidence implicates the increased use of food prepared outside the home as a risk for obesity. However, this phenomenon is largely limited to the United States and other Western countries, and may not be the case in Asian countries such as China.

18.4.2.3 *Increased frequency of drinking of soft beverages*

Soft drink consumption has increased by 300% in the past 20 years, and 56–85% of children in school consume at least one soft drink daily. The risk of becoming obese among children increases 1.6 times for each additional can or glass of sugar-sweetened drink consumed beyond their usual daily intake of the beverage (Harrington 2008). Sugar-sweetened drinks, classified as GI liquids, increase postprandial blood glucose levels and decrease insulin sensitivity. Additionally, high-GI drinks submit to a decreased satiety level and subsequent overeating (Harrington 2008). Sugar consumed in liquids may induce less satiety as compared with the same foods in a solid form because of the rapid transit of liquids through the stomach and intestines that may lead to reduced stimulation of satiety signals, differences in the regulation of thirst and hunger, and lower perceived energy content (van Dam & Seidell 2007).

18.4.3 Environmental factors contributable to obesity

The increasing 'obesogenicity' of the environments surrounding individuals is likely to be the major driving force for the increasing obesity epidemic (Swinburn *et al.* 2004). The environments in which people live are complex and their individual and combined elements have a marked influence on people's food behaviors. Individuals interact with a variety of environmental elements and settings such as schools, homes, restaurants and fast food outlets (Swinburn *et al.* 1999).

18.4.3.1 *Socioeconomic status*

The relationship between socioeconomic status and obesity is complicated, and the patterns are more exaggerated in women than in men and children. In developed countries, there is a higher rate of obesity in low socioeconomic families than in high socioeconomic families. In these countries, high socioeconomic class provides certain protection from obesity. Individuals with high socioeconomic status have higher education levels and a better understanding of the benefits of healthy foods. They are paying more attention to food choices and practicing better eating habits. They live in an environment with greater opportunities for healthy eating, and greater capacity to manipulate their environment to suit their needs. By contrast, people living in low socioeconomic environments generally take food by the default choices which are offered, and have a higher concentration of fast food outlets in their communities (Swinburn *et al.* 2004). The opposite is true in developing countries. The change in obesity prevalence patterns can be easily seen in countries that have gone through a significant socioeconomic transition period such as Brazil (Monteiro *et al.* 2001) and China (Reynolds 2007).

18.4.3.2 *School environment*

School is a key environmental setting that affects childhood obesity. The elements that contribute to the overall school food environment include available food choices, food policies, health education for teachers and service staff, guidelines for food and drink choices, and the curriculum on food and nutrition. Additionally, the food available at school impacts children and adolescents substantially on what they eat throughout the day. In the later half of the last century, the increase in the number of soft drink vending machines was linked to a fast-growing prevalence of obesity and overweight (Ludwig *et al.* 2001). Thus, certain types of soft drinks have recently been removed from most schools in North America.

18.4.3.3 *Home environment*

Undoubtedly, home is the most important environment where childhood eating behavior is influenced. The availability, accessibility and exposure to a range of different variety of foods, vegetables, fruits and drinks influence to a large extent the food and drink preference in children and adults. This is particularly true as compared between Western and Asian populations in terms of their food choices. At home, peers and parents provide a strong influence on a child's eating behaviors. Parental knowledge on nutrition, attitudes and food behaviors are significant predictors of consumption of different varieties of foods and drinks by children (Gibson *et al.* 1998). The continued trend for eating more food prepared outside the home reduces the exposure of children to cooking skills learned at home. Restricted availability of food choices increases the selectiveness of eating, and a high degree of parental control of a child's food intake is associated with a lower ability of the child to self-regulate energy intake (Swinburn *et al.* 2004).

18.4.3.4 *Food marketing sector (food processing and restaurants)*

Generally, food supplied in restaurants has high fat content compared to food prepared at home (Swinburn *et al.* 2004). Fast food restaurants and energy-dense foods and drinks are among the most advertised products on television, and children are often the targeted age group. The fat, sugar and energy content of foods advertised to children is very high compared to their daily needs, and most of the foods advertised fall into the 'eat least' or 'eat occasionally' sections of the recommended dietary guidelines. The prevalence of marketing messages can overwhelm the public with products that provide the opposite nutritional value than what is recommended for a healthy diet.

The food industry invests heavily in mass media marketing, mainly through television advertisements. Unfortunately, most of those pertain to fast food restaurants and manufacturers producing high-fat or high-sugar foods and drinks (Swinburn *et al.* 2004). The high volume of advertising for energy-dense foods and beverages is undoubtedly fueling the increasing consumption of these products since advertised products are more often requested for purchase and consumed by children (Lewis & Hill 1998). The prevalence of overweight and obesity is higher among children who watch more television, and the increased energy intakes may be partly responsible in these children (Taras *et al.* 1989; Lewis & Hill 1998).

18.4.3.5 *Portion size*

The influence of portion size on the development of obesity and overweight may not be well recognized by the general public; however, along with high fat and high sugar, portion sizes in prepacked, ready-to-eat and restaurant foods are also increasing in many world markets. Increased portion size is built on the basis of the consumers' desire for 'value for money'. Portion size has ballooned by 12–20% since the 1970s both inside and outside the home, while the largest increases have occurred at fast food establishments. Concurrently, obesity rates have skyrocketed since the same period. In recent years, the number of restaurants offering 'supersize' options on their menu has rapidly risen, and other food items, especially snack foods, have also increased package weight. The increased size of packaging indicates lower unit cost and encourages use of more product than a smaller package size. Supersize portions can potentially lead to increased energy intake not only at the time of consumption but over the course of the day as well, and could therefore be a significant contributor to obesity, particularly in populations with high consumption of meals prepared outside the home. There is a strong epidemiological evidence of a concurrent increase in portion size and obesity prevalence in countries such as the United States (Howarth *et al.* 2001).

18.4.4 Physical activity

The human body consumes energy in the form of resting energy expenditure (basic metabolic rate), thermal effect of food, and energy expenditure resulting from physical activity. The resting energy expenditure and thermal effect of food are relatively constant and account for 70–80% of total energy expenditure (Luis Griera 2007). In contrast, the contribution of physical activity is highly variable within and between individuals and therefore greatly influences energy balance. Under normal circumstances, energy balance fluctuates from day to day, but the human body keeps energy stores and weight stable through multiple regulatory mechanisms. Development of overweight and obesity only occurs when positive energy balance prevails for a considerable period (Lakka & Bouchard 2005). In addition to energy intake and energy expenditure, nutrient partitioning is emerging as another important determinant of long-term energy balance. Under positive energy balance conditions, individuals who are more likely to gain weight will partition more energy for storage in adipose tissue, which results in adipocyte hypertrophy, while those who are less likely to gain weight tend to partition relatively more for fat oxidation by skeletal muscle and other tissues (Ravussin & Smith 2002). Physical activity accounts for 20–30% of total daily energy expenditure in sedentary individuals, but it may represent up to 50% of all energy expended in persons who engage in heavy manual work or in demanding exercise training (Livingstone 1991). Physical activity accounts for most of the variation in total energy expenditure. The contribution of physical activity to total energy expenditure depends on the amount and intensity of physical activity and other factors such as body mass (Ravussin *et al.* 1986). Total physical activity can be divided into (1) spontaneous activities such as movement of arms, legs and head, taking small steps, fidgeting, and even mastication, (2) work-related activities, such as office work, construction and agriculture fieldwork, (3) the activities of daily living such as climbing stairs, walking or cycling to and from work, household work, and yard work and (4) conditioning exercise such as walking, running, cycling, skiing, dancing, swimming, sports and other fitness activities (Lakka & Bouchard 2005). In modern society, the contribution of work-related activity to total energy expenditure is much smaller than in the past. The activity of daily living accounts for most of the energy cost of physical activity in individuals who do not engage in regular exercise and who represent the majority of the population in developed countries (Lakka & Bouchard 2005). In addition, physical activity influences substrate balance; i.e. individuals who are physically active utilize rather than store excess fat and tolerate high-fat diets better than sedentary persons (Lakka & Bouchard 2005).

Fast development of high technology in the late 1980s has dramatically changed the nature of jobs and the degree of labor involvement in many areas, resulting in a significant increase of the population working in a sedentary environment. Consequently, labor density and energy expenditure have considerably decreased. Industrialization and modernization cause a similar shift in time allocation and physical effort in home and leisure activities. In the last century, the evolution of household technology has accelerated and facilitated the processing, storage and preparation of food by replacing manual labor with mechanical or electronic tools. These food-preparation technologies, together with home electrification, washing machines and dryers, vacuum cleaners, piped water, and so forth, have transformed home management from a time-consuming, often exhausting, full-time occupation to merely a minor hassle. Although home production still requires time and energy, purchased technology is widely accessible to expedite home production activities.

Possibly an even more astounding shift has come in leisure activities. In addition to computer games and television programs, other sedentary forms of entertainment have become available to almost every family, thus increasing the sedentary time of the entire population. The rapid shift in television ownership and the equally important provision of cable linkages which offer a wide array of choices to each household are key elements. In the past, leisure activities for children often meant active play, but leisure today may well mean a sedentary activity such as viewing television or playing a computer game (Popkin 2001). This change of lifestyle is more dramatic in children and adolescents than in adults. Therefore, overweight and obesity have rapidly reached epidemic levels in children in recent years, although the overall rate is still lower in children than in adults due to physiological conditions and developmental stages that are related to weight gain. There is no doubt in the considerable benefits high technology has brought to the general population, but meanwhile, it has indirectly contributed to weight gain and other health problems. Unfortunately, our body is not ready to adapt to all these sudden changes in lifestyle, and we have not developed a set of strategies to overcome the impact of high technology on human health. In facing this change, increasing physical activity is still one of the key strategies to prevent or treat obesity.

18.4.5 Genetic evolution and familial history

Certainly, evolution has also occurred in the human genome; however, this change equates to only a 1.6% difference between modern humans and most developed primates. In spite of this overall stability of the human genome, a number of minor alterations/mutations in gene structures have been observed and have attributed to the racial differences of various traits. It has been found that several gene mutations are associated with childhood obesity, such as congenital leptin deficiency, leptin receptor deficiency, pro-opiomelanocortin mutations and melanocortin-4 receptor mutations (ORahilly *et al.* 2003).

Quantitative genetic analyses have shown significant heritability. An individual's chances of being obese are increased when having obese relatives with the estimates of heritability ranging from 30 to 70% (Bell *et al.* 2005; Farooqi 2005; Hunter 2005). It is however difficult to reconcile the genetic etiology of common human obesity with the markedly variable prevalence of the disease in individuals of different socioeconomic backgrounds and demographic origins (Hill & Peters 1998). One of the hypotheses based on mouse studies is that susceptibility of obesity is determined largely by genetic factors, but the environment determines phenotypic expression (West *et al.* 1995). Genetic research of obesity has gone through several different stages, with increased efforts beginning in the late 1980s. The new research tools and techniques developed since the late 1990s allow the detection of gene polymorphisms, especially single-nucleotide polymorphism from a blood sample. This advancement opens a new era of research work that is devoted to studying the interaction among diets, metabolic variables, disease risk factors and gene polymorphisms. To date, over 60 genes have been identified to be strongly associated with obesity. People who have familial history or genetic traits of obesity should be more cautious and pay close attention to their weight change. Preventive diets and being physically active are highly recommended.

18.4.6 Virus infection

Recent studies have demonstrated that the infection of certain viruses is also linked to the development of obesity. As the rapid spread of obesity resembles epidemiologically the spread of infectious diseases, the relationships between the viruses and obesity have been increasingly investigated since the last decade (Atkinson 2007; Vasilakopoulou & le Roux 2007). Numerous studies have validated the relationship between obesity and virus infections in some cases. Consequently, it has been found that five animal viruses and three human viruses cause obesity in animals, and one human adenovirus, Ad-36, is linked to obesity in humans. Although the principles of weight gain after infection with specific viruses have been demonstrated and some epidemiological evidence in humans has been reported, a plausible mechanism is still awaited. The two most likely mechanisms are proposed to be either a peripheral effect on fat cell differentiation and storage or a central nervous system effect on appetite and energy expenditure. It is possible for a virus to affect specific neuronal pathways involved in the control of appetite and energy balance, without any other obvious detrimental effects or behavioral changes. Infection of certain viruses stimulates a rapid differentiation response in preadipocytes with significantly faster appearance of enzymes of fat storage and differentiation factors, differentiation into mature adipocytes, and accumulation of triacylglycerides. Viral infection also changes insulin and leptin secretion and insulin sensitivity (Vangipuram 2007). The development of overweight and obesity resulting from viral infection should be classified as pathological process and treated as an infectious disease where the obesity or weight gain is the appearance or symptoms of infection, which do not cause severe pathological problem but rather changes the central nervous system function and peripheral adipose tissue differentiation and function of fat storage tissues. Further work to unravel the mechanism is of importance to provide a solid explanation on how these viruses can cause weight gain or obesity. It should be noted however that the portion of obesity caused by viruses is considered small.

18.5 Obesity and cardiovascular disease

18.5.1 Association between obesity and various cardiovascular disease

The traditional risk factors for CVD consisted of smoking, hypertension, diabetes and hyperlipidemia (Assmann 1999). However, obesity has recently emerged as an important independent risk factor for the development of CVD (Rabkin *et al.* 1997). A close association between obesity and CVD has been observed in different ethnic groups in different locations and in both industrialized and non-industrialized countries (Stamler *et al.* 1978; Cooper *et al.* 1999; Suk 2003). Obesity markedly reduces lifespan, especially in young adults, and, in addition, is an independent predictor of premature cardiovascular death (Manson 1995; Stevens 1998; Rosengren *et al.* 1999). The pattern of fat distribution in the body is gaining increasing prominence as a more potent risk factor for CVD than body weight itself (Han *et al.* 1995). Abdominal obesity is a stronger risk factor for CVD than overall obesity in both men and women (Lakka & Bouchard 2005). CVDs associated with obesity include mainly coronary heart disease, ischemic stroke, peripheral artery disease, congestive heart failure and deep venous thromboembolism (Samama 2000; White *et al.* 2000; Suk 2003; Lakka & Bouchard 2005).

Deep venous thrombosis is a serious and potentially life-threatening condition where blood clots develop in deep veins within idle leg muscles. This problem has long been known to be caused by prolonged sitting due to varied reasons such as extended automobile journeys and flying on airplanes, and even during excessive TV viewing or computer and video game use (Hamilton *et al.* 2007). Recent epidemiological evidence has demonstrated that obesity is a risk factor for deep venous thrombosis that is also referred to as venous thromboembolism (Samama 2000; White *et al.* 2000). In a population-based prospective cohort study, an overweight group with a BMI of 25–29 and an obese group with a BMI of 30–35 had significantly higher risk of deep venous thrombosis when compared with a reference group of BMI of less than 25 (Tsai 2002). In another population-based cohort study, it was found that risk for venous thromboembolism sharply increased in middle-aged men of 50 years or older who surpassed the threshold waist circumference (Hansson *et al.* 1999).

18.5.2 Mechanisms involved in the development of cardiovascular disease due to obesity

Obesity raises the risk of developing CVD partially through its effects on metabolic syndrome, characterized by hyperinsulinemia, glucose intolerance and diabetes, hyperlipidemia, hypertension and chronic inflammation (Despres 2001; Douketis & Sharma 2005). Obesity causes endothelial dysfunction, which is an early marker of generalized atherosclerotic vascular disease (Arcaro 1999). It may promote platelet activation and aggregation through increased lipid peroxidation, inflammation, hyperviscosity and reduced red blood cell deformability (Solerte 1997). Obese people are at increased risk of developing congestive heart failure because of the continuous pressure overload that is exerted on the left ventricle from an increased body mass (Must 1999). Obesity is also associated with chronic volume overload and a hyperdynamic, high-cardiac output circulation, which can lead to systolic and diastolic ventricular dysfunction (Kuch 1998). The mechanisms of obesity on hypertension are proposed to include cardiovascular dysfunction, renal structural and hemodynamic changes, activation of the renin–angiotensin–aldosterone system, natriuretic peptide effects, changes in hormonal and biochemical mediators, and nervous system changes (Kolatkar *et al.* 2006).

Obesity is associated with alterations in lipid metabolism including elevations in serum concentrations of total cholesterol, low-density lipoprotein cholesterol, and triacylglycerides and a reduction of serum high-density lipoprotein cholesterol (Schulte *et al.* 1999). Central obesity and high waist-to-hip circumference ratio, in particular, account for much of the variance in elevated triacylglycerides and low high-density lipoprotein cholesterol (Robinson *et al.* 2006). It is well documented that dyslipidemia is a very important factor that is associated with CVD through the development of arteriosclerosis. Obesity, especially central obesity, is associated with an increased concentration of circulating free fatty acids, which contribute to dyslipidemia in obese people, and is independently associated with CVD mortality of subjects with coronary heart disease. Elevated free fatty acids can induce endothelial dysfunction, resulting in suppressed endothelial nitric oxide synthase and prostacyclin synthase activities, which are the two important antiatherogenic enzymes. In addition, free fatty acids may have lipotoxic effects on the heart (Bodary *et al.* 2007).

It is unquestionable that white adipose tissue is a highly metabolic active endocrine organ whose main products of adipokines have essential roles in energy homeostasis, glucose and lipid metabolism, cell viability, control of feeding, thermogenesis, neuroendocrine function, reproduction, immunity, and importantly, cardiovascular function (Trujillo & Scherer 2006). Obesity is a pro-inflammatory state in which hyperplasia and hypertrophy of fat cells induce severe alterations in the secretion of adipokines that lead to insulin resistance and inflammation (Skurk *et al.* 2007). The intensive research on adipokines in the past two decades has provided a plausible link between obesity, especially visceral obesity, the metabolic syndrome, inflammation and CVD (Despres & Lemieux 2006). White adipose tissue releases adiponectin, leptin, resistin, visfatin, apelin, omentin, chemerin, and others, that are shared with other systems, such as interleukin-6 (IL-6), tumor necrosis factor-α (TNF-α), monocyte chemoattractant protein-1 (MCP-1) and plasminogen activator inhibitor-1 (PAI-1). All these factors are altered under the condition of obesity and linked to the increase of type 2 diabetes and CVD in obese and overweight subjects (Gualillo *et al.* 2007; Lago *et al.* 2007).

18.6 Obesity and type 2 diabetes

18.6.1 Epidemic of diabetes

According to the American Diabetes Association diagnostic criteria (Report of the Expert Committee 1997), the prevalence of diagnosed and undiagnosed diabetes in the United States in adults of 40–74 years increased by 48% in a 12-year period between 1976–1980 and 1988–1994 (Harris 1998). During the same time period, the prevalence of impaired fasting glucose increased by 49%. More recent data indicate that the prevalence of diabetes (diagnosed and undiagnosed) in the total population of adults of 20 years and older was

8.6% in 1999–2000. An additional 6.1% of adults have impaired fasting glucose, increasing to 14.6% for adults 60 years and older. Overall, in 1999–2000, an estimated 14.5% of US adults at the age of 20 years and over and 33.9% of those 60 years and over had either diabetes or impaired fasting glucose (Centers for Disease Control and Prevention 2003). The prevalence of diabetes in Canada increased from 4.3% in 1997–1998 to 5.1% in 2000–2001 (Health Canada 2003), and up to 5.5% in 2004–2005 (Public Health Agency of Canada). The prevalence of diabetes rate in adults aged 20 years or over increased from 5.5% in 2000–2001 to 7.1% in 2004–2005 (Public Health Agency of Canada).

Diabetes is not only increasing at an alarming rate in North America, it has become a world epidemic. Based on available information and the trend of diabetes development in the past decades, it is estimated that the global prevalence of diabetes (including type 1 and type 2) in adults aged 20 years and over will increase by 39%, from 4.6% in 2000 to 6.4% in 2030. The prevalence is higher in developed countries than in developing countries, but the relative increase will be greater in the developing countries where the prevalence of diabetes is estimated to rise by 46% (from 4.1 to 6.0%). In contrast, the prevalence of diabetes in the developed countries will increase by 33% (from 6.3 to 8.4%). The total number of adults of 20 years and older with diabetes is projected to approximately double between 2000 and 2030 and the number will increase from 171 to 366 million (Wild *et al.* 2004).

Diabetes is a major complication of obesity and overweight and has been increasing in parallel with the prevalence of obesity all over the world. Type 2 diabetes is a disease that previously occurred predominantly in adults. However, it has grown to become a health problem in children and adolescents, which is thought to be associated with the prevalence of obesity and overweight in children and adolescents (Larsson & Wolk 2006). Excess weight plays a critical role in the etiology of type 2 diabetes mellitus, which is an important risk factor for premature coronary heart disease and stroke, as well as the main cause of kidney failure, limb amputations and new-onset blindness in adults.

18.6.2 Mechanism of obesity on the development of type 2 diabetes

Obesity is nearly invariably associated with insulin resistance and strongly associated with the increased prevalence of diabetes, but obesity is not sufficient to cause diabetes, meaning not all obese subjects will develop the disease. In fact, diabetes develops in a subset of genetically predisposed people whose pancreatic insulin secretion fails to meet insulin requirements set by the individual's insulin sensitivity (Kahn 2003). The mechanism through which obesity increases insulin resistance is currently thought to be related to the increased circulating free fatty acids, triacylglycerides, altered levels of adipokines, altered body fat distribution, chronic inflammation and other factors (Soodini & Hamdy 2006).

18.6.2.1 *Circulating free fatty acids and type 2 diabetes*

Intercommunication between adipocytes and pancreatic β-cells is mediated by free fatty acids and adipocyte-secreted adipokines (Cnop 2008). Blood free fatty acids are often higher in obese subjects, and it is reported that elevated circulating free fatty acids stimulate hepatic gluconeogenesis, induce hepatic and muscle insulin resistance, and impair insulin secretion (Boden & Chen 1995; Paolisso 1995). Constant high concentration of free fatty acids in the blood stream increases the accumulation of triacylglycerides in both liver and muscle, which are correlated with the degree of insulin resistance in these tissues. In addition, since triacylglycerides are in a state of constant turnover, their metabolites – acyl-coenzyme A, ceramides and diacylglycerides – contribute to both impaired hepatic and peripheral insulin action. This sequence of events is frequently referred to as lipotoxicity (Unger 1995), which accumulating evidence suggests to be an important contributor to the pancreatic β-cell dysfunction in patients with type 2 diabetes (Shimabukuro *et al.* 1998). High levels of circulating free fatty acids, particularly saturated fatty acids and low levels of adiponectin, are predictive of diabetes development (Cnop 2008).

18.6.2.2 *Role of adiponectin in the development of diabetes*

Adipocytes express and secrete a variety of bioactive peptides, termed adipokines including adiponectin, leptin, TNF-α, IL-6, MCP-1 and PAI-1 (Kralisch 2007). With the exception of adiponectin, which is decreased, all other adipokines are increased in overweight and obesity subjects. Adiponectin is an endogenous insulin sensitizer through its effect on adenosine monophosphate-dependent kinase (AMPK) (Kralisch 2007). Binding of adiponectin to its receptors activates AMPK, which phosphorylates and inhibits acetyl coenzyme A carboxylase (ACC) (Wu 2003). ACC is a key enzyme involved in fatty acid synthesis and acts as a 'control switch' between fatty acid synthesis and oxidation. The inhibition of ACC results in a decrease of fatty acid synthesis and reciprocally an increase in fatty acid oxidation. This change in fatty acid metabolism leads to reductions of circulating free fatty acids, and thus increases in insulin sensitivity. Furthermore, AMPK activation induces an increase of proteins involved in fatty acid transport and oxidation in myocytes and reduces the expression of gluconeogenic enzymes in hepatocytes (Yamauchi 2002).

The role of adiponectin on insulin sensitization has been proved in several studies using genetically modified animal models that lack one or two adiponectin receptors (Kralisch 2007). Circulating adiponectin levels are inversely proportional to obesity and insulin resistance, and increase with weight loss and with the use of insulin-sensitizing drugs (Maeda 2001; Kadowaki & Yamauchi 2005). It is suggested that inflammation contributes to hypoadiponectinemia in insulin resistance and obesity (Lago *et al.* 2007); and by contrast, physical activity increases circulating adiponectin levels and its receptor expression in muscles, which may mediate the improvement of insulin resistance in response to exercise (Bluher 2006). It is convincing that increased adiponectin levels may protect against later development of type 2 diabetes, whereas decreased adiponectin may predispose to diabetes (Lindsay 2002).

18.6.2.3 *TNF-α and type 2 diabetes*

In contrast to adiponectin, elevation of several potent pro-inflammatory cytokines in the plasma is implicated in the development of insulin resistance, type 2 diabetes and atherosclerosis. TNF-α is proposed to link insulin resistance with obesity. TNF-α impairs insulin sensitivity in hepatocytes, myocytes and adipocytes (Hotamisligil *et al.* 1993). TNF-α expression in adipocytes is increased in most animal models and human subjects with obesity and insulin resistance. In contrast, TNF-α expression in fat tissue decreases in obese patients after weight loss (Kern 1995). The role of TNF-α in the induction of insulin resistance has been demonstrated in rodent models that are either TNF-α deficient (Uysal *et al.* 1997) or administered with a TNF-α antibody (Hotamisligil *et al.* 1993). Convincing evidence shows that TNF-α infusion in healthy lean male volunteers potently induces insulin resistance similar to the results in rodents (Rask-Madsen 2003). Interestingly, the correlation between insulin resistance and plasma concentration of TNF-α is weak in obese subjects with and without diabetes although the circulating TNF-α levels are increased in both groups (Zinman *et al.* 1999). It has been postulated that tissue insulin resistance may be more strongly related to the local tissue TNF-α levels than to its plasma levels, and that the circulating levels of TNF-α receptors may reflect more accurately the status of activation of the TNF-α system. Increased release of TNF-α from adipocytes may play a major and direct role in the impairment of insulin action. TNF-α influences insulin signaling through serine phosphorylation of the insulin receptor and insulin receptor substrate-1, thus inhibiting insulin action at the organ level through autocrine and paracrine mechanisms (Soodini & Hamdy 2006).

18.6.2.4 *Interleukin-6 and type 2 diabetes*

Similar to TNF-α, IL-6 is another systemic pro-inflammatory cytokine, which regulates the hepatic production of cholesterol-reactive protein (CRP) and other acute-phase proteins. In humans, plasma IL-6 concentrations are positively correlated with BMI, the severity of inflammation as indicated by serum levels of CRP, and glucose intolerance and type 2 diabetes (Pickup *et al.* 2000; Pradhan *et al.* 2001; Bastard 2002). IL-6 induces insulin resistance and diabetes by affecting insulin signaling and glucose metabolism in hepatocytes, adipocytes and muscle, as well as β-cell apoptosis (Sandler *et al.* 1990; Shimabukuro *et al.* 1998; Fasshauer 2004).

18.6.2.5 *Leptin and type 2 diabetes*

Leptin is exclusively expressed in adipose tissue, especially subcutaneous fat. Leptin may exert a direct effect in metabolically important tissues and/or indirect effects by activating specific centers in the hypothalamus to decrease food intake and increase energy expenditure, thus influencing glucose and fat metabolism (Ahima & Flier 2000). In accordance with an appetite-suppressive effect, leptin-deficient ob/ob mice and leptin receptor-deficient db/db mice develop severe hyperphagia and excessive morbid obesity associated with insulin resistance, hyperinsulinemia and diabetes mellitus (Ahima & Flier 2000). In contrast, administration of recombinant leptin to ob/ob mice reduces food intake, body weight and fat mass, suggesting that leptin is responsible for the development of the obese phenotype in leptin-deficient mice (Pelleymounter 1995). Leptin-deficient human subjects show a similar phenotype of severe obesity, glucose intolerance and insulin resistance, and treatment of patients with leptin mutation or having no functional leptin with leptin leads to a sustained reduction in weight, predominantly as a result of loss of adipose tissue (Clement 1998; Farooqi 1999; Mantzoros & Flier 2000). Leptin treatment decreases glycosylated hemoglobin in patients with lipodystrophy and low serum leptin levels due to improved insulin-stimulated hepatic and peripheral glucose metabolism, as well as a marked reduction in muscle and hepatic triacylglyceride content (Petersen 2002). Food intake is decreased by systemic administration of leptin in normal weight animals, but the response is reduced in obese animals (Van Heek 1997). Similarly, in humans, leptin levels are increased in common obesity and further administration of leptin does not alter appetite and body weight (Considine 1996; Savage & ORahilly 2002). Obesity is a condition of relative leptin resistance such as insulin resistance seen in type 2 diabetes mellitus (El-Haschimi & 2000). Saturation of leptin transporters across the blood–brain barrier and impaired ability of leptin to activate hypothalamic signaling in diet-induced obesity are potential mechanisms for leptin resistance in obesity (El-Haschimi *et al.* 2000; Banks 2003).

18.6.2.6 *PAI-1 and type 2 diabetes*

PAI-1 is expressed mainly in endothelial and liver cells and also in adipose tissue (Alessi 1997). PAI-1 is involved in weight gain and insulin resistance (Ma 2004) and markedly upregulated in the conditions of obesity and insulin resistance (Alessi 1997). Accordingly, disruption of the PAI-1 gene leads to reduced adiposity in ob/ob mice (Ma 2004). The resting metabolic rates, total energy expenditure and insulin sensitivity are increased in PAI-1 null mice when compared with wild-type controls (Schafer *et al.* 2001).

18.6.2.7 *Other adipokines and type 2 diabetes*

In addition to the aforementioned adipokines, several other adipokines such as resistin, apelin, retinol-binding protein and visfatin are affected by obesity. Plasma concentration of these adipokines is associated with insulin resistance.

18.6.2.8 *Pattern of obesity and type 2 diabetes*

The pattern of body fat distribution rather than total fat mass is suggested to be an important predictor of insulin resistance. Individuals with upper-body fat accumulation or higher visceral fat mass are more resistant to insulin than those with a predominantly lower-body fat accumulation or more subcutaneous fat (Soodini & Hamdy 2006). The association between visceral fat accumulation and insulin resistance has been attributed to the increased sensitivity of visceral fat to lipolytic stimuli. This increases the flux of free fatty acids into the portal and systemic circulations (Motoshima 2002). In contrast to subcutaneous fat, visceral fat cells also produce excessive amounts of pro-inflammatory adipokines such as TNF-α, IL-6 and PAI-1 and decreased amounts of insulin-sensitizing adipokines such as adiponectin (Soodini & Hamdy 2006). Moreover, organ-specific deposition of body fat is a better predictor of insulin resistance in a given organ than visceral adiposity. For example, increased intramyocellular triacylglycerides correlate closely with muscle insulin resistance, and intrahepatic fat accumulation is associated with hepatic insulin resistance (Soodini & Hamdy 2006). Excessive storage of ingested fat in skeletal muscle and liver leads to the development of severe insulin resistance in these organs. Conversely, replacement therapy with leptin and surgical transplantation of adipose tissue in lipodystrophy animals resulted in mobilization of fat out of the liver and muscle and improved insulin sensitivity in these animals (Soodini & Hamdy 2006).

18.6.3 Potential biomarkers for diagnosis of prediabetes

Type 2 diabetes is generally a preventable chronic metabolic disease that is closely related to lifestyle. It is believed that some metabolic alterations occur before the onset and during the progression of type 2 diabetes. Recent studies have found several blood biomarkers associated with the predisease state of type 2 diabetes, which may be used to diagnose the risk and accordingly apply preventive procedures to delay or even prevent the development of the condition.

Many studies have demonstrated that increased blood insulin and glucose concentrations and obesity predict the development of type 2 diabetes (Sicree *et al.* 1987; Bergstrom 1990; Haffner *et al.* 1990b; Lillioja 1993). In addition, increased blood triacylglycerides, decreased high-density lipoprotein cholesterol and increased systolic blood pressure are predisease indicators (Haffner *et al.* 1990a). High levels of circulating free fatty acids, particularly saturated fatty acids, and low levels of adiponectin are also predictive of diabetes development (Cnop 2008).

Inflammatory biomarkers are increasingly found to be related to the development of type 2 diabetes. Chronic or subclinical inflammation as indicated by elevated CRP levels in blood appears to be associated with insulin resistance and features related to the insulin resistance syndrome that is characteristic of a prediabetic state (Haffner 2003b). In patients with type 2 diabetes, treatment with insulin sensitizer, rosiglitazone, significantly reduced serum levels of CRP (Jialal 2001). Another inflammatory biomarker that has been shown to be associated with the development of type 2 diabetes is PAI-1 (Haffner 2003a). Certainly, more research is needed in this area in order to select a few or a set of sensitive measures that are able to provide a better indication of risk for developing type 2 diabetes.

18.6.4 Association of diabetes with cardiovascular disease

In recent years, emerging evidence has demonstrated the association between diabetes and CVD, and it has gradually become clear that the cardiovascular complications of diabetes are due, at least in part, to its inflammatory effect mediated by a variety of adipokines and cytokines (Gualillo *et al.* 2007; Mehta & Farmer 2007). Although not inclusive, low circulating levels of adiponectin are suggested to be involved in the pathogenesis of hypertension (Gualillo *et al.* 2007). In addition, dyslipidemia is associated with low circulating adiponectin levels, even in the absence of other metabolic syndrome risk factors (Trujillo & Scherer 2006). Obesity-related hypoadiponectinemia is associated with subclinic inflammation, and there is a general consensus about a putative protective role of adiponectin from an inflammatory state. Low levels of adiponectin have been linked to inflammatory atherosclerosis in humans, suggesting that normal levels of the adipokine are required to maintain a non-inflammatory phenotype in the vascular wall (Fontana *et al.* 2007; Lago *et al.* 2007). Many *in vivo* animal models and *in vitro* studies have shown that adiponectin is of critical importance in atherosclerosis physiopathology (Gualillo *et al.* 2007). High plasma adiponectin levels are associated with a lower risk of myocardial infarction in humans, reduced coronary heart disease in diabetics and lower risk of acute coronary syndrome. In contrast, plasma adiponectin levels rapidly decline following acute myocardial infarction and are closely associated with coronary lesion complex in humans with coronary artery disease (Gualillo *et al.* 2007).

Adiponectin synthesis is not restricted to adipose tissue, but also occurs in bone-forming cells, cultured myotubes from skeletal muscle, liver cells and cardiomyocytes, where it induces glucose and fatty acid uptake and oxidation (Gualillo *et al.* 2007). In addition, adiponectin downregulates the secretion of other adipokines involved in inflammation and insulin resistance, such as IL-6, IL-8, macrophage inflammatory protein-1α/β and MCP-1 (Sell *et al.* 2006), and inhibits PAI-1 activity (Mertens 2005). Adiponectin has significant antiatherogenic effects by improving insulin sensitivity and endothelial function, reducing the expression of adhesion molecules and increasing angiogenesis. It also acts through anti-inflammation pathways (Mehta & Farmer 2007).

Leptin gene expression is mainly regulated by food intake, energy status and various hormones, but also by inflammatory mediators. Increased serum leptin levels in humans are associated with myocardial infarction and stroke, independently from traditional cardiovascular factors and obesity status, with insulin resistance, inflammation and disturbance in homeostasis, and with the extent of coronary artery calcification (Gualillo *et al.* 2007). Elevated plasma leptin levels have been found in patients with hypertension (Beltowski 2006b). Leptin is an efficient vasodilator in humans with coronary artery disease and also in rat aorta by increasing nitric oxide release (Momin 2006; Rodriguez 2007). In leptin-deficient mice, endothelial function is impaired (Beltowski 2006a). Leptin may play a role in pathogenesis of atheromatous plaques, acting synergistically with other inflammatory mediators, such as IL-6, CRP, soluble vascular cell adhesion molecules 1, MCP-1 and PAI-1 (Gualillo *et al.* 2007). Leptin has direct effects at the cardiac level and regulates cardiac contractile function, metabolism, cell size and the extracellular matrix components in cardiomyocytes.

18.7 Prevention of obesity

A moderate excess of weight confers a noticeable reduction in lifespan, and as the degree of overweight increases, a striking and steady contraction of lifespan occurs (Fontaine *et al.* 2003; Peeters 2003). Whether obese, at risk of becoming obese, or have a healthy body weight, the strategies for preventing obesity are much the same as losing weight, except the extremely obese, for whom prescribed medications or surgical treatment is required. As obesity is caused by an excess intake of energy relative to energy expenditure, changes in diet, lifestyle and food behavior are essential to succeed in an effort to prevent or treat obesity. Although socioeconomic status can affect the capacity to carry out changes in lifestyle, the necessary adaptations should be taken regardless of factors such as income and convenience. With regard to food habits, despite any factors which may contribute to differences in the understanding of the relationship between nutrition and health, healthier food behavior can be achieved through education and increasing self-awareness of the impact of food choices on body weight, body composition and the health consequences of overweight and obesity. Education at school, home, and through various forms of media on how to manage diet and lifestyle is critically important for all ethnic and socioeconomic groups to maintain their body weights within the healthy range. Emphasized in the following are the several factors that are commonly considered important in affecting body weight and obesity, and accordingly some advice has been provided.

18.7.1 Enjoy healthy food

Energy density and composition of macronutrients in food may be considered the primary focus in the management of body weight. Food energy is contributed by the three major macronutrients – carbohydrates, fat and protein – all of which have different energy densities. Typically, fat has twice the amount of calories as carbohydrate and protein. To prevent obesity or incur weight loss, diets should have low calories by reducing the percentage of fat and increasing the amount of fiber. Fiber can provide a feeling of fullness without contributing significant amount of energy as most types of fiber are not digestible. Dietary fat has the highest conversion rate to body fat or can be directly stored in fat tissues. Carbohydrates are primarily used as fuel to provide energy despite having less energy density than fats because the conversion of carbohydrates into energy occurs at a much quicker rate. Accordingly, fat is more efficient in storing energy than in providing energy for the use of body activity. In addition, fat increases palatability of foods and is less effective in producing a satiation feeling. Thus, high-fat diets lead to overconsumption of energy as compared to foods that have higher ratios of other macronutrients. Sugar in food and drinks are another important contributor of energy overconsumption. Similarly with fat, sugars or sweeteners increase palatability of foods, leading to overintake of energy. Therefore, foods or diets should contain low amounts of fat by replacing it with carbohydrates and fiber. However, it should be noted that no single food can provide all necessary or a balanced supply of nutrients required for body function and the maintenance of good health. It is strongly recommended that individuals eat a variety of foods to get a sufficient and balanced supply of all nutrients and especially specific nutrients needed for improving the quality of life. Whole grains, vegetables and fruits should be considered as a primary base for food choice. Meats, due to their high fat content, should be considered secondarily.

The choice of dairy products should always be linked to their fat content and preparation process. Low-fat and fermented dairy products such as yoghurt are generally considered healthy. The consumption of dairy products should also be differentiated between gender and age groups. Milk consumption may be important in children and those who need more calcium or have difficulty with eating solid foods. Few studies have investigated the association of milk consumption with BMI in preschool-aged children; however, a large cross-sectional school-based study in Italy found a statistically significant inverse relationship with milk consumption and BMI. The higher levels of milk consumption associated with the lowest mean BMI in children aged 3–11 years (Barba *et al.* 2005). Other studies have also demonstrated a similar inverse association between body fat in children and adolescents and calcium intake (Andersen *et al.* 1998; Martinez-Gonzalez *et al.* 1999). The new USDA Dietary Guidelines recommend that children 2–8 years of age consume two cups of low-fat milk per day (http://www.healthierus.gov/dietaryguidelines).

18.7.2 Avoid food traps

There are many situations which are not well recognized or even unknown to consumers, but can unconsciously trigger overeating. Individuals have their own special traps for overeating and must identify these traps in order to avoid or have better control of them. Fast food, for example, can be associated with several traps. Increased portion size or 'value meals' with high fat content generally provide an abundance of energy above what is required, which can be especially detrimental to children. Soft drinks with high sugar content not only provide additional energy but also increase appetite, leading to increased food intake. Recently, many fast food restaurants have started to provide salads to compensate for the nutrition deficiencies of many of their menu items. In general, such salads offered at fast food establishments provide healthy alternatives to the regular choices. However, some salads with extras, such as toppings containing bacon, certain dressings, and cheese and meat can be a caloric nightmare. The term 'couch potato' is another example of a food trap that can lead to consumption of extra energy. Moreover, being one is confounded with increased time spent watching television or movies, which in another words, means an increased sedentary lifestyle. The high energy content of soft drinks has added further to this problem. High frequency of eating out, eating foods with low nutritional values and drinking high caloric beverages are general food traps that potentially promote excessive energy intake and weight gain although they may not be tied to obesity in every individual. This phenomenon is magnified in children when compared to adults, which can be attributed to differences in lifestyle, understanding of health impact of food/nutrition, and self-control. Although these are the most common, each individual should identify his/her own in order to avoid over food or energy intake due to food traps.

18.7.3 Increase physical activity

It is well understood that weight gain is a result of positive energy balance between intake and expenditure. Energy intake is determined by total food intake and the energy density of food which is determined by its composition of macronutrients. Although the composition of food influences metabolic rate and thus energy expenditure, energy loss is influenced substantially by physical activity. Due to a general decrease in physical activity in most populations, energy expenditure has significantly reduced for most individuals.

Spontaneous activity can increase energy expenditure considerably in some individuals, but purposeful conditioning and training routines are the most important determinants of the energy expenditure due to physical activity for active individuals (Lakka & Bouchard 2005). This also applies to those who perform labor-intensive work. Regular physical activity increases the capacity of the body to use lipid substrates rather than carbohydrates as a source of energy during low- and moderate-intensity exercise, especially when maintained over a long period (Hurley 1986). Thus, physically active individuals burn rather than store excess fat and tolerate high-fat diets better than sedentary persons. Cross-sectional studies have shown that physically active adults and children are leaner and have less abdominal fat than sedentary individuals (Andersen *et al.* 1998; Martinez-Gonzalez *et al.* 1999). It has been estimated that about 30% of new cases of obesity could be prevented by adopting a relatively active lifestyle, which includes more than 30 minutes of brisk walking and fewer than 10 hours of TV watching per week (Hu *et al.* 2003). Evidence shows that a sedentary lifestyle is a better population-level predictor of weight gain than increased caloric or fat intake (Prentice & Jebb 1995). Epidemiological studies suggest that moderate-intensity physical activity of 45–60 minutes per day is needed to prevent unhealthy weight gain and obesity. Brisk walking is effective in the prevention of obesity, but the low-intensity activities typical of daily living also appear to be beneficial. Vigorous exercise provides additional benefits beyond low-intensity and moderate-intensity physical activity in the prevention of weight gain (Lakka & Bouchard 2005).

18.7.4 Monitor body weight regularly and stick to the plan

Those who monitor their body weight regularly are more successful in keeping the pounds off. While weekly weigh-ins will tell whether the weight control or weight-loss efforts work, it may also be important to watch your waist circumference as it is a better indicator of fatness and measures the pattern of fat distribution, which is more relevant to the complications of obesity. The monitoring of body weight and waist circumference can help individuals stay on or adjust the plan to achieve the goals set up. It may also serve as a reminder to keep weight-gaining foods away and reduce the consumption of energy-dense foods or drinks, and increase exercise. Another benefit is its helpfulness in keeping consistently to the adopted weight loss plan during the week, on the weekends, and amid vacation and holidays to increase chances of long-term success.

In sum, several approaches are involved in weight control and prevention of obesity. For long-term success, the best way is to adopt lifestyle and food behavior changes to keep physically active and avoid excessive energy intake. Certainly, the combination of both strategies works more effectively than either alone.

18.8 Treatment of obesity

Except for the cases caused by genetic factors or viral infection, obesity is classified as a preventable metabolic disorder. Because all treatments currently available have certain side effects and entail some risk, it is important to evaluate the benefits over risks for any drug therapy. It may be essential to assess the risks associated with total fat and fat distribution, metabolic fitness and complicating factors. Once relative risk is evaluated, the appropriateness of various treatments can be determined based on BMI and comorbidities. Generally, medical treatments may be considered in patients with BMI over 30, and surgical treatment may be performed if BMI is over 40 (Mantzoros 2006).

18.8.1 Diet, lifestyle and behavioral changes

Similar to the prevention of obesity and overweight, diet, exercise and behavioral changes should be the three main considerations in any attempts to lose weight and maintain weight after weight loss. The principles and strategies of losing weight using these approaches are described in detail in the prevention of overweight and obesity, and are not described in this section.

18.8.2 Medical treatment of obesity

It is important to understand that drug therapy is the last resort rather than the best option for the treatment of obesity, and all medicinal treatments of obesity currently available involve some risk. Therefore, before initiating drug therapy, patient consultations are essential. Patients must understand at least three things before taking any medicinal treatment for weight loss: (1) The amount of weight reduction is strongly related to the degree of behavioral change. Thus, assessing readiness to change lifestyle is the first step. The medical treatment requires food restriction and eating behavior change. If patients are to maximize weight reduction during the active weight-loss phase, they must be prepared to change their entrenched food behavior patterns. (2) Patients need to set up a realistic or achievable goal for their weight loss. It is unrealistic for a very obese patient to expect to achieve the ideal BMI of 25. However, a loss of 5–10% of the initial body weight for obese patients can translate into significant health benefits. Therefore, for these patients, the aim of losing weight is to stay healthy instead of gaining aesthetic value. Once the first goal is reached, the patient can then set up a new target. By doing so, the patient will eventually achieve a healthy body weight while at the same time possibly attaining the aesthetic benefit as well. Establishing a small but achievable goal is critically important in preventing failure and inducing the patients to continue on the weight-loss program and maintain the changes of lifestyle and food behavior. (3) Patients should also be informed as to the side effects of drug therapy in order to prevent failure and to achieve the targeted weight loss (Mantzoros 2006). During treatment, it is important for physicians to regularly monitor the health status for any manifestation of side effects in their patients.

Table 18.2 Medications used in the treatment of obesity

Drug	Mechanism of action	Weight-loss benefits (kg)	Side effects
Medications approved for obesity treatment			
Long-term therapy			
Orlistat[a,b]	Absorption inhibitor	2.54	Gastrointestinal
Sibutramine[a,b]	Appetite suppression	4.5	Blood pressure, heart rate, sympathomimetic
Rimonabant[b]	Appetite suppression	4.9	Psychiatric, neurologic
Short-term therapy			
Phentermine[a]	Appetite suppression	3.6	Sympathomimetic
Diethylpropion[a]	Appetite suppression	3.0	Sympathomimetic
Medications associated with weight loss but not approved for obesity treatment			
Fluoxetine	Appetite suppression	Variable (–14 to 0.4)	Agitation, nervousness
Bupropion	Appetite suppression	2.77	Insomnia, dry mouth
Topiramate	Appetite suppression	6.5% of body weight	Neurologic
Metformin	Unknown	1–2	Gastrointestinal
Exenatide	Appetite suppression	2.8	Nausea
Pramlintide	Appetite suppression	1.2	Nausea

Source: Karam *et al.* (2007).
[a]Approved by Food and Drug Administration.
[b]Approved by the European Medical Agency.

Despite the increased prevalence of obesity and its related comorbidities, few medications have been approved in North America and Europe for its treatment. Only orlistat and sibutramine are approved by the Food and Drug Administration to be used, in conjunction with diet and exercise, for the long-term therapy of obesity in patients with BMI of more than 30 or BMI of more than 27 with obesity-related comorbidity. An additional medication, rimonabant, is only approved in Europe. Old medications approved more than 30 years ago for the short-term use lack data on the efficacy and safety beyond 6 months of treatment. Weight-loss medications can be grouped into three categories based on their mechanisms of action: (1) appetite suppressors, (2) absorption inhibitors and (3) energy expenditure promoters (Karam *et al.* 2007). Some of these products are listed in Table 18.2.

18.8.3 Surgical treatment of obesity

Those with a BMI over 40 or 35 with comorbidities may consider having surgery to remove a portion of the excessive body fat. The details regarding such surgery are beyond the scope of this chapter.

18.9 Natural products for obesity prevention and intervention

Due to the high prevalence of overweight and obesity, development and marketing of non-drug products for preventing weight gain or for weight loss have been expanding during the last decade. In 2006, around 41% of US adults and 30% of European adults attempted to lose weight. Sales of diet products for weight control were estimated to be US$54 billion in Europe and US$50 billion in the United States in 2007. It is predicted that diet products for weight loss will increase by 3.4% annually until 2010. Diet products for weight loss are actually traditional diets that have low fat or low carbohydrates. Unfortunately, the success rate of using diet products for weight loss is quite low, and therefore, the market is changing. Consumers have recently become more interested in weight management products and techniques that target 'how' to suppress appetite, promote satiety, inhibit fat or carbohydrate digestion and absorption, and/or increase fat oxidation (Humphries 2007). Accordingly, a variety of natural products, mainly functional foods, have been introduced to the market. Based on market survey and analysis (Humphries 2007), natural products for weight management covers four categories:

(1) Fat burners
(2) Fat and carbohydrate blockers
(3) Appetite suppressants
(4) Satiety promoters

Between 2004 and 2006, the number of new weight-loss products launched around the world increased by 50%. A large portion of these products are herbal extracts. Despite the negative view of herbs by the mainstream medical community, botanical dietary supplements for weight loss remain extremely popular with the general public (Heber 2003). Although the safety and efficacy of available herbal products for weight reduction remain major issues that need to be resolved through research, herbal products for weight management may ultimately prove to be helpful in overcoming clinically significant obesity when combined with healthy diet and lifestyle changes.

18.9.1 Fat burners

Fat burners are the most significant part of the new generation of weight control products. Several botanical ingredients have been used either alone or combined together with others, and they are generally delivered in foods or drinks. Japan and the United States are the leading countries in producing weight control products in this category (Humphries 2007). The following ingredients are marketed as dietary supplements by companies mainly in these two countries.

18.9.1.1 *Citrus aurantium*

This plant is commonly named bitter orange, sour orange or Seville orange. Components of the fruit are sometimes used as a food, but the plant is more widely used as a medicinal or dietary supplement. The extract has been used in dietary supplements as an aid to fat loss and also as an appetite suppressant although it is not recommended for use in isolation. In traditional Chinese medicine, *Citrus aurantium* is always prescribed in concert with other support herbs. *C. aurantium* contains a number of phytochemicals, including *p*-octopamine and synephrine alkaloids, which are adrenergic agonists and often cited as its active ingredients. Synephrine alkaloids may increase energy expenditure through thermogenesis and reduce food intake by decreasing gut motility.

Although *C. aurantium* extracts have been used in a variety of cultures for thousands of years, they have not been used traditionally for long periods, or specifically for weight loss (Preuss *et al.* 2002). Currently, there is little if any basis for making definitive statements about the safety or risk of weight-loss products containing this plant (Haaz 2006). While a similar substance, ephedra, was banned from sale in the US market because of concerns about adverse effects, *C. aurantium* has been suggested as a safe alternative (Haaz 2006). However, reservations surround potentially adverse cardiovascular and cerebrovascular effects.

18.9.1.2 *Green tea catechins*

Green tea prepared by heating or steaming the leaves of *Camelia sinensis* is widely consumed on a regular basis throughout Asia. Green tea contains both caffeine and catechins, which make up approximately 30% of the dry matter of green tea leaves. Epigallocatechin gallate is one of the most powerful polyphenols and is the most abundant catechin in tea, especially green tea. Green tea extract protects from free radicals and suppresses inflammation in addition to promoting energy expenditure and fat oxidation. Because caffeine occurs naturally in green tea extract, it has been difficult to separate the effects of catechins from caffeine in humans drinking green tea. In fact, it has been shown that green tea extracts containing both catechins and caffeine are more potent in stimulating brown adipose tissue thermogenesis than equimolar concentration of caffeine alone. It is possible that there exists a synergistic interaction between catechins and caffeine in increasing energy expenditure by promoting thermogenesis (Haaz 2006).

18.9.1.3 *Capsaicin*

Capsaicin comes from red peppers and chili peppers. Capsaicin stimulates fat oxidation and thermogenesis and reduces food intake. Capsaicin may increase thermogenesis by enhancing catecholamine secretion from the adrenal medulla involving capsaicin-sensitive neurons. It is also implied that capsaicin induces thermogenesis through β-adrenergic stimulation. The long-term use of capsaicin may be limited by its strong pungency. A possible solution for this may be using the fruit of a non-pungent cultivar of pepper, such as CH-19 Sweet (Diepvens *et al.* 2007).

18.9.1.4 *Garcinia cambogia extract*

Garcinia cambogia is a subtropical species of *Garcinia* native to Indonesia. Hydroxycitric acid is one of 16 isomers of citric acid in the extract of *G. cambogia* and is the only one that inhibits citrate lyase, the enzyme that catalyzes the first step in fatty acid synthesis outside the mitochondrion. It is suggested that hydroxycitric acid suppresses *de novo* fatty acid synthesis and food intake (Heymsfield 1998). Overall, the evidence of this extract for weight control is encouraging. Few and mild adverse effects were reported for both *G. cambogia* and its active component, hydroxycitric acid (Pittler *et al.* 2005). However, further studies are needed to determine their efficacy and safety on weight loss.

18.9.1.5 *Fucoxanthin*

Fucoxanthin is a characteristic carotenoid of brown seaweeds and has been shown to have antiobesity effects. Fucoxanthin promotes thermogenic fat burning in white adipose tissue through the expression of uncoupling protein-1 at both mRNA and protein levels. Uncoupling protein-1 is usually expressed in brown adipose tissue and functions to release chemical energy and induce heat production. However, it has been found that feeding of fucoxanthin increases uncoupling protein-1 expression in white adipose tissue, which in turn leads to fatty acid oxidation and energy expenditure, and ultimately reduces fat mass (Maeda *et al.* 2007). Xanthigen is an extract from deep seawater-cultivated marine vegetables standardized to fucoxanthin, neoxanthin and violaxanthin with stronger thermogenic and weight control properties (Humphries 2007).

18.9.1.6 *Yohimbe*

Yohimbe (*Pausinystalia yohimbe*) is a tall evergreen tree, which is native to Central Africa. Yohimbine, an α-2 receptor antagonist, is the main active constituent of the ground bark of yohimbe. It is often promoted as a weight-loss supplement. The adverse effects associated with the use of yohimbine include headache, hypertension, anxiety and agitation (Pittler *et al.* 2005).

18.9.1.7 *Raspberry ketones*

Raspberry (European red raspberry, *Rubus idaeus*) is one of the oldest fruits recorded and has been used throughout the centuries for nutritional and medicinal purposes. Like its popular relatives the strawberry and blueberry, raspberry contains an abundance of sugars, vitamins, minerals and polyphenols. Raspberry ketone is a major aromatic compound of red raspberry and widely used as a fragrance in cosmetics and foodstuffs. The chemical structure of raspberry ketones resembles that of the capsaicin of red hot peppers and synephrine of *C. aurantium*. Raspberry ketones promote the breakdown of subcutaneous fat by increasing norepinephrine-induced lipolysis (Morimoto 2005).

18.9.2 Fat and carbohydrate blockers

As fat and carbohydrates are the main energy supply for the body and the cause of over energy intake and obesity in most cases, inhibition of fat and carbohydrate absorption is beneficial to weight management. However, few fat-blocker products have been made available to consumers. The most dynamic area is in carbohydrate blockers, in particular the best-known ones – *Phaseolamin vulgaris* and the branded ingredient Svetol (Humphries 2007).

18.9.2.1 *Phaseolamin vulgaris*

Phaseolamin vulgaris is an extract of white kidney beans (*Phaseolus vulgaris*) and is an inhibitor of α-amylase (which is an enzyme that helps the breakdown of starches). Therefore, phaseolamin may reduce the rate at which starches are broken down into sugars in the digestive tract. The extracts are potential ingredients in foods for increasing carbohydrate tolerance in diabetics, decreasing starch digestion and energy intake for weight control (Obiro *et al.* 2008).

18.9.2.2 *Chitosan*

Chitosan is a mucopolysaccharide component of the shells of crab, lobster, shrimp and other marine organisms, and is one of the more popular weight-loss supplements. It is reported that chitosan binds to fat in the gastrointestinal tract and thus decreases fat digestion and absorption (Wydro *et al.* 2007). It is also reported that chitosan oligosaccharides exert antiobesity effects by inhibiting adipocyte differentiation through downregulating the expression of adipogenic transcription factors and other specific genes (Cho 2008).

18.9.2.3 *Green coffee bean extract*

A possible mechanism for the efficiency of green coffee bean extract against weight gain and fat accumulation is by the inhibition of fat absorption in the intestine and the activation of fat metabolism in the liver. Caffeine is a known suppressor of fat absorption, but chlorogenic acid has also been found to be partially involved in the suppressive effect of green coffee bean extract on fat absorption, which results in the reduction of hepatic triglyceride level. Phenolic compounds such as neochlorogenic acid and feruloylquinic acid mixture, except chlorogenic acid, can enhance hepatic carnitine palmitoyltransferase activity (Shimoda *et al.* 2006), which is important in fatty acid transport from cytosol into mitochondria for β-oxidation.

18.9.2.4 *Banaba extract*

The leaves of *Lagerstroemia speciosa* (Lythraceae), a Southeast Asian tree more commonly known as banaba, have been traditionally consumed in various forms in the Philippines for treatment of diabetes and kidney-related diseases. In the 1990s, the popularity of this herbal medicine began to attract the attention of scientists worldwide. Gallotannins are identified in the banaba extract as the component responsible for the activity, and penta-*O*-galloyl is the most potent gallotannin. These compounds not only stimulate glucose uptake but also exhibit antiadipogenic effects (Klein *et al.* 2007). The combination of glucose uptake and antiadipogenic activity may suggest that banaba extract could be developed into a therapeutic agent for the prevention and/or treatment of both diabetes and obesity.

18.9.3 Appetite suppressants

Appetite suppressants are an exciting area in the new generation of weight management products. Current market growth is mainly focusing on appetite suppression products. Several products have been marketed in the United States, Europe and South Africa (Humphries 2007).

18.9.3.1 *Caralluma fimbriata*

Caralluma fimbriata is an edible, succulent cactus used by tribal Indians to suppress appetite and enhance endurance. It belongs to the family Asclepiadaceae and is well-known as a famine food, appetite suppressant and thirst quencher among tribal populations. It grows wild all over India and is also planted as a roadside shrub and boundary marker in gardens. Native Indian diets over many centuries have included these edible, wild, succulent cacti, with claims in folklore about its appetite-suppressant activity. A study has shown that caralluma extract suppresses appetite and reduces waist circumference (Kuriyan 2007). It is believed that this plant extract acts through the appetite control in the brain to block the activity of several enzymes involved in the formation of fat, forcing fat reserves to be burned (Humphries 2007).

18.9.3.2 *Hoodia gordonii*

Hoodia gordonii is a member of the *Hoodia* genus of stem succulents. It has long been in use by the indigenous populations of Southern Africa. P57 has been isolated and claimed as active ingredient in *H. gordonii*, responsible for its appetite-suppressant effect (Hoodia 2006; Lee & Balick 2007; van Heerden 2007).

18.9.3.3 *Oat and palm oils*

Oat and palm oil mixture is one of the new categories of weight-loss products that have been developed recently. The mechanism is based on appetite control and feelings of fullness. Oat and palm oils in combination have an important effect on satiety. Polar lipids from the oat oil extract enclose palm oil droplets, preventing digestion of the palm oil until it reaches the small intestine. When the palm oil reaches the ileum, it is interpreted as undigested fat, which triggers the brain to release satiety hormones and suppresses the hunger signals that would normally be sent. The branded ingredient, Fabuless (once called Olibra and also offered in a consumer product called Slimthru), is taking advantage of functional oils for weight loss, and is currently the market leader with its patent-protected combination of palm and oat oils (both naturally occurring dietary lipids). Fabuless has been clinically shown to create and maintain a feeling of satiety by slowing absorption (Burns 2001; Burns *et al.* 2002).

18.9.3.4 *Pinolenic acid*

Pinolenic acid is an appetite suppressant extracted from the Korean pine nut (*Pinus koraiensis*). Korean pine nut oil consists of more than 92% of poly- and monounsaturated fatty acids such as pinolenic acid (C18:3), linoleic acid (C18:2) and oleic acid (C18:1) (Alper & Mattes 2002). It is unique in that it contains approximately 15% pinolenic acid (C18:3) (Asset 1999). Korean pine nut may work as an appetite suppressant through an increasing effect on satiety hormones such as cholecystokinin and glucagon-like peptide-1 and a reduced prospective food intake (Hughes 2008; Pasman 2008).

18.9.3.5 *Yerba mate*

Yerba mate is prepared from *Iles paraguariensis*, an evergreen tree native to subtropical South America. Yerba mate tea, prepared by steeping the dry leaves and twigs of yerba mate in hot water, is widely consumed as a non-alcoholic beverage in South America. Yerba mate has been rapidly introduced into the world market, either as tea itself or as ingredient in formulated foods or dietary supplements. The indigenous people have used it for centuries as a social and medicinal beverage. Yerba mate has been shown to be hypocholesterolemic, hepatoprotective, central nervous system stimulant, diuretic, and to benefit the cardiovascular system. Yerba mate has been suggested to induce weight loss through suppressing appetite (Heck & de Mejia 2007) and enhancing the expression of uncoupling proteins and elevating AMPK phosphorylation (Pang *et al.* 2008). Increases of AMPK phosphorylation lead to upregulation of fatty acid oxidation and downregulation of fatty acid synthesis. Enhanced expression of uncoupling proteins promotes heat production and/or energy expenditure.

18.9.4 Satiety promoters

Satiety promoter is a subset of the appetite suppressants and refers to the feeling of fullness and disappearance of appetite after a meal. The feeling is controlled by the hypothalamus and is triggered by various hormones. Satiety is a weight management weapon that reduces caloric intake by managing hunger. Apart from the products mentioned above for appetite suppression which also work through satiety promotion, dietary fiber is a typical ingredient that functions as a satiety promoter to reduce food intake.

18.9.4.1 *Dietary fiber*

There are two different types of fibers – soluble and insoluble fibers – which are commonly used in diet and weight management. Fibers are beneficial to weight control through promoting satiety, decreasing absorption of macronutrients such as fat and glucose, and changing the secretion of gut hormones. A variety of foods and drinks containing fiber have been marketed by different companies. Evidence has demonstrated that dietary fiber increases fullness and decreases hunger, leading to lower food intake, especially in overweight or obese subjects. Fiber intake is associated with weight loss in both children and adults, and males and females. The inclusion of fiber in low-calorie diets helps to reduce the feeling of hunger. Soluble fiber appears to be more efficient than insoluble fiber. The most commonly used soluble dietary fibers include guar gum and glucomannan, while insoluble fibers include mainly bran fiber. Although weight loss associated with dietary fiber is less than 5% of the initial body weight, it is believed that this small loss of body weight confers clinically significant health benefits (Heber 2003).

18.10 Conclusion

Obesity and overweight have been rapidly increasing health problems during the last two decades and have become a pandemic health problem. The cause of obesity is complicated and involves a number of factors, including genetic traits, environments, lifestyle changes and behavioral changes. Of the many factors and risks involved in the development of overweight and obesity, excessive energy intake relative to energy expenditure is the key. Except for cases that are due to expression of weight gain as phenotypes or phenotypes, obesity is generally classified as a preventable health problem rather than a disease until its severity reaches the point of causing serious complications. The impact of obesity on the quality of life, as well as the economy and health care system, is predominantly through its associated complications such as morbidity, CVD and diabetes. Therefore, to minimize the societal impact of the condition, prevention of obesity is the crucial step and should be considered the primary focus, while the management and treatment of the complications of obesity are done in parallel to supplement prevention.

As mentioned earlier, the industrial revolution and high technology have brought quick and dramatic changes to our lifestyles; however, our body has not yet evolved to adapt to these sudden changes. The contradiction between excessive energy intake versus energy expenditure due to an increasing sedentary lifestyle has become a common situation in the general population. To date, the most feasible approach for combating the increased chance of developing overweight and obesity is to control food/energy intake while in the meantime increasing physical activity. Due to an increase of sedentary environments in both the workplace and home, self-initiated exercise is becoming increasingly important. Although there are natural products available that may be able to help maintain or control a healthy body weight, the effect is minimal. Furthermore, medical treatment or surgical removal of excessive fat mass should always be the last solution rather than the best option due to the risk associated with these procedures and to the difficulty in maintaining the weight during the postsurgery period. Instead of resorting to these measures, the most efficient way to prevent obesity for individuals is to change their lifestyle by eating healthier and being more physically active.

References

Agus, M.S., Swain, J.F., Larson, C.L., Eckert, E.A. & Ludwig, D.S. (2000). Dietary composition and physiologic adaptations to energy restriction. *American Journal of Clinical Nutrition*, **71**, 901–907.

Ahima, R.S. & Flier, J.S. (2000). Leptin. *Annual Review Physiology*, **62**, 413–437.

Alessi, M.C., Peiretti, F., Morange, P., Henry, M., Nalbone, G. & Juhan-Vague, I. (1997). Production of plasminogen activator inhibitor 1 by human adipose tissue: possible link between visceral fat accumulation and vascular disease. *Diabetes*, **46**, 860–867.

Alper, C.M. & Mattes, R.D. (2002). Effects of chronic peanut consumption on energy balance and hedonics. *International Journal of Obesity and Related Metabolic Disorders*, **26**, 1129–1137.

Andersen, R.E., Crespo, C.J., Bartlett, S.J., Cheskin, L.J. & Pratt, M. (1998). Relationship of physical activity and television watching with body weight and level of fatness among children: results from the Third National Health and Nutrition Examination Survey. *The Journal of the American Medical Association*, **279**, 938–942.

Anderson, G.H. (1995). Sugars, sweetness, and food intake. *American Journal of Clinical Nutrition*, **62**, 195S–201S; discussion 201S–202S.

Arcaro, G., Zamboni, M., Rossi, L., Turcato, E., Covi, G., Armellini, F., Bosello, O. & Lechi, A. (1999). Body fat distribution predicts the degree of endothelial dysfunction in uncomplicated obesity. *International Journal of Obesity and Related Metabic Disorders*, **23**, 936–942.

Asset, G, Staels, B., Wolff, R.L., Bauge, E., Madj, Z., Fruchart, J.C., & Dallongeville, J. (1999). Effects of Pinus pinaster and *Pinus koraiensis* seed oil supplementation on lipoprotein metabolism in the rat. *Lipids*, **34**, 39–44.

Assmann, G, Carmena, R., Cullen, P., Fruchart, J.C., Jossa, F., Lewis, B., Mancini, M. & Paoletti, R. (1999). Coronary heart disease: reducing the risk: a worldwide view. International Task Force for the Prevention of Coronary Heart Disease. *Circulation*, **100**, 1930–1938.

Atkinson, R.L. (2007). Viruses as an etiology of obesity. *Mayo Clinic Proceedings*, **82**, 1192–1198.

Banks, W.A. (2003). Is obesity a disease of the blood-brain barrier? Physiological, pathological, and evolutionary considerations. *Current Pharmaceutical Design*, **9**, 801–809.

Barba, G., Troiano, E., Russo, P., Venezia, A. & Siani, A. (2005). Inverse association between body mass and frequency of milk consumption in children. *The British Journal of Nutrition*, **93**, 15–19.

Bastard, J.P., Maachi, M., Van Nhieu, J.T., Jardel, C., Bruckert, E., Grimaldi, A., Robert, J.J., Capeau, J. & Hainque, B. (2002). Adipose tissue IL-6 content correlates with resistance to insulin activation of glucose uptake both in vivo and in vitro. *Journal of Clinical Endocrinology & Metabolism*, **87**, 2084–2089.

Bell, C.G., Walley, A.J. & Froguel, P. (2005). The genetics of human obesity. *Nature Reviews Genetics*, **6**, 221–234.

Beltowski, J. (2006a). Leptin and atherosclerosis. *Atherosclerosis*, **189**, 47–60.

Beltowski, J. (2006b). Role of leptin in blood pressure regulation and arterial hypertension. *Journal of Hypertension*, **24**, 789–801.

Bergstrom, R.W, R.W., Newell-Morris, L.L., Leonetti, D.L., Shuman, W.P., Wahl, P.W. & Fujimoto, W.Y. (1990). Association of elevated fasting C-peptide level and increased intra-abdominal fat distribution with development of NIDDM in Japanese-American men. *Diabetes*, **39**, 104–111
Birmingham, C.L., Muller, J.L., Palepu, A., Spinelli, J.J. & Anis, A.H. (1999). The cost of obesity in Canada. *Canadian Medical Association Journal*, **160**, 483–488.
Bluher, M, Bullen, J.W., Jr, Lee, J.H., Kralisch, S., Fasshauer, M., Kloting, N., Niebauer, J., Schon, M.R., Williams, C.J. & Mantzoros, C.S. (2006). Circulating adiponectin and expression of adiponectin receptors in human skeletal muscle: associations with metabolic parameters and insulin resistance and regulation by physical training. *Journal of Clinical Endocrinology and Metabolism*, **91**, 2310–2316.
Bodary, P.F., Iglay, H.B. & Eitzman, D.T. (2007). Strategies to reduce vascular risk associated with obesity. *Current Vascular Pharmacology*, **5**, 249–258.
Boden, G. & Chen, X. (1995). Effects of fat on glucose uptake and utilization in patients with non-insulin-dependent diabetes. *The Journal of Clinical Investigation*, **96**, 1261–1268.
Brand-Miller, J.C., Holt, S.H., Pawlak, D.B. & McMillan, J. (2002). Glycemic index and obesity. *American Journal of Clinical Nutrition* **76**, 281S–285S.
Bray, G.A. (2004). The epidemic of obesity and changes in food intake: the fluoride hypothesis. *Physiology and Behaviour*, **82**, 115–121.
Buchholz, A.C. & Schoeller, D.A. (2004). Is a calorie a calorie? *American Journal of Clinical Nutrition*, **79**, 899S–906S.
Burns, A.A., Livingstone, M.B., Welch, R.W., Dunne, A. & Rowland, I.R. (2002). Dose-response effects of a novel fat emulsion (Olibra) on energy and macronutrient intakes up to 36 h post-consumption. *European Journal of Clinical Nutrition*, **56**, 368–377.
Burns, A.A., Livingstone, M.B., Welch, R.W., Dunne, A., Reid, C.A. & Rowland, I.R. (2001). The effects of yoghurt containing a novel fat emulsion on energy and macronutrient intakes in non-overweight, overweight and obese subjects. *International Journal of Obesity and Related Metabolic Disorders*, **25**, 1487–1496.
Centers for Disease Control and Prevention (2003). Prevalence of diabetes and impaired fasting glucose in adults–United States, 1999–2000. *MMWR Morbidity and Mortality Weekly Report*, **52**, 833–837.
Cho, E.J., Rahman, M.A., Kim, S.W., Baek, Y.M., Hwang, H.J., Oh, J.Y., Hwang, H.S., Lee, S.H. & Yun, J.W. (2008). Chitosan oligosaccharides inhibit adipogenesis in 3T3-L1 adipocytes. *Journal of Microbiology and Biotechnology*, **18**, 80–87.
Clement, K., Vaisse, C., Lahlou, N., Cabrol, S., Pelloux, V., Cassuto, D., Gourmelen, M., Dina, C., Chambaz, J., Lacorte, J.M., Basdevant, A., Bougneres, P., Lebouc, Y., Froguel, P. & Guy-Grand, B. (1998). A mutation in the human leptin receptor gene causes obesity and pituitary dysfunction. *Nature*, **392**, 398–401.
Cnop, M. (2008). Fatty acids and glucolipotoxicity in the pathogenesis of type 2 diabetes. *Biochemical Society Transactions*, **36**, 348–352.
Considine, R.V., Sinha, M.K., Heiman, M.L., Kriauciunas, A., Stephens, T.W., Nyce, M.R., Ohannesian, J.P., Marco, C.C., McKee, L.J., Bauer, T.L. & Caro, J.F. (1996). Serum immunoreactive-leptin concentrations in normal-weight and obese humans. *The New England Journal of Medicine*, **334**, 292–295.
Cooper, R.S., Rotimi, C.N. & Ward, R. (1999). The puzzle of hypertension in African-Americans. *Scientific American*, **280**, 56–63.
Cordain, L., Miller, J.B., Eaton, S.B., Mann, N., Holt, S.H., Speth, J.D. (2000). Plant-animal subsistence ratios and macronutrient energy estimations in worldwide hunter-gatherer diets. *American Journal of Clinical Nutrition*, **71**, 682–692.
Despres, J.P. (2001). Health consequences of visceral obesity. *Annals of Medicine*, **33**, 534–541.
Despres, J.P. & Lemieux, I. (2006). Abdominal obesity and metabolic syndrome. *Nature*, **444**, 881–887.
Diepvens, K., Westerterp, K.R. & Westerterp-Plantenga, M.S. (2007). Obesity and thermogenesis related to the consumption of caffeine, ephedrine, capsaicin, and green tea. *American Journal of Physiology, Regularty, Integrative and Comparative Physiology*, **292**, R77–R85.
Douketis, J.D. & Sharma, A.M. (2005). Obesity and cardiovascular disease: pathogenic mechanisms and potential benefits of weight reduction. *Seminars in Vascular Medicine*, **5**, 25–33.
Drewnowski, A. (2000). Nutrition transition and global dietary trends. *Nutrition*, **16**, 486–487.
Drewnowski, A., Kurth, C., Holden-Wiltse, J. & Saari, J. (1992). Food preferences in human obesity: carbohydrates versus fats. *Appetite*, **18**, 207–221.
El-Haschimi, K., Pierroz, D.D., Hileman, S.M., Bjorbaek, C. & Flier, J.S. (2000). Two defects contribute to hypothalamic leptin resistance in mice with diet-induced obesity. *The Journal of Clinical Investigation*, **105**, 1827–1832.

Farooqi, I.S. (2005). Genetic and hereditary aspects of childhood obesity. *Best Practice Research. Clinical Endocrinology and Metabolism*, **19**, 359–374.

Farooqi, I.S., Jebb, S.A., Langmack, G., Lawrence, E., Cheetham, C.H., Prentice, A.M., Hughes, I.A., McCamish, M.A. & O'Rahilly, S. (1999). Effects of recombinant leptin therapy in a child with congenital leptin deficiency. *The New England Journal of Medicine*, **341**, 879–884.

Fasshauer, M, Kralisch, S., Klier, M., Lossner, U., Bluher, M., Klein, J. & Paschke, R. (2004). Insulin resistance-inducing cytokines differentially regulate SOCS mRNA expression via growth factor- and Jak/Stat-signaling pathways in 3T3-L1 adipocytes. *Journal of Endocrinology*, **181**, 129–138.

Ferreira, I.M., Verreschi, I.T., Nery, L.E., Goldstein, R.S., Zamel, N., Brooks, D. & Jardim, J.R. (1998). The influence of 6 months of oral anabolic steroids on body mass and respiratory muscles in undernourished COPD patients. *Chest*, **114**, 19–28.

Finkelstein, E.A., Fiebelkorn, I.C. & Wang, G. (2003). National medical spending attributable to overweight and obesity: how much, and who's paying? *Health Affairs (Millwood), (*Suppl Web Exclusives*)*, W3-219–W3-226.

Flegal, K.M., Carroll, M.D., Ogden, C.L. & Johnson, C.L. (2002). Prevalence and trends in obesity among US adults, 1999–2000. *The Journal of the American Medical Association*, **288**, 1723–1727.

Fontaine, K.R., Redden, D.T., Wang, C., Westfall, A.O. & Allison, D.B. (2003). Years of life lost due to obesity. *The Journal of the American Medical Association*, **289**, 187–193.

Fontana, L., Eagon, J.C., Trujillo, M.E., Scherer, P.E. & Klein, S. (2007). Visceral fat adipokine secretion is associated with systemic inflammation in obese humans. *Diabetes*, **56**, 1010–1013.

Gibson, E.L., Wardle, J. & Watts, C.J. (1998). Fruit and vegetable consumption, nutritional knowledge and beliefs in mothers and children. *Appetite*, **31**, 205–228.

Gualillo, O., Gonzalez-Juanatey, J.R. & Lago, F. (2007). The emerging role of adipokines as mediators of cardiovascular function: physiologic and clinical perspectives. *Trends in Cardiovascular Medicine*, **17**, 275–283.

Gurr, M.I. (1991). Diet, nutrition and the prevention of chronic diseases (WHO, 1990). *European Journal of Clinlical Nutrition*, **45**, 619–623.

Haaz, S., Fontaine, K.R., Cutter, G., Limdi, N., Perumean-Chaney, S. & Allison, D.B. (2006). *Citrus aurantium* and synephrine alkaloids in the treatment of overweight and obesity: an update. *Obesity Reviews*, **7**, 79–88.

Haffner, S.M. (2003a). Insulin resistance, inflammation, and the prediabetic state. *American Journal of Cardiology*, **92**, 18J–26J.

Haffner, S.M. (2003b). Pre-diabetes, insulin resistance, inflammation and CVD risk. *Diabetes Research and Clinical Practice*, **61** (Suppl 1), S9–S18.

Haffner, S.M., Stern, M.P., Hazuda, H.P., Mitchell, B.D. & Patterson, J.K. (1990a). Cardiovascular risk factors in confirmed prediabetic individuals. Does the clock for coronary heart disease start ticking before the onset of clinical diabetes? *The Journal of the Amwrican Medical Association*, **263**, 2893–2898.

Haffner, S.M., Stern, M.P., Mitchell, B.D., Hazuda, H.P. & Patterson, J.K. (1990b). Incidence of type II diabetes in Mexican Americans predicted by fasting insulin and glucose levels, obesity, and body-fat distribution. *Diabetes*, **39**, 283–288.

Hamilton, M.T., Hamilton, D.G. & Zderic, T.W. (2007). Role of low energy expenditure and sitting in obesity, metabolic syndrome, type 2 diabetes, and cardiovascular disease. *Diabetes*, **56**, 2655–2667.

Han, T.S., van Leer, E.M., Seidell, J.C. & Lean, M.E. (1995). Waist circumference action levels in the identification of cardiovascular risk factors: prevalence study in a random sample. *British Medical Journal*, **311**, 1401–1405.

Hansson, P.O., Eriksson, H., Welin, L., Svardsudd, K. & Wilhelmsen, L. (1999). Smoking and abdominal obesity: risk factors for venous thromboembolism among middle-aged men: 'the study of men born in 1913'. *Archives of Internal Medicine*, **159**, 1886–1890.

Harnack, L., Stang, J. & Story, M. (1999). Soft drink consumption among US children and adolescents: nutritional consequences. *Journal of American Dietetic Association*, **99**, 436–441.

Harrington, S. (2008). The role of sugar-sweetened beverage consumption in adolescent obesity: a review of the literature. *The Journal of School Nursing*, **24**, 3–12.

Harris, M.I., Flegal, K.M., Cowie, C.C., Eberhardt, M.S., Goldstein, D.E., Little, R.R., Wiedmeyer, H.M. & Byrd-Holt, D.D. (1998). Prevalence of diabetes, impaired fasting glucose, and impaired glucose tolerance in U.S. adults. The Third National Health and Nutrition Examination Survey, 1988–1994. *Diabetes Care*, **21**, 518–524.

Health Canada (2003). *Canadian Guidelines for Body Weight Classification in Adults*. Health Canada, Ottawa, p. 3.
Heber, D. (2003). Herbal preparations for obesity: are they useful? *Primary Care*, **30**, 441–463.
Heck, C.I. & de Mejia, E.G. (2007). Yerba mate tea (*Ilex paraguariensis*): a comprehensive review on chemistry, health implications, and technological considerations. *Journal of Food Science*, **72**, R138–R151.
Heymsfield, S.B. (2004). The weight debate: balancing food composition and physical activity. Preface. *American Journal of Clinical Nutrition*, **79**, 897S–898S.
Heymsfield, S.B., Allison, D.B., Vasselli, J.R., Pietrobelli, A., Greenfield, D. & Nunez, C. (1998). *Garcinia cambogia* (hydroxycitric acid) as a potential antiobesity agent: a randomized controlled trial. *The Journal of the American Medical Association*, **280**, 1596–1600.
Hill, J.O. & Peters, J.C. (1998). Environmental contributions to the obesity epidemic. *Science*, **280**, 1371–1374.
Hoodia Supreme Dietary Supplement (2006). Hoodia: lose weight without feeling hungry? *Consumer Reports*, 71, 49.
Hotamisligil, G.S., Shargill, N.S. & Spiegelman, B.M. (1993). Adipose expression of tumor necrosis factor-alpha: direct role in obesity-linked insulin resistance. *Science*, **259**, 87–91.
Howarth, N.C., Saltzman, E. & Roberts, S.B. (2001). Dietary fiber and weight regulation. *Nutrition Reviews*, **59**, 129–139.
Hu, F.B., Li, T.Y., Colditz, G.A., Willett, W.C. & Manson, J.E. (2003). Television watching and other sedentary behaviors in relation to risk of obesity and type 2 diabetes mellitus in women. *The Journal of the American Medical Association*, **289**, 1785–1791.
Hughes, G.M., Boyland, E.J., Williams, N.J., Mennen, L., Scott, C., Kirkham, T.C., Harrold, J.A., Keizer, H.G. & Halford, J. C. (2008). The effect of Korean pine nut oil (PinnoThin) on food intake, feeding behaviour and appetite: a double-blind placebo-controlled trial. *Lipids in Health and Disease*, **7**, 6.
Humphries, G. (2007). *NPD in Satiety and Weight Control Food and Drinks: Next Generation Fat Burners, Blockers and Appetite Suppressants*, Business Insights Ltd., London.
Hunter, D.J. (2005). Gene-environment interactions in human diseases. *Nature Reviews Genetics*, **6**, 287–298.
Hurley, B.F., Nemeth, P.M., Martin, W.H. III, Hagberg, J. M., Dalsky, G.P. & Holloszy, J.O. (1986). Muscle triglyceride utilization during exercise: effect of training. *Journal of Applied Physiology*, **60**, 562–567.
Jialal, I., Stein, D., Balis, D., Grundy, S.M., Adams-Huet, B. & Devaraj, S. (2001). Effect of hydroxymethyl glutaryl coenzyme a reductase inhibitor therapy on high sensitive C-reactive protein levels. *Circulation*, **103**, 1933–1935.
Kadowaki, T. & Yamauchi, T. (2005). Adiponectin and adiponectin receptors. *Endocrine Reviews*, **26**, 439–451.
Kahn, S.E. (2003). The relative contributions of insulin resistance and beta-cell dysfunction to the pathophysiology of type 2 diabetes. *Diabetologia*, **46**, 3–19.
Karam, J.G., El-Sayegh, S., Nessim, F., Farag, A. & McFarlane, S.I. (2007). Medical management of obesity: an update. *Minerva Endocrinology*, **32**, 185–207.
Katzmarzyk, P.T. & Janssen, I. (2004). The economic costs associated with physical inactivity and obesity in Canada: an update. *Canadian Journal of Applied Physiology*, **29**, 90–115.
Kern, P.A., Saghizadeh, M., Ong, J.M., Bosch, R.J., Deem, R. & Simsolo, R.B. (1995). The expression of tumor necrosis factor in human adipose tissue. Regulation by obesity, weight loss, and relationship to lipoprotein lipase. *The Journal of Clinical Investigation*, **95**, 2111–2119.
Klein, G., Kim, J., Himmeldirk, K., Cao, Y. & Chen, X. (2007). Antidiabetes and Anti-obesity Activity of *Lagerstroemia speciosa*. *Evidence-Based Complementary and Alternative Medicine*, **4**, 401–407.
Kolatkar, N.S., Thomas, A. & Williams, G.H. (2006). Obesity and hypertension. In: *Obesity and Cardiovascular Disease*. Robinson, M.K. & Thomas, A. (eds), Taylor & Francis Group, Boca Raton, FL, USA, pp 33–50.
Kralisch, S., Sommer, G., Deckert, C.M., Linke, A., Bluher, M., Stumvoll, M. & Fasshauer, M. (2007). Adipokines in diabetes and cardiovascular diseases. *Minerva Endocrinologica*, **32**, 161–171.
Kuch, B., Muscholl, M., Luchner, A., Doring, A., Riegger, G.A., Schunkert, H. & Hense, H.W.. (1998). Gender specific differences in left ventricular adaptation to obesity and hypertension. *Journal of Human Hypertension*, **12**, 685–691.
Kuriyan, R., Raj, T., Srinivas, S.K., Vaz, M., Rajendran, R. & Kurpad, A.V. (2007). Effect of *Caralluma fimbriata* extract on appetite, food intake and anthropometry in adult Indian men and women. *Appetite*, **48**, 338–344.

Lago, F., Dieguez, C., Gomez-Reino, J. & Gualillo, O. (2007). The emerging role of adipokines as mediators of inflammation and immune responses. *Cytokine and Growth Factor Reviews*, **18**, 313–325.
Lakka, T.A. & Bouchard, C. (2005). Physical activity, obesity and cardiovascular diseases. *Handbook of Experimental Pharmacology*, **170**, 137–163.
Larsson, S.C. & Wolk, A. (2006). Epidemiology of obesity and diabetes. In: *Obesity and Diabetes*. Mantzoros, C.S. (ed.), Human Press, Totowa, NJ, pp. 15–38.
Lee, R.A. & Balick, M.J. (2007). Indigenous use of *Hoodia gordonii* and appetite suppression. *Explore (New York)* **3**, 404–406.
Lewis, M.K. & Hill, A.J. (1998). Food advertising on British children's television: a content analysis and experimental study with nine-year olds. *International Journal of Obesity and Related Metabolic Disorders*, **22**, 206–214.
Lillioja, S., Mott, D.M., Spraul, M., Ferraro, R., Foley, J.E., Ravussin, E., Knowler, W.C., Bennett, P.H. & Bogardus, C. (1993). Insulin resistance and insulin secretory dysfunction as precursors of non-insulin-dependent diabetes mellitus. Prospective studies of Pima Indians. *The New England Journal of Medicine*, **329**, 1988–1992.
Lindsay, R.S., Funahashi, T., Hanson, R.L., Matsuzawa, Y., Tanaka, S., Tataranni, P.A., Knowler, W.C. & Krakoff, J. (2002). Adiponectin and development of type 2 diabetes in the Pima Indian population. *Lancet*, **360**, 57–58.
Livingstone, M.B., Strain, J.J., Prentice, A.M., Coward, W.A., Nevin, G.B., Barker, M.E., Hickey, R.J., McKenna, P.G. & Whitehead, R.G. (1991). Potential contribution of leisure activity to the energy expenditure patterns of sedentary populations. *British Journal of Nutrition*, **65**, 145–155.
Ludwig, D.S. (2000). Dietary glycemic index and obesity. *Journal of Nutrition*, **130**, 280S–283S.
Ludwig, D.S., Peterson, K.E. & Gortmaker, S.L. (2001). Relation between consumption of sugar-sweetened drinks and childhood obesity: a prospective, observational analysis. *Lancet*, **357**, 505–508.
Ludwig, D.S., Majzoub, J.A., Al-Zahrani, A., Dallal, G.E., Blanco, I. & Roberts, S.B. (1999). High glycemic index foods, overeating, and obesity. *Pediatrics*, **103**, E26.
Luis Griera, J., Maria Manzanares, J., Barbany, M., Contreras, J., Amigo, P. & Salas-Salvado, J. (2007). Physical activity, energy balance and obesity. *Public Health Nutrition*, **10**, 1194–1199.
Ma, L.J., Mao, S.L., Taylor, K.L., Kanjanabuch, T., Guan, Y., Zhang, Y., Brown, N.J., Swift, L.L., McGuinness, O.P., Wasserman, D.H., Vaughan, D.E. & Fogo, A.B. (2004). Prevention of obesity and insulin resistance in mice lacking plasminogen activator inhibitor 1. *Diabetes*, **53**, 336–346.
Maeda, H., Hosokawa, M., Sashima, T., Funayama, K. & Miyashita, K. (2007). Effect of medium-chain triacylglycerols on anti-obesity effect of fucoxanthin. *Journal of Oleo Science*, **56**, 615–621.
Maeda, N., Takahashi, M., Funahashi, T., Kihara, S., Nishizawa, H., Kishida, K., Nagaretani, H., Matsuda, M., Komuro, R., Ouchi, N., Kuriyama, H., Hotta, K., Nakamura, T., Shimomura, I. & Matsuzawa, Y. (2001). PPARgamma ligands increase expression and plasma concentrations of adiponectin, an adipose-derived protein. *Diabetes*, **50**, 2094–2099.
Manson, J.E., Willett, W.C., Stampfer, M.J., Colditz, G.A., Hunter, D.J., Hankinson, S.E., Hennekens, C.H. & Speizer, F.E. (1995). Body weight and mortality among women. *The New England Journal of Medicine*, **333**, 677–685.
Mantzoros, C.S. (2006). *Obesity and Diabetes*. Human Press Inc., Totowa, NJ, pp. 457–469.
Mantzoros, C.S. & Flier, J.S. (2000). Editorial: leptin as a therapeutic agent–trials and tribulations. *Journal of Clinical Endocrinology and Metabolism*, **85**, 4000–4002.
Marti, A. & Martinez, J.A. (2006). Genetics of obesity: gene x nutrient interactions. *International Journal of Vitamin and Nutrition Research*, **76**, 184–193.
Martinez, J.A. (2000). Body-weight regulation: causes of obesity. *The Proceedings of the Nutrition Society*, **59**, 337–345.
Martinez-Gonzalez, M.A., Martinez, J.A., Hu, F.B., Gibney, M.J. & Kearney, J. (1999). Physical inactivity, sedentary lifestyle and obesity in the European Union. *International Journal of Obesity and Related Metabolic Disorders*, **23**, 1192–1201.
Mehta, S. & Farmer, J.A. (2007). Obesity and inflammation: a new look at an old problem. *Current Atherosclerosis Reports*, **9**, 134–138.
Mertens, I., Ballaux, D., Funahashi, T., Matsuzawa, Y., Van Der Planken, M., Verrijken, A., Ruige, J.B. & Van Gaal, L.F. (2005). Inverse relationship between plasminogen activator inhibitor-I activity and adiponectin in overweight and obese women. Interrelationship with visceral adipose tissue, insulin resistance, HDL-chol and inflammation. *Thrombosis Haemostasis*, **94**, 1190–1195.

Momin, A.U., Melikian, N., Shah, A.M., Grieve, D.J., Wheatcroft, S.B., John, L., El Gamel, A., Desai, J.B., Nelson, T., Driver, C., Sherwood, R.A. & Kearney, M.T. (2006). Leptin is an endothelial-independent vasodilator in humans with coronary artery disease: evidence for tissue specificity of leptin resistance. *European Heart Journal*, **27**, 2294–2299.

Monteiro, C.A., Conde, W.L. & Popkin, B.M. (2001). Independent effects of income and education on the risk of obesity in the Brazilian adult population. *Journal of Nutrition*, **131**, 881S–886S.

Morimoto, C., Satoh, Y., Hara, M., Inoue, S., Tsujita, T. & Okuda, H. (2005). Anti-obese action of raspberry ketone. *Life Sciences*, **77**, 194–204.

Motoshima, H., Wu, X., Sinha, M.K., Hardy, V.E., Rosato, E.L., Barbot, D.J., Rosato, F.E. & Goldstein, B.J. (2002). Differential regulation of adiponectin secretion from cultured human omental and subcutaneous adipocytes: effects of insulin and rosiglitazone. *Journal of Clinical Endocrinology and Metabolism*, **87**, 5662–5667.

Must, A., Spadano, J., Coakley, E.H., Field, A.E., Colditz, G. & Dietz, W.H. (1999). The disease burden associated with overweight and obesity. *The Journal of the American Medicine Association*, **282**, 1523–1529.

O'Rahilly, S., Farooqi, I.S., Yeo, G.S. & Challis, B.G. (2003). Minireview: human obesity-lessons from monogenic disorders. *Endocrinology*, **144**, 3757–3764.

Obiro, W.C., Zhang, T. & Jiang, B. (2008). The nutraceutical role of the *Phaseolus vulgaris* alpha-amylase inhibitor. *British Journal of Nutrition*, **100**, 1–12.

Ogden, C.L., Carroll, M.D., Curtin, L.R., McDowell, M.A., Tabak, C.J. & Flegal, K.M. (2006). Prevalence of overweight and obesity in the United States, 1999–2004. *The Journal of the American Medicine Association*, **295**, 1549–1555.

Pang, J., Choi, Y. & Park, T., (2008). Ilex paraguariensis extract ameliorates obesity induced by high-fat diet: potential role of AMPK in the visceral adipose tissue. *Archives of Biochemistry and Biophysics*, **476** (2), 178–185.

Paolisso, G., Tataranni, P.A., Foley, J.E., Bogardus, C., Howard, B.V. & Ravussin, E. (1995). A high concentration of fasting plasma non-esterified fatty acids is a risk factor for the development of NIDDM. *Diabetologia*, **38**, 1213–1217.

Pasman, W.J., Heimerikx, J., Rubingh, C.M., van den Berg, R., O'Shea, M., Gambelli, L., Hendriks, H.F., Einerhand, A.W., Scott, C., Keizer, H.G. & Mennen, L.I. (2008). The effect of Korean pine nut oil on in vitro CCK release, on appetite sensations and on gut hormones in post-menopausal overweight women. *Lipids in Health and Disease*, **7**, 10.

Peeters, A., Barendregt, J.J., Willekens, F., Mackenbach, J.P., Al Mamun, A. & Bonneux, L. (2003). Obesity in adulthood and its consequences for life expectancy: a life-table analysis. *Annals of Internal Medicine*, **138**, 24–32.

Pelleymounter, M.A., Cullen, M.J., Baker, M.B., Hecht, R., Winters, D., Boone, T. & Collins, F. (1995). Effects of the obese gene product on body weight regulation in ob/ob mice. *Science*, **269**, 540–543.

Petersen, K.F., Oral, E.A., Dufour, S., Befroy, D., Ariyan, C., Yu, C., Cline, G.W., DePaoli, A.M., Taylor, S.I., Gorden, P. & Shulman, G.I. (2002). Leptin reverses insulin resistance and hepatic steatosis in patients with severe lipodystrophy. *The Journal of Clinical Investigation*, **109**, 1345–1350.

Pickup, J.C., Chusney, G.D., Thomas, S.M. & Burt, D. (2000). Plasma interleukin-6, tumour necrosis factor alpha and blood cytokine production in type 2 diabetes. *Life Sciences*, **67**, 291–300.

Pi-Sunyer, F.X. (1991). Health implications of obesity. *American Journal of Clinical Nutrition*, **53**, 1595S–1603S.

Pittler, M.H., Schmidt, K. & Ernst, E. (2005). Adverse events of herbal food supplements for body weight reduction: systematic review. *Obesity Reviews*, **6**, 93–111.

Popkin, B.M. (2001). The nutrition transition and obesity in the developing world. *Journal of Nutrition*, **131**, 871S–873S.

Popkin, B.M., Conde, W., Hou, N. & Monteiro, C. (2006). Is there a lag globally in overweight trends for children compared with adults? *Obesity (Silver Spring)* **14**, 1846–1853.

Pradhan, A.D., Manson, J.E., Rifai, N., Buring, J.E. & Ridker, P.M. (2001). C-reactive protein, interleukin 6, and risk of developing type 2 diabetes mellitus. *The Journal of the American Medicine Association*, **286**, 327–334.

Prentice, A.M. (2006). The emerging epidemic of obesity in developing countries. *International Journal of Epidemiology*, **35**, 93–99.

Prentice, A.M. & Jebb, S.A. (1995). Obesity in Britain: gluttony or sloth? *British Medical Journal*, **311**, 437–439.

Preuss, H.G., DiFerdinando, D., Bagchi, M. & Bagchi, D. (2002). *Citrus aurantium* as a thermogenic, weight-reduction replacement for ephedra: an overview. *Journal of Medicine*, **33**, 247–264.

Raben, A., Macdonald, I. & Astrup, A. (1997). Replacement of dietary fat by sucrose or starch: effects on 14d ad libitum energy intake, energy expenditure and body weight in formerly obese and never-obese subjects. *International Journal of Obesity and Related Metabolic Disorders*, **21**, 846–859.

Rabkin, S.W., Chen, Y., Leiter, L., Liu, L. & Reeder, B.A. (1997). Risk factor correlates of body mass index. Canadian Heart Health Surveys Research Group. *Canadian Medical Association Journal*, **157**(Suppl 1), S26–S31.

Rask-Madsen, C., Dominguez, H., Ihlemann, N., Hermann, T., Kober, L. & Torp-Pedersen, C. (2003). Tumor necrosis factor-alpha inhibits insulin's stimulating effect on glucose uptake and endothelium-dependent vasodilation in humans. *Circulation*, **108**, 1815–1821.

Ravussin, E., Lillioja, S., Anderson, T.E., Christin, L. & Bogardus, C. (1986). Determinants of 24-hour energy expenditure in man. Methods and results using a respiratory chamber. *The Journal of Clinical Investigation*, **78**, 1568–1578.

Ravussin, E. & Smith, S.R. (2002). Increased fat intake, impaired fat oxidation, and failure of fat cell proliferation result in ectopic fat storage, insulin resistance, and type 2 diabetes mellitus. *Annals of the New York Academy of Sciences*, **967**, 363–378.

Report of the Expert Committee (1997). Report of the Expert Committee on the Diagnosis and Classification of Diabetes Mellitus. *Diabetes Care*, **20**, 1183–1197.

Reynolds, K., Gu, D., Whelton, P.K., Wu, X., Duan, X., Mo, J. & He, J. (2007). Prevalence and risk factors of overweight and obesity in China. *Obesity (Silver Spring)*, **15**, 10–18.

Robinson, M.K., Thomas, A. & MyiLibrary. (2006). *Obesity and Cardiovascular Disease*. Taylor & Francis, New York, p. 389.

Rodriguez, A., Fortuno, A., Gomez-Ambrosi, J., Zalba, G., Diez, J. & Fruhbeck, G. (2007). The inhibitory effect of leptin on angiotensin II-induced vasoconstriction in vascular smooth muscle cells is mediated via a nitric oxide-dependent mechanism. *Endocrinology*, **148**, 324–331.

Rosengren, A., Wedel, H. & Wilhelmsen, L. (1999). Body weight and weight gain during adult life in men in relation to coronary heart disease and mortality. A prospective population study. *European Heart Journal*, **20**, 269–277.

Samama, M.M. (2000). An epidemiologic study of risk factors for deep vein thrombosis in medical outpatients: the Sirius study. *Archives of Internal Medicine*, **160**, 3415–3420.

Sandler, S., Bendtzen, K., Eizirik, D.L. & Welsh, M. (1990). Interleukin-6 affects insulin secretion and glucose metabolism of rat pancreatic islets in vitro. *Endocrinology*, **126**, 1288–1294.

Saris, W.H. (2003). Sugars, energy metabolism, and body weight control. *American Journal of Clinical Nutrition*, **78**, 850S–857S.

Savage, D.B. & O'Rahilly, S. (2002). Leptin: a novel therapeutic role in lipodystrophy. *The Journal of Clinical Investigation*, **109**, 1285–1286.

Schafer, K., Fujisawa, K., Konstantinides, S. & Loskutoff, D.J. (2001). Disruption of the plasminogen activator inhibitor 1 gene reduces the adiposity and improves the metabolic profile of genetically obese and diabetic ob/ob mice. *The FASEB Journal*, **15**, 1840–1842.

Schenck, C.H. & Mahowald, M.W. (1994). Review of nocturnal sleep-related eating disorders. *International Journal of Eating Disorders*, **15**, 343–356.

Schulte, H., Cullen, P. & Assmann, G. (1999). Obesity, mortality and cardiovascular disease in the Munster Heart Study (PROCAM). *Atherosclerosis*, **144**, 199–209.

Seidell, J.C. & Flegal, K.M. (1997). Assessing obesity: classification and epidemiology. *British Medical Bulletin*, **53**, 238–252.

Sell, H., Dietze-Schroeder, D., Eckardt, K. & Eckel, J. (2006). Cytokine secretion by human adipocytes is differentially regulated by adiponectin, AICAR, and troglitazone. *Biochemical and Biophysical Resarch Communications*, **343**, 700–706.

Shari, S. & Bassuk, J.E.M. (2006). Overview of the obesity epidemic and its relationship to cardiovascular disease. In: *Obesity and Cardiovascular Disease*. Robinson, M.K. & Thomas A. (eds), Taylor & Francis, New York, pp. 1–32.

Shimabukuro, M., Zhou, Y.T., Levi, M. & Unger, R.H. (1998). Fatty acid-induced beta cell apoptosis: a link between obesity and diabetes. *Proceedings of the National Academy of Sciences United State of America*, **95**, 2498–2502.

Shimoda, H., Seki, E. & Aitani, M. (2006). Inhibitory effect of green coffee bean extract on fat accumulation and body weight gain in mice. *BMC Complementary and Alternative Medicine*, **6**, 9.

Sicree, R.A., Zimmet, P.Z., King, H.O. & Coventry, J.S. (1987). Plasma insulin response among Nauruans. Prediction of deterioration in glucose tolerance over 6 yr. *Diabetes*, **36**, 179–186.
Sikaris, K.A. (2004). The clinical biochemistry of obesity. *Clinical Biochemistry Reviews*, **25**, 165–181.
Skurk, T., Alberti-Huber, C., Herder, C. & Hauner, H. (2007). Relationship between adipocyte size and adipokine expression and secretion. *Journal of Clinical Endocrinology and Metabolism*, **92**, 1023–1033.
Solerte, S.B., Fioravanti, M., Pezza, N., Locatelli, M., Schifino, N., Cerutti, N., Severgnini, S., Rondanelli, M. & Ferrari, E. (1997). Hyperviscosity and microproteinuria in central obesity: relevance to cardiovascular risk. *International Journal of Obesity Related Metabolic Disorders*, **21**, 417–423.
Soodini, G.R. & Hamdy, O. (2006). Pathophysiology of diabetes in obesity. In: *Obesity and Diabetes*. Mantzoros, C.S. (ed.), Humana Press, Totowa, NJ, pp. 117–125.
Stamler, R., Stamler, J., Riedlinger, W.F., Algera, G. & Roberts, R.H. (1978). Weight and blood pressure. Findings in hypertension screening of 1 million Americans. *The Journal of the American Medical Association*, **240**, 1607–1610.
Stevens, J., Cai, J., Pamuk, E.R., Williamson, D.F., Thun, M.J. & Wood, J.L. (1998). The effect of age on the association between body-mass index and mortality. *The New England Journal of Medicine*, **338**, 1–7.
Sturm, R. & Wells, K.B. (2001). Does obesity contribute as much to morbidity as poverty or smoking? *Public Health*, **115**, 229–235.
Suk, S.H., Sacco, R.L., Boden-Albala, B., Cheun, J.F., Pittman, J.G., Elkind, M.S. & Paik, M.C. (2003). Abdominal obesity and risk of ischemic stroke: the Northern Manhattan Stroke Study. *Stroke*, **34**, 1586–1592.
Swinburn, B., Egger, G. & Raza, F. (1999). Dissecting obesogenic environments: the development and application of a framework for identifying and prioritizing environmental interventions for obesity. *Preventive Medicine*, **29**, 563–570.
Swinburn, B.A., Caterson, I., Seidell, J.C. & James, W.P. (2004). Diet, nutrition and the prevention of excess weight gain and obesity. *Public Health Nutrition*, **7**, 123–146.
Taras, H.L., Sallis, J.F., Patterson, T.L., Nader, P.R. & Nelson, J.A. (1989). Television's influence on children's diet and physical activity. *Journal of Developmental & Behavioral Pediatrics*, **10**, 176–180.
Trujillo, M.E. & Scherer, P.E. (2006). Adipose tissue-derived factors: impact on health and disease. *Endocrine Reviews*, **27**, 762–778.
Tsai, A.W., Cushman, M., Rosamond, W.D., Heckbert, S.R., Polak, J.F. & Folsom, A.R. (2002). Cardiovascular risk factors and venous thromboembolism incidence: the longitudinal investigation of thromboembolism etiology. *Archives of Internal Medicine*, **162**, 1182–1189.
Unger, R.H. (1995). Lipotoxicity in the pathogenesis of obesity-dependent NIDDM. Genetic and clinical implications. *Diabetes*, **44**, 863–870.
Uysal, K.T., Wiesbrock, S.M., Marino, M.W. & Hotamisligil, G.S. (1997). Protection from obesity-induced insulin resistance in mice lacking TNF-alpha function. *Nature*, **389**, 610–614.
van Dam, R.M. & Seidell, J.C. (2007). Carbohydrate intake and obesity. *European Journal of Clinical Nutrition*, **61** (Suppl 1), S75–S99.
Van Heek, M., Compton, D.S., France, C.F., Tedesco, R.P., Fawzi, A.B., Graziano, M.P., Sybertz, E.J., Strader, C.D., Davis, H.R., Jr (1997). Diet-induced obese mice develop peripheral, but not central, resistance to leptin. *The Journal of Clinical Investigation*, **99**, 385–390.
van Heerden, F.R., Marthinus Horak, R., Maharaj, V.J., Vleggaar, R., Senabe, J.V. & Gunning, P.J. (2007). An appetite suppressant from Hoodia species. *Phytochemistry*, **68**, 2545–2553.
Vangipuram, S.D., Yu, M., Tian, J., Stanhope, K.L., Pasarica, M., Havel, P.J., Heydari, A.R. & Dhurandhar, N.V. (2007). Adipogenic human adenovirus-36 reduces leptin expression and secretion and increases glucose uptake by fat cells. *International Journal of Obesity (London)*, **31**, 87–96.
Vasilakopoulou, A. & le Roux, C.W. (2007). Could a virus contribute to weight gain? *International Journal of Obesity (London)*, **31**, 1350–1356.
Wang, Y., Mi, J., Shan, X.Y., Wang, Q.J. & Ge, K.Y. (2007). Is China facing an obesity epidemic and the consequences? The trends in obesity and chronic disease in China. *International Journal of Obesity (London)*, **31**, 177–188.
West, D.B., Waguespack, J. & McCollister, S. (1995). Dietary obesity in the mouse: interaction of strain with diet composition. *American Journal of Physiology*, **268**, R658–R665.
White, R.H., Gettner, S., Newman, J.M., Trauner, K.B. & Romano, P.S. (2000). Predictors of rehospitalization for symptomatic venous thromboembolism after total hip arthroplasty. *The New England Journal of Medicine*, **343**, 1758–1764.
Wilborn, C., Beckham, J., Campbell, B., Harvey, T., Galbreath, M., La Bounty, P., Nassar, E., Wismann, J. & Kreider, R. (2005). Obesity: prevalence, theories, medical consequences, management, and research directions. *Journal of International Society of Sports Nutrition*, **2**, 4–31.

Wild, S., Roglic, G., Green, A., Sicree, R. & King, H. (2004). Global prevalence of diabetes: estimates for the year 2000 and projections for 2030. *Diabetes Care*, **27**, 1047–1053.
Willett, W.C., Dietz, W.H. & Colditz, G.A. (1999). Guidelines for healthy weight. *The New England Journal of Medicine*, **341**, 427–434.
Wu, X., Motoshima, H., Mahadev, K., Stalker, T.J., Scalia, R. & Goldstein, B.J. (2003). Involvement of AMP-activated protein kinase in glucose uptake stimulated by the globular domain of adiponectin in primary rat adipocytes. *Diabetes*, **52**, 1355–1363.
Wydro, P., Krajewska, B. & Hac-Wydro, K. (2007). Chitosan as a lipid binder: a langmuir monolayer study of chitosan-lipid interactions. *Biomacromolecules*, **8**, 2611–2617.
Yamauchi, T., Kamon, J., Minokoshi, Y., Ito, Y., Waki, H., Uchida, S., Yamashita, S., Noda, M., Kita, S., Ueki, K., Eto, K., Akanuma, Y., Froguel, P., Foufelle, F., Ferre, P., Carling, D., Kimura, S., Nagai, R., Kahn, B.B. & Kadowaki, T. (2002). Adiponectin stimulates glucose utilization and fatty-acid oxidation by activating AMP-activated protein kinase. *Nature Medicine*, **8**, 1288–1295.
Yanovski, S.Z. & Sebring, N.G. (1994). Recorded food intake of obese women with binge eating disorder before and after weight loss. *International Journal of Eating Disorders*, **15**, 135–150.
Zinman, B., Hanley, A.J., Harris, S.B., Kwan, J. & Fantus, I.G. (1999). Circulating tumor necrosis factor-alpha concentrations in a native Canadian population with high rates of type 2 diabetes mellitus. *Journal of Clinical Endocrinology and Metabolism*, **84**, 272–278.
Zizza, C., Siega-Riz, A.M. & Popkin, B.M. (2001). Significant increase in young adults' snacking between 1977–1978 and 1994–1996 represents a cause for concern! *Preventive Medicine*, **32**, 303–310.

19 Omega-3, 6 and 9 fatty acids, inflammation and neurodegenerative diseases

Cai Song

19.1 Introduction

19.1.1 Neuroinflammation and neurodegenerative diseases

In the last two decades, strong evidence suggests that inflammatory abnormalities are involved in the onset and progression of a number of psychiatric and neurodegenerative diseases, including depression, Alzheimer's disease (AD) and Parkinson's disease (PD) (Zorrilla *et al.* 2001; Ringheim & Conant 2004; Ashdown *et al.* 2006). Depression is often associated with and develops into neurodegenerative diseases.

In depressed patients, increases in macrophage activity and the production of pro-inflammatory cytokines and some acute-phase proteins have been consistently reported (Song *et al.* 1994; Ringheim & Conant 2004). Conversely, more than 70% of volunteers (non-depressed patients) showed severe depressive symptoms after receiving tumor necrosis factor-α (TNF-α) or interferon-γ (IFN-γ) treatment. Furthermore, animal experiments have demonstrated that pro-inflammatory cytokines, such as interleukin (IL)-1β; IL-6 and TNF-α, can stimulate the hypothalamus to release corticotrophin-releasing factor that, via adreno-corticotropic hormone, induces glucocorticoid (GC) secretion (Smith 1991; Ringheim & Conant 2004). Excessive secretion of GCs can cause stress and depressive symptoms, and downregulate GC receptors in the hippocampus, which impairs the GC feedback system. Similar neuroendocrine changes also occur in depressed patients. On the other hand, excessive secretion of GCs can induce neuron death and atrophy of the hippocampus, which occur in several neurodegenerative diseases (Hoschl & Hajek 2001; Rasmuson *et al.* 2002). From the neurotransmitter perspective, pro-inflammatory cytokines have been found to reduce both serotonin and noradrenaline availability to the brain because inflammation can activate indoleamine 2,3-dioxygenase enzyme that breaks down these neurotransmitter precursors (Song *et al.* 1994; Ringheim & Conant 2004). Deficiencies in serotonergic and noradrenergic neurotransmitters are well accepted as a theory of depression in the past 50 years. Thus, inflammation and cytokines may play an important role in depression.

In the blood of AD patients, increased autoimmune and inflammatory responses have been found, which include increased lymphocyte function and production of pro-inflammatory cytokines (Song *et al.* 1999). Pro-inflammatory cytokines, such as IL-1β, can activate glial cells in the brain. Glial cells (microglia and astrocytes) play a major role in the neuroinflammation (Griffin *et al.* 2006) in which (1) the activation of glial cells leads to enhancements in the production of inflammatory mediators, such as TNF-α, IL-1β and prostaglandin E2

(PGE2) (Block & Hong 2005); (2) inflammation rapidly releases toxic reactive oxygen species (ROS) and turns on inducible nitric oxide to produce nitric oxide, which is mediated by phospholipase A2 enzymes (Zhu *et al.* 2006) – these oxidants are highly toxic to cells and cause apoptosis; and (3) IL-1β can induce the expression of amyloid precursor proteins (APP). Indeed, in the brains of patients with AD, activated microglia, increased expression of phospholipase A2 and production of pro-inflammatory cytokines, and increased production of nitric oxide and other oxidants have been consistently reported (Gebicke-Haerter 2001).

In the brain of PD patients, pro-inflammatory cytokines and oxidants are also regulated by microglia and astrocytes. The substantia nigra is the brain region with the highest density of microglial cells (Lawson *et al.* 1990). Robust microglial activation and reduced antioxidant capacity of dopamine (DA) neurons have been reported in the brain of PD patients (Czlonkowska *et al.* 2002). This implies that DA neurons are particularly susceptible to microglia-mediated inflammation and oxidative toxicity in PD because both inflammation and oxidation can directly damage DA neurons. Evidence shows that GCs may also be associated with PD neuropathology (van Craenenbroeck *et al.* 2005). GC receptors are widely distributed in the striatum, frontal cortex and subtantia nigra. Due to stimulation by inflammation, the overproduction of GCs may downregulate GC receptors. It has been reported that the GC receptor deficiency in transgenic (Tg) mice expressing GC receptor antisense RNA from early embryonic life has a dramatic impact on 'programming' the vulnerability of DA neurons to 1-methyl-4-phenyl-1,2,3,6-tetrahydropyridine (MPTP), a model of PD (Morale *et al.* 2004). Dexamethasone protects against DA neuron damage in a mouse model of PD (Kurkowska-Jastrzebska *et al.* 2004). The above evidence strongly suggests that anti-inflammatory treatments should have beneficial effects on these diseases. Increasing evidence indicates that omega (*n*)-3 and 9 fatty acids may improve and even treat these diseases by anti-inflammatory and antioxidative effects. However, excessive *n*-6 fatty acids may induce inflammation.

19.1.2 Food sources of omega-3, 6 and 9 fatty acids, and their structures

Dietary consumption of fatty acids (FAs) by humans from different natural sources such as fish, meat, vegetable seeds and olives provides omega (*n*)-3, 6, and 9 fatty acids. The *n*-3 and *n*-6 fatty acids are two groups of essential unsaturated fatty acids (EFAs). EFAs are essential to human health but cannot be made in the body. For this reason, they must be obtained from food, and only synthesized from dietary precursors such as α-linolenic (*n*-3), and linoleic (*n*-6) fatty acids, while *n*-9 fatty acids are semi-essential because the body can manufacture in a limited amount. Fat fishes, flaxseeds and green leaves contain *n*-3 fatty acids; meat and vegetable seeds contain *n*-6 fatty acids, and olives and walnuts contain *n*-9 fatty acids. The *n*-3 fatty acids include eicosapentaenoic acid (EPA) and docosahexaenoic acid (DHA); the *n*-6 fatty acids include dihomo-γ-linolenic acid (DGLA) and arachidonic acid (AA), and *n*-9 are oleic, nervonic acids and eicosatrienoic acid (ETrA). These unsaturated FAs are important components of membrane phospholipids in neurons, glial and immune cells, and are involved in many functions of the immune and central nervous systems (CNS) for the following reasons (Peet *et al.* 2003). First, changes in membrane FA components may change function of receptors, enzymes and peptides in the CNS and immune system. It is known that the quaternary structures of proteins and the final modeling and folding often depend on the precise nature of the lipid environment of the proteins because a high proportion of proteins in the cell is actually embedded in the membrane. Second, FAs can influence signal transduction molecules. Neurotransmitters, hormones and cytokines hit a target and induce functional changes by activating phospholipases that then generate a wide range of cell signaling or signal transduction. Third, FAs and other lipids can switch on and off many different genes. In particular, by binding to peroxisome proliferator-activated receptors or nuclear receptors, FAs can switch on and off the whole genetic programs. Peroxisomal enzymes are essential for the synthesis of plasmalogen, which is used for myelin formation (Mazza *et al.* 2007). Lipids or carbohydrates also modulate heat shock proteins that aid the expression of mRNA and the synthesis of proteins. Fourth, FAs influence ongoing metabolic regulation. Several studies have revealed that phospholipids undergo constant remodeling, with key FA components having half-lives of a few minutes. Some evidence also suggests *n*-9 FAs as a new class of biological signal molecule and a modulator of membrane homeoviscosity (Gobbi *et al.* 1999; Bourre and Dumont 2003; Ross *et al.* 2003). Recent evidence has suggested that dietary fatty acids may play an important role in the etiology of neurodegenerative diseases, such as AD, PD and depression. In this chapter (1) functions of these fatty acids in the brain and immune system are introduced; (2) changes in concentrations and ratios of these fatty acids in neurodegenerative diseases are reviewed; and (3) the therapeutic mechanisms by which these fatty acids to treat different neurodegenerative diseases are discussed. At the end, the application and weakness of current research in this area, and the future research direction are raised.

19.2 The functions of omega-3, 6, 9 fatty acids in the brain and in the immune system

19.2.1 Fatty acids and brain functions

Free FAs that are released into the blood, and then across the blood–brain barrier, can act at specific binding sites in the brain. Changes in the phospholipid content of neuronal membranes directly affect membrane viscosity and fluidity, which may cause abnormalities in basic physiological functions such as neurotransmitter binding and reuptake, membrane enzyme binding, receptor density and affinity, ion channels, and hormone secretion. In the CNS, *n*-3 and *n*-6 fatty acids fulfill different roles. AA enhances the release of glutamate neurotransmitter, inhibits neurotransmitter uptake, stimulates stress hormone secretion and enhances synaptic transmission (Maes *et al.* 1999; Song *et al.* 2003). AA may trigger microglial activity and induce inflammatory responses. As a consequence of glial activation and inflammatory responses, oxidants are produced, which may be a major cause for neurodegeneration. Therefore, AA plays an important role for neurotransmission but may also contribute to inflammatory and oxidative toxicity in neurodegenerative diseases (Kim *et al.* 2002). In contrast, *n*-3 FAs have been found to compete with *n*-6 FAs. Both EPA and DHA have been found to protect neurons from inflammation and oxidants. *n*-3 FA deficiency has been associated with abnormal monoamine neurotransmissions (Chalon 2006). *n*-3 FAs also modulate the expression of many genes in the aging brain (Barcelo-Coblijn *et al.* 2003). Furthermore, accumulated evidence suggests that *n*-3 or *n*-6 FA precursors or derivatives may have different functions from their ending products. For example, EPA, a precursor of DHA, can effectively treat depression and schizophrenia but not DHA; GLA, a precursor of AA, can inhibit an inflammatory response (Hibbeln 1998; Maes *et al.* 1999; Nemets *et al.* 2002; Kapoor & Huang 2006).

The *n*-9 FA functions in the brain are less understood. As mentioned above, *n*-9 FA nervonic acid is an important FA that comprises the myelin sheath, especially in early development. Loss of myelin is a neuropathological marker for several neurodegenerative diseases. Some evidence also suggests that *n*-9 FAs are involved in brain development (Winniczek *et al.* 1975; Matheson *et al.* 1981). In aging rats, increased oleic acid and decreased DHA were found in the hippocampus and frontal cortex, which has been associated with oxidative stress during aging (Favrelere *et al.* 2000).

19.2.2 Fatty acids and immune function

In the immune system, the onset of autoimmune and inflammatory diseases has been related to an unbalanced intake of *n*-3 and *n*-6 FAs (James *et al.* 2000). Inflammation is an important component of the early immunological responses, while inappropriate or dysfunctional immune responses underlie chronic inflammatory and autoimmune diseases. AA, an *n*-6 FA, is the precursor of eicosanoids that produce PGE2, leukotrienes, thromboxanes and related compounds that activate macrophages to produce pro-inflammatory cytokines, and shift the response of Th1 and Th2. Th1 cells trigger pro-inflammatory responses, while Th2 cells suppress Th1 responses. In contrast to AA, high intake of long-chain *n*-3 FAs, such as EPA (in fish oils), inhibits certain immune functions, e.g. antigen presentation, adhesion molecule expression, Th1 responses, and the production of eicosanoids and pro-inflammatory cytokines (Doshi *et al.* 2004). Clinical studies have also reported that oral fish oil supplementation (contains both EPA and DHA) has beneficial effects on rheumatoid arthritis and asthma (James *et al.* 2000; Darlington & Stone 2001), supporting the idea that the *n*-3 FAs are anti-inflammatory. However, EPA and DHA may have different effects on inflammatory response. There are investigations that compared EPA and DHA effects on immune system. DHA has been reported to reduce some immune cellular functions but enhance Th1 (inflammation) type of response, while EPA shifts Th1 to Th2 (anti-inflammation) type (Song *et al.* 2004; Lynch *et al.* 2007; Sierra *et al.* 2008). Compared to DHA, EPA has greater effects on infection and macrophage activity (Sierra *et al.* 2008). These studies indicate that DHA and EPA combined together to treat cognitive deficit or mental diseases may reduce DHA disadvantages. Many researches have reported that *n*-6 precursor GLA is an anti-inflammatory FA. The mechanism is that GLA further metabolized to DGLA, which undergoes oxidative metabolism by cyclooxygenases and lipoxygenases to produce anti-inflammatory eicosanoids (prostaglandins of series 1 and leukotrienes of series 3) (Kapoor & Huang 2006). In addition, an increase in *n*-9 FAs has been associated with the inhibition of the eicosanoid generation from AA and with the attenuation of experimentally induced inflammation (Schreiner *et al.* 1989; Doshi *et al.* 2004). An inverse relationship between olive oil (enriched with *n*-9 FAs) intake and rheumatoid arthritis was found (Linos *et al.* 1999). Several *in vitro* studies have also demonstrated that dietary enrichment with *n*-9 FAs suppresses the generation of leukotrienes in rat neutrophils, without inducing *n*-6 FA deficiency (Cleland *et al.* 1996a). A recent *in vivo* experiment has shown that an *n*-9 FA ETrA can suppress leukotriene production and leukocyte accumulation, which was similar to the effects of *n*-3 FAs, while this *n*-9 FA attenuated galactosamine/lipopolysaccharide-induced liver injury more effectively than the *n*-3 polyunsaturated fatty acid-rich diet (Cleland *et al.* 1996b). These data may suggest that different FAs play different roles in the modulation of inflammation, and not all *n*-6 fatty acids are pro-inflammatory.

19.3 Changes in concentrations and ratios of these fatty acids in neurodegenerative diseases

Several epidemiologic investigations have reported that the consumption of fish (rich in EPA and DHA) is associated with a slower onset of psychiatric and neurodegenerative diseases, such as depression, AD and PD. For example, the intake of fatty fish more than twice per week has been associated with a reduction in the risk of dementia by 28% and AD by 41% in comparison to those who eat fish less than once per month (Huang *et al.* 2005). Furthermore, clinical investigations have reported that a significant decrease in *n*-3 fatty acids and increase in *n*-6/*n*-3 ratio have been found in the blood of patients with AD (De Lau *et al.* 2005). In another clinical investigation, decreased concentration of *n*-9 fatty acid, oleic and AA, was found in postmortem hippocampus (Prasad *et al.* 1998). However, other *n*-9 fatty acids have not been studied in AD patients.

Regard to the role of EFAs in PD, a longitudinal investigation has shown that long-chain *n*-3 fatty acid intake is associated with a lower risk of PD. However, there is no significant difference in the profile of fatty acids between postmortem brains of PD patients and healthy controls (Julien *et al.* 2006).

Depression is the most widely studied area, where most investigations have shown a direct association between the drop in *n*-3 fatty acids and the risk of depression (Maes *et al.* 1999). The negative correlations have been demonstrated between fish consumption and rates of depression (Hibbeln 1998; Nemets *et al.* 2002; Su *et al.* 2003). Some epidemiological reports also support the connection that dietary fish intake can protect against bipolar disorder (Noaghiul & Hibbeln 2003) and seasonal depression (Cott & Hibbeln 2003). Depletion of red blood cell EPA has been related to suicide attempts (Huan *et al.* 2004). Reduced levels of several *n*-9 fatty acids, such as oleic and nervonic acids, were also found in erythrocyte membranes of depressed patients, which was associated with a decrease in *n*-3 fatty acids (Assies *et al.* 2004). In depressed patients with sleep disorder, decreased oleic acid seems to play a role, which may be due to their function as precursors of the sleep-inducing oleamide (Irmisch *et al.* 2007). The evidence suggests that EFAs may be used to treat depression and neurodegenerative diseases by the correction of dysfunction of membrane components.

19.4 The therapeutic effects in clinical investigations

19.4.1 In the treatment of depression and bipolar disorder

Generally, clinical studies indicate that *n*-3 fatty acid EPA improves almost all the symptoms of depression. Peet and Horrobin (2002) conducted a 12-week study testing a range of 1, 2, or 4 g/day of EPA in 70 patients with major depression and currently taking prescribed antidepressants. As the optimal dose, 1 g/day of EPA significantly improved the scores of depression, sleep and anxiety in the fourth week. Su *et al.* (2003) also found that the improvement occurred after treatment for 4 weeks. They also noted that combining EPA and DHA together may provide a better treatment option. By contrast, a recent clinical study has shown no effect of DHA and EPA combination on depressed mood and cognitive function in patients with mild to moderate depression (Rogers *et al.* 2008). Nemets *et al.* (2002) treated 20 depressed patients with ethyl-EPA over a period of 4 weeks based on scores on the Depression Rating Scale, compared with a placebo-controlled group. All patients had

Hamilton Depression Rating Scale (HDRS) scores above 18 before the treatment; at the end of the fourth week, the placebo group had a mean score of 20.0, whereas the group treated with EPA had a mean score 11.6. Puri *et al.* (2001) used 4 g/day EPA on a patient of remittent, treatment-resistant depression. The patient's suicidal ideation disappeared at 1 month and all other symptoms declined in the subsequent 9 months.

EPA has also been used to treat bipolar patients. The results show that their depression scores on the HDRS and the CGI (Clinical Global Impressions-Severity of Illness Scale) were improved greatly (Frangou & Lewis 2002). However, mania, as scored by the Young Mania Rate Score, did not change significantly. Subsequently, in an open-label study, depressed patients was treated with 2 g/day of EPA for 6 months and the 24-item HDRS (HAM-D-24) was rated every month (Huang *et al.* 2005). At 1 month, seven of ten patients already had a reduction in the HAM-D-24 score. Two patients who completed the study had an end score of 0. Although the study was small, it established that EPA is an effective treatment and does not induce mania.

19.4.2 Antiaging and improvement of cognitive functions

Even though many reviews indicated that DHA could be a good treatment for AD or other aging-associated cognitive deficient, there are only few clinical trials that investigate DHA or combine with other EFAs as a treatment. A study in human amnesic patients, including 21 mild cognitive dysfunctions, 10 organic brain lesions (organic) and 8 ADs, was carried out for the evaluation of a treatment with both DHA and AA. The cognitive functions were evaluated using Japanese version of repeatable battery for assessment of neuropsychological status at two time points: before and 90 days after the supplementation of 240 mg/day AA and DHA, or 240 mg/day of olive oil, respectively. Mild cognitive dysfunction group showed a significant improvement of the immediate memory and attention score. In addition, organic group showed a significant improvement of immediate and delayed memories. However, there were no significant improvements of each score in AD when compared to the placebo group. It is suggested from these data that AA and DHA supplementation can improve the cognitive dysfunction due to organic brain damages or aging (Kotani *et al.* 2006). In 2004, a pilot study by Boston and coinvestigators was carried on 19 outpatients with mild to moderate AD, as indicated by a score on the Mini-Mental State Examination (MMSE) from 10 to 24 (Boston *et al.* 2004). CT scans show that the majority of patients had cerebral atrophy. After EPA treatment for 12 weeks at a dose of 500 mg twice daily, MMSE, AD Assessment Scale-Cognitive Subscale (ADAS-Cog) and visual analogue were carried out in these patients. Except for a small weight increase and significant improvement of the carer's visual analogue, there was a little difference between treatment and baseline periods. The researchers suggested that a longer period of EPA treatment might be necessary because (1) a small improvement in ADAS-Cog was observed in some patients, and this benefit lasted 9–15 months, and (2) EPA treatment slightly increased DHA, AA and EPA concentrations in the blood, while changes did not reach statistical significance. In 2006, a randomized, double-blind trial evaluated the combined effect of DHA (1.7 g) and EPA (0.6 g) on 174 patients with mild to moderate AD (Freund-Levi *et al.* 2006). Before the treatment, the mean values for MMSE and ADAS-Cog scales were similar between placebo and fatty acid-treated groups. After 6-month treatment, the decline in cognitive functions as assessed by the two scales did not differ between the groups. However, if patients with very mild AD were separated from other patients, DHA and EPA combination significantly improved AD testing scales in patients with mild AD. This study suggests that the fatty acids may be effective in the treatment of mild but not severe AD. However, the newest research in 2008 on aging people treated with DHA and EPA combination has shown no significant benefits to cognitive function when compared to placebo (van de Rest *et al.* 2008). In this study, the treatment time was not long enough though. Other FAs, such as *n*-9 FA, have not been tried in a clinical trial.

19.4.3 Improvement of motor abilities

So far, there are no clinical trials to investigate the effect of any EFA on the symptoms or the progress of PD even though some evidence has suggested that *n*-3 fatty acids may effectively improve PD symptoms or delay the progress of PD (Schmidt & Dyerberg 1989; Das 2000; De Lau *et al.* 2005). Indirect evident that supports that *n*-3 fatty acids may improve motor behavior was from a clinical study in children with phenylketonuria. These children had lower blood concentrations of EPA and DHA with poor motometric Rostock-Oseretzky scale (ROS). After fish oil treatment, the increase in blood *n*-3 fatty acid concentration and decrease in *n*-6/*n*-3 ratio were associated with the improvement of ROS (Beblo *et al.* 2007). Therefore, fish oil seems to improve motor skill.

19.5 Mechanism by which EFAs treat different diseases

19.5.1 DHA-reduced Aβ and inflammation

DHA alone or combination with EPA failed in treatment of depression and schizophrenia. The mechanism is unclear. However, DHA has been studied to improve cognitive decline or impairment in AD or aging models of animals and cells. In AD models, DHA has been reported to reduce the accumulation of APP, which is correlated with the memory improvement and reduction of inflammatory response (Lim *et al.* 2005). Furthermore, cellular and molecular studies have shown more mechanisms by which DHA improves cognitive functions. For example, a neuronal sorting protein LR11 that can reduce APP trafficking to secretases that generate Aβ was lower in AD patients and AD models. DHA treatment increases LR11 in multiple systems, including primary rat neurons, aged non-Tg mice and an aged DHA-depleted APPsw AD mouse (Ma *et al.* 2007). These results suggest that DHA may treat AD through increase in this protein. In another cell model of AD, DHA administration stimulated non-amyloidogenic APP processing and reduced levels of Aβ, providing a mechanism for the reported beneficial effects of DHA *in vivo* (Sahlin *et al.* 2007). In the brains of 18-month-old APP/PS1 mice, 0.5% DHA treatment also improved blood circulation and reduced Aβ deposition (Hooijmans *et al.* 2007). In an AD model induced by Aβ infusion, the memory impairment was attenuated by DHA treatment, which was associated with an increase in membrane fluidity (Hashimoto *et al.* 2006).

In neuroinflammation, brain injury and stroke, decreased long-term potentiation (LTP), an important process in the formation of memory, was reported. The most important finding to support that DHA may benefit to memory in neurodegenerative diseases and other conditions is DHA enhancement of LTP (Fujita *et al.* 2001). The other mechanism by which DHA improves memory could be the anti-inflammatory function, such as reducing antigen-related activities, reducing productions of pro-inflammatory cytokine IL-1, IL-6 and TNF-α and suppressing microglia activity in the periphery and the CNS, respectively (Calder 2007; De Smedt-Peyrusse *et al.* 2008). Thus, DHA neuroprotection may result partially from its anti-inflammatory effects. However, DHA has been found to decrease IL-10 but increase IFN-γ concentrations, suggesting a Th1-like response (Maes *et al.* 2007). Because there is no strong clinical evidence to support that DHA may significantly improve AD symptoms, more research needs to be carried out to explore DHA effects on cognitive functions.

19.5.2 EPA, depression and cognitive function

A previous research revealed that EPA could fortify cell membranes and thus help with second messenger systems, and induce brain regeneration. In animal experiments, Lynch's group has reported that EPA increases anti-inflammatory cytokine IL-4 and reverses the inhibition of the LTP induced by Aβ1–40 in aging rats (Lynch *et al.* 2007). My research team has reported that EPA can reverse memory impairment and anxiety-like behavior induced by IL-1β as a neuroinflammation model (Song *et al.* 2003; Song & Horrobin 2004; Song *et al.* 2004). EPA also increases the expression of nervous growth factor (NGF) (Song 2006), attenuates IL-1β-induced changes in noradrenaline in the hippocampus and other limbic regions, and decreases pro-inflammatory (PGE2) but increases anti-inflammatory (IL-10) mediators (Song & Horrobin 2004; Song *et al.* 2004). In a depression model, olfactory bulbectomized (OB) rats, chronic EPA treatment normalized animal behavior in the 'open field', increased serotonin and noradrenaline concentrations in the limbic brain regions, reduced GCs, increased the expression of NGF, and reduced IL-1β concentration (Song 2005). In both the IL-1-indiced model and the OB depression model, EPA treatment for 7–8 weeks markedly improved animal learning and memory in a Morris water maze (spatial memory), and 8-arm radial maze (working memory) (Song & Horrobin 2004; Song *et al.* 2004). These results demonstrated the anti-inflammation and neuroprotective effects of EPA in treatment of depression.

19.5.3 *n*-3 fatty acids and Parkinson's disease

Even though there are few clinical investigations that have explored the therapeutical effects of *n*-3 fatty acids on symptoms of PD. In the recent years, experimental studies have achieved some progress. In a rat model of PD induced by 6-hydroxydopamine (6-OHDA), rotational behavior and dopaminergic markers were evaluated after uridine-5′-monophosphate (UMP) and DHA treatment. UMP/DHA treatment reduced ipsilateral rotations by 57% and significantly elevated striatal dopamine, tyrosine hydroxylase (TH) activity, TH protein and synapsin-1 on the lesioned side. Therefore, UMP and DHA may partially restore dopaminergic neurotransmission in this model of PD (Cansev *et al.* 2008). In a neurotoxin MPTP-induced PD model, Song's team found that ethyl-EPA treatment improved animal muscle ability in the pole test and "open field". In a cellular model of PD induced by MPP+ (a MPTP metabolite), EPA protected DA-like neurons against the neurotoxin-induced reduction in cell viability and upregulated antiapoptosis gene expressions (unpublished data). Thus, EPA should be tried by clinical doctors to treat PD. In addition, a study investigated the effect of DHA on levodopa-induced dyskinesias in Parkinsonian MPTP-treated monkeys. The study explored the effect of DHA in two paradigms. First, a group of MPTP monkeys was primed with levodopa for several months before introducing DHA. A second group of MPTP monkeys (*de novo*) was exposed to DHA before levodopa therapy. DHA administration reduced dyskinesias in both paradigms without alteration of the anti-Parkinsonian effect of levodopa, indicating that DHA can reduce the severity or delay the development of levodopa-induced dyskinesias in a non-human primate model of PD (Samadi *et al.* 2006).

These results suggest that DHA can reduce the severity or delay the development of dyskinesias in a non-human primate model of PD. DHA may represent a new approach to improve the quality of life of PD patients.

The possibility that oleic acid might affect dopaminergic function in primary neurons was studied in two cell lines derived from the striatum and mesencephalic origin. Oleic acid can increase dopamine levels and dopaminergic phenotype, which may suggest that this *n*-9 fatty acid intrigues with respect to possible therapeutic approaches to the treatment of dopaminergic cell loss in PD (Heller *et al.* 2005). Because recent researches have reported that inflammation also plays an important role in PD, future studies should focus on the effects of *n*-3 and *n*-9 fatty acids on inflammation and related oxidative stress in PD.

19.6 Weakness of current treatments and researches, and the future research direction

19.6.1 Treatment doses and durations

From both clinical and experimental studies, inconsistent treatment doses and duration can be noticed, which may contribute to conflicting efficacy, especially observed in the clinical trials. For the treatment of depression, 1–2 g/day and more than 4 weeks of EPA treatment has been demonstrated as optimal doses and duration. However, for other diseases, there are not enough clinical or experimental results to suggest effective doses and duration for a specific disease. In the Brain Lipid Conference 2008, Dr Manku and Dr Puri have reported that a much longer duration (up to 6 months) is needed to treat Huntington's disease with significant improvement but not 2–3 months. Thus, different doses and duration should be investigated according to the different diseases and severity.

19.6.2 Different EFAs target different diseases

Using fatty acids as treatments or nutrition for neurodegenerative diseases was based on evidence from epidemiological investigations on different type of food consumptions. Therefore, it is unclear the function of each single fatty acid in the brain and immune system, and also unknown whether the same type of fatty acids has the same effects on certain diseases. The reason to choose DHA for the treatment of cognitive impairment was also unclear except of DHA as a major component of neuron membrane. It is also unknown why EPA has better therapeutic effects in treatment of depression. There are few clinical and experimental studies that compare DHA and EPA effects on symptoms or pathological changes of AD or depression. With regard to *n*-6 fatty acids, no comparison between different *n*-6 fatty acids has been carried out either. In addition, the understanding of functions of *n*-9 fatty acids in the brain is the most limited. Therefore, these research directions need to be carried out before randomly taking them as treatments or food supplements.

19.6.3 Interaction between different EFAs and optimal combination

It has been reported that *n*-3 FA administrations can significantly change concentrations of *n*-6 or *n*-9 FAs in the brain (Contreras & Rapoport 2002). Thus, the long-term treatment with single FAs may cause imbalance of the composition of other FAs in the membrane of brain or immune cells, which may induce other health concerns. A balanced diet and treatment need to be understood and developed. In addition, these FAs interact with each other in antagonistic and/or synergistic manners. In clinical studies, for example, that EPA and AA work together and did not antagonize each other has been reported when examining the effects of EPA and AA in the treatment of schizophrenia (Horrobin *et al.* 2002). In schizophrenia patients, decreased levels of both EPA and AA were found. After EPA (1–2 g/day) treatment, the blood concentration of *n*-3 FAs was not significantly increased but the blood level of AA was markedly normalized (Peet *et al.* 2001). Therefore, the clinical effects of EPA are much more closely related to the increase in AA concentration than to change in *n*-3 FA concentration. Several clinical studies have also reported that the EPA/DHA combined with GLA or AA can improve both cognitive and motor functions in AD and HD children. In animal experiments, we and other investigators have explored the best ratio between *n*-3 and *n*-6, or *n*-6 precursor, which can exert stronger anti-inflammation and antistress effects. For example, it has been reported that combinations of EPA and GLA at a ratio of 1:1 have stronger anti-inflammatory effect, such as reducing IL-1-induced elevation of PGE2, and blocking IL-1-induced noradrenalin and DA changes (Song *et al.* 2003, 2004). A mixed diet containing α-linolenic (*n*-3) (0.92 g/mL) and linoleic acids (*n*-6) (0.90 g/mL) for 3 weeks can reverse increases in blood concentration of cortisol and cholesterol induced by cortisol, or stress induced by cold water. Other publications have shown that the EPA and GLA at ratio 1:1 have greater anti-inflammation and neuroprotective effects than EPA alone in endotoxin-induced inflammatory response and LTP impairment in the hippocampus (Kavanagh *et al.* 2004). A treatment with a combination of canola and fish oil (*n*-9 and *n*-3) changed immune function and resulted in improved allograft survival in cyclosporine-treated and donor-specific, transfusion-treated rats (Alexander *et al.* 1998). However, in terms of providing benefits to the functionality and overall health of the CNS, optimal ratios of dietary FA combinations for best effects on various psychiatric and neurodegenerative diseases are unknown. These results strongly suggest that the combinations of different FAs at right ratios will provide a new and better generation of FA products for inflammatory and mental diseases.

19.6.4 EPA effects on other neurodegenerative diseases

Currently, there seems no English publication to show the effects of EPA on other neurodegenerative diseases, such as amyotrophic lateral sclerosis (ALS) and multiple sclerosis. Our pilot experiment has found that EPA cannot delay the neurodegenerative progress but can prolong animal lifespan in ALS mice (unpublished). Since EPA can suppress inflammation and protect neurons, new exploration to this direction may be necessary.

References

Alexander, J.W., Valente, J.F., Greenberg, N.A., Custer, D.A., Ogle, C.K., Gibson, S.W. & Babcock, G.F. (1998). Dietary omega-3 and omega-9 fatty acids uniquely enhance allograft survival in cyclosporine-treated and donor-specific transfusion-treated rats. *Transplantation*, **65**, 1304–1309.

Ashdown, H., Dumont, Y., Ng, M., Poole, S., Boksa, P. & Luheshi, G.N. (2006). The role of cytokines in mediating effects of prenatal infection on the fetus: implications for schizophrenia. *Molecular Psychiatry*, **11**, 47–55.

Assies, J., Lok, A., Bockting, C.L., Weverling, G.J., Lieverse, R., Visser, I., Abeling, N.G., Duran, M. & Schene, A.H. (2004). Fatty acids and homocysteine levels in patients with recurrent depression: an explorative pilot study. *Prostaglandins Leukotrienes and Essential Fatty Acids*, **70**, 349–356.

Barcelo-Coblijn, G., Hogyes, E., Kitajka, K., Puskas, L.G., Zvara, A., Hackler, L., Jr, Nyakas, C., Penke, Z. & Farkas, T. (2003). Modification by docosahexaenoic acid of age-induced alterations in gene expression and molecular composition of rat brain phospholipids. *Proceedings of the National Academy of Sciences of the United States of America*, **100**, 11321–11326.

Beblo, S., Reinhardt, H., Demmelmair, H., Muntau, A.C. & Koletzko, B. (2007). Effect of fish oil supplementation on fatty acid status, coordination, and fine motor skills in children with phenylketonuria. *Journal of Pediatrics*, **150**, 479–484.

Block, M.L. & Hong, J.S. (2005). Microglia and inflammation-mediated neurodegeneration: multiple triggers with a common mechanism. *Progress in Neurobiology*, **76**, 77–98.

Boston, P.F., Bennett, A., Horrobin, D.F. & Bennett, C.N. (2004). Ethyl-EPA in Alzheimer's disease – a pilot study. *Prostaglandins Leukotrienes and Essential Fatty Acids*, **71**, 341–346.

Bourre, J.M. & Dumont, O. (2003). Dietary oleic acid not used during brain development and in adult in rat, in contrast with sciatic nerve. *Neuroscience Letters*, **336**, 180–184.

Calder, P.C. (2007). Immunomodulation by omega-3 fatty acids. *Prostaglandins Leukotrienes and Essential Fatty Acids*, **77**, 327–335.

Cansev, M., Ulus, I.H., Wang, L., Maher, T.J. & Wurtman, R.J. (2008). Restorative effects of uridine plus docosahexaenoic acid in a rat model of Parkinson's disease. *Neuroscience Research*, **62**, 206–209.

Chalon, S. (2006). Omega-3 fatty acids and monoamine neurotransmission. *Prostaglandins Leukotrienes and Essential Fatty Acids*, **75**, 259–269.

Cleland, L.G., Gibson, R.A., Neumann, M.A., Hamazaki, T., Akimoto, K. & James, M.J. (1996a). Dietary (n-9) eicosatrienoic acid from a cultured fungus inhibits leukotriene B4 synthesis in rats and the effect is modified by dietary linoleic acid. *Journal of Nutrition*, **126**, 1534–1540.

Cleland, L.G., Neumann, M.A., Gibson, R.A., Hamazaki, T., Akimoto, K. & James, M.J. (1996b). Effect of dietary n-9 eicosatrienoic acid on the fatty acid composition of plasma lipid fractions and tissue phospholipids. *Lipids*, **31**, 829–837.

Contreras, M.A. & Rapoport, S.I. (2002). Recent studies on interactions between n-3 and n-6 polyunsaturated fatty acids in brain and other tissues. *Current Opinion of Lipidology*, **13**, 267–272.

Cott, J. & Hibbeln, J.R. (2003). Lack of seasonal mood change in Icelanders. *American Journal of Psychiatry*, **158**, 328.

Czlonkowska, A., Kurkowska-Jastrzebska, I., Czlonkowski, A., Peter, D. & Stefano, G.B. (2002). Immune processes in the pathogenesis of Parkinson's disease-a potential role for microglia and nitric oxide. *Medical Science Monitor*, **8**, 165–177.

Darlington, L.G. & Stone, T.W. (2001). Antioxidants and fatty acids in the amelioration of rheumatoid arthritis and related disorders. *British Journal of Nutrition*, **85**, 251–269.

Das, U.N. (2000). Beneficial effect(s) of n-3 fatty acids in cardiovascular diseases: but, why and how? *Prostaglandins Leukotrienes and Essential Fatty Acids*, **63**, 351–362.

De Lau, L.M., Bornebroek, M. & Witteman, J.C. (2005). Dietary fatty acids and the risk of Parkinson disease: the Rotterdam study. *Neurology*, **64**, 2040–2045.

De Smedt-Peyrusse, V., Sargueil, F., Moranis, A., Harizi, H., Mongrand, S. & Layé, S. (2008). Docosahexaenoic acid prevents lipopolysaccharide-induced cytokine production in microglial cells by inhibiting lipopolysaccharide receptor presentation but not its membrane subdomain localization. *Journal of Neurochemistry*, **105**, 296–307.

Doshi, M., Watanabe, S., Niimoto T., Kawashima, H., Ishikura, Y., Kiso, Y. & Hamazaki, T. (2004). Effect of dietary enrichment with n-3 polyunsaturated fatty acids (PUFA) or n-9 PUFA on arachidonate metabolism in vivo and experimentally induced inflammation in mice. *Biological and Pharmaceutical Bulletin*, **27**, 319–323.

Favrelere, S., Stadelmann-Ingrand, S., Huguet, F., De Javel, D., Piriou, A., Tallineau, C. & Durand, G. (2000). Age-related changes in ethanolamine glycerophospholipid fatty acid levels in rat frontal cortex and hippocampus. *Neurobiology of Aging*, **21**, 653–660.

Frangou, S. & Lewis, M. (2002). The Maudsley Bipolar Project: a double-blind, randomised, placebo-controlled study of ethyl-EPA as an adjunct treatment of depression in bipolar disorder. *Bipolar Disorders*, **4**, 123.

Freund-Levi, Y., Eriksdotter-Jonhagen, M. & Cederholm, T. (2006). Omega-3 fatty acid treatment in 174 patients with mild to moderate Alzheimer disease: Omeg AD study: a randomized double-blind trial. *Archives of Neurology*, **63**, 1402–1408.

Fujita, S., Ikegaya, Y., Nishikawa, M., Nishiyama, N. & Matsuki, N. (2001). Docosahexaenoic acid improves long-term potentiation attenuated by phospholipase A(2) inhibitor in rat hippocampal slices. *British Journal of Pharmacology*, **132**, 1417–1422.
Gebicke-Haerter, P.J. (2001). Microglia in neurodegeneration: molecular aspects. *Microscopy Research and Technique*, **54**, 47–58.
Gobbi, M., Mennini, T., Valle, D.F., Cervo, L., Salmona, M. & Diomede, L. (1999). Oleamide-mediated sleep induction does not depend on perturbation of membrane homeoviscosity. *FEBS Letters*, **463**, 281–284.
Griffin, W.S., Liu, L., Li, Y., Mrak, R.E. & Barger, S.W. (2006). Interleukin-1 mediates Alzheimer and Lewy body pathologies. *Journal of Neuroinflammation*, **3**, 5.
Hashimoto, M., Hossain, S., Shimada, T. & Shido, O. (2006). Docosahexaenoic acid-induced protective effect against impaired learning in amyloid beta-infused rats is associated with increased synaptosomal membrane fluidity. *Clinical and Experimental Pharmacology and Physiology*, **33**, 934–939.
Heller, A., Won, L., Bubula, N., Hessefort, S., Kurutz, J.W., Reddy, G.A. & Gross, M. (2005). Long-chain fatty acids increase cellular dopamine in an immortalized cell line (MN9D) derived from mouse mesencephalon. *Neuroscience Letters*, **376**, 35–39.
Hibbeln, J.R. (1998). Fish consumption and major depression. *Lancet*, **351**, 1213.
Hooijmans, C.R., Rutters, F., Dederen, P.J., Gambarota, G., Veltien, A., van Groen, T., Broersen, L.M., Lütjohann, D., Heerschap, A., Tanila, H. & Kiliaan, A.J. 2007. Changes in cerebral blood volume and amyloid pathology in aged Alzheimer APP/PS1 mice on a docosahexaenoic acid (DHA) diet or cholesterol enriched Typical Western Diet (TWD). *Neurobiology of Disease*, **28**, 16–29.
Horrobin, D.F., Jenkins, K., Bennett, C.N. & Christie, W.W. (2002). Eicosapentaenoic acid and arachidonic acid: collaboration and not antagonism is the key to biological understanding. *Prostaglandins Leukotrienes and Essential Fatty Acids*, **66**, 83–90.
Hoschl, C. & Hajek, T. (2001). Hippocampal damage mediated by corticosteroids – a neuropsychiatric research challenge. *European Archives of Psychiatry and Clinical Neuroscience*, **251**, S81–S88.
Huan, M., Hamazaki, K. & Sun, Y. (2004). Suicide attempt and n-3 fatty acid levels in red blood cells: a case control study in China. *Biological Psychiatry*, **56**, 490–496.
Huang, T.L., Zandi, P.P. & Tucker, K.L. (2005). Benefits of fatty fish on dementia risk are stronger for those without APOE epsilon4. *Neurology*, **65**, 1409–1414.
Irmisch, G., Schläfke, D., Gierow, W., Herpertz, S. & Richter, J. (2007). Fatty acids and sleep in depressed inpatients. *Prostaglandins Leukotrienes and Essential Fatty Acids*, **76**(1): 1–7.
James, M.J., Gibson, R.A. & Cleland, L.G. (2000). Dietary polyunsaturated fatty acids and inflammatory mediator production. *American Journal of Clinical Nutrition*, **71**, 343S–348S.
Julien, C., Berthiaume, L., Hadj-Tahar, A., Rajput, A.H., Bédard, P.J., Di Paolo, T., Julien, P. & Calon, F. (2006). Postmortem brain fatty acid profile of levodopa-treated Parkinson disease patients and parkinsonian monkeys. *Neurochemistry International*, **48**, 404–414.
Kapoor, R. & Huang, Y.S. (2006). Gamma linolenic acid: an antiinflammatory omega-6 fatty acid. *Current Pharmaceutical Biotechnology*, **7**, 531–534.
Kavanagh, T., Lonergan, P.E. & Lynch, M.A. (2004). Eicosapentaenoic acid and gamma-linolenic acid increase hippocampal concentrations of IL-4 and IL-10 and abrogate lipopolysaccharide-induced inhibition of long-term potentiation. *Prostaglandins Leukotrienes and Essential Fatty Acids*, **70**, 391–397.
Kim, E.J., Kwon, K.J., Park, J.Y., Lee, S.H., Moon, C.H. & Baik, E.J. (2002). Neuroprotective effects of prostaglandin E2 or cAMP against microglial and neuronal free radical mediated toxicity associated with inflammation. *Journal of Neuroscience Research*, **70**, 97–107.
Kotani, S., Sakaguchi, E., Warashina, S., Matsukawa, N., Ishikura, Y., Kiso, Y., Sakakibara, M., Yoshimoto, T., Guo, J. & Yamashima, T. (2006). Dietary supplementation of arachidonic and docosahexaenoic acids improves cognitive dysfunction. *Neuroscience Research*, **56**, 159–164.
Kurkowska-Jastrzebska, I., Litwin, T., Joniec, I., Ciesielska, A., Przybylkowski, A., Czlonkowski, A. & Czlonkowska, A. (2004). Dexamethasone protects against dopaminergic neurons damage in a mouse model of Parkinson's disease. *International Immunopharmacology*, **4**, 1307–1318.
Lawson, L.J., Perry, V.H., Dri, P. & Gordon, S. (1990). Heterogeneity in the distribution and morphology of microglia in the normal adult mouse brain. *Neuroscience*, **39**, 151–170.
Lim, G.P., Calon, F., Morihara, T., Yang, F., Teter, B., Ubeda, O., Salem, N., Jr, Frautschy, S.A. & Cole, G.M. (2005). A diet enriched with the omega-3 fatty acid docosahexaenoic acid reduces amyloid burden in an aged Alzheimer mouse model. *Journal of Neuroscience*, **25**, 3032–3040.

Linos, A., Kaklamani, A.G., Kaklamani, E., Koumantaki, Y., Giziaki, E., Papazoglou, S. & Mantzoros, C.S. (1999). Dietary factors in relation to rheumatoid arthritis: a role for olive oil and cooked vegetables. *American Journal of Clinical Nutrition*, **70**, 1077–1082.

Lynch, A.M., Loane, D.J., Minogue, A.M., Clarke, R.M., Kilroy, D., Nally, R.E., Roche, O.J., O'connell, F. & Lynch, M.A. (2007). Eicosapentaenoic acid confers neuroprotection in the amyloid-beta challenged aged hippocampus. *Neurobiology of Aging*, **28**, 845–855.

Ma, Q.L., Teter, B., Ubeda, O.J., Morihara, T., Dhoot, D., Nyby, M.D., Tuck, M.L., Frautschy, S.A. & Cole, G.M. (2007). Omega-3 fatty acid docosahexaenoic acid increases SorLA/LR11, a sorting protein with reduced expression in sporadic Alzheimer's disease (AD): relevance to AD prevention. *Journal of Neuroscience*, **27**, 14299–14307.

Maes, M., Christophe, A., Delanghe, J., Altamura, C., Neels, H. & Meltzer, H.Y. (1999). Lowered omega3 polyunsaturated fatty acids in serum phospholipids and cholesteryl esters of depressed patients. *Psychiatry Research*, **85**, 275–291.

Maes, M., Mihaylova, I., Kubera, M. & Bosmans, E. (2007). Why fish oils may not always be adequate treatments for depression or other inflammatory illnesses: docosahexaenoic acid, an omega-3 polyunsaturated fatty acid, induces a Th-1-like immune response. *Neuro Endocrinology Letters*, **28**, 875–880.

Matheson, D.F., Oei, R. & Roots, B.I. (1981). Effect of dietary lipid on the acyl group composition of glycerophospholipids of brain endothelial cells in the developing rat. *Journal of Neurochemistry*, **36**, 2073–2079.

Mazza, M., Pomponi, M., Janiri, L., Bria, P. & Mazza, S. (2007). Omega-3 fatty acids and antioxidants in neurological and psychiatric diseases: an overview. *Progress in Neuro-psychopharmacology and Biological Psychiatry*, **31**, 12–26.

Morale, M.C., Serra, P.A., Delogu, M.R., Migheli, R., Rocchitta, G., Tirolo, C., Caniglia, S., Testa, N., L'Episcopo, F., Gennuso, F., Scoto, G.M., Barden, N., Miele, E., Desole, M.S. & Marchetti, B. (2004). Glucocorticoid receptor deficiency increases vulnerability of the nigrostriatal dopaminergic system: critical role of glial nitric oxide. *FASEB Journal*, **18**, 164–166.

Nemets, B., Stahl, Z. & Belmaker, R.H. (2002). Addition of omega-3 fatty acid to maintenance medication treatment for recurrent unipolar depressive disorder. *American Journal of Psychiatry*, **159**, 477–479.

Noaghiul, S. & Hibbeln, J.R. (2003). Cross-national comparisons of seafood consumption and rates of bipolar disorders. *American Journal of Psychiatry*, **160**(12), 2222–2227.

Peet, M., Brind, J., Ramchand, C.N., Shah, S. & Vankar, G.K. (2001). Two double-blind placebo-controlled pilot studies of eicosapentaenoic acid in the treatment of schizophrenia. *Schizophrenia Research*, **49**, 243–251.

Peet, M., Glen, I. & Horrobin, D.F. (eds) (2003). *Phospholipid Spectrum Disorders in Psychiatry and Neurology*, 2nd edn. Marius Press, Lancashire, UK.

Peet, M. & Horrobin, D.F. (2002). A dose-ranging study of the effects of ethyl-eicopentaenoate in patients with ongoing depression despite apparently adequate treatment with standard drugs. *Archives of General Psychiatry*, **59**, 913–919.

Prasad, M.R., Lovell, M.A., Yatin, M., Dhillon, H. & Markesbery, W.R. (1998). Regional membrane phospholipid alterations in Alzheimer's disease. *Neurochemical Research*, **23**, 81–88.

Puri, B.K., Counsell, S.J. & Hamilton, G. (2001). Eicosapentaenoic acid in treatment-resistant depression associated with symptom remission, structural brain changes and reduced neuronal phospholipid turnover. *International Journal of Clinical Practice*, **55**, 560–563.

Rasmuson, S., Nasman, B., Carlstrom, K. & Olsson, T. (2002). Increased levels of adrenocortical and gonadal hormones in mild to moderate Alzheimer's disease. *Dementia and Geriatric Cognitive Disorders*, **13**, 74–79.

Ringheim, G.E. & Conant, K. (2004). Neurodegenerative disease and the neuroimmune axis (Alzheimer's and Parkinson's disease, and viral infections). *Journal of Neuroimmunology*, **147**, 43–49.

Rogers, P.J., Appleton, K.M., Kessler, D., Peters, T.J., Gunnell, D., Hayward, R.C., Heatherley, S.V., Christian, L.M., McNaughton, S.A. & Ness, A.R. (2008). No effect of n-3 long-chain polyunsaturated fatty acid (EPA and DHA) supplementation on depressed mood and cognitive function: a randomised controlled trial. *British Journal of Nutrition*, **99**, 421–431.

Ross, J.A., Maingay, J.P., Fearon, K.C., Sangster, K. & Powell, JJ. (2003). Eicosapentaenoic acid perturbs signalling via the NFkappaB transcriptional pathway in pancreatic tumour cells. *International Journal of Oncology*, **23**, 1733–1738.

Sahlin, C., Pettersson, F.E., Nilsson, L.N., Lannfelt, L. & Johansson, A.S. (2007). Docosahexaenoic acid stimulates non-amyloidogenic APP processing resulting in reduced Abeta levels in cellular models of Alzheimer's disease. *European Journal of Neuroscience*, **26**, 882–889.

Samadi, P., Grégoire, L., Rouillard, C., Bédard, P.J., Di Paolo, T. & Lévesque, D. (2006). Docosahexaenoic acid reduces levodopa-induced dyskinesias in 1-methyl-4-phenyl-1,2,3,6-tetrahydropyridine monkeys. *Annals of Neurology*, **59**, 282–288.

Schmidt, E.B. & Dyerberg, J. (1989). n-3 fatty acids and leucocytes. *Journal of International of Medicine*, **731**, 151–158.

Schreiner, G.F., Rovin, B., Lefkowith, J.B. (1989). The antiinflammatory effects of essential fatty acid deficiency in experimental glomerulonephritis. The modulation of macrophage migration and eicosanoid metabolism. *Journal of Immunology*, **143**, 3192–3199.

Sierra, S., Lara-Villoslada, F., Comalada, M., Olivares, M. & Xaus, J. (2008). Dietary eicosapentaenoic acid and docosahexaenoic acid equally incorporate as decosahexaenoic acid but differ in inflammatory effects. *Nutrition*, **24**, 245–254.

Smith, R.S. (1991). The macrophage theory of depression. *Medical Hypotheses*, **35**, 298–306.

Song, C. (2005). EPA alone or combination with GLA modulate gene expression related to stress and inflammation in a rodent model of depression. *Brain, Behavior, and Immunity*, **14**, 77–140.

Song, C. (2006). The therapeutic mechanism of n-3 fatty acid EPA: anti-inflammation and neuroprotection. 58th American Academy Neurology, San Diego, CA. *Neurology* 66:P05.097.

Song, C., Dinan, T. & Leonard, B.E. (1994). Changes in immunoglobulin, complement and acute phase protein levels in the depressed patients and normal controls. *Journal of Affective Disorders*, **30**, 283–288.

Song, C. & Horrobin, D.F. (2004). Omega-3 fatty acid ethyl-eicosapentaenoate, but not soybean oil, attenuates memory impairment induced by central IL-1beta administration. *Journal of Lipid Research*, **45**, 1112–1121.

Song, C., Li, X., Leonard, B.E. & Horrobin, D.F. (2003). Effects of dietary n-3 or n-6 fatty acids on interleukin-1beta-induced anxiety, stress, and inflammatory responses in rats. *Journal of Lipid Research*, **44**, 1984–1991.

Song, C., Phillips, A.G., Leonard, B.E. & Horrobin, D.F. (2004). Ethyl-eicosapentaenoic acid ingestion prevents corticosterone-mediated memory impairment induced by central administration of interleukin-1beta in rats. *Molecular Psychiatry*, **9**, 630–638.

Song, C., Vandewoude, M., Stevens, W., De Clerck, L., Van Der Planken, M., Whelan, A., Anisman, H., Dossche, A. & Maes, M. (1999). Alterations in immune functions during normal aging and Alzheimer's disease. *Psychiatry Research*, **85**, 71–80.

Su, K.P., Huang, S.Y., Chiu, C.C. & Shen, W.W. (2003). Omega-3 fatty acids in major depressive disorder. A preliminary double-blind, placebo-controlled trial. *European Neuropsychopharmacology*, **13**, 267–271.

van Craenenbroeck, K., De Bosscher, K., Vanden Berghe, W., Vanhoenacker, P. & Haegeman, G. (2005). Role of glucocorticoids in dopamine-related neuropsychiatric disorders. *Molecular and Cellular Endocrinology*, **245**, 10–22.

van de Rest, O., Geleijnse, J.M., Kok, F.J., van Staveren, W.A., Dullemeijer, C., Olderikkert, M.G., Beekman, A.T. & de Groot, C.P. (2008). Effect of fish oil on cognitive performance in older subjects: a randomized, controlled trial. *Neurology*, **71**, 430–438.

Winniczek, H., Go, J. & Sheng, S.L. (1975). Essential fatty acid deficiency: metabolism of 20:3(n-9) and 22:3(n-9) of major phosphoglycerides in subcellular fractions of developing and mature mouse brain. *Lipids*, **10**, 365–373.

Zhu, D., Lai, Y., Shelat, P.B., Hu, C., Sun, G.Y. & Lee, J.C. (2006). Phospholipases A2 mediate amyloid-beta peptide-induced mitochondrial dysfunction. *Journal of Neuroscience*, **26**, 11111–11119.

Zorrilla, E.P., Luborsky, L., McKay, J.R., Rosenthal, R., Houldin, A., Tax, A., McCorkle, R., Seligman, D.A. & Schmidt, K. (2001). The relationship of depression and stressors to immunological assays: a meta-analytic review. *Brain, Behavior, and Immunity*, **15**, 199–226.

20 Functional food in child nutrition

Martin Gotteland, Sylvia Cruchet and Oscar Brunser

20.1 Maternal milk: The gold standard of functional food for infants

The nutritional requirements of the newborn during the first months of life are more important and specific than at any other time in life. The rapid growth of infants, who double their weight within only 4–5 months after birth, depends on the supply of large amounts of nutrients by maternal milk that provides the best nutrition during this period (Oddy 2002; Picciano & McDonald 2006). Breast milk has a unique nutritional composition that changes over time to support newborns in their physiological adaptations to extrauterine life; at the same time, breast milk responds to their growth requirements (Picciano & McDonald 2006). Until now, more than 200 components have been identified in maternal milk, and new constituents are regularly described which further contribute to its health benefits. Breast milk may be considered as the best example of functional foods due to the natural presence of such a great number of bioactive compounds (Table 20.1) (Lönnerdal 2000). Most of these are synthesized by the mother's body, while some others mainly phytochemicals, some long-chain polyunsaturated fatty acids (LC-PUFAs), minerals and vitamins are provided by her diet.

The bioactive compounds present in breast milk exert two main functions: the protection of the newborn against pathogenic microorganisms from the environment, and the preferential stimulation and maturation of the digestive, immune and neuroendocrine systems. In general, protective factors such as lactoferrin, lysozyme, immunoglobulins and oligosaccharides are found in higher concentrations in colostrum and transitional milk than in mature milk; this represents an advantage because newborns are more fragile during the immediate postnatal period (Picciano & McDonald 2006). In addition, nutritional and bioactive factors present in breast milk, including growth factors, hormones and cytokines, exert metabolic programming, i.e. long-term effects modulating the risk of developing obesity, diabetes, hypertension, hypercholesterolemia and other diseases during adult life (Turck 2007).

20.2 Infant formulas

Infants should be breastfed for as long as possible, exclusively for at least 4 months and if possible until 6 months of age. The definition here for exclusive is that no other food

Table 20.1 Bioactive compounds in breastmilk

Breast milk components	Functional properties
Lactoferrin (and its bioactive peptide lactoferricin)	Microbicidal ↗ Iron absorption Antioxidant Modulation and maturation of the immune system Anti-inflammatory
α-Lactalbumin (and its bioactive peptides)	↗ Mineral absorption (calcium, zinc, etc.) Antitumoral (oligomeric form) Inhibitor of ACE activity Opioid-like Microbicidal Modulation of the immune system
Glycomacropeptide (from κ-casein digestion)	Microbicidal Modulation of the immune system Bifidogenic
Immunoglobulins	Microbicidal Modulation of the immune system
Nucleotides	Trophic activity on the intestinal mucosa Modulation of the immune system Bifidogenic ↗ LC-PUFA content Regulation of intestinal blood flow
Lactic acid bacteria	Antibacterial Anti-inflammatory Modulation and maturation of the immune system
Oligosaccharides	Antibacterial Bifidogenic Modulation of colonic microbiota Modulation of the immune system Anti-inflammatory ↗ Calcium absorption Regulation of blood lipids Regulation of appetite
LC-PUFAs	Maturation of neuroendocrine system Development of cognitive functions Modulation and maturation of the immune system
Other bioactive compounds: lysozyme	Antibacterial
Casein encrypted peptides	Opioid-like, analgesic, ACE inhibitor
Polyamine	Trophic activity
Growth factors, cytokines, hormones	Maturation of the intestinal mucosa Modulation and maturation of the immune system Anti-inflammatory Metabolic activities
Enzymes (lipases, lactoperoxidase, catalase)	Antibacterial Antioxidant Anti-inflammatory
Protease inhibitors	Inhibition of inflammatory proteases
Free fatty acids	Microbicidal
Vitamins (A, C, E)	Antioxidant Anti-inflammatory
Immune cells	Microbicidal Maturation of the intestinal mucosa Modulation and maturation of the immune system

↗, increase.

or fluid is provided, except for breast milk and small amounts of medicines or vitamin and mineral supplements (Kramer & Kakuma 2004; Picciano & McDonald 2006). The functional capabilities of human milk cannot be reproduced by molecules originating from cow's milk, which is the raw material most commonly used in formula production; for this reason, some components are currently provided from other sources. Formulas have to provide adequate nutrition for infants who, for any reason, cannot be breastfed. Furthermore, if intrauterine nutrition has been unsatisfactory, the nutrients provided by the formulas must be able to compensate as far as possible for these deficiencies, and assure the restoration of any altered functions. Many of the bioactive compounds present in breast milk such as antibodies, LC-PUFAs, nucleotides, prebiotic oligosaccharides, probiotic bacteria, vitamins, minerals and some bioactive peptides are currently added to infant formulas to make their composition as close as practically possible to that of maternal milk and to try to mimic its functionalities (Carver 2003; Eshach Adiv *et al.* 2004). These same compounds are also used to develop functional foodstuffs destined for older children and adults.

In the following paragraphs, we briefly describe the functional properties of the main nutritional components of maternal milk, which are also used to improve modern milk formulas.

20.3 Main bioactive compounds in breast milk and their use in infant formulas

20.3.1 α-Lactalbumin

α-Lactalbumin is a major human milk protein, with concentrations of 2–3 g/L (20–25% of total protein compared with 2–5% in bovine milk); it contains unusually high concentrations of the essential amino acids lysine (11%), cysteine (6%) and tryptophan (4–5%), which help to satisfy the high requirements of amino acid by infants (Lönnerdal & Lien 2003). α-Lactalbumin in human milk binds calcium, but the importance of this process for calcium nutrition of the newborn is not known, probably because the total amount of calcium bounds is low compared to the total calcium content in the milk (Segewa & Sugai 1983). α-Lactalbumin also binds manganese, zinc and cobalt, but its significance in the metabolism of these elements is not clear (Segewa & Sugai 1983).

The formation of oligomeric forms of α-lactalbumin conjugated with oleic acid, called 'human α-lactalbumin made lethal to tumor cells' and specific for human milk, has been described at low pH conditions similar to those prevailing in the stomach. It has the capacity of inducing apoptosis in several human and murine tumoral cell lines; however, there is not clear evidence of the formation of this compound in the infant gastrointestinal tract (Mok *et al.* 2007).

The release of bioactive peptides encrypted in the human α-lactalbumin sequence has been observed *in vitro* during its digestion by gastrointestinal proteases (Lönnerdal & Lien 2003). Some of these peptides exert angiotensin-1-converting enzyme (ACE) inhibitory activity (α-lactorphin), while others have opioid-like activity or are bactericidal, mainly against gram-positive bacteria (Pelligrini *et al.* 1999; Sipola *et al.* 2002). Finally, other bioactive peptides, which include a characteristic GLF (Gly-Leu-Phe) sequence, have been isolated from the intestine of infants after breastfeeding; it is possible that they could exert immunostimulating activities of biological significance, either directly by acting on breast milk phagocytes or, after absorption, on the macrophages in the intestinal mucosa of newborns (Migliore-Samour *et al.* 1992). As the molecular structure of bovine α-lactalbumin is very close to that of humans, the peptides derived from its enzymatic digestion are very similar and exert comparable systemic and local effects in the gastrointestinal tract.

Infant formulas were made originally with whole cow's milk (casein–whey ratio 80:20); this implicated the administration of large amounts of protein to satisfy the essential amino acid requirements of growing infants. Currently, there is consensus that these amounts exceed those needed by infants. The possibility of amino acid overloads has been suggested by some studies as representing an unnecessary burden to their immature metabolism (Schmidt *et al.* 2004); however, the long-term consequences of this overload have not been established clearly. As a result, the casein–whey ratio of most infant formulas is currently modified to approximately 60:40 to provide an amino acid profile closer to that of human milk (Räihä *et al.* 2002). This is accomplished through the addition of whey.

The development of methods such as tangential flow filtration allows the isolation of concentrated whey protein fractions of nutritional interest at an industrial scale, among these α-lactalbumin, while reducing simultaneously the amounts of β-lactoglobulin, which is highly allergenic (Krissansen 2007). As a result, concentrated α-lactalbumin of very high biological quality may now be added to the standard whey-predominant preparations to produce α-lactalbumin-enriched formulas, and allowing the replacement of greater amounts of casein. Despite providing lower amounts of total protein, this addition makes the plasma amino acid profile of the infants even more similar to that of breastfed infants, with higher concentrations of tryptophan and lower levels of the insulinogenic amino acids, threonine (valine, leucine and isoleucine). In a recent study, infants fed a standard formula tended to gain weight more rapidly than the group fed an α-lactalbumin-enriched (25% α-lactalbumin) who grew at a rate comparable to breastfed infants (Lien *et al.* 2004). This greater weight gain by the infants fed the standard formula may be explained by their higher protein intakes or the insulinogenic effects of the excess branched-chain amino acids. Infants fed α-lactalbumin-enriched formulas also had lower blood urea nitrogen levels, suggesting less amino acid oxidation and decreased metabolic stress (Sandström *et al.* 2008).

20.3.2 Lactoferrin

Lactoferrin is a glycosylated protein highly resistant to proteolytic digestion in the infant digestive tract and displays a broad spectrum of physiological properties. It is a major component of the innate immune system, and is present in concentrations ranging from about 6–8 g/L in colostrum to 2–4 g/L in mature milk (Lönnerdal 2000; Picciano & McDonald 2006; Krissansen 2007). The passing of this amount of lactoferrin from the mother to the newborn suggests that this molecule must play an important role in the defense mechanisms against pathogenic bacteria. This idea is supported by the presence of lactoferrin in other external secretions in the human body such as tears, saliva, seminal and synovial fluids, in concentrations ranging from 10 μg/mL to about 2 mg/mL.

The fully processed lactoferrin molecule has two lobes, each formed in turn by two domains, which generate a cleft where a ferric ion (Fe^{3+}) is tightly bound in cooperation with a bicarbonate anion (Legrand *et al.* 2008). Lactoferrin in maternal milk is essentially unsaturated with regard to iron and is called apolactoferrin; its affinity for this metal is very high and equivalent to about 260 times that of blood transferrin.

The binding of iron-loaded lactoferrin to its receptor, intelectin, activates its endocytosis and absorption (Shin *et al.* 2008); this process is probably important for other biological activities such as the stimulation of enterocyte proliferation and differentiation. In addition, lactoferrin acts as a microbicidal agent against pathogens such as coagulase-negative staphylococci, *Candida albicans* and *Entamoeba histolytica* (León-Sicairos *et al.* 2006; Venkatesh & Rong 2008; Ochoa & Cleary 2009). Lactoferrin has a high affinity for lipopolysaccharide, a phenomenon which may explain its bactericidal activity against gram-negative bacteria (Elass-Rochard *et al.* 1995). In addition, lactoferrin is active against viruses including herpes simplex, cytomegalovirus, rotavirus, hepatitis C and HIV (Marr *et al.* 2009).

The bactericidal activity of lactoferrin has also been associated with its considerable iron-binding capacity, which makes this element less available for pathogen growth in the intestinal lumen of the newborn (Krissansen 2007, Legrand *et al.* 2008). Iron binding could also decrease iron-associated oxidative reactions in the intestinal and colonic lumen. On the other hand, *in vitro* digestion of lactoferrin by gastrointestinal proteases releases lactoferricin, a peptide which also displays a potent bactericidal activity and inhibits the attachment of some pathogens to the intestinal epithelium (León-Sicairos *et al.* 2006, Marr *et al.* 2009). Additionally, lactoferrin exerts immunostimulating and anti-inflammatory activities, possibly through its binding to pro-inflammatory unmethylated bacterial DNA CpG motifs (Mulligan *et al.* 2006; Krissansen 2007). More recently, lactoferrin has been proposed to have a key role in the development of the immune system by acting as a maturation factor for dendritic cells (Spadaro *et al.* 2008).

The availability of human recombinant and purified bovine lactoferrin has opened the possibility of its current clinical assay when added to infant formulas, fermented milks or oral rehydration solutions (ORS). A study carried out with ORS in Peru by Zavaleta *et al.* demonstrated that the addition of lactoferrin and lysozyme was associated with a decrease of diarrhea duration and volume, and the possible prevention of repeat episodes (Zavaleta *et al.* 2007). However, the possible long-term effects and the optimal duration of treatment and dosages were not established. Another point of interest is that as diarrheal morbidity and mortality are much higher in the less developed countries, a cost/benefit should be calculated for the use of this compound. Another study was conducted in the United States in bottle-fed infants who received a formula supplemented with lactoferrin up to 850 mg/L or a control formula (102 mg/L). The infants receiving the supplemented formula for 12 months experienced fewer episodes of lower respiratory tract infections and had higher hematocrit levels than the controls (King *et al.* 2007). This suggests that orally administered lactoferrin may exert systemic effects. In another randomized, double-blind, placebo-controlled study conducted in children in Peru, the incidence of diarrhea was not decreased but *Giardia* colonization rates were lower and infant growth was improved (Ochoa *et al.* 2008). A study by Egashira *et al.* in Japan showed that the frequency of intense episodes of diarrhea associated with rotavirus infections was decreased without affecting their incidence (Egashira *et al.* 2007). However, the dose of lactoferrin administered (100 mg/day) was low, and the product was not pure and included probiotics and possibly other active molecules; in addition, the dropout rate was very high both in the experimental and control groups. In contrast, a study carried out in Sweden did not show any apparent benefits from the addition of lactoferrin and/or nucleotides to infant formulas (Hernell & Lönnerdal 2002).

In conclusion, with respect to lactoferrin, clinical trials in humans have been few; the dosages used have been variable and some definitely very low, and the molecule was sometimes administered together with other bioactive compounds, making it difficult to interpret the results. Furthermore, some studies were not blinded.

20.3.3 Glycomacropeptide

Glycomacropeptide (GMP) is a large acidic glycosylated polypeptide (64 amino acids) released from humans and bovine κ-casein by peptic digestion in the stomach. It is also formed during the curdling of bovine casein by chymosin during cheese production, being therefore available in large amounts for human nutrition (Brody 2000). GMP is rich in threonine and may be the main source of this amino acid for formula-fed infants; this may explain their increased rates of weight gain compared to breastfed infants (Rigo *et al.* 2001). On the other hand, GMP contains low levels of phenylalanine (2.5–5.00 mg Phe/g protein), and for this reason it has been proposed that GMP-containing food products may represent a palatable alternative source of protein for the nutritional management of children with phenylketonuria (Ney *et al.* 2009).

GMP has been shown to exert antithrombotic, antibacterial and antiviral activities *in vitro* (Zimecki & Kruzel 2007), and to increase the proliferation and phagocytic activity of macrophages *in vitro*; this immunostimulatory activity is related to both its peptidic and glycosidic moieties (Li & Mine 2004). Interestingly, it has been shown that GMP may act as a bifidogenic factor; i.e. it may stimulate the growth of bifidobacteria *in vitro* and *in vivo*. For example, in a clinical study carried out in 4- to 8-week-old healthy term infants, the administration of a GMP-enriched formula up to 6 months increased fecal bifidobacteria (evaluated by fluorescent *in situ* hybridization) compared with the infants fed the control formula. This bifidogenic effect, however, was mainly observed in subjects with low initial counts of bifidobacteria, i.e. in those who had not been previously breastfed (Brück *et al.* 2003).

20.3.4 Immunoglobulins

Immunoglobulins G and M and secretory IgA are found in colostrum and mature breast milk and are transferred from the mother to her infant (Lönnerdal 2000; Picciano & McDonald 2006). These antibodies protect passively the infant against infections by bacteria, viruses and parasites that may gain access to the gastrointestinal lumen (Bogsted *et al.* 1996; Loureiro *et al.* 1998). More especially, they influence the immune repertoire of T and B lymphocytes, a phenomenon which affects the quality of the immune responses later in life, reducing the risk of allergic reactions and autoimmune processes (Kelleher & Lönnerdal 2001).

Several publications summarize clinical studies conducted using bovine immunoglobulin concentrates against gastrointestinal pathogens (*Streptococcus mutans*, *C. albicans*, *Helicobacter pylori*, rotavirus, *Cryptosporidium parvum* and enteropathogenic *Escherichia coli*) in children (Brunser *et al.* 1992; Weiner *et al.* 1999). However, there are problems associated with the effects of bile salts and digestive enzymes on the immunoglobulins, besides those resulting from the shelf-life of the carrier products and their palatability, and their possible contamination by microbial and endogenous proteolytic enzymes. A number of studies show that close to 25% of the antibodies survive passage along the gastrointestinal tract in a biologically active form (Roos *et al.* 1995); this limitation could be resolved by microencapsulation. However, the use of these immunoglobulin concentrates is associated with uncertainties as to dosage and timing of administration, and the results of some assays have been disappointing (Brunser *et al.* 1992).

20.3.5 Nucleotides

The nucleotide concentration in breast milk reported in the literature is highly variable (4–70 mg/L, i.e. 10–20% of the non-protein nitrogen) (Picciano & McDonald 2006) because different studies have measured different compounds (free nucleotides, nucleosides, free bases, nucleic acids or/and adducts such as uridine diphosphate galactose). Some differences exist between the profiles in bovine milk and human milk as, for example, orotic acid is present in the former but not in the latter. Human milk nucleotides are transformed by intestinal alkaline phosphatase to nucleosides, which are absorbed and thus used to stimulate the growth and maturation of the newborn intestinal mucosa and to improve hepatic functions (as observed in animal models) (Uauy *et al.* 1994). In just weaned rats, dietary nucleoside supplementation increased intestinal villus height and disaccharidase activity, compared with control animals; this trophic activity is probably potentiated by polyamines (mainly spermine and spermidine) which are found in substantial amounts in breast milk (Picciano & McDonald 2006). In addition to the anti-inflammatory property of nucleotides, they probably contribute to the recovery of the intestinal mucosa after gut injury; this should be facilitated by the fact that nucleotides also increase the velocity of intestinal blood flow in infants (Carver *et al.* 2004). The need for an external supply of nucleotides through milk is explained by the limited capacity of the mucosa of the gastrointestinal tract to synthesize them *de novo*.

The supplementation of infant formulas with nucleotides has been practiced now for some 25 years. There is evidence for stimulatory effects on immune function and mucosal intestinal repair after episodes of acute diarrhea. In a randomized, double-blind, placebo-controlled study carried out in Chile (Brunser *et al.* 1994), 141 infants below 1 year of age were fed a formula supplemented with nucleotides (Group 1) while 148 infants of comparable age received the same formula but unsupplemented (Group 2). The children in Group 1 experienced less episodes of diarrhea including less first episodes; 45.0 and 31.1% of infants in Groups 1 and 2, respectively, never developed episodes of diarrhea. These results reveal a protective effect of nucleotides for episodes of acute diarrhea, including their absence in a proportion of these infants, who were living in a contaminated environment. The role of specific and nonspecific immune boosting in these responses in acute diarrhea remains unknown.

A role of dietary nucleotides on infant immune responses has been proposed. It is suggested that they act on the T-helper/inducer lymphocyte populations, preferably in the initial phase of antigen processing. Their proliferation results in more efficient T-cell responses and, in some cases, in increased antibody responses (Pickering *et al.* 1998; Buck *et al.* 2004; Schaller *et al.* 2004). Infants fed breast milk or nucleotide-supplemented milk formulas have higher natural killer cell activity and higher levels of IL-2 released by activated mononuclear cells compared with the non-supplemented formula; this suggests that nucleotides are one of the bioactive compounds that contribute to improve the immune function of breastfed infants (Carver *et al.* 1991). On the other hand, it has been observed that feeding preterm newborns with a nucleotide-supplemented formula partially inhibits the decrease of docosahexaenoic acid (DHA) and eicosapentaenoic acid (EPA) levels in red blood phospholipids occurring during their first month of life (Pita *et al.* 1987). An explanation proposed for this observation is that dietary nucleotides may increase the processing of LC-PUFAs by stimulating the activities of the intestinal and hepatic desaturases.

Finally, it has also been shown that dietary nucleotides incorporated into infant formulas may act as prebiotics by decreasing the *Bacteroides/Bifidobacterium* ratio in the infant colon, thus contributing to the typical microbiota of breastfed infants (Singhal *et al.* 2008).

20.3.6 Oligosaccharides

Oligosaccharides are present in human milk at concentrations ranging from 7 to 12 g/L, much higher than those found in bovine milk, which only contains traces (Kunz *et al.* 2000; Picciano & McDonald 2006). They are synthesized in the mammary gland through the attachment of galactose and *N*-acetylglucosamine to lactose by β-glucosidic linkages and the incorporation of fucose and sialic acid by α-glucosidic linkages. The presence of fucose and/or sialic acid determines their classification into neutral and acidic oligosaccharides and explains the great variety of their structures (more than 130 described) (Miller 1999). The presence of *N*-acetylglucosamine and fucose differentiates human milk oligosaccharides from galacto-oligosaccharides (GOS). As the human milk oligosaccharides are not digested and absorbed in the small intestine, they can be considered as dietary fiber that reaches the infant colon where they are fermented by the microbiota. The neutral oligosaccharidic fraction is a relevant factor, which contributes to the development of the *Bifidobacterium*-rich microbiota characteristic of breastfed infants, and it is a phenomenon considered as the basis of the concept of 'prebiotic' (Gibson & Roberfroid 1995). On the other hand, the acidic fraction could prevent the adhesion of pathogenic bacteria to the intestinal epithelium (Morrow *et al.* 2005). This property is due to the fact that because this fraction displays specific moieties, acting as analogues to host cell surface receptors to which enteropathogens may bind. This results in their fecal elimination, preventing the development of infectious episodes. In consequence, the great structural diversity of these oligosaccharides may be considered as an adaptive response on the part of the mother to the diversity of pathogens in the environment.

The original definition of prebiotic proposed by Gibson and Roberfroid defined these as dietary components with a specific fermentation pathway that stimulates populations of intestinal bacteria that are considered beneficial for human health (Gibson & Roberfroid 1995). In 2007, the FAO Technical Meeting on Prebiotics defined prebiotics as non-viable components present in food that confer benefits to the host in association with the modulation of host microbiota (Piñeiro *et al.* 2008). It is interesting to note that according to this new definition other breast milk components such as nucleotides and the glycomacropeptide, which also exert bifidogenic effects on the infant colonic microbiota, could be considered as prebiotics. The main prebiotics of plant origin are fructo-oligosaccharides (FOS) such as inulin and oligofructose; as these are easily obtained they have been used preferentially, singly or as mixtures, in infant formulas in early studies (Kolida & Gibson 2007). Currently, with the increased availability of GOS, the growing tendency is to use a mixture of 90% GOS and 10% FOS, as it has been shown that this is more effective in stimulating the growth of bifidobacteria.

The use of prebiotics in infant formula has been widely reviewed (Boehm *et al.* 2004; Fanaro *et al.* 2005; Veereman 2007). Some recent studies (Knol *et al.* 2005; Bakker-Zierikzee *et al.* 2006; Brunser *et al.* 2006a,b; Mihatsch *et al.* 2006; Savino *et al.* 2006; Scholtens *et al.* 2006; Arslanoglu *et al.* 2008) in this respect are summarized in Table 20.2; the results show that from the clinical point of view, the main changes induced by prebiotics added to infant formulas are related to the characteristics of feces (increased volume and water content, increased secretory IgA content, lower pH, shorter transit time and modifications of short-chain fatty acid excretion). Infants fed prebiotic-containing formulas experienced less episodes of colic (Savino *et al.* 2006), maybe due to a lower visceral hypersensitivity. Other effects may occur in distant organs, such as decreases of atopic dermatitis and allergic urticaria, and less recurrent wheezing and upper respiratory infections (Arslanoglu *et al.* 2008), probably through the modulation of the immune system.

Another aspect worth mentioning is the capacity of prebiotics to modulate the colonic microbiota (Knol *et al.* 2005; Brunser *et al.* 2006a,b; Scholtens *et al.* 2006); this is particularly significant after the administration of antibiotics. In a study in Chile in infants between 1 and 2 years of age suffering from acute bronchitis, a 1-week amoxicillin treatment induced a significant decrease of the total counts of fecal bacteria and an increase of *E. coli* (Brunser

Table 20.2 Results of recent clinical trials carried out in infants with prebiotic-supplemented formulas

Type of study	Population	Experimental treatment	Results	Author/reference
R – DB – PC	68 infants 7–8 weeks + one group of breastfed	GOS/FOS 8 g/L Control formula for 6 weeks	↗ % bifidobacteria ↘ pH ↗ Acetate and ↘ propionate	Knol *et al.* (2005)
R – DB – PC	20 preterm infants in full enteral nutrition	GOS/FOS 10 g/L Control formula for 14 days	↘ Stool viscosity ↘ Gastrointestinal transit time	Mihatsch *et al.* (2006)
R – PC	267 formula-fed infants <4 months	GOS/FOS 8 g/L Control formula + simeticona 4 g/L for 14 days	↘ Colic episodes ↘ Crying duration	Savino *et al.* (2006)
R – DB	35 infants 4–6 months at weaning	GOS/FOS 4.5 g/L	↗ % bifidobacteria	Scholtens *et al.* (2006)
R – DB – PC	57 infants at birth	GOS/FOS 6 g/L Probiotic formula Control formula	↗ Fecal secretory IgA	Bakker-Zierikzee *et al.* (2006)
R – DB – PC	90 infants at 4 months + one group of breastfed infants	FOS 4.5 g/L Probiotic formula Control formula for 3 weeks	↗ % bifidobacteria	Brunser *et al.* (2006a)
R – DB – PC	140 infants 1–2 years of age, after 1 week amoxicillin	FOS 4.5 g/L Control formula for 3 weeks	↗ % bifidobacteria	Brunser *et al.* (2006b)
R – DB – PC	152 infants at risk of atopy	GOS/FOS 8 g/L Control formula for 6 months + follow-up for 2 years	↘ Incidence of atopic dermatitis, recurrent wheezing, allergic urticaria ↘ IRA episodes ↘ Antibiotic prescription	Arslanoglu *et al.* (2008)

R, randomized; DB, double-blind; PC, placebo-controlled; ↘, decrease; ↗, increase.

et al. 2006b). Administration of a formula containing a mixture of inulin and oligofructose (70:30%, 4.5 g/L) for 3 weeks stimulated the reestablishment of a normal microbiota and increased the *Bifidobacterium* population, compared with the control formula.

Other physiological effects related to the colonic fermentation of prebiotics, such as the modulation of calcium absorption, the synthesis of folate and of other vitamins of the B complex, the maintenance of the normality of the multiplication, differentiation, restitution and apoptosis in the colonic epithelium, the decrease of blood cholesterol and triglycerides, and possibly the control of appetite and food intake, may have repercussions later in life; unfortunately, to date, most of these effects have not been explored in infants and children.

20.3.7 Long-chain polyunsaturated fatty acids

Lipids represent an important component of the dietary energy supply (Koletzko *et al.* 1997); at the same time they largely determine the flavor, texture and aroma of foods. Some of the most important LC-PUFAs are derived from the ω-3 and ω-6 essential fatty acids: α-linolenic and linoleic acids, respectively. In the organism, these are subjected to sequential elongation and desaturation to give origin to EPA and DHA from α-linolenic acid, and arachidonic acid (ARA) from linoleic acid. The ω-3 and ω-6 LC-PUFAs are naturally present in human milk, while they are absent in many infant formulas. DHA content in the breast milk of mothers from Western countries is about 0.4% and the ARA/DHA ratio is approximately 1.0–1.5. In consequence, plasma and red blood cell LC-PUFA status is higher in breastfed infants than in infants fed non-supplemented formulas (Gil *et al.* 2003). LC-PUFA content in maternal milk is an important factor to take into account in infant nutrition, because during the neonatal period there is a rapid increase of ARA and/or DHA in the body of the newborn, and more specifically in his/her brain and retina (Heird & Lapillonne 2005). It has been proposed that human milk LC-PUFAs contribute to the improved visual acuity, cognitive development, immune responses and motor functions of breastfed infants compared with infants fed formula without LC-PUFA (Heird & Lapillonne 2005; Innis & Friesen 2008; Rudnicka *et al.* 2008). As described later in the text, adding LC-PUFAs in adequate quantity and quality to infant formulas improves the LC-PUFA status for the infants consuming it, with beneficial effects on their visual acuity, cognitive and psychomotor parameters at 18 months of age (Lucas & Morley 1999; Innis *et al.* 2002; Lapillonne *et al.* 2003; Clandinin *et al.* 2005).

Industrial sources of high-quality LC-PUFAs are now available that allow their incorporation into infant formulas; in the case of ARA, one of the main sources is the filamentous fungus *Mortierella alpina*, in which this essential fatty acid accounts for 25% of the total cell dry weight, and 69–79% of the total fatty acids (Totani & Oba 1987). As for sources of ω-3 PUFAs and DHA, the microalga *Crypthecodinium cohnii* contains more than 30% of them (de Swaaf *et al.* 2003). The exact pathways for their synthesis in these organisms are not known.

Term and preterm infants have the capacity to synthesize DHA and ARA from their precursors, α-linolenic and linoleic acids, but the resulting amounts cannot be calculated with precision. Furthermore, there is poor correlation between intakes of precursors and plasma levels of DHA and ARA. Supplementation of the formulas of preterm infants seems to yield functional advantages in retinal and brain functions, although there are discrepancies between the publications reporting data on these matters; it seems that studies with larger numbers of participating infants tend to yield less clear differences between supplemented and non-supplemented groups (Clandinin *et al.* 2005). This may be explained by the individual differences in endogenous synthesis rates, which cause a wide dispersion of results. Another possibility is that the methods used for assessment of the results are not capable of discerning subtle differences in these aspects. The ample availability of formulas providing ARA and DHA has shown that their use is safe, that they have no negative effects on growth, and that, on the contrary, they may stimulate growth in preterm infants (Lucas & Morley 1999; Innis *et al.* 2002; Lapillonne *et al.* 2003; Clandinin *et al.* 2005). The positive effects of DHA and ARA may also become evident when infants are weaned and their diet may not provide optimal amounts of the former (Lucas & Morley 1999; Innis *et al.* 2002; Lapillonne *et al.* 2003; Clandinin *et al.* 2005; Innis & Friesen 2008; Rudnicka *et al.* 2008).

20.3.8 Lactic acid bacteria and probiotics

Studies have shown that the secretion of milk during the breastfeeding period promotes both an acidic environment on the mother's nipple and the development of a bacterial biofilm rich in lactic acid bacteria (Martín *et al.* 2003). This biofilm would cover the surface of galactophores, allowing the lactic acid bacteria to be released into the breast milk when the newborn is suckling. A great variety of lactic acid bacteria have been described, belonging to the genus *Streptococcus*, *Leuconostoc*, *Lactococcus*, *Lactobacillus*, *Weissella* and *Bifidobacterium*, usually found in fermented foods (Martin *et al.* 2007). Some of the strains of *Lactobacillus* isolated from fresh human breast milk have been shown to exert antibacterial activity against *Salmonella*, as well as anti-inflammatory and immunostimulatory properties, suggesting that they could act as protective factors for the neonate (Olivares *et al.* 2006; Diaz-Ropero *et al.* 2007). Probably these bacteria, mainly those belonging to the *Bifidobacterium* genus, contribute to the process of colonization of the newborn's gastrointestinal tract, their growth in the colon being stimulated by neutral milk oligosaccharides, nucleotides and/or the glycomacropeptide (Grönlund *et al.* 2007). In consequence, breast milk may be considered a symbiotic mixture as it provides prebiotics and probiotics to the newborn. Interestingly, the lactic acid bacteria present in human milk may represent a new source of probiotics for human consumption.

The Food and Agricultural Organization of the United Nations and the World Health Organization, in a joint statement, defined probiotics as 'live microorganisms which, when consumed in adequate amounts as part of food, confer a health benefit on the host' (Araya *et al.* 2002). The microorganisms used most frequently as probiotics in infant formulas belong to the *Bifidobacterium* or *Lactobacillus* genus and the yeast *Saccharomyces boulardii*. It is important to note that their functional effects are strain-specific and, therefore, not every strain of *Lactobacillus* or *Bifidobacterium* is a probiotic. Furthermore, the vehicle in which the probiotic is suspended for administration may play a role in its efficacy as it may contribute to their survival in the gastrointestinal tract by buffering gastric acidity, avoiding contact with bile salts or providing nutrients for these bacteria. Many of the benefits associated with the administration of probiotics would require that these agents colonize permanently the gastrointestinal tract of the host. However, most studies indicate that this does not happen and that probiotics rapidly disappear from stools when their intake is stopped (Garrido *et al.* 2005); for this reason, they must be consumed periodically to exert their health-promoting effects. The single example of more permanent gut colonization by a probiotic strain in humans was the administration of *Lactobacillus* GG to pregnant mothers; in this case the bacteria colonized the newborn's gut after delivery and persisted for up to 2 years in some individuals (Schultz *et al.* 2004).

20.3.8.1 *Probiotics and diarrhea*

The use of single or associated strains of probiotics to modify the resident microbiota and/or gut function has been widely studied in infants and children. Many of these studies have evaluated their efficacy in the management or prevention of acute or persistent diarrhea. Different primary and secondary outcomes were evaluated including the incidence and duration of the diarrhea, stool output, systemic and extraintestinal symptoms, anthropometric parameters, antibiotic use and volume of oral rehydration solutions required. A number of meta-analyses support the use of some strains of probiotics as a useful tools in the management of these conditions and in the prevention of antibiotic-associated diarrhea (Van Niel *et al.* 2002; Johnston *et al.* 2006; Szajewska *et al.* 2007a). The effects are generally more significant in rotavirus-associated diarrhea than in cases associated with bacterial agents (Szajewska *et al.* 2007b). From the point of view of public health, the shortening of rotavirus episodes by approximately 24 hours is important in the less developed countries where the outpatient and hospital facilities tend to be overloaded and difficult to reach. Furthermore, the nutrition status of these children is precarious in many instances.

A considerable number of studies in infants and children have also been carried out using fermented foods containing lactic acid bacteria to evaluate their protective effect in acute diarrhea (Brunser *et al.* 2007). In a randomized and controlled study carried out in 82 weaned infants less than 12 months of age, the administration of an acidified milk with *Lactobacillus helveticus* and *Streptococcus thermophilus* for 6 months exerted a preventive effect on the incidence of diarrhea and reduced the number of days during which children were affected, as well as the duration of the episodes. No differences in the enteropathogens detected were observed, but the carriage rate of enteropathogens was lower with the acidified milk (Brunser *et al.* 1989).

There is evidence that the lactobacilli and bifidobacteria resident in the colon may exert some of their antimicrobial activities either directly or by influencing both local and systemic immunities. The lower incidence of infections in breastfed infants may be related to the special quality of their microbiota. Probiotic strains belonging to these genera also inhibit the growth and cellular adhesion of potential pathogens through the production of molecules with antibacterial activities (bacteriocins, H_2O_2, organic acids). In addition, probiotics modulate innate immunity by enhancing natural killer cell activity, as well as the phagocytic activity of macrophages and neutrophils; furthermore, they also stimulate specific immune responses through an increase in antibody production. These properties may explain in part the protective effect of probiotics against infectious infantile diarrhea. This is illustrated by the study of Kila *et al.* (1992) who observed that *Lactobacillus* GG administration to children with rotavirus diarrhea was associated with the enhancement of total immunoglobulin-secreting cell numbers during the acute phase of the infection and in a higher IgA-specific secreting cell response to rotavirus during convalescence, compared with the placebo group. On the other hand, the antibacterial properties of some strains of probiotics, in addition to their potential anti-inflammatory activities, may explain why enteral supplementation of probiotics reduced the risk of severe necrotizing enterocolitis and mortality in preterm infants of more than 1000 g at birth (Alfaleh & Bassler 2008).

20.3.8.2 *Probiotics, immunity and allergy*

It has been proposed that the presence of probiotics in the intestinal lumen may enhance a more balanced immune response, with a shift of the T-helper cellular response from the Th2 pattern, characteristic of newborn infants and of infants with allergy, toward a Th1 pattern (Ouwehand 2007). This process is probably associated with changes in cytokine production and the stabilization of the gastrointestinal barrier function, thus improving the protection of the mucosal surfaces against pathogens and antigens. Probiotics may provide a safe microbial stimulation for the developing immune system in infants and for the management of allergy and atopic diseases such as eczema, allergic rhinitis and asthma, whose prevalence has increased considerably during the past 30 years. It is estimated that food allergies affect approximately 8–10% of children, the main allergens being cow's milk proteins, egg, soy, peanuts, fish, wheat, nuts, shellfish and kiwi fruit. According to the hygiene hypothesis, there is an inverse association between the prevalence of some infections early in life and atopy (Garn & Renz 2007). The higher frequency of allergies may also be related to changes in food consumption patterns, industrial treatments resulting in the disappearance of lactic acid bacteria and alterations in the intestinal microbiota. This is supported by the fact that allergic infants have an aberrant microbiota with higher numbers of clostridia and fewer bifidobacteria; in addition, their *Bifidobacterium* population has higher counts of *B. adolescentis* and *B. longum* and lower counts of *B. bifidum*, i.e. a more adult-like composition. This microbiota is associated with increased *in vitro* synthesis of TNF-α and IL-12 by macrophage-like cells (Kalliomäki *et al.* 2001a; Ouwehand 2007).

Randomized, double-blind, placebo-controlled clinical trials have been carried out to evaluate whether probiotic intake alleviates atopic eczema in children. The administration of *Lactobacillus* GG prenatally to atopic mothers and postnatally for 6 months to their at-risk newborns decreased the incidence of atopy in the infants (23% vs 46% in the probiotic and placebo groups, respectively) (Kalliomäki *et al.* 2001b). This protective effect persisted until 7 years of age (Kalliomäki *et al.* 2007). On the other hand, hydrolyzed infant formula supplemented with *Lactobacillus rhamnosus* GG or *Bifidobacterium lactis* Bb-12 significantly improved atopic eczema in infants, compared with the control formula, without probiotic and decreased biomarkers of inflammation such as fecal TNF-α and α1-antitrypsin, plasmatic sCD4 and the eosinophil protein X in urine while increasing fecal sIgA (Isolauri *et al.* 2000; Viljanen *et al.* 2005). However, a meta-analysis evaluating the benefits of probiotics in the prevention and management of allergy has concluded that the reduction in clinical eczema by probiotics is not consistent between the different studies included in the analysis and that methodological concerns exist regarding some of them (Osborn & Sinn 2007). In consequence, the evidence recommending the addition of probiotics to infant formulas for prevention of allergic disease or food hypersensitivity is as yet insufficient, and further studies are needed to determine the reproducibility of the findings.

20.3.8.3 *Probiotics and H. pylori colonization*

H. pylori is a highly prevalent pathogen that colonizes the surface of the human gastric mucosa and is considered the etiological factor for gastroduodenal ulcers and a risk factor for gastric cancer. Gastric colonization by this agent begins early in life and affects a high proportion of the pediatric population, including infants, in the developing countries (Langat *et al.* 2006). The early colonization process is probably an important factor in determining the development of gastric cancer later in life. Although all colonized individuals develop chronic gastritis, most remain asymptomatic and should not receive antibiotic treatment. *H. pylori* treatment has a high cost; it is not highly effective as a result of antibiotic resistance, and it induces adverse symptoms that affect compliance. Furthermore, when treated, children are rapidly recolonized.

For these reasons, probiotics have been proposed as a tool for the dietary management of *H. pylori* colonization in at-risk populations. Some of the probiotics used in infant formulas and in child foodstuffs have been shown to interfere with *H. pylori* and/or to exert gastroprotective activities. These effects are strain-specific and are related to the inhibition of *H. pylori* growth and adhesion by probiotic bacteriocins, as well as to their antioxidant and anti-inflammatory properties (Gotteland *et al.* 2006). In a randomized, double-blind and controlled trial carried out in asymptomatic colonized children, it was shown that a 4-week administration of *Lactobacillus johnsonii* NCC533 significantly decreased *H. pylori* colonization (determined by the ^{13}C-urea breath test), compared with the placebo. Interestingly, the decrease in colonization induced by the probiotic correlated with the levels of basal colonization before treatment (Cruchet *et al.* 2003). In another study, the same probiotic strain, alone or combined with cranberry juice, was shown to eradicate *H. pylori* in 14.9 and 22.9%, respectively, of colonized children (Gotteland *et al.* 2008). Similar rates of eradication were observed with *S. boulardii* in another trial (Gotteland *et al.* 2005). On the other hand, the administration of a yogurt with *Lactobacillus gasseri* OLL 2716 (LG21) to

H. pylori-positive children did not affect the density of colonization at the end of the intake period, but decreased the severity of the gastric inflammation, as reflected by the lower pepsinogen I/II ratio (Sakamoto *et al.* 2001).

Probiotics have also been used in combination with *H. pylori* therapy with the aim of increasing its efficiency. In a multicenter, prospective, randomized, double-blind, controlled study carried out in symptomatic children colonized by *H. pylori*, a 7-day treatment with antibiotics and omeprazole was compared with the same regime supplemented with a fermented milk containing *Lactobacillus casei* DN-114001 for 14 days (Sýkora *et al.* 2005). The fermented product significantly increased the eradication rate from 57.5 to 84.6%. However, no differences in the rate of eradication were observed in another study carried out in 65 children, using the standard triple therapy (for 7 days) supplemented with a yogurt containing *Bifidobacterium animalis* and *L. casei* for 3 months (Goldman *et al.* 2006).

These studies suggest that some probiotics may be useful to maintain low levels of *H. pylori* and to decrease the chronic inflammatory processes in the gastric antrum of colonized children. However, better-designed studies with adequate sample sizes are necessary to obtain more solid conclusions.

20.4 Conclusions

Breast milk may be considered as the gold standard of functional foods because it contains a great number of bioactive compounds with protective and maturational functions. Modern infant formulas have been trying to apply the knowledge derived from the evaluation of these compounds for infant nutrition. This includes LC-PUFAs, nucleotides, oligosaccharides, probiotics, immunoglobulins and some whey proteins. Many clinical trials have been carried out to evaluate the effect of such formula supplementation on infant growth and health; evidence has been accumulating supporting their use, although more studies are required. As children grow, they begin to partake of the family diet and then other functional components are incorporated into their diets depending on factors such as their socioeconomic stratum and educational level, country of residence, market conditions, season, ethnic group, and religious traditions, among other factors. This makes it possible for children to incorporate into their diet components such as phytosterols, polyphenols and other phytochemicals, as well as many other molecules with health-promoting effects.

References

Alfaleh, K. & Bassler, D. (2008). Probiotics for prevention of necrotizing enterocolitis in preterm infants. *Cochrane Database Systematic Reviews*, (1), CD005496.

Araya, M., Morelli, L., Reid, G., Sanders, M.E., Stanton, C. & Pineiro, M. (2002). *Joint FAO/WHO Working Group Report on Drafting Guidelines for the Evaluation of Probiotics in Food*, London, Ontario, Canada. Available online ftp://ftp.fao.org/es/esn/food/wgreport2.pdf (accessed 8 January 2010).

Arslanoglu, S., Moro, G.E., Schmitt, J., Tandoi, L., Rizzardi, S. & Boehm, G. (2008). Early dietary intervention with a mixture of prebiotic oligosaccharides reduces the incidence of allergic manifestations and infections during the first two years of life. *Journal of Nutrition*, **138**(6), 1091–1095.

Bakker-Zierikzee, A.M., Tol, E.A., Kroes, H., Alles, M.S., Kok, F.J. & Bindels, J.G. (2006). Faecal sIgA secretion in infants fed on pre- or probiotic infant formula. *Pediatric Allergy and Immunology*, **17**(2), 134–140.

Boehm, G., Jelinek, J., Stahl, B., van Laere, K., Knol, J., Fanaro, S., Moro, G. & Vigi, V. (2004). Prebiotics in infant formulas. *Journal of Clinical Gastroenterology*, **38**(6 Suppl), S76–S79.

Bogsted, A.K., Johansen, K., Hatta, R., Kim, M., Casswall, M., Svensson, L. & Hammerström, L. (1996). Passive immunity against diarrhea. *Acta Paediatrica*, **85**(2), 125–128.

Brody, E.P. (2000). Biological activities of bovine glycomacropeptide. *British Journal of Nutrition*, **84**(1 Suppl), S39–S46.
Brück, W.M., Kelleher, S.L., Gibson, G.R., Nielsen, K.E., Chatterton, D.E. & Lönnerdal, B. (2003). rRNA probes used to quantify the effects of glycomacropeptide and alpha-lactalbumin supplementation on the predominant groups of intestinal bacteria of infant rhesus monkeys challenged with enteropathogenic *Escherichia coli*. *Journal of Pediatric Gastroenterology and Nutrition*, **37**(3), 273–280.
Brunser, O., Araya, M., Espinoza, J., Guesry, P.R., Secretin, M.C. & Pacheco, I. (1989). Effect of an acidified milk on diarrhoea and the carrier state in infants of low socio-economic stratum. *Acta Paediatrica Scandinavia*, **78**(2), 259–264.
Brunser, O., Espinoza, J., Araya, M., Cruchet, S. & Gil, A. (1994). Effect of dietary nucleotide supplementation on diarrhoeal disease in infants. *Acta Paediatrica*, **83**(2), 188–191.
Brunser, O., Espinoza, J., Figueroa, G., Araya, M., Spencer, E., Hilpert, H., Link-Amster, H. & Brüssow, H. (1992). Field trial of an infant formula containing anti-rotavirus and anti-*Escherichia coli* milk antibodies from hyperimmunized cows. *Journal of Pediatric Gastroenterology and Nutrition*, **15**(1), 63–72.
Brunser, O., Figueroa, G., Gotteland, M., Haschke-Becher, E., Magliola, C., Rochat, F., Cruchet, S., Palframan, R., Gibson, G., Chauffard, F. & Haschke, F. (2006a). Effects of probiotic or prebiotic supplemented milk formulas on fecal microbiota composition of infants. *Asia Pacific Journal of Clinical Nutrition*, **15**(3), 368–376.
Brunser, O., Gotteland, M. & Cruchet, S. (2007). Functional fermented milk products. *Nestlé Nutrition Workshop Series Pediatric Program*, **60**, 235–250.
Brunser, O., Gotteland, M., Cruchet, S., Figueroa, G., Garrido, D. & Steenhout, P. (2006b). Effect of a milk formula with prebiotics on the intestinal microbiota of infants after an antibiotic treatment. *Pediatric Research*, **59**(3), 451–456.
Buck, R.H., Thomas, D.L., Winship, T.R., Cordle, C.T., Kuchan, M.J., Baggs, G.E., Schaller, J.P. & Wheeler, J.G. (2004). Effect of dietary ribonucleotides on infant immune status. Part 2: Immune cell development. *Pediatric Research*, **56**(6), 891–900.
Carver, J.D. (2003). Advances in nutritional modifications of infant formulas. *American Journal of Clinical Nutrition*, **77**(6): 1550S–1554S.
Carver, J.D., Pimentel, B., Cox, W.I. & Barness, L.A. (1991). Dietary nucleotide effects upon immune function in infants. *Pediatrics*, **88**(2), 359–363.
Carver, J.D., Sosa, R., Saste, M. & Kuchan, M. (2004). Dietary nucleotides and intestinal blood flow velocity in term infants. *Journal of Pediatric Gastroenterology and Nutrition*, **39**(1), 38–42.
Clandinin, M.T., Van Aerde, J.E., Merkel, K.L., Harris, C.L., Springer, M.A., Hansen, J.W. & Diersen-Schade, D.A. (2005). Growth and development of preterm infants fed infant formulas containing docosahexaenoic acid and arachidonic acid. *Journal of Pediatrics*, **146**(4), 461–468.
Cruchet, S., Obregon, M.C., Salazar, G., Diaz, E. & Gotteland, M. (2003). Effect of the ingestion of a dietary product containing *Lactobacillus johnsonii* La1 on *Helicobacter pylori* colonization in children. *Nutrition*, **19**(9), 716–721.
de Swaaf, M.E., de Rijk, T.C., Van Der Meer, P., Eggink, G. & Sijtsma, L. (2003). Analysis of docosahexaenoic acid biosynthesis in *Crypthecodinium cohnii* by ^{13}C labeling and desaturase inhibitor experiments. *Journal of Biotechnology*, **103**(1), 21–29.
Diaz-Ropero, M.P., Martın, R., Sierra, S., Lara-Villoslada, F., Rodrıguez, J.M., Xaus, J. & Olivares, M. (2007). Two *Lactobacillus* strains, isolated from breast milk, differently modulate the immune response. *Journal of Applied Microbiology*, **102**(2), 337–343.
Egashira, M., Takayanagi, T., Moriuchi, M. & Moriuchi, H. (2007). Does daily intake of bovine lactoferrin-containing products ameliorate rotaviral gastroenteritis? *Acta Paediatrica*, **96**(8), 1242–1244.
Elass-Rochard, E., Roseanu, A., Legrand, D., Trif, M., Salmon, V., Motas, C., Montreuil, J. & Spik, G. (1995). Lactoferrin-lipopolysaccharide interaction: involvement of the 28–34 loop region of human lactoferrin in the high-affinity binding to *Escherichia coli* 055B5 lipopolysaccharide. *Biochemical Journal*, **312**(3), 839–845.
Eshach Adiv, O., Berant, M. & Shamir, R. (2004). New supplements to infant formulas. *Pediatric Endocrinology Reviews*, **2**(2), 216–224.
Fanaro, S., Boehm, G., Garssen, J., Knol, J., Mosca, F., Stahl, B. & Vigi, V. (2005). Galacto-oligosaccharides and long-chain fructo-oligosaccharides as prebiotics in infant formulas: a review. *Acta Paediatrica Supplement*, **94**(449), 22–26.
Garn, H. & Renz, H. (2007). Epidemiological and immunological evidence for the hygiene hypothesis. *Immunobiology*, **212**(6), 441–452.

Garrido, D., Suau, A., Pochart, P., Cruchet, S. & Gotteland, M. (2005). Modulation of the fecal microbiota by the intake of a *Lactobacillus johnsonii* La1-containing product in human volunteers. *FEMS Microbiology Letters*, **248**(2), 249–256.

Gibson, G.R. & Roberfroid, M.B. (1995). Dietary modulation of the human colonic microbiota — introducing the concept of prebiotics. *Journal of Nutrition*, **125**(6), 1401–1412.

Gil, A., Ramirez, M. & Gil, M. (2003). Role of long-chain polyunsaturated fatty acids in infant nutrition. *European Journal of Clinical Nutrition*, **57**(1 Suppl), S31–S34.

Goldman, C.G., Barrado, D.A., Balcarce, N., Rua, E.C., Oshiro, M., Calcagno, M.L., Janjetic, M., Fuda, J., Weill, R., Salgueiro, M.J., Valencia, M.E., Zubillaga, M.B. & Boccio, J.R. (2006). Effect of a probiotic food as an adjuvant to triple therapy for eradication of *Helicobacter pylori* infection in children. *Nutrition*, **22**(10), 984–988.

Gotteland, M., Andrews, M., Toledo, M., Muñoz, L., Caceres, P., Anziani, A., Wittig, E., Speisky, H. & Salazar, G. (2008). Modulation of *Helicobacter pylori* colonization with cranberry juice and *Lactobacillus johnsonii* La1 in children. *Nutrition*, **24**(5), 421–426.

Gotteland, M., Brunser, O. & Cruchet, S. (2006). Systematic review: are probiotics useful in controlling gastric colonization by *Helicobacter pylori*? *Alimentary Pharmacology and Therapeutics*, **23**(8), 1077–1086.

Gotteland, M., Poliak, L., Cruchet, S. & Brunser, O. (2005). Effect of regular ingestion of *Saccharomyces boulardii* plus inulin or *Lactobacillus acidophilus* LB in children colonized by *Helicobacter pylori*. *Acta Paediatrica*, **94**(12), 1747–1751.

Grönlund, M.M., Gueimonde, M., Laitinenz, K., Kociubinskiz, G., Grönroosz, T., Salminen, S. & Isolauri, E. (2007). Maternal breast-milk and intestinal bifidobacteria guide the compositional development of the *Bifidobacterium* microbiota in infants at risk of allergic disease. *Clinical and Experimental Allergy*, **37**(12), 1764–1772.

Heird, W.C. & Lapillonne, A. (2005). The role of essential fatty acids in development. *Annual Review of Nutrition*, **25**, 549–571.

Hernell, O. & Lönnerdal, B. (2002). Iron status of infants fed low-iron formula: no effect of added bovine lactoferrin or nucleotides. *American Journal of Clinical Nutrition*, **76**(4), 858–864.

Innis, S.M., Adamkin, D.H., Hall, R.T., Kalhan, S.C., Lair, C., Lim, M., Stevens, D.C., Twist, P.F., Diersen-Schade, D.A., Harris, C.L., Merkel, K.L. & Hansen, J.W. (2002). Docosahexaenoic acid and arachidonic acid enhance growth with no adverse effects in preterm infants fed formula. *Journal of Pediatrics*, **140**(5), 547–554.

Innis, S.M. & Friesen, R.W. (2008). Essential n-3 fatty acids in pregnant women and early visual acuity maturation in term infants. *American Journal of Clinical Nutrition*, **87**(3), 548–557.

Isolauri, E., Arvola, T., Sütas, Y., Moilanen, E. & Salminen, S. (2000). Probiotics in the management of atopic eczema. *Clinical and Experimental Allergy*, **30**(11), 1604–1610.

Johnston, B.C., Supina, A.L. & Vohra, S. (2006). Probiotics for pediatric antibiotic-associated diarrhea: a meta-analysis of randomized placebo-controlled trials. *Canadian Medical Association Journal*, **175**(4), 377–383.

Kaila, M., Isolauri, E., Soppi, E., Virtanen, E., Laine, S. & Arvilommi, H. (1992). Enhancement of the circulating antibody secreting cell response in human diarrhea by a human *Lactobacillus* strain. *Pediatric Research*, **32**(2), 141–144.

Kalliomäki M, Kirjavainen P, Eerola E, Kero P, Salminen S & Isolauri E. (2001a). Distinct patterns of neonatal gut microflora in infants in whom atopy was and was not developing. *Journal of Allergy and Clinical Immunology*, **107**(1), 129–134.

Kalliomäki, M., Salminen, S., Arvilommi, H., Kero, P., Koskinen, P. & Isolauri, E. (2001b). Probiotics in primary prevention of atopic disease: a randomized placebo-controlled trial. *Lancet*, **357**(9262), 1076–1079.

Kalliomäki, M., Salminen, S., Poussa, T. & Isolauri, E. (2007). Probiotics during the first 7 years of life: a cumulative risk reduction of eczema in a randomized, placebo-controlled trial. *Journal of Allergy and Clinical Immunology*, **119**(4), 1019–1021.

Kelleher, S.L. & Lönnerdal, B. (2001). Immunological activities associated with milk. *Advances in Nutritional Research*, **10**, 39–65.

King, J.C., Jr, Cummings, G.E., Guo, N., Trivedi, L., Readmond, B.X., Keane, V., Feigelman, S. & de Waard, R. (2007). A double-blind, placebo-controlled, pilot study of bovine lactoferrin supplementation in bottle-fed infants. *Journal of Pediatric Gastroenterology and Nutrition*, **44**(2), 245–251.

Knol, J., Scholtens, P., Kafka, C., Steenbakkers, J., Gro, S., Helm, K., Klarczyk, M., Schöpfer, H., Böckler, H.M. & Wells, J. (2005). Colon microflora in infants fed formula with galacto- and fructo-oligosaccharides: more like breast-fed infants. *Journal of Pediatric Gastroenterology and Nutrition*, **40**(1), 36–42.

Koletzko, B., Tsang, R., Zlotkin, S.H., Nichols, B.L. & Hansen, J.W. (1997). Importance of dietary lipids. In: *Nutrition During Infancy: Principles and Practice*. Tsang, R.C. (ed), Digital Educational Publishing, Cincinnati, OH, pp. 123–153.

Kolida, S. & Gibson, G.R. (2007). Prebiotic capacity of inulin-type fructans. *Journal of Nutrition*, **137**(11 Suppl), 2503S–2506S.

Kramer, M.S. & Kakuma, R. (2004). The optimal duration of exclusive breastfeeding: a systematic review. *Advances in Experimental Medicine and Biology*, **554**, 63–77.

Krissansen, G.W. (2007). Emerging health properties of whey proteins and their clinical implications. *Journal of the American College of Nutrition*, **26**(6), 713S–723S.

Kunz, C., Rudloff, S., Baier, W., Klein, N. & Strobel, S. (2000). Oligosaccharides in human milk: structural, functional, and metabolic aspects. *Annual Review of Nutrition*, **20**, 699–722.

Langat, A.C., Ogutu, E., Kamenwa, R. & Simiyu, D.E. (2006). Prevalence of *Helicobacter pylori* in children less than three years of age in health facilities in Nairobi Province. *East African Medical Journal*, **83**(9), 471–477.

Lapillonne, A., Clarke, S.D. & Heird, W.C. (2003). Plausible mechanisms for effects of polyunsaturated fatty acids on growth. *Journal of Pediatrics*, **143**(4 Suppl), S9–S16.

Legrand, D., Pierce, A., Elass, E., Carpentier, M., Mariller, C. & Mazurier, J. (2008). Lactoferrin structure and functions. *Advances in Experimental Medicine and Biology*, **606**, 163–194.

León-Sicairos, N., Reyes-López, M., Ordaz-Chipardo, C. & de la Garza, M. (2006). Microbicidal action of lactoferrin and lactofericin and their synergistic effect with metronidazole in *Entamoeba histolytica*. *Biochemistry and Cell Biology*, **84**(3), 327–336.

Li, E.W. & Mine, Y. (2004). Immunoenhancing effects of bovine glycomacropeptide and its derivatives on the proliferative response and phagocytic activities of human macrophage like cells, U937. *Journal of Agricultural and Food Chemistry*, **52**(9), 2704–2708.

Lien, E.L., Davis, A.M. & Euler, A.R. (2004). Growth and safety in term infants fed reduced-protein formula with added bovine alpha-lactalbumin. *Journal of Pediatric Gastroenterology and Nutrition*, **38**(2), 170–176.

Lönnerdal, B. (2000). Breast milk: a truly functional food. *Nutrition*, **16**(7/8), 509–511.

Lönnerdal, B. & Lien, E.L. (2003). Nutritional and physiologic significance of α-lactalbumin in infants. *Nutrition Reviews*, **61**(9), 295–305.

Loureiro, I., Frankel, G., Adu-Bobie, J., Dougan, G., Trabulsi, L.R. & Carneiro-Sampaio, M.M. (1998). Human colostrum contains IgA antibodies reactive to enteropathogenic *Escherichia coli* virulence-associated proteins: intimin, BfpA, EspA, and EspB. *Journal of Pediatric Gastroenterology and Nutrition*, **27**(2), 166–171.

Lucas, A. & Morley, R. (1999). Efficacy and safety of long-chain polyunsaturated fatty acid supplementation of infant-formula milk: a randomized trial. *Lancet*, **354**(9209), 1948–1954.

Marr, A.K., Jenssen, H., Moniri, M.R., Hancock, R.E. & Panté, N. (2009). Bovine lactoferrin and lactoferricin interfere with intracellular trafficking of Herpes simplex virus-1. *Biochimie*, **91**(1), 160–164.

Martin, R., Heilig, H., Zoetendal, E.G., Jimenez, E., Fernandez, L., Smidt, H. & Rodrıguez, J.M. (2007). Cultivation-independent assessment of the bacterial diversity of breast milk among healthy women. *Research in Microbiology*, **158**(1), 31–37.

Martín, R., Langa, S., Reviriego, C., Jimenez, E., Marın, M., Xaus, J., Fernandez, L. & Rodrıguez, J.M. (2003). Human milk is a source of lactic acid bacteria for the infant gut. *Journal of Pediatrics*, **143**(6), 754–758.

Migliore-Samour, D., Roch-Arveiller, M., Tissot, M., Jazziri, M., Keddad, K., Giroud, J.P. & Jollès, P. (1992). Effects of tripeptides derived from milk proteins on polymorphonuclear oxidative and phosphoinositide metabolisms. *Biochemical Pharmacology*, **44**(4), 673–680.

Mihatsch, W.A., Hoegel, J. & Pohlandt, F. (2006). Prebiotic oligosaccharides reduce stool viscosity and accelerate gastrointestinal transport in preterm infants. *Acta Paediatrica*, **95**(7), 843–848.

Miller, J.B. (1999). Human milk oligosaccharides: 130 reasons to breast feed. *British Journal of Nutrition*, **82**(5), 333–335.

Mok, K.H., Pettersson, J., Orrenius, S. & Svanborg, C. (2007). HAMLET, protein folding, and tumor cell death. *Biochemical and Biophysical Research Communications*, **354**(1), 1–7.

Morrow, A.L., Ruiz-Palacios, G.M., Jiang, X. & Newburg, D.S. (2005). Human-milk glycans that inhibit pathogen binding protect breast-feeding infants against infectious diarrhea. *Journal of Nutrition*, **135**(5), 1304–1307.

Mulligan, P., White, N.R.J., Monteleone, G., Wang, P., Wilson, J.W., Ohtsuka, Y. & Sanderson, I. (2006). Breast milk lactoferrin regulates gene expression by binding bacterial DNA CpG motifs but not genomic DNA promoters in model intestinal cells. *Pediatric Research*, **59**(5), 656–661.

Ney, D.M., Gleason, S.T., van Calcar, S.C., Macleod, E.L., Nelson, K.L., Etzel, M.R., Rice, G.M. & Wolff, J.A. (2009). Nutritional management of PKU with glycomacropeptide from cheese whey. *Journal of Inherited Metabolic Diseases*, **32**(1), 32–39.

Ochoa, T.J., Chea-Woo, E., Campos, M., Pecho, I., Prada, A., McMahon, R.J. & Cleary, T.G. (2008). Impact of lactoferrin supplementation on growth and prevalence of Giardia colonization in children. *Clinical Infectious Diseases*, **46**(12), 1881–1883.

Ochoa, T.J. & Cleary, T.G. (2009). Effect of lactoferrin on enteric pathogens. *Biochimie*, **91**(1), 1–7.

Oddy, W.H. (2002). The impact of breast milk on infant and child health. *Breastfeeding Review*, **10**(3), 5–18.

Olivares, M., Dıaz-Ropero, M.P., Martın, R., Rodrıguez, J.M. & Xaus, J. (2006). Antimicrobial potential of four *Lactobacillus* strains isolated from breast milk. *Journal of Applied Microbiology*, **101**(1), 72–79.

Osborn, D.A. & Sinn, J.K. (2007). Probiotics in infants for prevention of allergic disease and food hypersensitivity. *Cochrane Database Systematic Reviews*, (4), CD006475.

Ouwehand, A.C. (2007). Antiallergic effects of probiotics. *Journal of Nutrition*, **137**(2 Suppl), 794S–797S.

Pelligrini, A., Thomas, U., Bramaz, N., Hunziker, P. & Von Fellenberg, R. (1999). Isolation and identification of three bactericidal domains in the bovine α-lactalbumin molecule. *Biochimica et Biophysica Acta*, **1426**(3), 439–448.

Picciano, M.F. & McDonald, S.S. (2006). Lactation. In: *Modern Nutrition in Health and Disease*. Shils, M.E., Shike, M., Ross, A.C., Caballero, B. & Cousins, R.J. (eds), Lippincott Williams and Wilkins, Baltimore, pp. 784–796.

Pickering, L.K., Granoff, D.M., Erickson, J.R., Masor, M.L., Cordle, C.T., Schaller, J.P., Winship, T.R., Paule, C.L. & Hilty, M.D. (1998). Modulation of the immune system by human milk and infant formulae containing nucleotides. *Pediatrics*, **101**(2), 242–249.

Piñeiro, M., Asp, N.G., Reid, G., Macfarlane, S., Morelli, L., Brunser, O. & Tuohy, K. (2008). FAO Technical Meeting on prebiotics. *Journal of Clinical Gastroenterology*, **42**(3 Suppl Pt 2), S156–S159.

Pita, M.L., Fernández, M.R., De-Lucchi, C., Medina, A., Martínez-Valverde, A., Uauy, R. & Gil, A. (1987). Effect of dietary nucleotides on the fatty acid composition of erythrocyte membrane lipids in term infants. *Journal of Pediatric Gastroenterology and Nutrition*, **6**(4), 568–574.

Räihä, N.C., Fazzolari-Nesci, A., Cajozzo, C., Puccio, G., Monestier, A., Moro, G., Minoli, I., Haschke-Becher, E., Bachmann, C., Van't Hof, M., Carrié Fässler, A.L. & Haschke, F. (2002). Whey-predominant, whey-modified infant formula with protein/energy ratio of 1.8 g/100 kcal: adequate and safe for term infants from birth to four months. *Journal of Pediatric Gastroenterology and Nutrition*, **35**(3), 275–281.

Rigo, J., Boehm, G., Georgi, G., Jelinek, J., Nyambugabo, K., Sawatzki, G. & Studzinski, F. (2001). An infant formula free of glycoymacropeptide prevents hyperthreoninemia in formula-fed preterm infants. *Journal of Pediatric Gastroenterology and Nutrition*, **32**(2), 127–130.

Roos, N., Mahé, S., Benamouzig, R., Sick, H., Rautureau, J. & Tomé D. (1995). ^{15}N-labelled immunoglobulins from bovine colostrums are partially resistant to digestion in human intestine. *Journal of Nutrition*, **125**(5), 1238–1244.

Rudnicka, A.R., Owen, C.G., Richards, M. & Wadsworth, M.E.J. (2008). Effect of breastfeeding and sociodemographic factors on visual outcome in childhood and adolescence. *American Journal of Clinical Nutrition*, **87**(5), 1392–1399.

Sakamoto, I., Igarashi, M., Kimura, K., Takagi, A., Miwa, T. & Koga, Y. (2001). Suppressive effect of *Lactobacillus gasseri* OLL 2716 (LG21) on *Helicobacter pylori* infection in humans. *Journal of Antimicrobial Chemotherapy*, **47**(5), 709–710.

Sandström, O., Lönnerdal, B., Graverholt, G. & Hernell, O. (2008). Effects of α-lactalbumin-enriched formula containing different concentrations of glycomacropeptide on infant nutrition. *American Journal of Clinical Nutrition*, **87**(4), 921–928.

Savino, F., Palumeri, E., Castagno, E., Cresi, F., Dalmasso, P., Cavallo, F. & Oggero, R. (2006). Reduction of crying episodes owing to infantile colic: a randomized controlled study on the efficacy of a new infant formula. *European Journal of Clinical Nutrition*, **60**(11), 1304–1310.

Schaller, J.P., Kuchan, M.J., Thomas, D.L., Cordle, C.T., Winship, T.R., Buck, R.H., Baggs, G.E. & Wheeler, J.G. (2004). Effect of dietary ribonucleotides on infant immune status. Part 1: Humoral responses. *Pediatric Research*, **56**(6), 883–890.

Schmidt, I.M., Damgaard, I.N., Boisen, K.A., Mau, C., Chellakooty, M., Olgaard, K. & Main, K.M. (2004). Increased kidney growth in formula-fed versus breast-fed healthy infants. *Pediatric Nephrology*, **19**(10), 1137–1144.

Scholtens, P.A., Alles, M.S., Bindels, J.G., Van Der Linde, E.G., Tolboom, J.J. & Knol, J. (2006). Bifidogenic effects of solid weaning foods with added prebiotic oligosaccharides: a randomised controlled clinical trial. *Journal of Pediatric Gastroenterology and Nutrition*, **42**(5), 553–559.

Schultz, M., Göttl, C., Young, R.J, Iwen, P. & Vanderhoof, J.A. (2004). Administration of oral probiotic bacteria to pregnant women causes temporary infantile colonization. *Journal of Pediatric Gastroenterology and Nutrition*, **38**(3), 293–297.

Segewa, T. & Sugai, S. (1983). Interactions of divalent metal ions with bovine, human, and goat alpha-lactalbumins. *The Journal of Biochemistry*, **93**(5), 1321–1328.

Shin, K., Wakabayashi, H., Yamauchi, K., Yaeshima, T. & Iwatsuki, K. (2008). Recombinant human intelectin binds bovine lactoferrin and its peptides. *Biological and Pharmacological Bulletin*, **31**(8), 1605–1608.

Singhal, A., Macfarlane, G., Macfarlane, S., Lanigan, J., Kennedy, K., Elias-Jones, A., Stephenson, T., Dudek, P. & Lucas, A. (2008). Dietary nucleotides and fecal microbiota in formula-fed infants: a randomized controlled trial. *American Journal of Clinical Nutrition*, **87**(6), 1785–1792.

Sipola, M., Finckenberg, P., Vapaatalo, H., Pihlanto-Leppälä, A., Korhonen, H., Korpela, R. & Nurminen, M.L. (2002). Alpha-lactorphin and beta-lactorphin improve arterial function in spontaneously hypertensive rats. *Life Sciences*, **71**(11), 1245–1253.

Spadaro, M., Caorsi, C., Ceruti, P., Varadhachary, A., Forni, G., Pericle, F. & Giovarelli, M. (2008). Lactoferrin, a major defense protein of innate immunity, is a novel maturation factor for human dendritic cells. *Journal of the Federation of American Societies for Experimental Biology*, **22**(8), 2747–2757.

Sýkora, J., Valecková, K., Amlerová, J., Siala, K., Dedek, P., Watkins, S., Varvarovská, J., Stozický, F., Pazdiora, P. & Schwarz, J. (2005). Effects of a specially designed fermented milk product containing probiotic *Lactobacillus casei* DN-114 001 and the eradication of *H. pylori* in children: a prospective randomized double-blind study. *Journal of Clinical Gastroenterology*, **39**(8), 692–698.

Szajewska, H., Skórka, A. & Dylag, M. (2007a). Meta-analysis: *Saccharomyces boulardii* for treating acute diarrhoea in children. *Alimentary Pharmacology and Therapeutics*, **25**(3), 257–264.

Szajewska, H., Skórka, A., Ruszczyński, M. & Gieruszczak-Białek, D. (2007b). Meta-analysis: *Lactobacillus* GG for treating acute diarrhoea in children. *Alimentary Pharmacology and Therapeutics*, **25**(8), 871–881.

Totani, N. & Oba, K. (1987). The filamentous fungus *Mortierella alpina*, high in arachidonic acid. *Lipids*, **22**(12), 1060–1062.

Turck, D. (2007). Later effects of breastfeeding practice: the evidence. *Nestle Nutrition Workshop Series Pediatric Program*, **60**, 31–39.

Uauy, R., Quan, R. & Gil, A. (1994). Role of nucleotides in intestinal development and repair: implications for infant nutrition. *Journal of Nutrition*, **124**(8 Suppl), 1436S–1441S.

Van Niel, C.W., Feudtner, C., Garrison, M.M. & Christakis, D.A. (2002). *Lactobacillus* therapy for acute infectious diarrhea in children: a meta-analysis. *Pediatrics*, **109**(4), 678–684.

Veereman, G. (2007). Pediatric applications of inulin and oligofructose. *Journal of Nutrition*, **137**(11 Suppl), 2585S–2589S.

Venkatesh, M.P. & Rong, L. (2008). Human recombinant lactoferrin acts synergistically with antimicrobial compounds used in neonatal practice against coagulase-negative staphylococci and *Candida albicans* causing neonatal sepsis. *Journal of Medical Microbiology*, **57**(9), 1113–1121.

Viljanen, M., Kuitunen, M., Haahtela, T., Juntunen-Backman, K., Korpela, R. & Savilahti, E. (2005). Probiotic effects on faecal inflammatory markers and on faecal IgA in food allergic atopic eczema/dermatitis syndrome infants. *Pediatric Allergy and Immunology*, **16**(1), 65–71.

Weiner, C., Pan, Q., Hurtif, M., Boren, T., Bostwick, E. & Hammerström, L. (1999). Passive immunity against human pathogens using bovine antibodies. *Clinical and Experimental Immunology*, **116**(2), 193–205.

Zavaleta, N., Figueroa, D., Rivera, J., Sánchez, J., Alfaro, S. & Lönnerdal, B. (2007). Efficacy of rice-based oral rehydration solution containing recombinant human lactoferrin and lysozyme in Peruvian children with acute diarrhea. *Journal of Pediatric Gastroenterology and Nutrition*, **44**(2), 258–264.

Zimecki, M. & Kruzel, M.L. (2007). Milk-derived proteins and peptides of potential therapeutic and nutritive value. *Journal of Experimental Therapeutics and Oncology*, **6**(2), 89–106.

21 Functional foods and bone health: Where are we at?

Wendy E. Ward, Beatrice Lau, Jovana Kaludjerovic and Sandra M. Sacco

21.1 Osteoporosis is a significant public health issue

Osteoporosis is defined as 'a skeletal disorder resulting from compromised bone strength predisposing a person to an increased risk of fracture' (NIH Consensus Development Panel on Osteoporosis Prevention Diagnosis and Therapy 2001). The definition also identifies that 'bone strength reflects the integration of bone density and bone quality' (NIH Consensus Development Panel on Osteoporosis Prevention Diagnosis and Therapy 2001). At the present time, classifying whether an individual has osteoporosis is based on his or her bone mineral density (BMD), measured by dual-energy x-ray absorptiometry (DEXA) according to criteria established by the World Health Organization (WHO). Thus, an individual with a low BMD, in comparison to young healthy controls, is diagnosed as having osteoporosis (WHO 1994). However, the definition of osteoporosis includes bone strength and makes specific reference to bone quality that goes beyond BMD – there is much discussion among clinicians and scientists striving to more accurately classify an individual's bone health.

An unfortunate reality of osteoporosis is that fragility fractures lead to significant morbidity and mortality among North Americans. Estimates from the United States indicate that an elderly individual who experiences a broken hip is four times more likely to die within 3 months of fracture, and that one in five individuals who fracture a hip will be living at a nursing home within 1 year (U.S. Surgeon General Department of Health and Human Services 2004). A particularly startling statistic is that one in two adults will be at risk for osteoporosis-related fractures by the year 2020 (U.S. Surgeon General Department of Health and Human Services 2004), and bone health will continue to worsen as the population ages.

There is no ideal drug treatment for osteoporosis. Specific drugs are effective at slowing the loss of BMD and result in decreased risk of fragility fractures, but side effects are a valid concern (Rossouw *et al.* 2002; Brandao *et al.* 2008). Moreover, prevention of osteoporosis is the ideal approach, and as such, dietary strategies that maximize bone health throughout the life cycle are urgently needed to help prevent bone loss, preserve bone structure, and ultimately reduce the risk of fragility fractures that result from a low BMD and poor bone quality. Thus, there is a tremendous potential for the development of functional food products that can assist in prevention, and possibly treatment, of osteoporosis.

21.2 Bone is a dynamic tissue throughout the life cycle

Osteoporosis and overall bone health is often perceived as a disease of aging or even a natural phenomenon of aging. And while it is most often diagnosed during aging, it is important to understand that bone is a dynamic tissue and that lifelong bone health influences bone health during aging. Thus, it is imperative to assess bone health throughout the life cycle, and understand that the timing of dietary interventions may be critical to achieve optimal bone health, i.e. bones that are very resistant to fracture. Figure 21.1 depicts the relative changes in bone mineral content (BMC) throughout the life cycle and identifies specific times in the life cycle when foods and/or food components may favorably influence either the acquisition of peak bone mass (the maximal amount of mineral the skeleton will ever have) or slow the loss of bone mineral during aging. It is hypothesized that attaining a higher peak bone mass may help protect against the natural deterioration of bone such that an increased risk of fracture will occur at a later stage of life. Foods that contain dietary estrogens, i.e. soy, flaxseed (FS), or *n*-3 fatty acids such as α-linolenic acid (ALA) and longer chain *n*-3 fatty acids such as eicosapentaenoic acid (EPA) and docosahexaenoic acid (DHA) may enhance bone development and/or slow bone loss during aging (Figure 21.1). This chapter discusses findings to date regarding these novel foods and food components with respect to bone health. While Figure 21.1 focuses on changes in BMC, it is imperative to recognize that changes in bone matrix and overall changes in bone structure also contribute to overall bone health. Some of the outcome measures discussed in Section 21.3 and Table 21.1 evaluate these aspects in addition to quantifying BMC.

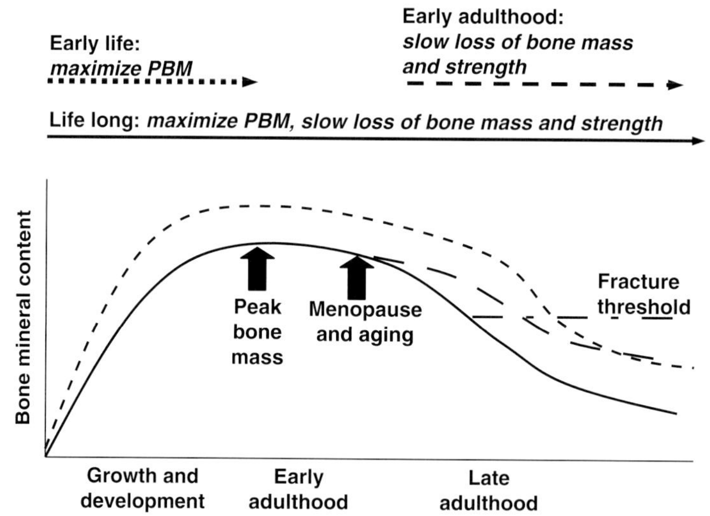

Fig. 21.1 The solid black line illustrates the relative changes in bone mineral content (BMC) throughout the life cycle. During early growth and development, BMC steadily increases until peak bone mass (PBM), the maximal amount of bone the body will ever attain, is reached. At menopause (in women), BMC declines as sex steroid levels decline. Men also experience a decline in BMC due to lower production of sex steroids. The timing of exposure to foods and/or food components may be critical in determining the biological effect to bone health, and interventions can be loosely grouped into one of the following categories: early-life exposure, later life exposure and life-long exposure. Interventions with a specific food and/or food component may result in greater acquisition of BMC throughout development and into early adulthood (dotted line) and/or slow the loss of BMC during aging (dashed line) such that an individual's risk of fragility fracture occurs at a later age.

Table 21.1 Summary of outcomes used in human or animal studies to understand how bone responds to a dietary intervention

Outcome of bone health	Human or animal studies	What information the outcome measure provides
BMD by DEXA	Human or animal	Measures the BMC of the region of interest. BMD is a derived measurement: BMD = BMC (g)/area (cm^2) There is much discussion that BMD may not be an ideal measure of fracture risk, but it is the most widely used measure for assessing fracture risk at present. Low exposure to radiation.
Biochemical markers of bone turnover	Human or animal	Measured in serum, plasma or urine, depending on the specific marker studied. There are two types of markers: *Bone formation markers* include AP, OC and type I procollagen *Bone resorption markers* include DPyr, N-telopeptide of type I collagen (NTx), Pyr and tartrate-resistant alkaline phosphatase
Biomechanical strength testing	Animal	Provides a measurement of how much force a specific skeletal site can withstand before fracture – this is called the 'peak load'. Common tests and sites: i. Compression test of an individual lumbar vertebra – mimics a compression fracture in humans ii. Femur neck – mimics a hip fracture in humans iii. Femur midpoint – not a common site of fragility fracture but does provide information about cortical bone, which predominates at this site Each of these tests is destructive.
Histomorphometry	Mostly animal[a]	Measures the levels of bone cell activity (osteoblast, osteoclasts). Rates of bone formation and resorption can be directly measured.
Microcomputed tomography (animals) or quantitative computed tomography (humans)	Human or animal	Provides a three-dimensional image of bone structure. Trabecular network and cortical bone can be evaluated at key skeletal sites: lumbar vertebra, femur. In animals, excised bones are analyzed. In humans, this measure is limited to extremities (i.e. forearm, ankle) to ensure a minimal exposure to radiation.
Incidence of fragility fracture	Human	Mostly used in studies evaluating drug interventions that are at least two years in duration. Not an ideal outcome measure as the goal is to prevent fragility fractures from occurring.
BMD by ultrasound	Human	Non-invasive technique. How measurements of BMD by ultrasound relate to BMD measurements obtained by DEXA and ultimately to risk of fragility fracture is an ongoing area of study

AP, alkaline phosphatase; BMD, bone mineral density; BMC, bone mineral content; DEXA, dual-energy x-ray absorptiometry; DPyr, deoxypyridinoline; OC, osteocalcin; Pyr, pyridinoline.
[a] In humans, histomorphometrical analysis can be performed on samples collected through a bone biopsy, an invasive procedure.

21.3 Assessment of bone health

Assessment of bone health can involve many different outcomes depending on whether humans or animals are being studied, and whether the specific question being asked relates directly to mineral content, bone strength and fracture risk, or mechanisms of action (Table 21.1). Studies using animal models, particularly rodents, have the potential to measure outcomes that are too invasive to measure in humans. Moreover, investigators may use a combination of these different outcomes to adequately understand how a dietary intervention alters bone health. Table 21.1 summarizes the outcomes of bone health that may be measured, and the information that the specific outcome measure provides regarding bone health, with emphasis on the outcomes that are presented in Tables 21.2 through 21.9.

21.3.1 BMD

Arguably, the most common outcome measure of bone health is BMD. As discussed previously, the WHO bases the classification of osteoporosis on an individual's BMD measurement (WHO 1994). BMD is measured by DEXA, and while it does expose the body to an x-ray dose, the dose of radiation is only roughly equivalent to the exposure an individual receives from an airplane flight from Toronto to Vancouver. In humans, regional measurements of BMD at sites susceptible to fracture are commonly studied: lumbar spine, hip (femur) and forearm. Animal studies also tend to focus on these sites, particularly lumbar spine and femur, to model the human scenario. Human studies (in adults) would typically measure BMD at intervals of 1 year or more to allow adequate time for remodeling to occur in response to treatment. Because of the more rapid growth rate in mice and rats, intervals of weeks or months are adequate to observe responses to dietary interventions. The length of the interval depends on the stage of the life cycle – changes in BMD in developing animals can be detected after weeks, while months are required in older animals, particularly after ovariectomy or orchidectomy (ovariectomy, removal of the ovaries, and orchidectomy, removal of testes, provide useful animal models for studying the effectiveness of an intervention at attenuating the deterioration of bone tissue that occurs after cessation of gonadal production of estrogen/testosterone).

21.3.2 Biochemical markers of bone turnover

Measurements of biochemical markers of bone turnover are particularly useful in human studies as they can be measured in serum, plasma or urine by radioimmunoassay and/or enzyme-linked immunoassays. It is ideal to include at least one measure of bone formation, reflective of osteoblastic activity, and one biomechanical measure of bone resorption, reflective of osteoclastic activity. Thus, serial measurements of a formation and resorption marker will reveal whether osteoblast and/or osteoclast activities are modulated by a specific dietary intervention. An advantage is that changes in osteoblastic and osteoclastic activity can be detected in weeks rather than months and years. Many commercially available kits can also be used in mouse or rat samples, and some species-specific kits are available.

Table 21.2 Epidemiological studies of soy isoflavones and bone

Population	Method used to assess isoflavone intake	Measured outcomes	Outcome of bone health	Reference
Pre- and postmenopausal Chinese women (19–86 years of age) (*n* = 650)	Self-reported food frequency questionnaire	• Femur and lumbar spine BMD • Biochemical markers: PTH, OC, Ntx, AP, serum calcium and phosphate	• Highest isoflavone (53.3 mg/day) intake was associated with ↑ BMD at the lumbar spine and femur, ↓ PTH, ↓ OC, ↓ urinary NTx • No association between isoflavone intake and bone was seen in premenopausal women	Mei *et al.* (2001)
Postmenopausal Chinese women (mean age 60 years) (*n* = 24 403)	Interviewed food frequency questionnaire, prospective cohort study of 4.5 years	• Risk of fracture	• There was a correlation between soy consumption and a ↓ in fracture incidence particularly among women who began to use soy in the early years following menopause	Zhang *et al.* (2005)
Post-menopausal Chinese women (48–63 years of age) (*n* = 454)	Self-reported food frequency questionnaire	• BMC and BMD of the whole body, lumbar spine and hip	• BMD of the whole body, lumbar spine and hip was 4–8% different between the first and fourth soy protein intake quartiles with the fourth quartile having the ↑ BMD • Soy protein/isoflavones intake had a modest but significant association with hip BMD as well as total body BMC	Ho *et al.* (2003)
Pre- and post-menopausal white, Hispanic, African American, Chinese and Japanese women (42–52 years of age) (*n* = 3302)	Self-reported block food frequency questionnaire	• BMD of lumbar spine, femur neck and total hip	• Consumption of isoflavones was significantly ↑ among Chinese and Japanese women compared to other ethnicities • Genistein intake was associated with ↑ BMD only in Japanese women	Greendale *et al.* (2002)
Post-menopausal Japanese women (52–83 years of age) (*n* = 85)	Dietary intake was recorded over 3 days and dieticians used these data to calculate energy, protein, calcium and soy intake	• Lumbar vertebrae (LV2–4) BMD • Serum AP, serum OC, urinary Pyr and urinary DPyr	• Isoflavone intake was positively associated with bone mass relative to age (Z-score BMD of LV 2–4) and negatively associated with urinary DPyr • Isoflavone intake did not have a significant effect on AP, OC or Pyr	Horiuchi *et al.* (2000)

(*continued*)

Table 21.2 *Continued*

Population	Method used to assess isoflavone intake	Measured outcomes	Outcome of bone health	Reference
Post-menopausal Japanese women (44–80 years of age) (*n* = 478)	Self-reported food frequency questionnaires were obtained and urinary genistein, daidzein and equol were measured by high-performance liquid chromatography	• Lumbar spine (LV2–4) BMD	• Weight and years since menopause were significantly independent predictors of BMD • BMD adjusted to years since menopause and weight were significantly higher in women who consumed 50–65 mg of isoflavones per day than in women who consume <50 mg of isoflavones per day	Somekawa *et al.* (2001)
Postmenopausal Southern California women (45–74 years of age) (*n* = 208)	Self-reported block food frequency questionnaire	• BMD of the lumbar spine and hip • Urinary NTx, Pyr, AP	• Women who consumed ↑ amounts of isoflavones (~45 mg of genistein/day) had an 18% ↓ in urinary NTx compared to women who consumed no genistein	Kritz-Silverstein & Goodman-Gruen (2002)
Post-menopausal Australian women (51–62 years of age) (*n* = 354)	Self-reported food frequency questionnaire	• Femur neck BMD • Waist and hip measurements • BMI	• Isoflavone intake in the cohort was 17 mg/day • Women who consumed isoflavones ate healthier foods, were more likely to exercise, had ↓ BMI, and ↑ BMD of the femur neck	Guthrie *et al.* (2000)

AP, alkaline phosphatase; BMC, bone mineral content; BMD, bone mineral density; BMI, body mass index; OC, osteocalcin; Pyr, pyridinoline; DPyr, deoxypyridinoline; NTx, *N*-telopeptide; PTH, parathyroid hormone.

21.3.3 Biomechanical strength properties

Biomechanical strength properties at skeletal sites that are common sites of fracture can be determined using a materials testing system by performing the following tests: compression test of an individual lumbar vertebra (LV) and femur neck fracture that mimics a hip fracture in humans. These sites contain a higher proportion of metabolically active trabecular bone – thus skeletal sites with more trabecular compared to cortical bone are more prone to fragility fracture as bone is lost at a faster rate at these sites. Since the femur midpoint is largely composed of cortical bone, measurement at this site reflects the strength of cortical bone. Of greatest clinical relevance is the 'peak load' measurement, the maximum force a bone withstands before it fractures. Other strength properties are also determined during a fracture test. These properties include yield load, ultimate stiffness and Young's modulus, among others (Ward & Thompson 2006).

Table 21.3 Effect of soy isoflavones on bone in feeding intervention trials in humans

Population	Study design and duration	Intervention	Measured outcomes	Outcome of bone health	Reference
Peri-menopausal Japanese women (49–53 years of age) (*n* = 23)	Randomized, placebo controlled 4 weeks	• 62.8 mg of isoflavones per day was administered as soybean isoflavones extract through two daily capsules • Placebo capsule contained dextrin	• Urinary excretion of isoflavones • Pyr, DPyr • Bone stiffness • Serum cholesterol and triglyceride	• Isoflavone excretion was ↑ at 2 and 4 weeks following treatment • Excretion of bone marks, total serum cholesterol and LDL was ↓ with isoflavone treatment	Uesugi *et al.* (2002)
Post-menopausal Japanese women living in Brazil (45–59 years of age) (*n* = 40)	Randomized, placebo-controlled 10 weeks	• 37.3 mg of isoflavones per day was given as a mixture of roasted germ (hypocotyl) of soybean and sesame (1:2 w/w) • Placebo – sesame	• Bone resorption markers: Pyr, DPyr • Bone stiffness • BMI	• Isoflavone excretion was ↑ at 3 and 10 weeks • Excretion of bone resorption markers was ↓ in the isoflavone-treated group but bone stiffness was unchanged	Yamori *et al.* (2002)
Post-menopausal Japanese women (*n* = 102)	Randomized, double-blind, placebo-controlled 4 weeks	• Capsule containing 40 mg of isoflavones per day • Capsule containing 25 mg of vitamin C + 5 mg of vitamin E • Placebo capsule	• Bone resorption marker: DPyr • Bone formation marker: BGP	• Isoflavones ↓ DPyr but had no effect on BGP	Mori *et al.* (2004)

(*continued*)

Table 21.3 *(Continued)*

Population	Study design and duration	Intervention	Measured outcomes	Outcome of bone health	Reference
Post-menopausal Chinese women with low BMD (48–62 years of age) ($n = 238$)	Randomized, double-blind, placebo-controlled 52 weeks	• 40 mg of isoflavones per day in 0.5 g of soy germ extract • 80 mg of isoflavones per day as 1.0 g of soy extract • Placebo – corn starch	• BMC and BMD of the whole body, lumbar spine (LV1–4) and hip	• Isoflavone effects on bone were observed only among women whose BMD was low initially • 80 mg of isoflavones per day for 1 year ↑ total hip BMC and trochanter BMC	Chen *et al.* (2003b)
Post-menopausal Chinese women (45–60 years of age) ($n = 90$)	Randomized, placebo-controlled 26 weeks	• 84 or 126 mg of isoflavones per day was administered as soy germ isoflavone extract powder • Placebo – starch	• BMD of the lumbar spine and hip • Bone formation markers: AP, OC • Bone resorption markers: DPyr	• Isoflavones ↑ BMD at the lumbar spine and femur neck in a dose-dependent manner 3 months posttreatment	Ye *et al.* (2006)
Pre-menopausal American women (21–25 years of age) ($n = 25$)	Randomized, double-blind, placebo-controlled 12 weeks	• Soy protein supplement containing 90 mg of isoflavones per day • Placebo – soy protein minus the isoflavones	• Whole body BMC and BMD • Fat, lean mass, BMI	• After 12 months there were no changes in BMC and BMD among groups	Anderson *et al.* (2002)

Peri-menopausal American women (41–62 years of age) ($n = 69$)	Randomized, double-blind, placebo-controlled/26 weeks	• 4.4 or 80.4 mg of isoflavones per day was consumed as soy protein isolate • Placebo – whey protein	• Lumbar spine BMC and BMD • NTx, AP	• 80.4 mg isoflavones per day attenuates BMC and BMD loss from the lumbar spine and increases AP in a dose-dependent manner	Alekel *et al*. (2000)
Post-menopausal American women (47–61 years of age) ($n = 87$)	Randomized, controlled/52 weeks	• 25 g of soy protein (60 mg of isoflavones per day) was consumed from snack bar, drink mix or cereal • Placebo food	• BMD and BMC of whole body, lumbar vertebrae (LV1–4), total hip • Bone formation markers: AP, IGF-I, OC • Bone resorption markers: DPyr	• Treatment did not induce a significant effect on BMC and BMD at the lumbar spine or hip • Isoflavones ↑ bone formation markers (AP, IGF-I, OC) but had no effect on bone resorption markers DPyr	Arjmandi *et al*. (2005)
Post-menopausal hypercholesterolemic American women (39–83 years of age) ($n = 66$)	Randomized, double-blind, placebo-controlled/26 weeks	• 40 g of soy protein containing 90 mg of isoflavones per day • 40 g of soy protein containing 56 mg of isoflavones per day • Placebo – casein and non-fat dry milk	• BMC and BMD of the lumbar spine (LV1–4), total cholesterol, HDL, LDL	• 90 mg isoflavone group had ↑ BMC and BMD at the lumbar spine than placebo group	Potter *et al*. (1998)

(*continued*)

Table 21.3 *(Continued)*

Population	Study design and duration	Intervention	Measured outcomes	Outcome of bone health	Reference
Post-menopausal Dutch women (60–75 years of age) ($n = 175$)	Randomized, double-blind, placebo-controlled/52 weeks	• 99 mg of isoflavones per day was consumed in 25.6 g of soy protein • Placebo – milk protein	• BMD of the hip and lumbar spine • Lipid assessment: lipoprotein, HDL, LDL, cholesterol, triglycerides	• Isoflavones ↑ BMD at the lumbar spine and femur neck in a dose-dependent manner 3 months posttreatment	Kreijkamp-Kaspers *et al.* (2004)
Post-menopausal Australian women (50–75 years of age) ($n = 78$)	Randomized, double-blind, crossover/12 weeks	• 118 mg of isoflavones per day was consumed in 40 g of soy protein • Placebo – casein	• Urinary isoflavone excretion • Bone resorption markers: Pyr, DPyr • HDL, LDL	• Women in the soy group had ↑ urinary isoflavone excretion than the control group • Soy isoflavones caused a ↓ HDL and LDL, but had no effect on Pyr and DPyr	Dalais *et al.* (2003)

AP, alkaline phosphatase; BGP, bone gamma-carboxyglutamic acid-containing protein; BMC, bone mineral content; BMD, bone mineral density; BMI, body mass index; HDL, high-density lipoprotein; IGF-I, insulin-like growth factor-I; LDL, low-density lipoprotein; OC, osteocalcin; Pyr, pyridinoline; DPyr, deoxypyridinoline; NTx, *N*-telopeptide; PTH, parathyroid hormone; SPI, soy protein isolate.

Table 21.4 Selected studies investigating the effects of long-chain PUFA on bone mineral and biomechanical strength in aging or ovariectomized animal models

Study subject and age	PUFA source	Amount of PUFA	Control group	Treatment duration	Bone mineral	Biomechanical strength	Reference
OVX mice (25–37 weeks)	Menhaden oil	7% menhaden oil and 3% safflower oil (*n*-6/*n*-3 ratio calculated to be ~0.9)	AIN-93G with 10% safflower oil	12 weeks	• Higher femur and lumbar spine BMD	• Higher yield load, peak load and stiffness at femur midpoint • No difference in vertebra peak load, but higher peak load if combined menhaden oil with soy isoflavones	Ward and Fonseca (2007)
OVX rats (9–21 weeks)	Menhaden oil plus safflower oil	*n*-6/*n*-3 ratio: 5.6	Safflower oil *n*-6/*n*-3 ratio: 344	12 weeks	• Higher tibia BMC • No difference in femur	• Not tested	Watkins *et al.* (2005)
OVX rats (3–7 months)	DHA (DHASCO oil) with AIN93-G	*n*-6/*n*-3 ratio: 5.3	Safflower oil *n*-6/*n*-3 ratio: 10.5	4 months	• Higher femur BMC • No difference in tibia and lumbar spine	• Not tested	Watkins *et al.* (2006)
OVX mice (8–33 weeks)	Fish oil	5% fish oil with 0.5% corn oil *n*-6/*n*-3 ratio: 0.4	5% corn oil *n*-6/*n*-3 ratio: 44.1	25 weeks	• Higher distal femur and lumbar spine BMD • No difference in BMD in central femur	• Not tested	Sun *et al.* (2003)

(continued)

Table 21.4 *(Continued)*

Study subject and age	PUFA source	Amount of PUFA	Control group	Treatment duration	Bone mineral	Biomechanical strength	Reference
Female mice (12–18 months)	Fish oil	10% fish, oil *n*-6/*n*-3 ratio: 0.2	10% corn oil *n*-6/*n*-3 ratio: 44.1	6 months	• Higher BMD in distal femoral metaphysis, proximal tibial metaphysis and tibia diaphysis • No difference in lumbar spine	• Not tested	Bhattacharya *et al*. (2007)
Male rats (12–17 months)	*n*-3 group (menhaden oil plus corn oil), *n*-6+*n*-3 group (menhaden oil plus safflower oil)	*n*-6/*n*-3 ratio: 0.2 for *n*-3 group, 10.1 for *n*-6+*n*-3 group	*n*-6 group (safflower oil) *n*-6/*n*-3 ratio: 242	5 months	• Higher femur BMC, cortical and subcortical bone density in *n*-3 group	• Not tested	Shen *et al*. (2006)
OVX rats (32–41 weeks)	EPA	Low group (0.1 g/kg body weight), high group (1.0 g/kg body weight)	4% corn oil (AIN-93M) *n*-6/*n*-3 ratio: 45.3	9 weeks	• No difference in lumbar spine BMD • Lower BMD in high group	• Not tested	Poulsen and Kruger (2006)
OVX rats (17–22 weeks)	EPA	0.32% EPA plus 2.88% corn oil (*n*-6/*n*-3 ratio: ~5.1)	3.2% corn oil	5 weeks	• Not tested	• Inhibit decrease in peak load due to OVX • Inhibit decrease in peak load due to lack of calcium intake	Sakaguchi *et al*. (1994)

Male rats (12–17 months)	*n*-3 group (menhaden oil plus corn oil), *n*-6+*n*-3 group (menhaden oil plus safflower oil)	*n*-6/*n*-3 ratio: 0.2 for *n*-3 group, 10.1 for *n*-6+*n*-3 group	*n*-6 group (safflower oil), *n*-6/*n*-3 ratio: 242	5 months	• Not tested	• Higher peak load, ultimate stiffness, Young's modulus at femur midpoint in *n*-3 group • Higher tibia trabecular bone volume, trabecular number in *n*-3 group and *n*-6/*n*-3 group • Lower tibia trabecular separation than *n*-6 group • Lower eroded surface/bone surface ratio of *n*-3 group • Higher tibia cortical bone area and lower marrow area in *n*-3 group	Shen *et al.* (2007)

BMD, bone mineral density; BMC, bone mineral content; OVX, ovariectomized; DHA, docosahexaenoic acid; DHASCO, DHA-rich single-cell oil; EPA, eicosapentaenoic acid.

Table 21.5 Effect of flaxseed or its components on bone health *in vitro*

Cell line and experimental period	Source	Amount	Control group	Treatment duration	Outcome	Reference
MG-63 human osteosarcoma cell line	Enterolactone or enterodiol	0.01–10 mg/mL	0.0 mg/mL	2–7 days	• Higher cell viability, alkaline phosphatase activity, mRNA of osteonectin and collagen I at low doses • Lower cell viability, alkaline phosphatase activity, mRNA of osteonectin and type I collagen at high doses	Feng *et al.* (2008)
MC3T3-E1 mouse calvaria osteoblastic cell line	α-linolenic acid	10^{-6}–10^{-4} M	0 M	2–4 days	• Higher DNA content at 10^{-5} and 10^{-4} M versus control	Fujimori *et al.* (1989)

DNA, deoxyribonucleic acid; mRNA, messenger ribonucleic acid.

Table 21.6 Effect of flaxseed or its components on bone health in developing animal models

Study subject and experimental period (age)	Source	Amount	Control group	Treatment duration	Bone site	BMD, BMC, strength or biochemical markers	Reference
Female rats (3–11 weeks)	Ground flaxseed or defatted flax meal	5% or 10% of diet, 6.2% of diet	0% (basal diet)	8 weeks	Non-specific	• Lower plasma alkaline phosphatase in the 5%, 10% FS diets, and 6.2% FS meal diet versus control	Babu *et al.* (2000)
Gilts (12–30 wks)	Whole flaxseed or flaxseed meal or flaxseed oil	10% of diet 6.5% of diet 3.5% of diet	0% (basal diet)	18–19 weeks	Non-specific	• Trend for higher serum NTx in all groups versus control	Farmer *et al.* (2007)
Female mice (0–19 weeks)	Secoisolariciresinol diglycoside	5% or 10% of flaxseed diet equivalent	0% (basal diet)	PND 21, PND 50 or PND 132	Femur	• Higher peak load at femur midpoint at PND 50, but benefit was lost by PND 132	Ward *et al.* (2001b)
Male mice (0–19 weeks)	Flaxseed or secoisolariciresinol diglycoside	10% of flaxseed diet or equivalent	0% (basal diet)	PND 21, PND 50 or PND 132	Femur	• Lower peak load at femur midpoint at PND 50, but no differences by PND 132	Ward *et al.* (2001a)
Male and female mice (4–13 weeks)	Flaxseed oil	10% of diet	Corn oil	9 weeks	Femur, lumbar spine	• No differences in BMD, BMC, or strength in femur or lumbar spine	Cohen and Ward (2005)
Male mice (4–13 weeks)	Flaxseed oil	10% of diet	Corn oil	9 weeks	Femur, lumbar spine	• No differences in BMD, BMC, or strength in femur or lumbar spine	Cohen *et al.* (2005)
Female rats (1–30 weeks)	Flaxseed oil (*n*-6:*n*-3 = 0.4)	7% of diet	Soybean oil (*n*-6/*n*-3 = 9) or sunflower seed oil (*n*-6/*n*-3 = 216)	4 weeks (diets given to pregnant dams from late gestation until pups were weaned)	Femur	• Lower cortical BMC versus soybean oil group • Lower cortical BMD versus sunflower seed oil group • No differences in trabecular BMD	Korotkova *et al.* (2004)

BMC, bone mineral content; BMD, bone mineral density; FS, flaxseed; PND, postnatal day.

Table 21.7 Effect of flaxseed or its components on bone health in aging or ovariectomized animal models

Study subject and experimental period (age)	Source	Amount	Control group	Treatment duration	Bone site	BMD, BMC, strength or biochemical markers	Reference
OVX mice (6–38 weeks)	Ground flaxseed	10% of diet	0% (basal diet)	25 weeks	Femur, lumbar spine	• Higher peak load and stiffness at femur midpoint versus control group	Power *et al.* (2007)
OVX rats (12–26 weeks)	Ground flaxseed	10% of diet	0% (basal diet)	12 weeks	Femur, tibia, lumbar spine	• No differences in BMD, BMC or strength at femur, tibia and lumbar spine versus control	Sacco *et al.* (2009)
OVX mice (6–35 weeks)	Enterolactone or enterodiol subcutaneous injections	10 mg/kg body weight	Vehicle injection	22 weeks	Femur, lumbar spine	• No differences in BMD, BMC or strength at femur and lumbar spine versus control	Power *et al.* (2006)
Female mice (0–19 weeks)	Secoisolariciresinol diglycoside	5% or 10% of flaxseed diet equivalent	0% (basal diet)	PND 132	Femur	• No differences in BMD or femur strength	Ward *et al.* (2001b)
Male mice (0–19 weeks)	Flaxseed or secoisolariciresinol diglycoside	10% of flaxseed diet or equivalent	0% (basal diet)	PND 132	Femur	• No differences in BMD or femur strength	Ward *et al.* (2001a)

BMC, bone mineral content; BMD, bone mineral density; OVX, ovariectomized; PND, postnatal day.

Table 21.8 Effect of flaxseed or its components on bone health in humans

Study type	Study subjects and age	Food source	Amount	Control group	Treatment duration	Outcome of bone health	Reference
Cohort study	642 men (72.9 years), 564 women not on HRT (74.0 years), 326 women on HRT (69.4 years)	Dietary LA:ALA	N/A	N/A	N/A	• Negatively associated with BMD at hip	Weiss *et al.* (2005)
Cross-sectional study	75 women (57.6 years)	Urinary enterolactone	N/A	N/A	N/A	• Positively associated with BMD at vertebra, femur neck, Ward's triangle	Kim *et al.* (2002)
Cohort study	67 women (63.1 years)	Urinary enterolactone	N/A	N/A	N/A	• Not associated with BMD at radius • Positively associated with rate of bone loss	Kardinaal *et al.* (1998)
Randomized, double-blind crossover trial	23 men and women (49.3 years)	Walnut oil + flaxseed oil combination	3.6 or 6.5% ALA	0.8% ALA	6 weeks	• Lower NTx in ALA group versus control group • NTx positively associated with TNF-α	Griel *et al.* (2007)
Randomized, double-blind, placebo-controlled trial	101 menopausal women (54.0 years)	Flaxseed muffin	40 g/day	98 menopausal women on wheat germ placebo (55.4 years)	12 months	• No differences in BMD at lumbar vertebrae and femur neck	Dodin *et al.* (2005)

(*Continued*)

Table 21.8 *(Continued)*

Study type	Study subjects and age	Food source	Amount	Control group	Treatment duration	Outcome of bone health	Reference
Randomized, double-blind, placebo-controlled trial	16 menopausal women (52.6 years)	Flaxseed muffin	25 g/day	15 menopausal women on wheat flour placebo (52.7 years)	16 weeks	• No differences in serum alkaline phosphatase or urinary de-oxypyridinoline	Brooks *et al.* (2004)
Randomized, double-blind placebo-controlled trial	20 menopausal women (54 years)	Ground flaxseed	40 g/day	16 menopausal women on wheat-based comparative control	3 months	• No differences in serum alkaline phosphatase, tartrate resistant acid phosphatase, urinary de-oxypyridinoline or helical peptide	Lucas *et al.* (2002)

BMD, bone mineral density; HRT, hormone replacement therapy; LA, linoleic acid; ALA, α-linolenic acid; NTx, cross-links of *N*-telopeptides; TNF-α, tumor necrosis factor-α; N/A, not applicable.

Table 21.9 Effect of flaxseed or its components on bone health in animal models of disease states

Study subject and experimental period (age)	Source	Amount	Control group	Treatment duration	Bone site	BMD, BMC, strength or biochemical markers	Reference
Male mice Han:SPRD-cy (3–11 weeks)	Flaxseed oil	5% of diet	Corn oil	8 weeks	Femur and non-specific	• No differences in BMD • No differences in plasma osteocalcin, urinary *N*-telopeptides	Weiler *et al.* (2002)
Male and female mice Han:SPRD-cy (3–15 weeks)	Flaxseed oil or flaxseed oil + SDG	7% of diet flaxseed oil (7%) + SDG (20 mg/kg diet)	Corn oil	12 weeks	Femur	• Higher BMD and BMC in rats fed flaxseed oil • Lower BMC in flaxseed oil + SDG supplemented females	Weiler *et al.* (2007)
Male mice IL-10 KO (4–13 weeks)	Flaxseed oil	10% of diet	IL-10 KO and control groups fed corn oil BD	9 weeks	Femur lumbar spine	• Higher femur BMC versus IL-10 KO mice fed BD; similar BMC to wild-type mice • Intermediate femur BMD and stiffness at femur midpoint versus IL-10 KO and controls • Intermediate lumbar spine BMD versus IL-10 KO and controls	Cohen *et al.* (2005)
Male rats *fa/fa* Zucker (6–15 weeks)	Flaxseed oil	6% of diet	*n*-6 group (safflower oil) and *n*-3 LCPUFA group (menhaden oil)	9 weeks	Femur and non-specific	• No differences in femur BMC or BMD • No differences in plasma C-terminal telopeptides of type I collagen or plasma OC	Mollard *et al.* (2005)

BD, basal diet; BMC, bone mineral content; BMD, bone mineral density; SDG, secoisolariciresinol diglycoside; LCPUFA, long-chain polyunsaturated fatty acid; IL-10 KO, interleukin-10 knockout; OC, osteocalcin.

21.3.4 Histomorphometry

Histomorphometry measures the levels of osteoblastic and osteoclastic activity, and the rates of bone formation and resorption can be directly measured. In humans, histomorphometrical analysis can be performed on samples collected through a bone biopsy. Data from these analyses provide mechanistic information regarding how dietary interventions modulate bone health.

21.3.5 Microcomputed tomography/peripheral quantitative tomography

Microcomputed tomography (used in animal studies) and peripheral quantitative computed tomography (used in human studies) provide measures of BMD as well as overall bone geometry and structure in three dimensions. The strength of this technology is that specific measurements provide surrogate measures of fracture risk, and thus relate to biomechanical strength testing.

21.3.6 Other outcome measures

Other outcome measures such as incidence of fragility fracture and BMD by ultrasound are available, but have not been evaluated in the studies investigating dietary interventions discussed in this chapter. Some information pertaining to these measures is provided in Table 21.1.

21.4 Foods and dietary components that may modulate bone metabolism throughout the life cycle

Data suggest that foods or food components such as soy and its isoflavones – fish oil and its *n*-3 long-chain polyunsaturated fatty acids (LCPUFA) – and FS, a rich source of the *n*-3 fatty acid ALA and the lignan secoisolariciresinol diglycoside (SDG), may favorably modulate bone health throughout the life cycle.

21.5 Soy and its isoflavones

A balanced diet provides sufficient nutrients to meet the metabolic requirements of an individual and may also contain an array of naturally occurring bioactive molecules, such as soy isoflavones, that can confer significant health benefits. Soy isoflavones, found in soy-rich foods, have been extensively studied for their role in preventing age-related diseases including cardiovascular disease, cancer and osteoporosis (Setchell & Cassidy 1999). Health benefits associated with soy protein consumption have driven the rapidly growing industry to push governing food agencies in establishing health claim labels for advertising soy-based foods. In 1999, the U.S. Food and Drug Administration (FDA) gave food manufacturers permission to put health claims labels on food products high in soy protein indicating that these foods may help to reduce the risk of coronary heart disease (FDA 1999). In order to qualify for this health claim, a food must contain a minimum of 6.25 g of soy protein per serving, such that 25 g of soy protein can be consumed in four servings of the food (FDA

1999). In every 25 g of soy protein, there is approximately 60 mg of isoflavones (Watanabe *et al.* 2000). Consequently, a large portion of the population susceptible to coronary heart disease may consume soy isoflavones. Moreover, many health-conscious women are opting for soy-based foods rich in isoflavones in anticipation of potential health benefits (Reinwald & Weaver 2006). This is because soy isoflavones are widely advertised for their ability to reduce the severity of menopausal symptoms, including attenuation of bone loss that is associated with a decline in estrogen production.

Over the past 8 years, consumption of isoflavones has also increased substantially among grade-school children because the U.S. Department of Agriculture approved legislation for the use of soy in school lunch programs in 2000 (Reinwald & Weaver 2006). As a result, grade-school children are often consuming soy-based food products even if their parents are not. Consumption of isoflavones is also common among infants fed soy protein-based infant formula. Soy formulas have been used for more than 70 years and reports of adverse effects are minimal (Setchell & Cassidy 1999; Strom *et al.* 2001). As a result, the American Academy of Pediatrics recommends soy protein-based formula as a safe and effective alternative for providing adequate nutrition for normal growth and development for term infants who are lactose intolerant, or whose nutritional needs are not being met by maternal breast milk (American Academy of Pediatrics: Committee on Nutrition 1998). Individuals in all age groups are being exposed to soy-based foods that are rich in soy isoflavones, thus, protective effects may be observed at all stages of the life cycle.

Soybean-derived isoflavones, such as genistein, daidzein and glycitein, may have positive effects on bone health because they are structurally similar to 17β-estradiol (Knight & Eden 1996). The phenolic ring shared by all the isoflavones increases their affinity for the estrogen receptors in mammalian cells and allows them to function as both receptor agonists and antagonists (Setchell 1998). X-ray crystallography studies examining the dimerization of the estrogen receptor complex indicate that folding of estrogen receptor helix 12 plays a crucial role in recruitment of various coactivators and corepressors, which modulate estrogen receptor activity (Kalkhoven *et al.* 1998).

There is evidence that soy isoflavones exert the greatest biological effects on bone metabolism when endogenous levels of sex steroids are low. There are two critical stages of the life cycle when endogenous hormones are low: (i) early postnatal life and (ii) during aging (Reinwald & Weaver 2006). As a result, exposure to soy isoflavones during these stages of life may compensate for the naturally low estrogen or testosterone levels to provide protection against age-associated diseases including osteoporosis. The following sections summarize the findings from epidemiological studies and human and animal feeding intervention studies on the use of soy isoflavones for prevention and treatment of osteoporosis at various stages of the life cycle.

21.5.1 Epidemiological evidence

Epidemiological studies indicate that the incidence of osteoporosis-related fractures is significantly lower in Asian women than in Western women (Table 21.2) (Cooper *et al.* 1992; Lau & Cooper 1996; Lauderdale *et al.* 1997; Tham *et al.* 1998; Greendale *et al.* 2002). However, this health advantage diminishes as Asian women adopt a Western lifestyle (Adlercreutz & Mazur 1997). Soy-based foods such as tofu, miso and tempeh, which are rich sources of soy isoflavones, are common dietary staples in many Asian countries including China, Japan and

Indonesia, but not in the Western world. Thus, it is hypothesized that the lower incidence of osteoporosis among Asian women may in part be explained by their high intake of soy protein. Work by Mai *et al.* revealed, using a food frequency questionnaire, that postmenopausal women who consumed 53.3 mg of isoflavones each day had a significantly higher BMD at the femur and lumbar spine than women who consumed negligible amounts (2.1 mg/day) of soy isoflavones (Mei *et al.* 2001). In contrast to postmenopausal women, there was no association between soy isoflavone intake and bone outcomes among pre-menopausal women. This finding reinforces the concept that specific stages of the life cycle are more responsive to isoflavone interventions, i.e. when endogenous 17-β estradiol is low.

The dose of isoflavones that modulates bone metabolism has not been identified because different diets can vary in the availability of isoflavones. Non-fermented soy foods, such as roasted soy beans and/or soy powder, have two to three times more isoflavones than fermented foods, such as miso and tempeh (Atmaca *et al.* 2008). Moreover, there are large interindividual differences in isoflavone metabolism due to different compositions of intestinal microflora, which may modulate the biological effects of these compounds. Individuals who can metabolize daidzein into equol are classified as equol producers (Setchell *et al.* 1984; Atmaca *et al.* 2008) and may exhibit the greatest biological effects (Setchell *et al.* 2002). The frequency of equol producers is much higher among vegetarian and Asian adults (59%) than among non-vegetarian adults (25%) (Setchell & Cole 2006), suggesting that dietary components consumed in conjunction with soy isoflavones may modulate its biological effects.

21.5.2 Soy, isoflavones and bone metabolism

Many feeding intervention trials in humans and animals have investigated the relationship between soy isoflavones and bone metabolism. The majority of these studies have demonstrated that consumption of soy protein and/or soy isoflavones upregulates bone formation and downregulates bone resorption. However, the methods of experimentation vary significantly and therefore drawing definitive conclusions across numerous studies is challenging. Different studies used different forms of isoflavones (i.e. isoflavones present in soy protein isolate, mixtures of soy isoflavones – usually containing genistein and daidzein, or individual purified isoflavones – i.e. genistein or daidzein), different negative controls (i.e. casein, semipurified diets or conjugated estrogen) and different routes of administration (i.e. subcutaneous injections or oral administration). Moreover, the frequency of isoflavone exposure varies between different studies. Some studies have examined the effects of once daily isoflavones administration on bone metabolism, while others have assessed how consuming isoflavones multiple times a day modulates bone health. One interesting general conclusion from these studies is that consuming the same quantity of soy isoflavones several times a day rather than once a day results in greater bioavailability.

21.5.2.1 *Feeding intervention trials in humans*

The first study to investigate the effects of soy isoflavones on bone was done by Potter *et al.* (1998) who reported that in hypercholesterolemic postmenopausal American women, 90 mg of isoflavones/day (as 40 g of soy protein isolate) significantly increased BMD of the lumbar spine after a 24-week treatment period (Table 21.3). In contrast, subjects consuming 56 mg of isoflavones/day (as 40 g of soy protein isolate) or the placebo treatment, containing casein and non-fat dry milk, did not experience any benefits to bone health. This finding suggests that a significant level of isoflavones is needed to observe bone-preserving effects in postmenopausal women.

Randomized, placebo-controlled trials investigating isoflavone effects on bone health in postmenopausal Asian women showed that isoflavones increase BMD and decrease excretion of biochemical markers of bone resorption (i.e. pyridinoline and deoxypyridinoline) (Table 21.3) (Uesugi *et al.* 2002; Yamori *et al.* 2002; Chen *et al.* 2003b; Mori *et al.* 2004; Ye *et al.* 2006). However, studies conducted on postmenopausal American women have provided conflicting results – some studies report no effect on bone metabolism (Anderson *et al.* 2002; Arjmandi *et al.* 2005), while others report a positive effect (Table 21.3) (Potter *et al.* 1998; Alekel *et al.* 2000).

Anderson *et al.* (2002) revealed that treatment with 90 mg of isoflavones per day for 12 weeks had no significant effect on BMC or BMD 12 months post-treatment in young, menstruating American women. However, this study was conducted using a very short time frame and small number of subjects in a pre-menopausal state. Estradiol may compete with soy isoflavones, inhibiting their effect on bone metabolism. In a 1-year soy protein supplementation study, consumption of 60 mg of isoflavones per day by postmenopausal American women did not induce a significant effect on BMC or BMD (Arjmandi *et al.* 2005). However, the isoflavone treatment had a significant positive effect on bone formation markers including alkaline phosphatase, osteocalcin and insulin growth factor-1, suggesting that isoflavones can modulate bone metabolism. Alekel *et al.* (2000) studied the effects of isoflavones on BMD in 69 peri-menopausal women in a randomized, double-blind, placebo-controlled trial. Subjects were randomly assigned to one of three treatment groups: an isoflavone-rich soy protein (80.4 mg of aglycones/day), an isoflavone-poor soy protein (4.4 mg of aglycones/day) or whey protein group (0 mg of aglycones/day). Lumbar spine BMC and BMD were assessed at baseline and after 24 weeks of treatment. Only the isoflavone-rich soy protein diet had a positive effect on lumbar spine BMC and BMD, suggesting that substantial quantities of isoflavones are needed to attenuate the deterioration of bone tissue. Current guidelines do not support isoflavones as a treatment for osteoporosis (Brown & Josse 2002). In summary, soy isoflavones induce either positive effects (Potter *et al.* 1998; Alekel *et al.* 2000; Morabito *et al.* 2002; Uesugi *et al.* 2002; Yamori *et al.* 2002; Chen *et al.* 2003b; Kreijkamp-Kaspers *et al.* 2004; Mori *et al.* 2004; Ye *et al.* 2006) or no effects (Anderson *et al.* 2002; Arjmandi *et al.* 2005) on bone metabolism in pre-menopausal and postmenopausal women. To our knowledge, no human feeding intervention studies have reported any negative effects of soy isoflavone consumption on bone metabolism.

21.5.2.2 *Feeding intervention trials using animal models*

Many animal feeding intervention studies have demonstrated that consuming soy protein and/or soy isoflavones has a positive effect on bone mass and strength (Blair *et al.* 1996; Fanti *et al.* 1998; Ishimi *et al.* 2000). Most of these studies have been conducted using the ovariectomized (ovx) rodent model, an accepted model for human osteoporosis as it results in increased bone resorption and a loss of bone mass.

The first animal studies conducted in rodent models examined the effects of soy milk or soy protein isolate compared to casein diets and reported that BMD was higher among the soy-fed rats compared to the casein-fed rats (Omi *et al.* 1994; Blum *et al.* 2003; Taguchi *et al.* 2006). Thereafter, researchers investigated whether the isoflavones present in soy protein were mediating the positive effects. Studies have shown that isoflavones are more effective in conserving bone than the protein component of soy. Blaire *et al.* (1996) reported that oral exposure to 30 μmol of genistein per day for 30 days in ovx Sprague-Dawley rats increased dry femur bone ash weight by 12%. Similar findings were reported in a ddy mouse model in which 0.7 mg of genistein per day for 30 days reduced ovariectomy-induced bone loss in the femur (Ishimi *et al.* 2000). Female ddy mice exposed to low (0.7 mg/day) or high (5 mg/day) doses of genistein had a significantly higher BMD at the femur than sham-operated mice exposed to placebo. This study is in agreement with work by Anderson *et al.* (1998) that reported lower doses of genistein are more effective in preserving bone loss post-ovariectomy than higher doses of genistein. Consequently, the biphasic effects of genistein may ultimately modulate bone metabolism. Furthermore, delaying administration of genistein to several weeks after ovariectomy was less effective at conserving bone loss than introducing the same treatment immediately post-ovariectomy. Thus, findings from these animal studies have shown that consumption of isoflavones is most beneficial in an estrogen-depleted state – similar to human intervention trials.

It is important to note, however, that not all studies have shown soy protein or isoflavone interventions protect against loss of bone mass and strength. Breitman *et al.* (2003) examined whether isoflavones (1.6 g/kg of diet) attenuated deterioration of bone tissue in an ovx rat model after an 8-week intervention period. Isoflavone intervention did not preserve BMD or bone strength at the femur or lumbar spine. In another study, 6-month-old ovx Sprague-Dawley rats were randomly assigned to one of nine treatment groups and were pair-fed soy or casein-based diet with or without soy isoflavones for 8 weeks to characterize the effects of soy protein and soy isoflavones on calcium and bone metabolism (Cai *et al.* 2005). In this study, two casein diets were enriched with isoflavones (0.3 mg of isoflavones/g of diet or 0.8 mg of isoflavones/g of diet) and two soy protein-based diets contained isoflavones (0.2 mg of isoflavones/g of diet or 0.4 mg of isoflavones/g of diet). The results from this study indicate that the isoflavone treatments did not significantly affect total calcium balance or calcium absorption, and that isoflavones did not suppress bone remodeling in trabecular or cortical bone after ovariectomy. Because this was a large animal study that had many treatment groups, it was not possible to detect subtle differences induced by soy isoflavones. Similar studies should be repeated using a smaller number of treatment groups.

Soy isoflavones may also induce biological effects during the earliest stages of postnatal life. Our laboratory published the first study demonstrating that exposure to genistein during early-life programs bone metabolism in mice (Piekarz & Ward 2007). Male and female CD-1 mice were treated with genistein at a similar level to that of human infants consuming soy protein-based infant formula, from postnatal day 1 through 5, and then followed to 4 months of age (representing young adulthood). Compared to controls, mice treated with genistein had a higher BMD at the lumbar spine (LV1–4) and, importantly, this translated into individual vertebrae that were stronger and more resistant to compression fracture as demonstrated by the significantly higher peak load of LV3 (Piekarz & Ward 2007). Previous studies had shown that treatment with synthetic estrogen, diethylstilbestrol, was similar to that of genistein in females (Migliaccio *et al.* 1992, 2000). An unexpected finding was that treatment with diethylstilbestrol and genistein in males had divergent effects, suggesting that in males genistein was not acting via an estrogen-dependent mechanism. Based on another study conducted in our laboratory (Ward & Piekarz 2007), we are confident that neonatal, and not the *in utero*, period represents the critical window of development for favorably modulating bone metabolism in developing CD-1 mice. Whether the benefits observed in the mouse model translate to human infants consuming soy infant formula requires study.

21.5.3 Safety issues

Compounds with estrogenic activity can disrupt endocrine function if exposure occurs during critical periods of cellular differentiation. Thus, any study investigating biological effects of food components with potential estrogen-like compounds should consider effects on fertility and other reproductive indices. Early exposure to synthetic estrogen, diethylstilbesterol, is associated with an increase in genital tract abnormalities in female mice (Takamatsu *et al.* 1992; Jefferson *et al.* 2005) and an alteration of reproductive functions in male mice (Sato *et al.* 2006). In humans, adenocarcinoma of the vagina has been reported in young women whose mothers were exposed to diethylstilbestrol during pregnancy (Herbst 1999). This finding illustrates that estrogen-like compounds such as isoflavones reprogram tissues such that adverse effects may not be observed until later life, long after the exposure has occurred. Although diethylstilbestrol has a significantly greater estrogenicity compared with soy isoflavones (diethylstilbestrol is 50 000-fold more estrogenic than genistein) (Jefferson *et al.* 2000), isoflavones have the potential to induce estrogen-like effects.

Many rodent studies of reproductive health have reported that soy isoflavones, in particular genistein, disrupt the estrous cycle, impair fertility and cause a reduction in the number of live offspring (Faber & Hughes 1991; Strauss *et al.* 1998; Herbst 1999; Jefferson *et al.* 2000, 2002, 2005, 2006, 2007b; Delclos *et al.* 2001; Newbold *et al.* 2001, 2005; Adeoya-Osiguwa *et al.* 2003; Goyal *et al.* 2003; Padilla-Banks *et al.* 2006; Ryokkynen *et al.* 2006). However, these studies have several shortcomings when extrapolating study findings to humans. The doses of isoflavones used in these studies were often higher than can be attained by diet alone (Delclos *et al.* 2001; Newbold *et al.* 2001; Jefferson *et al.* 2002, 2005, 2007a; Padilla-Banks *et al.* 2006). In addition, isoflavones, primarily genistein, were often investigated in isolation (Delclos *et al.* 2001; Newbold *et al.* 2001; Jefferson *et al.* 2002, 2007a, 2005; Fielden *et al.* 2003; Padilla-Banks *et al.* 2006). Soy protein-based foods contain a mixture of genistein, daidzein and glycitein, which could have interactive effects. It is possible that daidzein and glycitein may mitigate the potential estrogenic activity of genistein to prevent some or all of its adverse effect on reproductive organs.

To date, only one retrospective cohort study has compared the growth and development of young men and women who were exposed to either soy formula or cow's milk formula during infancy (Strom *et al.* 2001). Findings from this study indicate no difference in adult height, body weight, body mass index, pubertal maturation or fertility as measured by pregnancy, miscarriage or ectopic pregnancy rates due to type of feeding. There were no differences in cancer, reproductive or hormonal disorders, sexual orientation, libido dysfunction or birth defects between infants fed soy or cow's milk formula. However, women fed soy formula reported slightly longer duration of menstrual bleeding and greater discomfort with menstruation, with no difference in severity of menstrual flow. Although these findings suggest that soy formulas are safe for human consumption, long-term effects have not been extensively studied and further research is warranted.

Another potential concern is whether women with breast cancer or survivors of breast cancer should consume isoflavones, either in foods or as supplements, because of the potential estrogen-like activity of isoflavones (Messina & Wood 2008). The American Cancer Society has stated that up to three servings of traditional soy foods per day are safe (Doyle *et al.* 2006). The Society does caution that purified isoflavones in powder form or as isolated supplements should not be consumed. A review by Messina and Wood concluded that 'while more research is required to definitely allay concerns, the existing data should provide some degree of assurance that isoflavone exposure at levels consistent with historical Asian soyfood intake does not result in adverse stimulatory effects on breast tissue (Messina & Wood 2008). Moreover, dietary levels of isoflavones are considered to be less than 100 mg isoflavones per day. The authors note that further study is required to more fully determine 'safe' levels for healthy individuals and whether isoflavones may improve prognosis or affect recurrence of breast cancer (Messina & Wood 2008).

21.5.4 Recommended intakes

Based on the available data from clinical studies, it is recommended that women in early menopause consume between 75 and 100 mg of isoflavones each day to attenuate the deterioration of bone tissue. This level of isoflavones can be obtained by consuming soy protein-based foods such as soy flour, soybean, natto, fried tofu, tempeh and miso two to three times a day. At this point in time, soy isoflavones are not recommended as a treatment for osteoporosis, but they may be considered as a second-line preventative therapy in postmenopausal women. The effects of isoflavones in pre-menopausal women and aging men have not been well characterized, and recommendations regarding their use in these populations cannot be made at this time. Consuming soy-containing foods may be the safest way to obtain adequate concentrations of isoflavones since soy-based foods have a long history of use in the Asian cultures. Pure isoflavones administered as supplements have not been well characterized with respect to long-term safety.

21.6 Fish oil and *n*-3 long-chain polyunsaturated fatty acids

n-3 LCPUFAs have been studied extensively for their potential health benefits over the past few decades. Most research into *n*-3 LCPUFA has focused on its role in the promotion of cardiovascular health and neuronal development. An increasingly health-conscious public has driven the rapidly growing industry in developing food products enriched or fortified with such fatty acids. Specifically, EPA (20:5*n*-3) and DHA (22:6*n*-3), which are *n*-3 LCPUFA found in fatty fish, have gained considerable attention for their bioactive properties. As the taste for fish is not shared by all interested consumers, the derivation of fish oils (from menhaden, salmon, tuna and anchovy) and algal oils rich in DHA (from *Crypthecodinium cohnii* and *Schizochytrium* sp.) allows for the consumption of *n*-3 LCPUFA without actual fish intake. Concerns associated with possible contaminants in fish oils are further eliminated with algal oils, which become an option for vegetarians.

The palatability of fish oils is enhanced by their microencapsulation into food products or biodelivery into livestock tissues as the fish odor becomes negligible (Whelan & Rust 2006). Further benefits of microencapsulation include the increased shelf-life and a reduction in oxidation potential (Whelan & Rust 2006). Recently developed processing techniques have enabled the incorporation of fish and algal oils, hence EPA and/or DHA, into novel products that previously contained little or no *n*-3 LCPUFA. Examples of such food items include eggs, breads, pasta, dairy products, baby food, milk, baby formula, juices, cereals, meats and salad dressings (Whelan & Rust 2006).

21.6.1 Metabolism of *n*-3 long-chain polyunsaturated fatty acids

EPA and DHA can be derived from ALA (18:3*n*-3) through the activity of desaturase and elongase enzymes. However, because the rate of conversion from ALA is low, an exogenous supply of dietary *n*-3 LCPUFA is considered to be the main source. Linoleic acid (LA; 18:2 *n*-6) is the essential *n*-6 PUFA that competes with ALA for the elongation and desaturation into arachidonic acid (AA; 20:4*n*-6). EPA and AA are substrates for eicosanoid production, and DHA can be metabolized to docosanoids. The excessive production of eicosanoids from *n*-6 PUFA exerts inflammatory effects, while the products of *n*-3 PUFA are anti-inflammatory. In addition, PUFA can affect phospholipid composition and the fluidity of cellular membranes, thereby influencing cellular processes such as ion channel modulation, endocytosis and exocytosis (Nakamura *et al.* 2001).

21.6.2 *n*-3 long-chain polyunsaturated fatty acids and bone metabolism

A major role of *n*-3 LCPUFA is postulated to be the attenuation of inflammation. Their anti-inflammatory properties have also been used to explain the effects on bone metabolism. In the inflammatory pathway, AA, the main *n*-6 PUFA cleaved from membrane phospholipids, becomes a substrate of cyclooxygenase (COX) for production of prostaglandin E_2 (PGE_2). PGE_2, along with pro-inflammatory cytokines, such as tumor necrosis factor-α (TNF-α), interleukin-1 (IL-1) and IL-6, can stimulate the expression of receptor activator for nuclear factor-κβ ligand (RANKL) by osteoblasts. Increased activity of RANKL ultimately leads to activation of osteoclasts and thus increased bone breakdown. The incorporation of EPA and DHA into cell membranes displaces AA from membrane phospholipids, thereby lowering the proportion of AA available for PGE_2 synthesis. Available data support the reduction of COX expression by *n*-3 LCPUFA (Curtis *et al.* 2000). Given PGE_2 production depends on substrate and enzyme availability, it is expected that PGE_2 production is reduced with fish oil intake.

Indeed, animal feeding studies have shown that fish oil intake resulted in lower *ex vivo* PGE_2 production in bone (Watkins *et al.* 2000; Green *et al.* 2004). Furthermore, lower pro-inflammatory cytokine levels were observed in mice that consumed fish oil diets (Sun *et al.* 2003; Bhattacharya *et al.* 2007). Since the reduction of cytokines and eicosanoids can lower RANKL expression, *n*-3 LCPUFA intake is hypothesized to also reduce RANKL expression. In a cell culture study, DHA treatment resulted in RANKL expression below detection limits, while AA treatment increased its expression (Coetzee *et al.* 2007). Feeding OVX mice fish oil for 25 weeks attenuated the increase of RANKL expression after ovariectomy (Sun *et al.* 2003).

In general, the proposed mechanism suggests that a diet rich in *n*-3 LCPUFA, which can be achieved through a higher intake of fish and/or algal oils, reduces bone resorption. There was a lowering trend in resorption markers in mice fed tuna oil as compared to mice fed corn oil, although significance was not reached (Kruger & Schollum 2005). A menhaden oil diet resulted in rats with a greater level of alkaline phosphatase, a bone formation marker (Watkins *et al.* 2000). In the same study, the bone formation rate in the tibia of rats fed menhaden oil was also higher than the rate in the rats fed a higher level of safflower oil (Watkins *et al.* 2000).

It has been proposed that *n*-3 LCPUFA may modulate calcium metabolism and related factors such as calcitriol and parathyroid hormone by increasing calcium absorption and decreasing excretion. Although the results ranged from none to modest beneficial effects, none reported adverse outcomes on calcium metabolism (Green *et al.* 2004; Kruger & Schollum 2005; Poulsen & Kruger 2006).

21.6.2.1 *Feeding intervention trials in developing animal models*

With respect to bone development, some but not all studies have shown that *n*-3 LCPUFAs are beneficial to bone mineral, biomechanical strength or both. In male quails aged 8 months, those fed menhaden oil for 7 months had greater tibia BMC and biomechanical strength than their counterparts that were fed soybean oil (Liu *et al.* 2003; Sirois *et al.* 2003). Menhaden oil also resulted in higher femur BMD in 9-week-old male rats compared to rats fed the standard laboratory feed, AIN-93G, with soybean oil (Green *et al.* 2004). However, one study did not demonstrate any benefits or harmful effects of a fish oil intake on BMD and biomechanical strength in male rats (Sirois *et al.* 2003).

Six-week-old male mice that were fed tuna oil for 6 weeks had greater BMD in femur and lumbar spine than mice that were fed corn oil (Kruger & Schollum 2005). Biomechanical strength was tested at femur midpoint; however, no significant differences were observed (Kruger & Schollum 2005). Interestingly, no effects on bone were observed in mice that were given fish oil (Kruger & Schollum 2005). Since tuna oil has an even greater DHA content than fish oil, it is probable that DHA is an important mediator in bone development. In another study, DHASCO, a DHA-rich algal oil, was given to rats that were depleted of *n*-3 PUFA. Rats fed algal oil were shown to have greater biomechanical strength in their tibia than rats that continued to be fed an *n*-3-depleted diet (Reinwald *et al.* 2004). Therefore, both fish oils and algal oils appear to play important, and perhaps essential, roles in bone development. DHA is especially of interest as it may be the primary fatty acid that modulates bone metabolism.

21.6.2.2 *Feeding intervention trials in aging and ovariectomized animal models*

As well as maximizing PBM and/or strength, osteoporosis can be delayed or prevented by minimizing the rate of bone loss after the establishment of PBM. One study fed a fish oil diet to gonad-intact female mice from 12 to 18 months of age (Table 21.4) (Bhattacharya *et al.* 2007). The study showed a greater BMD in distal femoral metaphysis, proximal tibial metaphysis and tibial diaphysis in these animals than mice that were fed corn oil for the same duration (Bhattacharya *et al.* 2007). Positive effects including greater BMC and cortical and subcortical femur BMD were also observed in 17-month-old gonad-intact male rats that were fed a menhaden oil diet for 5 months (Table 21.4) (Shen *et al.* 2006). A similar dietary approach was undertaken as reported in another paper (Shen *et al.* 2007). At 17 months of age, male rats fed a menhaden oil diet had greater peak load, ultimate stiffness, and Young's modulus at the femur midpoint than rats fed a safflower oil diet (Shen *et al.* 2007). They were also reported to have higher trabecular bone volume, trabecular number and cortical bone area, and lower trabecular separation and eroded surface/bone surface ratio in the tibia (Shen *et al.* 2007).

However, the majority of studies used OVX rodent models (Table 21.4). OVX mice fed a fish oil diet for 25 weeks from 8 to 33 weeks of age had higher BMD in distal femur and lumbar spine compared to mice fed a corn oil diet (Sun *et al.* 2003). Twenty-one-week-old

OVX rats that were fed a menhaden oil diet for 12 weeks had greater BMC in tibia than control rats fed a safflower oil diet, and no differences were observed in the femur (Watkins *et al.* 2005). Interestingly, opposite findings were noted in another study by the same group (Watkins *et al.* 2006). The femur BMC was greater in 3-month-old OVX rats fed a diet high in DHA (DHASCO oil) than their counterparts fed a safflower oil diet, while there were no differences in tibia BMC (Watkins *et al.* 2006). Despite these differences, which could simply be a consequence of the unavoidable variability between studies, a possible beneficial effect of long-chain *n*-3 PUFA on long bones remains. In another study, EPA was fed to 17-month-old OVX rats for 5 weeks. At 22 weeks of age, the decrease in peak load at the femur midpoint due to ovariectomy or low calcium intake was inhibited (Sakaguchi *et al.* 1994).

Only one study has shown negative results with *n*-3 LCPUFA feeding (Poulsen & Kruger 2006). Thirty-two-week-old OVX rats fed a diet rich in EPA for 9 weeks had lower femur BMD compared to rats that were fed corn oil (Poulsen & Kruger 2006). Although this finding may seem to contradict earlier experiments, an important point to note is that the dietary level of 1.0 g of EPA per kilogram of body weight may have been excessively high for normal function.

To our knowledge, only one study has examined the effect of fish oil on both the bone mineral and bone strength in mice (Ward & Fonseca 2007). Thirty-seven-week-old OVX mice that were fed a menhaden oil diet for 12 weeks had greater BMD at the femur and lumbar spine than mice that were fed a control AIN-93G diet with 10% safflower oil (Ward & Fonseca 2007). Moreover, this was accompanied by higher yield load, peak load and stiffness at the femur midpoint (Ward & Fonseca 2007).

21.6.2.3 *Studies in humans*

To date, only one cohort study has examined the relationship between serum fatty acid concentrations and BMD in humans (Hogstrom *et al.* 2007). In men, DHA was positively correlated with BMD of the whole body and the spine at 22 years of age, and with the change in BMD at the spine between 16 and 22 years of age (Hogstrom *et al.* 2007). The benefits of *n*-3 LCPUFA on bone development seem to be shared among animals and humans; however, it is clear that more research is necessary to confirm this hypothesis. Randomized, controlled trials are foreseeable in the near future to determine directly if fish and algal oils aid in maximizing human bone development; and if so, whether this translates into a preventative strategy to defend against the onset of osteoporosis and/or other bone diseases. Given that individuals are highly susceptible to developing diseases that lead to fragility fractures, human studies examining the relationship between fatty acids and BMD become imperative as oils may offer an attractive alternative strategy to bone-preserving drugs, or delay the use of such drugs until later stages of the life cycle.

There is a paucity of data from human feeding intervention studies. In one study, supplements were given to 69 women as capsules containing EPA and γ-linolenic acid (GLA) or coconut oil (Kruger *et al.* 1998). The decrease in lumbar spine BMD was attenuated in women who had taken capsules with EPA and GLA compared to those taking capsules with coconut oil (Kruger *et al.* 1998). Moreover, their femur BMD was higher than the control subjects (Kruger *et al.* 1998). Therefore, it is reasonable to suggest that the supplementation of *n*-3 LCPUFA may promote the maintenance of bone mineral. However, an epidemiological study has reported that fragility fracture risk was associated not with *n*-3 PUFA intake, but with *n*-6 PUFA intake. A possible explanation is that the study did not discriminate between the intakes of *n*-3 LCPUFA and ALA (ALA is discussed in detail in the section on flaxseed).

The findings are also far from conclusive as they are associative. Given the competitive nature between *n*-6 and *n*-3 PUFA, a diet rich in long-chain *n*-3 PUFA may prove to be beneficial, even if its consequence is the manipulation of *n*-6 PUFA availability. More clinical trials are needed to elucidate how *n*-3 LCPUFAs affect bone health in adults, especially at a time when bone loss normally dominates.

21.6.3 Safety issues

A major concern of the public surrounding fish oil is the potential contamination with mercury and other toxins. Five over-the-counter brands of fish oils that were tested showed that mercury content ranged from non-detectable to negligible, and they were similar to the basal concentration in human blood (Foran *et al.* 2003). Although only five were tested, this study has shown that fish oils, particularly those derived from mid-sized fish such as tuna and trout, pose a low threat to toxicity. However, it is important to note that swordfish and shark are large carnivorous fish that are typically exposed to higher levels of mercury; therefore, oils made from such fish may be more contaminated. More independent studies evaluating fish oils in the market should be carried out. At the present time, it appears that fish oil is a safe alternative to fish intake.

21.6.4 Recommended intakes

There is no unified recommendation for *n*-3 LCPUFA intake. A health claim was issued by the FDA stating that there is supportive, although not conclusive, research indicating EPA and DHA may reduce risk of coronary heart disease (U.S. Food and Drug Administration 2004). The FDA recommends that the combined daily intakes of EPA and DHA not exceed 3 g, and no more than 2 g from dietary supplements. Current dietary recommendations for EPA and DHA are 10% of total *n*-3 PUFA intake. The guidelines issued by the American Heart Association propose eating fatty fish at least twice a week. Although eating fatty fish is one way to acquire *n*-3 LCPUFA, fish and algal oils offer alternative ways for those who do not want to or cannot eat fish to meet these recommendations. A growing number of food products fortified with fish and algal oils are marketed to and consumed by children, many of whom are not fond of the taste of fish. Many adults are also taking fish oil supplements for the cardiovascular benefits without giving much thought to their potential effects on bone metabolism. Studies presented in this section suggest that the possible advantages of EPA and DHA intake go beyond neuronal and visual development and cardiovascular health, with potential benefits in optimizing bone development in children and maintaining bone health in adults.

21.7 Flaxseed and its components, secoisolariciresinol diglycoside and α-linolenic acid

FS is an oilseed that has become increasingly available in the Western diet, in part due to its potential to modulate a number of diseases such as cancers of the breast and prostate, cardiovascular disease, and diabetes. The identified components of FS that are thought to have the most bioactive potential in the body are the lignans and *n*-3 PUFA. FS is the richest source of the lignan SDG and one of the richest sources of ALA, an *n*-3 PUFA. Once flaxseed is consumed, SDG in flaxseed is released and the glucose moiety is removed, forming secoisolariciresinol. Secoisolariciresinol can be converted to lignan metabolites, such as enterodiol and enterolactone, by intestinal bacteria. Enterodiol and enterolactone can exert mild estrogenic and antiestrogenic effects in the body. The *n*-3 PUFA, ALA, is a substrate for the synthesis of *n*-3 LCPUFA including EPA and DHA, which can induce a number of effects in the body including inhibiting the production of pro-inflammatory cytokines. The physiological action of ALA is not yet established, and it is believed that treatment-induced effects of ALA-rich diets may likely be attributed to its longer chain metabolites.

21.7.1 Flaxseed, flaxseed lignan, flaxseed oil and bone metabolism

Using *in vitro* models, the mammalian lignans and ALA have been shown to exert direct effects on bone cells by modulating cell viability, DNA content, alkaline phosphatase activity, and gene expression of markers of bone formation, including osteonectin and type I collagen (Table 21.5) (Fujimori *et al.* 1989; Feng *et al.* 2008). Due to the estrogenic/antiestrogenic potential of the lignans, as well as the ability for the *n*-3 PUFA to modulate bone metabolism, research interest on the action of FS on bone health has expanded in recent years.

21.7.1.1 *Feeding intervention trials in developing animal models*

In a study by Babu *et al.* (2000), 3-week-old female rats were randomized to a 0, 5, 10% FS diet or a 6.2% defatted FS diet for 8 weeks, after which plasma alkaline phosphatase as well as a number of hematological indices was measured (Table 21.6) (Babu *et al.* 2000). While no differences were observed between the 5 and 10% FS diets, as well as the 6.2% defatted FS diet, all treatments significantly induced lower plasma alkaline phosphatase levels compared to the control (0%) group, suggesting that bone formation may be reduced by FS, and that this reduction is due to the defatted fraction (i.e. lignans and fiber) of FS. The consequences of lower plasma alkaline phosphatase levels on bone strength are unknown since no functional tests (i.e. biomechanical strength testing) were performed.

To determine the effect of lignan exposure on BMC and biomechanical bone strength during bone development, Ward *et al.* (2001b) exposed female rats to an SDG-rich diet from birth throughout suckling (birth to postnatal day 21), continuously to adolescence (birth to postnatal day 50), or continuously to adulthood (birth to postnatal day 132) (Table 21.6) (Ward *et al.* 2001b). Two doses of SDG were used to reflect the levels found in a 5 or 10% FS diet. A group fed basal diet throughout the study was included as a control group. Both doses of SDG significantly induced higher peak load at the femur midpoint compared to the control group at postnatal day 50, but this benefit to femur strength was lost by postnatal day 132. No differences in BMC at postnatal day 50 were observed. In addition, rats who received SDG (at a level equivalent to that in a 5% FS diet) from birth to adulthood (day 132), SDG (at a level equivalent to that in a 10% FS diet) from birth to adulthood (day 132), or SDG (at a level equivalent to that in a 10% FS diet) from birth to weaning (day 21) then switched to a basal diet until adulthood (day 132), had significantly lower BMC at day 132 compared to the control group that received a basal diet throughout the study (Ward *et al.* 2001b). These results suggest that continuous exposure to 5 and 10% SDG results in stronger bones at adolescence only. Although lower BMC was observed in adulthood, these results did not translate into weaker femurs (i.e. peak load at the femur midpoint) (Ward *et al.* 2001b). Ward and colleagues also examined the effect of 10% FS and 10% SDG exposure on BMD, BMC and bone strength in male rats from birth until adulthood (postnatal day 132) (Table 21.6) (Ward *et al.* 2001a). Using a similar study design as in their previous work (Ward *et al.* 2001b), newborn male rats were exposed to 10% FS or 10% SDG (equivalent to that found in a 10% FS diet) throughout suckling (birth to postnatal day 21), continuously to adolescence (birth to postnatal day 50), or continuously to adulthood (birth to postnatal day 132) (Ward *et al.* 2001a). A control group was fed basal diet throughout the study. Unlike mice receiving the 10% SDG diet, mice that received the 10% FS diet from birth until postnatal day 50 had significantly lower ultimate bending stress and Young's modulus. However, these differences were no longer observed at adulthood (day 132). No differences in BMD or BMC were observed throughout the study. Thus, continuous exposure to 10% FS results in weaker bones in male rats at adolescence only, which is due to a component other than its SDG, as evident by the lack of observed effects in the groups exposed to SDG. In addition, the negative effects of FS on bone strength in these male rats were not apparent in adulthood, suggesting that over the long term, FS is safe to bone health (Ward *et al.* 2001a). Thus, FS and SDG-rich diets induce gender-specific effects on bone development and bone strength in female and male rats (Ward *et al.* 2001a; b). It is possible that variations in levels of circulating sex steroids between genders throughout development result in altered bone sensitivity to FS and its lignan, SDG.

Farmer *et al.* (2007) investigated the effects of 10% FS, 6.5% FS meal and 3.5% FS oil on markers of bone resorption (*N*-telopeptides of type I collagen) in developing gilts in order to answer the question of whether effects can be attributed to the lignans or oil component of FS (Table 21.6) (Farmer *et al.* 2007). Following approximately 18 weeks of treatment, there was a trend for higher circulating concentrations of *N*-telopeptides of type I collagen in all FS groups compared to the control group, suggesting that FS and its components may slightly promote degradation of bone collagen – a process that can lead to weaker bones. However, these results did not reach statistical significance. Moreover, other markers of bone formation or bone resorption, as well as the effect of these treatments on bone mineralization and biomechanical strength, were not determined. Thus, FS, FS meal and FS oil at a 10% equivalency during prepuberty may have slightly negative effects on bone development in gilts. However, further investigation on functional measures of bone strength is needed to confirm this finding. A lack of significant effects on bone health by FS oil was also observed by Cohen *et al.* (2005) using male and female developing mice (Table 21.6) (Cohen & Ward 2005). This 9-week intervention measured BMD, BMC and biomechanical bone strength in both the femurs and lumbar spine as primary outcome measures. Serum levels of cytokines (interleukin-1β, interleukin-6, TNF-α) were also measured. Interestingly, feeding FS oil did not modulate bone metabolism or serum cytokines, suggesting that while FS oil does not confer a benefit on bone development, it also does not impede bone development.

In summary, the potential beneficial or adverse effects of FS and its components on bone development appear to be transient and gender-specific in some studies (Ward *et al.* 2001a,b). Other developmental models (Weiler *et al.* 2002, 2007; Cohen *et al.* 2005; Mollard *et al.* 2005) that are representative of disease states (polycystic kidney disease, inflammatory bowel disease, hyperinsulinemia) are discussed in Section 21.7.1.4 (Table 21.9).

21.7.1.2 *Feeding intervention trials in aging and ovariectomized animal models*

Only one animal study has investigated the effect of FS on bone metabolism in an animal model of postmenopausal osteoporosis (Table 21.7). Adult (3 months old) ovx rats fed a 10% ground FS diet for 12 weeks showed no differences in BMD or BMC at the femur, tibia and LV (Table 21.7) (Sacco *et al.* 2009). In addition, no differences in strength at the femur and tibia midpoints as well as the LV2 were observed. This suggests that FS does not modulate bone mineral or bone strength in this well-characterized and FDA-approved ovx model of postmenopausal osteoporosis. Using ovx athymic mice with human breast cancer, Power *et al.* (2007) observed higher peak load and stiffness at the femur midpoint compared to control mice after a 25-week feeding of a 10% ground FS diet, suggesting that FS induces beneficial effects on cortical-rich bone in this mouse model of postmenopausal breast cancer (Table 21.7) (Power *et al.* 2007). However, using the same experimental model, Power *et al.* (2006) did not observe similar benefits to the femur after 22 weeks of daily injections with enterolactone or enterodiol (Table 21.7) (Power *et al.* 2006). Thus, other components of FS (i.e. other lignan metabolites and/or ALA) may be responsible for the protective effects of FS on bone strength in ovx athymic mice with human breast cancer.

21.7.1.3 *Feeding intervention trials in humans*

To date, only one epidemiological study has examined the role of the lignan metabolite, enterolactone, in postmenopausal osteoporosis (Table 21.8) (Kim *et al.* 2002). In this study, greater urinary excretion of enterolactone was correlated with higher BMD at the vertebrae, femur neck and Ward's triangle in Korean postmenopausal women with existing osteoporosis. It is important to make the distinction that these women had osteoporosis as other studies of postmenopausal women that are discussed and included in Table 21.8 include women without osteoporosis or low BMD. In healthy Dutch postmenopausal women with low ($\leq$0.5%) or high ($\geq$2.5%) rates of cortical bone loss at the radius (but no diagnosis of osteoporosis), greater urinary enterolactone was correlated with greater rates of bone loss (Kardinaal *et al.* 1998). The reasons for discrepancies in the findings reported in these studies may involve a number of factors including variations in subject populations and methods of enterolactone analyses. Urinary enterolactone may also have been a marker of other unidentified dietary or lifestyle factors that are associated with BMD and the rate of cortical bone loss.

Using the database from the Rancho Bernardo Study, a higher dietary LA-to-ALA ratio was associated with a lower hip BMD in healthy older men ($\sim$72.9 years of age), menopausal women not using hormone replacement therapy ($\sim$74.0 years of age) and postmenopausal women using hormone replacement therapy ($\sim$69.4 years of age) (Weiss *et al.* 2005). This inverse relationship was also observed at the lumbar spine in women not using therapy. Thus, ALA may be useful in attenuating bone loss. However, future studies are needed to confirm whether ALA can indeed attenuate bone loss, or if its potential protective action on bone is mediated through conversion of ALA to its longer-chain products, EPA and DHA, or the resulting ratio of n-6 to n-3 PUFA.

Three randomized, double-blind, placebo-controlled trials have investigated whether FS can modulate BMD or biochemical markers of bone metabolism in healthy postmenopausal women (Table 21.8) (Lucas *et al.* 2002; Brooks *et al.* 2004; Dodin *et al.* 2005). The trial by Dodin *et al.* (2005) consisted of long-term (12 months) daily treatment of 40 g FS in the form of a muffin (Dodin *et al.* 2005). In this trial, no changes in BMD at the lumbar vertebrae or femur neck were observed from baseline or between the FS and wheat germ placebo groups. The other two studies consisted of short-term (3–4 months) daily treatment of 25 g or 40 g FS, in the form of a muffin (Brooks *et al.* 2004) or ground seed (Lucas *et al.* 2002), respectively. In these studies by Brooks *et al.* (2004) and Lucas *et al.* (2002), no changes in serum markers (i.e. alkaline phosphatase, tartrate-resistant acid phosphatase) or urinary markers (i.e. deoxypyridinoline) of bone metabolism were observed from baseline or between the FS and wheat flour/wheat-based placebo groups (Lucas *et al.* 2002; Brooks *et al.* 2004). Thus, these short- and long-term studies demonstrate that FS does not modulate bone metabolism or BMD in healthy postmenopausal women without low BMD. In addition to its effects on biochemical markers of bone metabolism, Brooks *et al.* (2004) observed a significant increase in urinary 2-hydroxyestrone to 16α-hydroxyestrone ratio with FS treatment from baseline (Table 21.8). The increase in 2-hydroxyestrone to 16α-hydroxyestrone ratio after FS treatment was also correlated with urinary lignan excretion suggesting that FS alters estrogen metabolism to favor a potentially antiestrogenic environment. However, no differences in urinary estrogen metabolites (2-hydroxyestrone, 16α-hydroxyestrone) were observed between the FS and placebo groups and no differences in serum hormones (estradiol, estrone, estrone sulfate) were observed (Lucas *et al.* 2002; Brooks *et al.* 2004), suggesting that the potential modulation of FS on estrogen metabolism is too weak to induce changes in hormone levels and markers of bone metabolism.

Griel *et al.* (2007) used a combination of walnut oil and FS oil to provide ALA to 20 overweight and hypercholesterolemic men (48.6 years of age) and 3 overweight and hypercholesterolemic women (58.3 years of age) (Table 21.8). Subjects were randomized to one of three diets containing different ratios of LA to ALA: LA/ALA = 9.5/1 (representing an average American diet); LA/ALA = 3.5/1; or LA/ALA = 1.6/1 (representing an ALA-rich diet) for 6 weeks in a double-blind, crossover trial (Griel *et al.* 2007). At the end of this study, serum *N*-telopeptides of type I collagen were significantly lower following the ALA diet compared to the average American diet, while no differences in bone-specific alkaline phosphatase were observed, suggesting that a high-ALA diet may support bone health by decreasing bone breakdown. In addition, the ALA diet resulted in significantly lower serum TNF-α, a pro-inflammatory cytokine and osteoclastogenic factor, relative to the average American diet suggesting that a high-ALA diet may attenuate osteoclastic activity and thus the breakdown of bone. However, this study was too short in duration to measure differences in BMD. Whether a longer trial results in higher BMD, a surrogate measure of fracture risk, is unknown. It is important to note that these human trials are limited by their small sample sizes. Thus, larger trials similar in design are warranted to determine the ability of FS to modulate bone metabolism and estrogen status.

21.7.1.4 *Feeding intervention trials in animal models of disease states*

The effects of FS oil or FS oil in combination with SDG have been investigated using animal models of kidney disease (Weiler *et al.* 2002, 2007), obesity (Mollard *et al.* 2005) and inflammatory bowel disease (Cohen *et al.* 2005) (Table 21.9). Compromised bone health is a secondary characteristic of these diseases. Moreover, phytoestrogens and *n*-3 PUFA are hypothesized to inhibit bone loss through their potential to act as estrogen agonists or to inhibit the production of pro-inflammatory cytokines such as PGE_2 and TNF-α (Table 21.9). Using a rat model of kidney disease, Weiler *et al.* (2002) compared the effects of high *n*-3 (5% FS oil) and *n*-6 (corn oil) PUFA diet on bone metabolism in weanling male rats (Table 21.9) (Weiler *et al.* 2002). Healthy, wild-type male rats matched for treatments were also included to compare the effects of treatments in both normal and diseased states. After 8 weeks of treatment, femur BMD and BMC were lower in the mice with kidney disease compared to their healthy counterparts. Markers of bone formation (plasma osteocalcin) and bone resorption (*N*-telopeptides of type I collagen) were also higher in the mice with kidney disease compared to wild-type mice. However, no differences in BMD, BMC and markers of plasma osteocalcin or *N*-telopeptides of type I collagen were observed between control and FS oil groups regardless of health status (Weiler *et al.* 2002). A similar study by the same authors (Weiler *et al.* 2007) included 3-week-old male and female mice that were fed for 12 weeks with a 7% FS oil diet, a diet containing a combination of 7% FS oil and SDG (20 mg/kg diet), or a control diet consisting of corn oil (Table 21.9). Since estrogen can prevent bone loss in chronic renal failure, SDG was included in this study for the estrogenic potential that its lignan metabolites might induce. Similar to their previous study (Weiler *et al.* 2002), the authors included healthy, wild-type male and female mice matched for treatments to compare treatment effects in both normal and diseased states. In both male and female mice, femur BMD and BMC were higher in rats fed FS oil, regardless of kidney disease status. Femur BMC was significantly lower in female mice that received the combination of FS oil and SDG, suggesting that this lignan and its metabolites do not act estrogenically on bone in this mouse model of kidney disease (Weiler *et al.* 2007). Thus, FS oil may be beneficial to bone mineral in kidney disease when fed at a 7% level. However, SDG may induce detrimental effects to bone in females only.

In 4-week-old male and female interleukin-10 knockout mice, which develop intestinal inflammation that resembles human inflammatory bowel disease, a FS oil diet equivalent to that found in a whole 10% FS diet was fed for 9 weeks to test whether ALA or its long-chain metabolites can inhibit inflammatory-induced bone loss (Table 21.9) (Cohen *et al.* 2005). BMD, BMC and biomechanical strength testing were performed in both the femurs and vertebrae. BMD, BMC and peak load at the femur as well as BMD at the LV were significantly lower in the IL-10 knockout mice compared to their healthy, wild-type counterparts. FS oil induced modest benefits to bone health of IL-10 knockout mice, resulting in femur BMD, BMC and peak load that was similar to wild-type mice fed either the control or FS oil diets. Similar benefits to the LV were observed with FS oil treatment in IL-10 knockout mice, resulting in BMD that did not differ from the wild-type mice fed either the control or FS oil diets. Additional analyses revealed lower serum TNF-α in the IL-10 knockout mice fed FS oil. Thus, *n*-3 PUFA from FS oil can benefit bone health in IL-10 knockout mice, potentially by reducing serum levels of TNF-α.

The *fa/fa* Zucker rat model represents hyperinsulinemic and obese states, which also induces compromised bone health. Using male 6-week-old lean and *fa/fa* Zucker rats, Mollard *et al.* (2005) determined the effects of high LA (sunflower oil mixture), ALA (FS oil mixture) and EPA/DHA (menhaden oil mixture) diets on BMD, BMC and markers of bone formation (osteocalcin) and bone resorption (C-terminal telopeptides of type I collagen) in the early stages of disease, which is also a representation of juvenile obesity (Table 21.9) (Mollard *et al.* 2005). After 9 weeks of treatment on their respective diets, significantly lower BMD and BMC were observed in the diseased rats compared to their healthy counterparts. However, no diet-induced effects on femur BMD, BMC or markers of bone metabolism were observed among groups. While a main effect on femur diaphysis width was observed, with menhaden oil inducing a significantly greater width compared to the control group that received the safflower oil mixture, femur diaphysis width was intermediate in the group receiving the FS oil mixture. This suggests that FS oil induces mild effects on bone size, but these effects do not translate into greater mineralization or higher levels of bone formation. Both the FS oil and menhaden oil groups induced significantly lower *ex vivo* release of PGE_2 release, but this did not result in higher BMD or favorable changes in biochemical markers of bone metabolism.

In summary, it is evident that FS and its components induce a wide range of effects on bone metabolism that can be in part attributed to the timing of feeding along with the variations in treated populations. Insufficient and conflicting data on the role of FS on the aging skeleton, combined with its widespread consumption in the Western population, reinforce the need for well-designed trials to determine how FS may modulate bone metabolism, particularly in postmenopausal osteoporosis.

21.7.2 Safety issues

The greatest concern of FS consumption is the potential for SDG and its lignan metabolites to induce estrogenic effects on tissues such as the uterus and breast, which may result in an increased risk for tumorigenesis. Moreover, FS accumulates cadmium (Angelova *et al.* 2004; Hocking & McLaughlin 2006), a heavy metal that can also induce estrogenic effects (Johnson *et al.* 2003). In the case of enterolactone, it has been shown to exert estrogen-like action in the uterus of transgenic estrogen-sensitive reporter mice (Penttinen *et al.* 2007) and in MCF-7 and T-47D estrogen-sensitive breast cancer cells (Welshons *et al.* 1987; Mousavi & Adlercreutz 1992). Cadmium has been reported to exert estrogen-like action in the uterus and mammary gland (Johnson *et al.* 2003). In fact, exposure to a 10% FS diet during pregnancy and lactation has been shown to shorten tumor latency and increase tumor multiplicity in 7,12-dimethylbenz[*a*]anthracene (DMBA)-induced female rat offspring (Khan *et al.* 2007). However, another study demonstrated that postnatal exposure to 10% flaxseed resulted in lower DMBA-induced mammary tumors (Chen *et al.* 2003a). Thompson *et al.* (2005) also demonstrated that daily intake of 25 g FS in the form of a muffin reduces biological markers of tumor aggressiveness (Ki-67 labeling index, c-erB2 expression) and increases tumor cell death (apoptosis) in female patients with newly diagnosed breast cancer, suggesting that FS is not estrogenic (Thompson *et al.* 2005). A number of animal studies have also reported that FS and its components inhibit mammary tumor formation and growth without inducing changes in crude markers of organ toxicity such as wet weights of uterus, liver and kidneys (Chen *et al.* 2003a, 2006, 2007). However, it is important to note that wet organ weight does not infer safety as histological examination of tissue is needed to more fully assess safety and toxicity of a food component.

21.7.3 Recommended intakes

There are currently no dietary recommendations for the intake of FS, its lignans or its oil. However, ALA has a current adult recommendation of 1.1–1.6 g/per day as set out in the Daily Reference Intakes by the Institute of Medicine (Food and Nutrition Board 2005). A 10% FS level is used in many feeding studies because it can be easily attained in the human diet and represents an intake of approximately 25–50 g or 2.5–5 tablespoons of FS per day. Thus, the ALA intake in a 10% FS diet is equivalent to 2.5–5.0 g/day, which exceeds current dietary recommendations. Indeed, FS consumption is very popular in Western societies. Flaxseed oil is one of the ten most commonly used natural health products in Canada (Singh & Levine 2006), and lignans such as SDG found in FS are ubiquitous in our food supply, as well as herbal and other non-vitamin and non-mineral natural health products (Mazur 1998; Mazur *et al.* 1998; Nurmi *et al.* 2003; Thompson *et al.* 2007). Since consumption of FS and its components has been reported to be relatively safe and inhibits a number of chronic diseases, it is likely that its use will continue to increase.

21.8 Summary – Where are we at?

In summary, there are many studies providing data from humans and clinically relevant animal models of human bone health that suggest foods or food components such as soy and its isoflavones, fish oil and its *n*-3 LCPUFAs and FS, a rich source of the *n*-3 fatty acid ALA and phytoestrogen lignan, may favorably modulate bone health throughout the life cycle.

Certainly there is much need for further research, particularly long-term human intervention studies in which safety aspects, in addition to benefits to bone, are studied.

Because foods can contain multiple healthful food components in addition to the fact that a healthful diet will combine multiple functional foods, there is tremendous potential for developing functional food products that combine multiple functional food components (i.e. food synergy) or combining functional foods with drugs (food–drug synergy) to protect against osteoporosis and related fragility fracture.

21.8.1 Food synergy and food–drug synergy

Food synergy can be defined as the interaction of two or more components within a food, or of two or more foods working together, such that the potential health benefit is greater than the effect of the single component or food (Ward & Thompson 2006). Food–drug synergy can be defined as the interaction of a food or food component and a specific drug that confers a greater health benefit than either the food or food component or drug alone (Ward & Thompson 2006). With respect to bone health, the field of food synergy and food–drug synergy is an emerging field. There are, however, a few examples in the literature.

21.8.2 Food synergy: Isoflavones, supplemental calcium, fish oil

The potential synergy of soy isoflavones in combination with supplemental calcium was demonstrated in an ovx rat model. This study showed that the combination of isoflavone (a mixture of genistein and daidzein) and supplemental calcium resulted in higher BMD at both the femur and the lumbar spine (Breitman *et al.* 2003). The synergy is believed to occur due to the favorable estrogen-like effect of isoflavones on bone turnover, with the supplemental calcium, ensuring that there is adequate calcium available for mineralization. This finding was consistent with the conclusions of a meta-analysis that concluded that combining hormone replacement therapy with Ca supplements resulted in higher BMD at three common sites of fragility fracture: spine, hip and forearm (Nieves *et al.* 1998).

A study investigating the synergy of isoflavones and fish oil in the ovx mouse model demonstrated that the combination of isoflavones and fish oil resulted in stronger vertebrae than when fish oil or isoflavones were provided alone (Ward & Fonseca 2007). The mechanisms by which the combination of isoflavones and fish oil mediated the benefit to bone are an area for further investigation. Furthermore, whether a dietary strategy providing fish oil and isoflavones attenuates the deterioration of bone tissue in postmenopausal women awaits investigation.

21.8.3 Food–drug synergy: Flaxseed, low-dose estrogen therapy

Potential food–drug synergy should also be investigated. One example is the combination of FS with estrogen replacement therapy (ERT) at doses lower than traditional ERT. The findings of the Women's Health Initiative (WHI) Study (Writing Group for the Women's Health Initiative Investigators 2002) identifying a higher risk of cardiovascular disease events and breast cancer changed how many clinicians and patients view ERT and ultimately resulted in a widespread decline in usage. It also provided a unique opportunity for dietary strategies to attenuate bone loss after cessation of endogenous estrogen production. The findings from the WHI study were the impetus for our research group to study whether combining lower doses of estrogen are effective in the presence of a healthful food such as FS in an ovx rat model. Ovx rats fed a 10% FS diet in combination with continuous low-dose estrogen experienced the greatest protection against ovariectomy-induced bone loss at the lumbar spine (Sacco *et al.* 2009). This is an area of further investigation.

21.9 Where do we go from here?

Where do we go from here? Without question, there need to be collaborative relationships between scientists and the food industry to further our understanding of key components that should be considered in functional food product development as it relates to bone health. Many aspects must be considered: targeting critical stages of the life cycle when consumption of a functional food product may be most beneficial to bone health; dose-response studies to formulate effective levels of food components; potential food synergy; potential food–drug synergy; and safety with consideration of whether a functional food should be consumed by 'healthy' individuals and/or 'individuals with preexisting disease'. An example would be soy isoflavones and whether consumption of high levels is safe in patients with or survivors of hormone-dependent cancers. As with all research areas, funding is often a limiting factor. It is critical to forge relationships among scientists, food industry and government to ensure that this research area will receive appropriate resources.

References

Adeoya-Osiguwa, S.A., Markoulaki, S., Pocock, V., Milligan, S.R. & Fraser, L.R. (2003). 17β-Estradiol and environmental estrogens significantly affect mammalian sperm function. *Human Reproduction*, **18**, 100–107.

Adlercreutz, H. & Mazur, W. (1997). Phyto-oestrogens and Western diseases. *Annals of Medicine*, **29**, 95–120.

Alekel, D.L., Germain, A.S., Peterson, C.T., Hanson, K.B., Stewart, J.W. & Toda, T. (2000). Isoflavone-rich soy protein isolate attenuates bone loss in the lumbar spine of perimenopausal women. *American Journal of Clinical Nutrition*, **72**, 844–852.

American Academy of Pediatrics: Committee on Nutrition (1998). Soy protein-based formulas: recommendations for use in infant feeding. *Pediatrics*, **101**, 148–153.

Anderson, J.J., Ambrose, W.W. & Garner, S.C. (1998). Biphasic effects of genistein on bone tissue in the ovariectomized, lactating rat model. *Proceedings of the Society Experimental Biology and Medicine*, **217**, 345–350.

Anderson, J.J.B., Chen, X., Boass, A., Symons, M., Kohlmeier, M., Renner, J.B. & Garner, S.C. (2002). Soy isoflavones: no effects on bone mineral content and bone mineral density in healthy, menstruating young adult women after one year. *Journal of the American College of Nutrition*, **21**, 388–393.

Angelova, V., Ivanova, R., Delibaltova, V. & Ivanov, K. (2004). Bio-accumulation and distribution of heavy metals in fibre crops (flax, cotton, hemp). *Industrial Crops and Products*, **19**, 197–205.

Arjmandi, B., Lucas, E., Khalil, D., Devareddy, L., Smith, B., Mcdonald, J., Arquitt, A., Payton, M. & Mason, C. (2005). One year soy protein supplementation has positive effects on bone formation markers but not bone density in postmenopausal women. *Nutrition Journal*, **4**, 8.

Atmaca, A., Kleerekoper, M., Bayraktar, M. & Kucuk, O. (2008). Soy isoflavones in the management of postmenopausal osteoporosis. *Menopause*, **15**, 748–757.

Babu, U.S., Mitchell, G.V., Wiesenfeld, P., Jenkins, M.Y. & Gowda, H. (2000). Nutritional and hematological impact of dietary flaxseed and defatted flaxseed meal in rats. *International Journal of Food Sciences and Nutrition*, **51**, 109–117.

Bhattacharya, A., Rahman, M., Sun, D. & Fernandes, G. (2007). Effect of fish oil on bone mineral density in aging C57BL/6 female mice. *The Journal of Nutritional Biochemistry*, **18**, 372–379.

Blair, H.C., Jordan, S.E., Peterson, T.G. & Barnes, S. (1996). Variable effects of tyrosine kinase inhibitors on avian osteoclastic activity and reduction of bone loss in ovariectomized rats. *Journal of Cellular Biochemistry*, **61**, 629–637.

Blum, S.C., Heaton, S.N., Bowman, B.M., Hegsted, M. & Miller, S.C. (2003). Dietary soy protein maintains some indices of bone mineral density and bone formation in aged ovariectomized rats. *Journal of Nutrition*, **133**, 1244–1249.

Brandao, C.M., Lima, M.G., Da Silva, A.L., Silva, G.D., Guerra, A.A., Jr & Acurcio Fde, A. (2008). Treatment of postmenopausal osteoporosis in women: a systematic review. *Cadernos de Saude Publica*, **24**(Suppl 4), 592S–606S.

Breitman, P.L., Fonseca, D., Cheung, A.M. & Ward, W.E. (2003). Isoflavones with supplemental calcium provide greater protection against the loss of bone mass and strength after ovariectomy compared to isoflavones alone. *Bone*, **33**, 597–605.

Brooks, J.D., Ward, W.E., Lewis, J.E., Hilditch, J., Nickell, L., Wong, E. & Thompson, L.U. (2004). Supplementation with flaxseed alters estrogen metabolism in postmenopausal women to a greater extent than does supplementation with an equal amount of soy. *American Journal of Clinical Nutrition*, **79**, 318–325.

Brown, J.P. & Josse, R.G. (2002). 2002 clinical practice guidelines for the diagnosis and management of osteoporosis in Canada. *Canadian Medical Association Journal*, **167**, 1S–34S.

Cai, D.J., Zhao, Y., Glasier, J., Cullen, D., Barnes, S., Turner, C.H., Wastney, M. & Weaver, C.M. (2005). Comparative effect of soy protein, soy isoflavones, and 17beta-estradiol on bone metabolism in adult ovariectomized rats. *Journal of Bone and Mineral Research*, **20**, 828–839.

Chen, J., Power, K.A., Mann, J., Cheng, A. & Thompson, L.U. (2007). Flaxseed alone or in combination with tamoxifen inhibits MCF-7 breast tumor growth in ovariectomized athymic mice with high circulating levels of estrogen. *Experimental Biology and Medicine (Maywood)*, **232**, 1071–1080.

Chen, J., Tan, K.P., Ward, W.E. & Thompson, L.U. (2003a). Exposure to flaxseed or its purified lignan during suckling inhibits chemically induced rat mammary tumorigenesis. *Experimental Biology and Medicine (Maywood)*, **228**, 951–958.

Chen, J., Wang, L. & Thompson, L.U. (2006). Flaxseed and its components reduce metastasis after surgical excision of solid human breast tumor in nude mice. *Cancer Letters*, **234**, 168–175.

Chen, Y.M., Ho, S.C., Lam, S.S.H., Ho, S.S.S. & Woo, J.L.F. (2003b). Soy isoflavones have a favorable effect on bone loss in Chinese postmenopausal women with lower bone mass: A double-blind, randomized, controlled trial. *Journal of Clinical Endocrinology and Metabolism*, **88**, 4740–4747.

Coetzee, M., Haag, M. & Kruger, M.C. (2007). Effects of arachidonic acid, docosahexaenoic acid, prostaglandin E(2) and parathyroid hormone on osteoprotegerin and RANKL secretion by MC3T3-E1 osteoblast-like cells. *The Journal of Nutritional Biochemistry*, **18**, 54–63.

Cohen, S.L., Moore, A.M. & Ward, W.E. (2005). Flaxseed oil and inflammation-associated bone abnormalities in interleukin-10 knockout mice. *The Journal of Nutrition Biochemistry*, **16**, 368–374.

Cohen, S.L. & Ward, W.E. (2005). Flaxseed oil and bone development in growing male and female mice. *Journal of Toxicology and Environmental Health, Part A*, **68**, 1861–1870.

Cooper, C., Campion, G. & Melton, L.J.R. (1992). Hip fractures in the elderly: a world-wide projection. *Osteoporosis International*, **2**, 285–289.

Curtis, C.L., Hughes, C.E., Flannery, C.R., Little, C.B., Harwood, J.L. & Caterson, B. (2000). n-3 fatty acids specifically modulate catabolic factors involved in articular cartilage degradation. *The Journal of Biological Chemistry*, **275**, 721–724.

Dalais, F.S., Ebeling, P.R., Kotsopoulos, D., Mcgrath, B.P. & Teede, H.J. (2003). The effects of soy protein containing isoflavones on lipids and indices of bone resorption in postmenopausal women. *Clinical Endocrinology*, **58**, 704–709.

Delclos, K.B., Bucci, T.J., Lomax, L.G., Latendresse, J.R., Warbritton, A., Weis, C.C. & Newbold, R.R. (2001). Effects of dietary genistein exposure during development on male and female CD (Sprague-Dawley) rats. *Reproductive Toxicology*, **15**, 647–663.

Dodin, S., Lemay, A., Jacques, H., Legare, F., Forest, J.C. & Masse, B. (2005). The effects of flaxseed dietary supplement on lipid profile, bone mineral density, and symptoms in menopausal women: a randomized, double-blind, wheat germ placebo-controlled clinical trial. *Journal of Clinical Endocrinology and Metabolism*, **90**, 1390–1397.

Doyle, C., Kushi, L.H., Byers, T., Courneya, K.S., Demark-wahnefried, W., Grant, B., Mctiernan, A., Rock, C.L., Thompson, C., Gansler, T. & Andrews, K.S. (2006). Nutrition and physical activity during and after cancer treatment: an American Cancer Society guide for informed choices. *CA: A Cancer Journal for Clinicians*, **56**, 323–353.

Faber, K.A. & Hughes, C.L., Jr (1991). The effect of neonatal exposure to diethylstilbestrol, genistein, and zearalenone on pituitary responsiveness and sexually dimorphic nucleus volume in the castrated adult rat. *Biology of Reproduction*, **45**, 649–653.

Fanti, P., Monier-Faugere, M.C., Geng, Z., Schmidt, J., Morris, P.E., Cohen, D. & Malluche, H.H. (1998). The phytoestrogen genistein reduces bone loss in short-term ovariectomized rats. *Osteoporosis International*, **8**, 274–281.

Farmer, C., Petit, H.V., Weiler, H. & Capuco, A.V. (2007). Effects of dietary supplementation with flax during prepuberty on fatty acid profile, mammogenesis, and bone resorption in gilts. *Journal of Animal Science*, **85**, 1675–1686.

Feng, J., Shi, Z. & Ye, Z. (2008). Effects of metabolites of the lignans enterolactone and enterodiol on osteoblastic differentiation of MG-63 cells. *Biological Pharmaceutical Bulletin*, **31**, 1067–1070.

Fielden, M.R., Samy, S.M., Chou, K.C. & Zacharewski, T.R. (2003). Effect of human dietary exposure levels of genistein during gestation and lactation on long-term reproductive development and sperm quality in mice. *Food and Chemical Toxicology*, **41**, 447–454.

Food and Drug Administration (1999). *FDA Approves New Health Claim for Soy Protein and Coronary Heart Disease*. Fishers Lane Rockville. Available online http://www.cfsan.fda.gov/~lrd/tpsoypr2.html (accessed 20 March 2008).

Food and Nutrition Board (2005). *Dietary Reference Intakes for Energy, Cabohydrate, Fiber, Fat, Fatty Acids, Cholesterol, Protein, and Amino Acids (Macronutrients)*. National Academy Press, Washington, DC.

Foran, S.E., Flood, J.G. & Lewandrowski, K.B. (2003). Measurement of mercury levels in concentrated over-the-counter fish oil preparations: is fish oil healthier than fish? *Archives of Pathology and Laboratory Medicine*, **127**, 1603–1605.

Fujimori, A., Tsutsumi, M., Yamada, H., Fukase, M. & Fujita, T. (1989). Arachidonic acid stimulates cell growth in an osteoblastic cell line, MC3T3-E1, by noneicosanoid mechanism. *Calcified Tissue International*, **44**, 186–191.

Goyal, H.O., Robateau, A., Braden, T.D., Williams, C.S., Srivastava, K.K. & Ali, K. (2003). Neonatal estrogen exposure of male rats alters reproductive functions at adulthood. *Biology of Reproduction*, **68**, 2081–2091.

Green, K.H., Wong, S.C. & Weiler, H.A. (2004). The effect of dietary n-3 long-chain polyunsaturated fatty acids on femur mineral density and biomarkers of bone metabolism in healthy, diabetic and dietary-restricted growing rats. *Prostaglandins Leukotrienes Essential Fatty Acids*, **71**, 121–130.

Greendale, G.A., Fitzgerald, G., Huang, M.-H., Sternfeld, B., Gold, E., Seeman, T., Sherman, S. & Sowers, M. (2002). Dietary soy isoflavones and bone mineral density: results from the study of women's health across the nation. *American Journal of Epidemiology*, **155**, 746–754.

Griel, A.E., Kris-Etherton, P.M., Hilpert, K.F., Zhao, G., West, S.G. & Corwin, R.L. (2007). An increase in dietary n-3 fatty acids decreases a marker of bone resorption in humans. *Nutrition Journal*, **6**, 1–8.

Guthrie, J.R., Ball, M., Murkies, A. & Dennerstein, L. (2000). Dietary phytoestrogen intake in mid-life Australian-born women: relationship to health variables. *Climacteric*, **3**, 254–261.

Herbst, A. (1999). Diethylstilbestrol and adenocarcinoma of the vagina. *American Journal of Obstetrics and Gynecology*, **181**, 1576–1578.

Ho, S.C., Woo, J., Lam, S., Chen, Y., Sham, A. & Lau, J. (2003). Soy protein consumption and bone mass in early postmenopausal Chinese women. *Osteoporosis International*, **14**, 835–842.

Hocking, P.J. & Mclaughlin, M.J. (2006). Genotypic variation in cadmium accumulation by seed of linseed and comparison with seed of some other crop species. *Australian Journal of Agricultural Research*, **52**, 427–433.

Hogstrom, M., Nordstrom, P. & Nordstrom, A. (2007). n-3 Fatty acids are positively associated with peak bone mineral density and bone accrual in healthy men: the NO2 Study. *American Journal of Clinical Nutrition*, **85**, 803–807.

Horiuchi, T., Onouchi, T., Takahashi, M., Ito, H. & Orimo, H. (2000). Effect of soy protein on bone metabolism in postmenopausal Japanese women. *Osteoporosis International*, **11**, 721–724.

Ishimi, Y., Arai, N., Wang, X., Wu, J., Umegaki, K., Miyaura, C., Takeda, A. & Ikegami, S. (2000). Difference in effective dosage of genistein on bone and uterus in ovariectomized mice. *Biochemical and Biophysical Research Communications*, **274**, 697–701.

Jefferson, W.N., Couse, J.F., Banks, E.P., Korach, K.S. & Newbold, R.R. (2000). Expression of estrogen receptor-β is developmentally regulated in reproductive tissues of male and female mice. *Biology of Reproduction*, **62**, 310–317.

Jefferson, W.N., Couse, J.F., Padilla-Banks, E., Korach, K.S. & Newbold, R.R. (2002). Neonatal exposure to genistein Induces estrogen receptor (ER)α expression and multioocyte follicles in the maturing mouse ovary: evidence for ERβ-mediated and nonestrogenic actions. *Biology of Reproduction*, **67**, 1285–1296.

Jefferson, W.N., Padilla-Banks, E. & Newbold, R.R. (2005). Adverse effects on female development and reproduction in CD-1 mice following neonatal exposure to the phytoestrogen genistein at environmentally relevant doses. *Biology of Reproduction*, **73**, 798–806.

Jefferson, W.N., Padilla-Banks, E. & Newbold, R.R. (2006). Studies of the effects of neonatal exposure to genistein on the developing female reproductive system. *Journal of AOAC International*, **89**, 1189–1196.

Jefferson, W.N., Padilla-Banks, E. & Newbold, R.R. (2007a). Disruption of the developing female reproductive system by phytoestrogens: genistein as an example. *Molecualr Nutrition & Food Research*, **51**, 832–844.

Jefferson, W.N., Padilla-Banks, E. & Newbold, R.R. (2007b). Disruption of the female reproductive system by the phytoestrogen genistein. *Reproductive Toxicology*, **23**, 308.

Johnson, M.D., Kenney, N., Stoica, A., Hilakivi-Clarke, L., Singh, B., Chepko, G., Clarke, R., Sholler, P.F., Lirio, A.A., Foss, C., Reiter, R., Trock, B., Paik, S. & Martin, M.B. (2003). Cȃdmium mimics the in vivo effects of estrogen in the uterus and mammary gland. *Nature Medicine*, **9**, 1081–1084.

Kalkhoven, E., Valentine, J.E., Heery, D.M. & Parker, M.G. (1998). Isoforms of steroid receptor co-activator 1 differ in their ability to potentiate transcription by the oestrogen receptor. *The EMBO Journal*, **17**, 232–243.

Kardinaal, A.F., Morton, M.S., Bruggemann-Rotgans, I.E. & Van Beresteijn, E.C. (1998). Phyto-oestrogen excretion and rate of bone loss in postmenopausal women. *European Journal of Clinical Nutrition*, **52**, 850–855.

Khan, G., Penttinen, P., Cabanes, A., Foxworth, A., Chezek, A., Mastropole, K., Yu, B., Smeds, A., Halttunen, T., Good, C., Makela, S. & Hilakivi-Clarke, L. (2007). Maternal flaxseed diet during pregnancy or lactation increases female rat offspring's susceptibility to carcinogen-induced mammary tumorigenesis. *Reproductive Toxicology*, **23**, 397–406.

Kim, M.K., Chung, B.C., Yu, V.Y., Nam, J.H., Lee, H.C., Huh, K.B. & Lim, S.K. (2002). Relationships of urinary phyto-oestrogen excretion to BMD in postmenopausal women. *Clinical Endocrinology (Oxford)*, **56**, 321–328.

Knight, D.C. & Eden, J.A. (1996). A review of the clinical effects of phytoestrogens. *Obstetrics and Gynecology*, **87**, 897–904.

Korotkova, M., Ohlsson, C., Hanson, L.A. & Strandvik, B. (2004). Dietary n-6:n-3 fatty acid ratio in the perinatal period affects bone parameters in adult female rats. *The British Journal of Nutrition*, **92**, 643–648.

Kreijkamp-Kaspers, S., Kok, L., Grobbee, D.E., De Haan, E.H.F., Aleman, A., Lampe, J.W. & Van Der Schouw, Y.T. (2004). Effect of soy protein containing isoflavones on cognitive function, bone mineral density, and plasma lipids in postmenopausal women: a randomized controlled trial. *The Journal of the American Medical Association*, **292**, 65–74.

Kritz-Silverstein, D. & Goodman-Gruen, D.L. (2002). Usual dietary isoflavone intake, bone mineral density, and bone metabolism in postmenopausal women. *Journal of Women's Health and Gender-Based Medicine*, **11**, 69–78.

Kruger, M.C., Coetzer, H., De Winter, R., Gericke, G. & Van Papendorp, D.H. (1998). Calcium, gamma-linolenic acid and eicosapentaenoic acid supplementation in senile osteoporosis. *Aging (Milano)*, **10**, 385–394.

Kruger, M.C. & Schollum, L.M. (2005). Is docosahexaenoic acid more effective than eicosapentaenoic acid for increasing calcium bioavailability? *Prostaglandins Leukotrienes Essential Fatty Acids*, **73**, 327–334.

Lau, E.M. & Cooper, C. (1996). The epidemiology of osteoporosis. The oriental perspective in a world context. *Clinical Orthopaedics and Related Research*, 65–74.

Lauderdale, D.S., Jacobsen, S.J., Furner, S.E., Levy, P.S., Brody, J.A. & Goldberg, J. (1997). Hip fracture incidence among elderly Asian-American populations. *American Journal of Epidemiology*, **146**, 502–509.

Liu, D., Veit, H.P., Wilson, J.H. & Denbow, D.M. (2003). Long-term supplementation of various dietary lipids alters bone mineral content, mechanical properties and histological characteristics of Japanese quail. *Poultry Science*, **82**, 831–839.

Lucas, E.A., Wild, R.D., Hammond, L.J., Khalil, D.A., Juma, S., Daggy, B.P., Stoecker, B.J. & Arjmandi, B.H. (2002). Flaxseed improves lipid profile without altering biomarkers of bone metabolism in postmenopausal women. *Journal of Clinical Endocrinology and Metabolism*, **87**, 1527–1532.

Mazur, W. (1998). Phytoestrogen content in foods. *Baillieres Clinical Endocrinology and Metabolism*, **12**, 729–742.

Mazur, W.M., Wahala, K., Rasku, S., Salakka, A., Hase, T. & Adlercreutz, H. (1998). Lignan and isoflavonoid concentrations in tea and coffee. *British Journal of Nutrition*, **79**, 37–45.

Mei, J., Yeung, S.S.C. & Kung, A.W.C. (2001). High dietary phytoestrogen intake is associated with higher bone mineral density in postmenopausal but not premenopausal women. *Journal Clinical Endocrinology and Metabolism*, **86**, 5217–5221.

Messina, M.J. & Wood, C.E. (2008). Soy isoflavones, estrogen therapy, and breast cancer risk: analysis and commentary. *Nutrition Journal*, **7**, 17.

Migliaccio, S., Newbold, R.R., Bullock, B.C., Mclachlan, J.A. & Korach, K.S. (1992). Developmental exposure to estrogens induces persistent changes in skeletal tissue. *Endocrinology*, **130**, 1756–1758.

Migliaccio, S., Newbold, R.R., Teti, A., Jefferson, W.J., Toverud, S.U., Taranta, A., Bullock, B.C., Suggs, C.A., Spera, G. & Korach, K.S. (2000). Transient estrogen exposure of female mice during early development permanently affects osteoclastogenesis in adulthood. *Bone*, **27**, 47–52.

Mollard, R.C., Gillam, M.E., Wood, T.M., Taylor, C.G. & Weiler, H.A. (2005). (n-3) fatty acids reduce the release of prostaglandin E2 from bone but do not affect bone mass in obese (fa/fa) and lean Zucker rats. *Journal of Nutrition*, **135**, 499–504.

Morabito, N., Crisafulli, A., Vergara, C., Gaudio, A., Lasco, A., Frisina, N., D'anna, R., Corrado, F., Pizzoleo, M.A., Cincotta, M., Altavilla, D., Ientile, R. & Squadrito, F. (2002). Effects of genistein and hormone-replacement therapy on bone loss in early postmenopausal women: a randomized double-blind placebo-controlled study. *Journal of Bone and Mineral Research*, **17**, 1904–1912.

Mori, M., Sagara, M., Ikeda, K., Miki, T. & Yamori, Y. (2004). Soy isoflavones improve bone metabolism in postmenopausal Japanese women. *Clinical and Experimental Pharmacology and Physiology*, **31**(Suppl 2), 44S–46S.

Mousavi, Y. & Adlercreutz, H. (1992). Enterolactone and estradiol inhibit each other's proliferative effect on MCF-7 breast cancer cells in culture. *The Journal of Steroid Biochemistry and Molecular Biology*, **41**, 615–619.

Nakamura, M.T., Cho, H.P., Xu, J., Tang, Z. & Clarke, S.D. (2001). Metabolism and functions of highly unsaturated fatty acids: an update. *Lipids*, **36**, 961–964.

Newbold, R.R., Banks, E.P., Bullock, B. & Jefferson, W.N. (2001). Uterine adenocarcinoma in mice treated neonatally with genistein. *Cancer Research*, **61**, 4325–4328.

Newbold, R.R., Elizabeth, P.B., Snyder, R.J. & Jefferson, W.N. (2005). Developmental exposure to estrogenic compounds and obesity. *Birth Defects Research. Part A, Clinical and Molecular Teratology*, **73**, 478–480.

Nieves, J.W., Komar, L., Cosman, F. & Lindsay, R. (1998). Calcium potentiates the effect of estrogen and calcitonin on bone mass: review and analysis. *American Journal of Clinical Nutrition*, **67**, 18–24.

NIH Consensus Development Panel on Osteoporosis Prevention Diagnosis and Therapy (2001). Osteoporosis prevention diagnosis and therapy. *The Journal of the American Medical Association*, **285**, 785–795.

Nurmi, T., Heinonen, S., Mazur, W., Deyama, T., Nishibe, S. & Adlercreutz, H. (2003). Lignans in selected wines. *Food Chemistry*, **83**, 303–309.

Omi, N., Aoi, S., Murata, K. & Ezawa, I. (1994). Evaluation of the effect of soybean milk and soybean milk peptide on bone metabolism in the rat model with ovariectomized osteoporosis. *Journal of Nutritional Science and Vitaminology (Tokyo)*, **40**, 201–211.

Padilla-Banks, E., Jefferson, W.N. & Newbold, R.R. (2006). Neonatal exposure to the phytoestrogen genistein alters mammary gland growth and developmental programming of hormone receptor levels. *Endocrinology*, **147**, 4871–4882.

Penttinen, P., Jaehrling, J., Damdimopoulos, A.E., Inzunza, J., Lemmen, J.G., Van Der Saag, P., Pettersson, K., Gauglitz, G., Makela, S. & Pongratz, I. (2007). Diet-derived polyphenol metabolite enterolactone is a tissue-specific estrogen receptor activator. *Endocrinology*, **148**, 4875–4886.

Piekarz, A.V. & Ward, W.E. (2007). Effect of neonatal exposure to genistein on bone metabolism in mice at adulthood. *Pediatric Research*, **61**, 48–53.

Potter, S.M., Baum, J.A., Teng, H., Stillman, R.J., Shay, N.F. & Erdman, J.W., Jr (1998). Soy protein and isoflavones: their effects on blood lipids and bone density in postmenopausal women. *American Journal of Clinical Nutrition*, **68**, 1375S–1379S.

Poulsen, R.C. & Kruger, M.C. (2006). Detrimental effect of eicosapentaenoic acid supplementation on bone following ovariectomy in rats. *Prostaglandins Leukotrienes Essential Fatty Acids*, **75**, 419–427.

Power, K.A., Ward, W.E., Chen, J.M., Saarinen, N.M. & Thompson, L.U. (2006). Genistein alone and in combination with the mammalian lignans enterolactone and enterodiol induce estrogenic effects on bone and uterus in a postmenopausal breast cancer mouse model. *Bone*, **39**, 117–124.

Power, K.A., Ward, W.E., Chen, J.M., Saarinen, N.M. & Thompson, L.U. (2007). Flaxseed and soy protein isolate, alone and in combination, differ in their effect on bone mass, biomechanical strength, and uterus in ovariectomized nude mice with MCF-7 human breast tumor xenografts. *Journal of Toxicology and Environmental Health, Part A*, **70**, 1888–1896.

Reinwald, S., Li, Y., Moriguchi, T., Salem, N., Jr & Watkins, B.A. (2004). Repletion with (n-3) fatty acids reverses bone structural deficits in (n-3)-deficient rats. *Journal of Nutrition*, **134**, 388–394.

Reinwald, S. & Weaver, C.M. (2006). Soy isoflavones and bone health: a double-edged sword? *Journal of Natural Products*, **69**, 450–459.

Rossouw, J.E., Anderson, G.L., Prentice, R.L., Lacroix, A.Z., Kooperberg, C., Stefanick, M.L., Jackson, R.D., Beresford, S.A., Howard, B.V., Johnson, K.C., Kotchen, J.M. & Ockene, J. (2002). Risks and benefits of estrogen plus progestin in healthy postmenopausal women: principal results from the Women's Health Initiative randomized controlled trial. *The Journal of the American Medical Association*, **288**, 321–333.

Ryokkynen, A., Kukkonen, J.V.K. & Nieminen, P. (2006). Effects of dietary genistein on mouse reproduction, postnatal development and weight-regulation. *Animal Reproduction Science*, **93**, 337.

Sacco, S.M., Jiang, J.M.Y., Reza-López, S., Ma, D.W.L., Thompson, L.U.& Ward, W.E. (2009). Flaxseed combined with low-dose estrogen therapy preserves bone tissue in ovariectomized rats. *Menopause*, **165**, 45–54.

Sakaguchi, K., Morita, I. & Murota, S. (1994). Eicosapentaenoic acid inhibits bone loss due to ovariectomy in rats. *Prostaglandins Leukotrienes Essential Fatty Acids*, **50**, 81–84.

Sato, K., Fukata, H., Kogo, Y., Ohgane, J., Shiota, K. & Mori, C. (2006). Neonatal exposure to diethylstilbestrol alters the expression of DNA methyltransferases and methylation of genomic DNA in the epididymis of mice. *Endocrine Journal*, **53**, 331–337.

Setchell, K.D.R. (1998). Phytoestrogens: the biochemistry, physiology, and implications for human health of soy isoflavones. *American Journal of Clinical Nutrition*, **68**, 1333S–1346S.

Setchell, K.D.R., Borriello, S.P., Hulme, P., Kirk, D.N. & Axelson, M. (1984). Nonsteroidal estrogens of dietary origin: possible roles in hormone- dependent disease. *American Journal of Clinical Nutrition*, **40**, 569–578.

Setchell, K.D.R., Brown, N.M. & Lydeking-Olsen, E. (2002). The clinical importance of the metabolite equol-a clue to the effectiveness of soy and its isoflavones. *Journal of Nutrition*, **132**, 3577–3584.

Setchell, K.D.R. & Cassidy, A. (1999). Dietary isoflavones: biological effects and relevance to human health. *Journal of Nutrition*, **129**, 758S–767S.

Setchell, K.D.R. & Cole, S.J. (2006). Method of defining equol-producer status and its frequency among vegetarians. *Journal of Nutrition*, **136**, 2188–2193.

Shen, C.L., Yeh, J.K., Rasty, J., Chyu, M.C., Dunn, D.M., Li, Y. & Watkins, B.A. (2007). Improvement of bone quality in gonad-intact middle-aged male rats by long-chain n-3 polyunsaturated fatty acid. *Calcified Tissue International*, **80**, 286–293.

Shen, C.L., Yeh, J.K., Rasty, J., Li, Y. & Watkins, B.A. (2006). Protective effect of dietary long-chain n-3 polyunsaturated fatty acids on bone loss in gonad-intact middle-aged male rats. *British Journal of Nutrition*, **95**, 462–468.

Singh, S.R. & Levine, M.A. (2006). Natural health product use in Canada: analysis of the National Population Health Survey. *The Canadian Journal of Clinical Pharmacology*, **13**, e240–e250.

Sirois, I., Cheung, A.M. & Ward, W.E. (2003). Biomechanical bone strength and bone mass in young male and female rats fed a fish oil diet. *Prostaglandins Leukotrienes Essential Fatty Acids*, **68**, 415–421.

Somekawa, Y., Chiguchi, M., Ishibashi, T. & Aso, T. (2001). Soy intake related to menopausal symptoms, serum lipids, and bone mineral density in postmenopausal Japanese women. *Obstetrics and Gynecology*, **97**, 109–115.

Strauss, L., Makela, S., Joshi, S., Huhtaniemi, I. & Santti, R. (1998). Genistein exerts estrogen-like effects in male mouse reproductive tract. *Molecular and Cellular Endocrinology*, **144**, 83–93.

Strom, B.L., Schinnar, R., Ziegler, E.E., Barnhart, K.T., Sammel, M.D., Macones, G.A., Stallings, V.A., Drulis, J.M., Nelson, S.E. & Hanson, S.A. (2001). Exposure to soy-based formula in infancy and endocrinological and reproductive outcomes in young adulthood. *The Journal of the American Medical Association*, **286**, 807–814.

Sun, D., Krishnan, A., Zaman, K., Lawrence, R., Bhattacharya, A. & Fernandes, G. (2003). Dietary n-3 fatty acids decrease osteoclastogenesis and loss of bone mass in ovariectomized mice. *Journal of Bone and Mineral Research*, **18**, 1206–1216.

Taguchi, H., Chen, H., Yano, R. & Shoumura, S. (2006). Comparative effects of milk and soymilk on bone loss in adult ovariectomized osteoporosis rat. *Okajimas Folia Anatomica Japonica*, **83**, 53–59.

Takamatsu, Y., Iguchi, T. & Takasugi, N. (1992). Effects of neonatal exposure to diethylstilbestrol on protein expression by vagina and uterus in mice. *In Vivo* **6**, 1–8.
Tham, D.M., Gardner, C.D. & Haskell, W.L. (1998). Potential health benefits of dietary phytoestrogens: a review of the clinical, epidemiological, and mechanistic evidence. *Journal of Clinical Endocrinology and Metabolism*, **83**, 2223–2235.
Thompson, L.U., Boucher, B.A., Cotterchio, M., Kreiger, N. & Liu, Z. (2007). Dietary phytoestrogens, including isoflavones, lignans, and coumestrol, in nonvitamin, nonmineral supplements commonly consumed by women in Canada. *Nutrition and Cancer*, **59**, 176–184.
Thompson, L.U., Chen, J.M., Li, T., Strasser-Weippl, K. & Goss, P.E. (2005). Dietary flaxseed alters tumor biological markers in postmenopausal breast cancer. *Clinical Cancer Research*, **11**, 3828–3835.
Uesugi, T., Fukui, Y. & Yamori, Y. (2002). Beneficial effects of soybean isoflavone supplementation on bone metabolism and serum lipids in postmenopausal Japanese women: a four-week study. *Journal of the American College of Nutrition*, **21**, 97–102.
U.S. Food and Drug Administration (2004). *FDA Announces Qualified Health Claims for Omega-3 Fatty Acids*. U.S. Department of Health and Human Services. Available online http://www.fda.gov/NewsEvents/Newsroom/PressAnnouncements/2004/ucm108351.htm (accessed 15 October 2008).
U.S. Surgeon General Department of Health and Human Services (2004). Bone health and osteoporosis: A report of surgeon general. Rockville, MD, 1–371.
Ward, W.E. & Fonseca, D. (2007). Soy isoflavones and fatty acids: effects on bone tissue postovariectomy in mice. *Molecular Nutrition and Food Research*, **51**, 824–831.
Ward, W.E. & Piekarz, A.V. (2007). Effect of prenatal exposure to isoflavones on bone metabolism in mice at adulthood. *Pediatric Research*, **61**, 438–443.
Ward, W.E. & Thompson, L.U. (2006). Understanding food and food/drug synergy. In: *Food-Drug Synergy and Safety*. Thompson, L.U. & Ward, W.E. (eds), CRC Press, Boca Raton, FL.
Ward, W.E., Yuan, Y.V., Cheung, A.M. & Thompson, L.U. (2001a). Exposure to flaxseed and its purified lignan reduces bone strength in young but not older male rats. *Journal of Toxicology and Environmental Health, Part A*, **63**, 53–65.
Ward, W.E., Yuan, Y.V., Cheung, A.M. & Thompson, L.U. (2001b). Exposure to purified lignan from flaxseed (*Linum usitatissimum*) alters bone development in female rats. *British Journal of Nutrition*, **86**, 499–505.
Watanabe, S., Terashima, K., Sato, Y., Arai, S. & Eboshida, A. (2000). Effects of isoflavone supplement on healthy women. *Biofactors*, **12**, 233–241.
Watkins, B.A., Li, Y., Allen, K.G., Hoffmann, W.E. & Seifert, M.F. (2000). Dietary ratio of (n-6)/(n-3) polyunsaturated fatty acids alters the fatty acid composition of bone compartments and biomarkers of bone formation in rats. *Journal of Nutrition*, **130**, 2274–2284.
Watkins, B.A., Li, Y. & Seifert, M.F. (2006). Dietary ratio of n-6/n-3 PUFAs and docosahexaenoic acid: actions on bone mineral and serum biomarkers in ovariectomized rats. *The Journal of Nutritional Biochemistry*, **17**, 282–289.
Watkins, B.A., Reinwald, S., Li, Y. & Seifert, M.F. (2005). Protective actions of soy isoflavones and n-3 PUFAs on bone mass in ovariectomized rats. *The Journal of Nutritional Biochemistry*, **16**, 479–488.
Weiler, H.A., Kovacs, H., Nitschmann, E., Bankovic-Calic, N., Aukema, H. & Ogborn, M. (2007). Feeding flaxseed oil but not secoisolariciresinol diglucoside results in higher bone mass in healthy rats and rats with kidney disease. *Prostaglandins Leukotrienes Essential Fatty Acids*, **76**, 269–275.
Weiler, H., Kovacs, H., Nitschmann, E., Fitzpatrick Wong, S., Bankovic-Calic, N. & Ogborn, M. (2002). Elevated bone turnover in rat polycystic kidney disease is not due to prostaglandin E2. *Pediatric Nephrology*, **17**, 795–799.
Weiss, L.A., Barrett-Connor, E. & Von Muhlen, D. (2005). Ratio of n-6 to n-3 fatty acids and bone mineral density in older adults: the Rancho Bernardo Study. *American Journal of Clinical Nutrition*, **81**, 934–938.
Welshons, W.V., Murphy, C.S., Koch, R., Calaf, G. & Jordan, V.C. (1987). Stimulation of breast cancer cells in vitro by the environmental estrogen enterolactone and the phytoestrogen equol. *Breast Cancer Research and Treatment*, **10**, 169–175.
Whelan, J. & Rust, C. (2006). Innovative dietary sources of N-3 fatty acids. *Annual Review of Nutrition*, **26**, 75–103.
World Health Organization (1994). Assessment of fracture risk and its application to screening for postmenopausal osteoporosis. Report of a WHO Study Group. World Health Organization technical report series 843, 1–129.

Writing Group For The Women's Health Initiative Investigators (2002). Risks and benefits of estrogen plus progestin in healthy postmenopausal women – principal results from the women's health initiative randomized controlled trial. *The Journal of the American Medical Association*, **288**, 321–333.
Yamori, Y., Moriguchi, E.H., Teramoto, T., Miura, A., Fukui, Y., Honda, K.-I., Fukui, M., Nara, Y., Taira, K. & Moriguchi, Y. (2002). Soybean isoflavones reduce postmenopausal bone resorption in female Japanese immigrants in Brazil: A ten-week study. *Journal of American College of Nutrition*, **21**, 560–563.
Ye, Y.B., Tang, X.Y., Verbruggen, M.A. & Su, Y.X. (2006). Soy isoflavones attenuate bone loss in early postmenopausal Chinese women. *European Journal of Nutrition*, **45**, 327–334.
Zhang, X., Shu, X.O., Li, H., Yang, G., Li, Q., Gao, Y.T. & Zheng, W. (2005). Prospective cohort study of soy food consumption and risk of bone fracture among postmenopausal women. *Archives of Internal Medicine*, **165**, 1890–1895.

Index